STOCHASTIC SIMULATIONS of CLUSTERS

Quantum Methods in Flat and Curved Spaces

STOCHASTIC SIMULATIONS of CLUSTERS

Quantum Methods in Flat and Curved Spaces

EMANUELE CUROTTO

CRC Press
Taylor & Francis Group
Boca Raton   London   New York

CRC Press is an imprint of the
Taylor & Francis Group, an **informa** business

CRC Press
Taylor & Francis Group
6000 Broken Sound Parkway NW, Suite 300
Boca Raton, FL 33487-2742

First issued in paperback 2017

ISBN 13: 978-1-138-11241-4 (pbk)
ISBN 13: 978-1-4200-8225-8 (hbk)

Library of Congress Cataloging-in-Publication Data

Curotto, Emanuele.
  Stochastic simulations of clusters : quantum methods in flat and curved spaces / Emanuele Curotto.
    p. cm.
  Includes bibliographical references and index.
  ISBN 978-1-4200-8225-8 (alk. paper)
  1. Microclusters--Statistical methods. 2. Quasimolecules. 3. Quantum statistics--Data processing. 4. Monte Carlo method. 5. Path integrals. I. Title.

QC173.4.M48C87 2010
539'.6015195--dc22   2009024334

Visit the Taylor & Francis Web site at
http://www.taylorandfrancis.com

and the CRC Press Web site at
http://www.crcpress.com

Contents

III   Methods in Curved Spaces   435

10 Introduction to Differential Geometry   437

Preface

Our current research objective is the study of a class of weakly bound substances, sometimes referred to as van der Waals clusters or complexes. Weakly bound matter is unusual in many ways. For instance, in many cases, we cannot store a bottle of this material on a shelf. Some complexes are only stable at extremely low temperatures, and usually "live" only long enough for us to be able to measure their physical properties. One way to produce complexes in the gas phase is to shoot gaseous matter with high backing pressure through a small jet into an evacuated chamber. The effusing gas, at the proper conditions, is capable of achieving speeds that exceed the speed of sound in air. The supersonic jet expansion is nearly adiabatic, and produces a very cold gas. If the temperature is sufficiently low, condensation takes place and clusters are formed. One can think of a cluster as a molecular-sized microscopic droplet of liquid or a very tiny chunk of solid at sufficiently cold temperatures. Other types of clusters can be produced in vacuum by laser ablation, as in the case of metal clusters, or by thermal evaporation, as in the case of carbon clusters. Clusters have been the subject of a growing interest for the past two decades. They are important in atmospheric chemistry, and they can be technologically important new materials, especially in the case of semiconductors or transition metal clusters. Clusters also hold the key to our understanding of intermolecular forces and how these affect the physical properties of bulk condensed matter. We are interested in simulating phase changes at finite temperature or energy in large heterogeneous systems, and the study of phase processes that take place in these. Interesting phase phenomena, tied fundamentally with physical properties, seem to occur in heterogeneous clusters at very low temperatures. In recent times, we have been particularly interested in simulating aggregates of small covalent molecules. There are multitudes of important practical applications, ranging from novel fuel and material research, to nanotechnology and computational biology. For the latter, for example, we have been studying prototypical hydrogen-bonded systems for a number of years.

Clusters present interesting theoretical challenges, specially when they consist of small covalent molecules. The usual normal mode analysis, for example, so successful in the treatment of small covalently bound molecules, does not yield a reasonable model for a molecular aggregate. Therefore, our group has concentrated on the theoretical study of heterogeneous van der Waals clusters, developing a hierarchy of classical and quantum approaches. The presence of hydrogen and other light atoms, and cold temperatures, is pervasive in our work. Therefore, the argument for quantum mechanical corrections is compelling. This particular field of study requires the development of numerical techniques that represent a marriage of "real-time algorithms" for the exact solution of the Schrödinger equation, and finite temperature statistical simulation tools. The most important properties obtained from simulations are the energy and the inherent structures. We make use of potential energy models to search minima and transition states. The energy and structure of these provide a first insight into their behavior. Temperature effects are obtained using Monte Carlo methods, and quantum effects are studied using ground state variational and diffusion Monte Carlo simulations. Finally, the path integral approach allows us to study quantum effects at finite temperatures. All these techniques have been implemented by a community of researchers over the past two decades. The development and implementation of quantum methods is easiest in Euclidean spaces mapped by Cartesian coordinates. However, if one is interested in molecular aggregates, Cartesian coordinates are not efficient. The timescale difference between stiff internal modes and the weak van der Waals modes produces large fluctuations in classical stochastic simulations, and nonuniform convergence properties in quantum stochastic algorithms. Years ago, our group embarked on a journey to discover more efficient and accurate methods to handle the timescale problem. The journey has taken us far, as we have found the need to combine diagonalization techniques, differential geometry (the mathematics at the foundation of Einstein's general relativity theory), the path integral method in statistical mechanics, and the diffusion Monte Carlo approach.

This book was created from a collection of notes and tutorial documents that I have developed over the years to train students to simulate, and to become involved in the development of novel algorithms. As such, it is far from a complete reference of all numerical methods in theoretical chemistry; rather, it is a compendium of those methods that the group has either used or developed during years of research. The field has experienced considerable growth in the past decade. Nevertheless, there are no other training texts available to researchers in these methods. Few "hands on" books on stochastic methods exist, and these are largely confined to classical simulations of liquids. A training manual on quantum stochastic methods, even in the simpler Euclidean spaces mapped by Cartesian coordinates, is long overdue. A good portion of this book is designed to train the reader to use special techniques that accelerate the convergence of quantum Monte Carlo methods and overcome timescale problems. These methods have been implemented successfully to unravel what would otherwise have been formidable problems.

Where necessary, an introduction to the physical concepts and to the applied mathematics that the algorithms are constructed from, is provided. The reader will find a great deal of linear algebra, Lie algebra, differential geometry applied to computations, and simulations in nonrelativistic wave mechanics and statistical thermodynamics. In these cases, the theory seeks to introduce the reader to the practical aspects of the subject as it applies to the problem at hand, and the material is presented rather informally.

This book is designed for the advanced undergraduate students in physics or chemistry with at least two semesters of calculus, some background in programing principles, and a sound course in calculus-based physics, or graduate students in theoretical chemistry. The material on stochastic path integration, variational, and diffusion Monte Carlo in curved spaces is very recent, and most of the essential tensor analysis background is developed from the particle physics prospective. Researchers in chemical physics at all levels who are interested in the timescale problem and the techniques used to deal with it for molecular aggregates will find the reading and training useful.

The majority of the content of Chapters 1–6 is introductory in nature. The theoretical background in Chapter 2 contains the fundamental concepts of classical mechanics. Chapter 3 contains a cursory introduction to classical statistical mechanics [1], Chapter 4 contains a brief introduction to vector spaces and linear algebra, and we dedicate several sections of Chapter 5 to the development and elementary applications of the postulates of quantum mechanics. These concepts should be familiar to graduate students, but I found it useful to include them when training advanced undergraduates. However, unlike a typical introductory book, each chapter develops the concepts with hands-on numerical [2] exercises designed to bring the student closer to current literature topics. In Chapter 3, for example, programs are presented to introduce the basic principles of stochastic simulations, such as rejection techniques, importance sampling, the Metropolis algorithm, and parallel tempering [3,4]. The tutorials of Chapter 3 are designed to teach the concept of important sampling and raise the issue of quasiergodicity. Most computational aspects of classical statistical thermodynamics are very well treated in the books of Allen and Tildesley [6] and Frenkel and Smit [7]. These texts develop the importance sampling concept thoroughly; however, parallel tempering is only briefly treated because quasiergodicity is not pervasive in the simulation of liquids, whereas it becomes crucial to suppress it when simulating clusters. Further information on the theory of minimization of multidimensional functions can be found in the *Numerical Recipes* text [8].

Unlike the content of a typical linear algebra course, Chapter 4 introduces Lie algebras and matrix representations of evolution operators and other members of unitary groups. These provide the necessary mathematical background to follow the development of path integral techniques in Chapter 7 and are generally fundamental for the development of deterministic quantum methods. Lie algebras are the fundamental tools to unravel linear coupled systems of differential equations, the commutation rules of quantum operators, and the generators of groups of continuous transformations. Subgroups of these are the generators of the curvilinear or curved spaces treated in Chapter 10. The solution of coupled systems of differential equations by Lie groups, parallels the way linear algebras are used to solve linear coupled systems of algebraic equations. In Chapter 4, an important class of Lie algebras fundamental to the symplectic split operator techniques [9–48] for the integration of Hamilton's equations, is introduced. The reader will be able to find most of the introductory material on matrices and vector spaces, presented in Chapter 4, in a sufficiently enlightened linear algebra textbook [49,50], whereas the tome by Gilmore is an authoritative source of knowledge of Lie algebras and their properties [51]. The theory of Chapter 4 is applied to solve problems in quantum mechanics in Chapter 5, where we introduce the discrete variable representation [52–174], some sparse matrix technology, Krylov spaces, and the Lanczos algorithm [175–177, 179–181] to handle multidimensional problems. More detail about the representation of sparse matrices and many more algorithms for computations with sparse matrices can be found in Pissanetzky [178]. Chapter 6 introduces the time variable in quantum mechanics and treats variational Monte Carlo [182–204] and diffusion Monte Carlo with several importance sampling methods and with methods used to accelerate convergence [205–439]. Chapter 7 presents several algorithms for the computation of the path integral [440–754] in statistical mechanics for Euclidean spaces. The books by Feynman and Hibbs [754] and Schulman [747] are authoritative sources for the reader interested in deepening his or her understanding of the subject. Chapters 8 and 9 present a first example for the simulation of the properties of a cluster: Ar_7. Here, the reader has a chance to grasp the power of the numerical techniques using realistic examples, and to get closer to the literature on Lennard–Jones systems [757–851]. In Chapter 8, in particular, we search for the global minimum and transition states. Chapter 9 applies all the simulation techniques developed in the preceding chapters of the book to this relatively small cluster.

Chapter 10 introduces the fundamental concepts of differential geometry and tensor analysis necessary to understand how holonomic constraints are used in quantum simulations of molecular matter. References on the material can be easily found. The books by Schutz [852] and Curtis and Miller [853] are good complements to the brief introduction presented in Chapter 10. Chapters 11 and 12 deal with modifications to the path integral and variational and diffusion Monte Carlo methods necessary to carry out simulations in curved manifolds and for clusters of rigid tops. A great deal of the underlying theory of path integration in manifolds with curvature and torsion can be found in the book by Kleinert [515]. Finally, the first chapter is a brief FORTRAN tutorial, designed to familiarize the reader with the essential aspects of the programing language. The material contained here is not inclusive, and I recommend that the reader complements my explanations with a good FORTRAN text [854]. Additionally, Chapter 1 covers some basics of numerical analysis.

The reader must have a FORTRAN compiler at his or her disposal, since this is a hands-on text. Mastery of the material in this book can only come with the *endless toil* of the programs contained within. Typing, compiling, and running the programs is not enough. The reader is urged to test each program and subroutine carefully, to add write statements and test programs with different inputs. However, modifying the programs by forming conjectures, testing, and verifying these numerically (then proving them for the purist in my audience) is the only way to learn the underlying concepts. Computers are merely tools that should not be allowed to rob us of the joy of learning. True, in the business of computation, some routines are best left as canned.

However, those who read this book and use the programs contained within as black boxes, are destined to wander in very dark labyrinths. The reader should approach each program in this text with the assumption that any source code with more than 20 lines contains an error. We made no attempt to optimize the code given in this book for "production" simulations. The objective for providing most of the code in this book is purely pedagogical.

Much of the text is written in a guided inquiry style in the hope of making the learning as efficient as possible. A number of suggestions for modifications are made, and the reader is advised to follow these, as each exercise is designed to provide a distinct learning experience and introduce key concepts. Perhaps the greatest reward for the intense labor that awaits the reader is to discover just how much is left to be discovered in the rich field of theoretical chemistry. Even though in the past half-century we have made great strides forward, we are far from the time and place where computer simulations are routinely integrated in medicinal and material discovery. These are outstanding challenges that the scientist of tomorrow is facing.

Finally, a word on computer platforms and software. At the time of writing, our laboratory is running Linux CentOS with gfortran as the FORTRAN compiler. A good reference for some LINUX shell basics is Chapter 3 of the tome by Christopher Negus, published by Wiley. For a short but good online tutorial, we find the material at http://www.ee.surrey.ac.uk/Teaching/Unix/ useful. Our operating system has been augmented by graphic programs, such as grace, xmakemol, rasmol, and gnuplot, and these were used to generate the figures contained in this book. The source code for these programs is free and the installation on a LINUX workstation is straightforward.

Acknowledgments

This book would not have been possible without the constant encouragement and support of a number of important people whom I would like to thank formally. My mentors, Professors Irvin Lipschitz, James R. Cross, and David L. Freeman, have introduced me to theoretical chemistry, and have trained me to develop and use most of the techniques discussed in this book. They shared with me their passion for theoretical and computational inquiry, and their deep love for teaching.

This book is intended to be a training tool for what is a complex, ever-evolving, and intellectually stimulating craft. Collaborations with Professors Freeman, Doll, and Mella have had an incommensurably positive impact on my professional career. Our synergistic and constant exchange of ideas continues to deepen my understanding of many aspects of our craft. I am grateful to my colleague, Professor Peter Campbell, for constant stimulating conversations. Our exchanges are always grounded in realism, a prospective that only a careful experimentalist like Peter can offer. I thank my colleague, Professor Stephen Huber, for providing access to his personal collection of books, journals, lectures, and monographs in theoretical physics. None of these are among the holdings at our library, and their availability was, and continues to be, an invaluable asset. Additionally, I am grateful to Professors P. Nightingale and H. Blöte for giving me permission to publish their random number generator, `ransi`.

The programs in this book have benefited from the hard work of numerous brave and talented undergraduates that I have had the privilege and pleasure to train over the years. Given that the students, with my guidance, through their collective efforts, constructed the concepts, tested the programs, and derived nearly all the equations in this book, I felt compelled to avoid using singular pronouns in much of the writing, even though I am the only author of this book. During the writing of this book, I was supported by a number of grants. Therefore, I am acknowledging the donors of the Petroleum Research Fund, administered by the ACS (grant numbers 36870-B6, 40946-B6, and 41846-B6), and the National Science Foundation (grant number CHE0554922).

I dedicate this book, and the collective hard work of all the acknowledged individuals that it represents, to my two sons, Nicolas and Christopher.

Author

Emanuele Curotto is a professor of chemistry at Arcadia University in Glenside, Pennsylvania. He immigrated to the United States in 1987. He married his wife Kathleen in 1998, graduated *summa cum laude* from the University of Massachusetts Lowell in 1992, and went on to do graduate work in the Cross group at Yale University, where he worked on developing theories for molecular inelastic scattering. In 1996, he received his doctoral degree from Yale University and joined the Freeman group at the University of Rhode Island, where he worked on simulations of metal clusters and the development of Monte Carlo Path Integral methods. In 1998, he joined the chemistry and physics department at Arcadia University. In 2001, he became a US citizen, and in 2009, he was inducted into the Phi Kappa Phi honor society. Professor Curotto has made several contributions to the development of Path Integral, Variational, and Diffusion Monte Carlo methods for curved spaces, and has authored a number of scientific articles on the subject.

Part I

Fundamentals

1

FORTRAN Essentials

1.1 Introduction

Modern programing languages, such as JAVA and C++, require the assimilation of complex concepts, such as classes and object-oriented design. Classes are user-defined variable types with customized functionality, and are the key to object-oriented programing. Object-oriented programing was developed out of the necessity to increase code portability. The availability of classes made it possible to quickly, and smoothly, join together the work of a team of programers to efficiently produce code with millions of lines. Such tasks would be nearly impossible for the old style top-down design, but can be accomplished routinely with the maximized portability that classes afford. On the other hand, it is rare that theoretical and computational chemistry research groups need to delve in to such gargantuan endeavors as a graphical user interface, for example. The majority of the programing is carried out still using top-down design and FORTRAN, as, the numerical tasks are relatively less complicated than an event-driven demon, for example. Most programs we use in our group require less than 10,000 lines of code. The complexity of our work lies in the theory behind the algorithms, rather than in their implementations.

FORTRAN is based on a simpler set of principles, like top-down design, therefore, mastery of object-oriented programing is not a prerequisite for the audience of this book. Furthermore, readers unfamiliar with programing principles can gain the programing skills necessary to be productive researchers in a short time. In fact, most trainees in theoretical chemistry seldom need to design a program from scratch. Rather, it is far more likely that they will be given a program to modify. We are often asked by students, why FORTRAN is still so heavily used by scientists and engineers. There are a number of reasons, but perhaps the principle one is economy. A great deal of resources would have to be invested to translate the large amount of source code that is available, often shared among groups of scientists and engineers. This book contains at least one example of old code that was produced, likely, in the late sixties. The code happens to be the heart of the Lanczos algorithm, an application of sparse matrix technology heavily employed as an important and modern research tool in molecular physics.

Scientific and engineering research and development groups possess archives of this material, affectionately known as "spaghetti code" in the community. These algorithms were generated at a time when storage space and memory were much more scarce than today. Legibility had to be sacrificed, in order to minimize the number of punch cards needed. These codes are found embedded inside subroutines that can be readopted after proper testing. The added benefits of translating these into other programing languages are marginal and do not justify the effort. For these reasons, we structured this chapter to provide an overview of FORTRAN that is broader (but at the same time less didactical) than in modern textbooks, and to expose the reader to some older constructs, such as the **computed goto** statement and the **arithmetic if** statement. Other features characteristic of FORTRAN, such as the implicit typing, are presented as well. Here, as well as in the rest of the book, the reader is guided through a number of practical exercises designed to force the construction of the concepts from first-hand experience. We assume that the student reading this chapter is familiar with fundamental programing principles. Most students that work in our group have taken Java or a similar programing class. The reader unfamiliar with programing principles should complement the present chapter with a good introductory text [854]. The second purpose of this chapter is to introduce some fundamental concepts of numerical analysis. Numerical derivatives and simple numerical integration algorithms are introduced, and the reader is guided to explore the sources of error associated with these simple numerical procedures.

1.2　What is FORTRAN?

FORTRAN is an acronym for FORmula TRANslation. FORTRAN is a higher level programing language that allows us to instruct the computer with commands very similar to the English language, as opposed to a list of hexadecimal or, even worse, binary numbers. The general idea of higher level computer languages is to make the formulation of commands a less error-prone process, and to make the code more legible to humans. The translation of the higher level programing language into assembly language (machine language) is carried out by an assembly language program, called the compiler. Work on the first FORTRAN compiler began in the early 1950s at IBM. From the beginning, a proliferation of the compiler took place, with a new version created for every new computer platform. Consequently, the issue that emerged early on was code portability. A poorly portable higher language code needs substantial revisions to work on a different compiler. The solution to this problem was the introduction of higher computing language standards by the American National Standard Institute (ANSI). The first standard version of FORTRAN was introduced in 1966. Many other versions have been introduced since.

- FORTRAN 66

- FORTRAN 77

- FORTRAN 90

- FORTRAN 95

- High Performance FORTRAN

Most of the exercises in this book are written in a mixture of FORTRAN 77 and FORTRAN 90 constructs.

1.3   FORTRAN Basics

A FORTRAN program is just a sequence of command lines. A program (or a subprogram) has the following basic structure:

```
program name
declarations
:   :
statements
:   :
end
```

The declarations are used to introduce variables (so that memory locations can be identified). The execution of the code then takes place in the statements section. Let us begin by discussing data types, since these are specified in the declarations part of the code.

1.4   Data Types

FORTRAN supports the following data types:

`integer`	A whole number: positive, negative, or zero.
`real`	A real number in decimal or exponential representation.
`double precision` or	
`real*8`	A real number with higher precision.
`complex`	A pair of real numbers $(a,b){:}(a + ib)$.
`logical`	A holder for the logical values .true. or .false.
`character`	A sequence of symbols.

To identify variables (really memory locations), constants, or even subprograms, FORTRAN uses identifiers, names composed of a string of characters, up to 30 in length. It is important to remember that identifiers need to begin with a letter and must not contain blank spaces. Let's look at several declarations:

```
program mixture
integer counter
real rho1,rho2,rho3, m1,m2,m3,mtot,v1,v2,v3,chiv1,chiv2
character(8) component1,component2,component3
```

The variable `counter` is an integer; the variables,

```
 rho1,rho2,rho3, dm1,dm2,dm3,dmtot,v1,v2,v3,chiv1,chim1
```

are declared as real; and `component1,component2,component3` are character strings each containing eight-characters.

1.5   The IMPLICIT Statement

A great deal of programs contained in this text, declare all the variables implicitly using the following statement,

```
implicit real*8 (a-h,o-z)
```

This statement declares all variables with an identifier beginning with a letter between a-h or between o-z as double precision type, and all the variables with an identifier that begins with either i, j, k, l, m, or n as integers. This statement is often used to eliminate the need to declare variables all together. Many programers believe it is not good practice to use identifiers without a declaration. Unlike C programing, a misspelled identifier will be implicitly typed, even if no implicit statement is made at the beginning of the program. Misspelled identifiers are, perhaps, the greatest source of bugs in programs. One can force FORTRAN to require declarations by using the `implicit none` statement. If the compiler finds any undeclared identifier, and `implicit none` is in the declarations, it will stop compilation and print out an error. As some of the examples that follow demonstrate, it is very time consuming to modify a program written with implicit typing into one that does not have implicit typing or vice-versa. The use of the implicit typing or otherwise is more a matter of style than good form. A good programer should be able to modify both kinds of programs, with or without implicit typing, since chances are, a researcher will not begin by building his or her own programs from scratch. Even without implicit typing, there are no guarantees that the program is free of bugs. Extensive testing will continue to be a higher language programing necessity for a foreseeable future.

1.6   Initialization and Assignment

Once a memory location has been named with an identifier, it can be used to store values. Initialization, just like typing, is best done explicitly, either through an initialization statement or an assignment statement. Compilers will behave differently depending on the platform and the compilation options. Some compilers will implicitly initialize all numerical variables to zero and all character variables to blanks, but it is a mistake to assume that always happens. In the following line, the real variable dm1 is assigned a value of zero,

```
dm1 = 0.0
```

It is possible to assign variables and constants in the declaration section of the program as well. The **parameter** statement is used for the purpose of defining constants in the declarations section of a program. For example,

```
implicit real*8 (a-h,o-z)
parameter (PI = 3.1415926d0)
```

declares a constant PI as a double precision real, and assigns to that memory location a value of 3.14159260000000. The d0 ending replaces the e0 in the notation of floating numbers. The integer that follows d or e is the power of 10 from the scientific notation of the number in both cases.

1.7   Order of Operations

For the most part, FORTRAN statements consist of assigning a memory location with the outcome of one or more operations. Arithmetic operations are executed within an expression, according to the following priority rules:

1. All expressions contained inside a pair of parenthesis ( ) are executed first. Only round brackets are accepted and they must appear in matching pairs.

2. Exponentiations (**) are performed next; consecutive exponentiations are executed from right to left.

3. Multiplications and divisions (*, /) are next. Consecutive multiplications and divisions are executed from left to right.

4. Additions and subtractions (+, -) are performed last.

1.8   Column Position Rules

We are almost ready to look at an example, but first a warning. FORTRAN is not a free-format language. It has a very strict set of rules for how the source code should be formatted. The most important of these are the column position rules.

- Col. 1   Blank, or a "c" or "!" for comments
- Col. 2–5   Statement label (optional)
- Col. 6   Continuation of previous line (optional)
- Col. 7–72   Statements
- Col. 73–80   Sequence number (optional, rarely used today)

Formatted source code lines are a feature that dates back from the days of punch cards, and are still with us today. Any line of code that begins in column 1 through 6 will cause the compiler to print error messages like

```
Non-numeric character at (^) in label field [info -f g77 M LEX]
```

The error messages from the compiler can be quite cryptic, however, in most cases the line number containing the offending statement will be printed on the screen. Any statement line that extends beyond column 72 should be continued in the next line by placing any character in the sixth-column of the line immediately below. Any part that extends beyond column 72 of a statement will be ignored by the compiler and cause run time errors or compilation errors. The lines that begin with a "c" in the first column are comments and have no purpose other than to make the program more readable. It is possible to add comments in the same line as a statement by using the ! character. Originally, all FORTRAN programs had to be written in uppercase letters. FORTRAN compilers are now case insensitive and most programers prefer lowercase statements, as these improve legibility.

1.9   A Typical Chemistry Problem Solved with FORTRAN

Consider a tertiary mixture problem, where the densities of each component ρ_1, ρ_2, and ρ_3 are known, together with the percent by mass composition for the first component, χ_{m_1}, the percent by volume composition for the first component, χ_{V_1}, and the total mass of the mixture m_t. We are going to write a program that takes all such data as input from the user and outputs the masses m_1, m_2, m_3, the volumes V_1, V_2, V_3, the density of the mixture ρ_t, and the percent by mass and by volume compositions of the other two components $\chi_{m_2}, \chi_{m_3}, \chi_{V_2}$, and χ_{V_3}.

As with the solution to any problem, we must begin by formulating an algorithm. The first step here is to solve a system of equations with all variables so that the program is as general as possible. We must solve the system of equations, formulated from various definitions and assuming ideal behavior,

$$
\begin{cases}
\rho_1 &= \dfrac{m_1}{V_1} \\[2mm]
\rho_2 &= \dfrac{m_2}{V_2} \\[2mm]
\rho_3 &= \dfrac{m_3}{V_3} \\[2mm]
\chi_{m_1} &= \dfrac{100\, m_1}{m_1 + m_2 + m_3} \\[2mm]
\chi_{V_1} &= \dfrac{100\, V_1}{V_1 + V_2 + V_3} \\[2mm]
\rho_t &= \dfrac{m_1 + m_2 + m_3}{V_1 + V_2 + V_3} \\[2mm]
m_t &= m_1 + m_2 + m_3.
\end{cases}
$$

It is straightforward to show that the solution of this system in terms of the given data is

$$
\rho_t = \rho_1 \frac{\chi_{V_1}}{\chi_{m_1}},
$$

$$
V_1 = \frac{m_t\, \chi_{m_1}}{100\, \rho_1},
$$

$$
V_3 = \frac{m_t}{(\rho_3 - \rho_2)} \left(1 - \frac{\chi_{m_1}}{\chi_{V_1}} + \frac{\rho_2\, \chi_{m_1}}{100\, \rho_1} - \frac{\rho_2\, \chi_{m_1}}{\rho_1\, \chi_{V_1}} \right),
$$

$$
V_2 = \frac{m_t}{\rho_1} \left(\frac{\chi_{m_1}}{\chi_{V_1}} - \frac{\chi_{m_1}}{100} \right) - V_3.
$$

Once the volumes are known, it is simple to find the masses and the percent compositions (both by mass and volume) for the other two components.

1.10   Free Format I/O

To implement the solution of our system into a FORTRAN program, we need some way of entering data into the program during execution. This is typically accomplished in two ways. Either the program opens a file, or reads data from the keyboard during execution. The latter is accomplished with the `read(5,*)` `identifier` statement. During execution, this statement will prompt the user for keyboard input and will store in the memory location, named `identifier`, the content of the entry. It is up to the user to make sure that the data entered is of the correct type. If the incorrect data type is entered at execution time, the operating system will throw an exception and print a message on the screen that may look something like this:

```
invalid number: incomprehensible list input
apparent state: unit 5 (unnamed)
last format: list IO
lately reading direct formatted external IO
Abort
```

It is just as important that the program communicates back to the user, data results and explanations. Programs can write data and information to a file or to the screen. The latter task is accomplished in free format by the statement `write(6,*)` `identifier`, where, during execution, the content of the memory location, named `identifier`, is printed on the screen. It is always good programing practice to add strings to write statements, such as `write(6,*)` `'identifier = '`, `identifier`

1.11   The FORTRAN Code for the Tertiary Mixtures Problem

Now inspect carefully the code lines below. First, we build a template so that the command lines are formatted properly. These start with a c in column one so that the compiler interprets them as comments.

```
c       1       2       3       4       5       6       7
c234567890123456789012345678901234567890123456789012345678901234567890123456789012
        program mixture
        implicit none
        real rho1,rho2,rho3,chim1,chiv1,rhot
        real v1,v2,v3,m1,m2,m3,mt
        real chim2,chiv2,chim3,chiv3
        real rhot2,vt
c This program calculates
c the masses, the volumes, the overall density,
c the percent by mass and by
c volume compositions of the other two components
c for a tertiary mixture given the densities of each component,
c the percent by mass composition for the first component, the
c percent by volume composition for the first component,
c and the total mass of the mixture.
c The program assumes densities are in g/cm^3 and that the total
c mass is in grams.
c i/0 section
```

```
      write(6,*) 'Enter the three densities in g/cm^3'
      read(5,*) rho1,rho2,rho3
      write(6,*) 'The densities you entered are:'
      write(6,*) rho1,rho2,rho3
      write(6,*) 'Enter the composition by mass and by volume of the'
      write(6,*) 'first component with density ',rho1,' g/cm^3'
      read(5,*) chim1,chiv1
      write(6,*) 'The mixture is', chim1,' % by mass', chiv1,' % by',
     & 'volume'
      write(6,*) 'Enter the total mass of the mixture'
      read(5,*) mt
c calculates the density of the mixture, and the total volume
      rhot = rho1*chiv1/chim1          ! derive this equation
      write(6,*) 'The density of this mixture must be',rhot
      vt = mt/rhot
c calculates the volumes and the masses of each component
      v1 = mt*chim1/(rho1*100.)
      m1 = mt*chim1/(100.)
      v3 = mt*(1 - chim1/100. + rho2*chim1/(100.*rho1) -
     & rho2*chim1/(chiv1*rho1))/(rho3- rho2)
      v2 = mt*((chim1/chiv1 - chim1/100.)/rho1) - v3
      m2 = rho2*v2
      m3 = rho3*v3
      write(6,*) 'Component 1 mass',m1,' g and volume',v1,' cm^3'
      write(6,*) 'Component 2 mass',m2,' g and volume',v2,' cm^3'
      write(6,*) 'Component 3 mass',m3,' g and volume',v3,' cm^3'
c calculates the compositions
      chim2 = 100.*m2/mt
      chim3 = 100.*m3/mt
      chiv2 = 100.*v2/(v1+v2+v3)
      chiv3 = 100.*v3/(v1+v2+v3)
      write(6,*) 'Compositions by',
     & 'mass',chim1,chim2,chim3,(chim1+chim2+chim3)
      write(6,*) 'Compositions by',
     & 'volume',chiv1,chiv2,chiv3,(chiv1+chiv2+chiv3)
c checks the calculations of the total density
      rhot2 = mt/(v1+v2+v3)
      write(6,*) 'Checking the density', rhot2,rhot
      write(6,*) 'Checking the volumes', vt,(v1+v2+v3)
      write(6,*) 'Checking the masses', mt,(m1+m2+m3)
      write(6,*) 'Checking all the densities', m1/v1,m2/v2,m3/v3
      end
```

Several blank lines increase the readability of the program, with each block headed by the comments directing the reader to the task performed by the code lines that follow. Unfortunately, editorial constraints force us to exclude such blank lines. However, we encourage the reader to test and verify the outcome of each command with a pocket calculator, and to introduce blank lines and additional comments to improve legibility in all the codes in this book. Four lines of code were longer than the 72 columns given by the strict formatting requirements. Each line is broken up and the compiler directive to continue to interpret the line that follows as a continuation of the line above is given, in each case, by the & character in the sixth-column. Any character can be used here, but & is less confusing than a + or a − character in that position.

The implicit none statement is used, and consequently all the variables had to be declared. It is always good practice to add comments to the program at the beginning. It would be very difficult for anyone to decipher what the program is doing exactly (except on a line by line basis) if it does not contain an ample explanation in the comment lines. The I/O session is also highly documented for additional clarity during execution time. Additionally, a number of checks to make sure the program works properly are made, for example, all three compositions are added to check if the sum is 100.

The program mixture has been thoroughly tested and a sample run follows.

```
 Enter the three densities in g/cm^3
0.990400016 0.789799988 12.323
 The densities you entered are:
  0.990400016  0.789799988  12.323
 Enter the composition by mass and by volume of the
 first component with density  0.990400016 g/cm^3
22.0954 22.5534
 The mixture is  22.0953999 % by mass  22.5534 % by volume
 Enter the total mass of the mixture
45.07
 The density of this mixture must be  1.01092935
 Component 1 mass  9.95839691 g and volume  10.054924 cm^3
 Component 2 mass  26.7330761 g and volume  33.847908 cm^3
 Component 3 mass  8.37852669 g and volume  0.679909647 cm^3
 Compositions by mass  22.0953999  59.3145676  18.5900307  100.
 Compositions by volume  22.5534  75.9215469  1.52505124  100.
 Checking the density  1.01092935  1.01092935
 Checking the volumes  44.5827408  44.5827408
 Checking the masses  45.0699997  45.0699997
 Checking all the densities  0.990400016  0.789799929  12.323
```

Exercises

1. Compile the code above and run it with the test data given. Then, run it with other input data as well.

2. Run mixture with $\rho_3 = \rho_2$. What happens? Physically, this situation corresponds to a binary mixture as opposed to a tertiary one. After we

look at branching, we will modify this example program so that this case is included and computed properly.

3. Remove the `implicit none` statement and all four declarations. Compile and run the `mixture` using the test data. What happens? Fix the problem without making any declarations. [Hint: change the identifiers for the masses from `mt`, `m1`, `m2`, `m3` everywhere in the code to `xmt`, `xm1`, `xm2`, `xm3`.]

4. Modify `mixture` so that the total density is part of the input and the percent by volume of the first component is unknown, then use the test data for `mixture` to test the modifications you have made.

1.12   Conditional Execution

1.12.1   Logical Expressions

Logical variables and expressions can only have the value .true. or .false. A logical expression can be formed by comparing numerical contents of memory using a number of relational operators.

FORTRAN operator	Mathematical symbol	Meaning
.lt.	$<$	(less than)
.le.	$\leq$	(less than or equal to)
.gt.	$>$	(greater than)
.ge.	$\geq$	(greater than or equal to)
.eq.	$=$	(equal to)
.ne.	$\neq$	(not equal to)

For example,

```
logical test1
test1 = 5 .lt. 3
```

At execution time, the line above stores a `.false.` in the memory location, called `test1`. Logical expressions can be combined by the logical conjunction `.and.`, disjunction `.or.`, and negation `.not.` operators. For example, the program `testlogic`,

```
program testlogic
implicit none
logical a, b, c, d, e, f, g, h
a = .true.
b = .false.
c =  a .and. b
```

```
d = a .or. b
e = a .and. .not. b
f = b .and. a
g = b .or. a
h = b .and. .not. a
write(6,*) a
write(6,*) b
write(6,*) c
write(6,*) d
write(6,*) e
write(6,*) f
write(6,*) g
write(6,*) h
end
```

produces the following output

```
T  F  F  T  T  F  T  F
```

Check the output against each expression and compare each with your expectations.

1.12.2 The Logical if Statement

It is important in programing to have the ability to execute parts of the code based on conditions that arise at execution time. The simplest form of conditional execution is the logical if statement:

```
if (logical expression) executable statement
```

This has to be written on one line. For example, the line below is used to stop the execution of a program if the condition is true.

```
if (n .eq. 0) stop 'Illegal value for a whole number'
```

The string `Illegal value for a whole number` will print on the screen during execution if $n = 0$ is true and the program execution will stop. If more than one statement is to be executed based on a particular condition, then the following syntax should be used:

```
if (logical expression) then
    statements
endif
```

The most general form of the if statement is as follows:

```
if (logical expression) then
    statements
```

```
elseif (logical expression) then
  statements
  :  :
else
 statements
  :  :
endif
```

The logical expressions are evaluated in sequence from top to bottom until one is found to be true. Then, the associated code is executed and the control jumps to the next statement after the endif. For example, the following program

```
program testif
implicit none
logical a, b, c
a = .true.
b = .false.
c =  a .and. b
c some examples of the general if statement
  if (a) then
     write(6,*) 'a is true'
   elseif (b) then
     write(6,*) 'b is true'
   elseif (c) then
     write(6,*) 'c is true'
   elseif (c) then
   endif
   write(6,*) 'execution continues from here'
c some examples of the general if statement
  if (c) then
     write(6,*) 'c is true'
   elseif (b) then
     write(6,*) 'b is true'
   elseif (a) then
     write(6,*) 'a is true'
   elseif (c) then
   endif
   write(6,*) 'done!'
  end
```

produces the output:

```
a is true
execution continues from here
a is true
done!
```

Change the program `testif` given above so that users can input data from
the keyboard for a and b. Then, test the program with several input values.
In particular, try inputting .true. for both a and b, and observe the behavior
of the program.

1.12.3   The Arithmetic if Statements

In older FORTRAN code, one may still find the arithmetic if statement where
the conditional execution is carried out based on the condition of an integer.
Integers can either be negative, positive, or zero. This construct must use
statement labels. Here is an example.

```
      program arithmif
      implicit none
      integer x
      write(6,*) 'Enter an integer'
      read(5,*) x
      if (x) 10, 20, 30
   10 write(6,*) 'x is negative'
      goto 40
   20 write(6,*) 'x is zero'
      goto 40
   30 write(6,*) 'x is positive'
   40 write(6,*) 'x = ',x
      stop
      end
```

Try the code above and test all three cases, then try removing the `goto`
statements.

1.12.4   The Computed goto Statement

Similarly, one may still find the computed goto construct in older code. Again,
the computed goto is a branching construct based on the condition of a positive
integer. Statement labels must be used here too. Here is an example.

```
      program compgoto
      implicit none
      integer x
      write(6,*) 'Enter an integer'
      read(5,*) x
      goto (10,20,30,40,50) x
    1 write(6,*) 'x is not good'
      goto 100
   10 write(6,*) 'x is 1'
      goto 100
```

```
20    write(6,*) 'x is 2'
      goto 100
30    write(6,*) 'x is 3'
      goto 100
40    write(6,*) 'x is 4'
      goto 100
50    write(6,*) 'x is 5'
100   write(6,*) 'x = ',x
      stop
      end
```

Try the code above and test all the cases, then try removing the goto statements. Both the arithmetic if and the computed goto have fallen out of use since the new FORTRAN 90 standard introduced the **case** statement.

1.12.5 The Case Statement (FORTRAN 90)

An example of the case construct is,

```
program casetest
implicit none
integer x
write(6,*) 'Enter an integer'
read(5,*) x
number: select case (x)
case default
write(6,*) 'illegal value for x'
  case (1)
    write(6,*) 'x is 1'
  case (2)
    write(6,*) 'x is 2'
  case (3)
    write(6,*) 'x is 3'
  case (4)
    write(6,*) 'x is 4'
end select number
cnd
```

Some compilers, like the version used to test most of the programs here, still do not support character types in the selector variable and do not support the **default** option. Most of the conditional execution the reader is likely to find in the programs in this text use the **if else endif** construct.

```
if (logical expression) then
  statements
else
  statements
endif
```

Exercises

1. Modify the program `mixture` given earlier, so that the $\rho_2 = \rho_3$ condition is tested and the program stops and prints an error message using a logical if construct.

2. Consider the $\rho_2 = \rho_3$ case in the the program `mixture` as a binary mixture. Show that if the densities ρ_1, ρ_2, the total mass of the mixture m_t, and the composition by volume of component $1, \chi_{v_1}$ are known, the system of equations

$$\begin{cases} \rho_1 &= \dfrac{m_1}{V_1} \\ \rho_2 &= \dfrac{m_2}{V_2} \\ \chi_{m_1} &= \dfrac{m_1\,100}{m_1 + m_2} \\ \chi_{v_1} &= \dfrac{V_1\,100}{V_1 + V_2} \\ \rho_t &= \dfrac{m_1 + m_2}{V_1 + V_2} \\ m_t &= m_1 + m_2, \end{cases}$$

is solved by

$$\rho_t = \frac{(\rho_1 - \rho_2)\,\chi_{v_1} + 100\,\rho_2}{100},$$

$$V_2 = \frac{m_t\,(100 - \chi_{v_1})}{(\rho_1 - \rho_2)\,\chi_{v_1} + 100\,\rho_2},$$

$$\chi_{m1} = \frac{100\rho_1\chi_{v_1}}{(\rho_1 - \rho_2)\,\chi_{v_1} + 100\,\rho_2}.$$

3. Type the program `mixture2` below, where the `if - then - else - endif` construct is used to treat the $\rho_2 = \rho_3$ case separately,

```
program mixture2
implicit none
real rho1,rho2,rho3,chim1,chiv1,rhot
real v1,v2,v3,m1,m2,m3,mt
real chim2,chiv2,chim3,chiv3
real rhot2,vt
real den,chim4
c This program calculates
c the masses, the volumes, the overall density,
c the percent by mass and by
c volume compositions of the other other two components
c for a tertiary mixture Given the densities of each component,
c the percent by mass composition for the first component, the
```

```
c percent by volume composition for the first component,
c and the total mass of the mixture.
c The program assumes densities are in g/cm^3 and that the total
c mass is in grams.
c The program check the input (in part) and treats
c the  rho2 .eq. rho3 as a binary mixture.
c i/0 section
      write(6,*) 'Enter the three densities in g/cm^3'
      read(5,*) rho1,rho2,rho3
      write(6,*) 'The densities you entered are:'
      write(6,*) rho1,rho2,rho3

c   check the condition rho2 .eq. rho3
      if (abs(rho2 - rho3) .lt. 1.e-6) then
c ++++++++++++++++ binary mixture case
      write(6,*) 'Interpreting the input as a binary mixture'
      write(6,*) 'Enter the composition by volume for the'
      write(6,*) 'first component with density ',rho1,' g/cm^3'
      read(5,*)  chiv1
      write(6,*) 'The mixture is', chiv1,' % by',
     & ' volume'
      write(6,*) 'Enter the total mass of the mixture'
      read(5,*) mt
      den =  (rho1-rho2)*chiv1 + 100.*rho2
      rhot = den/100.
      write(6,*) 'The density of this mixture must be',rhot
      vt = mt/rhot
c calculates the volumes and the masses of each component
      v2 = mt*(100. - chiv1)/den
      v1 = vt - v2
      m2 = rho2*v2
      m1 = rho1*v1
      write(6,*) 'Component 1 mass',m1,' g and volume',v1,' cm^3'
      write(6,*) 'Component 2 mass',m2,' g and volume',v2,' cm^3'
c calculates the compositions
      chim1 = 100.*rho1*chiv1/den
      chim2 = 100.*m2/mt
      chiv2 = 100.*v2/(v1+v2)
      write(6,*) 'Compositions by',
     & ' mass',chim1,chim2,(chim1+chim2)
      write(6,*) 'Compositions by',
     & ' volume',chiv1,chiv2,(chiv1+chiv2)
c checks the calculations of the total density
      rhot2 = mt/(v1+v2)
      write(6,*) 'Checking the density', rhot2,rhot
      write(6,*) 'Checking the volumes', vt,(v1+v2)
      write(6,*) 'Checking the masses', mt,(m1+m2)
      write(6,*) 'Checking all the densities', m1/v1,m2/v2
      else
c+++++++++++++++++++++++++ tertiary mixture case
      write(6,*) 'Enter the composition by mass and by volume of the'
      write(6,*) 'first component with density ',rho1,' g/cm^3'
      read(5,*) chim1,chiv1
      write(6,*) 'The mixture is', chim1,' % by mass', chiv1,' % by',
```

```
      & ' volume'
        write(6,*) 'Enter the total mass of the mixture'
        read(5,*) mt
c calculates the density of the mixture, and the total volume
        rhot = rho1*chiv1/chim1                      ! derive this equation
        write(6,*) 'The density of this mixture must be',rhot
        vt = mt/rhot
c calculates the volumes and the masses of each component
        v1 = mt*chim1/(rho1*100.)
        m1 = mt*chim1/(100.)
        v3 = mt*(1 - chim1/100. + rho2*chim1/(100.*rho1) -
      & rho2*chim1/(chiv1*rho1))/(rho3- rho2)
        v2 = mt*((chim1/chiv1 - chim1/100.)/rho1) - v3
        m2 = rho2*v2
        m3 = rho3*v3
        write(6,*) 'Component 1 mass',m1,' g and volume',v1,' cm^3'
        write(6,*) 'Component 2 mass',m2,' g and volume',v2,' cm^3'
        write(6,*) 'Component 3 mass',m3,' g and volume',v3,' cm^3'
c calculates the compositions
        chim2 = 100.*m2/mt
        chim3 = 100.*m3/mt
        chiv2 = 100.*v2/(v1+v2+v3)
        chiv3 = 100.*v3/(v1+v2+v3)
        write(6,*) 'Compositions by',
      & ' mass',chim1,chim2,chim3,(chim1+chim2+chim3)
        write(6,*) 'Compositions by',
      & ' volume',chiv1,chiv2,chiv3,(chiv1+chiv2+chiv3)
c checks the calculations of the total density
        rhot2 = mt/(v1+v2+v3)
        write(6,*) 'Checking the density', rhot2,rhot
        write(6,*) 'Checking the volumes', vt,(v1+v2+v3)
        write(6,*) 'Checking the masses', mt,(m1+m2+m3)
        write(6,*) 'Checking all the densities', m1/v1,m2/v2,m3/v3
      endif
      stop
      end
```

In the **if** statement, the logical expression is computed using the absolute value of the difference between ρ_2 and ρ_3 (obtained by calling the **abs** intrinsic function). When comparing real numbers, the outcome is more stable when one compares the absolute value of the difference against a number slightly larger than the machine precision. Note how the code lines between the **if (logical expression) then** and the **else**, and between the **else** line and the **endif** line are indented. It is good programing practice to use indentation to increase the readability of the code and find possible syntax errors or bugs. Compile the new version of **mixture** and test it with the following input.

```
0.990400016   0.789799988 0.789799988
26.7494144 22.5534
45.07
```

The program `mixture2` should produce the following output:

```
Interpreting the input as a binary mixture
Enter the composition by volume for the
first component with density   0.990400016 g/cm^3
The mixture is 26.7494144 % by volume
Enter the total mass of the mixture
The density of this mixture must be   0.843459308
Component 1 mass   14.1562538 g and volume   14.2934713 cm^3
Component 2 mass   30.9137459 g and volume   39.1412354 cm^3
Compositions by mass   31.409483   68.5905151   100.
Compositions by volume   26.7494144   73.2505875   100.
Checking the density   0.843459308   0.843459308
Checking the volumes   53.4347038   53.4347076
Checking the masses   45.0699997   45.0699997
Checking all the densities   0.990399957   0.789799929
```

4. Modify `mixture2` as you did earlier, so that the density of the mixture is part of the input and the compositions are calculated instead. Make sure you modify both cases.

5. Perform additional tests, for example, check for the $\rho_1 = \rho_2$ or $\rho_1 = \rho_3$ condition.

1.13 Loops

1.13.1 Counter Controlled do-loop

Repeated execution of identical lines of code can be made most efficient with loops. FORTRAN 77 has two loop constructs, called the counter controlled do-loop and the goto loop. The form of the counter controlled do-loop is

```
do (statement label) counter = integer-value1 , integer-value2, step
     statement
(statement label)  continue
```

where `counter`, `integer-value1`, `integer-value2`, and `step` must be integers. The value of `step` can be negative, in which case the statements inside the loop will be executed only if the value in `integer-value1` is greater than that in `integer-value2`. The specification of the step size can be omitted, in which case the value in `counter` is incremented by one each time control goes back to the line containing the `do` statement. The example below is the typical FORTRAN 77 construct.

```
      program testloop
      integer i
      do 10 i = 1, 10
         write(*,*) 'i =', i
 10   continue
      end
```

Try running the code above, then play with the specification of the step. For example, try running the code with a do 10 i = 1, 10,2 statement and a do 10 i = 10, 1,-1 statement.

1.13.2   Do–enddo Loops (FORTRAN 90)

The FORTRAN 90 standard eliminated the need of statement labels in the do–enddo loop construct.

```
      do counter = integer-value1 , integer-value2,step
         statements
      enddo
```

Additionally, it is possible to exit conditionally a do–enddo loop without employing a goto statement. This is accomplished with the exit statement,

```
      do counter = integer-value1 , integer-value2,step
         if (logical expression) exit
         statements
      enddo
```

Note that with the do–enddo loop construct it is still possible to change the behavior of the counter as with the older do–continue construct. Try the code below,

```
      program testloop
      integer i
      do  i = 10, 1,-2
         write(*,*) 'i =', i
      enddo
      end
```

then try inserting an if (i .lt. 5) exit inside the loop and see how the program behaves. The majority of the programs in this text use the more modern do–enddo construct, but older examples are common in the technical literature; and it is important that the reader is familiar with the majority of the loop constructs.

1.14   Intrinsic Functions

The FORTRAN language contains a number of mathematical functions used to simplify computations. Among these, the most common algebraic and transcendental functions can be found.

Function Call	Description	Argument type	value type
abs(x)	absolute value	real	real
min(x1,x2,x3)	minimum value among x1,x2,x3	real or integer	real or integer
max(x1,x2,x3)	maximum value among x1,x2,x3	real or integer	real or integer
sqrt(x)	square root of x	real	real
sin(x)	sine of x radians	real	real
asin(x)	arcsine of x, in radians	real	real
cos(x)	cosine of x radians	real	real
tan(x)	tangent of x (radians)	real	real
atan(x)	arctangent of x, in radians (-pi/2 to + pi/2 range)	real	real
atan2(y,x)	arctangent of x, in radians (-pi to + pi range)	real*8	real*8
exp(x)	exponential (natural base)	real	real
log(x)	logarithm (natural base)	real	real
log10(x)	logarithm (base 10)	real	real
int(x)	integer part of x	real	integer
float(x)	convert x to real	integer	real

Many of these mathematical functions have a double precision version, for example, these functions are called inside an assignment statement, such as y = asin(x).

Function Call	Description	Argument type	value type
dabs(x)	absolute value	real*8	real*8
dsqrt(x)	square root of x	real*8	real*8
dsin(x)	arcsine of x in radians	real*8	real*8
dasin(x)	arcsine of x in radians	real*8	real*8
dcos(x)	cosine of x radians	real*8	real*8
dtan(x)	tangent of x (radians)	real*8	real*8
datan(x)	arctangent of x, in radians (-pi/2 to + pi/2 range)	real*8	real*8
datan2(y,x)	arctangent of x, in radians (-pi to + pi range)	real*8	real*8
dexp(x)	exponential (natural base)	real*8	real*8
dlog(x)	logarithm (natural base)	real*8	real*8
dlog10(x)	logarithm (base 10)	real*8	real*8
dfloat(x)	convert x to real	integer	real*8

1.15 User-Defined Functions

The pseudo code showing the proper syntax for an external user-defined function is as follows,

```
program program-name
declarations
  :   :
```

```
        statements
        :  :
        end
        (type-identifier) function function-name(argument-list)
        declarations
          statements
          function-name = expression
        end function function-name
```

The type identifier is necessary if implicit typing is not used. The following example demonstrates a few features of external user-defined functions in FORTRAN.

```
        program testfunction
        implicit none
        real*8 x,y,z,distance
        real*8 myfunction
c This program calculates the distance from the origin
c to a point with coordinates x,y,z
        write(6,*) 'Enter the coordinates of a point'
        read(5,*) x,y,z
        write(6,*) 'calling myfunction  with', x,y,z
        distance = myfunction(x,y,z)
        write(6,*) 'distance from the origin:',distance
        write(6,*) 'myfunction called with', x,y,z
        end
        real*8 function myfunction(xin,yin,zin)
        implicit none
        real*8 xin,yin,zin
         write(6,*) 'myfunction called with', xin,yin,zin
         myfunction = sqrt(xin*xin + yin*yin + zin*zin)
        end function myfunction
```

The user-defined external function `myfunction` calculates the distance, as a function of its calling parameters x, y, z

$$f(x, y, z) = \sqrt{x^2 + y^2 + z^2}.$$

With no implicit typing, the function identifier has to be declared in the main program declaration segment. The parameters in the calling statement `distance = myfunction(x,y,z)` do not have the same identifier as the function declaration statement `real*8 function myfunction(xin,yin,zin)`. The order of the calling parameters is crucial, of course. Running the program with the input,

1 1 1

produces the following output,

```
calling myfunction  with  1.  1.  1.
myfunction called with  1.  1.  1.
distance from the origin:  1.73205081
myfunction called with  1.  1.  1.
```

Subprograms, in general, have been proposed to deal with the complexity of programs when a real engineering or scientific problem is solved. With user-defined functions and subroutines, it is possible to break down the overall task into a set of smaller tasks, the combinations of which yield the solution of the problem. The benefit of subprograms is that they can be turned into portable modules (canned) and reused in many programs. Unfortunately, there are two issues that can potentially decrease the portability of a subprogram in FORTRAN. The first one is the issue of passing the calling parameters by reference. Consider the following modification to the program testfunction above,

```
program testfunction2
implicit none
real*8 x,y,z,distance
real*8 myfunction
write(6,*) 'Enter the coordinates of a point'
read(5,*) x,y,z
write(6,*) 'calling myfunction  with', x,y,z
distance = myfunction(x,y,z)
write(6,*) 'distance from the origin:',distance
write(6,*) 'myfunction called with', x,y,z
end
real*8 function myfunction(xin,yin,zin)
implicit none
real*8 xin,yin,zin
 write(6,*) 'myfunction called with', xin,yin,zin
 xin = 22.d0
 myfunction = sqrt(xin*xin + yin*yin + zin*zin)
end function myfunction
```

All we did here is add an innocent-looking assignment statement, xin = 22.d0. The value of the function should now change, but a surprise awaits. Running the program with the same input as before now produces the following output

```
calling myfunction  with  1.  1.  1.
myfunction called with  1.  1.  1.
distance from the origin:  22.0454077
myfunction called with  22.  1.  1.
```

To our horror, the value of the calling parameter is changed inside the function execution, and that change is reflected in the main function, as confirmed in the last line of output. In fact, the two identifiers x and xin, even if different, point to the same memory address, and that memory location can become compromised inside a function. This feature of FORTRAN subprogram is known as *passing by reference*. The second practice that can potentially decrease the portability of a subprogram in FORTRAN is the use of common blocks. Variables in common blocks are accessible by the main program and any subprogram that contains a matching common statement. The use of common variables is essentially another way of passing arguments to a subprogram. Again, the variables in common blocks are passed by reference. When porting a subprogram with common statements to another program, substantial revisions and extensive testing are usually needed. The C language did away with calling by reference. Pointers are used instead when passing by reference is needed in C. However, the greatest leap forward in modular programing languages is object-oriented programing, like C++, C#, Java, and the like. The main feature that makes these higher level languages object oriented is the availability of classes, user-defined data types with embedded user-defined functionality. Only classes are truly portable subunits of code.

1.16   Subroutines

In the last example, as a side effect, the function returns two values to the main program, the value of the function itself and the value of the changed argument. The proper subprogram to use when more than one value needs to be returned to the main program is a subroutine, a true subprogram that is able to perform many tasks and can be used to change all of its calling parameters. Here is the syntax for a subroutine code and its call

```
program program-name
declarations
:   :
statements
:   :
call subroutine-name(argument-list)
:   :
statements
:   :
end
subroutine subroutine-name(argument-list)
declarations
:   :
```

```
statements
:   :
return
end
```

In the example below, the calculation of the distance of a point from the origin is handled by the subroutine get_distance. The subroutine output is now part of the argument list as well. It is always a good idea to document in the comment lines, the list of calling parameters that are output and those that are input.

```
program testsubroutine
implicit none
real*8 x,y,z,dis,sdis
write(6,*) 'Enter the coordinates of a point'
read(5,*) x,y,z
write(6,*) 'calling get_distance  with', x,y,z
call get_distance(x,y,z,dis,sdis)
write(6,*) 'distance from the origin:',dis
write(6,*) 'distance from the origin squared:',sdis
write(6,*) 'get_distance called with', x,y,z
end
subroutine get_distance(xin,yin,zin,dis,sdis)
implicit none
real*8 xin,yin,zin,dis,sdis
 write(6,*) 'get_distance called with', xin,yin,zin
 sdis = xin*xin + yin*yin + zin*zin
 dis = sqrt(sdis)
return
end
```

Running this program with 1 1 1 produces the following output,

```
calling get_distance  with  1.  1.  1.
get-distance called with  1.  1.  1.
distance from the origin:  1.73205081
distance from the origin squared:  3.
get_distance called with  1.  1.  1.
```

Here is the same program using the common statement

```
program testsubroutine2
implicit none
real*8 x,y,z,dis,sdis
common x,y,z,dis,sdis
```

```
write(6,*) 'Enter the coordinates of a point'
read(5,*) x,y,z
write(6,*) 'calling get_distance  with', x,y,z
call get_distance
write(6,*) 'distance from the origin:',dis
write(6,*) 'distance from the origin squared:',sdis
write(6,*) 'get_distance called with', x,y,z
end
subroutine get_distance
implicit none
real*8 xin,yin,zin,dis,sdis
common  xin,yin,zin,dis,sdis
  write(6,*) 'get-distance called with', xin,yin,zin
  sdis = xin*xin + yin*yin + zin*zin
  dis = sqrt(sdis)
return
end
```

1.17   Numerical Derivatives

The reader will recall from introductory calculus that the derivative of a function $f(x)$ is defined as a limit

$$\frac{df}{dx} = \lim_{h \to 0} \frac{f(x+h) - f(x)}{h}$$

where the R.H.S. is understood as the value the sequence

$$\frac{f(x+h_i) - f(x)}{h_i}$$

converges to, if the sequence $h_1, h_2, h_3, \ldots$ converges to zero. The example that follows is designed to explore the numerical issues associated with the computations of derivatives by finite difference methods. The truncation error derives from the fact that the finite difference sequence is ended at some power, in practice. The rounding error is the consequence of using a finite precision in the decimal representation of real numbers. In the example below, the numerical derivative of a function is compared against the analytical value. The function $x^{5/2}$ is used, and the numerical derivative of this function is calculated at $x = 2$.

In the program `derivative` that follows, `fx,fxph,dfx` are the value of $f(x), f(x + h_i)$, and the analytical derivative of $f(x)$, respectively, whereas `ndfx` stores the values of the numerical derivative, and `error` stores values of

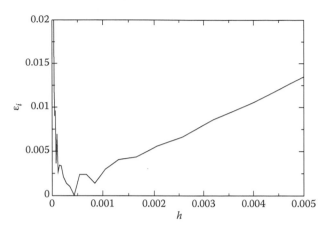

FIGURE 1.1
The numerical error of the first order finite difference numerical derivative.

the error sequence ϵ_i generated from

$$\epsilon_i = \left| \frac{f(x + h_i) - f(x)}{h_i} - \frac{df}{dx} \right|$$

```
program derivative
implicit none
integer i
real x,h,fx,fxph,dfx
real ndfx, error
x = 2
h = 0.005
do   i = 1,100
  fx = x**(2.5)
  dfx = 2.5*x**(1.5)
  fxph = (x + h)**(2.5)
  ndfx = (fxph - fx)/h
  error = abs(ndfx - dfx)
  write(6,*) i,h,error
  h = 0.8*h
  if (error .gt. 0.02) exit
enddo
end
```

The program is used to produce the graph in Figure 1.1, where the error ϵ_i is plotted as a function of h. In theory, ϵ_i in Figure 1.1 should converge to zero. In practice, however, the values of ϵ_i decrease for a while, as h decreases from right to left. In the region $(0.0005 < h < 0.005)$, the largest contribution to ϵ_i is the truncation error. At values of h smaller than 0.0005, ϵ_i increases rapidly because of the rounding error. There is a great deal of loss of precision in the computed difference (`fxph` - `fx`) when `h` becomes small, since (`fxph` - `fx`) becomes a small difference between two relatively large and relatively close numbers.

1.18   The Extended Trapezoid Rule to Evaluate Integrals

The second example deals with numerical integration. Consider the integral

$$\int_a^b f(x)dx,$$

its numerical value is the area under the graph of $f(x)$ between the points a and b on the x-axis. Let us first consider approximating the area in question with a single trapezoid

$$\int_a^b f(x)dx \approx (b-a)\,f(a) + \frac{1}{2}\,(b-a)\,(f(b) - f(a))\,.$$

The first term is the area under the parallelogram that is $(b-a)$ wide and $f(a)$ high. The second term is the area of the right triangle that has as one side, the width $(b-a)$ and as the other side, the height $(f(b) - f(a))$. With trivial algebra, the R.H.S. becomes,

$$\int_a^b f(x)dx \approx \frac{1}{2}\,(b-a)\,(f(b) + f(a))\,.$$

To generate a convergent sequence, the so-called extended trapezoid algorithm is developed using the result in the last equation. Imagine subdividing the integration range into n identical subintervals of equal width h. Then, the integral is approximated as the sum of the area of all the trapezoids. Let $x_1, x_2, \ldots$ be a n term sequence of values of x with $x_1 = a$ and $x_{n+1} = b$, such that,

$$x_{i+1} = x_i + h.$$

The area of the trapezoid between x_i and x_{i+1} is,

$$\frac{h}{2}\,(f(x_{i+1}) + f(x_i))\,.$$

So, the integral calculated from the n-segment partition of the range is

$$\int_a^b f(x)dx \approx \frac{h}{2}\sum_{i=1}^{n}(f(x_{i+1}) + f(x_i))\,.$$

We could translate this formula directly into code, but there is a minor improvement that is very useful when $f(x)$ is particularly expensive to calculate. Consider the $n = 2$ case,

$$\int_a^b f(x)dx \approx \frac{h}{2}\left(f(a+h) + f(a)\right) + \frac{h}{2}\left(f(b) + f(a+h)\right),$$

where $h = (b-a)/2$. We can combine the two terms on the R.H.S. into a single term with some trivial algebra,

$$\int_a^b f(x)dx \approx \frac{h}{2}\left(f(b) + 2f(a+h) + f(a)\right).$$

Similarly, the $n = 3$ case gives,

$$\int_a^b f(x)dx \approx \frac{h}{2}\left(f(b) + 2f(a+h) + 2f(a+2h) + f(a)\right),$$

where $h = (b-a)/3$. This process generalizes into an algorithm known as the extended trapezoid rule

$$\int_a^b f(x)dx \approx h \sum_{i=1}^{n+1} w_i\, f(x_i),$$

where w_i is either $1/2$ (when $i = 1$ or $n+1$) or 1 otherwise.

In the example that follows, the integral

$$\int_0^2 6\,x^5\,dx = 64,$$

is calculated numerically using the extended trapezoid algorithm for a chosen value of n.

```fortran
program trapezoid
implicit none
integer i,n,j
real h,x,area,sum,fx
real a,b,w,error
  n = 10
 do j = 1, 100
  n = n + 10
  a = 0.0
  b = 2.0
  sum = 0.0
  h = (b-a)/float(n)
  x = a
  do  i = 1,n+1
```

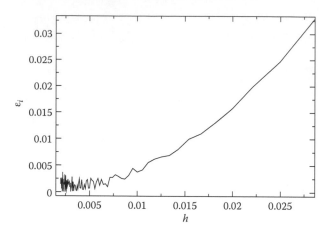

FIGURE 1.2
The numerical error of the composite trapezoid rule.

```
      w = 1.0
      if (i .eq. 1 .or. i .eq. n+1) w = 0.5
      fx = 6.*x**5
      sum = sum + h*w*fx
      x = x + h
   enddo
     error = abs(sum - 64)
     write(6,*) n,h,sum,error
   enddo
   end
```

Most variables have obvious meaning. The reader should focus on how the memory content **sum** is updated inside the loop. The reader should also note that **sum** is initialized to zero just before the execution of the loop begins.

The trapezoid rule is much less susceptible to rounding error and is far more stable than the central formula for derivatives. In Figure 1.2, the graph of the error as a function of h fluctuates at small values of h, but it never rises abruptly as it does in Figure 1.1.

Exercises

1. Many numerical analysis textbooks claim that the central difference formula

    $$\frac{f(x + h_i) - f(x - h_i)}{2\,h_i}$$

    produces a much better behaved estimate of the derivative and reaches the highest precision possible at significantly larger values of h. Verify this statement by modifying the **derivative** program above. Initialize h at 0.1.

2. In the program **derivative** change both declaration lines from **real** to **real*8** and see what happens.

3. Inside **trapezoid**, change the **real** declarations in the program to **real*8**, change the function **float** (used to convert the integer **n** into a floating number) to **dfloat** in the calculation of **h**, increase the outer loop counter from 100 to 500, rerun, and observe the behavior of the error. How does the rounding error behave?

4. Modify the program **trapezoid** so that the integral

$$\int_0^{1/2} \sin(x) \, dx$$

is computed. Use a $n = 10,000$ quadrature. The analytical value of the integral can be calculated with $1 - \cos(0.5)$. Compare the accuracy of the integral with single precision and double precision. With all real variables declared as double precision, use calls to **sin(x)** and compare the behavior of the algorithm with calls to the **dsin(x)** intrinsic function. When using double precision, the analytical value of the integral (used to obtain the numerical error) should be calculated with $- \text{dcos}(0.5d0) + 1.0d0$, where the argument of the double precision cosine function $(0.5d0)$ is an exponential format that declares 0.5 as a double precision floating point number. $(0.5d0)$ is the double precision version of the $(0.5e0)$. A call such as **dcos(0.5)** will cause most compilers to complain about the argument type. With a $n = 10,000$ quadrature, the numerical error should drop by two orders of magnitude by changing **sin(x)** into **dsin(x)** for the evaluation of the integrand, if you did everything correctly.

5. Perform the same test between **sin(x)** and **dsin(x)** using the central difference formula to evaluate the derivative.

2

Basics of Classical Dynamics

2.1 Introduction

For a chemist, the typical first introduction to the Hamiltonian function takes place in an undergraduate class, perhaps physical chemistry, in the context of quantum mechanics. Historically, however, the concepts are introduced into classical mechanics first, and are linked to two principles more profound than Newton's three laws. The two principles are the **conservation of energy** principle for **autonomous** mechanical systems and the **least action** principle. The latter is expressed typically using a Legendre transform of the Hamiltonian function, called the **Lagrangian**. In formulating a quantum theory, it is necessary to embed Heisenberg's uncertainty principle into the equations of motion. Feynman derives the path integral formalism by departing from the least action principle. Learning about the path integral as the element of the time evolution Lie group is easier if the reader is familiar with the principle of least action and the Lagrangian concept. With the exception of some special applications in quantum field theory, path integrals can be formulated using configuration space and the Lagrangian function. The conservation of energy principle has survived the physics revolution of the last century, and generalizes to other situations, such as the laws of thermodynamics.

In this chapter, after introducing the most frequently used dynamic variables of classical physics, we construct two important **state functions**: the lagrangian and the Hamiltonian. From these, we formulate the least action principle and the equations of motion in a general form. Analytic solutions of the equations of motion are only attainable for a few simple systems. Among these are the two-body problem with central isotropic potential, where the trajectory can be written with a single integral. The rigid rotor is a simple example that allows us to introduce holonomic constraints. Holonomic constraints occupy us a great deal in later chapters, therefore it is very beneficial to see a simple introduction of them here. Numerical integration methods and a class of symplectic integrators for Hamilton's equations are introduced. Finally, we discuss dissipative systems and the Fourier transform of dynamic variables. From the study of dissipative systems, we obtain an efficient minimization algorithm used to explore multidimensional potential energy surfaces. From the Fourier transform, we obtain the "characteristic frequencies of the system."

2.2   Some Important Variables of Classical Physics

For a system of N particles in three-dimensional Euclidean space, the vector $\mathbf{x}(t)$ has $3N$ components,

$$\mathbf{x}(t) = \{x_1(t), y_1(t), z_1(t), \ldots, z_N(t)\}. \tag{2.1}$$

We are using Cartesian coordinates, and the subscript is the particle number. The derivative of the position function is the velocity $\mathbf{v}(t)$,

$$\mathbf{v}(t) = \frac{d\mathbf{x}}{dt} = \dot{\mathbf{x}}, \tag{2.2}$$

and the second derivative of the position is the acceleration $\mathbf{a}(t)$,

$$\mathbf{a}(t) = \frac{d\mathbf{v}}{dt} = \frac{d^2\mathbf{x}}{dt^2} = \ddot{\mathbf{x}}, \tag{2.3}$$

with SI units of m/s and m/s², respectively. The time derivative applies for each component. In Equations 2.1 and 2.2, we introduce the dot notation. Each dot on top of a physical quantity symbolizes a time derivative. This is a standard notation used in many classical physics texts. Even for a single particle in one dimension, the position, the velocity, and the acceleration are **vector** quantities, as their signs indicate the direction of motion (either positive or negative); e.g., a velocity of -5.3 m/s indicates that a particle in one dimension is moving in the negative direction with a speed of 5.3 m/s. The size of the velocity vector for a particle in three-dimensional space, for example, is the speed v:

$$v_i = |\mathbf{v}| = \sqrt{\dot{x}_i^2 + \dot{y}_i^2 + \dot{z}_i^2}. \tag{2.4}$$

Another important vector quantity is the momentum. In Cartesian coordinates, in the absence of external force fields, and for velocity independent potentials, the momentum takes the following simple form:

$$\mathbf{p}(t) = \{m_1\dot{x}_1(t), m_1\dot{y}_1(t), m_1\dot{z}_1(t), \ldots, m_N\dot{z}_N(t)\}. \tag{2.5}$$

The kinetic energy K is a useful scalar quantity,

$$K = \frac{1}{2}\sum_{i=1}^{N} m_i \left(\dot{x}_i^2 + \dot{y}_i^2 + \dot{z}_i^2\right). \tag{2.6}$$

The potential energy (the energy available to the body or system of bodies by virtue of their position) is the work performed on the system to bring all its particles from an infinite separation limit to its present configuration $\mathbf{x}(t)$,

$$V(\mathbf{x}) = -\int dx_1 \int dy_1 \cdots \int dz_N F(\mathbf{x}) = -\int d\mathbf{x}\, F(\mathbf{x}). \tag{2.7}$$

For example, consider two particles with charges q and q'. The force between them only depends on the separation distance r, and follows an inverse square law:

$$F(r) = \frac{qq'}{4\pi\epsilon_0 r^2}.$$ (2.8)

The Coulomb potential for two charges at distance r is,

$$V(r) = \int_r^{+\infty} \frac{qq'}{4\pi\epsilon_0 s^2} ds = \frac{qq'}{4\pi\epsilon_0 r}.$$ (2.9)

In three dimensions, the angular momentum $\mathbf{L}$ is an important quantity,

$$\mathbf{L} = \mathbf{x} \times \mathbf{p}.$$ (2.10)

This vector is useful to describe rotational motion for a mechanical system. The angular momentum vector is perpendicular to the plane of procession and is related to the rotational energy of the mechanical system. The SI units for momentum are kg m/s; those for the kinetic and potential energy are Joules $(1 \text{ J} = 1 \text{ kg m}^2/\text{s}^2)$.

2.3 The Lagrangian and the Hamiltonian in Euclidean Spaces

Consider a particle of mass m in three-dimensional space subjected to a potential $V(q_1, q_2, q_3)$, where q_1, q_2, q_3 are general coordinates. The Lagrangian is a function of the generalized velocity and coordinate

$$\mathcal{L}(\dot{q}_1, \dot{q}_2, \dot{q}_3, q_1, q_2, q_3) = K(\dot{q}_1, \dot{q}_2, \dot{q}_3) - V(q_1, q_2, q_3),$$ (2.11)

$$\mathcal{L}(\dot{q}_1, \dot{q}_2, \dot{q}_3, q_1, q_2, q_3) = \frac{1}{2}m\left(\dot{q}_1^2 + \dot{q}_2^2 + \dot{q}_3^2\right) - V(q_1, q_2, q_3).$$ (2.12)

We are assuming the potential energy depends only on the coordinates. The Lagrangian then, is the difference between the kinetic and potential energy. Later on, we will see that the Lagrangian enters in the formulation of the least action principle and the derivation of the Euler–Lagrange equations of motion. The Lagrangian is the center piece of classical mechanics and is an important theoretical tool. Hamilton's contribution to classical mechanics is, however, equally important and begins with the formulation of the Legendre transform of the Lagrangian from a function of velocities and coordinates, to a function of coordinates, and the conjugate **momenta**.

The total differential of $\mathcal{L}$ for a particle in three-dimensional space is:

$$d\mathcal{L} = \frac{\partial \mathcal{L}}{\partial \dot{q}_1} d\dot{q}_1 + \frac{\partial \mathcal{L}}{\partial \dot{q}_2} d\dot{q}_2 + \frac{\partial \mathcal{L}}{\partial \dot{q}_3} d\dot{q}_3 + \frac{\partial \mathcal{L}}{\partial q_1} dq_1 + \frac{\partial \mathcal{L}}{\partial q_2} dq_2 + \frac{\partial \mathcal{L}}{\partial q_3} dq_3, \quad (2.13)$$

$$d\mathcal{L} = p_1 d\dot{q}_1 + p_2 d\dot{q}_2 + p_3 d\dot{q}_3 + \frac{\partial \mathcal{L}}{\partial q_1} dq_1 + \frac{\partial \mathcal{L}}{\partial q_2} dq_2 + \frac{\partial \mathcal{L}}{\partial q_3} dq_3. \quad (2.14)$$

Here, we define the canonical momentum

$$p_i = \frac{\partial \mathcal{L}}{\partial \dot{q}_i}. \quad (2.15)$$

p_i is also known as the conjugate momentum of coordinate q_i. The beauty of Equations 2.13 through 2.15 is that they are independent of choice of coordinates, can be easily extended to systems of many particles, and are, therefore, particularly useful to generalize theories based on classical mechanics. Let us define a new function $H(p_1, p_2, p_3, q_1, q_2, q_3)$ such that

$$H = -\mathcal{L} + p_1 \dot{q}_1 + p_2 \dot{q}_2 + p_3 \dot{q}_3. \quad (2.16)$$

That is, H is the Legendre transform of the Lagrangian. Then, using the product rule

$$dH = -d\mathcal{L} + p_1 d\dot{q}_1 + p_2 d\dot{q}_2 + p_3 d\dot{q}_3 + \dot{q}_1 dp_1 + \dot{q}_2 dp_2 + \dot{q}_3 dp_3. \quad (2.17)$$

Inserting this expression into Equation 2.14 and canceling yields,

$$dH = \dot{q}_1 dp_1 + \dot{q}_2 dp_2 + \dot{q}_3 dp_3 - \frac{\partial \mathcal{L}}{\partial q_1} dq_1 - \frac{\partial \mathcal{L}}{\partial q_2} dq_2 - \frac{\partial \mathcal{L}}{\partial q_3} dq_3. \quad (2.18)$$

The formalism that we are briefly touching upon leads to Hamilton's equations of motion, another useful theoretical tool in classical mechanics. Continuing with the single particle example, we have

$$p_i = \frac{\partial \mathcal{L}}{\partial \dot{q}_i} = m\dot{q}_i, \quad (2.19)$$

or

$$\dot{q}_i = \frac{p_i}{m}, \quad (2.20)$$

and

$$-\frac{\partial \mathcal{L}}{\partial q_i} = \frac{\partial V}{\partial q_i}. \quad (2.21)$$

$$dH = \frac{1}{m}(p_1 dp_1 + p_2 dp_2 + p_3 dp_3) + \frac{\partial V}{\partial q_1} dq_1 + \frac{\partial V}{\partial q_2} dq_2 + \frac{\partial V}{\partial q_3} dq_3, \quad (2.22)$$

$$dH = \frac{1}{m}(p_1 dp_1 + p_2 dp_2 + p_3 dp_3) + dV. \quad (2.23)$$

Direct integration of this **state function** gives

$$H = \frac{1}{2m} \left(p_1^2 + p_2^2 + p_3^2 \right) + V = K + V. \tag{2.24}$$

The result, $H = K + V$, with K expressed as a function of the canonical momenta, is quite general even though we derive it for a single particle. The Hamiltonian H is the total energy. The Hamiltonian is an important object used throughout the text. The convenience of the Hamiltonian is easily appreciated if one considers that phase space (the collection of all coordinates and conjugate momenta, also known Liouville space) has special properties. The phase space volume differential

$$dp_1 dp_2 \cdots dp_{3N} dx_1 dy_1 \cdots dz_N, \tag{2.25}$$

has the special property of being an invariant quantity under coordinate changes. This invariance, together with the energy conservation principle for autonomous systems, made the use of Hamiltonians appealing for the development of nineteenth century statistical mechanics theories, such as the Gibbs ensemble and the kinetic theory of matter. Theoretical physicists spend a great deal of time dealing with the subtleties of coordinate changes. This is especially true for modern physics in general relativity and string theory. Therefore, it should not be surprising that one aims to derive expressions that are generally valid in any system of coordinates. We will look at phase space in some detail in this chapter; however, the proof that the Liouville space volume element is invariant under coordinate change is in Chapter 11, after some of the powerful tools of differential geometry and tensor analysis are introduced.

Exercises

1. Consider the one-dimensional harmonic oscillator, i.e., a particle of mass m, attached to a spring with constant k. The Lagrangian is a function of two independent variables, x and $\dot{x}$,

$$\mathcal{L} = \frac{1}{2} m \dot{x}^2 - \frac{1}{2} k x^2. \tag{2.26}$$

Evaluate

$$\frac{\partial \mathcal{L}}{\partial \dot{x}}, \quad \frac{\partial \mathcal{L}}{\partial x}, \tag{2.27}$$

where the symbol ∂ stands for partial derivative. The partial derivative is defined in a straightforward way. Suppose f is a function of x and y, then

$$\frac{\partial f}{\partial x} = \lim_{h \to 0} \frac{f(x+h, y) - f(x, y)}{h}. \tag{2.28}$$

The partial derivative of $f(x, y)$ with respect to x is the slope of the tangent of the graph of $f(x, y)$ obtained by keeping y constant evaluated at the point x, y.

2. The total derivative of a function f of x and y represents the total (infinitesimal) change to f that results from changing both x and y infinitesimally,

$$df = \frac{\partial f}{\partial x}\, dx + \frac{\partial f}{\partial y}\, dy. \qquad (2.29)$$

Find an expression for $d\mathcal{L}$ using Equation 2.26.

3. A function f of x and y is analytic (also known as a state function) if the mixed second partials satisfy Euler's condition,

$$\frac{\partial^2 f}{\partial y \partial x} = \frac{\partial^2 f}{\partial x \partial y}, \qquad (2.30)$$

everywhere in their domain. Namely, the second order mixed partials can be evaluated in any order. Show that $\mathcal{L}$ in Equation 2.26 is analytic. This property of $\mathcal{L}$ is of fundamental importance for the proof of the existence and uniqueness of the equations of motion.

4. Define $H = -\mathcal{L} + p\dot{x}$, and show that for the harmonic oscillator, the Hamiltonian total derivative becomes

$$dH = kx dx + \frac{p}{m} dp. \qquad (2.31)$$

Show that dH is analytic.

5. Given that

$$dH = kx dx + \frac{p}{m} dp, \qquad (2.32)$$

is analytic (i.e., a state function), the line integral in phase space

$$\int_C dH, \qquad (2.33)$$

is path independent. Show that the Hamiltonian from the previous problem becomes

$$H(p, x) = \frac{1}{2m} p^2 + \frac{1}{2} kx^2. \qquad (2.34)$$

2.4   The Least Action Principle and the Equations of Motion

The laws of motion in classical Newtonian mechanics are not the most fundamental axioms of physics. In fact, the "law" label is really a misnomer. The laws of motion can be derived from a more fundamental principle, called the least action principle, which states:

The trajectory $\mathbf{q}(t)$ *for a mechanical system is one for which the action*

$$I\left[\dot{\mathbf{q}}(t), \mathbf{q}(t)\right] = \int_0^t \mathcal{L}\left[\dot{\mathbf{q}}(u), \mathbf{q}(u)\right] du, \tag{2.35}$$

is minimized.

The minimum of the functional $I\left[\dot{\mathbf{q}}(t), \mathbf{q}(t)\right]$ is the trajectory $\mathbf{q}(t)$ that satisfies the following set of differential equations:

$$\frac{d}{dt}\frac{\partial \mathcal{L}}{\partial \dot{q}_i} - \frac{\partial \mathcal{L}}{\partial q_i} = 0. \tag{2.36}$$

These are the Euler–Lagrange equations, typically a coupled set of second order differential equations. Rather than proving this far-reaching statement, we show that for a single particle in Cartesian coordinates, one recovers Newton's equations.

The Lagrangian in this case is simply,

$$\mathcal{L}\left(\dot{x}, \dot{y}, \dot{z}, x, y, z\right) = \frac{1}{2}m\left(\dot{x}^2 + \dot{y}^2 + \dot{z}^2\right) - V\left(x, y, z\right). \tag{2.37}$$

The first term of the Euler–Lagrange equations boils down to

$$\frac{d}{dt}\frac{\partial \mathcal{L}}{\partial \dot{q}_1} = \frac{d}{dt}\frac{\partial \mathcal{L}}{\partial \dot{x}} = m\ddot{x}, \tag{2.38}$$

whereas the second term becomes

$$\frac{\partial \mathcal{L}}{\partial q_1} = -\frac{\partial V}{\partial x} = F_x. \tag{2.39}$$

Therefore, one obtains Newton's equations

$$m\ddot{x} - F_x = 0. \tag{2.40}$$

Equivalent expressions apply for the y and z components. The action integral $I\left[\dot{\mathbf{q}}(t), \mathbf{q}(t)\right]$, with appropriate modifications, is a possible point of departure for the development of a quantum theory.

Imagine a particle of mass m attached to a spring, with spring constant k. The potential function is obtained from Hooke's law:

$$V = \frac{1}{2}kx^2. \tag{2.41}$$

Therefore, Newton's equation becomes

$$-kx = m\frac{d^2x}{dt^2}. \tag{2.42}$$

Given the initial conditions:

$$x(t = 0) = x_0, \qquad v(t = 0) = v_0, \tag{2.43}$$

we may integrate the differential equation and find a unique solution for the position as a function of time. The trial function

$$x(t) = A \, \cos(\omega t) + B \, \sin(\omega t), \tag{2.44}$$

is substituted into the differential equation, and this determines the parameter ω,

$$\frac{d^2 x}{dt^2} = -\omega^2 A \cos(\omega t) - \omega^2 B \sin(\omega t) = -\omega^2 x. \tag{2.45}$$

Therefore,

$$-kx = -m\omega^2 x, \tag{2.46}$$

or

$$\omega = \sqrt{\frac{k}{m}}. \tag{2.47}$$

ω is known as the natural frequency of the oscillator. Setting $t = 0$ in the trial function for x yields $x_0 = A$, and setting $t = 0$ in dx/dt yields $B = v_0/\omega$.

Finally, the general solution is:

$$x(t) = x_0 \cos(\omega t) + \frac{v_0}{\omega} \sin(\omega t). \tag{2.48}$$

2.5   The Two-Body Problem with Central Potential

The center of mass of N bodies located by the position vector $\mathbf{r}_i$ and with mass m_i, $i = 1, 2, \ldots, N$, is the mass-weighted average

$$\mathbf{R}_{\mathrm{CM}} = \frac{\displaystyle\sum_{i=1}^{N} m_i \mathbf{r}_i}{\displaystyle\sum_{i=1}^{N} m_i}, \tag{2.49}$$

where $\mathbf{R}_{\mathrm{CM}}$ and $\mathbf{r}_i$ are three-dimensional vectors. For the x component, this equation is

$$x_{\mathrm{CM}} = \frac{\displaystyle\sum_{i=1}^{N} m_i x_i}{\displaystyle\sum_{i=1}^{N} m_i}, \tag{2.50}$$

and similar equations apply for the y and z components. The rotation of the system of particles takes place without changing the location of the center of mass, only translations can change its location. If there is no net force on the center of mass, a situation that is considered often in molecular mechanics, then introducing the location of the center of mass as the origin of coordinates, simplifies the equations of motion. Rather than concerning ourselves with the equations of motion in the general case, as is often done in a classical physics course, we focus on two special cases that are useful later on. These are the two-body problem with a central potential, and the classical rigid rotor, a system of two point masses held at a fixed distance by a rigid weightless rod. We prove that in both cases, the two-body problem can be separated in the motion of the center of mass with a constant velocity (since we assume central forces only) and the motion of a reduced mass body.

Consider two point particles with masses m_1 and m_2, respectively, free to move in three-dimensional space. Let x_1, y_1, z_1 be the coordinates for particle one and x_2, y_2, z_2 the coordinates for particle two. Let us further assume that the potential energy depends only on the differences $x_1 - x_2$, $y_1 - y_2$, $z_1 - z_2$. These conditions define the central potential. Many two-body problems can be found to satisfy the central potential assumption, at least approximately, e.g., the earth–moon system, two charged bodies, the hydrogen atom, the nuclei of a diatomic molecule in the gas phase, etc.

The Lagrangian is

$$\mathcal{L} = \frac{1}{2}m_1\left(\dot{x}_1^2 + \dot{y}_1^2 + \dot{z}_1^2\right) + \frac{1}{2}m_2\left(\dot{x}_2^2 + \dot{y}_2^2 + \dot{z}_2^2\right) - V. \tag{2.51}$$

Consider the following change of coordinates:

$$x_1' = x_1 - X \qquad x_2' = x_2 - X, \tag{2.52}$$

where X is the coordinate of the center of mass.

$$X = \frac{m_1 x_1 + m_2 x_2}{m_1 + m_2}. \tag{2.53}$$

We omit writing equations for the y and z components in the derivation that follows, since these can be obtained simply by mimicking those for x. We wish to rewrite the Lagrangian with six different degrees of freedom. Instead of using $\dot{x}_1, \dot{y}_1, \dot{z}_1, \dot{x}_2, \dot{y}_2, \dot{z}_2$, we choose the following six coordinates,

$$\dot{X}, \dot{Y}, \dot{Z}, (\dot{q}_1 = \dot{x}_1' - \dot{x}_2'), (\dot{q}_2 = \dot{y}_1' - \dot{y}_2'), (\dot{q}_3 = \dot{z}_1' - \dot{z}_2'). \tag{2.54}$$

To make any progress, we must consider an expression for x_1' and x_2' in terms of x_1 and x_2, obtained by using the definition of X,

$$x_1' = x_1 - \frac{m_1 x_1 + m_2 x_2}{m_1 + m_2}, \tag{2.55}$$

$$x_2' = x_2 - \frac{m_1 x_1 + m_2 x_2}{m_1 + m_2}. \tag{2.56}$$

Solving for x_1 in Equation 2.55, and inserting it into Equation 2.56 gives,

$$m_2 x_2' = -m_1 x_1', \qquad (2.57)$$

from which it follows trivially that

$$m_2 \dot{x}_2' = -m_1 \dot{x}_1'. \qquad (2.58)$$

At this point, we are ready to rewrite the Lagrangian for the two-body problem. Starting with the definition of the primed coordinate system, using Equation 2.58, $m_2 \dot{x}_2' = -m_1 \dot{x}_1'$ and simplifying, we arrive at

$$m_1 \dot{x}_1^2 + m_2 \dot{x}_2^2 = m_1 (\dot{x}_1')^2 + m_2 (\dot{x}_2')^2 + (m_1 + m_2) \dot{X}^2. \qquad (2.59)$$

Finally, defining a new coordinate

$$q_1 = x_1' - x_2', \qquad (2.60)$$

taking the time derivative, and using Equation 2.58 again, gives

$$\dot{q}_1 = \frac{m_1 + m_2}{m_2} \dot{x}_1'. \qquad (2.61)$$

In a similar manner, it is possible to show that

$$\dot{q}_1 = -\frac{m_1 + m_2}{m_1} \dot{x}_2'. \qquad (2.62)$$

Noting that we can invert Equation 2.61 and Equation 2.62 to solve for $\dot{x}_1'$ and $\dot{x}_2'$, we can set up the sum $m_1 (\dot{x}_1')^2 + m_2 (\dot{x}_2')^2$ that occurs in the Lagrangian, and simplify to

$$m_1 (\dot{x}_1')^2 + m_2 (\dot{x}_2')^2 = \left[\frac{m_1 m_2}{m_1 + m_2} \right] \dot{q}_1^2, \qquad (2.63)$$

$$m_1 (\dot{x}_1')^2 + m_2 (\dot{x}_2')^2 = \mu \dot{q}_1^2. \qquad (2.64)$$

Equation 2.64 defines the reduced mass μ. Putting this result into Equation 2.59, we obtain

$$m_1 \dot{x}_1^2 + m_2 \dot{x}_2^2 = \mu \dot{q}_1^2 + (m_1 + m_2) \dot{X}^2. \qquad (2.65)$$

When the corresponding y and z parts are added to this last result, one arrives at the desired Lagrangian expression,

$$\mathcal{L} \left(\dot{X}, \dot{Y}, \dot{Z}, \dot{q}_1, \dot{q}_2, \dot{q}_3 \right) = \frac{1}{2} (m_1 + m_2) \left(\dot{X}^2 + \dot{Y}^2 + \dot{Z}^2 \right)$$
$$+ \frac{1}{2} \mu \left(\dot{q}_1^2 + \dot{q}_2^2 + \dot{q}_3^2 \right) - V \left(q_1, q_2, q_3 \right). \qquad (2.66)$$

If the potential energy only depends on q_1, q_2, q_3, we can remove the center of mass from the picture and we can reduce the two-body problem to a pseudo-one-body problem. To see this, note that if V does not depend on X, Y, and Z, the Euler–Lagrange equations for the center of mass components yield

$$m\ddot{X} = 0, \tag{2.67}$$

with similar equations applying for the y and z components. Therefore, the velocity of the center of mass is constant, the related momentum is conserved, and the coordinates X, Y, and Z are called **cyclic**. Cyclic coordinates and conjugate momenta of cyclic coordinates can be ignored in the Lagrangian, Hamiltonian, and in the equations of motion. Of course, this leads to a great deal of simplification for the two-body problem, reducing its most general case to a single one-dimensional integral when the central potential is isotropic.

2.6 Isotropic Potentials and the Two-Body Problem

For many applications, it is convenient to change coordinates for the pseudo-particle of mass μ. One case of great importance is the set of two-body problems with **isotropic** potentials. Isotropic simply means that the potential energy between the two bodies depends only on the distance $r = \left[q_1^2 + q_2^2 + q_3^2\right]^{1/2}$. In the isotropic case, it is simpler to work with spherical polar coordinates

$$\begin{aligned}
q_1 &= r \, \cos\phi \, \sin\theta, \\
q_2 &= r \sin\phi \sin\theta, \\
q_3 &= r \cos\theta.
\end{aligned} \tag{2.68}$$

The time derivatives of these expressions are

$$\begin{aligned}
\dot{q}_1 &= (\cos\phi\sin\theta)\,\dot{r} - (r\sin\phi\sin\theta)\,\dot{\phi} + (r\cos\phi\cos\theta)\,\dot{\theta}, \\
\dot{q}_2 &= (\sin\phi\sin\theta)\,\dot{r} + (r\cos\phi\sin\theta)\,\dot{\phi} + (r\sin\phi\cos\theta)\,\dot{\theta}, \\
\dot{q}_3 &= (\cos\theta)\,\dot{r} - (r\sin\theta)\,\dot{\theta}.
\end{aligned} \tag{2.69}$$

To transform the Lagrangian again, we must square all three of the last expressions and add them up. The result is

$$\dot{q}_1^2 + \dot{q}_2^2 + \dot{q}_3^2 = r^2\,\dot{\theta}^2 + r^2\sin^2\theta\,\dot{\phi}^2 + \dot{r}^2. \tag{2.70}$$

The complete Lagrangian for the two-body problem becomes

$$\mathcal{L} = \frac{1}{2}(m_1 + m_2)\left(\dot{X}^2 + \dot{Y}^2 + \dot{Z}^2\right) + \frac{1}{2}\mu\left(r^2\,\dot{\theta}^2 + r^2\sin^2\theta\,\dot{\phi}^2 + \dot{r}^2\right) - V(r). \tag{2.71}$$

Exercises

1. Derive Equation 2.59 using Equation 2.58.

2. Provide the missing steps to derive Equation 2.61 and Equation 2.62 from the definition of q_1.

3. Use the result in Equation 2.61 and Equation 2.62 to derive Equation 2.63.

4. Derive Equation 2.70 from Equation 2.69.

5. Find the Hamiltonian for the two-body problem using the X, Y, Z, r, θ, ϕ coordinate system. Follow these directions:

   i. Find the expressions for the conjugate momenta using

   $$p_i = \frac{\partial \mathcal{L}}{\partial \dot{q}_i},$$

   where q_i is any one of the six coordinates X, Y, Z, r, θ, ϕ. $\mathcal{L}$ is given in Equation 2.71.

   ii. Show that when the potential is added, the Hamiltonian takes the form,

   $$H = \frac{1}{2(m_1 + m_2)} \left(p_X^2 + p_Y^2 + p_Z^2\right) + \frac{1}{2\mu r^2} p_\theta^2$$
   $$+ \frac{1}{2\mu r^2 \sin^2 \theta} p_\phi^2 + \frac{1}{2\mu} p_r^2 + V(r). \qquad (2.72)$$

6. Use the expression for the trajectory of a harmonic oscillator in Equation 2.48 to show that, unlike the Hamiltonian, the Lagrangian $\mathcal{L} = K - V$ is not a conserved quantity. Obtain an expression for the action $I\left[v(t), x(t)\right]$ after a single vibration period, $\tau = 2\pi/\omega$.

2.7   The Rigid Rotor

Now imagine a system where the two particles are held at fixed distance R by a weightless rod. In the absence of external fields, there is no potential energy, only kinetic energy. This is one example where the spherical polar coordinates are essential. It is difficult to write down the Lagrangian and the Hamiltonian in Cartesian coordinates for the rigid rotor case. In spherical polar coordinates, all we need to do is set $r = R$ and $\dot{r} = 0$, therefore, the constraint leads to a smoothly conserved quantity.

The type of constraints that can be formulated with an *analytical one-to-one* map, $q_1, q_2, q_3 \rightarrow \theta, \phi$, are called **holonomic constraints**. The one-to-one condition requires that for each point on the sphere of radius R with coordinates θ, ϕ, there corresponds one and only one point with coordinates q_1, q_2, q_3, and that the converse must also be true. The analytical condition dictates that

$$\frac{\partial^2 q_i}{\partial \phi \partial \theta} = \frac{\partial^2 q_i}{\partial \theta \partial \phi}, \tag{2.73}$$

for $i = 1, 2, 3$.

The Lagrangian and the Hamiltonian are simply,

$$\mathcal{L} = \frac{1}{2}(m_1 + m_2)\left(\dot{X}^2 + \dot{Y}^2 + \dot{Z}^2\right) + \frac{1}{2}\mu\left(R^2\dot{\theta}^2 + R^2\sin^2\theta\,\dot{\phi}^2\right), \tag{2.74}$$

$$H = \frac{1}{2(m_1 + m_2)}\left(p_X^2 + p_Y^2 + p_Z^2\right) + \frac{1}{2\mu R^2}p_\theta^2 + \frac{1}{2\mu R^2\sin^2\theta}p_\phi^2. \tag{2.75}$$

Since we are assuming no net force from external fields, the angular momentum is conserved. **L** can point along any arbitrary axis, but we can always rotate our system of coordinates so that **L** points along the q_3 direction. Since **L** is conserved, this direction is a constant of motion. One can quickly verify that fixing the angular momentum along the q_3 axis leads to the result

$$\dot{\theta} = 0; \qquad \theta = \frac{\pi}{2}. \tag{2.76}$$

The Lagrangian and the Hamiltonian simplify further,

$$\mathcal{L} = \frac{1}{2}(m_1 + m_2)\left(\dot{X}^2 + \dot{Y}^2 + \dot{Z}^2\right) + \frac{1}{2}\mu R^2\,\dot{\phi}^2, \tag{2.77}$$

$$H = \frac{1}{2(m_1 + m_2)}\left(p_X^2 + p_Y^2 + p_Z^2\right) + \frac{1}{2\mu R^2}L^2, \tag{2.78}$$

where L in Equation 2.78 is the size of the orbital angular momentum. The quantity, μR^2, in the denominator of the last term is known as the moment of inertia of the rigid rotor.

Exercises

1. Show that the map, $q_1, q_2, q_3 \leftarrow \theta, \phi$, in Equation 2.69 satisfies Equation 2.73.

2. Invert the map in Equation 2.69 and prove that the following relationships hold:

$$\frac{\partial^2 \phi}{\partial q_i \partial q_j} = \frac{\partial^2 \phi}{\partial q_j \partial q_i},$$

$$\frac{\partial^2 \theta}{\partial q_i \partial q_j} = \frac{\partial^2 \theta}{\partial q_j \partial q_i}.$$

3. Prove that for a particle of mass μ, rotating in a circle of radius R, the size of the angular momentum is

$$|\mathbf{L}| = \mu R^2 \omega, \tag{2.79}$$

where $\omega = d\theta/dt$ is the angular velocity. Follow the steps below. First, using trigonometry, show that

$$\begin{aligned} x &= R\cos\theta \\ y &= R\sin\theta, \end{aligned} \tag{2.80}$$

therefore, this proves that

$$\mathbf{x} = (R\cos\theta\,\mathbf{i} + R\sin\theta\,\mathbf{j} + 0\,\mathbf{k}),$$

where $\mathbf{i}, \mathbf{j}, \mathbf{k}$ are unit vectors along the x, y, and z direction, respectively. Take the derivative with respect to t of x and y in Equation 2.80, holding R constant, and show that

$$\mathbf{p} = \mu\omega\left(-R\sin\theta\,\mathbf{i} + R\cos\theta\,\mathbf{j} + 0\,\mathbf{k}\right).$$

The cross product formula for two three-dimensional vectors $\mathbf{a} = (a_x\,\mathbf{i} + a_y\,\mathbf{j} + a_z\,\mathbf{k})$ and $\mathbf{b} = (b_x\,\mathbf{i} + b_y\,\mathbf{j} + b_z\,\mathbf{k})$, is

$$\mathbf{a}\times\mathbf{b} = \left[(a_y b_z - a_z b_y)\,\mathbf{i} + (a_z b_x - a_x b_z)\,\mathbf{j} + (a_x b_y - a_y b_x)\,\mathbf{k}\right].$$

Use this formula for $\mathbf{x}\times\mathbf{p}$ to show that

$$\mathbf{L} = \mu R^2 \omega\,\mathbf{k},$$

and use the result to obtain the following expression,

$$L = |\mathbf{L}| = \mu R^2 \omega.$$

Recall that the size of a vector $\mathbf{a}$ in Euclidean three-dimensional space is $|\mathbf{a}| = \left(a_x^2 + a_y^2 + a_z^2\right)^{1/2}$.

4. Use the result of the last exercise together with the expression of the kinetic energy of a rigid rotor,

$$K = \frac{1}{2}\mu R^2 \dot{\theta}^2,$$

to show that an equivalent expression for this quantity is

$$K = \frac{L^2}{2\mu R^2}.$$

This expression shows that the angular momentum for a rigid rotor is analogous to the linear momentum for a single particle. The denominator, $2\mu R^2$, appears in place of the 2μ term of the kinetic energy for a single particle expressed in terms of the linear momentum.

2.8 Numerical Integration Methods

Let us use the harmonic oscillator example to develop a simple numerical procedure to integrate Equation 2.42. If one assumes that the time step Δt, through which one propagates the solution from point x_n to point x_{n+1}, is sufficiently small, then the acceleration is approximately constant,

$$\frac{d^2x}{dt^2} \approx -\omega^2 x_n, \qquad (2.81)$$

and the velocity can be integrated trivially,

$$v_{n+1} \approx v_n - \omega^2 x_n \Delta t. \qquad (2.82)$$

This equation can be integrated once more to obtain the position update

$$x_{n+1} = x_n + v_n \Delta t - \frac{1}{2}\omega^2 x_n (\Delta t)^2 \qquad O\left(\Delta t^2\right). \qquad (2.83)$$

The recursion in Equation 2.83 is an application of the first order point-slope method, and can be converted to code easily. To implement this integrator, one simply sets x_0, v_0 to the desired initial values and chooses a sufficiently small value for Δt. Here is an implementation:

```
      program point_slope
      implicit real*8 (a-h,o-z)
c point - slope integrator for the harmonic oscillator
      parameter (dm = 2.34d0 ) ! mass in kilograms
      parameter (dk = 0.23d0 ) ! force constant in N/m
      parameter (omega = dsqrt(dk/dm)) ! frequency (Hz)
c initial conditions
      x0 = 0.2d0
      v0 = 0.2d0
      xn = x0
      vn = v0
      dt = 0.2    ! the time step in seconds
      t = 0.d0
      do k=1,500
       vnp1 = vn - omega**2*xn*dt
       xnp1 = xn + vn*dt - 0.5d0*xn*(omega*dt)**2
       t = t + dt
```

```
c compare the numerical solution with the analytical one
      xa = x0*cos(omega*t) + (v0/omega)*sin(omega*t)
      write(8,*) t,xnp1,xa
c reset for the next propagation
      xn = xnp1
      vn = vnp1
   enddo
   end
```

Exercises

1. Show that Equation 2.48 satisfies the differential equation in Equation 2.42 and the correct initial conditions by direct substitution.

2. Use Equation 2.48 to find an expression for the Hamiltonian as a function of time, then show that

$$\frac{\partial H}{\partial t} = 0. \tag{2.84}$$

   This means that the energy is conserved at all times. This is always the case in autonomous systems.

3. When the content in fort.8, obtained by running point_slope, is graphed, the reader should see two plots similar to those in Figure 2.1. Notably, the numerical solution becomes of increasingly poor quality as the simulation time goes on. This is the result of the truncation error generated by approximating the acceleration as a constant. Equation 2.83

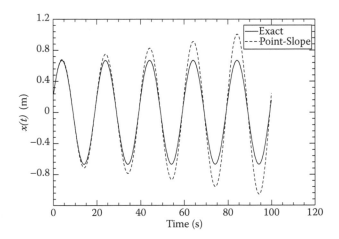

FIGURE 2.1
The exact and numerical trajectory for the harmonic oscillator.

can be interpreted as a Taylor expansion of the position about t_n, at which time $x(t_n) = x_n$. One notes that the term dropped is cubic in Δt (leaving a quadratic term) for the position and quadratic in Δt for the velocity. However, because the recursions rely on the value of position and velocity from the previous step, the error propagates through and grows, as evident in Figure 2.1. The error after the first step is known as the **local error**, whereas the error at the end time is the **global error**. The propagation of error analysis is straightforward; however, we guide the reader to determine the error orders empirically.

i. Compute the local and global error by making appropriate modifications to the program `point_slope` for a number of values of Δt from 0.4 to 0.0125 s. The local error is obtained by comparing the exact solution with the numerical solution after one propagation step. The global error is computed at the last integration step. Including an outer loop in the program is probably the most efficient way, rather than changing the code and recompiling every time. However, care must be taken to reinitialize the position and the velocity after each trajectory.

ii. Set the number of time points to 10,000 for all the runs.

iii. Plot the logarithm of the local and global error vs. the logarithm of Δt to determine the power laws for both errors. The local error should have a quadratic dependence, and the global error should depend only linearly on Δt. This is usually denoted $O(\Delta t^2)$, which should explain the symbol in Equation 2.83.

iv. Change the number of runs so that the end time $t_N = N\Delta t$ is always 100 s, then compare the global error for several values of Δt from 0.4 to 0.025. Are the global errors different?

4. Repeat the error analysis for the velocity.

5. The Verlet algorithm,

$$x_{n+1} = 2x_n - x_{n-1} - \omega^2 x_n (\Delta t)^2 \qquad O(\Delta t^3), \qquad (2.85)$$

$$v_{n+1} = \frac{x_{n+1} - x_{n-1}}{2\Delta t} \qquad O(\Delta t^2), \qquad (2.86)$$

converges cubically (global error) in the position and linearly in the velocity.

```
      program Verlet
      implicit real*8 (a-h,o-z)
c Verlet integrator for the harmonic oscillator
      parameter (dm = 2.34d0 ) ! mass in kilograms
      parameter (dk = 0.23d0 ) ! force constant in N/m
      parameter (omega = dsqrt(dk/dm)) ! frequency (Hz)
```

```
c initial conditions
      x0 = 0.2d0
      v0 = 0.2d0
      dt = 0.2   ! the time step in seconds
      xn = x0
      xnm1 =  x0 - v0*dt - 0.5d0*x0*(omega*dt)**2
      t = 0.d0
      do k=1,500
       xnp1 = 2.d0*xn - xnm1 - xn*(omega*dt)**2
       t = t + dt
c compare the numerical solution with the analytical one
       xa = x0*cos(omega*t) + (v0/omega)*sin(omega*t)
       write(8,*) t,xnp1,xa
c update for the next propagation
       xnm1 = xn
       xn = xnp1
      enddo
      end
```

After testing the program `Verlet`, modify it appropriately and verify these claims by performing the error analysis on the position, following the same empirical procedure outlined earlier.

6. Modify both `point_slope` and `Verlet` so that the total energy is computed as a function of time from

$$E(t) = \frac{1}{2}mv(t)^2 + \frac{1}{2}kx(t)^2. \tag{2.87}$$

Plot $|E - E_0|/E_0$ as a function of time, where $E_0 = E(t = 0)$. Given that the energy should be conserved theoretically, how do the two integrators behave? To what order in Δt is the error of $|E - E_0|/E_0$? Increase the final time of the Verlet integrator to 1000 s, then 10,000 s, keeping Δt at a constant value. Does $|E - E_0|/E_0$ get progressively worse?

7. Modify the program `Verlet` so that the spring is twice as stiff (k is twice as large). Compute the global error at 10,000 s for several values of Δt. Compare the magnitude of these errors with those obtained with the less stiff case, then comment on your observations. Repeat the computation of the global error after 10,000 simulation seconds with 10 times the stiffness. Your observations are crucial for the understanding of the timescale problem that occurs very often in simulations of chemically interesting systems. For example, an assembly of water molecules contains stiff degrees of freedom along the covalent bonds, and relatively softer ones along the intermolecular coordinates. The integration of these systems is challenging because the presence of stiff constants requires relatively

small values of Δt, while the "looser" van der Waals interactions require much longer integration times. The greater the frequency scale difference is, the more challenging and inefficient the integration becomes.

8. Modify `Verlet` so that the more general problem

$$-\frac{\partial V}{\partial x} = m\frac{d^2x}{dt^2}, \tag{2.88}$$

is integrated. Use a subroutine called `pot`, where the potential energy and the gradient are calculated for an input value of x and then returned to the main program. The position update is simply,

$$x_{n+1} = 2x_n - x_{n-1} + a\left(\Delta t\right)^2, \tag{2.89}$$

where

$$a = -\frac{1}{m}\frac{\partial V}{\partial x}, \tag{2.90}$$

and x_{n-1} is initiated with $x_{n-1} = x_0 - v_0\Delta t + a\left(\Delta t\right)^2/2$. Test your program with the quartic potential

$$V = 0.23x^4, \tag{2.91}$$

by computing $|E - E_0|$ at 1,000 s, then 10,000 s. Does changing the potential shape affect the behavior of the Verlet algorithm? In particular, how does the global error (estimated from the conservation of energy) behave as a function of Δt?

2.9 Hamilton's Equations and Symplectic Integrators

It is relatively simple to combine Equations 2.15 and 2.23, and derive a system of coupled first order differential equations that involve the Hamiltonian rather than the Lagrangian.

$$\dot{q}_i = \frac{\partial H}{\partial p_i}, \tag{2.92}$$

$$\dot{p}_i = -\frac{\partial H}{\partial q_i}. \tag{2.93}$$

For n particles in three-dimensional space, there are a total of $6n$ linear differential equations. Hamilton's equations satisfy special symmetry properties that make their integration technically less difficult. The propagation of the solution can be understood as the operation of a member of a special group, called the symplectic group. The discussion on symplectic groups is presented after some classical Lie algebras are introduced in Chapter 4. For the moment, we introduce a class of integrators called symplectic integrators. This class is important because it can be shown that the energy is rigorously conserved if the integration algorithm preserves the symplectic structure of Hamilton's equation. The energy is conserved within a global error similar to the local error. This important feature of symplectic algorithms can be used to explore questions that require exceedingly long trajectories as, for example, the evolution of a protein conformation in water, or to determine if Pluto's orbit is chaotic.

The code that follows is an application of the Candy–Rozmus algorithm [46], one of the many symplectic algorithms that have been proposed in the literature. The position and the momenta are updated in four stages recursively. The positions and momenta at $t + \Delta t$ are given at the end of the four stages:

$$p_n = p_{n-1} + \sum_{i=1}^{4} b_i f_{n-1} \Delta t, \qquad (2.94)$$

$$x_n = x_{n-1} + \sum_{i=1}^{4} a_i \frac{p_n}{m} \Delta t, \qquad (2.95)$$

where the a_i and b_i coefficients are optimized to eliminate terms in the expansion that are third order in Δt

$$a_1 = a_4 = \frac{2 + 2^{1/3} + 2^{-1/3}}{6}, \quad a_3 = a_2 = \frac{1 - 2^{1/3} - 2^{-1/3}}{6}, \qquad (2.96)$$

$$b_1 = 0, \qquad b_2 = b_4 = \frac{1}{2 - 2^{1/3}}, \quad b_3 = \frac{1}{2 - 2^{2/3}}, \qquad (2.97)$$

and f_n is the force at x_n.

```
      program Candy_Rozmus
      implicit real*8 (a-h,o-z)
c Forth order symplectic integrator for Hamilton's
c equations for a Hamiltonian H = T(p) + V(q)
      parameter (dk = 0.23d0 ) ! force constant in N/m
      parameter (dm = 2.34d0 ) ! mass in kilograms
      parameter (omega = dsqrt(dk/dm)) ! frequency (Hz)
      real*8 a(4),b(4)
c initialize the propagator parameters
      a(1) = 1.d0/3.d0 + (2.d0**(0.3333333333))/6.d0
     & + (2.d0**(-0.333333333))/6.d0
      a(2) = 1.d0/6.d0 - (2.d0**(0.3333333333))/6.d0
     & - (2.d0**(-0.33333333333))/6.d0
      a(3) = a(2)
      a(4) = a(1)
      b(1) = 0.d0
```

```
      b(2) = 1.d0/(2.0d0 - 2.0d0**(0.33333333333))
      b(3) = 1.d0/(1.d0 -2.0d0**(0.666666666666))
      b(4) = b(2)
c initial conditions
      xnm1 = 0.2d0
      pnm1 = 0.2d0
      x0 = xnm1
      p0 = pnm1
      call pot(xnm1,vnm1,fnm1)
      e0 = 0.5d0*pnm1*pnm1/dm + vnm1
      dt = 0.8d0    ! the time step in seconds
      t = 0.d0
      do k=1,1000000
        do i=1,4
          call pot(xnm1,vnm1,fnm1)
          pn = pnm1 + b(i)*fnm1*dt
          xn = xnm1 + a(i)*pn*dt/dm
c reset for the next stage of the propagation
          pnm1 = pn
          xnm1 = xn
        enddo
        t = t + dt
        e = 0.5d0*pn*pn/dm + vnm1
c compare the numerical solution with the analytical one
        xa = x0*cos(omega*t) + (p0/omega/dm)*sin(omega*t)
        if (k.eq. 1) write(6,*) t,(xn-xa),100.d0*(e-e0)/e0
        if (k.eq. 500) write(6,*) t,(xn-xa),100.d0*(e-e0)/e0
        write(8,*) t,pn,xn,100.d0*(e-e0)/e0
      enddo
      end
      subroutine pot(x,v,f)
      implicit real*8 (a-h,o-z)
      parameter (dk = 0.23d0 ) ! force constant in N/m
c on input: x position
c on output: v = potential energy
c            f = force (-dv/dx)
      v = 0.5d0*dk*x*x
      f = -dk*x
      return
      end
```

Exercises

1. Test the Candy–Rozmus algorithm, and verify that the local and global errors in p and x are fourth order with Δt. Show that $|E - E_0|$ is small,

stable, and that its error is first order in Δt, just as with the Verlet algorithm. With the parameters as given in the program, the values of Δt should be no smaller than 0.1 s, in order to capture the order of convergence. Rounding errors in the difference between the numerical and analytical solutions creep into our estimate of the error, and reduce the convergence to second order or worse when small values of Δt are used.

2. Graph $p(t)$ against $x(t)$. Change the initial value of the momentum, and graph the same two variables again. These graphs are the phase space representations of the trajectory at a given energy.

3. Show that for the harmonic oscillator

$$\dot{x} = \frac{p}{m}, \tag{2.98}$$

$$\dot{p} = -kx. \tag{2.99}$$

Then, using these results, take the derivative with respect to t of $\dot{x}$ and show that it yields

$$\ddot{x} = -\frac{1}{m}kx. \tag{2.100}$$

This is Newton's equation for the harmonic oscillator.

2.10 The Potential Energy Surface

The simplest theory of molecular physics begins with the assumption that atoms are point-like masses that interact with one another. The forces between atoms are caused by the electrons they share. The electron dynamics are essentially "hidden" under a phenomenological **potential energy surface** for the nuclei. The potential energy is constructed using the following procedure. The nuclei are clamped in place, the Schrödinger equation is solved (approximately) for the electrons, and these are used to fill the energy levels, following the Pauli exclusion principle as for the ground electronic state of atoms. The ground state electronic configuration of the molecule has a specific value of the energy, obtained by adding the energy of each electron in the configuration, and the Coulomb repulsion from all the nuclei. The configuration energy depends parametrically on the position of the nuclei. The nuclei are released, moved to a new configuration, clamped again, and the process is repeated. In principle, one can repeat the process of finding the ground state configuration of the molecule(s), and the electronic ground state energy associated with all the possible configurations, obtaining a function of the nuclei coordinates. The electronic ground state energy, therefore, is a function of the nuclei positions, and we call this function the potential energy surface. In this textbook, with few exceptions, we deal with nuclei, and the potential energy among them, which results from the electrons they share in their ground electronic state. The electrons are always assumed to be in the ground state, regardless of the position of the nuclei. This treatment is called the adiabatic or Born–Oppenheimer approximation, and it is justified by the large mass difference between the typical nucleus and the electrons. The Born–Oppenheimer approximation rests on the assumption that electrons move at a much shorter time scale, and essentially a readjustment of their equilibrium state takes place instantaneously with every infinitesimal nuclear movement. Solving the Schrödinger equation for a generic multifermion system remains a problem, especially when simulations of condensed matter are desired. These first principle electronic computations are approximate and quite expensive, though the approximations can be systematically improved. In molecular simulations, it is more practical to make use of analytical models constructed empirically, or by fitting a number of electronic computations. It is not an exaggeration to call the potential energy surface "the holy grail" of theoretical chemistry. A large number of theoretical chemistry groups around the world are engaged in the main stream research with the goal of finding ever faster and more accurate electronic ground states and potential energy surfaces that can be included in simulations of matter. When simulations are performed on complex systems, such as clusters, or molecular liquids, where millions of potential energy values are needed, the first principal treatment is not feasible with modern computers.

Many excellent empirical models for potential energy surfaces exist. Several potential energy models are presented in this book. In Chapter 8, for example, we introduce the Lennard-Jones potential as a model for atomic clusters. In Chapter 12, we present a large family of potential models for the simulation of clusters of linear molecules. An excellent empirical model for the covalent bond is the Morse potential,

$$V(r) = V_0 \left\{ 1 - \exp\left[-a\left(r - r_\mathrm{e}\right) \right] \right\}^2. \tag{2.101}$$

The Morse potential is a two-body isotropic potential. r is the distance between the two bodies. The potential in Equation 2.101 is frequently used to interpret spectroscopic data of diatomic molecules. V_0 is the well depth, the classical energy necessary to dissociate the bond. If $r \to \infty$, then $V \to V_0$. On the other hand, if $r \to r_\mathrm{e}$, then $V \to 0$, the minimum. This result can be easily verified by differentiation. a is a parameter related to the rigidity of the bond, a type of force constant. The Morse model for a diatomic molecule is quite simple to analyze. One just needs a graph and some simple calculus. However, even the simplest models for clusters of atoms or molecules have complicated potential energy surfaces that are functions of numerous variables. Finding minima and **transition states** of different ranks in many dimensions is challenging, and the location of these is often necessary to interpret the outcome of simulations.

A good portion of Chapter 8 is dedicated to the presentation of methods used to characterize multidimensional potential energy surfaces. To initiate the training in these methods, we find it best to use as simple a model as possible at first. The quartic potential in Equation 2.102 in particular, is useful to mimic the situation often found in clusters, where two minima are close in energy, and they are separated by a barrier much larger than the difference in energy between the two minima. The quartic potential in Equation 2.102 has been proposed by Frantz [839] to study the phenomena named quasiergodicity. The program `quartic` calls the subroutine `pot`, where the following quartic potential is computed,

$$V(x) = \frac{3}{2\alpha+1}x^4 + \frac{4\alpha-4}{2\alpha+1}x^3 - \frac{6\alpha}{2\alpha+1}x^2 + 1, \qquad (2.102)$$

along with its first and second derivative for several values of the "reaction coordinate" x. By setting the first derivative to zero and solving the resulting cubic equation, one can verify that the potential has a global minimum at $x = 1$, where $V = 0$, and a local minimum at $x = -\alpha$, where

$$V(-\alpha) = 1 - \frac{\alpha^3(\alpha+2)}{2\alpha+1} = 1 - \gamma. \qquad (2.103)$$

The second derivatives at the two minima are

$$V''(-\alpha) = \frac{12\alpha(\alpha+1)}{2\alpha+1}, \qquad (2.104)$$

and

$$V''(1) = \frac{12(\alpha+1)}{2\alpha+1}. \qquad (2.105)$$

Whereas, the second derivative at the transition state, $x = 0$ is

$$V''(0) = -\frac{12\alpha}{2\alpha+1}. \qquad (2.106)$$

The main subroutine initializes the coordinate at -1.5 and steps through 50 equally spaced values.

```
program quartic
implicit real*8 (a-h,o-z)
write(6,*) 'Enter gamma between 0 and 1'
read(5,*) gamma
write(6,*) 'Gamma is: ',gamma
x = -1.5
dx = 0.06
do k=1,50
    call pot(gamma, x,v,dv,d2v)
```

```
      write(8,*) x,v,dv,d2v
      x = x + dx
   enddo
   end
   subroutine pot(gamma, x,v,dv,d2v)
c This subroutine calculates the quartic potential
c proposed by Frantz et. al. J. Chem. Phys. 93, 2769 (1990).
c to study quasiergodicity.

c On input:  x = coordinate
c            gamma = parameter

c On output: v =  the value of the potential
c            dv = the value of the first derivative
c            d2v = the second derivative

   implicit real*8 (a-h,o-z)
   alpha = 0.d0
    do k=1,20
      alpha = (gamma*(2.*alpha + 1)/(alpha + 2))**(1./3.)
    enddo
   term1 = 3./(2.*alpha + 1.)
   term2 = (4.*alpha -4.)/(2.*alpha + 1)
   term3 = -6.*alpha/(2.*alpha + 1.)
   v = term1*x**4 + term2*x**3 + term3*x**2 + 1.
   dv = 4.*term1*x**3 + 3.*term2*x**2 + 2.*term3*x
   d2v = 12.*term1*x**2 + 6.*term2*x + 2.*term3
   return
   end
```

The line `write(8,*) x,v,dv,d2v` places the output in a device file called `fort.8` instead of the screen. The parameter γ is used to call the subroutine. In the body of the subroutine, the relationship between the parameters α and γ is inverted using a recursive algorithm. By simple iteration, the equation

$$\gamma = \alpha^3 \left(\frac{\alpha + 2}{2\alpha + 1} \right) \tag{2.107}$$

is solved for α.

Exercises

1. Run the program `quartic` with $\gamma = 0.9$ as input. Graph the first two columns of `fort.8` and compare it with the one in Figure 2.2. The value of x at the peak in V between the two minima is the transition state of the quartic potential.

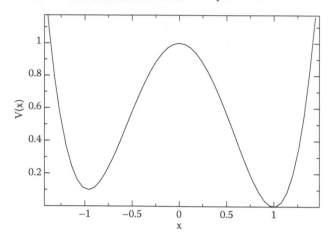

FIGURE 2.2
The quartic double well potential.

2. Place a `stop` statement right before the `return` statement inside `pot`. Use a `write(6,*)` statement inside `pot` to verify that the sequence,

$$\alpha_{n+1} = \sqrt[3]{\gamma \left(\frac{2\alpha_n + 1}{\alpha_n + 2} \right)} \tag{2.108}$$

converges to machine precision with 20 terms, and confirm that Equation 2.107 is satisfied by the recursion limit.

3. Obtain a graph of the first and second derivative of $V(x)$ using the data in fort.8 for $\gamma = 0.9$, then calculate the value of the second derivative of $V(x)$ at the location of the two minima and the transition state $x = 0$.

4. Derive the expressions for the first and second derivative of $V(x)$. Set the first derivative to zero, and verify that the roots of the first derivative are $x = -\alpha, 0, 1$. Then verify Equations 2.103 through 2.106.

5. Modify the program `Candy_Rozmus` to integrate the equations of motion over the quartic potential. Note that the subroutine `pot` in `quartic` returns the derivative of the potential, not the force, so the body of `Candy_Rozmus` should be modified accordingly. Use a loop to initiate a set of 20 trajectories at $x = 1$, for a range of energies from 0.8 to 1.4 units. The energy value is used to determine the initial value of p. Inspect the phase space diagrams (the graph of p vs x) for all the trajectories. Repeat by beginning with $x = -\alpha$. Note that the initial value of V has to be subtracted from the energy to compute the initial value of p when starting at $x = -\alpha$.

6. Starting with the modified program `Candy_Rozmus` from the previous exercise, use a loop to initiate a set of 20 trajectories at $x = 0$, for a range of initial values of p from 0.001 to 1.0×10^{-5} units. Inspect the phase space diagrams (the graph of p vs x) for all the trajectories. Graph the force on the particle for each trajectory as a function of time, and describe what is happening physically to the particle.

2.11 Dissipative Systems

For the quartic potential of the previous section, we can inspect the graph and deduce its minima and maxima, and any other feature of interest, e.g., confining vs. nonconfining, etc. Earlier, we explained that in clusters of atoms and molecules, the exploration of the potential energy surface is a complicated task. In this section, we introduce an efficient algorithm to locate the nearest minimum from any **configuration** (the set of coordinates for the atoms of a cluster). The algorithm amounts to adding to the equations of motion a phenomenological term proportional to the velocity. In one dimension, the phenomenological differential equation for x is

$$\ddot{x} = -\frac{1}{m}\frac{dV}{dx} - \gamma_d \dot{x}, \tag{2.109}$$

where γ_d is a coefficient of drag. The last term on the R.H.S. can be interpreted as the collective effect of the collisions between the simulated particle and a continuous medium composed of much smaller particles. The net effect of the collisions between the simulated particle and its environment is known as drag. The main outcome of adding a drag to the equations of motion is that there is no longer an overall state function, like the Hamiltonian, that can be used to prove conservation of energy. In fact, the drag term with $\gamma_d > 0$ causes the energy to drop. Consequently, the system becomes confined to regions closer to the minimum. This process continues as the dynamics progress in time, until the system stops at some local minimum. This phenomenon can be easily observed in a laboratory with, for example, a free oscillating pendulum. The air drag (and friction at the pivot) eventually brings the system to an halt at the lowest potential energy state, the vertical position. Since the energy is no longer conserved, integrations of the differential equation for dissipative systems can be done reasonably accurately with the point-slope method. Again, if one assumes that the time step Δt through which one propagates the solution from point x_n to point x_{n+1} is sufficiently small, then the acceleration is approximately constant

$$a \approx -\frac{1}{m}\frac{dV(x_n)}{dx} - \gamma_d v_n, \tag{2.110}$$

and the velocity can be found with a trivial integral,

$$v_{n+1} \approx v_n + a\Delta t. \tag{2.111}$$

This equation can be integrated once more to obtain the recursion relation for x,

$$x_{n+1} = x_n + v_{n+1}\Delta t. \tag{2.112}$$

The following code calls the quartic potential (which needs to be appended before compilation).

```
      program Brown
      implicit real*8 (a-h,o-z)
c point slope integrator for dissipative systems
c in one dimension
      parameter (dm = 2.34d0 ) ! mass
c initial conditions
      x0 = 2.0d0
      v0 = 0.0d0
      gamma = 0.9d0    ! potential parameter
      gammad = 1.0d0   ! drag coefficient
      xn = x0
      vn = v0
      dt = 0.02d0    ! the time step in seconds
      t = 0.0d0
      do k=1,50000
      call pot(gamma,xn,v,dv,d2v)
       a = -dv/dm - gammad*vn
c        write(6,*) t,a, xn,vn
      vnp1 = vn + a*dt
      xnp1 = xn + vnp1*dt
      t = t + dt
      write(8,*) t,vnp1,xnp1,v
c reset for the next propagation
      xn = xnp1
      vn = vnp1
      enddo
      end
```

2.12   The Fourier Transform and the Position Autocorrelation Function

We have seen that, in the case of the harmonic oscillator, there exists a fundamental property related to the period in the dynamics, the characteristic frequency $\omega = \sqrt{k/\mu}$, where k is the spring constant. What is the characteristic frequency associated with a nonharmonic potential? It is possible to answer this question by performing a Fourier analysis of the trajectory or any dynamic variable. The fundamental concept of the Fourier analysis can be introduced with simple dynamic variables, such as the position as a function of time.

The Fourier transform of a function of t, $f(t)$ is a function of ω, defined as

$$F(\omega) = \int_{-\infty}^{\infty} f(t) \exp(i\omega t), \tag{2.113}$$

where $i = \sqrt{-1}$, and ω is the frequency in rad/s. If one uses Euler's formula, the integral in Equation 2.113 can be split up into a real and imaginary part,

$$F(\omega) = \int_{-\infty}^{\infty} f(t) \cos(\omega t) + i \int_{-\infty}^{\infty} f(t) \sin(\omega t). \tag{2.114}$$

Normally, one is interested in the *power spectrum* of the dynamic variable $f(t)$; this is, literally, the size of $F(\omega)$ squared,

$$P(\omega) = |F(\omega)|^2 = \left| \int_{-\infty}^{\infty} f(t) \cos(\omega t) \right|^2 + \left| \int_{-\infty}^{\infty} f(t) \sin(\omega t) \right|^2. \tag{2.115}$$

In the code that follows, we modify **Verlet** to evaluate the Fourier transform of the position. To implement a numerical procedure for the Fourier transform, one can use the extended trapezoid rule to evaluate the two integrals in Equation 2.115. This takes place in the subroutine **Fourier**. The interval of integration cannot be infinite for obvious practical reasons. Simply truncating the integration between some time t_0 and t, no matter how large $t - t_0$ is, produces satellite peaks alongside the main features of the spectrum. The technical reason for the appearance of satellite peaks has to do with the abrupt cutoff (a square window) that the regular trapezoid rule produces. The square window is **convoluted** with $f(t)$, producing spurious features. A number of remedies have been proposed to alleviate the windowing problem. In the subroutine **Fourier**, the Bartlett window is employed:

$$w_i = 1 - \left| \frac{i - n_p/2}{n_p/2} \right|, \tag{2.116}$$

where n_p is the number of time points at which $f(t)$ is known. Another important aspect of the numerical implementation of the Fourier transform is the Nyquist frequency,

$$\omega_N = \frac{2\pi}{\Delta t}, \tag{2.117}$$

where Δt is the sampling time interval between points. The Nyquist frequency is of high importance; it is a limiting frequency in the numerical Fourier transform that can be obtained with the given sampling interval Δt. A sine wave sampled at its Nyquist frequency is a constant, which has no transform other than that of the window used. If $f(t)$ is composed of a function that oscillates at frequencies greater than the Nyquist frequency, spurious features appear in the spectrum in the range $0, \omega_N$, a phenomenon called aliasing.

```
      program Verlet_FT
      implicit real*8 (a-h,o-z)
c Verlet integrator and Fourier transform
      parameter (dm = 2.34d0 ) ! mass
      real*8 traj(200000)
c initial conditions
      x0 = 0.0d0
      v0 = 0.001d0
      dt = 0.01    ! the time step in seconds
      xnm1 = x0
      call pot(x0,pe,dv)
      xn =   x0 + v0*dt - 0.5d0*dv*(dt)**2/dm
      t = 0.d0
      np = 10000
      do k=1,np
       call pot(xn,pen,dv)
       xnp1 = 2.d0*xn - xnm1 - dv*(dt)**2/dm
       v = (xnp1 - xnm1)/(2.d0*dt)
       pnp1 = dm*v
c store the value of x relative to its initial value in an array
       traj(k) = (xnp1- x0)
       t = t + dt
c compare the numerical solution with the analytical one
      write(8,*) t,xnp1,pnp1,xa
c update for the next propagation
       xnm1 = xn
       xn = xnp1
      enddo
c now perform the Fourier Transform of the trajectory
      fny = 2.d0*3.1415926/(dt)
      npp = 1000
      df = 0.01*fny/dfloat(npp)
      f =   0.d0
      write(6,*) 'Nyquist frequency (rad/s) = ',fny,omega
      do k=1,npp
       call Fourier(traj,f,p,np,dt)
       write(9,*) f,p
       f = f + df
      enddo
      end
```

```
      subroutine pot(x,pe,dv)
      implicit real*8 (a-h,o-z)
      parameter (dk = 0.23d0 ) ! force constant
       pe = dk*x*x
       dv = dk*x
      return
      end
      subroutine Fourier(traj,f,p,np,dt)
      implicit real*8 (a-h,o-z)
      real*8 traj(200000)
      sum1 = 0.d0
      sum2 = 0.d0
      t = 0.d0
      do  i = 1,np
        w = 1.0 - abs(dfloat(i-np/2)/dfloat(np/2))  ! Bartlett window
c       w = 1.0                                     ! Square window
c         if (i .eq. 1 .or. i .eq. np) w = 0.5      !  "        "
      t = dfloat(i)*dt
        ft1 = traj(i)*cos(f*t)
        ft2 = traj(i)*sin(f*t)
        sum1 = sum1 + dt*w*ft1
        sum2 = sum2 + dt*w*ft2
      enddo
       p = sum1*sum1 + sum2*sum2
      return
      end
```

Note the new type of declaration made in the line

```
      real*8 traj(200000)
```

The variable `traj` is an array or a subscripted variable. In FORTRAN, to access the k entry in the array, one uses the expression `traj(k)`, as in the statement `write(6,*) traj(k)`, for example. In FORTRAN, all array variables need to be declared, even with implicit typing, and their size must be specified either by using numbers or by using the identifier of a previously assigned constant parameter.

Exercises

1. Run **Brown** and reproduce the results in Figures 2.3 and 2.4. By zooming on the graphs near the local attractors, you should note that the phase space trajectories exhibit fractal properties, in the sense that they look identical at several scales.

2. Run **Verlet_FT**, produce a graph from `fort.9` of the spectrum $P(\omega)$, and confirm that the frequency at the peak is the expected value.

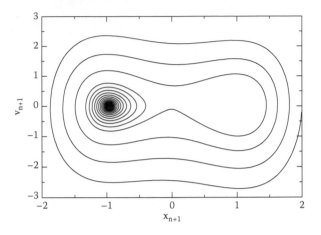

FIGURE 2.3

Phase space trajectory for a particle on the $\gamma = 0.9$ quartic potential with 2.3 units of mass, and with a drag coefficient of 0.1. The particle starts at $x = 2$, and crosses the barrier seven times before spiraling down to the minimum at $x = -\alpha$.

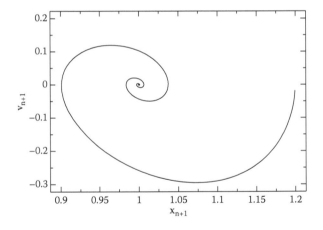

FIGURE 2.4

Phase space trajectory for a particle on the $\gamma = 0.9$ quartic potential with 2.3 units of mass, and with a drag coefficient of 1.0. The particle starts at $x = 1.2$, and spirals down to the minimum at $x = 1$.

3. Inside the subroutine `Fourier`, comment out (by typing a c on the first column) the line marked "Bartlett window" and uncomment the two lines marked "Square window." Recompile and rerun the program. Reproduce the graph of $P(\omega)$ and compare it with its graph obtained in the previous exercise.

4. Reset the integration weights to the Bartlett window weights. Increase the number of time points in the Verlet integration to 100,000, then compare $P(\omega)$ with its graph obtained in the previous exercises.

5. Evaluate the Fourier transform and the power spectrum of a square pulse analytically

$$f(t) = \begin{cases} 0 & t < 0 \\ 1 & 0 \leq t \leq 1 \\ 0 & t > 1, \end{cases}$$

then write a program that computes the Fourier transform numerically.

6. To see the aliasing effects, run the Verlet algorithm for 100,000 time points with the same Δt, so that the trajectory is accurate. However, change the sampling interval for the function $f(t)$, stored in the array `traj`, to $10\Delta t$, $100\Delta t$, ..., until aliasing occurs. This can be accomplished by changing the line

```
traj(k) = (xnp1- x0)
```

in the code to the lines

```
if (mod(k,10) .eq. 0) then
  traj(icc) = (xnp1- x0)
  icc = icc + 1
endif
```

for the $10\Delta t$ case. Note that the calling parameter `np` for `Fourier` needs to be recalculated and `icc` needs to be initiated to zero above the loop over k.

7. Modify `Verlet_FT` so that the potential is the quartic double well with $\gamma = 0.9$. Begin the trajectory at $x = 1$, with a small initial velocity. Compute the expected frequency ω_e by using the local curvature of the potential

$$\omega_e \approx \sqrt{\frac{12(\alpha+1)}{m(2\alpha+1)}}$$

[c.f. Equation 2.105]. Repeat with $x = -\alpha$.

3

Basics of Stochastic Computations

3.1 Introduction

The laws of classical mechanics, which we reviewed in the last chapter, are both incomplete and not suitable for applications to problems of chemical interest. Matter is composed of electrons and nuclei, which are both extremely small and extremely light on the laboratory scale. Light particles have been shown to obey the laws of quantum mechanics, a more complete theory from which classical mechanics emerges in the large mass limit. We introduce quantum mechanics later in the text; however, the typical experiment in chemical physics yields thermodynamic properties, like temperature, pressure, and volume. These variables can be understood as averages of dynamic variables, like the velocity and kinetic energy of samples of matter composed of a large number (on the order of Avogadro's number) of particles. It is impossible to integrate the laws of motion for such large systems. Instead, one constructs a statistical theory based upon models for small fractions of the sample of matter (e.g., for a gas, a single molecule, more generally this fraction is called a system), and seeks probability distributions for the dynamic variables. These fractions are much more convenient to handle theoretically and numerically. Once such a model is constructed, and the probability distributions derived, statistical averages can be obtained by integration. In this manner, gases, liquids, solids, and clusters can be simulated to obtain experimentally observable properties.

The **classical canonical ensemble** is a theoretical construct that allows us to develop a convenient statistical approach to the problem of finding distributions of mechanical variables, and to calculate thermodynamic variables. Our introduction to the classical canonical ensemble is rather brief and informal. Imagine a set of closed systems with a fixed volume, V, and number of particles, N, that are connected with one another. Energy exchange is allowed from system to system. However, since the systems are closed, no matter can cross the boundaries. We also assume that the volume of the systems is constant. If we allow an infinite amount of time, and we assume an infinity of such systems in contact with one another, the ensemble equilibrates to a finite temperature, $T > 0$. The quantities that one calculates from the canonical ensemble are averages, and these have the following simple interpretation. The average calculated corresponds to the average that one would measure experimentally, if a sufficiently large sample of such systems were assembled and allowed to equilibrate for a sufficiently long amount of time. The classical canonical ensemble is also known as the constant N, V, T ensemble. Other variants exist. For instance, the **microcanonical** or constant N, V, E, the **isothermal–isobaric** (constant N, P, T), and the grand canonical ensembles, are convenient for the stochastic study of liquids and solids. The canonical ensemble is the most frequently used tool to study clusters. Therefore, we limit the discussion to the canonical ensemble in this chapter, and we refer the reader to one of the many good texts in statistical mechanics for additional details and alternative ensembles [1]. The numerical methods we present in this chapter can be adapted to alternative ensembles without much difficulty.

The central concept for the theoretical developments of statistical mechanics is the **partition function**. The partition function for the canonical ensemble is introduced, after a brief review of the calculus of continuous random variables and their distributions. The central concept for the numerical developments of statistical mechanics is the **importance sampling** function. Because clusters undergo **phase changes** over relatively broad ranges of temperatures, they can be considered as systems that are solid-like for a significant fraction of the ensemble, and liquid-like for the remaining fraction. The reader will appreciate this more in later chapters, when atomic and molecular clusters are simulated. The unusual phase coexistence at finite temperatures produces additional challenges that are absent in traditional simulations of solids or liquids; these are numerical in nature, and are usually given the collective name **quasiergodicity**. The reader is guided in this chapter to experience quasiergodicity from the Metropolis algorithm, and a numerical strategy to overcome quasiergodicity, **parallel tempering**.

3.2   Continuous Random Variables and Their Distributions

We need to introduce a few basic statistical concepts. The continuous counterpart of a histogram for the random event x, is a function $f(x)$, which has the following interpretation. The probability of locating the event between x and $x + dx$ is the infinitesimal area $f(x)\,dx$. For continuous variables, the distribution must be normalized by integration, i.e.,

$$\int_{\text{over all space}} f(x)\,dx = 1. \tag{3.1}$$

Once the normalization condition is met, the average random event x, also known as the mean of the distribution, can be calculated using

$$\langle x \rangle = \int_{\text{over all space}} x f(x) dx. \tag{3.2}$$

The mean is a measure of central tendency of a distribution; the root mean square event is another measure of central tendency,

$$x_{\mathrm{rms}} = \sqrt{\int_{\text{over all space}} x^2 f(x) dx}. \tag{3.3}$$

Note that

$$\sqrt{\langle x^2 \rangle} \neq \left\langle \sqrt{x^2} \right\rangle, \tag{3.4}$$

since the square root is a nonlinear operator. The most probable event, x_{p}, is yet another measure of central tendency useful for nonsymmetric distributions;

$$\frac{df}{dx}\Big|_{x=x_{\mathrm{p}}} = 0. \tag{3.5}$$

Clearly, a distribution with more than one peak has more than one value of x_{p}. The standard deviation is a measure of the spread of x associated with the distribution,

$$\sigma_x = \sqrt{\langle x^2 \rangle - \langle x \rangle^2}, \tag{3.6}$$

where

$$\langle x^2 \rangle = \int_{\text{over all space}} x^2 f(x) dx. \tag{3.7}$$

These concepts are important for much of the work in this book. We showcase a few simple examples. Let us consider the random event to be a real number limited between x_1 and x_2 with uniform distribution. Then, let us obtain a distribution, normalize it, and use it to find an expression for the mean and standard deviation of x. The distribution is uniform, therefore, $f(x) = C$, where C is a constant. The normalization condition becomes:

$$\int_{x_1}^{x_2} C dx = 1, \tag{3.8}$$

$$C(x_2 - x_1) = 1, \tag{3.9}$$

$$C = \frac{1}{x_2 - x_1}. \tag{3.10}$$

Therefore, the normalized distribution $f(x)$ is simply

$$f(x) = \frac{1}{x_2 - x_1} \qquad (x_1 \leq x \leq x_2). \tag{3.11}$$

The average x can now be calculated,

$$\langle x \rangle = \int_{x_1}^{x_2} \frac{x dx}{x_2 - x_1}, \tag{3.12}$$

$$\langle x \rangle = \frac{x_2 + x_1}{2}. \tag{3.13}$$

The result in Equation (3.13) should not be surprising. Let us calculate the standard deviation predicted by the uniform model between x_1 and x_2. First, we calculate $\langle x^2 \rangle$,

$$\langle x^2 \rangle = \frac{1}{x_2 - x_1} \int_{x_1}^{x_2} x^2 dx = \frac{x_1^2 + x_1 x_2 + x_2^2}{3}. \tag{3.14}$$

The standard deviation for x, predicted by the uniform distribution between x_1 and x_2, is

$$\sigma_x = \frac{x_2 - x_1}{2\sqrt{3}}. \tag{3.15}$$

3.3 The Classical Statistical Mechanics of a Single Particle

The basic assumption of statistical mechanics is that the probability of finding a mechanical system at a point in the neighborhood of a phase space (p, q) with energy $\mathcal{H}(p, q)$, decreases exponentially as the energy of the state increases. For an ensemble of systems containing a particle moving in one dimension, the probability of it being between (p, q) and $(p + dp, q + dq)$ is postulated to be

$$P(p, q)\, dpdq = \frac{\exp\left\{-\beta \mathcal{H}(p, q)\right\} dpdq}{Q}, \tag{3.16}$$

where Q is a normalization constant, known as the **partition function**.

$$Q = \int_{-\infty}^{+\infty} dp \int_V dq \, \exp\left\{-\beta \mathcal{H}(p, q)\right\}. \tag{3.17}$$

The probability distribution in Equation (3.16) is known as the Boltzmann distribution. The average energy is calculated from the Boltzmann distribution using the continuous random variable formalism,

$$\langle \mathcal{H} \rangle = \frac{1}{Q} \int_{-\infty}^{+\infty} dp \int_V dq \, \mathcal{H}(p, q) \, \exp\left\{-\beta \mathcal{H}(p, q)\right\}. \tag{3.18}$$

Note that if the partition function is known as a function of β, one can obtain analytical expressions for the average energy without any additional integration;

$$\langle \mathcal{H} \rangle = -\frac{1}{Q}\frac{\partial Q}{\partial \beta} = -\frac{\partial \ln Q}{\partial \beta}. \tag{3.19}$$

It is instructive to see how this formalism works for some simple systems. Consider a single particle in three-dimensional space. Recall that the Hamiltonian for such a system is,

$$\mathcal{H} = \frac{1}{2m} \left(p_x^2 + p_y^2 + p_z^2 \right). \tag{3.20}$$

The partition function is,

$$Q = \int_{-\infty}^{+\infty} dp_x \int_{-\infty}^{+\infty} dp_y \int_{-\infty}^{+\infty} dp_z$$
$$\times \int dx \int dy \int dz \exp\left\{ -\frac{\beta}{2m} \left(p_x^2 + p_y^2 + p_z^2 \right) \right\}. \tag{3.21}$$

Since $\mathcal{H}$ does not depend on x, y, z, the integral over the three variables yields the volume of the container, V,

$$Q = V \int_{-\infty}^{+\infty} dp_x \int_{-\infty}^{+\infty} dp_y \int_{-\infty}^{+\infty} dp_z \exp\left\{ -\frac{\beta}{2m} \left(p_x^2 + p_y^2 + p_z^2 \right) \right\}, \tag{3.22}$$

and using the group properties of the exponential function, we can write,

$$Q = V \left[\int_{-\infty}^{+\infty} ds \exp\left\{ -\frac{\beta}{2m} s^2 \right\} \right]^3, \tag{3.23}$$

where s is an integration variable. The integral is standard and yields,

$$Q = V \left[\frac{2m\pi}{\beta} \right]^{3/2}. \tag{3.24}$$

The average energy becomes,

$$\langle \mathcal{H} \rangle = -\frac{1}{V} \left[\frac{\beta}{2m\pi} \right]^{3/2} \frac{\partial}{\partial \beta} V \left[\frac{2m\pi}{\beta} \right]^{3/2}, \tag{3.25}$$

$$\langle \mathcal{H} \rangle = \frac{3}{2\beta}. \tag{3.26}$$

This result should be compared with the kinetic theory of matter result for a single particle,

$$\langle K \rangle = \frac{3}{2} k_{\mathrm{B}} T, \tag{3.27}$$

where, k_{B} is Boltzmann's constant. Therefore, $\beta = 1/k_{\mathrm{B}} T$, and we use either symbol interchangeably from now on. The idea behind the classical canonical ensemble is to relate the average properties to thermodynamic variables. The first one we encounter is the energy. The following connection between statistical mechanics and thermodynamics is known as Gibbs' hypothesis,

$$U = \langle \mathcal{H} \rangle. \tag{3.28}$$

The ensemble of single particle systems yields the equations for the energy of a single particle as an ideal monoatomic gas.

3.4   The Monoatomic Ideal Gas

If we permit N noninteracting point particles in each system, then we have a slightly better model for the ideal gas, since we gain knowledge of how extensive thermodynamic variables change linearly with N. The Hamiltonian becomes,

$$\mathcal{H} = \sum_{i=1}^{N} \frac{1}{2m_i} \left(p_{x,i}^2 + p_{y,i}^2 + p_{z,i}^2 \right), \tag{3.29}$$

and the partition function takes the form,

$$Q = \int_{-\infty}^{+\infty} dp_{x,1}, \ldots, \int dz_N \, \exp\left\{ -\sum_{i=1}^{N} \frac{\beta}{2m_i} \left(p_{x_i}^2 + p_{y,i}^2 + p_{z,i}^2 \right) \right\}. \tag{3.30}$$

The integral over all $3N$ configuration coordinates is trivial, and yields the volume of the system raised to the Nth power,

$$Q = V^N \int_{-\infty}^{+\infty} dp_{x,1}, \ldots, \int_{-\infty}^{+\infty} dp_{z,N} \, \exp\left\{ -\sum_{i=1}^{N} \frac{\beta}{2m_i} \left(p_{x_i}^2 + p_{y,i}^2 + p_{z,i}^2 \right) \right\}; \tag{3.31}$$

and making use of the group properties of the exponential function once more, we can simplify the integration further,

$$Q = V^N \prod_{i=1}^{N} \left(\int_{-\infty}^{+\infty} ds \, \exp\left\{ -\frac{\beta s^2}{2m_i} \right\} \right)^3. \tag{3.32}$$

The integrals are all analytical, and the partition function is

$$Q = V^N \left(\frac{2\pi}{\beta} \right)^{3N/2} \prod_{i=1}^{N} m_i^{3/2} = V^N \left(2\pi k_B T \right)^{3N/2} \prod_{i=1}^{N} m_i^{3/2}. \tag{3.33}$$

Q is a function of N, V, and T. The average energy can be calculated by taking the derivative of Q with respect to the temperature, rather than using β. Using the chain rule, it is simple to show that

$$\frac{\partial Q(V, N, T)}{\partial \beta} = \frac{\partial Q(V, N, T)}{\partial T} \frac{\partial T}{\partial \beta}. \tag{3.34}$$

Equation 3.34 requires that T is given as a function of β,

$$T = \frac{1}{k_B \beta}, \qquad \frac{\partial T}{\partial \beta} = -\frac{1}{k_B \beta^2} = -k_B T^2, \tag{3.35}$$

therefore, the average energy can be expressed as a derivative of Q with respect to T,

$$\langle \mathcal{H} \rangle = k_{\mathrm{B}}T^2 \frac{\partial \ln Q}{\partial T} = k_{\mathrm{B}}T^2 Q^{-1} \frac{\partial}{\partial T} V^N (2\pi k_{\mathrm{B}})^{3N/2} T^{3N/2} \prod_{i=1}^{N} m_i^{3/2}, \quad (3.36)$$

$$\langle \mathcal{H} \rangle = k_{\mathrm{B}}T^2 \frac{(3N/2)T^{3N/2-1} V^N (2\pi k_{\mathrm{B}})^{3N/2} \prod_{i=1}^{N} m_i^{3/2}}{V^N (2\pi k_{\mathrm{B}})^{3N/2} T^{3N/2} \prod_{i=1}^{N} m_i^{3/2}}, \quad (3.37)$$

$$U = \langle \mathcal{H} \rangle = \frac{3N}{2} k_{\mathrm{B}}T.$$

This last derivation shows us an important result that we can easily generalize. Note that there are a total of $3N$ terms in the Hamiltonian for N **non interacting** particles. Each such term contributes $k_{\mathrm{B}}T/2$ to the total energy. This result generalizes to what is known as the equipartition theorem.

3.5 The Equipartition Theorem

The average energy for a system containing N square terms in the Hamiltonian is $N k_{\mathrm{B}}T/2$. For the monoatomic ideal gas, there is no potential and there are only three terms for each particle in the kinetic energy, which represent the translation along x, y, and z. Therefore, U is

$$U = \frac{3N}{2} k_{\mathrm{B}}T. \quad (3.38)$$

Consider the rigid dumbbell model for a gas of diatomic molecules. Let there be N such noninteracting rigid rotors inside our thermodynamic systems. In Chapter 2, we derived the Lagrangian and the Hamiltonian for a single rigid rotor,

$$\mathcal{L} = \frac{1}{2} m_T \left(\dot{X}^2 + \dot{Y}^2 + \dot{Z}^2 \right) + \frac{1}{2} \mu r^2 \left(\dot{\theta}^2 + \sin^2 \theta \dot{\phi}^2 \right), \quad (3.39)$$

$$H = \frac{1}{2m_T} \left(p_x^2 + p_y^2 + p_z^2 \right) + \frac{1}{2\mu} \left(\frac{p_\theta^2}{r^2} + \frac{p_\phi^2}{r^2 \sin^2 \theta} \right), \quad (3.40)$$

where $m_T = m_1 + m_2$ is the sum of the two masses, μ is the reduced mass, and r is the fixed distance. The equipartition theorem applies to this Hamiltonian even if a mixture of Cartesian and spherical polar coordinates is used. The partition function for an ensemble of single rigid rotors becomes

$$Q = \int_{-\infty}^{+\infty} dp_X \int_{-\infty}^{+\infty} dp_Y \int_{-\infty}^{+\infty} dp_Z \int dX \int dY \int dZ \int_{-\infty}^{+\infty} dp_\theta \int_{-\infty}^{+\infty} dp_\phi$$

$$\times \int_0^{\pi} d\theta \int_0^{2\pi} d\phi \, \exp\left\{ -\frac{\beta}{2m_T} \left(p_x^2 + p_y^2 + p_z^2 \right) + \frac{1}{2\mu} \left(\frac{p_\theta^2}{r^2} + \frac{p_\phi^2}{r^2 \sin^2 \theta} \right) \right\}.$$

$$(3.41)$$

The integral over X, Y, Z is trivial and yields the volume of the container as before,

$$Q = V \int_{-\infty}^{+\infty} dp_X \int_{-\infty}^{+\infty} dp_Y \int_{-\infty}^{+\infty} dp_Z \int_{-\infty}^{+\infty} dp_\theta \int_{-\infty}^{+\infty} dp_\phi$$
$$\times \int_0^\pi d\theta \int_0^{2\pi} d\phi \, \exp\left\{ -\frac{\beta}{2m_T}(p_x^2 + p_y^2 + p_z^2) + \frac{1}{2\mu}\left(\frac{p_\theta^2}{r^2} + \frac{p_\phi^2}{r^2 \sin^2\theta}\right)\right\}.$$
$$(3.42)$$

The integral over the momenta of the center of mass is the same as for the single particle and the ideal gas. Next, we break up the integration for the angular momenta using the group properties of the exponential function as usual. **Caution!** The integration over the angular momenta has to be carried out before that of the conjugate angles.

$$Q = V\left(\frac{2\pi\, m_T}{\beta}\right)^{3/2} \int_0^\pi d\theta \int_0^{2\pi} d\phi \int_{-\infty}^{+\infty} dp_\theta \, \exp\left\{ -\frac{1}{2\mu}\frac{p_\theta^2}{r^2}\right\}$$
$$\times \int_{-\infty}^{+\infty} dp_\phi \, \exp\left\{ -\frac{1}{2\mu}\frac{p_\phi^2}{r^2 \sin^2\theta}\right\}.$$
$$(3.43)$$

The integrals are all standard. Note that one of the expressions obtained after integrating over the angular momenta depends on the angular coordinate θ,

$$Q = V\left(\frac{2\pi\, m_T}{\beta}\right)^{3/2}\left(\frac{2\pi\mu r^2}{\beta}\right)\int_0^\pi \sin\theta d\theta \int_0^{2\pi} d\phi, \qquad (3.44)$$

and the last two integrals evaluate to 4π,

$$Q = 4\pi V\left(\frac{2\pi\, m_T}{\beta}\right)^{3/2}\left(\frac{2\pi\mu r^2}{\beta}\right) = 4\pi V \left(2\pi\, m_T\right)^{3/2}\left(2\pi\mu r^2\right) k_B^{5/2} T^{5/2}.$$
$$(3.45)$$

The average energy is

$$\langle \mathcal{H}\rangle = k_B T^2 \frac{\partial \ln Q}{\partial T} = \frac{5}{2}k_B T.$$

With this result, it is not hard to prove that for an ideal gas consisting of N such rigid rotors, one has

$$Q = (4\pi V)^N \left(2\pi\, m_T\right)^{3N/2}\left(2\pi\mu r^2\right)^N k_B^{5N/2} T^{5N/2}, \qquad (3.46)$$

$$U = \langle E\rangle = \frac{5N}{2}k_B T, \qquad (3.47)$$

from which it follows trivially that the molar heat capacity at constant volume, $\overline{C}_V$, and at constant pressure, $\overline{C}_p$ are

$$\overline{C}_V = \frac{5}{2}R, \tag{3.48}$$

$$\overline{C}_p = \frac{7}{2}R. \tag{3.49}$$

Where $R = N_A k_B$ is the ideal gas constant. We can compare $\overline{C}_p$ for N_2, O_2, and Cl_2, using the rigid diatomic model ($\overline{C}_p = 7R/2 = 29.099$ J mol^{-1} K^{-1}) to the experimental values found in thermodynamic tables at 298 K.

Molecule	$\overline{C}_p$ experimental $\left(\text{J mol}^{-1}\text{ K}^{-1}\right)$
N_2	29.125
O_2	29.355
Cl_2	33.91

Evidently, some diatomic molecules cannot be treated with the rigid rotor model. For Cl_2, for example, the stretching degree of freedom is relatively softer and contributes to the observed heat capacity. **But why?** And better yet, why are not all three values of $\overline{U}$ equal to $6RT$ (three translations, two rotations, one vibration, and six other square terms from the conjugate momenta), giving a value of $C_p = 58.198$ J mol^{-1} K^{-1}? The rigid rotor model works for N_2 and O_2. The observed heat capacity for most diatomic gases changes with temperature, suggesting that the rigid rotor model breaks down. For a diatomic molecule like H_2, for example, $\overline{C}_V$ is $3/2R$ between 25 and 100 K, it's $5/2R$ between 250 and 750 K, and climbs toward higher values. H_2 dissociates around 2500 K before the heat capacity plateau, corresponding to the equipartition value for the harmonically vibrating diatomic $\overline{C}_p = 58.198$ J mol^{-1} K^{-1}.

What is going on? This phenomenon was not at all understood by scientists in the late eighteen hundreds, and the answer to this nagging question had to wait the quantum revolution. As the temperature is lowered, the high frequency degrees of freedom (e.g., vibrations of N_2 and O_2) drop to the ground state, and the contribution to the heat capacity from those degrees of freedom approaches zero. If we lower the temperature further, the rotation degrees of freedom approach the ground state, and the heat capacity drops further. Eventually, at sufficiently cold temperatures, even the contribution from the translational degrees of freedom drops to the ground state, and the heat capacity drops to zero altogether. These are quantum effects; there is nothing in the classical ensembles that can account for the phenomena we describe. We will return to this point, since this set of phenomena occur frequently in molecular clusters. The fact that many internal stretching and bending degrees of freedom can be considered in the ground state is the crux of most of the advanced stochastic methods presented in the second half of this book. In this chapter, we continue with the classical ensemble, since the classical limit is always an important benchmark used to interpret more complicated simulations.

We make one last point about the rigid rotor example. Note that the differential volume of phase space for the rigid rotor in Equation (3.41) is simply

$$dX \, dY \, dZ \, dp_X \, dp_Y \, dp_Z \, dp_\theta \, dp_\phi \, d\theta \, d\phi.$$

As explained in Chapter 2, there is no Jacobian when changing coordinates for the points in phase space from

$$p_X, p_Y, p_Z, p_1, p_2, p_3, X, Y, Z, q^1, q^2, q^3,$$

to

$$p_X, p_Y, p_Z, p_\theta, p_\phi, X, Y, Z, \theta, \phi.$$

As a last example of the application of the equipartition theorem, consider a solid with regular simple cubic arrangement. In solids, a given atom has three translational degrees of freedom in the kinetic energy, and three vibrational degrees of freedom in the potential energy. Since the vibrations of the atoms are small and close to the equilibrium, we can assume a harmonic potential. Furthermore, if we assume a cubic arrangement, we obtain the following simple Lagrangian,

$$\mathcal{L} = \frac{1}{2}m\dot{x}^2 + \frac{1}{2}m\dot{y}^2 + \frac{1}{2}m\dot{z}^2 - \frac{1}{2}k_x x^2 - \frac{1}{2}k_y y^2 - \frac{1}{2}k_z z^2. \qquad (3.50)$$

The six degrees of freedom give rise to the following thermodynamic averages at equipartition,

$$\overline{U} = 3RT, \qquad \overline{C}_V = 3R, \qquad \overline{C}_p \approx 3R, \qquad (3.51)$$

where, for the constant pressure heat capacity, we use

$$\overline{C}_p - \overline{C}_V = \frac{\alpha^2 TV}{\kappa_T}, \qquad (3.52)$$

and we then neglect the cubic expansion coefficient α, a good approximation for a solid,

$$\overline{C}_p - \overline{C}_V \approx 0. \qquad (3.53)$$

The result, $C_p = 3R$, for simple solids is known as the Dulong and Petit law.

Exercises

1. Fill in the details in the derivation of Equations (3.13) and Equation (3.15).

2. Take the derivative of Equation (3.17) and derive Equation (3.18).

3. The lifetime of radioactive nuclei has the following distribution in $0 < t < \infty$:

$$f(t) = \exp\left[-\gamma t\right].$$

Calculate the mean lifetime of a radioactive nucleus and the standard deviation.

4. Use the probability in the last exercise to calculate the probability that the lifetime of a radioactive nucleus falls between plus or minus one standard deviation from the mean.

5. Given the distribution function,

$$\begin{cases} f(x) = A\sin^2\left(\dfrac{\pi x}{L}\right) & \text{for } 0 \leq x \leq L \\ f(x) = 0 & \text{for all other values of } x, \end{cases} \tag{3.54}$$

find $A, x_{\mathrm{av}}, x_{\mathrm{rms}}, \sigma$, and calculate the probability that a randomly picked value of x is inside the interval

$$x_{\mathrm{av}} - \sigma \leq x \leq x_{\mathrm{av}} + \sigma.$$

3.6 Basics of Stochastic Computation

In this section, the reader learns to use the FORTRAN 90 intrinsic random number generator routine. Random numbers are at the roots of stochastic algorithms (numerical procedures that rely on the laws of probability). In the first part, we generate a set of random numbers that have a uniform distribution. We write a program that allows us to statistically analyze the distribution of numbers.

```
      program random
      implicit real*8(a-h,o-z)
c
      parameter (maxn = 1)   ! Only draw one random number at the time
      real*8 xn(maxn)
      integer iseed(10)
       write(6,*) ' Uniform Random number in (0,1)    '
       write(6,*) '  This program explores a uniform distribution'
c Seeding procedure
c        iseed(1) =  secnds(0.)
c        call random_seed(put = iseed)
       write(6,*) 'How many random numbers do I draw?'
       read(5,*) n
       sum1 = 0.d0
       sum2 = 0.d0
```

```
do k=1,n
  call random_number(xn)
  write(6,*) xn(1)
   sum1 = sum1 + xn(1)
   sum2 = sum2 + xn(1)**2
  enddo
xav = sum1/dfloat(n)
std = sqrt(sum2/dfloat(n) - xav**2)
write(6,*) 'The average is:',xav
write(6,*) 'The standard deviation is:',std
end
```

Upon running **random** with, say $n = 6$, you should see that all n numbers are between zero and one. Run **random** again with the same input. You should see exactly the same n random numbers. This happens because exactly the same default seed gets fed into the intrinsic function **random_number**. The intrinsic random number generator, **random_number**, works by using congruent linear sequences

$$I_{n+1} = aI_n + b \quad \mathrm{mod}\,(m), \tag{3.55}$$

where a and m are large integers. The initial value of the sequence, I_0, is called the seed. The random numbers that are generated by the subroutine become reproducible if the sequence is initiated with the same seed. Reproducible random numbers are not truly random, and are called pseudorandom numbers. The reproducibility of pseudorandom numbers is an important feature of generators, which is useful when testing and debugging code. If the seed is a true random event, like the number of seconds since midnight (obtained from the call to **secnds(0.)** in the comment line), then the sequence of random numbers generated by the function **random_number** is truly random. This can be accomplished with the following lines of code

```
iseed(1) = secnds(0.)
call random_seed(put = iseed)
```

The reader should verify that upon uncommenting the lines, successive runs produce different random numbers.

Let us learn a little more about the distribution of numbers we just generated. The distribution is uniform in $(0,1)$, meaning that the probability that a number selected is between x and $x + dx$ is the same for every value of x between zero and one. The code calculates the average

$$x_{\mathrm{av}} = \frac{1}{n}\sum_{i=1}^{n} x_i, \tag{3.56}$$

and the standard deviation

$$s = \sqrt{\frac{1}{n}\sum_{i=1}^{n}(x_i - x_{\mathrm{av}})^2} = \sqrt{\frac{1}{n}\sum_{i=1}^{n}x_i^2 - x_{\mathrm{av}}^2}, \tag{3.57}$$

for the n numbers we have drawn. The average predicted from a uniform distribution in $[0,1]$ is,

$$\mu = \int_0^1 x\,dx = \frac{1}{2},\tag{3.58}$$

and the standard deviation is,

$$\sigma = \sqrt{\int_0^1 x^2\,dx - \left(\frac{1}{2}\right)^2} = \frac{1}{2\sqrt{3}} \approx 0.2886751.\tag{3.59}$$

See Equations (3.13) and Equations (3.15). The values of x_{av} and s should be close (because we use a finite sample) to μ and σ, respectively. The function `dfloat()` is a standard FORTRAN function that converts the integer `n` into a double precision floating point number.

3.7 Probability Distributions

So far, we have tested the random number generator by evaluating moments of what we think is a uniform distribution. Can one obtain an actual graph of the distribution of a set of random numbers? The answer is yes, if we are happy with a discredited version of a continuous distribution in one or two dimensions. The program we generate will be useful later on. The analysis is carried out by the program `histogram` that follows.

```
      program histogram
      implicit real*8 (a-h,o-z)
      real*8 hist(1000)
      character*72 filename
c Program for post analysis of data generated by
c Markovian chains
      ncat = 1000
      xmax = -1.d230
      xmin = 1.d230
      count = 0.d0
      sumx  = 0.d0
      sumx2  = 0.d0
      do i=1,1000
       hist(i) = 0.d0
      enddo
      write(6,1000)
      read(5,1002) filename
      open (2,file=filename,status='old')
```

```
1      read(2,*,end=100) x
        if (x .gt. xmax) xmax = x
        if (x .lt. xmin) xmin = x
        sumx = sumx + x
        sumx2 = sumx2 + x*x
        count = count + 1.d0
       goto 1
100    avx = sumx/count
       std = sqrt(sumx2/count - avx*avx)
       write(6,1101)
       write(6,1102)
       write(6,1103) count
       write(6,1104) xmax
       write(6,1105) xmin
       write(6,1106) avx
       write(6,1107) std
c prepare the histograms
       cmin = xmin - 0.05*dabs(xmin)
       cmax = xmax + 0.05*dabs(xmax)
       cwidth  = (cmax - cmin)/1000.d0
       rewind(2)
       do m=1,count
        read(2,*) x
        do k=1,1000
         blow = cmin + cwidth*dfloat(k-1)
         kin = k
         bhigh = cmin + cwidth*dfloat(k)
         if (x .gt. blow .and. x .lt. bhigh) then
         goto 110
         endif
        enddo
110    hist(kin) = hist(kin) + 1.d0
       enddo
c integrate the histograms
       sumh = 0.d0
       do i=1,1000
        sumh = sumh + hist(i)*cwidth
       enddo
c normalize the histogram
       totarea = sumh
       do i=1,1000
        hist(i) = hist(i)/totarea
       enddo
       write(6,1101)
       write(6,1108)
```

```
      write(6,1109) cmin
      write(6,1110) cmax
      write(6,1111) cwidth
      write(6,1112) sumh
      write(6,1201)
      write(6,1202)sumh
c The average, standard deviation, skewness and excess of
c the distribution. For comparison, the normal distribution
c has skewness and excess of 0, the exponential distribution
c has skewness and excess of 2 and 6 respectively.
      write(6,1300)
      write(6,1300)
      mu = 0.d0
      sigma = 0.d0
      skew = 0.d0
      excess = 0.d0
      sum0 = 0.d0
      sum = 0.d0
      do k=1,1000
       z = cmin + cwidth*(dfloat(k-1) + 0.5)
       dist = hist(k)
       sum0 = sum0 + dist
       sum = sum + z*dist
      enddo
c mean
      dmu = sum/sum0
      sum = 0.d0
      sum3 = 0.d0
      sum5 = 0.d0
      do k=1,1000
       z = cmin + cwidth*(dfloat(k-1) + 0.5)
       dist = hist(k)/sum0
       sum = sum + ((z - dmu)**2)*dist
       sum3 = sum3 + ((z - dmu)**3)*dist
       sum5 = sum5 + ((z - dmu)**4)*dist
      enddo
c standard deviation
      sigma = sqrt(sum)
c skewness
      skew = sum3/(sum*sigma)
c excess
      excess = sum5/(sum*sum) - 3.d0
c write a report
      write(6,1302) dmu
      write(6,1303) sigma
```

```
       write(6,1304) skew
       write(6,1305) excess
c
       write(6,1101)
       write(6,1108)
       do k=1,1000
        z = cmin + cwidth*(dfloat(k-1) + 0.5)
        write(12,1200) z,hist(k)
       enddo
       close(2)
       stop
1000 format('# Enter the file name')
1002 format(a72)
1101 format('#')
1102 format('# =============== Statistics Report =============')
1103 format('# The sample size is          ',f20.0)
1104 format('# The maximum value of x is ',f20.8)
1105 format('# The minimum value of x is ',f20.8)
1106 format('# The average is             ',f20.8)
1107 format( '# The standard deviation is',f20.8)
1108 format('# ====== Histogram data Report =========')
1109 format('# Beginning ',f14.8)
1110 format('# End',f14.8)
1111 format('# Class width',f14.8)
1112 format('# The normalization constant',f20.8)
1200 format(f12.6,11f12.8)
1201 format('# the relative normalization constants')
1202 format('#',11f11.4)
1300 format('# ====== Individual Histogram data Report =========')
1301 format('# Histogram number ',i5)
1302 format('# Mean          ',f20.8)
1303 format('# Std. dev. ',f20.8)
1304 format('# Skewness   ',f20.8)
1305 format('# Excess      ',f20.8)
       end
```

Assume a set of random numbers is stored as a column in a file. The program reads the data from the file (the file name is the user input) twice. The first time around, the data is analyzed to find the lowest and highest entry in the file. These are used to divide the range into 1000 classes by finding the boundaries for each class. After the program reaches the end of the file, it must read it again from the beginning. This is accomplished by the **rewind** (2) command. Then, the data is read again. This time, if a datum is within the class range specified by the boundaries, the tally for the class (stored in the array hist()) is incremented by one.

Exercises

1. Compile and run the program **random**. To test the program, comment out the line, **write(6,*) xn(1)**, then compile and run **random** with $n = 10^2$, 10^3, 10^4, 10^5, 10^6, and $n = 10^7$. Produce a plot of $\log |x_{av} - \mu|$ as a function of $\log(n)$. The graph should look similar to Figure 3.1. A linear regression analysis yields a slope of -0.60 ± 0.05. With this result, you have just demonstrated numerically, a theorem central to stochastic theory. The **statistical error**, ε, associated with a moment of a random distribution computed with a sample of size n, is proportional to $n^{-1/2}$.

2. Repeat the analysis above to show that the error in s is also proportional to n^{ξ}, and find ξ.

3. The linear transformation

$$y_r = a + (b - a)\, x_r,$$

where x_r is a uniform random number in $(0, 1)$, produces a distribution of y_r uniform in (a, b). Modify **random** and verify this claim for a few values of a and b.

4. Use **random** to evaluate the area of a circle of radius R with the strategy summarized in the pseudocode below.

   - Draw a large number of pairs of random x, y coordinates with uniform distribution between $-R$ and $+R$.

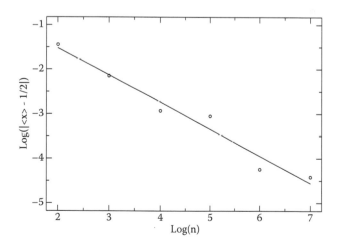

FIGURE 3.1
The common logarithm of the statistical error as a function of the common logarithm of the sample size.

- Count the number of pairs that are hits

$$x^2 + y^2 < R^2, \tag{3.60}$$

and the number of pairs that are misses

$$x^2 + y^2 > R^2. \tag{3.61}$$

- The area of the circle is estimated by the total number of hits divided by the total number of pairs drawn. The excluded area of the square is estimated by the total number of misses divided by the total number of pairs drawn. Calculate the area of the circle with

$$A_c = 4R^2 \frac{n_h}{n}, \tag{3.62}$$

where n_h is the number of hits, and n is the total number of pairs drawn. Compute also the excluded area of the square (of length $2R$) with,

$$A_e = 4R^2 - A_c = 4R^2 \frac{1 - n_h}{n}. \tag{3.63}$$

Run 10^2, 10^3, 10^4, 10^5, 10^6, 10^7, and 10^8 pairs with $R = 1.00$. Calculate the value of the absolute error

$$\left| A_c - \pi R^2 \right| = R^2 \left| 4\frac{n_h}{n} - \pi \right|, \tag{3.64}$$

for each run, and verify that it drops as $n^{-1/2}$.

Essentially, in this exercise, you carry out the numerical evaluation of the integral,

$$A_c = \int_{-R}^{+R} dy \int_{\sqrt{R^2-y^2}}^{\sqrt{R^2+y^2}} dx = \int_0^{2\pi} d\theta \int_0^R r\, dr. \tag{3.65}$$

5. Modify the program **random**, as in the previous exercise, to calculate the average value of e^{-x} in $(0,1)$. What is the theoretical value for this integral?

6. The last exercise demonstrates the use of random numbers to calculate integrals like,

$$\int_0^1 f(x)\, dx.$$

As shown in Exercise 2, it is simple to modify **random** to generate a uniform distribution in (a, b). If x_r is a uniform random number in $(0, 1)$, then y_r is in (a, b) if it is the following linear transformation of x_r,

$$y_r = a + (b - a)\, x_r.$$

Use this transformation to compute the following integral

$$\int_{-\infty}^{\infty} \exp\left(-\frac{x^2}{2}\right) dx.$$

Its value is $\sqrt{2\pi}$. The integrand is a Gaussian function with a unit standard deviation and zero mean. Since the Gaussian function is virtually zero at $x = \mu - 3\sigma$ and $x = \mu + 3\sigma$, the limits of integration can be safely changed to -3 and $+3$. Before attempting this exercise, however, you must realize that simply averaging $\exp\left(-x^2/2\right)$ over the uniform distribution of random numbers from -3 to +3 corresponds to a ratio of integrals

$$\left\langle \exp\left(-\frac{x^2}{2}\right) \right\rangle = \frac{\int_{-3}^{3} \exp\left(-\frac{x^2}{2}\right) dx}{\int_{-3}^{3} dx}.$$

7. Generate a file `fort.8`, which contains a large set of random numbers as follows. Modify the program `random` from the previous section by changing the line, `write(6,*) xn(k)`, in the main section of `random`, to `write(8,*) xn(k)`. Recompile and rerun `random` by drawing 10^6 random numbers. A file, called `fort.8`, should be in your directory. Compile and run `histogram` and enter `fort.8` as the input. Plot the data in `fort.12`. The graph should look like a noisy straight line at $y = 1$.

8. Random number generators may produce sets of numbers that are correlated sequentially. Such sequential correlations are undesirable features, as they introduce a bias in the sampling of a physical space. The sequential correlation coefficient,

$$C_{n,n+i} = \frac{\langle (I_n - \langle I_n \rangle)(I_{n+i} - \langle I_{n+1} \rangle) \rangle}{\langle (I_{n+i} - \langle I_{n+1} \rangle) \rangle \langle (I_n - \langle I_n \rangle) \rangle}, \tag{3.66}$$

is often used to test random number generators. Modify the program `histogram` to estimate $C_{n,n+i}$ for several values of i, then use it to test the intrinsic random number generator.

3.8   Minimizing V by Trial and Error

We are now going to combine the subroutine for the quartic double well potential, presented in the previous chapter, with the random number generator. Random trial-and-error moves are a quick and dirty technique to minimize functions in many dimensions. The bistable quartic potential, introduced in

Chapter 2, can be used to expose the main difficulty encountered when seeking to minimize potentials of molecules. The potential function of a cluster of atoms or molecules of sufficiently large size has an uncountable number of minima (isomers). Only one of these has the lowest possible potential energy, the **global minimum**. It is possible for uniform atomic clusters to have an optically active structure for the global minimum. Depending on the semantics, one can say that these clusters have two distinct global minima. The trial-and-error strategy applied to the particle in the bistable potential illustrates that trapping into a local minimum is a frequent event. The actual algorithm is quite simple: starting from an arbitrary position, a random move is attempted.

$$x'_{n+1} = x_n + \Delta \left(x_{\mathrm{r}} - \frac{1}{2} \right), \tag{3.67}$$

where x_{r} is a uniform random number in $(0, 1)$. If the attempted move lowers the energy, the move is accepted,

$$x_{n+1} \leftarrow x'_{n+1}, \tag{3.68}$$

otherwise it is rejected;

$$x_{n+1} \leftarrow x_n. \tag{3.69}$$

The cycle is repeated many times over, until the random walk converges to a minimum. We introduce the algorithm here for several reasons. Firstly, it is the simplest example of a random walk, and random walks are the heart of stochastic simulations. Secondly, it is interesting from the theoretical point of view, as it corresponds to the $T = 0$ limit of the configuration integral random walk. As we will see later, it is simple to modify it, and implement the Metropolis algorithm. For this reason, the trial-and-error algorithm we present here is sometimes called the $T = 0$ simulated annealing algorithm. Thirdly, the $T = 0$ simulated annealing algorithm can be more efficient than the $T = 0$ Brownian algorithm in Chapter 2, if the gradient of the potential is expensive to compute.

```
      program trial_error
      implicit real*8 (a-h,o-z)
      real*8 xn(1)
      integer iseed(10)
C this program demonstrates the simulated annealing
c algorithm on a bistable potential.
c I/O
      write(6,*) 'Enter gamma between 0 and 1'
      read(5,*) gamma
      write(6,*) 'Gamma is: ',gamma
c seed the generator
       iseed(1) =  secnds(0.)
       call random_seed(put = iseed)
```

```
c start at x = -2.0
      x = -2.0
      call pot(gamma,x,v,dv,d2v)
c dx is the size of the steps
      dx = 0.02
      do moves = 1,1000
c now try a move at random with a step between -dx/2 and + dx/2
      call random_number(xn)
      xt = x + dx*(xn(1) - 0.5)
c vt is the potential at the trial position
      call pot(gamma,xt,vt,dvt,d2vt)
c test to see if the potential energy is less then before
      if (vt .lt. v) then
c accept the move
      x = xt
      v = vt
      else
c reject the move
      endif
c keep track of the walk
      write(10,*)moves, x,v
      enddo
      end
```

Exercises

1. Append the subroutine **pot** with the quartic potential of Chapter 2, to the program **trial_error**, compile, and run the program with $\gamma = 0.9$. Obtain a graph of the data in fort.10 (first column as x and second column as y) and compare it with the graph in Figure 3.2. The y axis of this graph is the position x. Notice that the algorithm finds the minimum at $x = -\alpha$, which means that it is trapped into a local minimum and it failed to find the global minimum.

2. Change the starting position to see what it takes to find the global minimum at $x = 1$. Do this by starting from a random place between -2 and + 2, instead of using the assignment x = -2 before the moves loop, then use an outer loop to restart the simulated annealing from random several times. Graph the content of fort.10 (first two columns as before, use points, no lines between points or the graph will look confusing).

3. Set the initial position at -2 and systematically increase Δ using an outer loop again, until the algorithm once again finds the global minimum. Continue to loop until the global minimum is found from the -2 position.

4. Modify the program **trial_error** so that it finds the transition state(s) for the potential energy surface. One way to accomplish this is to

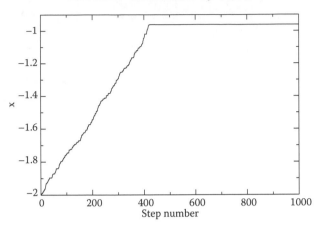

FIGURE 3.2
The random walk (trapped into the $x = -\alpha$ well) from a trial and error run.

minimize the force(s), but one has to perform a second derivative test to separate minima from transition states. The transition state for the quartic potential is at $x = 0$, where $V = 1$.

3.9 The Metropolis Algorithm

Let us return to the classical ensemble. So far, our discussion has been limited to some simple systems for which there are no interactions (i.e., $V = 0$), or the potential is harmonic. To study nonideal gases, liquids, imperfect solids, and clusters, we need to be able to develop tools for the integration of ensemble averages with nonharmonic potentials. Suppose that one would like to know the average energy for an ensemble of such systems,

$$E_{\mathrm{av}} = \frac{\displaystyle\int_{-\infty}^{\infty} d\mathbf{x}\,d\mathbf{p}\,\mathcal{H}(\mathbf{x},\mathbf{p})\exp(-\mathcal{H}(\mathbf{x},\mathbf{p})/k_{\mathrm{B}}T)}{\displaystyle\int_{-\infty}^{\infty} d\mathbf{x}\,d\mathbf{p}\,\exp(-\mathcal{H}(\mathbf{x},\mathbf{p})/k_{\mathrm{B}}T)}. \tag{3.70}$$

One has to consider both the kinetic and the potential energy. The Hamiltonian is

$$\mathcal{H}(\mathbf{x},\mathbf{p}) = \sum_{i=1}^{n} \frac{1}{2m_i}\mathbf{p}_i^2 + V(\mathbf{x}),$$

where $V(\mathbf{x})$ is the potential energy at position $\mathbf{x}$, and $\mathbf{p}_i^2$ is the size of the momentum vector squared for atom i. The momentum integration can be carried out analytically. The integration over all momenta yields at once,

$$E_{\mathrm{av}} = \frac{3k_{\mathrm{B}}T}{2}n + V_{\mathrm{av}}, \tag{3.71}$$

where V_{av} is the average potential energy defined as

$$V_{\mathrm{av}} = \frac{\displaystyle\int_{-\infty}^{\infty} d\mathbf{x}\, V(\mathbf{x}) \exp\left(-V(\mathbf{x})/k_{\mathrm{B}}T\right)}{\displaystyle\int_{-\infty}^{\infty} d\mathbf{x}\, \exp\left(-V(\mathbf{x})/k_{\mathrm{B}}T\right)}. \tag{3.72}$$

The ratio of integrals in Equation (3.72) are known as the configuration integrals. For a molecule of n atoms, V_{av} still requires a couple of $3n$ fold integrals. Computing them with a trapezoid rule requires computer resources that grow exponentially with n. The Metropolis algorithm works very well in many dimensions, contrary to other numerical integration schemes. The integral in the denominator of Equation (3.72) is known as the configuration integral. For now, we calculate the ensemble average of the potential energy for a particle trapped in the quartic double potential, since it is very instructive. Equation (3.72) becomes a ratio of two integrals in one dimension.

The trick of the Metropolis algorithm is to transform the series of random numbers from a uniform distribution into a series of random numbers (referred to as a random walk, or a Markov chain) that are distributed according to $f(\mathbf{x})$

$$f(\mathbf{x}) = \frac{\exp(-V(\mathbf{x})/k_{\mathrm{B}}T)}{\displaystyle\int_{-\infty}^{\infty} d\mathbf{x}\, \exp(-V(\mathbf{x})/k_{\mathrm{B}}T)}, \tag{3.73}$$

using a rejection technique. The numerical method used for the production of random numbers that are distributed according to some given continuous distribution $f(\mathbf{x})$, is called **importance sampling**.

To test the Metropolis algorithm, we need to obtain a graph of Equation (3.73) for the quartic double potential, by other means. For a problem in one dimension, the distribution can be easily obtained with the extended trapezoid rule. If we run a Metropolis walk, and analyze the distribution of x with histograms, and then compare these histograms with the distribution obtained by the extended trapezoid rule, we can learn a great deal. In the program exact_distribution, the integral in the denominator of Equation (3.73) is evaluated by an extended trapezoid quadrature.

```
program exact_distribution
implicit real*8 (a-h,o-z)
real*8 xn(10)
  write(6,*) 'Enter gamma'
```

```
      read(5,*) gamma
      write(6,*) 'Gamma is: ',gamma
c calculate the normalization constant
      write(6,*) 'Enter a temperature in K'
      read(5,*) temp
      write(6,*) 'T = ',temp
      x = -2
      sum = 0.d0
      dx = 0.02
      do k=1,200
        call pot(gamma, x,v,dv,d2v)
c the units for the potential are Kelvin
c so V/kT  = V/T
        sum = sum + dx*exp(-v/temp)
        x = x + dx
      enddo
c obtain the theoretical distribution
      x = -2
      dx = 0.02
      do k=1,200
        call pot(gamma, x,v,dv,d2v)
        write(1,*) x, exp(-v/temp)/sum
        x = x + dx
      enddo
      end
```

Append the quartic double well potential subroutine and run the program
with a temperature of 1 K, $\gamma = 0.9$, and look for a file called `fort.1` in your
directory. Graph the data in `fort.1`. As commented in the program, the units
we take for the potential are Kelvin. One can think of this choice of units for
the potential energy as follows. To obtain the parameters in `pot`, we divide
the potential energy by the Boltzmann constant. Assume this is done already,
and simply use `pot` as it is. The Kelvin is the preferred unit to use for Monte
Carlo calculations for obvious reasons. Look up how many J correspond to one
Kelvin in an energy conversion table, but do not add this conversion factor in
your code.

Rerun the program with a temperature of 0.05 K, $\gamma = 0.9$, and graph the
data in `fort.1` again. You should still see two peaks, one above $x = -\alpha$ and
one above $x = 1$, but at the lower temperature, the peak over $x = -\alpha$ should
be considerably smaller and the peaks much narrower.

For the stochastic simulation, the algorithm invented by Metropolis requires only a slight modification of the trial-and-error procedure, which you
learned in the last section. The idea of the rejection technique is very similar to
the idea of evaluating the area of a circle. Suppose we would like to draw random numbers distributed according to Equation (3.73). The algorithm takes
the following form,

i. Draw two random numbers ξ_1, ξ_2.

ii. Change x to

$$x' = x + \Delta \left(\xi_1 - \frac{1}{2}\right). \tag{3.74}$$

iii. Calculate

$$q = \frac{\exp[-V(x')/k_B T]}{\exp[-V(x)/k_B T]}. \tag{3.75}$$

iv. Compare q and ξ_2, if $\xi_2 < q$ accept x', otherwise reject x'.

The step size Δ has to be adjusted so that roughly 50% of the moves are accepted. Then, the average potential energy is simply

$$V_{\text{av}} = \frac{\displaystyle\int_{-\infty}^{\infty} dx\, V(x)\exp(-V(x)/k_B T)}{\displaystyle\int_{-\infty}^{\infty} dx\, \exp(-V(x)/k_B T)} \approx \frac{1}{N}\sum_{i-1}^{N} V(x_i), \tag{3.76}$$

where N is the number of Metropolis moves and $\{x_i\}_{i=1}^{N}$ are all the positions visited during the random walk. Let us calculate the average potential energy for a particle inside the quartic potential at different temperatures, and obtain the distributions from the random walks using the program metropolis.

```
      program metropolis
      implicit real*8 (a-h,o-z)
      real*8 xn(2)
      integer iseed(10)
      gamma = 0.9d0
      temp = 1.0d0
      nmoves = 1000000
      deltax = 3.5
c seed the random number generator
      iseed(1) =  secnds(0.)
      call random_seed(put = iseed)
c the units for the potential are Kelvin
c so V/kT  = V/T
      x = 1.0
      nrej = 0
      sumv = 0.d0
      call pot(gamma, x,v,dv,d2v)
      do moves=1,2*nmoves
```

```
c now try a metropolis move
          call random_number(xn)
          xt = x + deltax*(xn(1) - 0.5)
c vt is the potential at the trial position
          call pot(gamma, xt,vt,dvt,d2vt)
c test to see if it is a hit or a miss
          q = exp(-vt/temp)/exp(-v/temp)
          if (xn(2) .lt. q) then
c accept the move
              x = xt
              v = vt
          else
c reject the move
              nrej = nrej + 1
          endif
c keep track of the walk and
c accumulate the average energy.....
       if (moves .gt. nmoves) then
              write(11,*) x
              sumv = sumv + v
        endif
        enddo
          write(6,*) 'Average potential :'
          write(6,*) sumv/dfloat(nmoves),' Kelvin'
          write(6,*) 'Percent rejection:',
     &    100.*dfloat(nrej)/dfloat(2*nmoves)
          end
```

The subroutine pot for the quartic double well potential needs to be appended. Note how similar this code is to the code in trial_error of the previous section. We will see that the walk, however, is very different. It is also noteworthy to mention that the data points $\{V(x_i)\}_{i=1}^{N}$ for the average potential and the distribution of x are collected after the first nmoves. The random walk needs an "equilibration" period. The correct distribution is reached asymptotically.

Exercises

1. With the Metropolis algorithm, we can evaluate the moments of a continuous distribution $f(x)$ of random numbers using a sum over a random walk,

$$x_{\mathrm{av}} = \frac{1}{n}\sum_{i=1}^{n} x_i \approx \frac{\int x f(x) dx}{\int f(x) dx}, \qquad (3.77)$$

where the integration is over all possible values of x. Modify `metropolis` to compute the following integral

$$\frac{\displaystyle\int_{-\infty}^{\infty} x^2 \exp\left(-\frac{x^2}{2}\right) dx}{\displaystyle\int_{-\infty}^{\infty} \exp\left(-\frac{x^2}{2}\right) dx}.$$

Since the integrals are known analytically, this exercise represents a rigorous test of the program `metropolis`.

2. Append the quartic double well `pot`, compile and run the program `metropolis` with $\gamma = 0.9$, a temperature of 1.0 K, with 10^6 moves, and with $\Delta = 3.5$. Analyze the Metropolis walk by obtaining a histogram of the position distribution. Run `histogram` with `fort.11` as input. Run the program `exact_distribution` with a temperature of 1.0 K and $\gamma = 0.9$. Graph `fort.12`, which contains the histogram of the Metropolis walk together, and `fort.1`, which contains the exact distribution. The graph of the data in `fort.1` and `fort.12` should look like Figure 3.3.

3. Modify the program `exact_distribution` so that it computes V_{av}. Compute its value at 1.0 K and determine its error by running several computations with different values of n. Then, use the "exact" value of V_{av} to estimate the statistical error from `metropolis` runs with $n = 10^2, 10^3, 10^4, 10^5, 10^6, 10^7$. Confirm that the statistical error drops as $n^{-1/2}$, where n is the number of moves stored in `nmoves`.

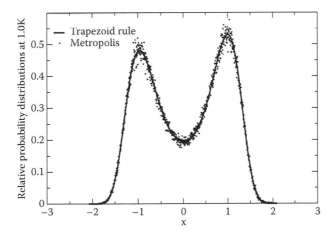

FIGURE 3.3
Exact (heavy line) and simulated probability $(\cdots)$ distributions for a particle in a quartic potential at 1.0 K.

4. Run `metropolis` with 10^6 moves at 1.0 K 10 times (each time with a new seed), average the values of V_{av} and obtain the standard deviation. The inverse of the standard deviation from these independent blocks is a rough measure of the efficiency of the algorithm. Redo the 10 simulations for 5 values of Δ between 0.3 and 14.3. Obtain the average percent rejections and the efficiency for each value of Δ, then graph the efficiency as a function of the percent rejections.

5. Run `metropolis` with $\gamma = 0.9$, a temperature of 0.05 K, with 10^6 Metropolis moves, and with $\Delta = 0.5$. Analyze the Metropolis walk by obtaining a histogram of the position distribution. Run `histogram` with `fort.11` as input. Run the program `exact_distribution` with a temperature of 0.05 K and $\gamma = 0.9$. Graph `fort.12`, which contains the histogram of the Metropolis walk together, and `fort.1`, which contains the exact distribution. The graph of the data in `fort.1` and `fort.12` should look like Figure 3.4.

Note that the Metropolis walk actually did not sample the distribution adequately, as the data labeled "metropolis" shows. Since we started the calculation on the well on the right, it never sampled the well on the left. Both minima are important at 0.05 K, however. Metropolis is trapped to sample only one! This phenomenon has been known for quite some time, and is called **quasiergodicity**. In the next section, we learn one of the ways that have been proposed to handle quasiergodicity, **parallel tempering**.

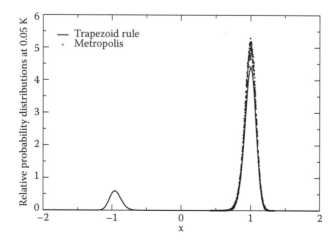

FIGURE 3.4

Exact (heavy line) and simulated probability ($\cdots$) distributions for a particle in a quartic potential at 0.05 K. The simulated probability visibly overestimates the lowest well and underestimates the second one.

6. Investigate the behavior at 0.001 K. Run `metropolis`, use `histogram`, and compare the distribution with the results from `exact_distribution` at the same temperature. Are the results in agreement? Repeat the Metropolis run at 0.001 K, but start the walk at $x = -1.2$. Examine the walk with `histogram` again. What do you observe?

7. Obtain the sequential correlation coefficients for x at 1.0 K,

$$C_{n,n+i} = \frac{\langle (x_n - \langle x_n \rangle)(x_{n+i} - \langle x_{n+i} \rangle) \rangle}{\langle (x_{n+i} - \langle x_{n+i} \rangle) \rangle \langle (x_n - \langle x_n \rangle) \rangle}, \tag{3.78}$$

for $i = 1, 2, 3, 10, 50, 100, 500, 1000$, by writing the positions during the random walk to a device file, and then running the modified version of `histogram` developed in a previous exercise. Compare the sequential correlation coefficients obtained with the Metropolis walk, with the one obtained from the intrinsic random number generator. What do you observe?

3.10   Parallel Tempering

The analysis in the last section demonstrates that simulations with potentials that have minima wells close in energy and separated by significant barriers, compared with $k_B T$, are problematic. Metropolis walk ergodicity issues are a cause for worry. One can only imagine the complexities of the multidimensional potential energy surface for a cluster of molecules, to cringe at the thought of using the Metropolis algorithm. The Metropolis algorithm has been used in many applications, but mainly for the simulations of liquids, where the well-to-well hopping is frequent and the barriers are not large, compared to $k_B T$. The parallel tempering scheme [3, 4], is one of several approaches that have been recently proposed in the literature to deal with quasiergodicity.

In `parallel_tempering`, we rewrite the Metropolis algorithm to run an array of walkers, each at a different temperature, simultaneously. The temperatures are between the interval $T_{\min}$ and $T_{\max}$, entered by the user and stored in the variables `tempm` and `tempx`. Then, for $n_W > 1$ walker (also user input stored in `nwalkers`), the temperature for the ith walker is,

$$T_i = \frac{(i-1)(T_{\max} - T_{\min})}{n_W - 1}.$$

The program is structured so that a move is attempted for all the walkers and not the other way around, meaning a single move is attempted for walker one, than a move for walker 2, and so on. The order is irrelevant for Metropolis simulations, but becomes important for the implementation of the parallel tempering algorithm. The updated walk is kept in the array `ym1`. The program writes to `fort.11`, the random walk at the highest temperature.

The trick to parallel tempering is to add a new kind of move that allows "communication" between walkers, so that the high temperature "view" of the potential surface can trickle down to the last walker. The move is a configuration swap between two adjacent walkers. We must do the move carefully, and either accept or reject the move using detailed balance to ensure that the walk remains Markovian. The probability for a swap between a walker at x_1 with temperature T_1, and a walker at x_2 with temperature T_2, is

$$p_s = \exp\left[-\left(\frac{1}{T_1} - \frac{1}{T_2}\right)(V(x_2) - V(x_1))\right]. \qquad (3.79)$$

A generalized version of Equation (3.79) is derived in Chapter 9, where the concept of detailed balance is introduced. For now, let's learn what these new types of moves do. We modify the move loop of `metropolis` and add the swap move as follows,

```
      program parallel_tempering
      implicit real*8 (a-h,o-z)
      real*8 xn(1),y(100),ym1(100),vk(100),temp(100)
      real*8 sumv(100),sumv2(100),deltax(100)
      integer nrej(100),nrejs(100), iseed(10),nrejt(100)
      gamma = 9.0d-1
      tempm = 5.0d-2
      tempx = 1.0d0
      nwalkers = 20
      nmoves = 1000000
      do nk = 1, nwalkers    ! initialize the arrays
       y(nk) = 1.d0
       ym1(nk) = 1.d0
       sumv(nk) = 0.d0
       sumv2(nk) = 0.d0
       deltax(nk) = 0.2
       x = y(nk)
       call pot(gamma, x,v,dv,d2v)
       vk(nk) = v
       temp(nk) = tempm + dfloat(nk-1)*(tempx-tempm)/dfloat(nwalkers-1)
       nrej(nk) = 0
       nrejs(nk) = 0
       nrejt(nk) = 0
      enddo
c seed the generator
       iseed(1) = secnds(0.)
      call random_seed(put=iseed)
      do moves=1, 2*nmoves
      do nk = 1, nwalkers        ! loop over all the walkers
       call random_number(xn)
       if (xn(1) .gt. 0.9 .and. nk .ne. nwalkers)  then ! 10 % of the time
c ************************** SWAP
          dv = (1.d0/temp(nk) - 1.d0/temp(nk+1))*(vk(nk+1) - vk(nk))
```

```
      q = exp(-dv)
      call random_number(xn)
      if (xn(1) .lt. q) then
        y(nk) = ym1(nk+1)            ! accept
        v = vk(nk+1)
        ym1(nk+1) = ym1(nk)
        vk(nk+1) = vk(nk)
        ym1(nk) = y(nk)
        vk(nk) = v
      else
        nrej(nk) = nrej(nk) + 1 ! reject
        nrejs(nk) = nrejs(nk) + 1
      endif
    else
c ************************** METROPOLIS move
      call random_number(xn)
      y(nk) = ym1(nk) + deltax(nk)*(xn(1) - 0.5)
      x = y(nk)
      call pot(gamma,x,v,dv,d2v)
      dv = (v - vk(nk))/temp(nk)
      q = exp(-dv)
      call random_number(xn)
      if (xn(1) .lt. q) then
c ************************** accept the move
      ym1(nk) = y(nk)
      vk(nk) = v
      else
c ************************** reject the move
      nrej(nk) = nrej(nk) + 1
      nrejt(nk) = nrejt(nk) + 1
      endif
      endif
      if (moves .gt. nmoves) then
      if (nk .eq. 1) write(11,*) ym1(nk)
      x = ym1(nk)
      call pot(gamma,x,v,dv,d2v)
      sumv(nk) = sumv(nk) + v/dfloat(nmoves)
      sumv2(nk) = sumv2(nk) + v*v/dfloat(nmoves)
      else
c adjust deltax automatically during the equilibration run
      if (mod(moves,1000) .eq. 0) then
        if (nrejt(nk) .gt. 550) deltax(nk) = deltax(nk)*0.9
        if (nrejt(nk) .lt. 450) deltax(nk) = deltax(nk)/0.9
        nrejt(nk) = 0
      endif
      endif
      enddo      ! walker loop
      enddo
      do nk=1,nwalkers
```

```
        e = 0.5*temp(nk) + sumv(nk)
        cv = 0.5d0 + (sumv2(nk) - sumv(nk)*sumv(nk))/(temp(nk)*temp(nk))
        write(6,1000) temp(nk),e,sumv(nk),cv,deltax(nk),nrej(nk),nrejs(nk)
        enddo
1000    format(5f15.8,2i8)
        end
```

Run the program with $\gamma = 0.9$, the lowest temperature 0.05 K, the highest 0.9 K, 20 walkers, and with `deltax = 1.2`. This time, the random walk for the walker at 0.05 K is written to the file `fort.11`. The agreement between the theoretical and the simulated distributions should now be excellent, even for the 0.05 K walker. Append the subroutine `pot`, before attempting to compile the program. The data in `fort.11` should be processed with `histogram`, and the simulated distribution at 0.05 K in file `fort.12` should match the content of `fort.1` obtained by running `exact_distribution` with $T = 0.05$ K.

Understanding quasiergodicity can only come with experience. The following exercises are designed to explore parallel tempering and the particle in a quartic double well. The thermodynamics of a particle in the quartic double well bear many similarities to those of atomic and molecular clusters.

3.11   A Random Number Generator

The intrinsic FORTRAN 90 random generator is sufficiently robust for training purposes. For research purposes, our group has relied on the robustness of a version of a random number generator, which originated from the Blöte group, and has been later modified by the Nightingale group. We reproduce their subroutine with their permission. We use the subroutine `ransi` later on.

```
        program test_ransi
        implicit real*8(a-h,o-z)
        real*8 xr(10)
        iseed = 11
        max = 1000000
         call ransi(iseed)
         do j=1,max
         call ransr(xr,1)
          write(8,*) xr(1)
         enddo
        end
c Date of last modification: Thu Aug 29 16:53:45 EDT 1996
        subroutine ransi(iseed)
c ransr: purpose: generate uniform random numbers (fast version)
c shift register random generator with very long period
c combined with linear congruential rn generator
```

```
c sequential version
c ransr and ransr_2 differ:
c the first is a shift register xor-ed with a linear congruential
c the second xor-es two shift registers
c::::::::::::::::::::::::::::::::::::::::::::::::::::::::::::::::c
c Copyright: H.W.J. Bloete and M.P. Nightingale, June 1995  c
c::::::::::::::::::::::::::::::::::::::::::::::::::::::::::::::::c
      implicit real*8(a-h,o-z)
c The following parameters for a linear congruential rn generator
c are from Numerical Recipes "An Even Quicker and Dirtier Generator."
c It relies on integer arithmetic that ignores overflows.
c The second set is the same rn generator iterated twice.
      parameter (MULT1=1664525   ,IADD1=1013904223)
      parameter (MULT2=389569705,IADD2=1196435762)
c In k'=mod(k*mult+iadd+2**31,2**32)-2**31
c mod and shift are assumed to be done implicitly by hardware
c the next multiplier is no longer used
c       parameter (mult=32781)
      parameter (TWO=2,TM32=two**(-32),HALF=1/TWO)
      dimension rn(1)
      parameter (LENR=9689,IFDB=471)
      common/ransrb/ irs(LENR),inxt(LENR)
      save ipoint,ipoinf,k
      k=3**18+1seed
      l=MULT1*k+IADD1
c initialize shift registers
      do i=1,LENR
        k=k*MULT2+IADD2
        l=l*MULT2+IADD2
        irs(i)=ieor(k,ishft(l,-16))
        inxt(i)=i+1
      enddo
      inxt(LENR)=1
      ipoint=1
      ipoinf=ifdb+1
      return
c
      entry ransr(rn,n)
c     calculate n random numbers
      do i=1,n
        l=ieor(irs(ipoint),irs(ipoinf))
        k=k*MULT1+IADD1
        rn(i)=ieor(k,l)*TM32+HALF
        irs(ipoint)=l
        ipoint=inxt(ipoint)
        ipoinf=inxt(ipoinf)
      enddo
      return
      end
```

Exercises

1. Change the parameter γ, systematically, to values closer and closer to 1, but keep all the other parameters the same, namely, $T_{max} = 1.0$ K, $T_{min} = 0.05$ K, and $n_W = 20$. Can you get `parallel_tempering` to fail as $\gamma \to 1$ with these parameters?

2. For $\gamma = 0.9$, $T_{min} = 0.05$ K, and $n_W = 20$, lower, systematically, the value of T_{max} until `parallel_tempering` visibly fails to yield the correct distribution at $T_{min} = 0.05$ K.

3. Are 20 walkers necessary for simulations at $\gamma = 0.9$, $T_{max} = 0.9$ K, and $T_{min} = 0.05$ K? Try running the algorithm with 16, 12, 8, 4, and 2 walkers, and find out.

4. Consult Figure 3.5. At 1.0 K, the ensemble composed of one particle in a double well system is "gaseous-like," and C_v/k_B tends toward 1/2 (the ideal monodimensional gas). At 0.01 K and below, it is solid-like and in the global minimum. The value of C_v/k_B approaches 1.0 (1/2 from the average kinetic energy and 1/2 from the harmonic well centered at $x = 1$). Between 0.02 and 0.1 K, the two minima are in coexistence in a solid-liquid exchange. At higher temperatures, the ensemble averages smaller and smaller contributions from the potential, compared to the kinetic energy, and the systems begin to approach the ideal gas behavior. All these "phase changes" are best seen by inspecting the heat capacity

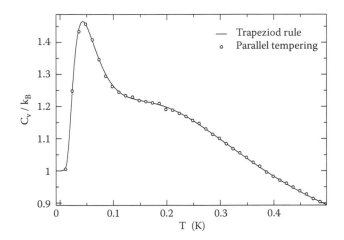

FIGURE 3.5
Constant volume heat capacity for a particle in a quartic double well potential from 0.001 K to 0.5 K.

obtained as the following average,

$$\frac{C_v}{k_B} = \frac{1}{2} + \frac{\langle V^2 \rangle - \langle V \rangle^2}{T^2}, \qquad (3.80)$$

assuming the potential V is in K. Modify exact_distribution and compute C_v/k_B for 1000 values of T, from 0.001 K up to 1 K. Then, use parallel tempering with $\gamma = 0.9$, $T_{min} = 0.01$, $T_{max} = 1.0$ K, and $n_W = 80$, to reproduce points in the same graph. These should look like Figure 3.5.

5. Change the line of code

```
if (xn(1) .gt. 0.9 .and. nk .ne. nwalkers)
then ! 10 % of the time
```

to

```
if (xn(1) .gt. 1.1 .and. nk .ne. nwalkers)   then ! never !
```

so that swaps are never attempted. Run the program with $\gamma = 0.9$, $T_{min} = 0.01$, $T_{max} = 1.0$ K, and $n_W = 80$, and compare the heat capacity with the results in Figure 3.5. Where is the Metropolis walk not **ergodic**?

6. Examine distributions of V at several temperatures in the solid, solid-liquid, and in the gas phase. Place all these in the same graph, then find two temperatures for which the distributions of V do not overlap. Use these two temperatures for a two-walker parallel tempering simulation. Monitor the number of rejected swaps.

7. Modify random to compute the two integrals involved in the computation of V_{av} at $T = 0.05$ K, using a uniform distribution from -2 to $+2$. Once modified, run the simulation 10 times with one million random numbers. Repeat 10 times with a new seed. Use these independent blocks to estimate the efficiency of the stochastic integration. What are your conclusions?

8. The **ergodic hypothesis** states that the **time average** of a physical property $\langle A(x) \rangle$, computed with

$$\langle A(x) \rangle = \lim_{t \to \infty} \frac{1}{t} \int_{t_0}^{t_0+t} A[x(\tau)] \, d\tau, \qquad (3.81)$$

where $x(\tau)$ is a **single** solution of Hamilton's equations, is the same as the average computed stochastically with

$$\langle A(x) \rangle = \frac{\int_{-\infty}^{\infty} d\mathbf{x}\, d\mathbf{p}\, A(x) \exp(-\mathcal{H}(\mathbf{x}, \mathbf{p})/k_B T)}{\int_{-\infty}^{\infty} d\mathbf{x}\, d\mathbf{p}\, \exp(-\mathcal{H}(\mathbf{x}, \mathbf{p})/k_B T)}. \qquad (3.82)$$

Note that a single long integration of Hamilton's equation is a constant energy simulation, the temperature is computed as an average in the microcanonical ensemble,

$$\langle T \rangle = \lim_{t \to \infty} \frac{2}{t} \int_{t_0}^{t_0+t} \frac{K}{k_B} d\tau, \qquad (3.83)$$

where K is the kinetic energy and k_B is Boltzmann's constant. Continue to use units of Kelvin for the potential and kinetic energy, and assume a unit mass in the quartic double well. Integrate Hamilton's equations using the Candy–Rozmus, as in Chapter 2. Compute the average temperature using

$$\langle T \rangle = \lim_{t \to \infty} \frac{1}{t} \int_{t_0}^{t_0+t} p^2 (\tau) \, d\tau, \qquad (3.84)$$

and the average potential using

$$\langle V(x) \rangle = \lim_{t \to \infty} \frac{1}{t} \int_{t_0}^{t_0+t} V[x(\tau)] \, d\tau. \qquad (3.85)$$

Start the trajectory at $x = 1$, and run the dynamic simulations for several energy values. Compare the average potential energy at $\langle T \rangle \approx 0.01$ K with that predicted by parallel tempering. Repeat at $\langle T \rangle \approx 0.05$ K. Are Hamilton's equations ergodic? Quasiergodicity can be fixed for molecular simulations by using an algorithm similar to parallel tempering, called **replica exchange**.

9. Test the routine **ransi** and verify that the distribution in $(0,1)$ is uniform, and that several sequential correlation coefficients are below statistical significance.

4

Vector Spaces, Groups, and Algebras

4.1 Introduction

In this chapter, we introduce the reader to several fundamental concepts. We begin by introducing matrices, and operations that pertain to them. Matrix addition and scalar multiplication, for example, allow us to demonstrate that all $n \times n$ matrices with entries from the number field $\mathbb{F}$ belong to a vector space, denoted $\mathcal{V}(n \times n, \mathbb{F})$. Since the definition of vector spaces requires fields and is constructed by adding structure to a group, we briefly introduce these concepts. In this chapter, the theory of Lie algebras and Lie groups over the real number field $\mathbb{R}$ will be introduced first, as these are easier to handle. However, vector space and group theory applications to quantum mechanics require the extension of these concepts to the complex field $\mathbb{C}$. The definition of matrix product allows us to add algebraic structure to the vector space $\mathcal{V}(n \times n, \mathbb{R})$ and generate the **general linear algebra** of order n. The reader is guided to learn the noncommutative nature of the matrix product from which antisymmetrized and symmetrized product definitions can be constructed. The antisymmetric commuted product and the Jacobi identity, are used to introduce two important Lie algebras: $so(n, \mathbb{R})$ and $sp(n, \mathbb{R})$. Using Taylor's theorem, we construct the corresponding **special orthogonal** groups, $SO(n, \mathbb{R})$, and the **symplectic** groups, $Sp(n, \mathbb{R})$. The complex field equivalent of the $so(n, \mathbb{R})$ algebra and the special orthogonal group are $su(n, \mathbb{C})$ and $SU(n, \mathbb{C})$, respectively, the special unitary Lie algebra and group. We revisit the general linear group $Gl(n, \mathbb{R})$ on the real number field and $Gl(n, \mathbb{C})$ on the complex number field, as groups of continuous linear transformations on two sets of vector spaces, $\mathcal{V}(1 \times n, \mathbb{R})$ and $\mathcal{V}(n \times 1, \mathbb{R})$. The Lie groups are subgroups of these general groups over the respective fields. Diagonalization is introduced as a way to produce subgroups of $SO(n, \mathbb{R})$ for symmetric real matrices and $SU(n, \mathbb{C})$ for complex Hermitian matrices.

The theory of continuous groups of transformations is fundamental in theoretical physics. The laws of nature must obey certain symmetry relations, often stated as "invariance" under transformations. For example, the laws of classical mechanics are invariant under Galilean frame transformations. Properties that are invariant under Galilean frame transformations are guaranteed to be measured by two observers, which may choose to set up different coordinate origins and orientations and obtain the same outcome from the same experiment. In fact, the laws of physics are unchanged by translations, rotations, space inversions, and time reversal. In special relativity, the Galilean group is extended to include transformations between Lorenzian frames. The Poincaré group is the proper group that contains all the possible continuous transformations of coordinates, including those between moving frames, under which the laws of physics are invariant. Groups additionally provide the proper method to extend the **canonical quantization rules** for quantum mechanics. However, there are more mundane reasons to study Lie groups. In particular, there are two main reasons why we are interested in Lie groups in this book.

i. The Lie group of transformations are the solutions of systems of coupled differential equations. The connection between groups and systems of coupled differential equations is merely demonstrated as an application of the theory. The rigorous development of the theorems of Lie, Taylor, and Magnus and a systematic classifications of all the classical groups, are outside the scope of this book [51]. Solutions of the Schrödinger equation of quantum mechanics belong to the $SO(n, \mathbb{R})$ or the $SU(n, \mathbb{C})$ groups of linear transformations, and their subgroups arising from eigen-analysis. Whereas, members of the $Sp(n, \mathbb{R})$ group can be used to solve Hamilton's equations analytically in some simple cases and in general, the theory can be used to produce efficient and stable numerical algorithms.

ii. Members of a group of transformations can be understood as changes of coordinates. The exponentiation of elements of certain Lie algebras produces maps that change coordinates from flat spaces to curved manifolds. These latter objects and the tensor analysis that ensues from them, occupy us for several chapters in this book. In Chapter 10, we make use of coordinate changes and the properties of maps, to introduce differential geometry, the concept of manifolds, flat and curved spaces, metric tensors, and the other items necessary to extend the theoretical tools at our disposal, and eventually apply these to the study of assembly of molecular systems.

Therefore, this chapter, while almost entirely devoid of physical content, is central to the developments and applications in all the chapters that follow. The intimate connection between the areas of differential geometry and Lie groups as solutions of differential equations is profound, and in our opinion's, is still not fully explored by the theoretical chemistry community. Particularly, this remains true for Lie superalgebras. Typically, chemists receive training to apply concepts, such as the time evolution operator as a solution of the Schrödinger equation, in advanced graduate courses. However, in most cases, this is done with no mention of the underlying Lie algebra and its structure. Differential geometry, tensor analysis, and path integration are not routinely included as part of the regular training in chemistry graduate programs. It is our hope that by the end of the training presented in this book, the interested reader will have been exposed to enough applications of Lie group and Lie algebra theory to initiate research into this rich area on his/her own.

4.2   A Few Useful Definitions

A large number of applications in chemistry employ matrices. One can think of a matrix as a table of numbers or as a doubly indexed variable. Most matrices that have direct physical applications are either square (n rows and n columns), column vectors (n rows and one column), or row vectors (one row and n columns). Suppose $\mathbf{A}$ is a $n \times n$ matrix. We call the entry or the number contained in row i of the matrix and column j of the matrix, A_{ij}. The main diagonal is the collection of n numbers that are in column i and row i for all the values of i. Here is an example of a matrix that is square and real (meaning it is composed of real numbers) and is symmetric about the main diagonal.

$$\mathbf{A} = \begin{pmatrix} 2 & 5 & 2 \\ 5 & 3 & 6 \\ 2 & 6 & 22 \end{pmatrix}. \tag{4.1}$$

One can verify that for this matrix, the following relationship holds:

$$A_{ij} = A_{ji}, \tag{4.2}$$

for all values of i and j between 1 and 3. The entries in the **main diagonal** are 2, 3, and 22. These numbers correspond to A_{11}, A_{22}, A_{33}, respectively. The **minor diagonal** of a $n \times n$ matrix is sometimes important. These are the entries along the line that begins from the bottom left and ends at the top right of the matrix.

$$A_{ij} \rightarrow \begin{cases} j = i & \text{Main diagonal elements} \\ j = n - i + 1 & \text{Minor diagonal elements.} \end{cases} \tag{4.3}$$

The elements on the minor diagonal from the top right, beginning with row 1, are A_{13}, A_{22}, A_{31}, equal to 2,3,2, respectively. The **transpose** of a matrix $\mathbf{A}$ is denoted $\mathbf{A}^T$, and is obtained (in $n \times n$ matrices) by exchanging entries across the main diagonal,

$$A_{ij}^T = A_{ji}. \tag{4.4}$$

Clearly, a better definition for real $n \times n$ symmetric matrices is that the matrix is equal to its transpose, i.e., $\mathbf{A}^T = \mathbf{A}$. A matrix $\mathbf{D}$ is called **diagonal**, if and only if

$$A_{ij} = 0 \quad \forall\, i,j \quad (i \neq j). \tag{4.5}$$

Therefore,

$$D = \begin{pmatrix} 7 & 0 & 0 \\ 0 & 3 & 0 \\ 0 & 0 & 6 \end{pmatrix},$$

is diagonal, however,

$$B = \begin{pmatrix} 7 & 0 & 1 \\ 0 & 3 & 0 \\ 0 & 0 & 6 \end{pmatrix},$$

is neither diagonal, nor is it symmetric. The **trace** of a $n \times n$ matrix is defined as the sum of all the elements along the main diagonal,

$$\text{trace}(\mathbf{A}) = \sum_{i=1}^{N} A_{ii}. \tag{4.6}$$

The **addition** of two matrices is well defined, provided they are both of the same size. Let $\mathbf{C} = \mathbf{A} + \mathbf{B}$, then the elements of $\mathbf{C}$ are

$$C_{ij} = A_{ij} + B_{ij}. \tag{4.7}$$

Therefore, if we use the 3×3 matrices in previous examples, we get

$$\mathbf{A} + \mathbf{B} = \begin{pmatrix} 2 & 5 & 3 \\ 5 & 3 & 6 \\ 3 & 6 & 22 \end{pmatrix} + \begin{pmatrix} 7 & 0 & 1 \\ 0 & 3 & 0 \\ 1 & 0 & 6 \end{pmatrix} = \begin{pmatrix} 9 & 5 & 4 \\ 5 & 6 & 6 \\ 4 & 6 & 28 \end{pmatrix}.$$

4.3   Groups

With one operation, we can begin to build a mathematical structure on a set. If we consider a set as the simplest mathematical object, then the next object more complex than the set is the **group**. The set of all $n \times n$ matrices is a legitimate example of an infinite set.

A group $\mathcal{G}$ is composed of:

1. A set $\mathbf{A}, \mathbf{B}, \mathbf{C}, \ldots$

2. A group operation $*$

The group operation must satisfy the following postulates.

i. Closure

$$\mathbf{A}, \mathbf{B} \in \mathcal{G}, \quad \mathbf{A} * \mathbf{B} \in \mathcal{G}. \tag{4.8}$$

ii. Associativity

$$(\mathbf{A} * \mathbf{B}) * \mathbf{C} = \mathbf{A} * (\mathbf{B} * \mathbf{C}). \tag{4.9}$$

iii. Existence of an identity element in $\mathcal{G}$

$$\mathbf{A} * \mathbf{1} = \mathbf{A}. \tag{4.10}$$

iv. Existence of an inverse

$$\mathbf{A} * (-\mathbf{A}) = \mathbf{1}. \tag{4.11}$$

If commutativity under the group operation is satisfied,

$$\mathbf{A} * \mathbf{B} = \mathbf{B} * \mathbf{A}, \tag{4.12}$$

for all $\mathbf{A}, \mathbf{B} \in \mathcal{G}$, then $\mathcal{G}$ is called an **Abelian** group. The majority of the groups we are interested in developing can be represented by $n \times n$ matrices with the matrix multiplication, which is defined later. However, the sets of $n \times n$ matrices constitute a group under matrix addition. It is possible to show that matrix addition, as defined by Equation 4.7, satisfies all the postulates and that the group of $n \times n$ matrices is Abelian. The reader is urged to prove these statements. The next level of complexity is achieved by introducing a field into a group and the scalar product; when a group is endowed with these additional elements, one has a vector space.

4.4   Number Fields

In discussing the group of $n \times n$ matrices, we have tacitly filled each entry of the matrix with real numbers. The field of real numbers is denoted with $\mathbb{R}$. There are several number fields that find applications in physics, therefore, we formalize the definition of a general number field $\mathbb{F}$ as a nonempty set of objects with two operations, the sum $(+)$ and the product $(\cdot)$ with the following properties:

i. Closure: if $x, y \in \mathbb{F}$, then $x + y \in \mathbb{F}$, and $x \cdot y \in \mathbb{F}$.

ii. Associativity: $(x + y) + z = x + (y + z)$, $(x \cdot y) \cdot z = x \cdot (y \cdot z)$.

iii. Existence of the identity: $x + 0 = x$, $x \cdot 1 = x$ $0, 1 \in \mathbb{F}$.

iv. Existence of the inverse: $x + (-x) = 0$, $x \cdot x^{-1} = 1$ $-x, x^{-1} \in \mathbb{F}$.

vi. Linearity: $x \cdot (y + z) = x \cdot y + x \cdot z$.

vii. Commutativity: $x + y = y + x$, $x \cdot y = y \cdot x$.

Three commonly used fields in physics and chemistry are the real number field $\mathbb{R}$, the complex number field $\mathbb{C}$, and the quaternion field $\mathbb{Q}$. The real and complex number fields are both associative and commutative under multiplication. Occasionally, number fields that have a noncommutative product find applications. One example is the field of quaternions, $\mathbb{Q}$, which we introduce later in this chapter.

4.5 Vector Spaces

The **scalar product of a matrix and a number** ($\cdot$) is defined as follows. Let $a \in \mathbb{F}$, and let $\mathbf{A}$ be a matrix built over the field $\mathbb{F}$. Then, the scalar product of a with a matrix $\mathbf{A}$, is the matrix that has every entry of $\mathbf{A}$ multiplied by a. In other words,

$$[a \cdot \mathbf{A}]_{ij} = aA_{ij}. \tag{4.13}$$

Here is an example to clarify the definition,

$$5 \cdot \mathbf{A} = 5 \cdot \begin{pmatrix} 2 & 5 & 3 \\ 5 & 3 & 6 \\ 3 & 6 & 22 \end{pmatrix} = \begin{pmatrix} 10 & 25 & 15 \\ 25 & 15 & 30 \\ 15 & 30 & 110 \end{pmatrix}.$$

The set of all real square $n \times n$ matrices is an example of a linear vector space, which we denote $\mathcal{V}(n \times n, \mathbb{R})$. The word vector here is used broadly, i.e., if a collection of mathematical objects can be shown to be a vector space, then each member of the collection is a vector.

A **linear vector space**, $\mathcal{V}$, consists of two sets of elements:

1. A collection of vectors $\mathbf{A}, \mathbf{B}, \mathbf{C}, \ldots$

2. A collection of numbers $a, b, c, \ldots$ that belong to a **field** $\mathbb{F}$.

And two operations:

1. Vector sum ($+$)

2. Scalar product ($\cdot$)

The following postulates regarding the operations apply:

1. Closure under vector addition and under scalar multiplication,

$$\mathbf{A}, \mathbf{B} \in \mathcal{V}, \qquad \mathbf{A} + \mathbf{B} \in \mathcal{V}, \tag{4.14}$$

$$a \in \mathbb{F}, \mathbf{A} \in \mathcal{V}, \qquad a \cdot \mathbf{A} \in \mathcal{V}. \tag{4.15}$$

2. Associativity of the vector sum and the scalar product,

$$(\mathbf{A} + \mathbf{B}) + \mathbf{C} = \mathbf{A} + (\mathbf{B} + \mathbf{C}), \tag{4.16}$$

$$a \cdot (b \cdot \mathbf{A}) = (a \cdot b) \cdot \mathbf{A} \qquad a, b \in \mathbb{F}, \mathbf{A} \in \mathcal{V}. \tag{4.17}$$

3. There exist an identity element under vector addition and scalar product,

$$\mathbf{A} + \mathbf{0} = \mathbf{A}, \tag{4.18}$$

$$a \cdot \mathbf{A} = \mathbf{A}. \tag{4.19}$$

4. There exist an additive inverse to each vector in $\mathcal{V}$,

$$\mathbf{A} + (-\mathbf{A}) = \mathbf{0}. \tag{4.20}$$

5. The vector sum is commutative,

$$\mathbf{A} + \mathbf{B} = \mathbf{B} + \mathbf{A}. \tag{4.21}$$

6. The scalar product is linear,

$$a \cdot (\mathbf{A} + \mathbf{B}) = a \cdot \mathbf{A} + a \cdot \mathbf{B}, \tag{4.22}$$

$$(a + b) \cdot \mathbf{A} = a \cdot \mathbf{A} + b \cdot \mathbf{A}. \tag{4.23}$$

Postulate 5 makes $\mathcal{V}$ an Abelian group under addition, whereas postulate 6 answers the linear property of $\mathcal{V}$. The reader should realize that this definition of vector spaces is far broader than the commonly understood concept of a vector as a mathematical entity with magnitude and direction. Of course, the familiar Euclidean three-dimensional space is a vector space under this broad definition. It is straightforward to show that the vector sum and the scalar products, defined earlier for matrices over the real number field, have all the properties postulated as requirements. Therefore, $\mathcal{V}(n \times n, \mathbb{R})$ is a genuine vector space. In fact, the set of $n \times n'$ matrices is a vector field and in particular, the sets of $n \times 1$ (or column vector space of dimension n) and $1 \times n$ (or row vector space of dimension n) are vector spaces. For example, we could arrange

$$v_1 = 5i + 3j - 7k,$$

as a vector that belongs to $\mathcal{V}(3 \times 1, \mathbb{R})$,

$$\begin{pmatrix} 5 \\ 3 \\ -7 \end{pmatrix}.$$

Therefore, Euclidean three-dimensional space is a vector space, as one would expect. It is common to find a different symbol for vector spaces $\mathcal{V}(n \times 1, \mathbb{R})$, especially when their matrix representation is not necessary. The alternate common symbol for $\mathcal{V}(n \times 1, \mathbb{R})$ is $\mathbb{R}^n$. It is useful to consider other, more abstract vector spaces. A collection of functions, f_n, over the real or complex field is a vector space. Here are some examples. The set of functions,

$$\{\exp(im\phi)\}_{m=-\infty}^{\infty},$$

over the circle $0 < \phi < 2\pi$, ($i = \sqrt{-1}$) is a vector space. The set of functions,

$$\left\{\sin\left(\frac{n\pi x}{a}\right)\right\}_{n=0}^{\infty},$$

for $0 < x < a$ is a vector space. The last two are examples of a special kind of vector space, since they have infinite dimensions. Such vector spaces are called **Hilbert spaces**. The reader familiar with introductory quantum mechanics will have recognized some of the simplest solutions of the time independent Schrödinger equation. It is true, in general, that the set of eigenfunctions of a Hamiltonian operator constitutes an infinite dimensional vector space, a Hilbert space. It is this powerful result that allows us to solve a large number of problems in quantum mechanics by numerical methods. Much of Chapter 5 focuses on such applications.

The vectors in the set $\mathbf{A}, \mathbf{B}, \mathbf{C}, \ldots$ are **linearly independent**, if and only if the sum

$$a \cdot \mathbf{A} + b \cdot \mathbf{B} + c \cdot \mathbf{C} + \cdots = 0, \tag{4.24}$$

is true when all the scalars $a, b, c, \ldots$ are zero. A **basis set** for a vector space is any set of linearly independent vectors in the space from which it is impossible to find any other linearly independent vector that belongs to the same space. **The number of linearly independent vectors is equal to the dimension of the vector space.** Let us return to the $\mathcal{V}(n \times n, \mathbb{R})$ vector space example. The dimension of $\mathcal{V}(n \times n, \mathbb{R})$ is n^2. Here is an example for the 3×3 case. A possibility for the basis is

$$E_{11}, E_{12}, E_{13}, \ldots, E_{33}$$

$$= \begin{pmatrix} 1 & 0 & 0 \\ 0 & 0 & 0 \\ 0 & 0 & 0 \end{pmatrix}, \begin{pmatrix} 0 & 1 & 0 \\ 0 & 0 & 0 \\ 0 & 0 & 0 \end{pmatrix}, \begin{pmatrix} 0 & 0 & 1 \\ 0 & 0 & 0 \\ 0 & 0 & 0 \end{pmatrix}, \ldots, \begin{pmatrix} 0 & 0 & 0 \\ 0 & 0 & 0 \\ 0 & 0 & 1 \end{pmatrix}. \tag{4.25}$$

It is trivial to verify that the nine matrices in Equation 4.25 are linearly independent, and that it is impossible to find an additional linearly independent vector. This example illustrates a profound truth about basis sets. In every vector space, there exists an infinite number of distinct basis sets. It is obvious, for example, that replacing the 1 entry in each of the nine matrices by any real number will produce a valid basis set. If we put the closure under vector sum postulate, together with the concept of basis sets, we have the following powerful result, known as the **linear superposition principle**. If $E_1, E_2, \ldots, E_n$ are the basis vectors of $\mathcal{V}_n$, an n-dimensional vector space, then,

$$\alpha_1 E_1 + \alpha_2 E_2 + \cdots + \alpha_n E_n \in \mathcal{V}_n, \tag{4.26}$$

for any real value of the scalar $\alpha_1, \alpha_2, \ldots, \alpha_n$. With this theorem in hand, we can write for our $\mathcal{V}(3 \times 3, \mathbb{R})$ example,

$$
\begin{pmatrix} 2 & 5 & 3 \\ 5 & 3 & 6 \\ 3 & 6 & 22 \end{pmatrix} = 2E_{11} + 5E_{12} + 3E_{13} + 5E_{21} + 3E_{22} + 6E_{23}
$$

$$
+ 3E_{31} + 6E_{32} + 22E_{33}.
$$

4.6 Algebras

An algebra, $\mathcal{A}$ is a vector space with one additional operation, the vector product ($\times$). The vector product must satisfy a closure and a linear property relation.

$$
\mathbf{A}, \mathbf{B} \in \mathcal{A}, \qquad \mathbf{A} \times \mathbf{B} \in \mathcal{A}, \tag{4.27}
$$
$$
\mathbf{A} \times (\mathbf{B} + \mathbf{C}) = \mathbf{A} \times \mathbf{B} + \mathbf{A} \times \mathbf{C}. \tag{4.28}
$$

The definition for the matrix product of $n \times n$ matrices allows us to construct an algebra: the **vector product** of two $n \times n$ matrices is defined symbolically as follows. If

$$
\mathbf{C} = \mathbf{A} \times \mathbf{B}, \tag{4.29}
$$

then

$$
C_{ij} = \sum_{k=1}^{n} A_{ik} B_{kj}. \tag{4.30}
$$

In other words, to calculate the entry of the product in row i and column j, we take the "dot" product of the i row vector of $\mathbf{A}$ with the column vector j of $\mathbf{B}$. An example should clarify.

$$
\begin{pmatrix} 2 & 5 & 2 \\ 5 & 3 & 6 \\ 2 & 6 & 1 \end{pmatrix} \times \begin{pmatrix} 7 & 1 & 0 \\ 0 & 3 & 0 \\ 1 & 2 & 6 \end{pmatrix} = \begin{pmatrix} 16 & 21 & 12 \\ 41 & 26 & 36 \\ 15 & 22 & 6 \end{pmatrix},
$$

where, to obtain the 1,1 element of the product, we isolate row 1 of the matrix to the left of the ($\times$) operator and column 1 of the matrix to the right of the ($\times$) operator,

$$
\begin{pmatrix} 2 & 5 & 2 \end{pmatrix} \times \begin{pmatrix} 7 \\ 0 \\ 1 \end{pmatrix} = 2 \times 7 + 5 \times 0 + 2 \times 1 = 16.
$$

The reader should check these results, and then should show that

$$\begin{pmatrix} 2 & 5 & 2 \\ 5 & 3 & 6 \\ 2 & 6 & 1 \end{pmatrix} \times \begin{pmatrix} 7 & 1 & 0 \\ 0 & 3 & 0 \\ 1 & 2 & 6 \end{pmatrix} \neq \begin{pmatrix} 7 & 1 & 0 \\ 0 & 3 & 0 \\ 1 & 2 & 6 \end{pmatrix} \times \begin{pmatrix} 2 & 5 & 2 \\ 5 & 3 & 6 \\ 2 & 6 & 1 \end{pmatrix}.$$

The matrix product is not commutative in general, i.e.,

$$\mathbf{AB} \neq \mathbf{BA}. \tag{4.31}$$

Most of the time, as we have done above, the matrix product operator ($\times$) and the scalar product operator ($\cdot$) are omitted. The two operations are readily distinguished by using normal fonts for scalars and bold type fonts for matrices and vectors. To distinguish members of the $\mathcal{V}(n \times n, \mathbb{R})$ vector space from those of the $\mathcal{V}(n \times 1, \mathbb{R})$ vector space, usually uppercase roman letters in bold font are used for matrices and lowercase roman letters in bold font are used for column vectors. Finally, elements of $\mathcal{V}(1 \times n, \mathbb{R})$, i.e., row vectors, are represented as lowercase roman letters in bold font with a T superscript, since the transpose operation maps elements of $\mathcal{V}(n \times 1, \mathbb{R})$ into elements of $\mathcal{V}(n \times 1, \mathbb{R})$, and vice-versa. This notation is used in most books as well as in most of the literature on the subject. Therefore, we use this notation throughout.

The **commutator** of two matrices is a matrix defined as the antisymmetric product of the two matrices.

$$[\mathbf{A}, \mathbf{B}] = \mathbf{AB} - \mathbf{BA}. \tag{4.32}$$

If

$$\mathbf{C} = [\mathbf{A}, \mathbf{B}], \tag{4.33}$$

then

$$C_{ij} = \sum_{k=1}^{n} \left(A_{ik} B_{kj} - B_{ik} A_{kj} \right). \tag{4.34}$$

The product of two matrices is a tedious operation for most but the smallest-sized matrices. It involves on the order of n^3 multiplications and additions. Multiplications of matrices of size $10^7 \times 10^7$ or larger are routine in modern applications of quantum mechanics. The number of repetitive operations is best handled by a computer. The program `matrix_product` and its modifications, will serve the reader in constructing a number of useful concepts in this chapter.

```
program matrix_product
implicit real*8 (a-h,o-z)
parameter (NMAX = 100)
real*8 a(NMAX,NMAX),b(NMAX,NMAX),c(NMAX,NMAX),d(NMAX,NMAX)
```

```
      do i=1,NMAX    ! initialize
       do j=1,NMAX
        a(i,j) = 0.d0
        b(i,j) = 0.d0
        c(i,j) = 0.d0
        d(i,j) = 0.d0
       enddo
      enddo
      read(5,*) N   ! input
      do i=1,N*N
      read(5,*,end=10) in,jn,x,y
      a(in,jn) = x
      b(in,jn) = y
      enddo
10    do i=1,N      ! computation
       do j=1,N       ! C is AB
        c(i,j) = 0.d0
         do k=1,N
          c(i,j) = c(i,j) + a(i,k)*b(k,j)
         enddo
        enddo
      enddo
      do i=1,N     ! D is [A,B]
       do j=1,N
        d(i,j) = 0.d0
        do k=1,N
         d(i,j) = d(i,j) + a(i,k)*b(k,j) - b(i,k)*a(k,j)
        enddo
       enddo
      enddo
      if (N .le. 10) then   !output
       write(6,*) 'AB'
       do i=1,N
        write(6,2000) (c(i,j),j=1,N)
       enddo
       write(6,*) '[A,B]'
       do i=1,N
        write(6,2000) (d(i,j),j=1,N)
       enddo
      else
       write(6,*) 'AB'
       do i=1,N
        do j=1,N
         write(6,*) i,j,c(i,j)
        enddo
       enddo
```

```
      write(6,*) '[A,B]'
      do i=1,N
       do j=1,N
        write(6,*) i,j,d(i,j)
       enddo
      enddo
      endif
2000  format(10f10.3)
      end
```

Exercises

1. Test the program `matrix_product` by entering the 3×3 matrices from the example in the text, and check the output by hand.

2. Using this program, the reader should verify the following theorem: the set of $n \times n$ antisymmetric matrices, i.e., all $n \times n$ matrices for which

$$\mathbf{A}^T = -\mathbf{A}, \tag{4.35}$$

is an algebra if the commutator is used as the vector product. To verify the theorem, we need to show first that the set of $n \times n$ antisymmetric matrices over the field of real numbers is a vector space, and then, that the commutator satisfies the linear property and closure in the set. The closure under the commutator product is less trivial to prove. To verify closure, if we choose two arbitrary antisymmetric matrices, say,

$$\mathbf{A} = \begin{pmatrix} 0 & 2 & 0 \\ -2 & 0 & -7 \\ 0 & 7 & 0 \end{pmatrix} \quad \mathbf{B} = \begin{pmatrix} 0 & -3 & 8 \\ 3 & 0 & 6 \\ -8 & -6 & 0 \end{pmatrix},$$

then one checks that

$$\mathbf{C} = [\mathbf{A}, \mathbf{B}],$$

is antisymmetric as well. Change the input file, then run the program `matrix_product`, and confirm. Verifying the closure part of the theorem is not the same as proving it, of course. This simple example demonstrates one of the uses of computers in mathematic research. The computer is only a tool used for the purpose of testing conjectures, which only after logical arguments, can be proved and turned into theorems. The reader is challenged to prove the theorem, in general, by showing that the set of $n \times n$ antisymmetric matrices is closed under the commuting product operation. Here are some hints on how to proceed. First, modify `matrix_product` to test the following conjecture about the matrix product and the transpose operation,

$$(\mathbf{AB})^T = \mathbf{B}^T \mathbf{A}^T, \tag{4.36}$$

then prove this conjecture generally, by using the definition of the matrix product in Equation 4.30. Finally, note that one consequence of this last theorem is that

$$[\mathbf{A}, \mathbf{B}]^T = - \left[\mathbf{A}^T, \mathbf{B}^T \right].$$

(4.37)

Therefore, if $\mathbf{A}$ and $\mathbf{B}$ are symmetric, then $[\mathbf{A}, \mathbf{B}]$ is antisymmetric. What happens if $\mathbf{A}$ and $\mathbf{B}$ are antisymmetric?

3. Use the program to test the conjecture that the set of diagonal $n \times n$ matrices is an Abelian algebra under the regular matrix product, then prove it, using the definition of the matrix product. The symbol $\mathbb{D}^n$ represents the vector space of diagonal $n \times n$ matrices.

4. Show that the set of $n \times n$ real matrices is an Abelian group under the matrix addition operation $(+)$.

5. Show that the set of $n \times n$ real matrices, for which a multiplicative inverse exists, is a non-Abelian group under the matrix multiplication operation $(\times)$. This group is called $Gl\,(n, \mathbb{R})$.

6. Prove that the set of functions, $x^n \ n = 1, 2, \dots$, is a vector space over the real number field. These constitute the basis upon which one can expand **any** analytical function on the field, and these are in the vector field. These expansions are familiar to the reader from calculus, as Taylor series.

7. A **Lie algebra** is an algebra with a vector product that satisfies

$$\mathbf{A} * \mathbf{A} = \mathbf{0},$$

(4.38)

and

$$\mathbf{A} * (\mathbf{B} * \mathbf{C}) + \mathbf{C} * (\mathbf{A} * \mathbf{B}) + \mathbf{B} * (\mathbf{C} * \mathbf{A}) = \mathbf{0}.$$

(4.39)

The last expression is known as the Jacobi identity. Show that the commuted product, defined in Equation 4.34, satisfies both criteria. The Jacobi identity for the commuted product reads,

$$[\mathbf{A}, [\mathbf{B}, \mathbf{C}]] + [\mathbf{C}, [\mathbf{A}, \mathbf{B}]] + [\mathbf{B}, [\mathbf{C}, \mathbf{A}]] = \mathbf{0}.$$

(4.40)

Therefore, we can construct a Lie algebra over the vector space $\mathcal{V}\,(n \times n, \mathbb{R})$ by using the commuted product. There are two important subalgebras of this general Lie algebra that we investigate closely in this chapter.

8. The set of $n \times n$ antisymmetric matrices under the commuted product is a Lie algebra. Test this conjecture using the following program for few values of n.

```
         program jacobi_identity
         implicit real*8 (a-h,o-z)
         parameter (NMAX = 100)
         real*8 a(NMAX,NMAX),b(NMAX,NMAX),c(NMAX,NMAX)
         real*8 c1(NMAX,NMAX),c2(NMAX,NMAX),c3(NMAX,NMAX)
         real*8 c4(NMAX,NMAX),c5(NMAX,NMAX),c6(NMAX,NMAX)
         do i=1,NMAX    ! initialize
          do j=1,NMAX
           a(i,j) = 0.d0
           b(i,j) = 0.d0
           c(i,j) = 0.d0
           c1(i,j) = 0.d0
           c2(i,j) = 0.d0
           c3(i,j) = 0.d0
           c4(i,j) = 0.d0
           c5(i,j) = 0.d0
           c6(i,j) = 0.d0
          enddo
         enddo
         read(5,*) N  ! input
         do i=1,N*N
         read(5,*,end=10) in,jn,x,y,z
         a(in,jn) = x
         b(in,jn) = y
         c(in,jn) = z
         enddo
 10   do i=1,N      ! computation
         do j=1,N      ! C1 is [B,C], C2 is [A,B], C3 is [C,A]
          c1(i,j) = 0.d0
          c2(i,j) = 0.d0
          c3(i,j) = 0.d0
          do k=1,N
            c1(i,j) = c1(i,j) + b(i,k)*c(k,j) - c(i,k)*b(k,j)
            c2(i,j) = c2(i,j) + a(i,k)*b(k,j) - b(i,k)*a(k,j)
            c3(i,j) = c3(i,j) + c(i,k)*a(k,j) - a(i,k)*c(k,j)
          enddo
         enddo
         enddo
 c C4 is [A,[B,C]], C5 is [C,[A,B]], C6 is [B,[C,A]]
         do i=1,N
         do j=1,N
          c4(i,j) = 0.d0
          c5(i,j) = 0.d0
          c6(i,j) = 0.d0
          do k=1,N
```

```
            c4(i,j) = c4(i,j) + a(i,k)*c1(k,j) - c1(i,k)*a(k,j)
            c5(i,j) = c5(i,j) + c(i,k)*c2(k,j) - c2(i,k)*c(k,j)
            c6(i,j) = c6(i,j) + b(i,k)*c3(k,j) - c3(i,k)*b(k,j)
          enddo
        enddo
      enddo
      if (N .le. 10) then   ! output
        write(6,*) '[A,[B,C]] + [C,[A,B]] + [B,[C,A]]'
        do i=1,N
          write(6,2000) ((c4(i,j)+c5(i,j)+c6(i,j)),j=1,N)
        enddo
      else
        write(6,*) '[A,[B,C]] + [C,[A,B]] + [B,[C,A]]'
        do i=1,N
        do j=1,N
          write(6,*) i,j,(c4(i,j)+c5(i,j)+c6(i,j))
        enddo
        enddo
      endif
 2000  format(10f10.3)
      end
```

The Lie algebra we just explored is denoted $so\,(n, \mathbb{R})$. There is a great deal of physical content in these Lie algebras. $so\,(3, \mathbb{R})$, for example, is fundamental in the theory of angular momentum.

9. Show that a viable basis set for $so\,(n, \mathbb{R})$ from the last exercise, is

$$X_i = E_{ij} - E_{ji} \quad (i, j = 1, 2, \ldots, n), \tag{4.41}$$

where the definition of E_{ij} is given in Equation 4.25. Then, use this result to show that the dimension of $so\,(n, \mathbb{R})$ is $n\,(n-1)\,/2$.

10. A set of objects that is of importance in quantum mechanics, and in the classification of Lie algebras, is the set of the structure constants for a Lie algebra. These are defined from the commutation relations among the basis of a Lie algebra.

$$[X_i, X_j] = C_{ij}^k X_k. \tag{4.42}$$

The superscript k in Equation 4.42 is not to be understood as a power, rather, it is a **contravariant** label. **Einstein's sum convention** is used in Equation 4.42. Whenever an upper and a lower index match, the index is understood as "summed over,"

$$C_{ij}^k X_k \equiv \sum_{k=1}^{r} C_{ij}^k X_k, \tag{4.43}$$

where r is the dimension of the algebra, $n(n-1)/2$ in the case of $so(n, \mathbb{R})$. Let $X_1 = E_{12}^{(n)} - E_{21}^{(n)}$, $X_2 = E_{13}^{(n)} - E_{31}^{(n)}$, and $X_3 = E_{23}^{(n)} - E_{32}^{(n)}$. Confirm that the structure constants for $so(3, \mathbb{R})$ are,

$$C_{12}^1 = 0, \quad C_{12}^2 = 0, \quad C_{12}^3 = -1, \tag{4.44}$$

$$C_{13}^1 = 0, \quad C_{13}^2 = +1, \quad C_{13}^3 = 0, \tag{4.45}$$

$$C_{23}^1 = -1, \quad C_{23}^2 = 0, \quad C_{23}^3 = 0. \tag{4.46}$$

To find the remaining ones, it is sufficient to note that

$$C_{ji}^k = -C_{ij}^k \quad C_{ii}^k = 0, \tag{4.47}$$

follow from the definition. Structure constants are fundamental in the classifications of Lie algebras. Furthermore, structure constants can be used to discover the numerous **isomorphisms** among many representations of the same Lie algebra and among different Lie algebras.

11. Structure constants can be used to build a representation of the basis of **simple Lie algebras** by noting that for every value of k, one can define

$$(X_k)_{ij} \Rightarrow \left(C^k\right)_{ij}. \tag{4.48}$$

Confirm these statements for $so(3, \mathbb{R})$. In particular show that,

$$\left(C^1\right)_{ij} = -(X_3)_{ij}, \tag{4.49}$$

$$\left(C^3\right)_{ij} = -(X_1)_{ij}, \tag{4.50}$$

$$\left(C^2\right)_{ij} = (X_2)_{ij}. \tag{4.51}$$

Simple Lie algebras are Lie algebras that do not contain **closed** subalgebras. Subalgebras are readily identified from the structure constants.

12. Use the structure constants for $so(3, \mathbb{R})$ to find a representation for the Killing's form

$$g_{ij} = -C_{il}^k C_{jk}^l, \tag{4.52}$$

another important $r \times r$ matrix. Note the implied double sum in Equation 4.52. The Killing's form has a symmetric representation,

$$g_{ji} = g_{ij}, \tag{4.53}$$

and this property follows from the definition in Equation 4.52.

13. The Casimir invariants are also of fundamental importance in quantum mechanics. These are composed of polynomials of the basis for the algebra. One can show that Lie algebras have a Casimir invariant of order 2, C_2,

$$C_2 = g^{ij} X_i X_j, \tag{4.54}$$

where g^{ij} is the inverse of the Killing form. Show that the Casimir operator, C_2, for $so\,(3, \mathbb{R})$ is

$$C_2 = X_1 X_1 + X_2 X_2 + X_3 X_3 = -2 \begin{pmatrix} 1 & 0 & 0 \\ 0 & 1 & 0 \\ 0 & 0 & 1 \end{pmatrix}, \tag{4.55}$$

and that

$$[C_2, X_i] = 0. \tag{4.56}$$

Some Casimir invariants are proportional to the Hamiltonian that correspond to a continuous group of transformation. Therefore, the concepts of simple algebra and Casimir invariants are central to both classical and quantum theory. The structure of the Casimir invariant allows us to extend the canonical quantization rule to manifolds. The typical example encountered in advanced quantum mechanics courses is angular momentum theory.

4.7 The Exponential Mapping of Lie Algebras

Most linear algebra books spend a great deal of effort in finding the multiplicative inverse of a matrix (if such inverse exists). The set of $n \times n$ matrices over a number field form a group under addition. The set of **invertible** $n \times n$ matrices form a group under the matrix product. This group is called the **general linear group** and is generally denoted $Gl\,(n, \mathbb{F})$. All Lie groups built over $\mathbb{F}$ are subgroups of $Gl\,(n, \mathbb{F})$. Therefore, the existence of an inverse and methods to find it are important. Finding the inverse of a given matrix $\mathbf{A}$ amounts to finding $\mathbf{A}^{-1}$, such that

$$\mathbf{A}^{-1}\mathbf{A} = \begin{pmatrix} 1 & 0 & 0 & \cdots \\ 0 & 1 & 0 & \cdots \\ 0 & 0 & 1 & \cdots \\ \cdots & \cdots & \cdots & \cdots \end{pmatrix} = 1. \tag{4.57}$$

A number of methods, including the Gauss–Jordan row elimination method, can be found in many introductory texts in linear algebra. We will interest ourselves in a group of matrices with a special property. A matrix $\mathbf{U}$ over the field of real numbers is said to be **orthogonal** if its transpose is equal to its inverse.

$$\mathbf{U}^{-1} = \mathbf{U}^T. \tag{4.58}$$

Note that if $\mathbf{U}$ is orthogonal, then

$$\mathbf{U}^T\mathbf{U} = \mathbf{U}\mathbf{U}^T = 1. \tag{4.59}$$

The set of all orthogonal matrices is a group under multiplication, denoted $O(n, \mathbb{R})$, and is a subgroup of $Gl(n, \mathbb{R})$. Orthogonal matrices are important in matrix quantum mechanics because the solution of a problem in matrix quantum mechanics is generally an orthogonal matrix, or a **unitary** matrix. The set of unitary matrices is the complex field equivalent of the set of orthogonal matrices. The unitary group is denoted $U(n, \mathbb{C}) \supset Gl(n, \mathbb{C})$. The columns of **U** are the basis of a special vector space comprised of column vectors, which is a subset of $V(n \times 1, \mathbb{R})$. The orthogonality property of **U** translates to the orthogonality and normality of the basis vector of $V(n \times 1, \mathbb{R})$. Consider the (i, j) element of the product $\mathbf{U}^T\mathbf{U}$. If we label $\mathbf{u}_i$ as the ith column of **U**, then we have the result

$$\mathbf{u}_i^T \mathbf{u}_j = \delta_{ij}, \tag{4.60}$$

where δ_{ij} is the Kronecker delta,

$$\delta_{ij} = \begin{cases} 1 & i = j \\ 0 & i \neq j \end{cases}. \tag{4.61}$$

One of the reasons why we are interested in the Lie algebra of $n \times n$ antisymmetric matrices is a consequence of the following profound and remarkably simple result to prove. **The exponential** function maps the Lie algebra of $n \times n$ antisymmetric matrices into a group of $n \times n$ orthogonal matrices. This powerful result led, first Dirac and later Feynman, to the development of a very elegant theory and a powerful tool, the path integral. The reader familiar with introductory quantum mechanics will recall that the set of functions that solves the Schrödinger equation is an orthogonal set; this is no accident.

What is the exponential mapping of a Lie algebra? It is a function whose domain is the set of matrices in the algebra, and the range is the exponential of those matrices. If $\mathbf{A} \in so(n, \mathbb{R})$, then

$$\exp(\mathbf{A}) = \sum_{n=0}^{\infty} \frac{(\mathbf{A})^n}{n!} \in SO(n, \mathbb{R}), \tag{4.62}$$

where $SO(n, \mathbb{R})$ is a subgroup of $Gl(n, \mathbb{R})$. The S part in the notation for $SO(n, \mathbb{R})$ stands for "special." The special linear group, $Sl(n, \mathbb{R})$ (also known as the volume preserving group), is a subgroup of $Gl(n, \mathbb{F})$ constructed over the set of all $n \times n$ matrices that have a determinant equal to $+1$. The $SO(n, \mathbb{F})$ group is the intersection of the special linear group and the orthogonal group, $SO(n, \mathbb{F}) = Sl(n, \mathbb{F}) \cap O(n, \mathbb{F})$. We elaborate on determinants in the next section.

The definition of the exponential mapping is a special case of Taylor's theorem. Here, the exponential of the matrix $\mathbf{A}$ is defined with the Taylor series, just like the exponential of a real number. Since the Taylor series of e^x converges for all values of x, there are no problems with defining the exponential of a matrix. Of course, we let the computer do the work, since the series may take many terms to converge. Before we present the next program, however, there is one last detail that we need to address. The exponential of a matrix can be computed by using a convenient recursion relation,

$$\mathbf{B}_n = \frac{1}{n} \mathbf{A} \mathbf{B}_{n-1}, \tag{4.63}$$

initiated with

$$\mathbf{B}_0 = \mathbf{1}. \tag{4.64}$$

It is simple to show that

$$\exp(\mathbf{A}) = \sum_{n=0}^{\infty} \mathbf{B}_n. \tag{4.65}$$

Here is the algorithm for a $n \times n$ exponential mapping.

```
      program matrix_exponentiation
      implicit real*8 (a-h,o-z)
      parameter (NMAX = 100)
      real*8 a(NMAX,NMAX),bn(NMAX,NMAX),bnp1(NMAX,NMAX),sum(NMAX,NMAX)
      do i=1,NMAX    ! initialize
       do j=1,NMAX
        a(i,j) = 0.d0
        bn(i,j) = 0.d0
        bnp1(i,j) = 0.d0
        sum(i,j) = 0.d0
        if (i.eq.j) then
         bn(i,j) = 1.d0
         sum(i,j) = 1.d0
        endif
       enddo
      enddo
      read(5,*) N, mtop ! input
      do i=1,N*N
      read(5,*,end=10) in,jn,x
      a(in,jn) = x
      enddo
10    do m = 1,mtop ! exponentiation
       do i=1,N        ! do no more than mtop iterations
       do j=1,N
        bnp1(i,j) = 0.d0
        do k=1,N
          bnp1(i,j) = bnp1(i,j) + a(i,k)*bn(k,j)/dfloat(m)
        enddo
        sum(i,j) = sum(i,j) + bnp1(i,j)    ! update the sum
       enddo
```

```
         enddo
         do i=1,N      ! set bnp1 to bn
           do j=1,N
             bn(i,j) = bnp1(i,j)
           enddo
         enddo
         enddo          ! iteration loop ends here
       if (N .le. 10) then  ! output
         write(6,*) 'U = EXP(A)'
         do i=1,N
           write(6,2000) (sum(i,j),j=1,N)
         enddo
       else
         write(6,*) 'U = EXP(A)'
         do i=1,N
         do j=1,N
           write(6,*) i,j,sum(i,j)
         enddo
         enddo
       endif
20     write(6,*) 'U**T U'
       do i=1,N     ! Testing the orthogonality property of sum
       do j=1,N
         check = 0.d0 ! check the ij element of U**T U
         do k=1,N
           check = check + sum(k,i)*sum(k,j)
         enddo
         write(6,*) i,j,check
       enddo
       enddo
2000   format(10f12.6)
       end
```

Note that the number of terms of the power series is part of the input (kept in the variable mtop). Testing the program with 50 iterations and with the input matrix

$$\mathbf{A} = \begin{pmatrix} 0 & -3 & 8 \\ 3 & 0 & 6 \\ -8 & -6 & 0 \end{pmatrix}, \tag{4.66}$$

produces the following output:

$$\mathbf{U} = \exp\left(\mathbf{A}\right) = \begin{pmatrix} -0.022784 & -0.428339 & -0.903331 \\ -0.916692 & 0.369517 & -0.152095 \\ 0.398944 & 0.824611 & -0.401074 \end{pmatrix}, \tag{4.67}$$

where $\mathbf{U}^T\mathbf{U}$ should be the unit matrix within the machine precision ($\approx 10^{-13}$). The proof of this result, as anticipated, is remarkably simple, since

$$\exp\left(\mathbf{A}\right)^T = \exp\left(\mathbf{A}^T\right) = \exp\left(-\mathbf{A}\right), \tag{4.68}$$

where the fact that $\mathbf{A}$ is antisymmetric, is used on the right-hand side. It follows that

$$\mathbf{U}^T\mathbf{U} = \exp\left(-\mathbf{A}\right)\exp\left(\mathbf{A}\right) = \exp\left(-\mathbf{A} + \mathbf{A}\right) = \exp\left(\mathbf{0}\right) = \mathbf{1}. \qquad (4.69)$$

Exercises

1. In Equation 4.68, we use the following result:

$$\exp\left(\mathbf{A}\right)^T = \exp\left(\mathbf{A}^T\right). \qquad (4.70)$$

   Use a few terms of the power series expansion of $\exp\left(\mathbf{A}\right)$ to demonstrate this result.

2. Show that if $\mathbf{U}$ is as in Equation 4.67, then

$$\mathbf{U}^T\mathbf{U} = \begin{pmatrix} 1 & 0 & 0 \\ 0 & 1 & 0 \\ 0 & 0 & 1 \end{pmatrix}.$$

3. Use the program `matrix_exponentiation` to test the conjecture

$$\exp\begin{pmatrix} 0 & \theta \\ -\theta & 0 \end{pmatrix} = \begin{pmatrix} \cos\theta & \sin\theta \\ -\sin\theta & \cos\theta \end{pmatrix}, \qquad (4.71)$$

   then prove it by matrix multiplications. You will need to make use of the power series expansions of the sine and cosine functions:

$$\cos\left(x\right) = \sum_{n=0}^{\infty} \frac{(-1)^n x^{2n}}{(2n)!}, \quad \sin\left(x\right) = \sum_{n=0}^{\infty} \frac{(-1)^n x^{2n+1}}{(2n+1)!}. \qquad (4.72)$$

4. Write the first four terms of Equation 4.63 in terms of $\mathbf{A}$, using Equation 4.64, to show that Equation 4.65 holds.

5. Show that

$$\left[\exp\begin{pmatrix} 0 & \theta \\ -\theta & 0 \end{pmatrix}\right]^T \exp\begin{pmatrix} 0 & \theta \\ -\theta & 0 \end{pmatrix} = \begin{pmatrix} 1 & 0 \\ 0 & 1 \end{pmatrix}. \qquad (4.73)$$

6. Using the program `matrix_exponentiation`, what can be learned, if anything, from the exponential mapping of a symmetric matrix? An orthogonal matrix? A diagonal matrix? An upper and lower triangular matrix? When is the exponential of the sum of two equidimensional matrices $\mathbf{A}$ and $\mathbf{B}$,

$$\exp\left(\mathbf{A} + \mathbf{B}\right),$$

   the same as the product of the exponentials of the two matrices,

$$\exp\left(\mathbf{A}\right)\exp\left(\mathbf{B}\right),$$

evaluated separately? Form conjectures, verify them by writing or modifying programs and input files and then see if you can prove them as theorems.

7. One important property of the orthogonal group is that elements of the group are trace and determinant preserving transformations. The **similarity transform B′** under **U** of a matrix **B** is,

$$\mathbf{B}' = \mathbf{U}^T \mathbf{B} \mathbf{U}. \tag{4.74}$$

The matrices **B′** and **B** are different **representations** of the same abstract object. One can think of a similarity transformation as a change of coordinates. The transforming matrix **U** preserves the trace of **B** if, and only if, **U** is orthogonal. The proof is relatively simple. Using the matrix product definition,

$$B'_{ij} = \sum_{k=1}^{N} \sum_{k'=1}^{N} U_{ki} B_{kk'} U_{k'j}, \tag{4.75}$$

the trace of **B′** is,

$$\text{trace}\,(\mathbf{B}') = \sum_{i=1}^{N} B'_{ii} = \sum_{i=1}^{N} \sum_{k=1}^{N} \sum_{k'=1}^{N} U_{ki} B_{kk'} U_{k'i}. \tag{4.76}$$

If we change the summation order on the right,

$$\text{trace}\,(\mathbf{B}') = \sum_{k=1}^{N} \sum_{k'=1}^{N} \sum_{i=1}^{N} U_{ki} B_{kk'} U_{k'i}, \tag{4.77}$$

we can take $B_{kk'}$ out of the sum on i, since it does not depend on i,

$$\text{trace}\,(\mathbf{B}') = \sum_{k=1}^{N} \sum_{k'=1}^{N} B_{kk'} \sum_{i=1}^{N} U_{ki} U_{k'i}. \tag{4.78}$$

However, if **U** is orthogonal,

$$\sum_{i=1}^{N} U_{ki} U_{k'i} = \delta_{k'k}, \tag{4.79}$$

one obtains,

$$\text{trace}\,(\mathbf{B}') = \sum_{k=1}^{N} B_{kk} = \text{trace}\,(\mathbf{B}), \tag{4.80}$$

since the $\delta_{k'k}$ symbol collapses the sum on k' down to one surviving term when $k' = k$. Write a program that verifies this theorem for an arbitrary antisymmetric matrix **A** and an arbitrary matrix **B**. The antisymmetric matrix **A** is used to construct the orthogonal matrix **U**, using the exponential mapping algorithm. Obviously, **A** and **B** have to be the same size.

4.8 The Determinant of a $n \times n$ Matrix and the Levi–Civita Symbol

For completeness, we define the determinant of a matrix. The method described here does not lend itself to efficient algorithms, and many better ways to calculate the determinant of a matrix do exist. However, the definition given here has theoretical value. Before defining the determinant, we must investigate the operation of permutation of a list of n objects. The set of permutations of a list of n objects is the group P_n. In particular, let's look at the set of permutations of

$$1, 2, 3, 4, 5, 6, \ldots, n. \tag{4.81}$$

There are $n!$ such permutations, for example, for $n = 2$, we have $1, 2$ and $2, 1$, and $2! = 2$. We call an **even** permutation of $1, 2, 3, 4, 5, 6, \ldots, n$, the permutation that requires an even amount of exchanges of two numbers in the sequence. An **odd** permutation of $1, 2, 3, 4, 5, 6, \ldots, n$, is the permutation that requires an odd number of exchanges of two numbers in the sequence. The Levi–Civita symbol for the set $1, 2, 3, 4, 5, 6, \ldots, n$, is defined as

$$\epsilon_{i_1, i_2, \ldots, i_n} = \begin{cases} +1, & \text{if } i_1, i_2, \ldots, i_n \text{ is an even permutation of } 1, 2, 3, \ldots, n \\ -1, & \text{if } i_1, i_2, \ldots, i_n \text{ is an odd permutation of } 1, 2, 3, \ldots, n \\ 0, & \text{otherwise.} \end{cases}$$
$$\tag{4.82}$$

Some examples should clarify. For $n = 2$, one has

i_1	i_2	$\epsilon_{i_1 i_2}$
1	2	$+1$
2	1	-1

for $n = 3$, one obtains the following table

i_1	i_2	i_3	$\epsilon_{i_1 i_2 i_3}$
1	2	3	$+1$
1	3	2	-1
2	1	3	-1
2	3	1	$+1$
3	2	1	-1
3	1	2	$+1$

We are now ready to define the determinant of a matrix $|\mathbf{A}|$.

$$|\mathbf{A}| = \sum_{i_1, i_2, \ldots, i_n \in P_n} \epsilon_{i_1 i_2, \ldots, i_n} a_{i_1 1} a_{i_2 2} a_{i_3 3}, \ldots, a_{i_n n}, \tag{4.83}$$

where the sum runs over all the possible permutations of $i_1, i_2, i_3, \ldots, i_n$. Here, again, examples should clarify. For a 2×2 we get,

$$|\mathbf{A}| = \epsilon_{1\,2} a_{1\,1} a_{2\,2} + \epsilon_{2\,1} a_{2\,1} a_{1\,2} = a_{1\,1} a_{2\,2} - a_{2\,1} a_{1\,2},$$

and for a 3×3, the formula yields,

$$|\mathbf{A}| = a_{1\,1} a_{2\,2} a_{3\,3} - a_{2\,1} a_{1\,2} a_{3\,3} - a_{3\,1} a_{2\,2} a_{1\,3} - a_{1\,1} a_{3\,2} a_{2\,3}$$
$$+ a_{3\,1} a_{1\,2} a_{2\,3} + a_{2\,1} a_{3\,2} a_{1\,3}.$$

The direct application of the definition to an algorithm requires $n!$ operations, and that makes the evaluation of determinants by this method very expensive. The determinant of $\mathbf{U} \in SO\,(n, \mathbb{R})$ is $+1$. For example,

$$\begin{vmatrix} \cos\theta & \sin\theta \\ -\sin\theta & \cos\theta \end{vmatrix} = \cos^2\theta - (-\sin^2\theta) = 1.$$

The special group plays a central role in coordinate transformations, as will be clarified in Chapter 11. The determinant of a product is the product of the determinants.

$$|\mathbf{AB}| = |\mathbf{A}| \cdot |\mathbf{B}|. \tag{4.84}$$

For example,

$$\begin{vmatrix} 1 & 2 \\ 2 & 0 \end{vmatrix} = -4,$$

$$\begin{vmatrix} 1 & 3 \\ 0 & 2 \end{vmatrix} = 2,$$

$$\left| \begin{pmatrix} 1 & 2 \\ 2 & 0 \end{pmatrix} \begin{pmatrix} 1 & 3 \\ 0 & 2 \end{pmatrix} \right| = \begin{vmatrix} 1 & 7 \\ 2 & 6 \end{vmatrix} = 6 - 14 = -4(2) = -8.$$

The determinant of a diagonal matrix $\mathbf{D}$ is the product of all its elements,

$$|\mathbf{D}| = d_{1\,1} d_{2\,2} d_{3\,3}, \ldots, d_{n\,n} = \prod_{i=1}^{n} d_{ii}. \tag{4.85}$$

This result follows trivially from the definition of the determinant. Any permutation of the $1, 2, 3, \ldots, n$ set introduces off-diagonal elements, making all the terms in the sum over the permutations zero, except for the first one. It is far superior to evaluate the determinant of a matrix by first transforming it (using similarity transformations) into a diagonal matrix, if it is possible, since that operation only requires, at most, n^3 operations. The Gauss–Jordan elimination is the most frequently used method to compute the determinant of matrices. Matrices with zero determinant do not have an inverse.

4.9 Scalar Product, Outer Product, and Vector Space Mapping

We switch our focus to vector spaces of column and row vectors. Using the definition of the matrix product, we begin by demonstrating that, for a $n \times n$ matrix $\mathbf{A}$ and vector $\mathbf{v} \in \mathcal{V}(n \times 1, \mathbb{R})$, the vector

$$\mathbf{u} = \mathbf{A}\mathbf{v} \in \mathcal{V}(n \times 1, \mathbb{R}). \tag{4.86}$$

To see this, consider an example in detail,

$$\mathbf{u} = \mathbf{A}\mathbf{v} = \begin{pmatrix} 2 & 5 & 3 & 18 \\ 5 & 3 & 6 & 7 \\ 3 & 6 & 22 & 8 \\ 18 & 7 & 8 & 10 \end{pmatrix} \begin{pmatrix} -3 \\ 1 \\ -1 \\ 7 \end{pmatrix} = \begin{pmatrix} 122 \\ 31 \\ 31 \\ 15 \end{pmatrix}.$$

The top row of $\mathbf{A}$ contracted with the column vector yields the top entry in the column of $\mathbf{u}$, $2 \cdot (-3) + 5 \cdot 1 + 3 \cdot (-1) + 18 \cdot (7) = -6 + 5 - 3 + 126 = 122$. The second row of $\mathbf{A}$ contracted with the column vector yields the second entry from the top in the column of $\mathbf{u}$, and so on. In general, the elements of $\mathbf{u}$ are

$$u_i = \sum_{k=1}^{n} A_{ik} v_k. \tag{4.87}$$

The $n \times n$ matrix $\mathbf{A}$ is a $\mathcal{V}(n \times 1, \mathbb{R}) \to \mathcal{V}(n \times 1, \mathbb{R})$ map. The matrix $\mathbf{A}$ is often referred to as an **operator** on the vector space $\mathcal{V}(n \times 1, \mathbb{R})$. If $\mathbf{A} \in Gl(n, \mathbb{R})$, then the map is invertible, since $\mathbf{A}$ has an inverse. Therefore,

$$\mathbf{v} = \mathbf{A}^{-1}\mathbf{u} \in \mathcal{V}(n \times 1, \mathbb{R}) \text{ if } \mathbf{A} \in Gl(n, \mathbb{R}). \tag{4.88}$$

The scalar product of two vectors is the map $[\mathcal{V}(1 \times n, \mathbb{R}), \mathcal{V}(n \times 1, \mathbb{R})] \to \mathbb{R}$. In other words, the scalar product maps from a column and a row vector to a scalar,

$$\mathbf{u}^T \mathbf{v} = \sum_i u_i v_i, \tag{4.89}$$

following the same rules that define the matrix product. The scalar product of a vector with itself is known as the **norm** of the vector. If $\mathbf{A} \in SO(n, \mathbb{R})$, then the map is invertible, since $\mathbf{A}$ has an inverse and additionally, the norm of the vector is preserved. This is simple to demonstrate.

$$\mathbf{u}^T \mathbf{u} = (\mathbf{A}\mathbf{v})^T \mathbf{A}\mathbf{v} = \mathbf{v}^T \mathbf{A}^T \mathbf{A}\mathbf{v} = \mathbf{v}^T \mathbf{v}. \tag{4.90}$$

We note that the product of a column vector, with a row vector in order from left to right, produces a matrix instead of a scalar. The operation is called the vector outer product,

$$\mathbf{O} = \mathbf{v}\mathbf{u}^T. \tag{4.91}$$

Using the definition of the matrix product, it is not difficult to show that

$$O_{ij} = v_i u_j. \tag{4.92}$$

Therefore, the vector outer product is a map $\{\mathcal{V}(n \times 1, \mathbb{R}), \mathcal{V}(1 \times n, \mathbb{R})\} \to \mathcal{V}(n \times n, \mathbb{R})$. The reader should note that all three different types of products are represented by omitted operators, and only careful attention to the notation can distinguish between them. The notation we are using here is the most commonly used in the literature.

4.10   Rotations in Euclidean Space

Consider a vector $\mathbf{x} = (x, y, z) \in \mathcal{V}(3 \times 1, \mathbb{R})$. Usually, one interprets such a vector as the coordinates of a particle relative to some origin of space. The group of continuous transformations that preserve the norm in $\mathcal{V}(3 \times 1, \mathbb{R})$ (the distance from the origin), can be interpreted as a group of rotations. If we construct the Lie algebra, $so(3, \mathbb{R})$, we can identify the generators directly with the angular momentum operators and we discover the same commutation rules for angular momentum that one finds in quantum mechanics. From Lie's first theorem, we can derive Euler's equations of motion for a rigid body, to which our rotation matrix $\mathbf{U}$ is simply a solution. We deal with the Lie algebra of angular momentum and later, the reader will rediscover it in the context of quantum mechanics. Let us present an example of a two-parameter rotation operator,

$$\mathbf{U} = \mathbf{R}_{-\phi}\mathbf{R}_{\theta} = \begin{pmatrix} \cos(\theta)\cos(\phi) & -\sin(\phi) & \sin(\theta)\cos(\phi) \\ \cos(\theta)\sin(\phi) & \cos(\phi) & \sin(\theta)\sin(\phi) \\ -\sin(\theta) & 0 & \cos(\theta) \end{pmatrix}, \tag{4.93}$$

where θ and ϕ are angles used to rotate the vector $\mathbf{x}$. To understand a little better what each angle corresponds to physically, consider the case when $\theta = 0$, then

$$\mathbf{x}' = \begin{pmatrix} x' \\ y' \\ z' \end{pmatrix} = \begin{pmatrix} x\cos\phi - y\sin\phi \\ x\sin\phi + y\cos\phi \\ z \end{pmatrix}. \tag{4.94}$$

Therefore, a rotation by ϕ takes place about the z axis, as the x' coordinate is a mixture of x and y, but not z, and y' is a mixture of x and y, but not z. Similarly, if we set ϕ to zero,

$$\mathbf{x}' = \begin{pmatrix} x' \\ y' \\ z' \end{pmatrix} = \begin{pmatrix} x\cos\theta + z\sin\theta \\ y \\ -x\sin\theta + z\cos\theta \end{pmatrix}. \tag{4.95}$$

Therefore, a rotation by θ takes place about the y axis, as the x' coordinate is a mixture of x and z, but not y, and z' is a mixture of x and z, but not y, which remains unchanged. The matrix $\mathbf{U}$ acts directly on the components of $\mathbf{x}$. We can also think of the effects of $\mathbf{U}$ as "rotating the Cartesian axis," by acting on the basis of the vector space instead. The latter is the passive interpretation, as opposed to the former active interpretation of the action of $\mathbf{U}$.

The program `rotate` that follows, rotates a whole set of (x, y, z) vectors.

```
        program rotate
        implicit real*8 (a-h,o-z)
        parameter (PI = 3.141592654)
        real*8 rot(3,3),un(3,3),an(2)
c
c       On input:
c             Vector xv,yv,zv are the coordinates of a point particle
c             alpha and phi are angles in radians
c          floating point theta and phi are the polar angles
        read(5,*) xv,yv,zv,theta,phi
c       On output:
c             Vector xp,yp,zp
c             after the rotation has taken place

c Begin by constructing the rotation matrix rot(3,3)
        write(6,*) 'vector size =',sqrt(xv*xv+yv*yv + zv*zv)
        rot(1,1) = cos(theta)*cos(phi)
        rot(1,2) = -sin(phi)
        rot(1,3) = sin(theta)*cos(phi)
        rot(2,1) = cos(theta)*sin(phi)
        rot(2,2) = cos(phi)
        rot(2,3) = sin(theta)*sin(phi)
        rot(3,1) = -sin(theta)
        rot(3,2) = 0.d0
        rot(3,3) = cos(theta)
c rotate  ...
          xp = rot(1,1)*xv + rot(1,2)*yv + rot(1,3)*zv
          yp = rot(2,1)*xv) + rot(2,2)*yv + rot(2,3)*zv
          zp = rot(3,1)*xv + rot(3,2)*yv + rot(3,3)*zv
          write(6,*) xp,yp,zp
          write(6,*) 'vector size =',sqrt(xp*xp+yp*yp + zp*zp)
        enddo
        return
        end
```

Exercises

1. Run the program using arbitrary values for x, y, z, θ, let's call these x_0, y_0, z_0, and with $\phi = 0$. Check that the program rotates the coordinates about the y axis and that the vector norm is conserved. Let's call

the outcome of this operation x_1, y_1, z_1. Then, run the program using the values for x_1, y_1, z_1, obtained after the first rotation with an arbitrary value for ϕ, and with $\theta = 0$. Let's call the outcome of this operation, x_2, y_2, z_2. Check that the program rotates the coordinates about the z axis and that the vector norm is conserved. Then, rotate x_0, y_0, z_0 with the same values of θ and ϕ, but simultaneously with a single rotation. Are the results what you expected? Modify the program so that the transpose of **U** is used to perform the rotations. Repeat the entire three-step procedure. What do you note and how do you explain it?

2. Confirm that $\mathbf{U}^T \mathbf{U} = 1$, where **U** is given in Equation 4.93, by hand first, then by modifying `rotate` accordingly.

3. Derive the following results using matrix multiplication.

$$R_{-\phi} = \begin{pmatrix} \cos\phi & -\sin\phi & 0 \\ \sin\phi & \cos\phi & 0 \\ 0 & 0 & 0 \end{pmatrix} = \exp\begin{pmatrix} 0 & -\phi & 0 \\ \phi & 0 & 0 \\ 0 & 0 & 0 \end{pmatrix}.$$

$$R_\theta = \begin{pmatrix} \cos\theta & 0 & -\sin\theta \\ 0 & 0 & 0 \\ -\sin\theta & 0 & \cos\theta \end{pmatrix} = \exp\begin{pmatrix} 0 & 0 & \theta \\ 0 & 0 & 0 \\ -\theta & 0 & 0 \end{pmatrix}.$$

Then, use these results to find an expression for $R_{-\phi}R_\theta$.

4. Use the expressions for $R_{-\phi}$ and R_θ to show that $R_{-\phi}^T = R_\phi$ and $R_\theta^T = R_{-\theta}$.

5. The matrices

$$\mathbf{J}_z = \begin{pmatrix} 0 & 1 & 0 \\ -1 & 0 & 0 \\ 0 & 0 & 0 \end{pmatrix}, \quad \mathbf{J}_y = \begin{pmatrix} 0 & 0 & -1 \\ 0 & 0 & 0 \\ 1 & 0 & 0 \end{pmatrix},$$

and

$$\mathbf{J}_x = \begin{pmatrix} 0 & 0 & 0 \\ 0 & 0 & 1 \\ 0 & -1 & 0 \end{pmatrix},$$

are the matrix representations of the components of angular momentum. Prove the following commutation rules by hand,

$$[\mathbf{J}_x, \mathbf{J}_y] = \mathbf{J}_z,$$
$$[\mathbf{J}_x, \mathbf{J}_z] = \mathbf{J}_y,$$
$$[\mathbf{J}_z, \mathbf{J}_y] = \mathbf{J}_x,$$
$$[\mathbf{J}^2, \mathbf{J}_x] = [\mathbf{J}^2, \mathbf{J}_y] = [\mathbf{J}^2, \mathbf{J}_z] = 0,$$

where $\mathbf{J}^2 = \mathbf{J}_x^2 + \mathbf{J}_y^2 + \mathbf{J}_z^2$, is the Casimir operator of the algebra.

6. Prove the Jacobi identity for $\mathbf{J}_x$, $\mathbf{J}_y$, and $\mathbf{J}_z$ by hand.

4.11 Complex Field Extensions

The application of the machinery of Lie groups as solutions of differential equations requires that we extend the concepts of transpose of a vector and a matrix, the concept of orthogonality and the concept of symmetric matrix, to members of vector spaces and algebras constructed over the field of complex numbers. There are differences in the definition of addition and multiplication in the complex number field, which the reader is already aware. The differences become more striking when algebraic structures are built to generate groups, vector spaces, and algebras over $\mathbb{C}$. We begin by extending the operation of transpose. Numbers in $\mathbb{C}$, normally represented with a pair, $z = a + ib$, $z \in \mathbb{C}$, $a, b \in \mathbb{R}$ with $i = \sqrt{-1}$, have a **complex conjugated** $z^* = a - ib$, formed by changing the sign of i. Members of a vector space built on $\mathbb{C}$ have the **adjoint** member formed by taking the transpose of the matrix representation and the complex conjugate of each entry. The adjoint operation is typically symbolized with †. For example,

$$\begin{pmatrix} 5 & 2+i3 & i5 \\ 2-i3 & 1 & 3 \\ -i5 & 3 & 2 \end{pmatrix}^\dagger = \begin{pmatrix} 5 & 2+i3 & i5 \\ 2-i3 & 1 & 3 \\ -i5 & 3 & 2 \end{pmatrix}.$$

A matrix $\mathbf{A}$ containing complex numbers is called **Hermitian** (or self-adjoint), if the operations of transposition and evaluation of the complex conjugate leave the matrix unchanged. The matrix in the example above is Hermitian. For a Hermitian matrix

$$\mathbf{A}^\dagger = \mathbf{A}, \tag{4.96}$$

whereas for an anti-Hermitian matrix,

$$\mathbf{A}^\dagger = -\mathbf{A}. \tag{4.97}$$

Note that a matrix with complex entries can be written as a sum of two matrices with real entries $\mathbf{A}^R$ and $\mathbf{A}^I$,

$$\mathbf{A} = \mathbf{A}^{(R)} + i\mathbf{A}^{(I)}. \tag{4.98}$$

A Hermitian matrix has a symmetric $\mathbf{A}^{(R)}$ and antisymmetric $\mathbf{A}^{(I)}$. An anti-Hermitian matrix has an antisymmetric $\mathbf{A}^{(R)}$ and symmetric $\mathbf{A}^{(I)}$. Note also that if $\mathbf{A}$ is Hermitian, then $i\mathbf{A}$ is anti-Hermitian. The vector space of all anti-Hermitian $n \times n$ matrices over the complex field $\mathbb{C}$ with the commuter as the vector product is an important Lie algebra, denoted $su(n, \mathbb{C})$ and it is the generalization of the Lie algebra $so(n, \mathbb{R})$ introduced earlier. The Lie group obtained by exponentiation of elements in $su(n, \mathbb{C})$ is denoted $SU(n, \mathbb{C})$ for the "**special unitary**" group of transformations. If $\mathbf{U} \in SU(n, \mathbb{C})$, then

$$\mathbf{U}^\dagger = \mathbf{U}^{-1}. \tag{4.99}$$

In Chapter 5, we will see that the solutions of the Schrödinger equation are subgroups of $SO(n, \mathbb{R})$ or $SU(n, \mathbb{C})$, often depending on the particular representation that is chosen. The unitary group of transformations is composed of similarity transformations

$$\mathbf{U}^\dagger \mathbf{A} \mathbf{U} = \mathbf{B}', \tag{4.100}$$

that preserve the trace of $\mathbf{A}$

$$\text{trace}(\mathbf{B}') = \text{trace}(\mathbf{U}^\dagger \mathbf{A} \mathbf{U}), \tag{4.101}$$

and the norm of vectors in $\mathcal{V}(n \times 1, \mathbb{C})$:

$$\mathbf{v}' = \mathbf{U} \mathbf{v}, \tag{4.102}$$

$$(\mathbf{v}')^\dagger \mathbf{v}' = (\mathbf{U} \mathbf{v})^\dagger \mathbf{U} \mathbf{v} = \mathbf{v}^\dagger \mathbf{U}^\dagger \mathbf{U} \mathbf{v} = \mathbf{v}^\dagger \mathbf{v}. \tag{4.103}$$

Taylor's theorem can be used to map members of the Lie algebra $su(n, \mathbb{C})$ into members of the $SU(n, \mathbb{C})$ group of transformations. In particular, if $\mathbf{A}$ is real and symmetric, then $i\mathbf{A} \in su(n, \mathbb{C})$, and

$$\exp(i\mathbf{A}) = \sum_{m=0}^{\infty} i^m \frac{\mathbf{A}^m}{m!} = \cos(\mathbf{A}) + i \sin(\mathbf{A}) \in SU(n, \mathbb{C}). \tag{4.104}$$

Since the radius of convergence of the sine and cosine function is infinite, there are no mathematical difficulties in defining these operations for matrices.

The program `cmatrix_exponentiation` that follows, can be used to test the conjecture contained in Equation 4.104. The terms $\mathbf{A}^m/m!$ are accumulated recursively, the same way as in the `matrix_exponentiation` program, however, the sine and cosine power series have alternating signs. For odd values of m, terms of the sine are added to the expansion, and these must have a negative sign if $(m-1)/2$ is odd, or a positive sign otherwise. Similarly, even values of m are terms of the cosine expansion, and these must have a negative sign if $m/2$ is odd, or a positive sign otherwise. To test the unitary condition, we begin by noting that $\mathbf{U}^\dagger \mathbf{U}$ may have a real part and an imaginary part, since $\mathbf{U}$ contains a real part and an imaginary part:

$$\mathbf{U} = \mathbf{R} + i\mathbf{I}, \tag{4.105}$$

therefore,

$$\mathbf{U}^\dagger \mathbf{U} = (\mathbf{R}^T - i\mathbf{I}^T)(\mathbf{R} + i\mathbf{I}), \tag{4.106}$$

$$\mathbf{U}^\dagger \mathbf{U} = (\mathbf{R}^T \mathbf{R} + \mathbf{I}^T \mathbf{I}) + i(\mathbf{R}^T \mathbf{I} - \mathbf{I}^T \mathbf{R}). \tag{4.107}$$

This result allows us to test the unitary of $\mathbf{U}$ without using the FORTRAN complex variable type.

```fortran
      program cmatrix_exponentiation
      implicit real*8 (a-h,o-z)
      parameter (NMAX = 100)
      real*8 a(NMAX,NMAX),bn(NMAX,NMAX),bnp1(NMAX,NMAX)
      real*8 sumr(NMAX,NMAX),sumi(NMAX,NMAX)
c This program computes exp(iA) where i = sqrt(-1), and A is real
c and symmetric. Note that the complex type for this problem is not
c strictly necessary and it is not used by the program. The real
c part is stored in the array sumr and the imaginary part in sumi
      do i=1,NMAX      ! initialize
       do j=1,NMAX
        a(i,j) = 0.d0
        bn(i,j) = 0.d0
        bnp1(i,j) = 0.d0
        sumr(i,j) = 0.d0
        sumi(i,j) = 0.d0
        if (i.eq.j) then
         bn(i,j) = 1.d0
         sumr(i,j) = 1.d0
        endif
       enddo
      enddo
      read(5,*) N, mtop ! input
      do i=1,N*N
      read(5,*,end=10) in,jn,x
      a(in,jn) = x
      enddo
10    do m = 1,mtop ! exponentiation
       do i=1,N         ! do no more than mtop iterations
       do j=1,N
        bnp1(i,j) = 0.d0
        do k=1,N
          bnp1(i,j) - bnp1(i,j) + a(i,k)*bn(k,j)/dfloat(m)
        enddo
       if (mod(m,2) .eq. 0) then
        sign = -1.d0
        if (mod(m/2,2) .eq. 0) sign = +1.d0
        sumr(i,j) = sumr(i,j) + sign*bnp1(i,j)   !the real sum
       else
        sign = -1.d0
        if (mod((m-1)/2,2) .eq. 0) sign = +1.d0
        sumi(i,j) = sumi(i,j) + sign*bnp1(i,j)   !the imaginary sum
       endif
       enddo
       enddo
       do i=1,N       ! set bnp1 to bn
        do j=1,N
          bn(i,j) = bnp1(i,j)
        enddo
```

```
      enddo
      enddo        ! iteration loop ends here
      if (N .le. 10) then   ! output
      write(6,*) 'Re(U = EXP(iA))'
      do i=1,N
       write(6,2000) (sumr(i,j),j=1,N)
      enddo
      write(6,*) 'Im(U = EXP(iA))'
      do i=1,N
       write(6,2000) (sumi(i,j),j=1,N)
      enddo
      else
      write(6,*) 'U = EXP(iA)'
      do i=1,N
      do j=1,N
       write(6,*) i,j,sumr(i,j),sumi(i,j)
      enddo
      enddo
      endif
20    write(6,*) 'U**dagger U'
      do i=1,N    ! Testing the orthogonality property of sum
      do j=1,N
        check1 = 0.d0 ! check the real ij element of U**T U
        check2 = 0.d0 ! check the imaginary ij element of U**T U
        do k=1,N
          check1 = check1 + sumr(k,i)*sumr(k,j) + sumi(k,i)*sumi(k,j)
          check2 = check2 + sumr(k,i)*sumi(k,j) - sumi(k,i)*sumr(k,j)
        enddo
        write(6,2010) i,j,check1,check2
      enddo
      enddo
2000  format(10f12.6)
2010  format(2i4,2f12.6)
      end
```

This program, tested with

$$\mathbf{A} = \begin{pmatrix} 1 & 2 & -1 \\ 2 & 0 & 6 \\ -1 & 6 & 2 \end{pmatrix},$$

yields

$$\mathbf{U} = \exp(i\mathbf{A}) = \begin{pmatrix} 0.017137 & -0.170079 & 0.227858 \\ -0.170079 & 0.755062 & -0.061379 \\ 0.227858 & -0.061379 & 0.686996 \end{pmatrix}$$

$$+ i \begin{pmatrix} 0.946572 & 0.101006 & -0.112510 \\ 0.101006 & 0.612610 & 0.108141 \\ -0.112510 & 0.108141 & 0.669326 \end{pmatrix}.$$

4.12   Dirac Bra–Ket Notation

In the 1920s, as the development of quantum theory unfolded, a useful notation for members of the vector field $V(n \times 1, \mathbb{C})$ was proposed. A member of $V(n \times 1, \mathbb{C})$ is denoted with a **ket** symbol, for example, if $\mathbf{u} \in V(n \times 1, \mathbb{C})$, then $\mathbf{u} \Rightarrow |u\rangle$. Letting $|i\rangle$ represent the basis, the linear superposition principle allows us to expand $|u\rangle$,

$$|u\rangle = \sum_{i=1}^{n} c_i |i\rangle, \tag{4.108}$$

where $c_i \in \mathbb{C}$. We remind the reader that n can be infinite in the case of Hilbert spaces. The notation was proposed by Dirac, and was almost immediately adopted because it is a very convenient shorthand for a number of vector space manipulations. Note that the bra-ket symbols are used freely for both basis vectors and for superpositions of these. The reader has to distinguish between these two possibilities from the context. For example, consider $V(2 \times 1, \mathbb{C})$. The basis can be chosen as follows.

$$|1\rangle \Rightarrow \begin{pmatrix} 1 \\ 0 \end{pmatrix}, \quad |2\rangle \Rightarrow \begin{pmatrix} 0 \\ 1 \end{pmatrix}.$$

And

$$\begin{pmatrix} 5 \\ -3 + i \end{pmatrix} = 5|1\rangle + (-3 + i)|2\rangle.$$

The notation has a slightly different symbol for members of the adjoint vector field, $V(1 \times n, \mathbb{C})$, the **bra** symbol. If $\mathbf{u} \in V(1 \times n, \mathbb{C})$, then $\mathbf{u} \Rightarrow \langle u|$. The adjoint vector space basis can be expressed as $\langle i|$, and the linear superposition principle allows us to expand $\langle u|$,

$$\langle u| = \sum_{i=1}^{n} c_i \langle i|, \tag{4.109}$$

where $c_i \in \mathbb{C}$. The adjoint operation can take any ket and turn it into the corresponding bra,

$$|u\rangle^{\dagger} = \langle u|, \tag{4.110}$$

and in terms of the basis,

$$\left(\sum_{i=1}^{n} c_i |i\rangle \right)^{\dagger} = \sum_{i=1}^{n} c_i^* \langle i|. \tag{4.111}$$

Returning to our example, we write,

$$\begin{pmatrix} 5, & -3+i \end{pmatrix}^{\dagger} = \begin{pmatrix} 5 & -3-i \end{pmatrix} = 5\langle 1| + (-3+i)\langle 2|.$$

The scalar product takes place between vectors in $\mathcal{V}(1 \times n, \mathbb{C})$ and those in $\mathcal{V}(n \times 1, \mathbb{C})$, and has a simple symbol, $\langle v|\, u \rangle$. If we make use of the basis expansion, we can rewrite this as,

$$\langle v|\, u \rangle = \left(\sum_{i=1}^{n} b_i \langle i| \right) \sum_{j=1}^{n} c_j |j\rangle, \tag{4.112}$$

$$\langle v|\, u \rangle = \sum_{i=1}^{n} \sum_{j=1}^{n} b_i c_j \langle i|\, j \rangle. \tag{4.113}$$

If, additionally, the basis are unitary (which is not always the case, e.g., consider the Hilbert space spanned by the basis of all complex analytical functions, z^n, $z \in \mathbb{C}$), then, $\langle i|\, j \rangle = \delta_{ij}$, the Kronecker delta, and the innermost sum collapses to a single term when $j = i$,

$$\langle v|\, u \rangle = \sum_{i=1}^{n} b_i c_i. \tag{4.114}$$

The inner product of a vector with its adjoint is

$$\langle u|\, u \rangle = \sum_{i=1}^{n} \sum_{j=1}^{n} c_i^* c_j \langle i|\, j \rangle, \tag{4.115}$$

and with unitary basis it becomes,

$$\langle u|\, u \rangle = \sum_{i=1}^{n} |c_i|^2, \tag{4.116}$$

where the $||$ delimiters for a complex number, $z = a + ib$, imply its size, $|z|^2 = a^2 + b^2$, $a, b \in \mathbb{R}$.

A very useful property of unitary base vectors in $\mathcal{V}(n \times 1, \mathbb{C})$ and its adjoint space is the **resolution of the identity**, which when expressed in Dirac bra-ket notation becomes

$$\sum_{i=1}^{n} |i\rangle \langle i| = \mathbf{1}. \tag{4.117}$$

Since the ket precedes the bra from left to right, the resolution of the identity is a sum of outer products among basis vectors. For the $\mathcal{V}(2 \times 1, \mathbb{C})$ example, and its adjoint vector space, we have

$$\sum_{i=1}^{n} |i\rangle \langle i| = \begin{pmatrix} 1 \\ 0 \end{pmatrix} \begin{pmatrix} 1 & 0 \end{pmatrix} + \begin{pmatrix} 0 \\ 1 \end{pmatrix} \begin{pmatrix} 0 & 1 \end{pmatrix}, \tag{4.118}$$

$$\sum_{i=1}^{n} |i\rangle \langle i| = \begin{pmatrix} 1 & 0 \\ 0 & 0 \end{pmatrix} + \begin{pmatrix} 0 & 0 \\ 0 & 1 \end{pmatrix}. \tag{4.119}$$

4.13   Eigensystems

Diagonalization is at the heart of computation methods in matrix quantum mechanics. We have finally accumulated enough knowledge to tackle this rich subject. Needless to say, the literature on diagonalization is vast, since the solution of the eigenvalue problem finds many applications in science and engineering. The problem can be formulated using the vector space concept. Given a Hermitian $n \times n$ matrix $\mathbf{A}$, we call the vector $\mathbf{u} \in V(n \times 1, \mathbb{C})$ an **eigenvector** of $\mathbf{A}$, if it obeys the equation

$$\mathbf{A}\mathbf{u} = \lambda \mathbf{u}, \tag{4.120}$$

where $\lambda \in \mathbb{R}$ is a scalar called the **eigenvalue**. There are n such linearly independent vectors and together they form the basis set for a vector space, which is an n-dimensional subspace of $\mathcal{V}(n \times 1, \mathbb{C})$. This special vector space is called the eigenvector space of $\mathbf{A}$, and we denote it with $\mathcal{V}(n \times 1, \mathbf{A}, \mathbb{C})$. The **operator** $\mathbf{A}$ always maps any vector in $\mathcal{V}(n \times 1, \mathbf{A}, \mathbb{C})$ to another vector in $\mathcal{V}(n \times 1, \mathbf{\Lambda}, \mathbb{C})$. The set of vectors of $\mathcal{V}(n \times 1, \mathbf{A}, \mathbb{C})$ are the basis vectors that satisfy the eigenvalue-eigenvector equation (Equation 4.120). The base vectors for $\mathcal{V}(n \times 1, \mathbf{A}, \mathbb{C})$ are not unique. Unlike the basis of $\mathcal{V}(n \times 1, \mathbb{C})$, those of the subspace $\mathcal{V}(n \times 1, \mathbf{A}, \mathbb{C})$ are not trivial to obtain. The basis vectors (often called the Ritz vectors) are the outcome of a diagonalization operation.

Alternatively, and perhaps more instructively, the eigen-analysis problem can be stated using the concept of similarity transformation. Solving the eigenvalue problem is equivalent to finding the unitary transformation $\mathbf{U}$, such that

$$\mathbf{U}^{\dagger} \mathbf{A} \mathbf{U} = \mathbf{D}, \tag{4.121}$$

is diagonal. The diagonal elements of $\mathbf{D}$ are the eigenvalues λ_i, and the columns of $\mathbf{U}$ are the corresponding eigenvectors $\mathbf{u}_i$. $\mathbf{U} \in SO(n, \mathbb{R})$ if $\mathbf{A}$ is real and symmetric, or $\mathbf{U} \in SU(n, \mathbb{C})$ if $\mathbf{A}$ is complex and Hermitian. The collection of the eigenvalues of $\mathbf{A}$ is also called the spectrum of $\mathbf{A}$. The set $\{\mathbf{u}_1, \mathbf{u}_2, \ldots, \mathbf{u}_n\}$ is orthogonal for a nondegenerate spectrum (i.e., $\lambda_i \neq \lambda_j$). We start from Equation 4.120, and we construct the following product

$$\lambda_j \mathbf{u}_i^{\dagger} \mathbf{u}_j = \mathbf{u}_i^{\dagger} \mathbf{A} \mathbf{u}_j. \tag{4.122}$$

The expressions in Equation 4.122 are scalar quantities and we can transpose on the left, since the transpose of a scalar is,

$$\lambda_j \mathbf{u}_i^{\dagger} \mathbf{u}_j = \left(\mathbf{u}_i^{\dagger} \mathbf{A} \mathbf{u}_j\right)^{\dagger}, \tag{4.123}$$

then,

$$\lambda_j \mathbf{u}_i^{\dagger} \mathbf{u}_j = (\mathbf{A} \mathbf{u}_j)^{\dagger} \mathbf{u}_i, \tag{4.124}$$

$$\lambda_j \mathbf{u}_i^{\dagger} \mathbf{u}_j = \mathbf{u}_j^{\dagger} \mathbf{A}^{\dagger} \mathbf{u}_i, \tag{4.125}$$

and using the symmetry property of $\mathbf{A}$, we arrive at:

$$\lambda_j \mathbf{u}_i^\dagger \mathbf{u}_j = \mathbf{u}_j^\dagger \mathbf{A} \mathbf{u}_i. \tag{4.126}$$

Now we use the eigenvalue equation, and we take the adjoint on the right again,

$$\lambda_j \mathbf{u}_i^\dagger \mathbf{u}_j = \lambda_i \mathbf{u}_i^\dagger \mathbf{u}_j, \tag{4.127}$$

or

$$(\lambda_j - \lambda_i) \mathbf{u}_i^\dagger \mathbf{u}_j = 0. \tag{4.128}$$

Therefore, if $\lambda_j \neq \lambda_i$, the associated eigenvectors are orthogonal. If $\mathbf{A}$ is Hermitian, all the eigenvalues $\{\lambda_1, \lambda_2, \ldots, \lambda_m\}$ are real. The **secular determinant** of $\mathbf{A}$ is defined as

$$\begin{vmatrix} a_{11} - \lambda & a_{12} & \cdot & a_{1m} \\ a_{21} & a_{22} - \lambda & \cdot & a_{2m} \\ \cdot & \cdot & \cdot & \cdot \\ a_{m1} & a_{m2} & \cdot & a_{mm} - \lambda \end{vmatrix} = 0. \tag{4.129}$$

The secular determinant yields a polynomial of order n in λ, the roots of which are the eigenvalues of $\mathbf{A}$. The secular determinant can be used to compute eigenvalues for small Hermitian matrices. Then, the Ritz vector, $\mathbf{u}_i$, can be obtained by seeking the **null space** of $\mathbf{A} - \lambda_i \mathbf{1}$,

$$(\mathbf{A} - \lambda_i \mathbf{1}) \mathbf{u}_i = 0. \tag{4.130}$$

An example should clarify. Let us seek the eigenvalues and eigenvector basis for,

$$\begin{pmatrix} 1 & \sqrt{6} \\ \sqrt{6} & 2 \end{pmatrix}.$$

The secular determinant,

$$\begin{vmatrix} 1 - \lambda & \sqrt{6} \\ \sqrt{6} & 2 - \lambda \end{vmatrix} = 0,$$

yields the **characteristic polynomial**, $\lambda^2 - 3\lambda - 4$. The roots are the eigenvalues of the matrix: $\lambda_1 = 4$ $\lambda_2 = -1$. To each one of these two eigenvalues, there corresponds an eigenvector, $\mathbf{u}_1, \mathbf{u}_2$. To find these, we need to find the null space of

$$\begin{pmatrix} 2 & \sqrt{6} \\ \sqrt{6} & 3 \end{pmatrix} \quad \text{and} \quad \begin{pmatrix} -3 & \sqrt{6} \\ \sqrt{6} & -2 \end{pmatrix},$$

respectively. The matrix on the left is obtained following Equation 4.130, $\mathbf{A} - (-1)\mathbf{1}$, the one on the right is $\mathbf{A} - 4\mathbf{1}$. To solve for $\mathbf{u}_1$, we must find the solution of the system

$$\begin{pmatrix} -3 & \sqrt{6} \\ \sqrt{6} & -2 \end{pmatrix} \begin{pmatrix} u_{11} \\ u_{21} \end{pmatrix} = 0.$$

Multiplication of $(\mathbf{A} - \lambda_1 \mathbf{1})\,\mathbf{u}_1$ produces two linearly dependent equations for the entries of the eigenvector,

$$-3u_{1\,1} + \sqrt{6}u_{2\,1} = 0,$$
$$\sqrt{6}u_{1\,1} - 2u_{2\,1} = 0.$$

The best one can do is set either entry to one, say, $u_{2\,1} = 1$, and obtain a number for the remaining entry from either equation. Using the top equation, we get

$$u_{1\,1} = \frac{\sqrt{6}}{3}.$$

It is customary to normalize the vector, by dividing it by the square root of its norm $\mathbf{u}_1^{\dagger}\mathbf{u}_1$, giving finally,

$$\mathbf{u}_1 = \begin{pmatrix} \sqrt{2/5} \\ \sqrt{3/5} \end{pmatrix}.$$

Forcing the eigenvector to be normalized does not remove the degree of freedom in the problem, which is a reflection of the fact that there are an infinite number of basis set for the Ritz vector space. With a similar procedure, we obtain the eigenvector associated with $\lambda = -1$,

$$\mathbf{u}_2 = \begin{pmatrix} -\sqrt{3/5} \\ \sqrt{2/5} \end{pmatrix}.$$

With the basis vector for the eigenspace of $\mathbf{A}$ as $|i\rangle \in \mathcal{V}\,(n \times 1, \mathbf{A}, \mathbb{C})$, one can expand any vector in the same space. If $|f\rangle \in \mathcal{V}\,(n \times 1, \mathbf{A}, \mathbb{C})$, then

$$|f\rangle = \sum_{i=1}^{n} c_i\,|i\rangle. \tag{4.131}$$

With this result, it is not difficult to show that $\mathbf{A}\,|f\rangle \in \mathcal{V}\,(n \times 1, \mathbf{A}, \mathbb{C})$.

4.14 The Connection between Diagonalization and Lie Algebras

One can reformulate the diagonalization process to solve the Schrödinger equation into that of seeking a Lie algebra whose exponential mapping yields the basis of the vector space $\mathcal{V}\,(n \times 1, \mathbf{A}, \mathbb{C})$. Important things happen when we reformulate the problem that way. One develops the Schrödinger and the Heisenberg picture, the WKBJ semiclassical approximation, the Dyson time ordered series, the Magnus expansion, the Lie–Trotter formula, and the process ultimately culminates with the derivation of the path integral. These advanced methods will be much more palatable to the reader after we deal with the practical aspects of the matrix solution of the time-independent Schrödinger equation in Chapter 5. We bring quantum mechanics to the surface because the strict requirement that the abstract operators of quantum mechanics, associated with measurable physical properties, must yield Hermitian matrices is central to the whole process (including the simpler diagonalization issue). If the Hermitian condition is violated, the exponential mapping of the operator, $i\mathbf{O}$, which propagates the state vector, is no longer unitary and, consequently, the probability is not conserved by the transformation. The only exceptions that one finds to this rule are those for the creation and annihilation operators of quantum field theory, where the process of creating or destroying a particle must necessarily not conserve the probability.

The reformulation of the diagonalization procedure into a search for the correct subgroup of $SO\,(n,\mathbb{R})$ or $SU\,(n,\mathbb{C})$, is the connection between Lie algebras and such operation, and this in turn, tells us that diagonalizing is a way of solving coupled systems of differential equations. Often, texts introduce a different analytical formulation to obtain an orthogonal $\mathbf{U}$ for a 2×2, to reflect the connection more directly for real symmetric matrices. The process consists of writing $\mathbf{R}$ as a plane rotation,

$$\mathbf{R} = \begin{pmatrix} \cos\theta & \sin\theta \\ -\sin\theta & \cos\theta \end{pmatrix}, \qquad (4.132)$$

and determining a value of θ that causes the nondiagonal element of $\mathbf{R}^T \mathbf{A} \mathbf{R}$ to vanish. This method, however, only works with 2×2 real symmetric matrices. For larger real symmetric matrices, the procedure outlined in this section has to be performed iteratively a finite number of times, and the resulting algorithm is known as the Jacobi method [8]. The crux of the method is to obtain an expression for the $1, 2$ element of $\mathbf{R}^T \mathbf{A} \mathbf{R}$ and set it to zero by an appropriate choice of the rotation angle,

$$\left(\mathbf{R}^T \mathbf{A} \mathbf{R}\right)_{12} = a_{12} \cos^2\theta - a_{21} \sin^2\theta + (a_{11} - a_{22}) \sin\theta \cos\theta. \qquad (4.133)$$

When we set this to zero to determine θ, we obtain

$$a_{12} \cos^2\theta - a_{21} \sin^2\theta + (a_{11} - a_{22}) \sin\theta \cos\theta = 0, \qquad (4.134)$$

$$a_{12}\left(\cos^2\theta - \sin^2\theta\right) + (a_{11} - a_{22}) \sin\theta \cos\theta = 0, \qquad (4.135)$$

$$(a_{11} - a_{22}) \sin\theta \cos\theta = -a_{12}\left(\cos^2\theta - \sin^2\theta\right). \qquad (4.136)$$

Rearranging and using a trigonometric identity, we get

$$\frac{a_{22} - a_{11}}{a_{12}} = \frac{\cos^2\theta - \sin^2\theta}{\sin\theta \cos\theta} = 2\cot 2\theta. \qquad (4.137)$$

Therefore,

$$\theta_a = \frac{1}{2} \cot^{-1}\left(\frac{a_{22} - a_{11}}{2a_{12}}\right), \qquad (4.138)$$

and

$$\mathbf{R} = \exp \begin{pmatrix} 0 & \theta_a \\ -\theta_a & 0 \end{pmatrix}, \tag{4.139}$$

diagonalizes $\mathbf{A}$,

$$\mathbf{A} = \begin{pmatrix} a_{11} & a_{12} \\ a_{12} & a_{22} \end{pmatrix}. \tag{4.140}$$

Exercises

1. Combine the eigenvectors of

$$\begin{pmatrix} 1 & \sqrt{6} \\ \sqrt{6} & 2 \end{pmatrix},$$

   in the example:

$$\mathbf{u}_1 = \begin{pmatrix} \sqrt{2/5} \\ \sqrt{3/5} \end{pmatrix}, \quad \mathbf{u}_2 = \begin{pmatrix} -\sqrt{3/5} \\ \sqrt{2/5} \end{pmatrix},$$

   into the columns of the matrix $\mathbf{U}$, and show that

$$\mathbf{U}^\dagger \mathbf{U} = \begin{pmatrix} 1 & 0 \\ 0 & 1 \end{pmatrix},$$

$$\mathbf{U}^\dagger \begin{pmatrix} 1 & \sqrt{6} \\ \sqrt{6} & 2 \end{pmatrix} \mathbf{U} = \begin{pmatrix} 4 & 0 \\ 0 & -1 \end{pmatrix}.$$

2. Let

$$\begin{pmatrix} \sqrt{2/5} \\ \sqrt{3/5} \end{pmatrix} \Rightarrow |1\rangle \quad \begin{pmatrix} -\sqrt{3/5} \\ \sqrt{2/5} \end{pmatrix} \Rightarrow |2\rangle.$$

   Prove the resolution of the identity for this example,

$$|1\rangle \langle 1| + |2\rangle \langle 2| = \begin{pmatrix} 1 & 0 \\ 0 & 1 \end{pmatrix}.$$

3. Use Equation 4.138 to find the value of θ, which upon insertion into Equation 4.133, yields the similarity transformation that diagonalizes

$$\begin{pmatrix} 1 & \sqrt{6} \\ \sqrt{6} & 2 \end{pmatrix}.$$

4. Let

$$\mathbf{A} = \begin{pmatrix} a_{11} & a_{12} \\ a_{21} & a_{22} \end{pmatrix},$$

   where $a_{11}, a_{12}, a_{21}, a_{22} \in \mathbb{C}$.

- Show that

$$\lambda_\pm = \frac{1}{2}\left[\text{trace}\left(\mathbf{A}\right) \pm \Delta\right],$$

where

$$\Delta^2 = \text{trace}\left(\mathbf{A}\right)^2 - 4\,\|\mathbf{A}\| = \left(a_{11} - a_{22}\right)^2 + 4a_{12}a_{21}.$$

- Letting

$$\delta = \frac{1}{2}\left(a_{22} - a_{11} - \Delta\right),$$

derive the normalized eigenvectors

$$\mathbf{u}_+ = \frac{1}{\sqrt{|a_{21}|^2 + |\delta|^2}}\begin{pmatrix} \delta \\ -a_{21} \end{pmatrix},$$

$$\mathbf{u}_- = \frac{1}{\sqrt{|a_{12}|^2 + |\delta|^2}}\begin{pmatrix} a_{12} \\ \delta \end{pmatrix}.$$

- Show that the scalar product between the vectors is,

$$\mathbf{u}_+^\dagger \mathbf{u}_- = \frac{\delta\left(a_{12}^* - a_{21}\right)}{\delta^2 + |a_{21}|^2},$$

then show that if $\mathbf{A}$ is Hermitian, δ is real and the product $\mathbf{u}_+^\dagger \mathbf{u}_-$ vanishes.

- The matrix $\mathbf{U}$ is,

$$\mathbf{U} = \begin{pmatrix} \delta/\sqrt{|\delta|^2 + |a_{21}|^2} & a_{12}/\sqrt{|\delta|^2 + |a_{12}|^2} \\ -a_{21}/\sqrt{|\delta|^2 + |a_{21}|^2} & \delta/\sqrt{|\delta|^2 + |a_{12}|^2} \end{pmatrix}.$$

Show that if $\mathbf{A}$ is Hermitian, we can simplify $\mathbf{U}$ slightly,

$$\mathbf{U} = \frac{1}{\sqrt{\delta^2 + a_{21}a_{12}}}\begin{pmatrix} \delta & a_{12} \\ -a_{21} & \delta \end{pmatrix}.$$

- Show that if $\mathbf{A}$ is Hermitian, $\mathbf{U} \in SU\left(2, \mathbf{C}\right)$, and that $\mathbf{U}^\dagger \mathbf{A} \mathbf{U}$ is diagonal with the eigenvalues of $\mathbf{A}$ as entries.

5. Use `cmatrix_exponentiation` to test the following conjecture ($a \in \mathbb{R}$)

$$\exp\left[i\begin{pmatrix} 0 & a \\ a & 0 \end{pmatrix}\right] = \begin{pmatrix} \cos a & i\sin a \\ i\sin a & \cos a \end{pmatrix},$$

then prove it by using matrix multiplications.

6. Prove that if **A** is Hermitian (not necessarily real), then

$$\exp(i\mathbf{A}),$$

is unitary.

7. Write a program that verifies the trace invariance theorem for an arbitrary symmetric matrix **A** and an arbitrary matrix **B**. The symmetric matrix **A** is used to construct the orthogonal matrix **U**, using $\mathbf{U} = \exp(i\mathbf{A})$.

8. The method of expanding the secular determinant would not be very useful to implement a general diagonalization algorithm. The reason is the number of operations involved with the evaluation of the determinant $(n!)$. For some of the simple applications of diagonalization to quantum mechanics, intended to train the uninitiated, we will make use of the popular Householder method. This is a very stable method that requires on the order of n^3 operations. It is based on a series of matrix operations (plane rotations) that first reduce the matrix to a **tridiagonal** form. Then, a second, efficient method performs the final step to obtain eigenvalues and eigenvectors. Chapter 11 of the *Numerical Recipes* book is a good source of detail for the Householder, and other popular eigenanalysis methods [8]. Many of the programs we write make use of canned routines from EISPACK [855], available on the web.* In the program `laplacian`, we make use of canned subroutines r, `tred2`, and `tql`, available on the EISPACK website. These are connected together into a subroutine called **eigen**. To perform the next exercise, the reader must download the routines r, `tred2`, and `tql` and follow the directions in the comments to put them together. These are necessary for most of the work on the next chapter as well.

We use **eigen** as a black box here, though the present exercises test the routine reasonably carefully. Consider a matrix constructed as,

$$\mathbf{A} = \begin{pmatrix} -2 & 1 & 0 & \dots & 0 & 0 & 0 \\ 1 & -2 & 1 & \dots & 0 & 0 & 0 \\ \vdots & & & & & \vdots & \\ 0 & 0 & 0 & \dots & 1 & -2 & 1 \\ 0 & 0 & 0 & \dots & 0 & 1 & -2 \end{pmatrix}.$$

The elements of **A** can be expressed in terms of Kronecker's delta,

$$A_{ij} = -2\delta_{ij} + \delta_{i,j+1} + \delta_{i,j-1}.$$

The matrix is quite sparse[†] and the pattern of nonzero entries has a name, since it is encountered frequently. Matrices that have nonzero

---

*http://www.netlib.org/eispack subroutines **reduc**, **tred2**, and **tql2** to find the eigenvalues and eigenvectors of a real symmetric matrix.

[†]A sparse matrix is a matrix in which most entries are zeros.

entries only along the main diagonal and the diagonals immediately above and below, are called **tridiagonal**. There is something else special about the matrix we have chosen here. **A** in the last equation is related to the lattice Laplacian, and it is an important object in matrix quantum mechanics. What are the eigenvalues and the eigenvectors of a 10×10?

```
      program laplacian
      implicit real*8 (a-h,o-z)
c This program demonstrates the use of the EISPACK routines
c for the diagonalization of a Lattice Laplacian
      parameter (nbas = 100)
      real*8 a(nbas,nbas)
      real*8 d(nbas),z(nbas,nbas),e(nbas)
      do i=1,nbas             ! initialize
       d(i) = 0.d0
       e(i) = 0.d0
       do j=1,nbas
        a(i,j) = 0.d0
        z(i,j) = 0.d0
       enddo
      enddo
      n_max = 10
      do i=1,n_max            ! set up the matrix
      do j=1,n_max
       if (i .eq. j) a(i,j) = -2.d0
       if (i .eq. j-1) a(i,j) = 1.d0
       if (i .eq. j+1) a(i,j) = 1.d0
      enddo
      enddo
      call eigen(nbas,n_max,a,d,z,e,ierr)  ! EIGEN CALL
      write(6,*) 'ierr',ierr    ! output
      do k=1,n_max
        write(6,*) k,d(k)
      enddo
      write(6,*) 'Z**T Z'
      do i=1,n_max  ! check the orthogonality of the transform Z
       do j=1,n_max
        check = 0.d0
        do k=1,n_max
         check = check + z(k,i)*z(k,j)
        enddo
        if(dabs(check) .gt. 1.d-8) write(6,1000) i,j,check
       enddo
      enddo
      write(6,*) 'Z**T A Z'
      do i=1,n_max  ! check the eigenvalues and if A is diagonal
       do j=1,n_max
        check = 0.d0
        do k=1,n_max
```

```
      do l=1,n_max
        check = check + z(k,i)*a(k,l)*z(l,j)
      enddo
      enddo
      x = 0.d0
      if (i. eq. j) x = d(i)
      if(dabs(check) .gt. 1.d-8) write(6,1000) i,j,check, x
      enddo
      enddo
1000  format(2i4,3f12.8)
      end
```

Upon return, the array **d** contains the eigenvalues, and the array **z** the basis eigenvector. The integer `ierr` is zero for a normal return and a positive number in case of error. The orthogonality $U^T U = 1$ and similarity transform $U^T A U = \mathrm{diag}\,(\lambda_1, \lambda_2, \ldots)$ properties of the basis eigenvectors (stored in the rows of **z**) are tested in the code commented with "check the orthogonality of the transform z" and "check the eigenvalues and check if A is diagonal," respectively. A run with a size of $n\_max = 10$ gives the following eigenvalues,

n	λ_n
1	-3.91898595
2	-3.68250707
3	-3.30972147
4	-2.83083003
5	-2.28462968
6	-1.71537032
7	-1.16916997
8	-0.690278532
9	-0.317492934
10	-0.0810140528

The program outputs the outcome of $U^\dagger U$, which should be the unit matrix and compares $U^\dagger A U$ with the diagonal matrix **D**. Perform additional tests with the program, using small symmetric matrix (e.g., a 2×2), and then check the result by hand.

4.15 Symplectic Lie Algebras and Groups

Members of the symplectic group are represented by $2n \times 2n$ matrices, and satisfy the following property

$$A^T J A = J, \tag{4.141}$$

where **J** is the matrix with signed unit entries along the minor diagonal as follows,

$$
\mathbf{J} = \left(\begin{array}{ccc|ccc}
0 & \cdots & 0 & 0 & \cdots & 1 \\
\vdots & \ddots & \vdots & \vdots & \ddots & \vdots \\
0 & \cdots & 0 & 1 & \cdots & 0 \\
\hline
0 & \cdots & -1 & 0 & \cdots & 0 \\
\vdots & \ddots & \vdots & \vdots & \ddots & \vdots \\
-1 & \cdots & 0 & 0 & \cdots & 0
\end{array}\right). \tag{4.142}
$$

J is called the **symplectic norm**. Each of the four blocks consists of a $n \times n$ matrix. The basis set for the symplectic algebra, $sp(2n, \mathbb{R})$, can be constructed with

$$
M_{ij} = E_{ij}^{(2n)} - \sigma(i)\sigma(j) E_{-j-i}^{(2n)} \quad i, j = \pm 1, \pm 2, \ldots, \pm n, \tag{4.143}
$$

$$
M_{i,-i} = E_{i,-i}^{(2n)}, \tag{4.144}
$$

where, $\sigma(i) = i/|i| = \text{sign}(i)$, and $E_{ij}^{(2n)}$ is the matrix with one unit entry in row i, column j. There is a symmetry among these base vectors,

$$
M_{ij} = -\sigma(i)\sigma(j) M_{-i,-j}, \tag{4.145}
$$

consequently Equations 4.143 and 4.144 produce $n(2n+1)$ independent basis. Basis of a Lie algebra are also known as the **generators** for the corresponding group of transformations. Consider, for example, the $n = 1$ case. In using Equation 4.143, it helps to enumerate the rows and the columns, starting with $-n$ up to $+n$, skipping over 0. The three generators of the $sp(2, \mathbb{R})$ algebra are,

$$
M_{-1-1} = \begin{pmatrix} 1 & 0 \\ 0 & -1 \end{pmatrix}, \quad M_{-11} = \begin{pmatrix} 0 & 1 \\ 0 & 0 \end{pmatrix}, \tag{4.146}
$$

$$
M_{1-1} = \begin{pmatrix} 0 & 0 \\ 1 & 0 \end{pmatrix}. \tag{4.147}
$$

The members of the symplectic Lie group associated with these elements are,

$$
R_\alpha = \exp(\alpha M_{-1-1}) = \begin{pmatrix} e^\alpha & 0 \\ 0 & e^{-\alpha} \end{pmatrix}, \tag{4.148}
$$

$$
R_\beta = \exp(\beta M_{-11}) = \begin{pmatrix} 1 & \beta \\ 0 & 1 \end{pmatrix}, \tag{4.149}
$$

and

$$
R_\gamma = \exp(\gamma M_{1-1}) = \begin{pmatrix} 1 & 0 \\ \gamma & 1 \end{pmatrix}. \tag{4.150}
$$

Exercises

1. Derive Equations 4.148 through 4.150.

2. Show that Equation 4.141 is satisfied for R_α, R_β, and R_γ matrices.

3. Show that

$$R_\alpha R_\beta R_\gamma = \begin{pmatrix} (1 + \beta\gamma)\, e^\alpha & \gamma e^\alpha \\ \beta e^{-\alpha} & e^{-\alpha} \end{pmatrix}, \tag{4.151}$$

then demonstrate that $R_\alpha R_\beta R_\gamma$ satisfies Equation 4.141.

4. Derive

$$\exp\left(\beta M_{-11} - \beta M_{1-1}\right) = \begin{pmatrix} \cos\beta & \sin\beta \\ -\sin\beta & \cos\beta \end{pmatrix}. \tag{4.152}$$

5. Show that both

$$\exp\left(\beta M_{-11} - \beta M_{1-1}\right)^T \mathbf{J} \exp\left(\beta M_{-11} - \beta M_{1-1}\right) = \mathbf{J}, \tag{4.153}$$

$$\exp\left(\beta M_{-11} - \beta M_{1-1}\right)^T \exp\left(\beta M_{-11} - \beta M_{1-1}\right) = \mathbf{1}, \tag{4.154}$$

hold. Therefore, there are groups of transformations that are interceptions of the orthogonal and symplectic groups.

6. Show that

$$\exp\left(\beta M_{-11} - \beta M_{1-1}\right) \neq \exp\left(\beta M_{-11}\right) \exp\left(-\beta M_{1-1}\right). \tag{4.155}$$

Then, use the result from the previous exercise and show that if β is small enough, the two exponential mappings agree. The operation on the l.h.s. is said to occur along a straight line in parameter space. The operation on the r.h.s. is known as an operation along a **broken geodesic**. The relations between groups of transformations formed by single and multiple exponentiation are called the Baker–Campbell–Hausdorff relations.

7. Derive

$$\exp\left(\gamma M_{-11} + \gamma M_{1-1}\right) = \begin{pmatrix} \cosh\gamma & \sinh\gamma \\ \sinh\gamma & \cosh\gamma \end{pmatrix}, \tag{4.156}$$

using products of matrices and the power series that define the hyperbolic sine and cosine,

$$\cosh(x) = \sum_{n=0}^\infty \frac{x^{2n}}{(2n)!}, \quad \sinh(x) = \sum_{n=0}^\infty \frac{x^{2n+1}}{(2n+1)!}. \tag{4.157}$$

8. Use the result from the previous exercise to show that

$$\exp\left(\beta M_{-11} + \beta M_{1-1}\right)^T \mathbf{J} \exp\left(\beta M_{-11} + \beta M_{1-1}\right) = \mathbf{J}, \qquad (4.158)$$

holds, but

$$\exp\left(\beta M_{-11} + \beta M_{1-1}\right)^T \exp\left(\beta M_{-11} + \beta M_{1-1}\right) = \mathbf{1}, \qquad (4.159)$$

does not. Therefore, $\exp\left(\beta M_{-11} + \beta M_{1-1}\right)$ is a member of the symplectic group, but not a member of the orthogonal group.

9. Show that

$$\det\left(R_\alpha\right) = \det\left(R_\beta\right) = \det\left(R_\gamma\right) = 1, \qquad (4.160)$$

and that,

$$\det\left(\exp\left(\beta M_{-11} - \beta M_{1-1}\right)\right) = \det\left(\exp\left(\beta M_{-11} + \beta M_{1-1}\right)\right) = 1. \qquad (4.161)$$

10. Using the generators for $sp\left(2, \mathbb{R}\right)$, $M_{-1-1}, M_{-11}, M_{1-1}$, find the structure constants for this Lie algebra and use them to determine if it is simple. Recall that if a Lie algebra is not simple, then it contains a subalgebra.

11. Use Equations 4.143 and 4.144 to derive the 10 generators for $sp\left(4, \mathbb{R}\right)$, i.e., the $n = 2$ case. Then, test your results using the following program. The program **symplectic** uses Equations 4.143 and 4.144 to produce the generators. The representations are constructed based upon the ordering $n, n-1, \ldots, 1, -1, \ldots, -n$, from left to right. To avoid redundancies, i is changed only from n to 1 in increments of -1. For each value of i, all values of j from n to $-n$, excluding zero, are used. The matrix representation of each generator is stored in a two-dimensional array of dimension $2n \times 2n$ by shifting the indeces to i', j', according to

$$i' = n - i + 1, \quad j' = \begin{cases} n - j + 1 & j > 0 \\ n - j & j < 0 \end{cases}.$$

In developing the algorithm without redundancies, we encounter three cases:

   i. $j = -i$, for which Equation 4.144 must be used. The cases that occur for negative values of i are obtained by forming the transpose of the generator produced for $i > 0$.

   ii. $\sigma\left(i\right)\sigma\left(i\right) < 0$, if $-j < i$, Equation 4.143 produces a matrix representation identical to one that is produced for $-j > i$ when the two values are exchanged. For $sp\left(4, \mathbb{R}\right)$, this only happens when $i = 2$ and $j = -1$. The generator produced when $i = 1$ and $j = -2$ is identical to the one with $i = 2$ and $j = -1$, as Equation 4.144 demonstrates. Fortunately, the missing generator, which requires negative values of i, is obtained by forming the transpose of the generator created at this step.

iii. The $\sigma(i)\,\sigma(i) > 0$ case, when $i \neq j$, does not produce redundant generators in the range of i and j used by the program. The algorithm has not been tested for $n > 2$.

```
      program symplectic
      implicit real*8 (a-h,o-z)
      integer i,j,k,c,n,index,jndex,indexp,jndexp,sigma
      real*8 generator(100,100,200)
      real*8 commuter(100,100)
c This program produces the generators for the sp(2,R)
c and sp(4,R) algebras. The algorithm needs to be tested
c carefully for larger values of n.
      read(5,*) n
      c = 1
      do i=n,1,-1
        do j=n,-n,-1
          if (j .ne. 0) then
            index = -i + n + 1
            jndex = -j + n
            if (j .gt. 0)  jndex = jndex +1
            if (j .eq. -i) then
c E_{-i,-j}
            generator(index,jndex,c) = 1
             write(6,*) 1,i,j,index,jndex
             c = c + 1
            generator(jndex,index,c) = 1
             c = c + 1
            else
c E_{i,j} - sigma* E_{-j,-i}
            sigma = (i/abs(i))*(j/abs(j))
            if (sigma .lt. 0) then
            if (abs(i) .ne. abs(j) .and. abs(j) .lt. abs(i)) then
            generator(index,jndex,c) = 1
            indexp = i + n
            jndexp = j + n
            if (i .lt. 0)  indexp = indexp +1
            if (j .lt. 0)  jndexp = jndexp +1
            generator(jndexp,indexp,c) = -sigma
            write(6,*) 2,i,j,index,jndex,indexp,jndexp
            c = c + 1
c if sigma < 0  and i =/= j, the equation  E_{i,j}-sigma* E_{-j,-i}
c produces two equal generators stepping
c through i=n,n-1,...,1, j=n,n-1,...1,-1,...,-n
c the missing generator would require a negative value of i,
c however, the missing generator is the transpose of the
c one just computed.
            generator(jndex,index,c) = 1
            generator(indexp,jndexp,c) = -sigma
            write(6,*) 2,i,j,jndex,index,jndexp,indexp
```

```
                  c = c + 1
                 endif
                else
                 generator(index,jndex,c) = 1
                 indexp = i + n
                 jndexp = j + n
                 if (i .lt. 0)  indexp = indexp +1
                 if (j .lt. 0)  jndexp = jndexp +1
                 generator(jndexp,indexp,c) = -sigma
                 write(6,*) 3,i,j,index,jndex,indexp,jndexp
                 c = c + 1
                endif  ! sigma > 0
               endif  ! i = -i
              endif ! j =/= 0
              enddo
             enddo
c
c   write out all the generators of the symplectic sp(2n,r) algebra
         do k=1,c-1
          write(6,*) 'Generator number',k
          do i=1,2*n
           write(6,1000) (generator(i,j,k),j=1,2*n)
          enddo
         enddo
c work out the commutation rules among the generators to
c characterize the Lie algebra
         do k=1,c-1
          do kp = k+1,c-1
           write(6,*) 'Commutator between',k,kp
c compute the commutator between generator k and kp
           do i=1,2*n
            do j=1,2*n
             commuter(i,j) = 0.d0
             do jj=1,2*n
              commuter(i,j) = commuter(i,j)
     & + generator(i,jj,k)*generator(jj,j,kp) -
     &   generator(i,jj,kp)*generator(jj,j,k)
             enddo
            enddo
           enddo
c output the commutator
           do i=1,2*n
            write(6,1000) (commuter(i,j),j=1,2*n)
           enddo
c loop over all values of kp > k
          enddo
         enddo
1000  format(12f5.0)
         end
```

12. Using the program `symplectic`, inspect the structure constants for the $sp\,(4,\mathbb{R})$ case. Is $sp\,(4,\mathbb{R})$ a simple algebra?

13. Modify the program `symplectic` to demonstrate that

$$X_i^T \mathbf{J} = -\mathbf{J}X_i, \tag{4.162}$$

where X_i is the matrix representation of generator i of $sp\,(4,\mathbb{R})$.

14. Modify the program `symplectic` to demonstrate that

$$\exp\,(aX_i)^T \mathbf{J} \exp\,(aX_i) = \mathbf{J}, \tag{4.163}$$

where X_i is the matrix representation of generator i of $sp\,(4,\mathbb{R})$ and $a \in \mathbb{R}$.

15. Show that

$$\exp \begin{pmatrix} 0 & ia \\ 0 & 0 \end{pmatrix} = \begin{pmatrix} 1 & ia \\ 0 & 1 \end{pmatrix}.$$

Then, show that

$$\left[\exp \begin{pmatrix} 0 & ia \\ 0 & 0 \end{pmatrix} \right]^T \begin{pmatrix} 0 & 1 \\ -1 & 0 \end{pmatrix} \exp \begin{pmatrix} 0 & ia \\ 0 & 0 \end{pmatrix} = \begin{pmatrix} 0 & 1 \\ -1 & 0 \end{pmatrix}.$$

Therefore, there exists symplectic algebras over the complex number field. The symplectic norm is the same as for symplectic algebra over the real field, and the matrix operation that preserves it is not the adjoint operation, but simply the transpose.

4.16 Lie Groups as Solutions of Differential Equations

As mentioned earlier, Lie algebras are fundamental for the solutions of differential equations, the same way that linear algebras are fundamental for systems of algebraic equations. Let us consider, for example, the following second order differential equation,

$$\frac{d^2u}{dt^2} - \omega^2 u = 0, \qquad u\,(t=0) = u_0, \qquad \frac{du}{dt}\,(t=0) = v_0. \tag{4.164}$$

Introducing $v = du/dt$, we can reduce the original equation to a set of first order coupled differential equations,

$$\frac{du}{dt} - v = 0, \tag{4.165}$$

$$\frac{dv}{dt} - \omega^2 u = 0. \tag{4.166}$$

This can be written in matrix form,

$$\begin{pmatrix} \dot{u} \\ \dot{v} \end{pmatrix} = \begin{pmatrix} 0 & 1 \\ \omega^2 & 0 \end{pmatrix} \begin{pmatrix} u \\ v \end{pmatrix}. \tag{4.167}$$

The 2×2 matrix on the right-hand side, acts on the state vector (u, v) and returns its time derivative. It is straightforward to verify that

$$\exp \begin{pmatrix} 0 & t \\ \omega^2 t & 0 \end{pmatrix} = \begin{pmatrix} \cosh \omega t & \omega^{-1} \sinh \omega t \\ \omega \sinh \omega t & \cosh \omega t \end{pmatrix}. \tag{4.168}$$

Therefore, the solution of the set of differential equations is

$$\begin{pmatrix} u \\ v \end{pmatrix} = \exp \begin{pmatrix} 0 & t \\ \omega^2 t & 0 \end{pmatrix} \begin{pmatrix} u_0 \\ v_0 \end{pmatrix}, \tag{4.169}$$

or

$$u = u_0 \cosh \omega t + \frac{v_0}{\omega} \sinh \omega t, \tag{4.170}$$

$$v = v_0 \cosh \omega t + u_0 \omega \sinh \omega t. \tag{4.171}$$

The member of the group that propagates in time, the solution of a system of coupled differential equations from their original value is called the **time evolution operator**. It can be shown that the time evolution operator is both volume preserving and, in this case, also symplectic, i.e.,

$$\det \begin{pmatrix} \cosh \omega t & \omega^{-1} \sinh \omega t \\ \omega \sinh \omega t & \cosh \omega t \end{pmatrix} = 1, \tag{4.172}$$

and,

$$\begin{pmatrix} \cosh \omega t & \omega^{-1} \sinh \omega t \\ \omega \sinh \omega t & \cosh \omega t \end{pmatrix}^T \mathbf{J} \begin{pmatrix} \cosh \omega t & \omega^{-1} \sinh \omega t \\ \omega \sinh \omega t & \cosh \omega t \end{pmatrix} = \mathbf{J}. \tag{4.173}$$

Remarkably, the solution to the original differential equation is obtained by exponentiation. Note, however, that the argument of the exponential function in Equation 4.169 is the time integral of the coupling matrix in Equation 4.167.

Exercises

1. Verify Equation 4.168 and Equations 4.170 through 4.173.

2. Check that the solutions in Equations 4.170 and 4.171 satisfy the set of coupled differential equations, Equations 4.165 and 4.166.

3. Show that the Hamilton equations for the monodimensional harmonic oscillator are,

$$\begin{pmatrix} \dot{x} \\ \dot{p} \end{pmatrix} = \mathbb{L} \begin{pmatrix} x \\ p \end{pmatrix}, \tag{4.174}$$

where $\mathbb{L}$ is known as the Liouville operator,

$$\mathbb{L} = \begin{pmatrix} 0 & m^{-1} \\ -k & 0 \end{pmatrix}, \tag{4.175}$$

with m the mass, and k the spring constant. Using the matrix product and power series, show that,

$$[\exp(\mathbb{L}t)] \begin{pmatrix} x_0 \\ p_0 \end{pmatrix}, \tag{4.176}$$

is the solution of the differential equations, that the symplectic norm is preserved,

$$[\exp(\mathbb{L}t)]^T \mathbf{J} [\exp(\mathbb{L}t)] = \mathbf{J}, \tag{4.177}$$

and that,

$$[\exp(-\mathbb{L}t)] [\exp(\mathbb{L}t)] = \mathbf{1}, \tag{4.178}$$

where the first operation is a propagation back in time from $t = 0$ to $-t$.

4.17 Split Symplectic Integrators

Consider a general monodimensional Hamiltonian system,

$$H(x, p) = \frac{p^2}{2m} + V(x). \tag{4.179}$$

It can be shown that the following Liouville operator:

$$\mathbb{L} = \begin{pmatrix} 0 & \dfrac{\partial H}{\partial p} \dfrac{\partial}{\partial p} \\ -\dfrac{\partial H}{\partial x} \dfrac{\partial}{\partial x} & 0 \end{pmatrix}, \tag{4.180}$$

acting on the phase space state vector, produces Hamilton's equations,

$$\begin{pmatrix} \dot{x} \\ \dot{p} \end{pmatrix} = \mathbb{L} \begin{pmatrix} x \\ p \end{pmatrix}. \tag{4.181}$$

Unlike the simple harmonic oscillator, generally, the Liouville operator is a function of x, p. As such, its entries depend implicitly on time. The exponentiation of $\int \mathbb{L} \Delta t$ is valid only for a local time evolution operator. However, one can use the local operator and derive a large family of efficient algorithms for numerical solutions that preserve the volume of phase space, and the symplectic nature of Hamilton's equations.

To derive a simple integrator based on the local time evolution operator, it is convenient to split up the Liouville operator into stages, using the generators of the symplectic Lie algebra. If we use only one stage split time evolution operator, we get the Euler–Cromer integrator. Here is how:

$$\exp\left[\int_t^{t+\Delta t} \mathbb{L}\,dt\right] = \exp\left[\int_t^{t+\Delta t} \mathbb{A}\,dt\right]\exp\left[\int_t^{t+\Delta t} \mathbb{B}\,dt\right], \qquad (4.182)$$

where,

$$\int_t^{t+\Delta t} \mathbb{B}\,dt = \begin{pmatrix} 0 & 1 \\ 0 & 0 \end{pmatrix}\int_t^{t+\Delta t}\frac{\partial H}{\partial p}\frac{\partial}{\partial p}\,dt. \qquad (4.183)$$

For sufficiently small values of Δt, the integral can be approximated,

$$\int_t^{t+\Delta t}\frac{\partial H}{\partial p}\frac{\partial}{\partial p}\,dt \simeq \frac{\Delta t}{p_n}\frac{\partial H}{\partial p}\bigg|_{x=x_n\ p=p_n}. \qquad (4.184)$$

Similarly,

$$\int_t^{t+\Delta t} \mathbb{A}\,dt \simeq -\frac{\Delta t}{x_n}\frac{\partial H}{\partial x}\bigg|_{x=x_n\ p=p_n}\begin{pmatrix} 0 & 0 \\ 1 & 0 \end{pmatrix}. \qquad (4.185)$$

With H as in Equation 4.179, we get

$$\exp\left[\int_t^{t+\Delta t}\mathbb{L}\,dt\right] = \exp\begin{pmatrix} 0 & \dfrac{\Delta t}{m} \\ 0 & 0 \end{pmatrix}\exp\begin{pmatrix} 0 & 0 \\ \dfrac{F_n\Delta t}{x_n} & 0 \end{pmatrix}, \qquad (4.186)$$

where F_n is the local force,

$$F_n = -\frac{\partial V}{\partial x}\bigg|_{x=x_n}. \qquad (4.187)$$

This local operator evolves the state (x_n, p_n) to the state (x_{n+1}, p_{n+1}) as follows,

$$\begin{pmatrix} x_{n+1} \\ p_{n+1} \end{pmatrix} = \begin{pmatrix} 1 & \dfrac{\Delta t}{m} \\ 0 & 1 \end{pmatrix}\begin{pmatrix} 1 & 0 \\ \dfrac{F_n\Delta t}{x_n} & 1 \end{pmatrix}\begin{pmatrix} x_n \\ p_n \end{pmatrix}. \qquad (4.188)$$

This set of equations becomes,

$$p_{n+1} = p_n + F_n\Delta t, \qquad (4.189)$$

$$x_{n+1} = x_n + p_{n+1}\frac{\Delta t}{m}. \qquad (4.190)$$

Here is a FORTRAN version of the Euler–Cromer algorithm, implemented with a quadratic potential. The Euler–Cromer algorithm, while quite similar to the point-slope method of Chapter 2, is much more stable because the local time evolution operator is both volume preserving and symplectic within rounding errors.

```
      program Euler_Cromer
      implicit real*8 (a-h,o-z)
c Euler-Cromer integrator for the harmonic oscillator
      parameter (dm = 2.34d0 ) ! mass in kilograms
      parameter (dk = 0.23d0 ) ! force constant in N/m
      parameter (omega = dsqrt(dk/dm)) ! frequency (Hz)
c initial conditions
      x0 = 0.2d0
      p0 = 0.2d0
      xn = x0
      pn = p0
      dt = 0.2    ! the time step in seconds
      t = 0.d0
      do k=1,500
       call pot(xn,v,f)
       pnp1 = pn + f*dt
       xnp1 = xn + pnp1*dt/dm
       t = t + dt
c compare the numerical solution with the analytical one
      xa = x0*cos(omega*t) + (p0/(dm*omega))*sin(omega*t)
      write(8,*) t,xnp1,pnp1,xa
c reset for the next propagation
      xn = xnp1
      pn = pnp1
      enddo
      end
      subroutine pot(x,v,f)
      implicit real*8 (a-h,o-z)
      parameter (dk = 0.23d0 ) ! force constant in N/m
      v = 0.5d0*dk*x*x
      f = -dk*x
      return
      end
```

To develop higher order integrators [47], one increases the number of stages for a given step,

$$\exp\left[\int_t^{t+\Delta t} \mathbb{L}\, dt\right] = \prod_{k=1}^{n} \exp\left[c_k \int_t^{t+\Delta t} \mathbb{A}_k\, dt\right] \exp\left[d_k \int_t^{t+\Delta t} \mathbb{B}_k\, dt\right]. \quad (4.191)$$

An n-stage integrator with a global error, $O\left(\Delta t^{n+1}\right)$, can be obtained by finding appropriate values of c_k and d_k, such that all the error terms in Δt^n and lower order vanish. The fourth order Candy–Rozmus integrator [46], of Chapter 2 is obtained that way. An eight stage, sixth order integrator for separable Hamiltonians has recently been proposed by Yoshida [48]. Volume preserving integrators for the Nosé Hoover bath can be found in the literature [28], while the symplectic nature of time evolution operators in quantum mechanics has been exploited to find solutions of the time-dependent Schrödinger equation [12, 13, 21, 24, 34, 37, 40]. Methods to solve the time-independent Schrödinger equation are in Chapter 5, whereas the time evolution operator is presented in detail in Chapter 6.

Exercises

1. Show that the Liouville operator in Equation 4.180, applied to the state in Equation 4.181, produces the correct Hamilton equations.

2. Derive the time evolution operator in Equation 4.188 by performing the two exponentiations in Equation 4.186 analytically.

3. Derive Equations 4.189 and 4.190 from Equation 4.188.

4. Test the Euler–Cromer integrator and discover its local error in position, global error in position, and its global energy conservation properties.

4.18   Supermatrices and Superalgebras

The reader should become convinced, by reviewing the definition of a field $\mathbb{F}$ given in section 4.4, that any associative algebra that contains the identity and the inverse elements under addition and under multiplication is a field. Therefore, we can take a subset of the general linear algebra $gl\,(n \times n, \mathbb{F})$ that satisfies the additional requirements just mentioned, and use elements of that field to build the elements of a $m \times m$ matrix, a supermatrix. These form sets, which under the usual operations of addition and multiplication, can form groups, vector spaces, and algebras with various interesting properties. The concept of supermatrices is useful when applications of the differential equations to multidimensional spaces are necessary and occasionally, when fields other than the real number field are necessary. For example, to extend the Candy–Rozmus or the Euler–Cromer to three dimensions, we can choose the following symplectic group to contain the propagator $\mathbf{U}$,

$$\mathbf{U} \in sp\left(2, \mathbb{D}^3\right).$$

Then, the symplectic property of $\mathbf{U}$ is expressed as $\mathbf{U}^T\mathbf{U} = \mathbf{J}$, where

$$\mathbf{J} = \left(\begin{array}{c|c} \mathbf{0} & \mathbf{1} \\ \hline -\mathbf{1} & \mathbf{0} \end{array}\right),$$

and each of the four blocks is a 3×3 matrix representation of elements of the $\mathbb{D}^3$ vector space, and $\mathbf{1}$ is the additive identity in $\mathbb{D}^3$, a diagonal 3×3 matrix with 1 along each element of the main diagonal. It is not difficult to see that the Euler–Cromer algorithm generalizes to the following for a particle of mass m in three-dimensional space,

$$\mathbf{p}_{n+1} = \mathbf{p}_n + \Delta t \, \mathbf{F}_n, \tag{4.192}$$

$$\mathbf{x}_{n+1} = \mathbf{x}_n + \frac{\Delta t}{m} \mathbf{p}_{n+1}, \tag{4.193}$$

where,

$$\mathbf{F}_n = - \begin{pmatrix} \dfrac{\partial V}{\partial x} & 0 & 0 \\ 0 & \dfrac{\partial V}{\partial y} & 0 \\ 0 & 0 & \dfrac{\partial V}{\partial z} \end{pmatrix} \in \mathbb{D}^3. \tag{4.194}$$

Similar developments can be used to generalize the symplectic algorithms in this text, to a system of n particles in Euclidean spaces, without difficulty. Additionally, as some of the exercises below demonstrate, one can develop matrix representations of the complex and quaternion fields. Placing these representations inside the matrix representation of a member of the $SU(n, \mathbb{C})$ or a Hermitian matrix, eliminates the need for a higher level language to have intrinsic complex or quaternion types. On the other hand, for the object-oriented junkies out there, the matrix representation of these fields opens up the possibility to inherit matrix-like functionalities into the development of classes for a field. The computer is essential to discover properties of Lie algebras and Lie groups constructed over fields like the quaternion and complex numbers, or other fields that one can dream up.

Exercises

1. Start with the Hamiltonian,

$$\mathcal{H} = \frac{1}{2m} \left(p_x^2 + p_y^2 + p_z^2 \right) + V(x, y, z),$$

   and construct a matrix representation of the Liouville operator. Use $\mathbb{D}^3$ as the field for the group. Then, derive Equations 4.192 and 4.193.

2. Show that

$$a \begin{pmatrix} 1 & 0 \\ 0 & 1 \end{pmatrix} + b \begin{pmatrix} 0 & 1 \\ -1 & 0 \end{pmatrix}, \tag{4.195}$$

   is a faithful representation of the complex number $a + ib$ with $i = \sqrt{-1}$. Proceed as follows:

   - Prove that the elements represented in Equation 4.195 constitute a group under matrix addition, a vector space over $\mathbb{R}$, an associative

Abelian linear algebra under the matrix product, and finally, a field.

- Prove that $i^2 = -1$, where obviously,

$$-1 \Rightarrow - \begin{pmatrix} 1 & 0 \\ 0 & 1 \end{pmatrix}.$$

- Show that

$$(a + ib)(c + id) = ac - db + i(ad + bc) \Rightarrow \begin{pmatrix} ac - db & ad + bc \\ -ad - bc & ac - db \end{pmatrix}.$$

- Use the representation to show that $(a + ib) + (c + id) = (a + c) + i(b + d)$. Evidently, the proposed representation behaves identically to the field of complex numbers under the addition and multiplication operations.

3. Use the matrix representation of the complex field $\mathbb{C}$, presented in the previous exercise and the matrix product to derive Euler's formula,

$$\exp(ix) = \cos x + i \sin x, \tag{4.196}$$

where x is a real number.

4. Use the representation of $\mathbb{C}$ from the previous two exercises and the program `matrix_exponentiation`, to evaluate

$$\mathbf{U} = \exp \begin{pmatrix} 0 & 2i & -i \\ 2i & 0 & -6 \\ -i & 6 & 0 \end{pmatrix},$$

then show that $\mathbf{U}^T \mathbf{U} = \mathbf{1}$. Finally, find the representation of $\mathbf{U}$ in the set of 3×3 matrices over the field of complex numbers and show that $\mathbf{U}^\dagger \mathbf{U} = \mathbf{1}$, using that representation.

5. The field of quaternions $\mathbb{Q}$ is a four-parameter field,

$$q = a\lambda_1 + b\lambda_2 + c\lambda_3 + d\lambda_4, \tag{4.197}$$

where $a, b, c, d \in \mathbb{R}$, and the λs can be represented with 4×4 matrices:

$$\lambda_1 \Rightarrow \begin{pmatrix} 1 & 0 & 0 & 0 \\ 0 & 1 & 0 & 0 \\ 0 & 0 & 1 & 0 \\ 0 & 0 & 0 & 1 \end{pmatrix}, \tag{4.198}$$

$$\lambda_2 \Rightarrow \begin{pmatrix} 0 & 0 & 0 & -1 \\ 0 & 0 & 1 & 0 \\ 0 & -1 & 0 & 0 \\ 1 & 0 & 0 & 0 \end{pmatrix}, \tag{4.199}$$

$$\lambda_3 \Rightarrow \begin{pmatrix} 0 & 0 & -1 & 0 \\ 0 & 0 & 0 & -1 \\ 1 & 0 & 0 & 0 \\ 0 & 1 & 0 & 0 \end{pmatrix}, \qquad (4.200)$$

$$\lambda_4 \Rightarrow \begin{pmatrix} 0 & -1 & 0 & 0 \\ 1 & 0 & 0 & 0 \\ 0 & 0 & 0 & 1 \\ 0 & 0 & -1 & 0 \end{pmatrix}. \qquad (4.201)$$

Discover the multiplication rules for quaternions. Then, show that the field is associative under multiplication, but it is not commutative.

6. Show that λ_2, λ_3, and λ_4 from the previous exercise, produce a Lie algebra **isomorphic** to $so\,(3, \mathbb{R})$, by studying the commutation rules among the members of both Lie algebras. If the structure constants are the same within a proportionality constant, then the algebras are isomorphic. Use the program `matrix_product` to expedite the otherwise tedious computations. It is the isomorphism just derived that allows us to handle rotations with groups over unimodular (i.e., $a^2 + b^2 + c^2 + d^2 = 1$) quaternions. These comments will become clearer after the group derived by exponentiation of members of $so\,(3, \mathbb{R})$ are used to rotate coordinates of atoms in clusters.

7. Use the program `matrix_product` to explore the properties of $so\,(2, \mathbb{Q})$, the algebra of 2×2 quaternion-valued asymmetric matrices with zero along the diagonal under the commuted product. Find all the generators and the structure constants. Identify any subalgebra, and for any subalgebra (the algebra itself in the absence of subalgebras) find the Killing's form and the Casimir operator C_2. Exponentiate the generators separately, and find the mathematical properties of each of the resulting sets. Is it a group? Is it volume preserving? Is it orthogonal? Is it symplectic? To evaluate the determinant of an exponential of a matrix, the following result, which we challenge you to prove, is most useful.

$$\det[\exp(\mathbf{A})] = \exp[\text{trace}(\mathbf{A})]. \qquad (4.202)$$

Therefore, the exponential of a traceless matrix has a determinant of 1.

5

Matrix Quantum Mechanics

5.1 Introduction

We assume that some readers have not had previous exposure to quantum mechanics, other than the cursory introductory undergraduate level discussion. As such, we develop the subject from scratch, and we follow, at least for the first few sections of the chapter, a typical introductory course. We give a quick overview of the problems that forced physicists in the early part of the last century to propose quantum mechanics, and we formally present its fundamental postulates. However, the rest of the chapter takes a different course compared to the typical introductory treatment. The harmonic oscillator is introduced, but not from the analytical point of view, rather, the machinery of matrix quantum mechanics is unveiled and brought to bear on the problem. The rest of the chapter develops three powerful deterministic tools to solve complex problems in quantum mechanics, the Discrete variable representation (DVR), the Lanczos algorithm for the tridiagonalization of sparse Hamiltonians, and the theory of angular momentum. In the typical introductory curriculum of quantum mechanics, the harmonic oscillator is solved by the method of power series. In advanced courses, the same problem is solved with creation and annihilation operators, making a later transition to quantum field theory easier for the advanced student of quantum mechanics. Our approach is somewhere in between. It is important to realize that the power series solution is no different from other approaches, in that a vector space is used to transform the differential equation into an algebraic problem. The power series method is important because, for the harmonic oscillator, the finiteness of the solution at both asymptotes is the condition that ultimately yields a quantized energy, and this is mathematically more transparent in the power series approach.

In a book about stochastic methods, one may come to question the presence of chapters dedicated to deterministic approaches in quantum mechanics. However, there are two very good reasons to include deterministic methods. Firstly, we have developed a number of quantum stochastic strategies over the years with the aid of students trained in deterministic computations. In the development of a new algorithm, it is always important to perform tests on a hierarchy of increasingly complex systems, for which solutions can be found with alternative strategies. Stochastic methods in quantum mechanics are presently at the cutting edge, but their development would not have been possible if it were not for the luxury to confirm their answers for simple test systems using deterministic methods, which do not suffer from the statistical error. Secondly, our group is feverishly engaged in the development of mixed deterministic-stochastic methods. Several approaches exist in the literature, however, these typically treat a relatively small system with exact quantum methods similar to those in this and the next chapter, and its surrounding by molecular mechanics similar to those in Chapter 2. The broad spectrum of these approaches is known under the acronym QM-MM. Our group is developing similar methods. However, our goal is to treat a relatively small system with exact quantum methods, and its surrounding by path integration, or diffusion Monte Carlo, similar to those we introduce in Chapters 6 and 7. If these techniques emerge as feasible and broadly applicable, our group will have found a useful class of tools for the simulation of realistic models of condensed matter that are presently beyond our reach. These approaches are in development, and in this book we take the first step toward that lofty goal when we introduce the path integral and the diffusion Monte Carlo in spaces generated with holonomic constraints. This is a necessary step, since we anticipate finding that the partitions of degrees of freedom seldom yield Euclidean spaces.

5.2   The Failures of Classical Physics

Near the end of the twentieth century, several laws of classical physics had been established, that explained most experimental observations. Newtonian mechanics could predict the motion of planets, the stability of macroscopic structures, and the dynamics of systems of particles. Classical thermodynamics, classical statistical mechanics, and the kinetic theory of matter, had been extremely successful in unraveling complex physical–chemical phenomena, such as heat transfer, diffusion, chemical kinetics, and chemical equilibrium. Maxwell's theory unified two seemingly separate sets of phenomena (electric and magnetic) into a classical electrodynamic theory. Maxwell's theory explains how electromagnetic radiation can be created by oscillating charges. There remained only a handful of unexplained phenomena. Four of these, spectroscopy, the photoelectric effect, the heat capacity at low temperatures, and the black body radiation energy density, are elaborated in many introductory quantum mechanic texts. Since it is the explanation of these phenomena that brought a revolution in the theories of physics during the last century, it is worth spending some time revisiting them. The reader is reminded, however, that scientists had no valid explanation for the periodic table in 1900. Understanding why elements have periodic chemical properties had been a fundamental problem of chemistry since 1869, when Dimitri Ivanovich Mendeleyev first proposed his now famous periodic table. Only the development of quantum atomic physics, initiated by Bohr in 1915, can shed light on the unique chemical–physical behavior of elements.

5.3 Spectroscopy

Atoms and molecules can be excited (energized) by placing them in flames, electric arcs, or similar devices. When the radiation generated by a pure element is resolved by wavelength, one observes a series of distinct narrow lines. If the radiation emitted by pure elements is visible or ultraviolet light, it can be resolved into its colors by using a grating or a prism and made to impress a film. Unlike white light, the light from a pure element does not contain all wavelengths. Rather, a distinct pattern of lines is observed that is a unique fingerprint for the element. Throughout the latter part of the 1800s, researchers made use of spectroscopy and the uniqueness of the line patterns to systematically categorize all the known naturally occurring elements. The pattern of lines is particularly simple for hydrogen. Let λ be the wavelength of emitted radiation from H atoms. Rydberg was able to fit a simple mathematical relationship for the wavelength of the lines emitted by hydrogen,

$$\frac{1}{\lambda} = R_{\text{H}} \left(\frac{1}{n_1^2} - \frac{1}{n_2^2} \right), \qquad (5.1)$$

where n_1 and n_2 are integers from 1 to infinity, and R_{H} is Rydberg's constant, $R = 109677.5856 \text{cm}^{-1}$. In 1885, Balmer recorded the $n_1 = 2$ series of lines ($n_2 = 3, 4, 5, \ldots$) as each member of the series, using photographic plates sensitive to visible and near ultraviolet light. Equation 5.1 is intimidatingly simple. How could it be that the radiation wavelength from excited hydrogen atoms are so few and follow such simple mathematical pattern? More importantly, what is the physical meaning of the integers n_1 and n_2? No feature of Maxwell's thery can account for these observations. The spectroscopy phenomenon remained unexplained throughout the 1800s. Even the discovery of the electron by J. J. Thompson did not help matters along these lines. After the discovery of the electron, and the finding from the scattering of alpha particle by thin gold foils, Drude and Lorentz attempted to formulate a model for the atom that would have electrons interacting with a nucleus in a harmonic way. The oscillations of the electrons (charged particles) could perhaps account for the radiation generated. However, the model fails to yield patterns in the wavelength of emitted radiation.

5.4 The Heat Capacity of Solids at Low Temperature

Statistical thermodynamics (classical) also had unresolved problems. At low temperatures, the heat capacity of pure solid substances is predicted by classical statistical mechanics to approach the equipartition limit. The oscillations of atoms about the equilibrium positions in the crystal approach the harmonic limit as the temperature decreases, since the amplitude of oscillations becomes smaller. As explained in Chapter 3, heat capacities had been shown to approach zero as the absolute temperature approaches zero. Measurements of the heat capacity of diatomic gases at room temperatures are also a mystery. Classical statistical mechanics predicts that, as the temperature is lowered, the stretching mode of the diatomic gas, like hydrogen, approaches the harmonic limit. Room temperature experiments suggest that diatomic gases, like N_2, only have $5RT/2$ units of energy per mole in direct contradiction to the equipartition limit. The behavior of the heat capacity and the internal energy at low temperature remained a mystery until the beginning of the 1900s.

5.5   The Photoelectric Effect

A beam of light is made to impinge on a surface of a conductor inside a vacuum tube. The conductor ejects electrons only if light with frequency greater than a threshold value, υ_0, is used. Furthermore, the kinetic energy of ejected electrons (photoelectrons) can be measured by applying a counteracting voltage. This is increased slowly until at V_a the current of ejected electrons drops to zero. The kinetic energy for a photoelectron is

$$K_e = eV_a. \tag{5.2}$$

The kinetic energy of photoelectrons was observed to be proportional to the excess frequency of the impinging electromagnetic radiation on the conductor

$$K_e \propto (\upsilon - \upsilon_0). \tag{5.3}$$

This phenomenon could not be explained by classical theories either. In the classical theory, the energy of the electromagnetic field increases when the intensity of light is increased. The photoelectric experiment clearly shows that increasing the intensity of light with frequency below υ_0 does not produce any photoelectrons, and increasing the intensity of light with frequency above υ_0 increases the current, but **does not increase the kinetic energy of the photoelectrons**.

5.6   Black Body Radiator

A perfect black body radiator (hohlraum) is an idealized system where matter is in thermal equilibrium with electromagnetic radiation. Imagine a hohlraum to be a hollow cube of size a, with a tiny hole from which the radiation inside can be detected. The electromagnetic waves must have nodes at the walls, because at the boundaries, radiation is absorbed and emitted. Let υ be the frequency of radiation and let λ be the wavelength. For standing waves we must have

$$\lambda = \frac{2a}{m}, \tag{5.4}$$

where $m = 1, 2, 3, \ldots$ The wavelength λ, and the frequency υ of electromagnetic radiation are related to the speed of light c,

$$\lambda\upsilon = c. \tag{5.5}$$

A quantity that can be easily measured is the radiation energy density $\rho(\upsilon)$ in J m^{-3}. This quantity is related to the average energy per mode $\langle U \rangle$,

$$\rho(\upsilon)d\upsilon = \frac{8\pi\upsilon^2 \langle U \rangle}{c^3} d\upsilon. \tag{5.6}$$

The equipartition theorem for $\langle U \rangle = k_B T$ (two degrees of freedom for electromagnetic radiation) yields the classical Rayleigh–Jeans law,

$$\rho(\upsilon)d\upsilon = \frac{8\pi\upsilon^2 k_B T}{c^3} d\upsilon, \tag{5.7}$$

where k_B is Boltzmann's constant.

The problem with Equation 5.7 is that it does not reproduce the experimental data, especially at high frequencies where it becomes infinite. Experimentally, $\rho(\upsilon)$ reaches a maximum energy density at $\upsilon_{max} = c/\lambda_{max}$ and drops to zero at higher frequencies. λ_{max} is inversely proportional to temperature,

$$\lambda_{max} = \frac{\alpha}{T}. \tag{5.8}$$

This empirical equation is known as Wien's displacement law, and has been known since the late nineteenth century. You should verify that the derivative of $8\pi\upsilon^2 k_B T/c^3$, with respect to υ, does not have any zeros for $\upsilon > 0$.

The total radiated power density, obtained by integrating the experimental radiated power $c\rho_{exp}/4$ over all the emitted frequencies, depends on the fourth power of the temperature,

$$\frac{c}{4} \int_0^\infty \rho_{exp}(\upsilon) \, d\upsilon = \sigma T^4. \tag{5.9}$$

Equation 5.9 is known as the Stephan–Boltzmann law. You can verify that the integral of Rayleigh–Jeans expression for $\rho(\upsilon)$ over all the frequencies diverges (becomes infinite). The Rayleigh–Jeans theory fails to reproduce the empirical results. The failure of the Rayleigh–Jeans law at high frequencies is so dramatic that it has been named the "ultraviolet catastrophe." Wien's displacement law and Stephan–Boltzmann's radiated power law are empirical formulas, just like Rydberg's. And just like Rydberg's equation, there was no theoretical explanation for them in 1900.

5.7   The Beginning of the Quantum Revolution

5.7.1   Planck's Distribution

Planck was a thermodynamicist in Berlin, who had been studying the black body radiation density problem. In 1900, he assumed that the oscillations of charges inside the black body are constrained to only special values of energy. This, in turn, produces electromagnetic waves with special values of energy,

$$E = nh\upsilon, \tag{5.10}$$

where n is a whole number, υ is the frequency of light, and h is an empirical parameter that today we call Planck's constant. **This is a radical departure from classical ideas. There is nothing in Newtonian mechanics that can produce such constraints on the energy of an oscillator.** This assumption changes the average energy per mode of the electromagnetic field in the cavity at a given frequency υ. If the energy only has certain discrete values for a given frequency, as Equation 5.10 suggests, then, the partition function of Chapter 3 has to be modified so that a sum, rather than an integral of all the Boltzmann factor is performed. The result is,

$$\langle U \rangle = \frac{\displaystyle\sum_{n=0}^{\infty} nh\upsilon \exp\left(-nh\upsilon/k_{\mathrm{B}}T\right)}{\displaystyle\sum_{n=0}^{\infty} \exp\left(-nh\upsilon/k_{\mathrm{B}}T\right)} = \frac{h\upsilon}{\exp\left(h\upsilon/k_{\mathrm{B}}T\right) - 1}. \tag{5.11}$$

Inserting this result into Equation 5.6 gives Planck's distribution

$$\rho(\upsilon)d\upsilon = \frac{8\pi\upsilon^3 h}{c^3 \left[\exp(h\upsilon/k_{\mathrm{B}}T) - 1\right]} d\upsilon. \tag{5.12}$$

Agreement between theory and experiment demands that $h = 6.62607556 \times 10^{-34}$ Js. Note that if we let $h \to 0$, we return to the classical limit

$$\lim_{h \to 0} \left\{ \frac{8\pi\upsilon^3 h}{c^3 \left[\exp\left(h\upsilon/k_{\mathrm{B}}T\right) - 1\right]} \right\} = \lim_{h \to 0} \frac{8\pi\upsilon^3}{c^3 \left[\upsilon/k_{\mathrm{B}}T\right]} = \frac{8\pi\upsilon^2 k_{\mathrm{B}}T}{c^3}. \tag{5.13}$$

The Planck distribution can be written as a function of radiation wavelength, λ, by using the dispersion relation, $\upsilon = c/\lambda$, then

$$\rho(\lambda)d\lambda = \rho\left(\frac{c}{\upsilon}\right) \left|\frac{d\upsilon}{d\lambda}\right| d\lambda, \tag{5.14}$$

$$\rho(\lambda)d\lambda = \frac{8\pi hc}{\lambda^5 \left[\exp\left(hc/\lambda k_{\mathrm{B}}T\right) - 1\right]} d\lambda. \tag{5.15}$$

5.7.2 Wien's and Stephan–Boltzmann's Law

If we take the derivative of $\rho(\lambda)$ in Equation 5.15 with respect to λ and set it to zero, we find the maximum energy density at λ_{max} for a given temperature,

$$\frac{d}{d\lambda}\left\{\frac{8\pi h}{\lambda^5\left[\exp\left(hc/\lambda k_B T\right)-1\right]}\right\}$$

$$=\frac{8\pi hc}{\lambda^6\left[\exp\left(hc/\lambda k_B T\right)-1\right]}\left[-5+\left(\frac{hc}{\lambda k_B T}\right)\frac{\exp\left(hc/\lambda k_B T\right)}{\exp\left(hc/\lambda k_B T\right)-1}\right]. \quad (5.16)$$

Setting this derivative to zero yields

$$5=\frac{hc}{\lambda k_B T}\frac{\exp\left(hc/\lambda k_B T\right)}{\exp\left(hc/\lambda k_B T\right)-1}. \quad (5.17)$$

This equation can be solved recursively,

$$\lambda_{n+1}=\frac{hc}{5k_B T}\frac{\exp\left(hc/\lambda_n k_B T\right)}{\exp\left(hc/\lambda_n k_B T\right)-1}, \quad (5.18)$$

using

$$\lambda_0=\frac{hc}{5k_B T}, \quad (5.19)$$

as the initial guess. Then,

$$\lambda_1=\frac{hc}{5k_B T}\frac{\exp(5)}{\exp(5)-1}\approx 1.007\frac{hc}{5k_B T}, \quad (5.20)$$

and

$$\lambda_2=\frac{hc}{5k_B T}\frac{\exp(4.965)}{\exp(4.965)-1}\approx 1.007025\frac{hc}{5k_B T}. \quad (5.21)$$

Convergence is achieved with three iterations to six places. This accuracy is more than sufficient for most applications.

$$\lambda_{max}T=1.007025\frac{hc}{5k_B}. \quad (5.22)$$

Equation 5.22 is Wien's displacement law.

If we integrate the rate at which energy is radiated from the pinhole [also called the exitance $(c/4)\rho(\upsilon)$] over all frequencies, we obtain

$$\frac{c}{4}\int_0^\infty\frac{8\pi\upsilon^3 h}{c^3\left[\exp\left(h\upsilon/k_B T\right)-1\right]}d\upsilon=\frac{2\pi h}{c^2}\int_0^\infty\frac{\upsilon^3}{\exp\left(h\upsilon/k_B T\right)-1}d\upsilon. \quad (5.23)$$

Changing variables $u=h\upsilon/k_B T$, $du=h\,d\upsilon/k_B T$, $\upsilon^3=(k_B T/h)^3$, yields

$$\frac{2\pi h}{c^2}\left(\frac{k_B T}{h}\right)^4\int_0^\infty\frac{u^3}{\exp(u)-1}du, \quad (5.24)$$

the integral evaluates to $\pi^4/15$, leaving for the total radiated power density

$$\frac{2\pi^5 k_{\mathrm{B}}^4}{15c^2 h^3} T^4, \tag{5.25}$$

which is the Stephan–Boltzmann radiation law. Therefore, Planck is able to resolve the clash between classical statistical thermodynamics and Maxwell's theory of electromagnetism.

5.7.3   Einstein's Explanation of the Photoelectric Effect

In 1905, Einstein uses Planck's quantum idea to explain the photoelectric effect. In his explanation, monochromatic light is postulated to be made up of energy bundles, each carrying a "quantum" of energy, $h\upsilon$. These bundles of energy are now known as photons. The energy of a single photon can be increased by increasing its frequency. The photoelectric event is simply a collision between an electron in a conduction band of a metal and a photon of energy, $h\upsilon$. If $h\upsilon$ is greater than the ionization energy (also known as the work function $\Phi = h\upsilon_0$ of a metal), then the collision of a single photon produces a single electron outside the metal with kinetic energy,

$$K_{\mathrm{e}} = h\upsilon - \Phi. \tag{5.26}$$

Increasing the intensity of the electromagnetic radiation increases only the number of electrons, but not their kinetic energy. This explains why radiation with $\upsilon < \upsilon_0$ cannot produce photoelectrons, and why radiation with $\upsilon > \upsilon_0$ increases the photocurrent, but not the kinetic energy of the ejected electrons.

5.7.4   Einstein's Equation for the Heat Capacity of Solids

In 1907, Einstein uses the same assumption made by Planck to explain the heat capacity behavior of solids at low temperatures. Suppose atoms vibrate harmonically along the three Cartesian axes about their equilibrium positions and their energy is restricted to quantized levels. Einstein assumes that one single vibrational frequency (υ_{E}) could represent the statistical behavior of the solid, and writes $E_n = nh\upsilon_{\mathrm{E}}$. Then, the average energy [derived the same way as Equation 5.11] is

$$\langle U \rangle = 3\frac{\sum\limits_{n=1}^{\infty} nh\upsilon_{\mathrm{E}} \exp\left(-nh\upsilon_{\mathrm{E}}/k_{\mathrm{B}}T\right)}{\sum\limits_{n}^{\infty} \exp\left(-nh\upsilon_{\mathrm{E}}/k_{\mathrm{B}}T\right)} = \frac{3h\upsilon_{\mathrm{E}}}{\exp\left(h\upsilon_{\mathrm{E}}/k_{\mathrm{B}}T\right) - 1}. \tag{5.27}$$

The constant volume heat capacity is

$$C_{\mathrm{V}} = \frac{\partial \langle U \rangle}{\partial T} = \frac{3h^2 \upsilon_{\mathrm{E}}^2 \exp\left(h\upsilon_{\mathrm{E}}/k_{\mathrm{B}}T\right)}{k_{\mathrm{B}}T^2 \left[\exp\left(h\upsilon_{\mathrm{E}}/k_{\mathrm{B}}T\right) - 1\right]^2}. \tag{5.28}$$

At temperatures $T \gg h\upsilon_E/k_B$, we can expand the exponential into a power series. To first order, we get

$$C_V \sim \frac{3h^2\upsilon_E^2}{k_B T^2 \left[h\upsilon_E/k_B T\right]^2} \sim 3k_B, \tag{5.29}$$

which is the Dulong–Petit result, while as $T \to 0$, one can show that $C_V \to 0$. Let $x = h\upsilon_E/k_B T$, then

$$C_V = 3k\frac{x^2 e^x}{\left(e^x - 1\right)^2}, \tag{5.30}$$

and taking the $x \to \infty$ limit yields,

$$\lim_{x \to \infty} \frac{x^2 e^x}{\left(e^x - 1\right)^2} = 0. \tag{5.31}$$

Therefore, the quantum limit at low temperatures is reproduced with Einstein's heat capacity expression, however, experimental data is in better agreement with theory when Debye's expression (1912) is uses. Debye used the vibrational spectrum of solids, instead of assuming a single frequency of vibration. Nevertheless, Einstein's result is an important step forward.

5.7.5 Bohr's Model for Hydrogen

In 1913, Bohr assumes that the **angular momentum** $\mathbf{L} = \mathbf{r} \times \mathbf{p}$ of the electron in the hydrogen atom is quantized in a manner analogous to the vibrations of the charges in the black body walls and the atoms in a crystal. He assumed that $L = |\mathbf{L}| = n\hbar$, where $n = 1, 2, 3, \ldots$, and $\hbar$ is Planck's constant divided by 2π, $\hbar - h/2\pi$. The potential energy of the electron around the proton (after the center of mass motion is removed) is

$$E = \frac{L^2}{2I} + V = \frac{n^2\hbar^2}{2\mu r^2} + V, \tag{5.32}$$

where the first term is the centrifugal energy, $I = \mu r^2$ is the moment of inertia of the proton–electon system, μ is the proton–electron reduced mass

$$\mu = \frac{m_p m_e}{m_p + m_e} \approx 9.095 \times 10^{-31} \text{ kg}, \tag{5.33}$$

and V is Coulomb's attractive potential energy between the electron and the proton,

$$V = -\frac{e^2}{4\pi\epsilon_0 r}. \tag{5.34}$$

The electron is subjected to a centrifugal force, F_L,

$$F_L = -\frac{d}{dr}\left(\frac{n^2\hbar^2}{2\mu r^2}\right) = \frac{n^2\hbar^2}{\mu r^3}, \tag{5.35}$$

and to a Coulomb force, F_C,

$$F_C = -\frac{d}{dr}\left(-\frac{e^2}{4\pi\epsilon_0 r}\right) = -\frac{e^2}{4\pi\epsilon_0 r^2}. \tag{5.36}$$

These forces oppose each other and a stable orbit is attained when the two are in balance,

$$F_L = -F_C, \tag{5.37}$$

which yields

$$\frac{n^2\hbar^2}{\mu r^3} = \frac{e^2}{4\pi\epsilon_0 r^2}. \tag{5.38}$$

Rearranging and solving for r, and omitting the $r = 0$ double root, we get

$$r_n = \frac{4\pi\epsilon_0 n^2\hbar^2}{\mu e^2}. \tag{5.39}$$

Stable orbits are obtained at each value of r_n and are a consequence of the assumption that the values of L are quantized.

For a given orbit, the energy is,

$$E_n = \frac{n^2\hbar^2}{2\mu r_n^2} - \frac{e^2}{4\pi\epsilon_0 r_n} = -\frac{e^4\mu}{8\epsilon_0^2 n^2 h^2}. \tag{5.40}$$

Note that since the energies are negative, as n increases, the energy increases also. E_1 is the lowest possible energy and is called **the ground state energy. The $n \to \infty$, $E_n = 0$ limit is called the ionization limit.**

The model as it stands, has a major flaw. According to Maxwell's theory, moving charges must radiate electromagnetic energy. If that were the case, as the electron orbits around the nucleus it would lose energy and eventually spiral down into the nucleus. Matter as we know it, would not exist. **Therefore, when the electron is in one of the stable orbits, it must not radiate. Bohr's model does not explain why that is so.** Nevertheless, Bohr assumes that the electron radiates only when it jumps from one orbit to a lower one, and emits a photon with a frequency matching the energy difference between the two levels,

$$h\upsilon = E_{n1} - E_{n2}, \tag{5.41}$$

while a photon is absorbed by the electron when it goes from a less to a more energetic orbit. This equation, now called Bohr's condition, is simply an energy conservation equation. The L.H.S. applies to the electromagnetic field and the R.H.S. to the "quantum states" of matter involved. Using Equation 5.40, Bohr derives

$$h\upsilon = \frac{hc}{\lambda} = \frac{e^4\mu}{8\epsilon_0^2 h^2}\left(\frac{1}{n_1^2} - \frac{1}{n_2^2}\right), \tag{5.42}$$

or rearranging slightly.

$$\frac{1}{\lambda} = \frac{e^4\mu}{8\epsilon_0^2 h^3 c}\left(\frac{1}{n_1^2} - \frac{1}{n_2^2}\right). \tag{5.43}$$

Equation 5.43 is Rydberg's equation.

Equation 5.1 and the combination of fundamental constants in Bohr's equation yields Rydberg's constant,

$$R_{\mathrm{H}} = \frac{e^4\mu}{8\epsilon_0^2 h^3 c} = 1.0952 \times 10^5\,\mathrm{cm}^{-1}. \tag{5.44}$$

Rydberg's constant is one of the physical constants known most precisely. The value of r_n for $n = 1$ is known as Bohr's radius (the smallest radius allowed in hydrogen),

$$a_0 = \frac{4\pi\epsilon_0\hbar^2}{\mu e^2} = 5.2977 \times 10^{-11}\,\mathrm{m}. \tag{5.45}$$

Exercises

1. Derive Equation 5.11. Start by deriving the partition function

$$Z = \sum_{n=0}^{\infty} \exp\left(-nh\upsilon/k_{\mathrm{B}}T\right).$$

Use the geometric series

$$\frac{1}{1-x} = \sum_{n=0}^{\infty} x^n,$$

where $x = \exp\left(-h\upsilon/k_{\mathrm{B}}T\right)$. To find $\langle U \rangle$, take the derivative of Z as in Equation 3.19.

2. Verify, using an appropriate power series expansion, that the Rayleigh–Jeans law can be obtained from Planck's distribution in the small frequency regime (υ around zero).

3. Derive Equation 5.39 from Equation 5.38.

4. Derive Equation 5.40 by inserting Equation 5.39 into Equation 5.32.

5.8 Modern Quantum Theory and Schrödinger's Equation

Bohr's model of the hydrogen atom gives us an explanation of the spectroscopy phenomenon that remained unexplained for so many years. It also begins to shed some light on the interaction between matter and radiation. The old quantum theory, which has preoccupied us up to this point, leaves a number of open questions behind.

- Why don't electrons radiate as they process about the proton in a stable orbit?

- Why is the orbital angular momentum of electrons quantized?

- Why are oscillations of charges in the walls of the black body and atoms in crystals quantized?

- Why are only some values of the energy allowed?

- Since Newtonian mechanics does not restrict the energy to special values, is Newtonian mechanics wrong?

A new quantum theory of physics has emerged that answers all these questions in a satisfactory way. The theory is capable of making predictions that can be tested and proved accurate by experiment. Furthermore, it contains Newtonian mechanics in the "classical" limit (recall the $h \to 0$ limit of the Planck distribution), since **Newtonian mechanics is not wrong, it is simply incomplete.** It would be preposterous to discard three centuries of experimental evidence that confirmed its validity. Schrödinger's and Heisenberg's theories explain all the questions left behind by the old quantum theory, and contain the classical limit.

5.8.1   de Broglie's Hypothesis and Heisenberg's Uncertainty Principle

In explaining the photoelectric effect as a collision between a photon and an electron, Einstein shows that the photon carries a momentum proportional to $h\upsilon$. Photons have no rest mass, and the expression for the momentum of a photon can only be obtained using relativity theory. Einstein's result is,

$$p = \frac{h\upsilon}{c} = \frac{h}{\lambda}. \tag{5.46}$$

In 1924, Leuis de Broglie proposes the same equation to describe matter. He postulates that the "wavelength" associated with a particle is related to its momentum by

$$\lambda = \frac{h}{p}. \tag{5.47}$$

Therefore, it is de Broglie's idea that matter at the atomic level behaves like a wave. The experiment that confirmed the validity of such assumption comes in 1928, when Davisson and Germer obtain wavelike interference patterns from electrons scattered by a nickel crystal. Many such experiments follow afterward, always confirming the wave nature of matter at the atomic and subatomic level.

The idea that matter has wave-like properties, answers **all** the questions that the early developments in the field left behind. **Many classical phenomena involving standing waves have quantized (that is special) energies associated with them. If electrons are waves "smeared out" around a stable orbit about the nucleus, then they do not need to radiate at all, since they aren't moving at all.** If de Broglie's equation is true, then why did we not observe matter wave phenomena before the Davisson–Germer experiment? The answer is in the size of Planck's constant, h. The wave properties of a body are important only if the associated wavelength is of the same size or larger than the body. Now, compare the wavelength of a baseball $(10^{-1}$ kg) traveling at 1 m s^{-1} $\lambda \approx 7 \times 10^{-33}$ m, with a hydrogen atom $(10^{-27}$ kg) moving with the same speed $\lambda \approx 7 \times 10^{-7}$ m. The former is 32 orders of magnitude smaller than the baseball, whereas the latter is many times larger than the size of the hydrogen atom. The technology to carry out experiments with atoms and molecules that would reveal the wave nature of matter (excluding spectroscopy) is not available to researchers before the turn of the last century.

A direct consequence of the wave nature of matter is a fundamental limitation in our ability to measure certain pairs of physical properties simultaneously. The most obvious example is the momentum–position pair. If a wave phenomenon is known to have a very precise frequency, the wave will extend over all space. Conversely, we can sum waves with different frequencies together to get destructive interferences all over space except for a specific location. **By doing this, we are actually measuring the position of the particle, however, we cannot tell its momentum because we have added up all possible frequencies to localize it.** Heisenberg summarizes these observations into a principle. Suppose the position of the particle is known within $\pm \Delta x$. Then, the momentum of the particle has a fundamental imprecision that satisfies the following inequality:

$$\Delta p \Delta x > \frac{\hbar}{2}, \tag{5.48}$$

where $\hbar/2$ is called the minimum uncertainty. At this point, all that is needed is a way of actually quantifying the matter wave idea into a theory that can be used for making further predictions. This is the task that both Schrödinger and Heisenberg took upon. They developed two separate theories that were later shown by Dirac to be identical.

5.8.2   Schrödinger's Equation

In dealing with wave phenomena, we seek functions that tell us how the amplitude of the wave varies with distance and time. For now, we focus on the time-independent Schrödinger equation. The time dimension is taken up in Chapter 6. Normally, the matter waves are represented with the Greek letter $\Psi(x)$ in quantum mechanics. Physical properties are extracted from wavefunctions by mathematically operating on them. Operators can have a variety of mathematical forms. Most of the operators are either differential operators or are functions of position. Operating on a wave Ψ with an operator $\widehat{D}$ means extracting a **precise value of a property** only if the following condition is satisfied:

$$\widehat{D}\Psi = d\,\Psi. \tag{5.49}$$

Equation 5.49 is known as the **eigenvalue equation**, Ψ is known as the **eigenfunction**, and d the **eigenvalue**. The resemblance of the eigenvalue equation (Equation 5.49) with the diagonalization problem in Chapter 4 should be striking. Later in this chapter, this similarity will come sharply into focus. For the moment, the operator $\widehat{D}$ is some mathematical operation on a space of functions; d must be a number, not a function. Solving the eigenvalue equation (typically a differential equation with boundary values) yields both Ψ and d. Some examples should clarify all these notions.

- e^{3x} is an eigenfunction of the derivative operator d/dx. What is the eigenvalue?

$$\frac{d}{dx}e^{3x} = 3e^{3x}.$$

  Note how e^{3x} is the only function on the left of the equation but to the right of the operator, and is the only function on the right (after the number 3). The eigenvalue is 3.

- The function $\sin(\alpha x)$ is an eigenfunction of the operator d^2/dx^2 with eigenvalue $-\alpha^2$.

- The function $\sin(\alpha x)$ is not an eigenfunction of the operator d/dx.

$$\frac{d}{dx}\sin \alpha x = \alpha \cos(\alpha x),$$

  the function on the right is different from the function on the left.

Let us consider Planck's energy expression for a photon once again, $E = h\upsilon$, to emphasize that waves are more energetic when their frequency is higher. If the frequency of a wave increases, its wavelength decreases and its curvature increases. The curvature at point x of Ψ, you may recall from calculus, is obtained by evaluating the second derivative of Ψ at x.

$$\frac{d^2\Psi}{dx^2}. \tag{5.50}$$

The curvature must be proportional to the kinetic energy of a particle K. Consider a simple sine wave,

$$\Psi = \sin(kx). \tag{5.51}$$

The wavelength is $\lambda = 2\pi/k$. Taking two derivatives gives us

$$\frac{d^2\Psi}{dx^2} = -k^2 \sin(kx) = -k^2\Psi. \tag{5.52}$$

We operated on Ψ, and retrieved it along with a physical property we were interested in. From the argument above, $d^2\Psi/dx^2$ is proportional to the kinetic energy associated with the wave,

$$\widehat{K}\Psi = A\frac{d^2\Psi}{dx^2} = -k^2 A\Psi, \tag{5.53}$$

where A is a constant of proportionality that we have to determine and $\widehat{K}$ is the **quantum mechanical kinetic energy operator**. If we use the classical expression for the kinetic energy, we get

$$\frac{p^2}{2m} = -k^2 A, \tag{5.54}$$

and using de Broglie's equation for p gives,

$$\frac{h^2}{2m\lambda^2} = -k^2 A. \tag{5.55}$$

Recall that $\lambda = 2\pi/k$, therefore,

$$\frac{h^2 k^2}{2m\,(2\pi)^2} = -k^2 A, \tag{5.56}$$

$$-\frac{\hbar^2}{2m} = A. \tag{5.57}$$

Therefore, the quantum kinetic energy operator becomes

$$\widehat{K} = -\frac{\hbar^2}{2m}\frac{d^2}{dx^2}. \tag{5.58}$$

It operates on Ψ, returning Ψ (if Ψ is its eigenfunction) and the kinetic energy. When the particle experiences a potential $V(x)$, then the total energy operator $\widehat{H}$, the Hamiltonian operator, is

$$\widehat{H} = \widehat{K} + \widehat{V} = -\frac{\hbar^2}{2m}\frac{d^2}{dx^2} + V(x), \tag{5.59}$$

and solving Schrödinger's equation is reduced to finding sets of E_n and Ψ_n, such that the eigenvalue equation

$$\left[-\frac{\hbar^2}{2m}\frac{d^2}{dx^2} + V(x)\right]\Psi = E\Psi, \tag{5.60}$$

is satisfied. More than a single function satisfies Equation 5.60 for a given potential $V(x)$. One labels each distinct solution pair $\Psi_n(x), E_n$ in the set with an index n, which corresponds to a **quantum number**. The energy E_n has discrete values only when the particle of matter is bound by a potential. The matter wave equation (Equation 5.60) has a continuous spectrum when the particle of matter is not bound by a potential. Unbound states are called **scattering states**. This more complex case is not the main focus of this book. The proper treatment of quantum scattering requires the introduction of the time evolution operator. We do this in Chapter 6.

While the discussion has been limited to one-dimensional problems, higher dimensional problems frequently occur in applications. As a general rule, for bound systems there is one index (quantum number) for every dimension. The hydrogen atom, for example, requires the solution of a three-dimensional Schrödinger equation after the center of mass degrees of freedom are removed. Three quantum numbers label each individual state of the electron. At this point, we have a mathematical tool that can help us calculate a wavefunction Ψ, at least for some elementary problems. From this wavefunction, we can obtain the physical observables we need, as long as Ψ is an eigenfunction of the quantum mechanical operator. We now need to address two fundamental questions.

- What is the physical meaning of Ψ?

- How do we get physical properties from Ψ when it is not an eigenfunction of the corresponding operator?

The answers to these two questions are normally presented as the postulates of quantum mechanics.

5.8.3   The Interpretation of Ψ: The First Postulate of Quantum Mechanics

The physical interpretation of $\Psi(x)$ is a probabilistic one. The probability $P(x)$ of finding a particle between x and $x + dx$ is given by

$$P(x)\,dx = \Psi^*(x)\,\Psi(x)\,dx, \tag{5.61}$$

where Ψ^* is the complex conjugate of Ψ. Often, the wavefunction is a complex quantity. Taking the complex conjugate in an expression can be accomplished by replacing i with $-i$ in the expression.

In order for the probabilistic interpretation to work, we must normalize Ψ so that,

$$\int_{\text{(all space)}} \Psi^*(x)\,\Psi(x)\,dx = 1, \tag{5.62}$$

which means that the probability of finding the particle anywhere in the space defined by the problem is 1. The probability of finding the particle between

x_1 and x_2, when Ψ is normalized, is given by

$$P\left(x_1 \leq x \leq x_2\right) = \int_{x_1}^{x_2} \Psi^*\left(x\right) \Psi\left(x\right) dx. \tag{5.63}$$

The set of eigenfunctions of the Hamiltonian operator is a Hilbert space that has, additionally, a nice property. The set of eigenfunctions are mutually **orthogonal** if the corresponding states are not **degenerate**. Two states, n and m, are called degenerate if $E_n = E_m$. This statement has been proven in Chapter 4, in the context of the diagonalization problem. After we derive the most elementary of the Hilbert spaces, we tie these two concepts together into one. By orthogonality between $\Psi_n\left(x\right)$ and $\Psi_m\left(x\right)$, we mean,

$$\int_{\text{(all space)}} \Psi_n^*\left(x\right) \Psi_m\left(x\right) dx = 0 \quad \text{iff} \quad E_n \neq E_m. \tag{5.64}$$

5.8.4 Expectation Values: The Second Postulate of Quantum Mechanics

When the operator $\widehat{O}$, corresponding to the physical property o, and the wavefunction of the system Ψ, **does not satisfy the eigenvalue equation** ($\widehat{O}\Psi \neq o\Psi$), then the best we can do is **obtain an expectation value** for the physical property in question (i.e., an average). This is represented by,

$$\langle o \rangle = \int_{\text{(all space)}} \Psi^*\widehat{O}\Psi \, dx. \tag{5.65}$$

Equation 5.65 must be interpreted in the following way. The average property o is obtained by first operating $\widehat{O}$ on Ψ, then multiplying the result by the complex conjugate of Ψ, and finally integrating over all space. N.B. Ψ must be normalized. If Ψ is **not normalized**, then Equation 5.65 must be written as follows,

$$\langle o \rangle = \frac{\displaystyle\int_{\text{(all space)}} \Psi^*\widehat{O}\Psi \, dx}{\displaystyle\int_{\text{(all space)}} \Psi^*\Psi \, dx}. \tag{5.66}$$

The most important consequence of the fact that the physical property, o, and the wavefunction of the system, Ψ, do not satisfy the eigenvalue equation ($\widehat{O}\Psi \neq o\Psi$) is that there is an uncertainty associated with the average value predicted by the theory. The uncertainty associated with $\langle o \rangle$ can be obtained in the usual way,

$$\Delta o = \sqrt{\langle o^2 \rangle - \langle o \rangle^2}, \tag{5.67}$$

where, for a normalized Ψ,

$$\langle o^2 \rangle = \int_{\text{(all space)}} \Psi^*\widehat{O}^2\Psi \, dx. \tag{5.68}$$

Consider the case when $\widehat{O}$ and Ψ do obey the eigenvalue equation. Then $\widehat{O}\Psi = o\Psi$ and Equation 5.65 yields

$$\langle o \rangle = \int_{\text{(all space)}} \Psi^* \widehat{O}\Psi \, dx = \int_{\text{(all space)}} \Psi^* o\Psi \, dx = o \int_{\text{(all space)}} \Psi^* \Psi \, dx,$$

the last integral on the right equals 1 by assumption. Therefore, $\langle o \rangle = o$. This is an important result since it demonstrates that Equation 5.65 works, even when the pair $\widehat{O}$ and Ψ obey the eigenvalue equation. However, when $\widehat{O}$ and Ψ obey the eigenvalue equation, our knowledge of o is sharp, since the spread Δo goes to zero,

$$\langle o^2 \rangle = \int_{\text{(all space)}} \Psi^* \widehat{O}^2\Psi \, dx = o \int_{\text{(all space)}} \Psi^* \widehat{O}\Psi \, dx = o^2$$

$$\times \int_{\text{(all space)}} \Psi^* \Psi \, dx = o^2,$$

and $\langle o^2 \rangle = \langle o \rangle^2$, therefore, $\Delta o = 0$, as we wanted to demonstrate.

5.8.5 A Brief List of Operators

We encountered two operators already. The position operator,

$$\widehat{x}\Psi = x\Psi, \tag{5.69}$$

is simply the operation of "multiplying Ψ by x." Note that if Ψ is a function of x, it cannot be an eigenfunction of the position operator. We have seen operators that are functions of the position operator (e.g., the potential energy operator $\widehat{V}$), which operates on Ψ by multiplying $V(x)$ to it. Related to the second derivative, we have the kinetic energy operator (which we derived, as you recall, from the curvature and de Broglie's equation),

$$\widehat{K} = -\frac{\hbar^2}{2m}\frac{d^2}{dx^2}. \tag{5.70}$$

If we write $\widehat{K}$ as momentum operator squared divided by $2m$, we obtain

$$\widehat{K} = \frac{\widehat{p}^2}{2m} = -\frac{\hbar^2}{2m}\frac{d^2}{dx^2}, \tag{5.71}$$

$$\widehat{p}^2 = -\hbar^2 \frac{d^2}{dx^2}. \tag{5.72}$$

This means that $\widehat{p}$ can be defined (up to a sign)

$$\widehat{p} = \frac{\hbar}{i}\frac{d}{dx}, \tag{5.73}$$

since $i^2 = -1$. The momentum cannot be imaginary, except when particles tunnel (another interesting quantum effect). In three-dimensional space mapped by Cartesian coordinates ($\mathbb{R}^3$), the position operator $\hat{\mathbf{r}}$ is a vectorial quantity,

$$\hat{\mathbf{r}} = x\mathbf{i} + y\mathbf{j} + z\mathbf{k}, \tag{5.74}$$

where $\mathbf{i}, \mathbf{j}, \mathbf{k}$ are the unit vectors along the x, y, z axis, respectively. The momentum is also a vector operator,

$$\hat{\mathbf{p}} = \frac{\hbar}{i}\frac{\partial}{\partial x}\mathbf{i} + \frac{\hbar}{i}\frac{\partial}{\partial y}\mathbf{j} + \frac{\hbar}{i}\frac{\partial}{\partial z}\mathbf{k}. \tag{5.75}$$

The kinetic energy is the dot product of the momentum with itself, divided by $2m$,

$$\hat{K} = \frac{\hat{\mathbf{p}} \cdot \hat{\mathbf{p}}}{2m} = -\frac{\hbar^2}{2m}\frac{\partial^2}{\partial x^2} - \frac{\hbar^2}{2m}\frac{\partial^2}{\partial y^2} - \frac{\hbar^2}{2m}\frac{\partial^2}{\partial z^2}, \tag{5.76}$$

and is a scalar quantity as expected. The sum

$$\frac{\partial^2}{\partial x^2} + \frac{\partial^2}{\partial y^2} + \frac{\partial^2}{\partial z^2}, \tag{5.77}$$

is called the Laplacian operator in $\mathbb{R}^3$ and is abbreviated with the symbol ∇^2. The Hamiltonian operator in $\mathbb{R}^3$ has the following form:

$$-\frac{\hbar^2}{2m}\nabla^2 + V(x, y, z). \tag{5.78}$$

If we take the cross product between $\hat{\mathbf{r}}$ and $\hat{\mathbf{p}}$, we obtain the components of the angular momentum operator $\hat{\mathbf{L}}$ in $\mathbb{R}^3$,

$$\hat{\mathbf{L}} = \hat{\mathbf{r}} \times \hat{\mathbf{p}}, \tag{5.79}$$

$$L_x = \frac{\hbar}{i}\left(y\frac{\partial}{\partial z} - z\frac{\partial}{\partial y}\right), \tag{5.80}$$

$$L_y = -\frac{\hbar}{i}\left(x\frac{\partial}{\partial z} - z\frac{\partial}{\partial x}\right), \tag{5.81}$$

$$L_z = \frac{\hbar}{i}\left(x\frac{\partial}{\partial y} - y\frac{\partial}{\partial x}\right). \tag{5.82}$$

For central potentials [$V(x, y, z) = V(r)$, where $r = \left(x^2 + y^2 + z^2\right)^{1/2}$], it is convenient to use spherical polar coordinates, r, θ, ϕ. The Laplacian in these is

$$\nabla^2 = \frac{1}{r^2}\frac{\partial}{\partial r}\left(r^2\frac{\partial}{\partial r}\right) + \frac{1}{r^2\sin\theta}\frac{\partial}{\partial\theta}\left(\sin\theta\frac{\partial}{\partial\theta}\right) + \frac{1}{r^2\sin^2\theta}\frac{\partial^2}{\partial\phi^2}. \tag{5.83}$$

We derive Equation 5.83 in Chapter 10.

5.8.6   The Particle in a Box

Consider a particle with mass m trapped inside a potential energy $V(x)$ well, which is zero inside an interval $(0, L)$ and is infinite outside,

$$V(x) = \begin{cases} \infty & x \leq 0 \\ 0 & 0 < x < L \\ \infty & x \geq L \end{cases} . \tag{5.84}$$

No particle can exist in regions of space with infinite potentials, therefore, the wavefunction Ψ cannot have any amplitude for x below zero and above L. Otherwise, there would be some probability of finding the particle there. Ψ must be continuous everywhere, therefore, it must be zero at both 0 and L,

$$\Psi(0) = \Psi(L) = 0. \tag{5.85}$$

So, the particle must be somewhere between 0 and L, where V is zero. Schrödinger's equation for this region of space is simple,

$$-\frac{\hbar^2}{2m} \frac{d^2 \Psi}{dx^2} = E\Psi \qquad (0 < x < L). \tag{5.86}$$

The standard approach to solve this differential equation is to let the solution Ψ be

$$\Psi = A \sin(kx) + B \cos(kx). \tag{5.87}$$

Plugging Equation 5.87 into the differential equation in Equation 5.86 gives

$$-k^2 \Psi + \frac{2mE}{\hbar^2} \Psi = 0. \tag{5.88}$$

Therefore, Ψ satisfies the eigenvalue equation. Ψ can now be dropped from both sides. Solving for k after dropping Ψ in Equation 5.88 yields,

$$k = \sqrt{\frac{2mE}{\hbar^2}}. \tag{5.89}$$

The boundary conditions determine the allowed values of k and E, as we now demonstrate. At $x = 0$, we have

$$A \sin(0) + B \cos(0) = 0 \quad B = 0, \tag{5.90}$$

and at $x = L$ (taking $B = 0$ in consideration), gives

$$A \sin(kL) = 0. \tag{5.91}$$

The only way that this expression can be zero is if k takes discrete values: $k = \pi/L, 2\pi/L, \ldots, n\pi/L$. The $n = 0$ case yields the trivial solution $\Psi = 0$, which is interpreted to mean that there is no particle in the box. The solution now becomes

$$\Psi = A \sin\left(\frac{n\pi x}{L}\right),\tag{5.92}$$

and the energy is restricted by the allowed values of k. Using Equation 5.89, one obtains,

$$E_n = \frac{\hbar^2 n^2 \pi^2}{2mL^2} = \frac{h^2 n^2}{8mL^2}, \quad n = 1, 2, 3, \ldots.\tag{5.93}$$

We find A by normalizing $[\Psi^* = A \sin(n\pi x/L)]$,

$$1 = A^2 \int_0^L \sin^2\left(\frac{n\pi x}{a}\right) dx.\tag{5.94}$$

The integral evaluates to $L/2$ and solving for A gives the set of normalized wavefunctions that solve the particle in a box,

$$\Psi_n = \sqrt{\frac{2}{L}} \sin\left(\frac{n\pi x}{L}\right).\tag{5.95}$$

Note that the energy E **cannot be zero** because the $n = 0$ state is not allowed as a solution, as discussed earlier. The ground state energy is therefore,

$$E_1 = \frac{h^2}{8mL^2}.\tag{5.96}$$

The ground state energy is kinetic in nature ($V = 0$) for the particle in a box, and is purely a quantum effect. One cannot generate a stable quantum state in a well that has zero energy, or energy below this value, for that matter. More generally, one cannot generate a stable quantum state in a potential well that has the minimum potential energy. Classically, one **can** place a particle at the minimum without any initial velocity (i.e., any kinetic energy). The ground state energy in excess of the lowest possible classical energy is called **the zero point energy**. Its value decreases with increasing mass and, for the particle in a box, with increasing size L. The ground state energy of an electron in a 0.2-nm well is 10^{-18} J. This may seem like a small number, but take one mole of noninteracting wells, each containing one electron (a crude model for a metal). The resulting zero point energy is 907 kJ mol^{-1}. On the other hand, the zero point energy of a baseball (≈ 0.1 kg) in a 1-m box is 5×10^{-67} J, an insignificant amount.

Exercises

1. Plug Equation 5.87 into Equation 5.86 and derive Equation 5.89.

2. Work through the details of the derivation of the expression for A in Equation 5.92.

3. Show that the set

$$\left\{ \sqrt{\frac{2}{L}} \sin\left(\frac{n\pi x}{L}\right) \right\}_{n=1}^{\infty},$$

is a Hilbert space.

4. Use the postulates of quantum mechanics to obtain some physical properties from the wavefunctions of the particle in the box. It is quite simple to show that $\sin(n\pi x/L)$ is not an eigenfunction of the position operator $\hat{x}$. Therefore, the best we can do is obtain an average value for it. Show that this is

$$\langle x \rangle_n = \frac{L}{2}.$$

The average position is in the middle of the box, regardless of the quantum state that the particle happens to be in.

5. Obtain an expression for the spread in position Δx. Start by showing that the average of the square of the position is

$$\langle x^2 \rangle_n = L^2 \left(\frac{1}{3} - \frac{1}{2n^2\pi^2} \right),$$

then show that

$$\Delta x = 2n\pi L \left(\frac{n^2\pi^2 - 6}{3} \right)^{1/2}.$$

6. With the same method, we can obtain an expression for $\langle p_x \rangle_n$,

$$\langle p_x \rangle_n = \int_0^L \left[\left(\frac{2}{L}\right)^{1/2} \sin\frac{n\pi x}{L} \right] \left\{ \left(-i\hbar\frac{d}{dx}\right) \left[\left(\frac{2}{a}\right)^{1/2} \sin\frac{n\pi x}{L} \right] \right\} dx.$$

Show that the average momentum is zero.

7. To find the average square of the momentum, note that $\sin(n\pi x/L)$ is an eigenfunction of the kinetic energy operator, therefore, no integral is needed. Use this fact, and the results from previous exercises to explore the uncertainty principle in some detail for the particle in a box in any state n. Show that Heisenberg's uncertainty relation between position and momentum for the particle in a box is

$$\Delta x \Delta p = \frac{\hbar}{2} \left(\frac{n^2\pi^2 - 6}{3} \right)^{1/2}.$$

Note that Heisenberg's principle is respected because for $n = 1$ (the state with least uncertainty),

$$\Delta x \Delta p \approx (1.136) \frac{\hbar}{2} > \frac{\hbar}{2},$$

and note that $\Delta x \Delta p$ is independent of L.

8. Take the limit

$$\lim_{n \to \infty} \langle x^2 \rangle$$

for the particle in a box, and compare the result with the same quantity derived from a uniform distribution of x between 0 and L:

$$\langle x^2 \rangle = \frac{\displaystyle\int_0^L x^2 \, dx}{\displaystyle\int_0^L dx}.$$

To understand what this result means, consider the following experiment. Assume that there are one million boxes that are narrow enough to hold a golf ball, and each are precisely $a + d$ in length. A golf ball, with diameter $d \ll a$ (say, a is at least 10 times greater than d), is present in each box, and each box is shaken a random amount of times back and forth, then placed on a perfectly level surface, so that any kind of gravitational force is eliminated. Now, assume that your patient research assistant opens each box without any shock and records the precise location of the ball.

Note that the wavefunction squared for the particle in a box in the limit $n \longrightarrow \infty$ is not the same as the function you would develop for the probability distribution for the location of the center of the golf ball, however, you just learned that it predicts exactly the same "moments" (higher order moments agree as well, you can check for yourself). Such equivalence property between distributions is called **weak convergence** in functional analysis. So, classical behavior, apparently, **is** a subset of solutions or behaviors that quantum theory allows. Note, however, that there are several much more precise, transparent, and beautiful ways to arrive at the classical limit without ever evoking weak convergence.

To elaborate on what is not precise about our interpretation, if you now interpret the wavefunction simply as a distribution resulting from our ignorance of the location of the golf balls prior to opening, and you think that each golf ball has, in fact, a precisely defined position prior to the assistant opening the boxes, then you are essentially saying that *the wavefunction is not an element of reality*, rather, it is a meaningless mathematical entity used to describe the statistics of the experiment, whereas the position of the particle is an element of reality. However, such is not the prevailing view. Your interpretation is akin to a hidden variable theory, and you have just joined Einstein's camp. We can only take the golf ball, in the box analogy so far. In order to measure the location of the golf ball, our assistant had to open the box and let light in, which presumably caused some error in the measurement, but such error is much smaller than d, and so it must also be much smaller than a. Therefore, the true wavefunction of the golf ball (its true element of reality according to the Copenhagen interpretation) is never spread uniformly inside the box as it would be for an electron. That makes Bohr's classical limit not a true correspondence principle. True correspondence principles can be derived from quantum dynamics because the classical laws are dynamic laws. Quantum dynamics requires that we introduce the time variable, and we take up the issue of measurement and time evolution in Chapter 6. We will see that, while time evolution removes the wrinkle in the classical limit of quantum theory, it adds its own conundrum.

Here is a bit of philosophy, a necessary ingredient for the physical interpretation of results from quantum mechanics. According to the positivist school of thought (Ernst Mach and the Vienna Circle), an element of reality is only that which can be directly measured. Einstein takes a strict positivist stance against the Copenhagen interpretation, his argument being that the wavefunction cannot be measured directly. The weak convergence principle prevents the measurement of all the moments of a distribution to be a direct measurement of the distribution itself. For a positivist therefore, the wavefunction is not real because it cannot be measured directly. However, Einstein, while greatly influenced by Ernst Mach's ideas, was not a strict positivist. His theory of general relativity, for example, makes use of space-time and its curvature as an element of reality, even though it cannot be measured directly. The Copenhagen interpretation has emerged victorious after years of attacks from hidden variable theories. Bell's theorem defeated any type of hidden variable theory, while the advent of experimental techniques such as Scanning Probe Microscopy (SPM) and Scanning Tunneling Microscopy (STM) have enabled us to measure the wavefunction (squared) for electrons trapped on metal surfaces. Quantum theory, while admittedly strange, is a highly successful theory for the explanation of phenomena at the atomic and molecular scale.

5.8.7   Atomic Units

Atomic units have been introduced by physicists to make the numerical representation measurements and theoretical predictions more convenient. Consequently, computer code is considerably easier to develop in atomic units when simulations at the electronic, nuclear, and molecular scale phenomena are implemented. The atomic units are defined by taking the mass of the electron as the unit of mass, the bohr radius as the unit of length, and the time-scale to be such that $\hbar = 1$. Note that, naturally, from the equation

$$a_0 = \frac{4\pi\epsilon_0 \hbar^2}{m_e e^2}, \tag{5.97}$$

in SI units, it follows that

$$1 = \frac{4\pi\epsilon_0}{e^2}, \tag{5.98}$$

in atomic units. The unit of energy is the hartree (E_{h}),

$$E_{\mathrm{h}} = \frac{\hbar^2}{\mu a_0^2} = 1 \text{ hartree } = 4.359748 \times 10^{-18} \text{ J}, \tag{5.99}$$

or 27.211396 eV.

In atomic units, the Schrödinger equation for hydrogenic atoms (Z protons in the nucleus and one single electron) becomes

$$-\frac{1}{2}\nabla^2\psi - \frac{Z}{r}\psi = E\psi, \tag{5.100}$$

with energy levels in hartree given by

$$E = -\frac{Z^2}{2n^2}. \tag{5.101}$$

The electron in the hydrogen atom ($Z = 1$) is bound by -0.5 hartree. As an exercise, using atomic units, calculate the atomic time-scale in terms of seconds. This will give you an idea of the order of magnitude of the time in which events involving electrons (such as ionizations) take place. From here on, and for the rest of this book, the reader may assume that atomic units are used unless otherwise specified.

5.9 Matrix Quantum Mechanics

We now move on to a much more powerful way of applying the laws of quantum physics. Schrödinger's equation can only be solved analytically for a handful of simple problems, like the particle in a box. If, by analytical solution, we mean a solution in **closed form**, meaning with a finite number of algebraic or transcendental terms, then, even the elementary solution of the harmonic oscillator is not attainable in closed form, since the elements of the Hilbert space require an infinite number of orthogonal polynomials to be expressed.

In other words, we can obtain a solution in terms of a power series for the Schrödinger equation in one dimension, for a particle of mass m in a smooth potential V,

$$-\frac{1}{2m}\frac{d^2}{dx^2}\psi + V(x)\psi = E\psi, \tag{5.102}$$

but the power series has an infinite number of terms. Matrix quantum mechanics is far more powerful than Schrödinger's theory, since some spaces, such as the spin space, are not the solution of a Schrödinger equation. Assuming we have a basis set for a Hilbert space $|n\rangle$, then, to find a matrix representation for $|\Psi\rangle$, where $|\Psi\rangle$ is an abstract vector in the Hilbert space,

$$|\Psi\rangle = \sum_n a_n |n\rangle. \tag{5.103}$$

Note that we can use the resolution of the identity

$$|\Psi\rangle = \left\{\sum_n |n\rangle \langle n|\right\} |\Psi\rangle, \tag{5.104}$$

since the sum inside the curly brackets on the R.H.S. is equal to the identity operator. This equation can be rearranged to read

$$|\Psi\rangle = \sum_n \langle n| \Psi\rangle |n\rangle. \tag{5.105}$$

Comparing Equation 5.105 with Equation 5.103 yields the following result:

$$a_n = \langle n| \Psi\rangle. \tag{5.106}$$

Therefore, we can represent an arbitrary element of the Hilbert space with a column vector (of infinite dimension).

$$|\Psi\rangle \Rightarrow \begin{pmatrix} \langle 1| \Psi\rangle \\ \langle 2| \Psi\rangle \\ \vdots \\ \langle n| \Psi\rangle \\ \vdots \end{pmatrix}. \tag{5.107}$$

Next, we demonstrate that operators in the same Hilbert space can be represented by matrices. We begin with the operator $\widehat{O}$ operating on a basis vector $|m\rangle$,

$$\widehat{O} |m\rangle = |f\rangle = \sum_{n'} o_{n'm} |n'\rangle. \tag{5.108}$$

The far R.H.S. of Equation 5.108 must apply, since the Hilbert space spanned by the basis $|m\rangle$ is complete. Multiplying on both sides from the left by $\langle n|$, yields,

$$\langle n| \, \widehat{\mathbf{O}} \, |m\rangle = \sum_{n'} o_{n' \, m} \, \langle n \, |n'\rangle. \tag{5.109}$$

Assuming the basis set is orthonormal $\langle n \, |n'\rangle = \delta_{n \, m}$, gives,

$$\langle n| \, \widehat{\mathbf{O}} \, |m\rangle = o_{n \, m}, \tag{5.110}$$

since each term of the sum for which $n' \neq n$ is zero. Equation 5.110 tells us that we can represent an arbitrary operator in the Hilbert space with a matrix of infinite dimension,

$$\widehat{\mathbf{O}} \Rightarrow \begin{pmatrix} \langle 1| \, \widehat{\mathbf{O}} \, |1\rangle & \langle 1| \, \widehat{\mathbf{O}} \, |2\rangle & \dots \\ \langle 2| \, \widehat{\mathbf{O}} \, |1\rangle & \langle 2| \, \widehat{\mathbf{O}} \, |2\rangle & \dots \\ \vdots & \vdots & \vdots \\ \langle n| \, \widehat{\mathbf{O}} \, |1\rangle & \langle n| \, \widehat{\mathbf{O}} \, |2\rangle & \dots \\ \vdots & \vdots & \vdots \end{pmatrix}. \tag{5.111}$$

Consider the eigenvalue equation equivalent of Schrödinger's equation,

$$\widehat{\mathbf{H}} \, |\Psi\rangle = E \, |\Psi\rangle, \tag{5.112}$$

we can use the resolution of the identity to expand the eigenvector $|\Psi\rangle$, and, at the same time, use the matrix representation of the Hamiltonian operator,

$$\begin{pmatrix} \langle 1| \widehat{\mathbf{H}} \, |1\rangle & \langle 1| \widehat{\mathbf{H}} \, |2\rangle & \dots \\ \langle 2| \widehat{\mathbf{H}} \, |1\rangle & \langle 2| \widehat{\mathbf{H}} \, |2\rangle & \dots \\ \vdots & \vdots & \vdots \\ \langle n| \widehat{\mathbf{H}} \, |1\rangle & \langle n| \widehat{\mathbf{H}} \, |2\rangle & \dots \\ \vdots & \vdots & \vdots \end{pmatrix} \begin{pmatrix} \langle 1| \, \Psi\rangle \\ \langle 2| \, \Psi\rangle \\ \vdots \\ \langle n| \, \Psi\rangle \\ \vdots \end{pmatrix} = E \begin{pmatrix} \langle 1| \, \Psi\rangle \\ \langle 2| \, \Psi\rangle \\ \vdots \\ \langle n| \, \Psi\rangle \\ \vdots \end{pmatrix}. \tag{5.113}$$

Equation 5.113 tells us two important things. Firstly, it says that the basis of the Hilbert space, in which we have chosen to represent the Hamiltonian operator, need not diagonalize it. However, the basis vectors do need to satisfy the same boundary conditions as the basis of the eigenspace of the Hamiltonian. Secondly, if the basis set does not diagonalize the Hamiltonian, we can find the eigenvalues and the eigenvectors that satisfy the secular determinant by applying standard diagonalization methods. The resulting technique is the heart of matrix quantum mechanics applicable to bound states problems with time-independent Hamiltonians. This is the subject of exploration for the remainder of the present chapter. This section should have convinced the reader of the power of Dirac's notation in quantum mechanics. All the results are derived without making any assumption about the basis set of the space.

Bound states are all those quantum states that have energy below the value that it takes to separate the parts of the system by an infinite amount. States with greater energy are allowed, and as mentioned earlier, do not have quantized level of energy. Scattering theory plays an important role in the intepretation and the prediction of outcomes of collision experiments. It is impossible to overstate the importance of scattering experiments in chemistry. One just needs to recall the scattering of alpha particles by thin gold sheets, the photoelectric effect, and the Davisson and Germer experiment, where electrons are made to scatter off nickel crystals, to appreciate their fundamental role. The most important contribution to theoretical physics in recent times, arguably, remains the quantum theory of fields, a purely scattering theory. Applications of quantum theory to scattering require more advanced techniques than those presented in this chapter, but these build on the methods of matrix quantum mechanics, and by the end of the next chapter, the reader will have the tools necessary to handle scattering problems as well. These are important for clusters as well, since a growing number of cross beams experiments are used to study the energetics of clusters.

Before we move on to the first computer application of matrix quantum mechanics, we need to introduce one particularly important representation. Arguments in analysis centered around the Cauchy sequences and the completeness of the number fields, prove that both $\mathbb{R}$ and $\mathbb{C}$ are genuine Hilbert spaces. Both of these can be built to produce the **position representation**. The basis set vectors of the position representation are all the values of the real number line. These are typically represented with $|x\rangle$. The wavefunction, $\Psi(x)$, is

$$|\Psi\rangle = \int_{\text{all space}} dx' \, \Psi(x') \, |x'\rangle. \tag{5.114}$$

Mutiplying by $\langle x|$ on both sides gives,

$$\langle x| \, \Psi\rangle = \int_{\text{all space}} dx' \, \Psi(x') \, \langle x \, |x'\rangle. \tag{5.115}$$

The position basis set is assumed to be orthogonal, namely,

$$\langle x \, |x'\rangle = \delta(x - x'). \tag{5.116}$$

The symbol $\delta(x - x')$ is the Dirac delta distribution, a probability distribution with the following property for any function $f(x)$ in $\mathbb{R}$ and $\mathbb{C}$:

$$\int_{\text{all space}} dx' \, f(x') \, \delta(x - x') = f(x). \tag{5.117}$$

Therefore, Equation 5.114 becomes,

$$\langle x| \, \Psi\rangle = \Psi(x). \tag{5.118}$$

For example, the Hilbert space basis set that solves the particle in a box in $(0, L)$ has the following position representation:

$$\langle x| \, n \rangle = \sqrt{\frac{2}{L}} \sin \left(\frac{n \pi x}{L} \right). \tag{5.119}$$

5.10 The Monodimensional Hamiltonian in a Simple Hilbert Space

The Schrödinger equation for Harmonic oscillator is

$$-\frac{1}{2m} \frac{d^2}{dx^2} \psi + \frac{1}{2} k x^2 \psi = E \psi. \tag{5.120}$$

The solution of the Schrödinger equation represents the states of a particle of mass m bound in a harmonic well with a spring of constant k. This time, the configuration space is $-\infty < x < \infty$. The common approach is to solve the differential equation in Equation 5.120 using power series methods. The energy quantization results from the boundary condition that restricts the shape of the wavefunction at both asymptotes. The solutions that are rejected on physical grounds are those that grow without bound as either limit is approached. Only special values of the energy produce functions ψ that satisfy the square integrability condition,

$$\int_{-\infty}^{\infty} |\psi|^2 \, dx < \infty. \tag{5.121}$$

It is a standard exercise in mathematical physics to solve the Schrödinger equation for the harmonic oscillator using power series, and to obtain Hermite's polynomials. The eigenvalues are

$$E_n = \left(n + \frac{1}{2} \right) \omega, \tag{5.122}$$

where $\omega = \sqrt{k/m}$ is the characteristic frequency of the system, and n is an integer ranging from 0 to ∞. We are using atomic units in Equation 5.122.

We now represent the Hamiltonian and its eigenfunctions in the Hilbert space generated by the solutions of the particle in the box. Since the wavefunction must satisfy square integrability, it must become zero as we move away from the potential minimum in either direction. Therefore, if we pick L sufficiently large, we have a Hilbert space with a basis set that satisfies the same boundary conditions as the solution set. Since the set of solutions of the particle in a box in $(0, L)$ is complete, provided all values of n from 1 to ∞

are included, we can assume that our solution, ψ, of the general Schrödinger equation (Equation 5.102) can be expanded in that space,

$$\psi(x) = \sqrt{\frac{2}{L}} \sum_{n=1}^{\infty} a_n \sin\left(\frac{n\pi x}{L}\right), \tag{5.123}$$

where a_n is the projection along the direction n of the vector ψ. Placing this expression into Equation 5.102 gives

$$-\frac{1}{2m} \sqrt{\frac{2}{L}} \sum_{n=1}^{\infty} a_n \left(-\frac{n^2\pi^2}{L^2}\right) \sin\left(\frac{n\pi x}{L}\right)$$

$$+ [V(x) - E] \sqrt{\frac{2}{L}} \sum_{n=1}^{\infty} a_n \sin\left(\frac{n\pi x}{L}\right) = 0. \tag{5.124}$$

Now, let us left multiply by $\sqrt{2/L}\sin(l\pi x/L)$ and integrate from 0 to L.

$$\sum_{n=1}^{\infty} a_n \left[\left(\frac{n^2\pi^2}{2mL^2}\right) \delta_{ln} + \langle l|\, V \,|n\rangle - E\delta_{ln} \right] = 0. \tag{5.125}$$

Where we make use of the orthogonality condition for the particle in a box solutions,

$$\frac{2}{L} \int_0^L \sin\left(\frac{l\pi x}{L}\right) \sin\left(\frac{n\pi x}{L}\right) dx = \delta_{ln}, \tag{5.126}$$

the symbol δ_{nl} is the Kronecker delta, and the symbol $\langle l|\, V \,|n\rangle$ is the potential matrix element,

$$\langle l|\, V \,|n\rangle = \frac{2}{L} \int_0^L V(x) \sin\left(\frac{l\pi x}{L}\right) \sin\left(\frac{n\pi x}{L}\right) dx. \tag{5.127}$$

Equation 5.125 has the form of an eigenvalue-eigenvector pencil

$$[\mathbf{H} - E\mathbf{I}]\, \mathbf{v} = 0, \tag{5.128}$$

which can be solved easily by the Householder routine introduced in Chapter 4. The matrix elements of the Hamiltonian are obtained by inspecting Equation 5.125,

$$\langle l|\, H \,|n\rangle = \left(\frac{n^2\pi^2}{2mL^2}\right) \delta_{ln} + \langle l|\, V \,|n\rangle. \tag{5.129}$$

Therefore, one can calculate the matrix elements $\langle l|\, V \,|n\rangle$, add the diagonal terms of the kinetic energy matrix, and submit the Hamiltonian to a Householder routine to solve the problem.

5.11   Numerical Solution Issues in Vector Spaces

Three issues remain to be discussed before the Hilbert space theory we derived can be used in practice: the cutoff, the truncation, and the numerical quadrature. The careful reader will have realized that the Hamiltonian matrix is expanded in an infinite space. In this space, the wavefunction behaves much like a convergent power series expansion of a transcendental function. In practical applications, the power series is terminated when the computed properties converge to the desired precision. Therefore, one approximates the wavefunction with a truncated (i.e., finite) rather than infinite sum,

$$\psi \approx \sum_{n=1}^{n_{\max}} a_n \sin\left(\frac{n\pi x}{L}\right). \tag{5.130}$$

A quick and dirty way to find out if $n_{\max}$ was chosen sufficiently large, is to run the calculation with $n_{\max}$ basis and then run it again with a slightly larger basis set. If the desired results have not changed appreciably, we say the solution has converged. The second issue is that of the configuration space. Consider the harmonic oscillator problem for which the configuration space, as discussed earlier, is $-\infty < x < \infty$. What should the value of L be? The value of L has to be chosen carefully. Its value depends on how high is the window of energies that are of interest and how closely spaced are the energy levels. Selecting a value for L is the same as selecting a top value for the potential energy. Thus, the name cutoff energy is explained. Virtually every numerical solution of Schrödinger's equation must deal with truncation and cutoff. The closer the energy levels get to the cutoff energy, the greater their error. Much has been written in the literature about truncation and cutoff errors, but nothing is better than a little practice in investigating these issues ourselves.

Finally, the quadrature rule for the numerical integration of the potential matrix elements has to be chosen carefully as well. In the most general case, it may not be possible to obtain an analytic expression for the matrix elements $\langle l|\, V\, |n\rangle$. There will be $n_{\max}^2$ such elements to calculate from a numerical quadrature, so the procedure must not be too time consuming. The computation of all these integrals is the bottleneck for the evaluation of the Hamiltonian matrix, and it is this reason that causes the particle in a box Hilbert space method to rapidly "run out of steam" in many dimensions. For d dimensions, n_{max} basis, and a quadrature of order p, the computation of the matrix requires $(p n_{\max})^{2d}$ potential points. For example, a 100-point quadrature in three dimensions with $n_{\max} = 100$ requires 10^{10} potential points. Better methods are available to handle the multidimensional Schrödinger equation, but none can eliminate the exponential growth in the size of the matrix that comes from the Cartesian product of the basis in each dimension.

For the problem that we treat here, the extended trapezoid rule converges sufficiently fast. In the extended trapezoid rule, the integration range is subdivided into p regularly spaced intervals of length Δx. Then, the integral of a function $f(x)$ is approximated by

$$\int_a^b f(x)dx = \Delta x \left[\frac{1}{2}f(a) + f(a+h) + f(a+2h) + \cdots + \frac{1}{2}f(b) \right]. \quad (5.131)$$

This procedure too, introduces some error into the Hamiltonian matrix elements, and into the resulting eigenvalues.

5.12   The Harmonic Oscillator in Hilbert Space

Having discussed the main issues regarding the numerology of matrix quantum mechanics with the particle in a box Hilbert space, let us implement the algorithm into an actual program that solves the Harmonic oscillator.

```
      program Hilbert
      implicit real*8 (a-h,o-z)
c This program demonstrates the use of the particle in a box
c Hilbert space to solve 1 - D Schroedinger equations with an arbitrary
c (well behaved) time and velocity independent 1 parameter confining
c potential. To change the potential you must re-edit the subroutine POT.
      parameter(pi = 3.1415926)
      parameter (nbas = 100)
      real*8 h(nbas,nbas)        ! h() is the Hamiltonian matrix
      real*8 d(nbas),z(nbas,nbas),e(nbas)
      common/box/ d_L
      write(6,*) '         HILBERT                    '
      write(6,*) ' This program solves the 1 - D Schroedinger equation'
      write(6,*) ' Using the solution of the particle in a box as basis'
      d_L = 4.d0    ! the size of the well L
      d_K = 1.d0    ! the force constant in atomic units
      n_max = 22
c Plot out the potential
      x = 0.d0
      dx = 2.d-2
      do k=1,200
      call pot(x,d_K,v)
      write(8,*) x,v
      x = x + dx
      enddo
c L = 4 a.u. and V  is in a.u.
c The mass is 207 a.u. (a muon)
      do n=1,n_max
c the diagonal elements of the Hamiltonian
         h(n,n) = dfloat(n*n)*pi*pi/(207.d0*2.d0*d_L*d_L)
         write(9,*) n,h(n,n)
```

```
      d(n) = 0.d0
      e(n) = 0.d0
      do m = n,n_max
         call vme(n,m,d_K,ve)
         z(n,m) = 0.d0
         h(n,m) = h(n,m) + ve
         h(m,n) = h(n,m)
c          write(6,*) n,m,h(n,m)
      enddo
      enddo
c calculate the matrix elements of h: http://www.netlib.org/eispack
      call eigen(nbas,n_max,h,d,z,e,ierr)
      write(6,*) 'ierr',ierr
      do k=1,7
        write(6,*) k,d(k)
      enddo
      end
      subroutine pot(y,gamma,v)
      implicit real*8 (a-h,o-z)
      common/box/ d_L
      x = y - d_L/2
c gamma is just a dummy parameter used here as the force constant
c for debugging purposes use the harmonic potential centered at x = 2
c inside a box 4 bohr wide and that has a force constant of 1 a.u.
      v= 0.5*gamma*x*x
      return
      end
c
c The subroutine vme calculates the potential energy
c matrix elements using the trapezoid rule.  The algorithm
c first determines the best step size by determining the
c minimum period of the wavefunction product.
c The fast oscillatory term is calculated separately from
c the slowly moving one.
      subroutine vme(n,m,gamma,ve)
      implicit real*8(a-h,o-z)
      parameter (pi = 3.1415926)
      common/box/ d_L
c h is 2000 times smaller than  the smallest period of oscillation
      h = min(0.0001,.0004d0/(dfloat(n+m)))
      nq = int(4.0d0/h) + 1
      sum1 = 0.d0
      sum2 = 0.d0
      vm1 = 0.d0
      x = 0.d0
      do i=1,nq
        w = 1.d0
        if (i.eq. 1 .or. i .eq. nq) w = 0.5d0
        call pot(x,gamma,v)
```

```
if (n .eq. m) then
    sum1 = sum1 + w*h*v
else
    sum1 = sum1 + w*h*v*cos((n-m)*pi*x/4.d0)
endif
sum2 = sum2 + w*h*v*cos((n+m)*pi*x/4.d0)
x = x  + h
enddo
ve = sum1/d_L - sum2/d_L
return
end
```

In the main function, the size of the well d_L is set at 4 bohr, the force constant is set to 1.0 hartree bohr^{-2}, and $n_{max} = 22$. To test the program, a harmonic well centered at $L/2$ is used as the potential energy. The atomic units of $\sqrt{k/m}$ are hartree, since $\hbar\omega$ has units of J and the unit of action in atomic units is 1. The potential between $x = 0$ and 4 is plotted next, using 200 points. The outcome is written to the device file fort.8. The main function calculates the elements of the Hamiltonian matrix. Note that the mass is 207 a.u. (the mass of a muon). The potential matrix elements are calculated by the subroutine vme, which is called with the two integers l, n (m and n in the program) and the parameter of the potential. The result is returned in ve.

A call to eigen generates the eigenvalues. The subroutine pot is quite simple. The first line executes a shift of origin so that the bottom of the harmonic well is at $L/2$. Because the force constant is one atomic unit, the cutoff potential is 2 hartree. We are interested in the first seven energy levels, which for this choice of mass are all below 0.5 hartree. The subroutine vme requires some further explanation. The potential matrix elements are slightly rearranged using trigonometric identities for sum-difference of the arguments

$$\langle l| V |n\rangle = \frac{1}{L} \int_0^L V(x) \cos\left[\frac{(l-n)\pi x}{L}\right] dx - \frac{1}{L} \int_0^L V(x) \cos\left[\frac{(l+n)\pi x}{L}\right] dx.$$
(5.132)

In the special case, $n = l$, the slow moving term reduces to

$$\frac{1}{L} \int_0^L V(x) dx.$$
(5.133)

For the harmonic potential ($V = kx^2/2$), all these integrals have analytical expressions and the integration routine can be easily tested to make sure the quadrature order is appropriate, and that vme is working properly.

$$\langle l| V |n\rangle = \frac{kL^2}{\pi^2} \left[\frac{\cos(l-n)\pi}{(l-n)^2} - \frac{\cos(l+n)\pi}{(l+n)^2}\right], \quad l \neq n.$$
(5.134)

$$\langle l| V |n \rangle = \frac{kL^2}{\pi^2} \left[\frac{1}{6} - \frac{\cos 2n\pi}{4n^2} \right], \quad l = n. \tag{5.135}$$

The integral $(1/L) \int_0^L V(x)dx$, for example, with the force constant $k = 1$ a.u., has a numerical value of 2.666 hartree. If everything is working properly, the energy levels in atomic units for the first seven energy levels should be close to ($\omega = 0.0695048$ a.u.)

n	E_n (hartree)
0	0.0347524
1	0.1042572
2	0.1737620
3	0.2432668
4	0.3127716
5	0.3822764
6	0.4517812

$$\tag{5.136}$$

Note that n in Equation 5.136 represents the harmonic oscillator quantum number in Equation 5.122.

The application of the particle in a box Hilbert space can be easily generalized to other potentials in $\mathbb{R}^1$. However, as we pointed out earlier, the method is inefficient when generalized to higher dimensions, since the numerical integration for the potential energy matrix elements quickly becomes intractable. A major advance in the matrix solution of the multidimensional Schrödinger equation has been made with the DVR. To demonstrate the DVR approach, we use again a one-dimensional problem for the sake of simplicity. The power of the numerical technique will be illustrated by using examples of higher dimensions later in the chapter. The main advantage of the DVR is that the matrix elements can be obtained without having to calculate integrals numerically. The example we use here is designed to point out the next bottleneck of the matrix quantum mechanics approach. The number of operations required to diagonalize the matrix still scales as n^3, where n is the number of basis. The Hamiltonian matrix in our simple DVR will be shown to be very sparse and the diagonalization bottleneck can be overcome as well. This motivates the unveiling of a powerful technique, called the Lanczos method. Sparse matrix technology has allowed us to tackle, numerically, the matrix solution of the multidimensional Schrödinger equation. Six-dimensional solutions are routine at the time of writing. These would not be possible without the Lanczos algorithm.

Exercises

1. Derive Equations 5.134 and 5.135.

2. Change $n_{\max}$ to 32 in the program `hilbert`, and verify that the higher energy levels become more accurate.

3. How does the truncation error of the ground state depend on n_{max}? Use values of n_{max} between 4 and 22.

4. For the given value of L, obtain the cutoff energy, then determine the energy level that is theoretically immediately below the cutoff energy. Evaluate the absolute difference between the theory and numerical results for several values of n_{max}. Then repeat with a slightly larger value of L.

5. The truncation error for the harmonic oscillator is usually of greater order than expected. Change the potential to a simple quartic function,

$$V = \left(x - \frac{L}{2} \right)^4 \qquad 0 < x < L,$$

and evaluate the truncation error of the ground state as a function of n_{max}. Use values of n_{max} between 4 and a value for which the error visibly drops below the ninth place. Use the ground state energy with the largest value of n_{max} as the "exact" value.

6. The operators in Equations 5.69 through 5.73 are in the position representation. Obtain the top left 4×4 corner of their representation in the particle in a box Hilbert space. Show that all these operators are Hermitian.

7. Use the position representation of $\hat{x}$ and $\hat{p}$ and operate on an arbitrary function $f(x)$ to demonstrate that

$$[\hat{x}, \hat{p}] = -i\hbar. \tag{5.137}$$

Noncommuting operators imply that the physical observables associated with them are noncompatible, namely, their sharp values cannot be measured simultaneously, and the sharp knowledge of one makes the knowledge of the other uncertain. The Heisenberg uncertainty principle extends to any pair, a, b, of incompatible measurements as follows,

$$\Delta a \Delta b \geq \frac{\hbar}{2} \left| \left[\hat{A}, \hat{B} \right] \right|. \tag{5.138}$$

5.13   A Simple Discrete Variable Representation (DVR)

Consider a problem such as the Harmonic oscillator in $\mathbb{R}^1$. As before, we take an open set in $\mathbb{R}^1$ of length L, symmetrically placed about the minimum of the potential. We then represent this subset with a finite set of lattice points,

$$x_i, \quad i = 1, 2, \dots. \tag{5.139}$$

It is necessary to keep the interval between points a constant,

$$\epsilon = x_i - x_{i-1} = \frac{L}{N}. \tag{5.140}$$

The wavefunction that solves the Schrödinger equation is represented as a set $\psi_n(x_i)$,

$$\vec{\psi} \approx \begin{pmatrix} \psi_n(x_1) \\ \psi_n(x_2) \\ \psi_n(x_3) \\ \vdots \\ \psi_n(x_N) \end{pmatrix}, \tag{5.141}$$

where each entry is the value of the wavefunction at the particular lattice point.

Consider the derivative of $\psi_n(x_i)$ expressed on the lattice,

$$\frac{d\psi}{dx} \approx \frac{\psi_n(x_i) - \psi_n(x_{i-1})}{\epsilon}, \tag{5.142}$$

and, more importantly, the second derivative of $\psi_n(x_i)$,

$$\frac{d^2\psi}{dx^2} \approx \frac{\psi_n(x_{i+1}) - 2\psi_n(x_i) + \psi_n(x_{i-1})}{\epsilon^2}. \tag{5.143}$$

This formula is the so-called three-point formula in numerical analysis [2], and can be derived easily from the definition of the derivative. It is simple to obtain the representation of the Laplacian operator $\widehat{\nabla}^2$ in this vector space. If we consider $\widehat{\nabla}^2$ as a linear map in the vector space

$$\sum_{j=1}^{N} \widehat{\nabla}_{ij}^2 \vec{\psi}_j, \tag{5.144}$$

and we make use of Equation 5.143, it follows that

$$\langle i| \widehat{\nabla}^2 |j\rangle = -\frac{2}{\epsilon^2}\delta_{ij} + \frac{1}{\epsilon^2}\delta_{i,j+1} + \frac{1}{\epsilon^2}\delta_{i,j-1}. \tag{5.145}$$

This is the lattice Laplacian matrix we introduced in Chapter 4. The lattice Laplacian is not diagonal in this representation. The potential energy matrix, on the other hand, is diagonal. Each entry of the potential energy matrix is the value of V at the particular lattice point. It follows that the Hamiltonian operator can be represented using a lattice vector space in the following manner:

$$\langle i|\, H\, |j\rangle = -\frac{\hbar^2}{2m\epsilon^2}
\begin{pmatrix}
-2 & 1 & 0 & \cdots & 0 \\
1 & -2 & 1 & \cdots & 0 \\
0 & 1 & -2 & \cdots & 0 \\
\vdots & \vdots & \vdots & \cdots & \vdots \\
0 & 0 & 0 & \cdots & -2
\end{pmatrix}$$

$$+
\begin{pmatrix}
V(x_1) & 0 & 0 & \cdots & 0 \\
0 & V(x_2) & 0 & \cdots & 0 \\
0 & 0 & V(x_3) & \cdots & 0 \\
\vdots & \vdots & \vdots & \cdots & \vdots \\
0 & 0 & 0 & \cdots & V(x_N)
\end{pmatrix}.
\qquad (5.146)$$

This representation is called the simple DVR. Note that there is no need for numerical integration to obtain the representation of the Hamiltonian. The second great advantage of the DVR is the pattern of sparsity of the Hamiltonian. In fact, in $\mathbb{R}^1$ the Hamiltonian projected in the simple DVR is tridiagonal. Diagonalization of the Hamiltonian matrix yields both the eigenvalues, E_n, and the eigenvectors that represent the wavefunctions on the lattice. Consider the following program, where the same harmonic oscillator with a mass of 207 a.u. and a force constant of 1.00 hartree bohr^{-2} is solved, this time using the simple DVR.

```
program simple_DVR
implicit real*8 (a-h,o-z)
c Diagonalization of a simple DVR Hamiltonian
      parameter (nbas = 1000)
      real*8 a(nbas,nbas)
      real*8 d(nbas),z(nbas,nbas),e(nbas)
      do i=1,nbas            ! initialize all the arrays
       d(i) = 0.d0
       e(i) = 0.d0
       do j=1,nbas
        a(i,j) = 0.d0
        z(i,j) = 0.d0
       enddo
      enddo
      n_max = 10
      dL = 5.D0
      epsilon = dL/dfloat(n_max-1)
      do i=1,n_max           ! set up the Hamiltonian matrix
      x = -0.5*dL + dfloat(i-1)*epsilon
      do j=1,n_max
       if (i .eq. j) a(i,j) = 2.d0/(2.*207*epsilon**2) + 0.5*x*x
       if (i .eq. j-1) a(i,j) = -1.d0/(2.*207*epsilon**2)
       if (i .eq. j+1) a(i,j) = -1.d0/(2.*207*epsilon**2)
      enddo
```

```
      enddo
      call eigen(nbas,n_max,a,d,z,e,ierr)
      write(6,*) 'ierr',ierr    ! output
      do k=1,n_max
       write(6,*) k,d(k)
      enddo
      write(6,*) 'Z**T Z'
      do i=1,n_max  ! checking the orthogonality of the transform Z
      do j=1,n_max
       check = 0.d0
       do k=1,n_max
        check = check + z(k,i)*z(k,j)
       enddo
       if(dabs(check) .gt. 1.d-8) write(6,1000) i,j,check
      enddo
      enddo
      write(6,*) 'Z**T A Z'
      do i=1,n_max  ! check the eigenvalues and check if A is diagonal
      do j=1,n_max
       check = 0.d0
       do k=1,n_max
       do l=1,n_max
        check = check + z(k,i)*a(k,l)*z(l,j)
       enddo
       enddo
       x = 0.d0
       if (i. eq. j) x = d(i)
       if(dabs(check) .gt. 1.d-6) write(6,1000) i,j,check, x
      enddo
      enddo
1000  format(2i4,3f12.8)
      end
```

The orthogonality and similarity transformation check can be carried out first for small values of n max, but then should be commented out when **n_max** is increased. The **converged** eigenvalues have to be compared with the theoretical values in Equation 5.136. A wavefunction plot, like in Figure 5.1, can be obtained by adding the following snippet at the end of the main function.

```
      do i=1,n_max
       x = -0.5*dL + dfloat(i-1)*epsilon
       write(8,*) x,z(i,10)
      enddo
```

Note how the wavefunction is converging in amplitude to the correct answer for $N \geq 400$ and that the phase of ψ changes for $300 > N > 400$. Therefore, the convergence of the numerical scheme is slower than with the particle in a box Hilbert space in our first example of numerical matrix quantum mechanics. The trade-off is in favor of the DVR, which is faster overall

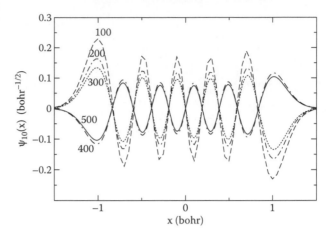

FIGURE 5.1

The convergence of $\psi_9(x)$ for the Harmonic oscillator as a function of the DVR vector space size N.

for a given accuracy level of the eigenvalues and eigenvectors, as the reader can verify.

5.14   Accelerating the Convergence of the Simple Discrete Variable Representation (DVR)

Can one improve upon the convergence of the simple DVR? If we return to a numerical analysis book, we can find a whole array of formulas for second derivatives. Consider the five-point formula for second derivatives, for example,

$$\frac{d^2\psi}{dx^2} \approx \frac{-\psi_n(x_{i+2}) + 16\psi_n(x_{i+1}) - 30\psi_n(x_i) + 16\psi_n(x_{i-1}t) - \psi_n(x_{i-2})}{12\epsilon^2}.$$

$$(5.147)$$

Equation 5.147 implies the following lattice representation for the Hamiltonian matrix:

$$\langle i|\,H\,|j\rangle = -\frac{\hbar^2}{24m\epsilon^2}
\begin{pmatrix}
-30 & 16 & -1 & \cdots & 0 \\
16 & -30 & 16 & \cdots & 0 \\
-1 & 16 & -30 & \cdots & 0 \\
\vdots & \vdots & \vdots & \cdots & \vdots \\
0 & 0 & 0 & \cdots & -30
\end{pmatrix}$$

$$+ \begin{pmatrix} V(x_1) & 0 & 0 & \cdots & 0 \\ 0 & V(x_2) & 0 & \cdots & 0 \\ 0 & 0 & V(x_3) & \cdots & 0 \\ \vdots & \vdots & \vdots & \cdots & \vdots \\ 0 & 0 & 0 & \cdots & V(x_N) \end{pmatrix}. \tag{5.148}$$

The program `simple_dvr` can be modified (in the construction of the Hamiltonian matrix part) as follows,

```
n_max = 100
dL = 5.D0
epsilon = dL/dfloat(n_max)
do i=1,n_max              !  set up the Hamiltonian matrix
x = -0.5*dL + dfloat(i-1)*epsilon
do j=1,n_max
  if (i .eq. j) a(i,j) = 30.d0/(24.*207*epsilon**2) + 0.5*x*x
  if (i .eq. j-1) a(i,j) = -16.d0/(24.*207*epsilon**2)
  if (i .eq. j+1) a(i,j) = -16.d0/(24.*207*epsilon**2)
  if (i .eq. j-2) a(i,j) = 1.d0/(24.*207*epsilon**2)
  if (i .eq. j+2) a(i,j) = 1.d0/(24.*207*epsilon**2)
enddo
enddo
```

We leave it to the reader to show that the five-point formula has better convergence properties than the simple DVR based on the three-point formula introduced earlier. The good news is that even though the Hamiltonian matrix is less sparse, there is still no need for numerical integrations. This begs the question: Can we continue to use higher order formulas to improve upon the convergence of the DVR indefinitely? The answer is no, of course. On a n-point lattice, we can use at most a n-point formula to represent the derivatives. This case has been considered by Colbert and Miller [156]. The n-point formula DVR in one dimension has a dense kinetic energy matrix and the potential energy is still diagonal. The Hamiltonian matrix elements in $\mathbb{R}^1$ are given by the formula,

$$\langle i | H | j \rangle = \begin{cases} \dfrac{\hbar^2}{2m\epsilon^2} \dfrac{\pi^2}{3} + V(x_i), & (i = j) \\[3mm] (-1)^{i-j} \dfrac{\hbar^2}{2m\epsilon^2} \dfrac{2}{(i-j)^2}, & (i \neq j). \end{cases} \tag{5.149}$$

It is straightforward to modify `simple_dvr` once more,

```
n_max = 100
dL = 5.D0
epsilon = dL/dfloat(n_max)
do i=1,n_max              !  set up the Hamiltonian matrix
x = -0.5*dL + dfloat(i-1)*epsilon
```

```
do j=1,n_max
  if (i .eq. j) then
    a(i,j) = 3.14159**2/(6.*207*epsilon**2) + 0.5*x*x
  else
    sign = -1.d0
    if (mod(abs(i-j),2) .eq. 0) sign = +1.d0
    a(i,j) = sign/(207*epsilon**2*(i-j)**2)
  endif
enddo
enddo
```

Exercises

1. Determine how the truncation error on the ground state of the harmonic oscillator depends on the basis set size n for the simple DVR, the five-point formula DVR, and the DVR of Colbert and Miller. Use five values of n between 4 and 40 for the simple and five-point DVR, and from 4 to 10 for the DVR of Colbert and Miller, to obtain the order. The convergence of the DVR of Colbert and Miller should be a factor of 10 better (that is, it should require a set about 10 times smaller) than the simple DVR, to achieve the same truncation error on the ground state. The actual numerical gain of the DVR of Colbert and Miller, with respect to the simple DVR, depends on the mass of the particle, the cutoff used (i.e., the size of dL), and the details of the potential energy. The trade-off is in the increase of the nonzero entries in the Hamiltonian.

2. Systematically increase the mass of the particle in the harmonic potential for masses ranging from 1800 to 80,000 a.u. Inspect graphs of the ground state wavefunction and verify that the probability of locating the particle at the minimum becomes sharper and sharper.

3. Change the potential energy by adding the subroutine **pot** for the quartic double well potential used in previous chapters. Use the resulting program to calculate the first 10 energy levels of the muon (207 atomic mass units) trapped in the asymmetric double well for at least 20 values of γ, from 0.05 to 0.95. Make a plot of the energy levels as a function of γ. Make sure the energy is converged to at least $\pm 10^{-5}$ hartree.

4. Use the program created in the previous example to calculate the first few energy levels of a mass trapped in the asymmetric double well for $\gamma = 0.9$, and for at least 10 values of its mass m between 10 and 240 a.u. Plot the ground state wavefunction for all values of m. Make sure the energy is converged to at least $\pm 10^{-5}$ hartree.

5. Use the DVR of Colbert and Miller to find the graphs of the first four wavefunctions for a hydrogen atom in a degenerate double well,

$$V = x^4 - 2x^2 + 1.$$

5.15 Elements of Sparse Matrix Technology

In more than one dimension, the DVR of Colbert and Miller, the simple DVR, and the five-point formula DVR are all very sparse, as shown in the following. The sparsity pattern in multidimensional DVR motivates the development of techniques to take advantage of the reduced number of computations. The number of basis sets, unfortunately, grows exponentially with the dimensions of the space d. The three-body problem, for example, requires special coordinates to eliminate the center of mass degrees of freedom and bring the total number of spacial dimensions to six. One possibility to derive the Jacobi coordinates for the three-body problem is discussed in Chapter 10. If one constructs a DVR with 20 lattice points per dimension, the Hamiltonian matrix is a 64 million by 64 million matrix containing 4.0×10^{15} entries and requiring on the order of 10^{45} operations. These staggering numbers make calculations of this caliber, using algorithms we have encountered so far, prohibitive. However, it can be shown that only on the order of 10^8 elements are nonzero in the DVR of the Hamiltonian. Diagonalization schemes exist that take full advantage of the sparsity of the Hamiltonian. These require only on the order of 10^8 operations to find few eignevalues in windows of interest. To see the sparsity pattern of the simple DVR in multidimensional problems, consider the $\mathbb{R}^3$ example,

$$H_{ijki'j'k'} = K_{ii'}\delta_{jj'}\,\delta_{kk'} + K_{jj'}\delta_{ii'}\,\delta_{kk'} + K_{kk'}\delta_{ii'}\,\delta_{jj'}$$
$$+ V\left(x_i, y_j, z_k\right)\delta_{ii'}\,\delta_{jj'}\delta_{kk'}, \tag{5.150}$$

where $K_{ii'}, K_{jj'}, K_{kk'}$ are representations of the second derivative operators with respect to x, y, and z. The following program produces the point plot of the Hamiltonian matrix in the three-point formula DVR. A point plot is a $x - y$ graph, where x is the column label and y is the row label. A point given by a x, y pair represents a nonzero entry for the Hamiltonian matrix.

```
program point_plot
implicit real*8(a-h,o-z)
c generates the point plot for the simple DVR in three
c dimensions.
   n_max = 50
   do i=1,n_max
    do j = 1, n_max
     do k= 1,n_max
      do ip = 1, n_max
       do jp = 1,n_max
        do kp = 1, n_max
          index = k + n_max*(j-1) + n_max*n_max*(i-1)
          indey = kp + n_max*(jp-1) + n_max*n_max*(ip-1)
```

```
c diagonal elements
            if (i .eq. ip .and. j .eq. jp .and. k .eq. kp) then
            write(8,*) index,-indey
            endif
c off diagonal elements
            if (j .eq. jp .and. k .eq. kp) then
            if (i. eq. ip+1) write(8,*) index,-indey
            if (i. eq. ip-1) write(8,*) index,-indey
            endif
            if (i .eq. ip .and. k .eq. kp) then
            if (j. eq. jp+1) write(8,*) index,-indey
            if (j. eq. jp-1) write(8,*) index,-indey
            endif
          if (i .eq. ip .and. j .eq. jp) then
            if (k. eq. kp+1) write(8,*) index,-indey
            if (k. eq. kp-1) write(8,*) index,-indey
            endif
        enddo
       enddo
      enddo
     enddo
    enddo
   enddo
  end
```

With n_max = 10, the program point_plot produces 6400 nonzero entries or $6.4n^3$ for n lattice point per dimension. A plot of the data points of the full range in fort.8 produces the graph in Figure. 5.2. Note that the negative sign of the y axis is removed so that the labels may represent the row number as increasing from the top down. For the five-point DVR, with $n = 50$ points per dimension, 11,200 or $N = 11.2n^3$ nonzero entries are obtained. The point plot for the five-point DVR in $\mathbb{R}^3$ with 10 lattice points per dimension is in Figure 5.3. Finally, the DVR of Colbert and Miller produces 30,000 nonzero entries in the Hamiltonian (or $N = 30n^3$), and the point plot with the same parameters is in Figure 5.4. Each one of the small black squares along the main diagonal is a filled 10×10 block. Each of the ten 100×100 blocks along the main diagonal are visibly sparse.

5.16   The Gram–Schmidt Process

Suppose we have m linearly independent vectors $\{\mathbf{x}_1, \mathbf{x}_2, \ldots, \mathbf{x}_m\}$ that are not orthonormal with each other. Start with the vector $\mathbf{u}_1 = \mathbf{x}_1/|\mathbf{x}_1|$ and construct another vector, $\mathbf{u}_2$, that is perpendicular to $\mathbf{u}_1$, by defining $\mathbf{u}_2$ in

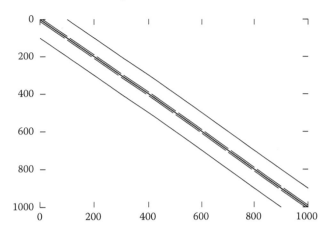

FIGURE 5.2
Point plot for the simple DVR in $\mathbb{R}^3$.

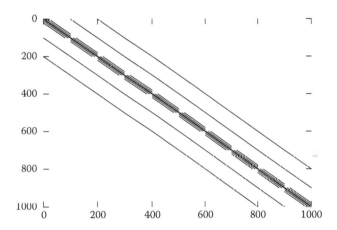

FIGURE 5.3
Point plot for the five-point DVR in $\mathbb{R}^3$.

the following way:

$$\mathbf{u}_2' = \mathbf{x}_2 - \left(\mathbf{u}_1^T \mathbf{x}_2\right) \mathbf{u}_1. \tag{5.151}$$

We "deflate" $\mathbf{u}_1$ from the $\mathbf{x}_2$ vector with the operation above. Next, normalize $\mathbf{u}_2'$,

$$\mathbf{u}_2 = \frac{\mathbf{u}_2'}{|\mathbf{u}_2'|}. \tag{5.152}$$

Stochastic Simulations of Clusters

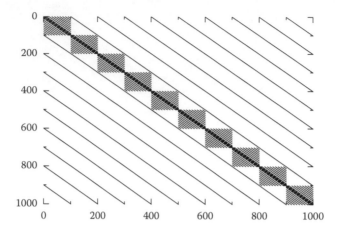

FIGURE 5.4
Point plot for the DVR of Colbert and Miller in $\mathbb{R}^3$.

The vector $\mathbf{u}_2'$ is perpendicular to $\mathbf{u}_1$ and it can be easily demonstrated by multiplying $\mathbf{u}_1^T$ from the left in Equation 5.152,

$$\mathbf{u}_1^T \mathbf{u}_2' = \mathbf{u}_1^T \mathbf{x}_2 - \mathbf{u}_1^T \left(\mathbf{u}_1^T \mathbf{x}_2\right) \mathbf{u}_1. \tag{5.153}$$

Now $(\mathbf{u}_1^T \mathbf{x}_2)$ is just a scalar, therefore,

$$\mathbf{u}_1^T \mathbf{u}_2' = \mathbf{u}_1^T \mathbf{x}_2 - (\mathbf{u}_1^T \mathbf{x}_2) \mathbf{u}_1^T \mathbf{u}_1, \tag{5.154}$$

and

$$\mathbf{u}_1^T \mathbf{u}_1 = \frac{\mathbf{x}_1^T \mathbf{x}_1}{|\mathbf{x}_1|^2} = \frac{\mathbf{x}_1^T \mathbf{x}_1}{\mathbf{x}_1^T \mathbf{x}_1} = 1. \tag{5.155}$$

This leaves

$$\mathbf{u}_1^T \mathbf{u}_2' = \mathbf{u}_1^T \mathbf{x}_2 - \left(\mathbf{u}_1^T \mathbf{x}_2\right) = 0, \tag{5.156}$$

which is the result we wanted to show.

The process just explained can be extended to all the vectors and, in general, it will be true that

$$\mathbf{u}_i^T \mathbf{u}_j = \delta_{ij}, \tag{5.157}$$

where δ_{ij} is the Kronecker delta symbol. Thus, this procedure converts m vectors $\{\mathbf{x}_1, \mathbf{x}_2, \ldots, \mathbf{x}_m\}$ into a set of m orthonormal vectors $\{\mathbf{u}_1, \mathbf{u}_2, \ldots, \mathbf{u}_m\}$. Given a set of m vectors $\{\mathbf{x}_1, \mathbf{x}_2, \ldots, \mathbf{x}_m\}$, a set of orthonormal vectors $\{\mathbf{u}_1, \mathbf{u}_2, \ldots, \mathbf{u}_m\}$ is constructed by setting $\mathbf{u}_1 = \mathbf{x}_1/|\mathbf{x}_1|$ and performing the following operation recursively for $j = 2, 3, \ldots, m$,

$$\mathbf{u}'_j = \mathbf{x}_j - \sum_{k=1}^{j-1} (\mathbf{u}_k^T \mathbf{x}_j) \mathbf{u}_k, \qquad (5.158)$$

$$\mathbf{u}_j = \frac{\mathbf{u}'_j}{|\mathbf{u}'_j|}. \qquad (5.159)$$

We have shown that the theorem holds for $j = 1$ and 2 in the discussion above. The general statement of the theorem can be proved by the induction principle. Assume the theorem holds for 1, 2, and up to some j, then form

$$|\mathbf{u}'_{j+1}|\mathbf{u}_{j+1} = \mathbf{x}_{j+1} - \sum_{k=1}^{j} (\mathbf{u}_k^T \mathbf{x}_{j+1}) \mathbf{u}_k. \qquad (5.160)$$

Now multiply from the left by $\mathbf{u}_n^T$, $n \leq j + 1$. If $n \leq j$, the orthogonality between $\mathbf{u}_n$ and $\mathbf{u}_k$ follows by assumption and the only surviving term in the sum on the right of the following equation is the $k = n$ term:

$$|\mathbf{u}'_{j+1}|\mathbf{u}_n^T \mathbf{u}_{j+1} = \mathbf{u}_n^T \mathbf{x}_{j+1} - \sum_{k=1}^{j} (\mathbf{u}_k^T \mathbf{x}_{j+1}) \mathbf{u}_n^T \mathbf{u}_k, \qquad (5.161)$$

$$|\mathbf{u}'_{j+1}|\mathbf{u}_n^T \mathbf{u}_{j+1} = \mathbf{u}_n^T \mathbf{x}_{j+1} - \sum_{k=1}^{j} (\mathbf{u}_k^T \mathbf{x}_{j+1}) \delta_{n\,k}. \qquad (5.162)$$

δ_{nk} is zero for all the terms of the sum above except for the $k = n$ term. The Kronecker delta collapses the sum into a single term, and the surviving term cancels exactly the first term on the L.H.S.,

$$|\mathbf{u}'_{j+1}|\mathbf{u}_n^T \mathbf{u}_{j+1} = \mathbf{u}_n^T \mathbf{x}_{j+1} - (\mathbf{u}_n^T \mathbf{x}_{j+1}) = 0. \qquad (5.163)$$

On the other hand, if $n = j + 1$, the sum on the R.H.S. is zero. As a result, we recover the normalization of $\mathbf{u}'_{j+1}$ by construction. Quod Erat Demonstrandum.

As shown in Chapter 4, one organizes the column vectors $\{\mathbf{u}_1, \mathbf{u}_2, \ldots, \mathbf{u}_m\}$ as the columns of a matrix $\mathbf{U}$. Then, the orthonormality of the set is concisely written as

$$\mathbf{U}^T \mathbf{U} = 1, \qquad (5.164)$$

from which it follows that $\mathbf{U}^T = \mathbf{U}^{-1}$.

5.17   The Krylov Space

In this section, we explore a special vector space for a real symmetric matrix (or a Hermitian matrix). The Krylov space for a $n \times n$ matrix is the subspace of $\mathcal{V}(n \times 1, \mathbb{R})$ that is spanned by the vectors

$$\mathbf{b}, \mathbf{A}\mathbf{b}, \mathbf{A}^2\mathbf{b}, \mathbf{A}^3\mathbf{b}, \ldots, \mathbf{A}^{N-1}\mathbf{b}. \tag{5.165}$$

The vector $\mathbf{b}$ is arbitrary. If these vectors are linearly independent, they represent a basis set for the Krylov space. However, a more convenient basis set can be formed from these by using the Gram Schmidt orthonormalization procedure. The matrix $\mathbf{A}$ is tridiagonal when represented in its own Krylov space. This is an important fact, since the tridiagonalization of a matrix is the first step for the diagonalization of a matrix. The Householder and the Givens methods are, in fact, tridiagonalization procedures. The task of diagonalizing a tridiagonal matrix can be carried out efficiently with well-established methods, such as the QL algorithm [8]. Performing the tridiagonalization by means of the Krylov space has some advantages when sparse matrices are involved, as we will soon see.

Consider the following example:

$$\mathbf{A} = \begin{pmatrix} 1 & 2 & 1 & 3 \\ 2 & 2 & 1 & 2 \\ 1 & 1 & 1 & 3 \\ 3 & 2 & 3 & 3 \end{pmatrix}.$$

Here is the set of four vectors in the Krylov space. Let us start with an arbitrary vector $\mathbf{b}$,

$$\mathbf{b} = \begin{pmatrix} 1 \\ 1 \\ 1 \\ 1 \end{pmatrix} = \mathbf{x}_1.$$

It is simple to verify the following results:

$$\mathbf{x}_2 = \mathbf{A}\mathbf{b} = \begin{pmatrix} 7 \\ 7 \\ 6 \\ 11 \end{pmatrix}, \quad \mathbf{x}_3 = \mathbf{A}^2\mathbf{b} = \begin{pmatrix} 60 \\ 56 \\ 53 \\ 86 \end{pmatrix}, \quad \mathbf{x}_4 = \mathbf{A}^3\mathbf{b} = \begin{pmatrix} 483 \\ 457 \\ 427 \\ 709 \end{pmatrix}.$$

It is easier to calculate these vectors recursively, i.e., $\mathbf{x}_3 = \mathbf{A}\mathbf{x}_2$, or in general, $\mathbf{x}_j = \mathbf{A}\mathbf{x}_{j-1}$. Now perform the Gram Schmidt process on these four vectors. Start with the normalization of the first one, $\mathbf{x}_1^T \mathbf{x}_1 = 4$, therefore, $|\mathbf{x}_1| = 2$,

$$\mathbf{u}_1 = \frac{1}{2} \begin{pmatrix} 1 \\ 1 \\ 1 \\ 1 \end{pmatrix}.$$

The calculations by hand are tedious, but can be carried out with a good pocket calculator. The projection of $\mathbf{x}_2$ onto $\mathbf{u}_1$, $\mathbf{u}_i^T \mathbf{x}_2$ is $= 31/2$. The deflated, unnormalized second vector becomes,

$$\mathbf{u}_2' = \begin{pmatrix} 7 \\ 7 \\ 6 \\ 11 \end{pmatrix} - \frac{31}{4} \begin{pmatrix} 1 \\ 1 \\ 1 \\ 1 \end{pmatrix} = \frac{1}{4} \begin{pmatrix} -3 \\ -3 \\ -7 \\ 13 \end{pmatrix}.$$

The size squared of $\mathbf{u}_2'$ is $\frac{9+9+49+169}{16}$. The normalized second vector becomes,

$$\mathbf{u}_2 = \begin{pmatrix} -0.1952 \\ -0.1952 \\ -0.4556 \\ 0.8462 \end{pmatrix}.$$

To form $\mathbf{u}_3'$, we must find $\mathbf{u}_2^T \mathbf{x}_3$ and $\mathbf{u}_1^T \mathbf{x}_3$. These are, $\mathbf{u}_1^T \mathbf{x}_3 = 137.5$ and $\mathbf{u}_2^T \mathbf{x}_3 = 25.9726$, therefore,

$$\mathbf{u}_3' = \begin{pmatrix} 60 \\ 56 \\ 53 \\ 86 \end{pmatrix} - 25.9726 \begin{pmatrix} -0.1952 \\ -0.1952 \\ -0.4556 \\ 0.8462 \end{pmatrix} - 137.5 \begin{pmatrix} 0.5 \\ 0.5 \\ 0.5 \\ 0.5 \end{pmatrix}.$$

After normalization, and with a similar process for the last vector, we obtain

$$\mathbf{u}_3 = \begin{pmatrix} 0.4145 \\ -0.8397 \\ 0.3401 \\ 0.0850 \end{pmatrix}, \quad \mathbf{u}_4 = \begin{pmatrix} -0.7348 \\ -0.0816 \\ 0.6531 \\ 0.1632 \end{pmatrix}.$$

If we assemble the four column vectors into a similarity transform matrix **U**,

$$\mathbf{U} = \begin{pmatrix} 0.5 & -0.1952 & 0.4145 & -0.7348 \\ 0.5 & -0.1952 & -0.8397 & -0.0816 \\ 0.5 & -0.4556 & 0.3401 & 0.6531 \\ 0.5 & 0.8462 & 0.0850 & 0.1632 \end{pmatrix},$$

the reader should show (perhaps by modifying `matrix_product`) that

$$\mathbf{U}^T \mathbf{U} = 1,$$

and that,

$$\mathbf{U}^T \mathbf{A} \mathbf{U} = \begin{pmatrix} 7.75 & 1.9202 & 0 & 0 \\ 1.9202 & -0.9872 & 0.8303 & 0 \\ 0 & 0.8303 & 0.1372 & 0.7681 \\ 0 & 0 & 0.7681 & 0.1 \end{pmatrix},$$

is tridiagonal and symmetric. The subroutine krylov, calculates first the N independent vectors $b, Ab, A^2b, \ldots$, stores them as the columns of the matrix z, and then applies the Gram–Schmidt process to orthonormalize the columns of z. The resulting vectors are stored in the columns of the matrix u.

```
      subroutine krylov(a,z,u,N,NMAX)
      implicit real*8 (a-h,o-z)
      real*8 a(NMAX,NMAX),z(NMAX,NMAX),u(NMAX,NMAX)
c On entry: a  = real symmetric matrix
c           z = scratch space
c        NMAX = size
c        N = number of used rows and columns
c On return:
c        z = contains the vectors b,a*b,a*a*b.... (a^N-1) b
c           arranged as columns
c        u = Krylov space basis arranged as columns
c This subroutine constructs a set of unit, mutually orthogonal
c vectors spanning the space av, aav, aaav, aaaav, .... (a^N) v
c where v is an arbitrary vector. Orthonormality is enforced
c by brute force deflation at every step of the recursion with
c a Gram Schmidt process.
c The similarity transform tridiagonalizes the input matrix a.
c calculation of  a*v,a*a*v, a*a*a*v, a*a*a*a*v, .... (a^N) v
c these vectors are calculated recursively and stored in the
c columns of the matrix u
      do i=1,N
       z(i,1) = 1.d0
      enddo
      do j=2,N
      do i=1,N
       z(i,j) = 0.d0
       do k=1,N
          z(i,j) = z(i,j) + a(i,k)*z(k,j-1)
       enddo
      enddo
      enddo
      if (N .lt. 20) then
        write(6,*) 'Z'
      do i=1,N
        write(6,999) (z(i,j), j=1,N)
      enddo
      endif
      do i=1,N                      ! Obtain u1
       u(i,1) = z(i,1)/sqrt(dfloat(N))
      enddo
c run the Gram-Schmidt orthogonalization
      do j=2,N
      do k=1,j-1
```

```
c dnkj is the dot product of vector u(*,k) and u(*,j) (k = 1 to j-1)
      dnkj = 0.d0
      do i=1,N
       dnkj = dnkj + u(i,k)*z(i,j)
      enddo
c  deflating vector z(*,j)
      do i=1,N
       z(i,j) = z(i,j) - dnkj*u(i,k)
      enddo
      enddo
c normalizing the vector in z(*,j)
      dnorm = 0.d0
      do i=1,N
      dnorm = dnorm + z(i,j)*z(i,j)
      enddo
c by dividing the vector z(*,j) by its norm we form a unit vector
      do i = 1,N
      u(i,j) = z(i,j)/sqrt(dnorm)
      enddo
      enddo  ! j loop
      if (N .lt. 20) then
        write(6,*) 'U'
      do i=1,N
        write(6,999) (u(i,j), j=1,N)
      enddo
      endif
      return
999   format(10f11.7)
      end
```

To test krylov, we call it from inside a main function where a 5×5 one-dimensional Colbert and Miller DVR matrix is constructed. The mass is 207 a.u. and the force constant for the harmonic potential is 1.0 hartree bohr^{-2}, as usual. Here is the output,

$$\mathbf{A} = \begin{pmatrix} 1.8187 & -0.0083 & 0.0020 & -0.0009 & 0.0005 \\ -0.0083 & 0.6635 & -0.0083 & 0.0020 & -0.0009 \\ 0.0020 & -0.0083 & 0.0859 & -0.0083 & 0.0020 \\ -0.0009 & 0.0020 & -0.0083 & 0.0859 & -0.0083 \\ 0.0005 & -0.0009 & 0.0020 & -0.0083 & 0.6635 \end{pmatrix}, \quad (5.166)$$

$$\mathbf{Z} = \begin{pmatrix} 1.0000 & 1.8120 & 3.2907 & 5.9818 & 10.8776 \\ 1.0000 & 0.6479 & 0.4137 & 0.2465 & 0.1132 \\ 1.0000 & 0.0734 & 0.0054 & 0.0048 & 0.0115 \\ 1.0000 & 0.0703 & -0.0003 & -0.0059 & -0.0080 \\ 1.0000 & 0.6568 & 0.4357 & 0.2905 & 0.1957 \end{pmatrix}, \quad (5.167)$$

$$\mathbf{U} = \begin{pmatrix} 0.4472 & 0.8163 & 0.3654 & 0.0047 & 0.0029 \\ 0.4472 & -0.0029 & -0.5512 & 0.5007 & 0.4953 \\ 0.4472 & -0.4073 & 0.3651 & -0.4975 & 0.5032 \\ 0.4472 & -0.4094 & 0.3649 & 0.4968 & -0.5023 \\ 0.4472 & 0.0033 & -0.5441 & -0.5047 & -0.4990 \end{pmatrix}, \qquad (5.168)$$

$$\mathbf{U}^T\mathbf{U} = \begin{pmatrix} 1 & 0 & 0 & 0 & 0 \\ 0 & 1 & 0 & 0 & 0 \\ 0 & 0 & 1 & 0 & 0 \\ 0 & 0 & 0 & 1 & 0 \\ 0 & 0 & 0 & 0 & 1 \end{pmatrix}, \qquad (5.169)$$

$$\mathbf{U}^T\mathbf{AU} = \begin{pmatrix} 0.6521 & 0.6354 & 0 & 0 & 0 \\ 0.6354 & 1.2372 & 0.5202 & 0 & 0 \\ 0 & 0.5202 & 0.6696 & 0.0001 & 0 \\ 0 & 0 & 0.0001 & 0.3930 & 0.2849 \\ 0 & 0 & 0 & 0.2849 & 0.3657 \end{pmatrix}. \qquad (5.170)$$

$\mathbf{U}^T\mathbf{AU}$ demonstrates that the projection of $\mathbf{A}$ in its own Krylov space is tridiagonal. The details of the programing of the main function are left to the reader as an exercise. The advantage of the Krylov space method for tridiagonalization is that only products of matrices with vectors are needed. This is an important feature that is exploited by the Lanczos algorithm [175].

5.18   The Row Representation of a Sparse Matrix

We are almost ready to introduce the Lanczos algorithm. There are now only two simple items of sparse matrix technology we need to work through [178]. The first is a new row representation scheme for sparse matrices. The idea of sparse matrix technology is to eliminate the need to manipulate the zeros present in a sparse matrix. Testing to see if the entry is zero before an operation is not a good strategy to gain significant numerical advantage. We need to rework the representation of the sparse matrix first, then we need to rework the algorithm for the sparse matrix–vector multiplication. In the row representation, instead of storing the matrix in a $n \times n$ array, three one-dimensional arrays are used. The first array keeps the values of the nonzero entries. The second array is a set of pointers (indices that point to the columns that contain nonzero entries to be found in the original matrix). Consider, for example, the sparse matrix,

$$\mathbf{A} = \begin{pmatrix} -1 & 0 & 0 & 3 \\ 0 & 0 & 3 & 1 \\ 0 & 3 & -2 & 0 \\ 3 & 1 & 0 & 1 \end{pmatrix}.$$

A contains nine nonzero entries and we store these sequentially from left to right, from top to bottom in a vector, $\mathbf{A}_n$,

$$\mathbf{A}_n = -1, 3, 3, 1, 3, -2, 3, 1, 1.$$

Next, define an array of pointers, call it $\mathbf{j}_a$, that contains the column number for each entry. For our example,

$$\mathbf{j}_a = 1, 4, 3, 4, 2, 3, 1, 2, 4.$$

Therefore, for each entry in $\mathbf{A}_n$ there is an integer in $\mathbf{j}_a$ that specifies the column associated with each entry in $\mathbf{A}_n$. Finally, the third array, say it is $\mathbf{i}_a$, is an array of pointers that specifies which entries in $\mathbf{j}_a$ are the beginning and the ending of each row of $\mathbf{A}$. The array $\mathbf{i}_a$ is arranged so that for a $n \times n$ sparse matrix there are $N + 1$ entries. To find the columns for row i in $\mathbf{j}_a$, one scans the entries of $\mathbf{j}_a$ and $\mathbf{A}_n$ between the values of $\mathbf{i}_a(i)$ and $\mathbf{i}_a(i+1) - 1$. For our example, the array $\mathbf{i}_a$ is

$$\mathbf{i}_a = 1, 3, 5, 7, 10.$$

The nonzero values of row one are in entries 1 through 2 of $\mathbf{A}_N(i)$, and the corresponding columns are in entries 1 through 2 of $\mathbf{j}_a(i)$. For row 2, look in entries 3 through 4, for row 3, in entries 5 through 6, and in row 4 in entries 7 through 9. This less friendly (to humans) representation is powerful in at least two ways. Firstly, the storage needed scales linearly with n and not n^2 for a sufficiently sparse matrix. Secondly, and more importantly, the number of operations for matrix multiplications can now be made to scale linearly with n instead of n^3. It is useful to make a program that produces the row representation of a sparse matrix. Consider the following lines of code:

```
      program rowrep
c ROWREP reads a sparse matrix from an input file
c stored with the usual format i,j,a(i,j)
c and converts the matrix to row representation
      implicit real*8 (a-h,o-z)
      parameter (maxdim = 2000)
      parameter (max = 60000)          !       Sparse BLAS vector
      real*8 an(max),b(maxdim),c(maxdim)
      integer*4 ia(max),ja(max)
      read(5,*) n        ! n is the size of the matrix
      k = 0
      k2 = 0
      ii_old = 0
      do i = 1,n*n
       read(5,*,end=10) ii,jj,ain
       k = k + 1
       an(k) = ain
```

```
        if (ii .eq. ii_old) then
          ja(k) = jj
        else
          k2 = k2 + 1
          ia(k2) = k
          ja(k) = jj
        endif
        ii_old = ii
      enddo
10    ia(k2+1) = k+1
c write out the row representation
      do i=1,k+2
        write(6,*) an(i),ja(i),ia(i)
      enddo
      call display(ia,ja,an,n,6)
      end
      subroutine display(ia,ja,an,n0,lu)
      implicit real*8 (a-h,o-z)
      parameter (max = 60000)
      real*8 an(max),ant(max)
      integer*4 ia(max),ja(max)
      integer*4 iat(max),jat(max)
      write(lu,*) '# MATRIX DISPLAYING ALGORITHM RR(C)O'
      irow = 1
      do j = 1,n0
       iaa = ia(j)
       iab = ia(j+1) -1
       do n=iaa,iab
         write(lu,*) irow,ja(n),an(n)
       enddo
       irow = irow + 1
      enddo
      return
      end
```

Only the nonzero entries of the matrix to be represented should be included in the input of rowrep. The subroutine display is designed to translate the row representation back to the familiar row, column, entry data format. The subroutine display is a great debugging tool when working with row representations of sparse matrices.

The reader should test the limit of the computer storage capacity by carefully increasing the size of max inside display and in the main program rowrep. This determines the largest matrix that can be handled with the storage capability available. Different compilers will complain differently when the fast storage (RAM) capacity is exceeded. When this happens, one has to resort to the much slower binary file storage.

Consider the matrix–vector product. The i element,

$$(\mathbf{Ab})_i = \sum_{k=1}^{N} A_{ik} b_k,$$

is the dot product of row i of $\mathbf{A}$ with the vector $\mathbf{b}$. It is not hard to envision an algorithm that takes advantage of the row representation of $\mathbf{A}$ to avoid multiplying by the zeros. All we need, for a given element of the matrix–vector product, is to find the right rows in $\mathbf{b}$ that correspond to the matching column number of $\mathbf{A}$. This is just $b\left[\mathbf{j}_{a(i)}\right]$. In fact, the row representation was developed with the purpose of expediting the matrix–vector product for sparse matrices. The following subroutine is reproduced with permission from page 252 of Ref. [178], and is the heart of the Lanczos method.

```
c Algorithm for the product of a general sparse matrix by a full
c column vector  O(N) operations
c Input:
c        n            Size of A
c        ia,ja,an     Matrix in RR(C)O format
c        b            Given full vector
c Output:
c        c            Output vector Ab
      subroutine pgsmcv(ia,ja,an,b,n,c)
      implicit real*8 (a-h,o-z)
      parameter (maxdim = 3000)
      parameter (max = 60000)
      real*8 an(max),b(maxdim),c(maxdim+1)
      integer*4 ia(max),ja(max)
      do 20 i=1,n
      u = 0.d0
      iaa = ia(i)
      iab = ia(i+1) - 1
      if (iab .lt. iaa) goto 20
      do 10 k=iaa,iab
10    u = u + an(k)*b(ja(k))
20    c(i) = u
      return
      end
```

Exercises

1. Write the main program that calls `krylov`, and reproduce the test data in section 5.17.

2. Modify `krylov` to find the representation of $\mathbf{A}$ in its own nonorthonormalized Krylov space, before the vectors are submitted to the Gram–Schmidt procedure.

3. Prove that the vectors of the Krylov space of $\mathbf{A}$ are linearly independent if $\mathbf{A}$ is nonsingular.

4. Prove that if $\mathbf{A}$ is nonsingular, it is tridiagonal in its own orthonormalized Krylov space.

5. Modify `rowrep` to read a $n = 4$ column vector and include the subroutine `pgsmcv`. Write out the result of $\mathbf{Ab}$ and check the answer by hand.

6. Modify the program `point_plot` to write the nonzero entries of an arbitrary Colbert and Miller Hamiltonian matrix $\mathbb{R}^3$ to a file, using the $i, j, a_{i,j}$ format. Then, use the file, as input to the `rowrep` program. Run both programs and systematically increase the size of the matrix. Keep track of the run time of `rowrep` on a quiet system (with no background jobs) and convince yourself that the time scales linearly with the size of the matrix n.

5.19   The Lanczos Algorithm

If one combines the recursion that calculates the Krylov space vectors with the Gram–Schmidt in one single step, one obtains the Lanczos algorithm. Something special happens when the Gram–Schmidt is carried out simultaneously with the computation of the Krylov space vectors. First, it can be shown [175] that the basis vector $j + 1$ generated in step j of the recursion is automatically orthogonal to the basis vectors $j - 1, j - 2, \ldots, 1$, so the newly formed Lanczos vector needs to be deflated of only two other vectors. However, the Lanczos vector in finite precision computations occasionally needs to be deflated of all the previously determined vectors. This is a rounding error problem that makes the Lanczos algorithm for tridiagonalization of dense matrices less stable, compared to the Householder or the Givens method. The real breakthrough for sparse matrix diagonalization came when Paige [176] realized that subsets of the full Krylov space, computed by the recursion relation proposed by Lanczos, are nearly invariant subspaces. That's what makes the Lanczos algorithm special. If a Krylov subspace is nearly invariant, it means that the set of $m \ll N$ Lanczos vectors can create a tridiagonal $m \times m$ matrix whose eigenvalues are a close approximation of the eigenvalues of the original sparse matrix. The simple Lanczos converges to eigenvalues at both ends of the spectrum first. If we are only interested in the ground state and the first few excited eigenvalues of a large and sparse Hamiltonian matrix, we can run a small number (much less than N) of Lanczos recursions and diagonalize by any method a significantly smaller matrix. Good approximations to the eigenvectors can also be constructed without much difficulty. The recursion relation proposed by Lancozos is

$$\mathbf{q}'_{j+1} = \mathbf{A}\mathbf{q}_j - \beta_{j-1}\mathbf{q}_{j-1} - \alpha_j\mathbf{q}_j, \tag{5.171}$$

where $\mathbf{q}'_{j+1}$ is the unnormalized Lanczos vector

$$\mathbf{q}'_{j+1} = \beta_j\mathbf{q}_{j+1}, \tag{5.172}$$

and β_j is defined as the size of $\mathbf{q}'_{j+1}$,

$$\beta_j = \left|\mathbf{q}'_{j+1}\right|.$$

Note that if we multiply both sides of Equation 5.171 by $\mathbf{q}_j^T$ from the left, and we use the orthonormality property of the Lanczos vectors, we obtain,

$$\alpha_j = \mathbf{q}_j^T \mathbf{A}\mathbf{q}_j, \tag{5.173}$$

whereas, if we multiply both sides of Equation 5.171 by $\mathbf{q}_{j+1}^T$ from the left, we obtain,

$$\beta_j = \mathbf{q}_{j+1}^T \mathbf{A}\mathbf{q}_j. \tag{5.174}$$

Therefore, the Lanczos recursion achieves three goals at once, it calculates the Krylov subspace vectors, orthogonalizes them, and constructs a $m \times m$ projection of the original sparse matrix $\mathbf{A}$ ready to be diagonalized. The following program uses the subroutine **pgsmcv** for sparse matrix–vector multiplication, the Lanczos algorithm, and the routines **tqli**, **eigsrt** from the *Numerical Recipes in FORTRAN* book [8]. For testing purposes, a call to the EISPACK routines is made as well.

```
      program lanczos
      implicit real*8 (a-h,o-z)
      parameter (maxdim = 1400)
      real*4 deltim1,deltim2,timar(2),tim0,etime
      real*8 a(maxdim,maxdim)
      real*8 w(0:maxdim,0:maxdim)
      real*8 eval(maxdim),subd(maxdim)
      real*8 alphaj(0:maxdim),betaj(0:maxdim)
      real*8 z(maxdim,maxdim),d(maxdim),e(maxdim)
      real*8 qmn1(maxdim),rmn1(maxdim)
      real*8 qm(maxdim,maxdim),rm(maxdim)
      parameter (max = 60000)            ! Choose this number wisely!
      real*8 an(max),b(maxdim),c(maxdim) !  Sparse BLAS vector
      integer*4 ia(max),ja(max)
c Diagonalization routine for real symmetric matrices
c                      ========================
c this program uses sparse BLAS to solve the eigenvalue
c problem  RR(C)U format is used for the real symmetric matrix H.
c This program implements the Lanczos algorithm with the selective
```

```
c reorthogonalization procedure of Simon.
c The mtop x mtop invariant subspace to be diagonalized
c (using a call to eigen for pedagogical reasons) is in the matrix a
c Curotto E.      07/14/2000
      do i=1,maxdim
       e(i) = 0.d0
       d(i) = 0.d0
       do j=1,maxdim
        a(i,j) = 0.d0
        z(i,j) = 0.d0
       enddo
      enddo
      w(1,0) = 0.d0
      w(1,1) = 1.d0
      w(0,0)= 1.d0
      tiny = 1.d-15    ! tiny is the machine precision
      tiny2 = 3.2d-8
      k = 0
      k2 = 0
      ii_old = 0
      read(5,*) mtop
      read(5,*) n,m
      if (n .gt. maxdim) stop 'TOO LARGE TO HANDLE'
      do i = 1,m
      read(5,*) ii,jj,ain
      k = k + 1
      an(k) = ain
      if (ii .eq. ii_old) then
      ja(k) = jj
      else
      k2 = k2 + 1
      ia(k2) = k
      ja(k) = jj
      endif
      ii_old = ii
      enddo
      ia(k2+1) = k+1
      icounta = m
c       call display(ia,ja,an,n,9)
2       nt3 = n
      tim0=etime(timar)
c Build the sequence of Krylov subspaces
      do i=1,nt3
      qmn1(i) = 0.d0
      rmn1(i) = 1.d0
      enddo
      betamn1 = sqrt(dfloat(nt3))
      betaj(0) = betamn1
      do m=1,mtop
```

```
c        write(6,*) 'step ',m,betamn1
c Step 1 Extend orthonormal basis
         do i=1,nt3
          qm(i,m) = rmn1(i)/betamn1
          b(i) = qm(i,m)
         enddo
c Step 2 Find partial residuals
         call pgsmcv(ia,ja,an,b,nt3,c)
         do i=1,nt3
          rm(i) = c(i) - betamn1*qmn1(i)
         enddo
c Step 3 Extend diagonal of Tm
         alpham = 0.d0
         do i=1,nt3
          alpham = alpham + qm(i,m)*rm(i)
         enddo
c Step 4 Complete the computation of the residuals
         sum = 0
         do i=1,nt3
          rm(i) = rm(i) - alpham *qm(i,m)
          sum = sum + rm(i)*rm(i)
         enddo
c Step 5 Compute magnitude of residuals
         betam = sqrt(sum)
c Step 6 set up the tridiagonal matrix
         alphaj(m) = alpham
         betaj(m) = betam
c Step 7a Recursion of the orthogonality loss in the Krylov space
         sum = 0.d0
         do k=1,nt3
          sum = sum + rm(k)*qm(k,m)
         enddo
         sum = sum/betaj(m)
         w(m,m+1) = sum
         w(m+1,m+1) = 1.d0
         wmax = tiny
         do k=1,m-1
          w(k,m+1) = betaj(k)*w(k+1,m)
          w(k,m+1) = w(k,m+1) + (alphaj(k) - alphaj(m))*w(k,m)
          w(k,m+1) = w(k,m+1) - betaj(m-1)*w(k,m-1)
          w(k,m+1) = w(k,m+1) + betaj(k-1)*w(k-1,m)
          w(k,m+1) = w(k,m+1)/betaj(m)
          if (dabs(w(k,m+1)) .gt. wmax) wmax = dabs(w(k,m+1))
         enddo
c Step 7b Reorthogonalization
         iflag = 0
         if (wmax .gt. tiny2) then
         do i1 = 1,m-3
```

```
c calculate the inner product of the residual with qm(,i1)
      sumk = 0.d0
      do j1=1,nt3
        sumk = sumk + rm(j1)*qm(j1,i1)
      enddo
c orthonormalize rm to qm(j1,i1)
      icr = icr + 1
      iflag = 1
      do j1=1,nt3
        rm(j1) = rm(j1) - sumk*qm(j1,i1)
      enddo
      w(i1,m+1) = tiny
      enddo
      endif
c recalculate the size of the residuals
      if (iflag .eq. 1) then
      sum = 0.d0
      do j1=1,nt3
        sum = sum +  rm(j1)*rm(j1)
      enddo
      betam = sqrt(sum)
      betaj(m) = betam
      endif
c Step 8 Reset rmn1,qmn1 and betamn1
      do i=1,nt3
        qmn1(i) = qm(i,m)
        rmn1(i) = rm(i)
      enddo
      betamn1 = betam
      enddo       ! m
c Step 9 Calculate some eigenvalues
      do k=1,mtop
        a(k,k) = alphaj(k)
        if (k. lt. mtop) then
          a(k,k+1) =  betaj(k)
          a(k+1,k) =  betaj(k)
        endif
c        eval(k) = alphaj(k)    ! more effient way
c        subd(k+1) = betaj(k)
c        z(k,k) = 1.d0
      enddo
c        call tqli(eval,subd,mtop,maxdim,z)   ! from numerical recipes
c        call eigsrt(eval,z,mtop,maxdim) ! from numerical recipes
c        do k=1,10
c          err = betam*dabs(z(mtop,k))
c          write(6,*) k, eval(k),err
c        enddo
900      format(2f20.8,2f20.8)
      deltim1=etime(timar)-tim0
```

```
write(6,*) icr,'  REORTHOGONALIZATIONS'
tim0=etime(timar)
call eigen(maxdim,mtop,a,d,z,e,ierr)
do i=1,10
write(6,*) i,d(i)
enddo
deltim2=etime(timar)-tim0
write(6,*) deltim1,deltim2
end
```

The first few lines of `lanczos` read a sparse matrix from a file in regular format and construct its row representation as in `rowrep`. Then, the vector $\mathbf{q}_0$ is initialized to zero. The program makes use of two residual vectors, $\mathbf{r}_{j-1}$ and $\mathbf{r}_j$. Every element of $\mathbf{r}_{j-1}$ is initialized to 1. The value of `betajm1` is initialized to the size of $\mathbf{r}_{j-1}$. Then, steps 1 through 8 are repeated `m` times, where `m` is the desired size of the Krylov subspace and is part of the input.

In step 1, the residual vector $\mathbf{r}_{j-1}$ is normalized, and the result is the jth Lanczos vector. j runs from 1 to `m`,

$$\mathbf{q}_j \leftarrow \frac{\mathbf{r}_{j-1}}{\beta_{j-1}}. \tag{5.175}$$

In step 2, a new partial residual vector is computed,

$$\mathbf{r}_j \leftarrow \mathbf{A}\mathbf{q}_j - \beta_{j-1}\mathbf{q}_{j-1}. \tag{5.176}$$

In step 3, the value of α_j is obtained,

$$\alpha_j \leftarrow \mathbf{q}_j^T \mathbf{r}_j. \tag{5.177}$$

In step 4, the computation of the residual vector (the unnormalized Lanczos vector) is completed,

$$\mathbf{r}_j \leftarrow \mathbf{r}_j - \alpha_j \mathbf{q}_j. \tag{5.178}$$

In step 5, the size of $\mathbf{r}_j$ is calculated,

$$\beta_j = |\mathbf{r}_j|. \tag{5.179}$$

The values of the main diagonal and subdiagonal for the $m \times m$ tridiagonal matrix are stored in separate arrays in step 6.

In step 7, the recursion procedure to detect loss of orthogonality, as suggested by Simon [177], is implemented. Let $\mathbf{W}$ be the "orthogonality" matrix

$$\mathbf{W} = \mathbf{Q}_m^T \mathbf{Q}_m, \tag{5.180}$$

where $\mathbf{Q}_j$ is a $N \times j$ matrix containing $\mathbf{q}_1, \mathbf{q}_2, \ldots, \mathbf{q}_j$ in its columns. The basis vector set is semiorthogonal if the maximum absolute value of an off-diagonal entry of $\mathbf{W}$ is less than $\epsilon^{1/2}$, where ϵ is the machine precision. The elements of **W** satisfy a simple recursion relation,

$$\beta_j W_{k,j+1} = \beta_k W_{k+1,j} + (\alpha_k - \alpha_j) W_{kj} - \beta_{j-1} W_{k,j-1} + \beta_{k-1} W_{k-1,j}. \quad (5.181)$$

Whenever an off-diagonal element of **W** exceeds $\epsilon^{1/2}$, the vector $\mathbf{q}_j$ is reorthogonalized against the entire basis set at hand. The program `lanczos` keeps track of the number of reorthogonalizations made during a run.

Finally, in step 8, the variables $\mathbf{r}_{j-1}, \beta_{j-1}$ and $\mathbf{q}_{j-1}$ are reset to continue the recursion.

The program keeps track of the execution time (user time) of the Lanczos recursion in seconds with the FORTRAN intrinsic function `etime()`.

To test the algorithm, we construct a 1000×1000 Colbert and Miller Hamiltonian matrix for a particle unit mass in a three-dimensional parabolic potential with unit force constant for all three dimensions. The matrix contains 28,000 nonzero entries. The value of `d_L` is set to $2\sqrt{10}$. A Krylov subspace size of 100 produces the following output:

i	E_i	
1	1.50041225	
2	2.50264195	
3	2.50264195	
4	3.50487166	
5	3.51814541	(5.182)
6	3.52423244	
7	4.50710136	
8	4.52037512	
9	4.58001226	
10	5.52260482	

The program `lanczos` performs around 1300 reorthogonalizations. The error in the eigenvalues can be estimated during the recursion. Note that the first three eigenvalues are converged to the exact ones* with many figures. The program `lanczos` finds other energy levels, but it finds difficulty in converging all the degenerate states. The lack of convergence for degenerate states is a problem that can be resolved by performing a spectral transformation of the matrix before subjecting it to the tridiagonalization process.

The Kryolv spaces find applications in several iterative methods for solving linear systems, $\mathbf{Ax} = \mathbf{b}$. Among these, one finds the generalized residual method (GMRES) [179], the quasiminimal residual method (QMR) [180], the conjugate gradient (CG) [181], and numerous others. The linear system, $\mathbf{Ax} = \mathbf{b}$, plays a role in numerical solutions of the Schrödinger equation when one wishes to have the Lanczos algorithm converge in a region of the spectrum of the Hamiltonian near an energy E. Then, the Lanczos procedure is performed for a spectral transform of **H**, such as $[\mathbf{H} - E\mathbf{1}]^{-1}$.

---

*Here exact means neglecting cutoff and truncation error. Of course, the eigenvalues obtained by EIGEN have cutoff and truncation error as the reader can appreciate from the output on this page.

Exercises

1. In a previous exercise, we asked you to modify the program `point_plot` to write the nonzero entries of an arbitrary Colbert and Miller Hamiltonian matrix in $\mathbb{R}^3$ to a file using the $i, j, a_{i,j}$ format. The potential energy should be the three-dimensional parabolic potential with unit force constant, and the mass should be 1 a.u. Use the output to run `lanczos` for several small values of the Krylov subspace size, and determine how the error in the ground state energy depends on this parameter.

2. Use the Lanczos algorithm and the DVR of Colbert and Miller to find the solutions to the following Schrödinger equation:

$$-\frac{\hbar}{2m}\left(\frac{\partial^2}{\partial x^2} + \frac{\partial^2}{\partial x^2} + \frac{\partial^2}{\partial x^2}\right)\Psi(x,y,z)$$

$$+\frac{1}{2}\left(k_x x^2 + k_y y^2 + k_z z^2\right)\Psi(x,y,z) = E\Psi(x,y,z).$$

Choose values of $k_x, k_y,$ and k_z that are convenient, but different from one another, so that the degeneracy is broken. Verify that the energy levels are given by the following equation by computing the first 10 energy levels,

$$E_{n,j,l} = \hbar\left(\frac{1}{2} + n\right)\sqrt{\frac{k_x}{m}} + \hbar\left(\frac{1}{2} + j\right)\sqrt{\frac{k_y}{m}} + \hbar\left(\frac{1}{2} + l\right)\sqrt{\frac{k_y}{m}},$$

$$j, n, l = 0, 1, 2, \ldots.$$

3. Turn off the automatic reorthogonalization of step 7 and rerun the previous example using a set of subspaces that span more than the size of the Hamiltonian matrix. The best way to turn off the automatic reorthogonalization is perhaps to set `tiny2` to some arbitrary large value. You should see numerous eigenvalues converging properly, as well as a number of spurious ones. The spurious eigenvalues can generally be identified and removed. However, the resulting method can be quite inefficient. A number of restarting strategies other than the selective reorganization approach used here can be found in the literature, each have their own merits and limitations.

5.20 Orbital Angular Momentum and the Spherical Harmonics

Let's begin by reviewing the definition of angular momentum in quantum mechanics [856, 857, 858]. For a point particle of mass m with a position vector $\mathbf{x}$ and a momentum $\mathbf{p}$, the angular momentum is given by

$$\mathbf{L} = \mathbf{x} \times \mathbf{p}. \tag{5.183}$$

The operators that correspond to the three components of the angular momentum vectors can be derived by applying the canonical quantization rules. Turning the variables x and p into operators, and then using the definition of the cross product yields the desired result. The operators for each component of the angular momentum vector are in Equations 5.80 through 5.82.

The angular momentum eigenstates are the solutions of the particle with mass μ on a two-sphere of radius R. The two-sphere is a space isomorphic to that of a rigid rotor after the removal of the center of mass coordinates. The Hamiltonian is simply

$$\mathcal{H} = \frac{\hat{L}^2}{2\mu R^2}. \tag{5.184}$$

It is a straightforward exercise in partial derivatives to derive the orbital angular momentum squared operator in spherical polar coordinates,

$$\hat{L}^2 = -\hbar^2 \left(\frac{\partial^2}{\partial \theta^2} + \frac{1}{\sin^2 \theta} \frac{\partial^2}{\partial \phi^2} + \frac{\cos \theta}{\sin \theta} \frac{\partial}{\partial \theta} \right). \tag{5.185}$$

The solution of Schrödinger's equation for the potential free particle, with mass μ on a two-sphere of radius R, gives states with well-defined values for the size of the orbital angular momentum l, and magnetic quantum number m,

$$Y_l^m (\theta, \phi), \quad E_l = \frac{\hbar^2 l (l + 1)}{2\mu R^2}, \tag{5.186}$$

where

$$Y_l^m = (-1)^m \left[\frac{(l - m)! (2l + 1)}{4\pi (l + m)!} \right]^{1/2} P_l^m (\cos \theta) \exp (im\phi), \tag{5.187}$$

are the spherical harmonics, and $P_l^m (x)$ is a set of orthogonal functions known as the associate Legendre polynomials,

$$P_l^m (x) = \frac{1}{2^l l!} \left(1 - x^2 \right)^{m/2} \frac{d^{l+m}}{dx^{l+m}} \left(x^2 - 1 \right)^l. \tag{5.188}$$

The quantum numbers, l, m, are integers that satisfy the following rules, $l = 0, 1, 2, \ldots, m = -l, \ldots, 0, 1, \ldots, l$.

5.21   Complete Sets of Commuting Observables

The following is designed to familiarize the reader with the computation of matrix elements of the Hamiltonian on a two-sphere for a spinless particle subjected to a potential V,

$$\mathcal{H} = \frac{\hat{L}^2}{2\mu R^2} + V, \tag{5.189}$$

and the subsequent solution with Hilbert space and diagonalization techniques. In order to develop the matter fully, we need to make use of two concepts. The first one, from quantum theory, is the concept of the total number of commuting observables. The other, from the theory of angular momentum, is the coupling of angular momentum vectors.

We begin with Equation 5.184. There are a number of physical observables that we can construct for a particle in a two-sphere. However, the state of the system is only specified with a **complete set of commuting observables**. If we decide that the energy is one such observable, then we must search for all the other operators that commute with the Hamiltonian, in order to construct a unique representation of the eigenstates. The operators associated with the physical observables we choose are isomorphic to a Lie algebra. Finding the basis set for the algebra and building the Casimir operator, C_2, is generally the way that we establish that we have obtained the complete set of mutually commuting observables and that we have quantized the system. We start by proving the following commutation rules among the components of the angular momentum vector,

$$\left[\hat{L}_x, \hat{L}_y\right] = i\hbar \hat{L}_z, \tag{5.190}$$

$$\left[\hat{L}_z, \hat{L}_x\right] = i\hbar \hat{L}_y, \tag{5.191}$$

$$\left[\hat{L}_y, \hat{L}_z\right] = i\hbar \hat{L}_x. \tag{5.192}$$

Note how these are isomorphic to the commutations for J_x, J_y, J_z we derive in Chapter 4. Let us derive Equation 5.190 in detail. Note that

$$\left[\hat{L}_x, \hat{L}_y\right] = [\hat{y}\hat{p}_z - \hat{z}\hat{p}_y, \hat{z}\hat{p}_x - \hat{x}\hat{p}_z], \tag{5.193}$$

$$\left[\hat{L}_x, \hat{L}_y\right] = [\hat{y}\hat{p}_z, \hat{z}\hat{p}_x] + [\hat{y}\hat{p}_z, -\hat{x}\hat{p}_z] + [-\hat{z}\hat{p}_y, \hat{z}\hat{p}_x] + [\hat{z}\hat{p}_y, \hat{x}\hat{p}_z], \tag{5.194}$$

and writing out each commutator explicitly in terms of differential operators, gives

$$[\hat{y}\hat{p}_z, \hat{z}\hat{p}_x] = -\hbar^2 \left(\hat{y}\frac{\partial}{\partial z}\hat{z}\frac{\partial}{\partial x} - \hat{z}\frac{\partial}{\partial x}\hat{y}\frac{\partial}{\partial z} \right), \tag{5.195}$$

where the differential operators can be moved to the right of each term after they operate on the term on the right in each factor. In some cases, the product rule has to be used,

$$[\hat{y}\hat{p}_z, \hat{z}\hat{p}_x] = -\hbar^2 \left(\hat{y}\frac{\partial}{\partial x} + \hat{y}\hat{z}\frac{\partial^2}{\partial z \partial x} - \hat{z}\hat{y}\frac{\partial^2}{\partial x \partial z} \right) = -\hbar^2 \hat{y}\frac{\partial}{\partial x}. \tag{5.196}$$

Note that the assumption made here is that the operators are applied to well-behaved functions f, for which the mixed partial derivative are equivalent,

$$\frac{\partial^2 f}{\partial x \partial z} = \frac{\partial^2 f}{\partial z \partial x}. \tag{5.197}$$

Following the same procedure, we can derive the following three results:

$$[\hat{y}\hat{p}_z, -\hat{x}\hat{p}_z] = \hbar^2 \left(\hat{y}\frac{\partial}{\partial z}\hat{x}\frac{\partial}{\partial z} - \hat{x}\frac{\partial}{\partial z}\hat{y}\frac{\partial}{\partial z} \right) = 0, \tag{5.198}$$

$$[-\hat{z}\hat{p}_y, \hat{z}\hat{p}_x] = -\hbar^2 \left(-\hat{z}\frac{\partial}{\partial y}\hat{z}\frac{\partial}{\partial x} + \hat{z}\frac{\partial}{\partial x}\hat{z}\frac{\partial}{\partial y} \right) = 0, \tag{5.199}$$

$$[\hat{z}\hat{p}_y, \hat{x}\hat{p}_z] = -\hbar^2 \left(-\hat{x}\frac{\partial}{\partial z}\hat{z}\frac{\partial}{\partial y} \right) = -\hbar^2 \left(-\hat{x}\frac{\partial}{\partial y} \right). \tag{5.200}$$

One uses the canonical quantization of momentum once again,

$$p_x = i\hbar\frac{\partial}{\partial x}, \quad p_y = i\hbar\frac{\partial}{\partial y},$$

etc. Putting these four results together, we establish Equation 5.190,

$$\left[\hat{L}_x, \hat{L}_y \right] = -\hbar^2 \left(\hat{y}\frac{\partial}{\partial x} - \hat{x}\frac{\partial}{\partial y} \right) = i\hbar\hat{L}_z. \tag{5.201}$$

The derivation of Equations 5.191 and 5.192 is left as an exercise.

It is obvious that the square of the angular momentum commutes with the Hamiltonian operator $\mathcal{H}$ in the potential free case,

$$\left[\hat{L}^2, \mathcal{H} \right] = \left[\hat{L}^2, \frac{\hat{L}^2}{2\mu R^2} \right] = \frac{1}{2\mu R^2} \left[\hat{L}^2, \hat{L}^2 \right] = 0. \tag{5.202}$$

Therefore, eigenstates of $\hat{L}^2$ are also eigenstates of the Hamiltonian in the potential free case. Note that $\hat{L}^2$ is the C_2 operator for the Lie algebra we study in Chapter 4. A set of observables, $A, B, C, \ldots$ is called a **complete set of commuting observables** if:

1. All observables $A, B, C, \ldots$ commute by pairs.

2. There exists a common orthonormal basis set of eigenvectors.

The last item in the definition assures us that once the values of the observables $A, B, C, \ldots$ are specified, the associated eigenvector in the Hilbert space is fully defined. For the potential free, spinless particle on a two-sphere, there are only two observables in the complete set of commuting observables, and these are (by convention) $\hat{L}^2$ and $\hat{L}_z$. To show this, we need to verify that $\hat{L}^2$ and $\hat{L}_z$ commute. We proceed as follows,

$$\left[\hat{L}^2, \hat{L}_z \right] = \left[\hat{L}_x^2 + \hat{L}_y^2 + \hat{L}_z^2, \hat{L}_z \right] = \left[\hat{L}_x^2, \hat{L}_z \right] + \left[\hat{L}_y^2, \hat{L}_z \right] + \left[\hat{L}_z^2, \hat{L}_z \right]. \tag{5.203}$$

The last commutator on the right is zero, since any operator commutes with itself. The other two we can tackle as follows:

$$\left[\hat{L}_x^2, \hat{L}_z\right] = \hat{L}_x \hat{L}_x \hat{L}_z - \hat{L}_z \hat{L}_x \hat{L}_x = \hat{L}_x \hat{L}_z \hat{L}_x - \hat{L}_z \hat{L}_x \hat{L}_x, +\hat{L}_x \left[\hat{L}_x, \hat{L}_z\right], \quad (5.204)$$

$$\left[\hat{L}_x^2, \hat{L}_z\right] = \left[\hat{L}_x, \hat{L}_z\right] \hat{L}_x + \hat{L}_x \left[\hat{L}_x, \hat{L}_z\right] = -i\hbar \left(\hat{L}_y \hat{L}_x + \hat{L}_x \hat{L}_y\right). \quad (5.205)$$

Similarly,

$$\left[\hat{L}_y^2, \hat{L}_z\right] = \left[\hat{L}_y, \hat{L}_z\right] \hat{L}_y + \hat{L}_y \left[\hat{L}_y, \hat{L}_z\right] = i\hbar \left(\hat{L}_x \hat{L}_y + \hat{L}_y \hat{L}_z\right). \quad (5.206)$$

It can be easily seen that $\left[\hat{L}_y^2, \hat{L}_z\right]$ and $\left[\hat{L}_x^2, \hat{L}_z\right]$ add up to zero, as one term is the negative of the other. This shows that

$$\left[\hat{L}^2, \hat{L}_z\right] = 0, \quad (5.207)$$

which implies that both the z component of the angular momentum and its size (squared) can be found simultaneously. The reader is urged to prove that

$$\left[\hat{L}^2, \hat{L}_x\right] = 0, \quad (5.208)$$

$$\left[\hat{L}^2, \hat{L}_y\right] = 0. \quad (5.209)$$

These results suggest that we could have chosen $\hat{L}^2, \hat{L}_y$ or $\hat{L}^2, \hat{L}_x$ as our complete sets of commuting observables. Note, however, that $\hat{L}^2, \hat{L}_x, \hat{L}_y$ does not constitute a valid set of commuting observables, since $\hat{L}_x$ and $\hat{L}_y$ do not commute and cannot be known simultaneously. The set of commuting observables used conventionally is $\hat{L}^2, \hat{L}_z$. Letting $l(l+1)$ and m represent the eigenvalues of the $\hat{L}^2$ and $\hat{L}_z$ operators, respectively, we can symbolize the state of a particle on a two-sphere by using the Dirac Bra–Ket notation as follows

$$|l\ m\rangle. \quad (5.210)$$

We are now ready to represent the operator,

$$\mathcal{H} = \frac{\hat{L}^2}{2\mu R^2}, \quad (5.211)$$

as a matrix in the Hilbert space spanned by the basis $|l\ m\rangle$,

$$\langle l'\ m'|\mathcal{H}|l\ m\rangle = \frac{1}{2\mu R^2} \langle l'\ m'|\hat{L}^2|l\ m\rangle = \hbar^2 \frac{l(l+1)}{2\mu R^2} \delta_{l'l}\delta_{m'm}. \quad (5.212)$$

In units of $\hbar^2/(\mu R^2)$ the Hamiltonian matrix is

$$\mathbf{H} = \begin{pmatrix} 0 & 0 & 0 & 0 & 0 & \cdots \\ 0 & 1 & 0 & 0 & 0 & \cdots \\ 0 & 0 & 1 & 0 & 0 & \cdots \\ 0 & 0 & 0 & 1 & 0 & \cdots \\ 0 & 0 & 0 & 0 & 3 & \cdots \\ \vdots & \vdots & \vdots & \vdots & \vdots & \end{pmatrix}. \quad (5.213)$$

The matrix is clearly diagonal, and along the main diagonal we can immediately read off energies. So far, we have done nothing more than re-express the rigid rotor problem and its solution in the language of angular momentum theory and Hilbert spaces. Adding a potential (that depends on the angles θ and ϕ) and calculating the matrix elements in the $|l\ m\rangle$ basis requires us to dive into angular momentum theory a little deeper and study how two angular momenta couple with one another. It will turn out (as one would expect) that the integrals of the form

$$\langle l'\ m'|\ V\left(\theta, \phi\right)|l\ m\rangle, \tag{5.214}$$

are not diagonal, implying that the potential energy matrix is not diagonal, in general. However, no numerical integration will be necessary for many cases of interest, since the matrix elements can be calculated quite conveniently using the **vector coupling coefficients**.

5.22 The Addition of Angular Momentum Vectors

Now imagine a physical system for which there are two sources of angular momentum. The Hamiltonian (or the orbital part of it) for a system with two sources of orbital angular momentum has the following form:

$$\mathcal{H} = \frac{\hat{L}_1^2}{2\mu R_1^2} + \frac{\hat{L}_2^2}{2\mu R_2^2}. \tag{5.215}$$

The reader will recall that a single particle may have two sources of angular momenta, however. For instance, the spin and the orbital angular momentum of an electron are two distinct angular momenta, which add according to the rules explained here. If we use $\mathbf{J}$ as a generic symbol for angular momentum (which may or may not be restricted to integer numbers), we write for the total angular momentum

$$\mathbf{J}_3 = \mathbf{J}_1 + \mathbf{J}_2. \tag{5.216}$$

We leave it to the interested reader to show that the nine commutations represented by the combination of the components of $\hat{\mathbf{J}}_1$ and $\hat{\mathbf{J}}_2$ are all zero, i.e.,

$$\left[\hat{\mathbf{J}}_1, \hat{\mathbf{J}}_2\right] = 0. \tag{5.217}$$

These commutations are easy to prove, since the coordinates of space 2 are constants for the derivatives with respect to space 1 coordinates and vice versa. The following commutations are less trivial to show, but are equally important.

$$\left[\hat{J}_1^2, \hat{J}_{3,z}\right] = 0, \qquad \left[\hat{J}_2^2, \hat{J}_{3,z}\right] = 0, \qquad \left[\hat{J}_3^2, \hat{J}_{3,z}\right] = 0, \tag{5.218}$$

$$\left[\hat{J}_1^2, \hat{J}_2^2\right] = 0, \qquad \left[\hat{J}_1^2, \hat{J}_3^2\right] = 0, \qquad \left[\hat{J}_2^2, \hat{J}_3^2\right] = 0. \tag{5.219}$$

Additionally, each angular momentum vector satisfies commutation rules identical to those for a single angular momentum, namely,

$$\left[\hat{J}_1^2, \hat{J}_{1,z}\right] = 0, \tag{5.220}$$

$$\left[\hat{J}_2^2, \hat{J}_{2,z}\right] = 0, \tag{5.221}$$

$$\left[\hat{J}_3^2, \hat{J}_{3,z}\right] = 0. \tag{5.222}$$

However, note that $\hat{J}_{1,z}$ and $\hat{J}_{2,z}$ do not commute with $\hat{J}_3^2$. All these commutation rules give rise to two distinct sets of commuting observables that are of particular importance for our discussion. One possible complete set of commuting observables is

$$\hat{J}_1^2, \hat{J}_{1,z}, \hat{J}_2^2, \hat{J}_{2,z}, \tag{5.223}$$

with a Dirac vector represented by the symbol,

$$|j_1 \ m_1 \ j_2 \ m_2\rangle. \tag{5.224}$$

This representation is commonly called the **uncoupled representation**. j_1 and j_2 are quantum numbers (integers or half integers) that are greater than zero and m_1 and m_2 are integers or half integers constrained between

$$-j_1, -j_1 + 1, \ldots \leq m_1 \leq \ldots, j_1 - 1, j_1, \tag{5.225}$$

and

$$-j_2, -j_2 + 1, \ldots \leq m_2 \leq \ldots, j_2 - 1, j_2, \tag{5.226}$$

respectively. These restrictions on the magnetic quantum numbers should be familiar to the reader. A second complete set of commuting observables can be formed with the following operators:

$$\hat{J}_1^2, \hat{J}_2^2, \hat{J}_3^2, \hat{J}_{3,z}. \tag{5.227}$$

The reader should check that all the operators in this list do indeed commute by pairs. The Dirac Bra–Ket representation for this set is

$$|j_1 \ j_2 \ j_3 \ m_3\rangle. \tag{5.228}$$

This new representation is known as the **coupled representation**. The values allowed for j_3 and m_3 follow the **vector addition rules**

$$j_3 = j_1 + j_2, j_1 + j_2 - 1, j_1 + j_2 - 2, \ldots, |j_1 - j_2|, \tag{5.229}$$

$$m_3 = m_1 + m_2. \tag{5.230}$$

At this point, we have two distinct representations for vectors in the same Hilbert space. It should be obvious (by inspecting a few commutation rules) that the state represented by one vector in the coupled representation is not equivalent to any other vector in the uncoupled representation, and the converse is also true. However, since both $|j_1\ m_1\ j_2\ m_2\rangle$ and $|j_1\ j_2\ j_3\ m_3\rangle$ are just different basis for the same vector space, we can expand a vector in one representation, in terms of linear combinations of vectors of the other. In other words, we can always write

$$|j_1\ j_2\ j_3\ m_3\rangle = \sum_{m_1}\sum_{m_2}\langle j_1\ m_1\ j_2\ m_2|j_1\ j_2\ j_3\ m_3\rangle\,|j_1\ m_1\ j_2\ m_2\rangle. \tag{5.231}$$

The quantities

$$\langle j_1\ m_1\ j_2\ m_2|j_1\ j_2\ j_3\ m_3\rangle, \tag{5.232}$$

are real (by convention) constant coefficients for given values of the quantum numbers $j_1, j_2, j_3, m_1, m_2, m_3$, and are known as the **vector coupling coefficients** or **Clebsch–Gordan coefficients**. Of course, the inverse relation is also possible,

$$|j_1\ m_1\ j_2\ m_2\rangle = \sum_{j_3}\sum_{m_3}\langle j_1\ j_2\ j_3\ m_3|j_1\ m_1\ j_2\ m_2\rangle\,|j_1\ j_2\ j_3\ m_3\rangle. \tag{5.233}$$

Since the vector coupling coefficients are real, it is true that

$$\langle j_1\ m_1\ j_2\ m_2|j_1\ j_2\ j_3\ m_3\rangle = \langle j_1\ j_2\ j_3\ m_3|j_1\ m_1\ j_2\ m_2\rangle. \tag{5.234}$$

It can be shown that the vector coupling coefficients have several important properties. They are elements of an orthogonal matrix that transforms vectors in the Hilbert space of two angular momenta from the coupled to the uncoupled representation, and vice versa. This translates into the following four equations:

$$\sum_{j_3}\sum_{m_3}\langle j_1\ m_1'\ j_2\ m_2'|j_1\ j_2\ j_3\ m_3\rangle\,\langle j_1\ j_2\ j_3\ m_3|j_1\ m_1\ j_2\ m_2\rangle$$
$$= \delta_{m_1'\,m_1}\delta_{m_2'\,m_2}, \tag{5.235}$$

$$\sum_{m_1}\sum_{m_2}\langle j_1\ j_2\ j_3'\ m_3'|j_1\ m_1\ j_2\ m_2\rangle\,\langle j_1\ m_1\ j_2\ m_2|j_1\ j_2\ j_3\ m_3\rangle$$
$$= \delta_{j_3'\,j_3}\delta_{m_3'\,m_3}\delta\,(j_1, j_2, j_3), \tag{5.236}$$

$$\sum_{j_3}\langle j_1\ m_1\ j_2\ m_3 - m_1'|j_1\ j_2\ j_3\ m_3\rangle\,\langle j_1\ j_2\ j_3\ m_3|j_1\ m_1\ j_2\ m_3 - m_1\rangle$$
$$= \delta_{m_1'\,m_1}, \tag{5.237}$$

$$\sum_{m_1}\langle j_1\ j_2\ j_3'\ m_3|j_1\ m_1\ j_2\ m_3 - m_1\rangle\,\langle j_1\ m_1\ j_2\ m_3 - m_1|j_1\ j_2\ j_3\ m_3\rangle$$
$$= \delta_{j_3'\,j_3}\delta\,(j_1, j_2, j_3), \tag{5.238}$$

where

$$\delta(j_1, j_2, j_3) = 1, \tag{5.239}$$

if j_1, j_2, j_3 satisfy the vector addition rules in Equation 5.229, and it is zero otherwise.

5.23 Computation of the Vector Coupling Coefficients

The computation of the vector coupling coefficients is carried out in the general case by a formula first derived by Racah in 1942.

$$\langle j_1 \, m_1 \, j_2 \, m_2 | j_1 \, j_2 \, j_3 \, m_3 \rangle$$

$$= \left[\frac{(2j_3 + 1)(j_1 + j_2 - j_3)!(j_1 - j_2 + j_3)!(-j_1 + j_2 + j_3)!}{(j_1 + j_2 + j_3 + 1)!} \right]^{1/2}$$

$$\times \left[(j_1 + m_1)!(j_1 - m_1)!(j_2 + m_2)!(j_2 - m_2)!(j_3 + m_3)!(j_3 - m_3)! \right]^{1/2}$$

$$\times \sum_z (-1)^z / [z!(j_1 + j_2 - j_3 - z)!(j_1 - m_1 - z)!(j_2 + m_2 - z)!$$

$$(j_3 - j_2 + m_1 + z)!(j_3 - j_1 - m_2 + z)!]. \tag{5.240}$$

The sum over z in Equation 5.240 is restricted to a set of positive integers (0 included) for which terms in the denominator of the summand are nonnegative.

```
c   calculate the Racah coefficient
      subroutine racah(j1,j2,j3,m1,m2,m3,c)
      implicit real*8(a-h,o-z)
      parameter(MAXN = 150)
      real*8 f(0:MAXN)
      c = 0.d0
c check that indeces comply with the vector addition rules
      if (j3 .gt. j1 + j2) return
      if (j3 .lt. abs(j1-j2))return
      if (m3 .ne. (m1 + m2))return
      if (abs(m1) .gt. j1) return
      if (abs(m2) .gt. j2) return
      if (abs(m3) .gt. j3) return
c check to see if Racah's formula can be used
      if ((j1 + j2 +j3 +1) .gt. 150) STOP 'J S ARE TOO LARGE FOR RACAH'
      call factorial(f)
      i10 =   j1 + j2 - j3
      i11 =   j1 - j2 + j3
      i12 = -j1 + j2 + j3
      i14 =   j1 + j2 + j3 + 1
```

```
      i20 = j1 + m1
      i21 = j1 - m1
      i22 = j2 + m2
      i23 = j2 - m2
      i33 = j3 + m3
      i34 = j3 - m3
      a = dfloat(2*j3+1)*f(i10)*f(i11)*f(i12)/f(i14)
      b = f(i20)*f(i21)*f(i22)*f(i23)*f(i33)*f(i34)
      c = 0.d0
      do iz=0,(j1 + j2 + j3 + 1)
       s = 1.d0
       if (mod(iz,2) .ne. 0) s = -1.d0
       i1 = j1 + j2 - j3 - iz
       i2 = j1 - m1 - iz
       i3 = j2 + m2 - iz
       i4 = j3 - j2 + m1 + iz
       i5 = j3 - j1 - m2 + iz
       if (i1 .lt. 0) goto 10
       if (i2 .lt. 0) goto 10
       if (i3 .lt. 0) goto 10
       if (i4 .lt. 0) goto 10
       if (i5 .lt. 0) goto 10
       c = c + s*sqrt(a*b)/(f(iz)*f(i1)*f(i2)*f(i3)*f(i4)*f(i5))
10     enddo
      c = c
      return
      end
c   calculate an array of factorials
      subroutine factorial(f)
      implicit real*8(a-h,o-z)
      parameter(MAXN = 150)
      real*8 f(0:MAXN)
      f(0) = 1.d0
      do i=1,MAXN
      f(i) = dfloat(i)*f(i-1)
      enddo
      return
      end
```

The subroutine **racah** is design to handle only integer values of the arguments $j_1, j_2, j_3, m_1, m_2, m_3$. The largest values of the orbital angular momentum should not exceed 75. We could toil with the powers of 10 for the factorials. However that would unnecessarily complicate the subroutine, a $j_{max} = 75$ is more than adequate for our purposes. The program checks internally to make sure that the calling arguments follow the vector addition rules. It is standard to test the algorithm before employing it anywhere. We guide the reader to test **racah** reasonably well in this section.

Several special values of the vector coupling coefficients can be obtained with a little thought. For instance, it is clear that when the quantum number for the angular momentum of one source is zero, the state vector in the coupled and uncoupled representation is the same. Therefore,

$$\langle j_1 \ m_1 \ 0 \ 0 | j_1 \ 0 \ j_1 \ m_1 \rangle = 1. \tag{5.241}$$

It is also relatively simple to derive the following special values from Racah's formula,

$$\langle j_1 \ 0 \ j_2 \ 0 | j_1 \ j_2 \ j_3 \ 0 \rangle = 0 \quad \text{if and only if } 2 \top (j_1 + j_2 + j_3), \tag{5.242}$$

i.e., if and only if $j_1 + j_2 + j_3$ is an odd integer. The symbol $\top$ is a number theory symbol that says "does not divide," meaning the division of the argument to the right of $\top$ by the argument to the left is not an integer. Other useful results are,

$$\langle j_1 \ m_1 \ j_1 \ -m_1 | j_1 \ j_1 \ 0 \ 0 \rangle = \frac{(-1)^{j_1-m_1}}{\sqrt{2j_1+1}}, \tag{5.243}$$

$$\langle j_1 \ j_1 \ j_2 \ j_3 - j_1 | j_1 \ j_2 \ j_3 \ j_3 \rangle = \left[\frac{(2j_1)! \, (2j_1+1)!}{(j_1+j_2+j_3+1)! \, (j_1-j_2+j_3)!} \right]^{1/2}. \tag{5.244}$$

Some of the values of the vector coupling coefficients can be found in tables, such as the one that follows.

j_1	j_2	j_3	m_1	m_2	m_3	$\langle j_1 m_1 j_2 m_2 | j_1 j_2 j_3 m_3 \rangle$
1	1	1	1	−1	0	0.707106781
1	1	1	1	0	0	0.000000000
1	1	1	−1	1	0	−0.707106781
1	1	1	1	0	1	0.707106781
1	1	1	0	1	1	−0.707106781
1	1	2	1	−1	0	0.408248290
1	1	2	−1	1	0	0.408248290
1	1	2	0	0	0	0.816496581
1	1	2	1	0	1	0.707106781
1	1	2	0	1	1	0.707106781
1	1	2	1	1	2	1.000000000

Write a program that calls **racah** and verifies the values in the preceeding table, and in Equations 5.241 through 5.244. Test **racah** with values of the argument that are not consistent with the vector addition rules (c.f. Equations 5.225, 5.226, 5.229, 5.230) and make sure that **racah** returns zero. Write a program that verifies the four orthogonality properties of the vector coupling coefficients expressed in Equations 5.235 through 5.238.

5.24 Matrix Elements of Anisotropic Potentials in the Angular Momentum Basis

Let 's evaluate matrix elements of the Hamiltonian in Equation 5.189.

$$
\langle l_1\, m_1|\, \mathcal{H}\,|\, l_2\, m_2\rangle = \langle l_1\, m_1|\, \frac{\hat{L}^2}{2\mu R^2} + V\,|\, l_2\, m_2\rangle
$$

$$
= \frac{\hbar^2\, l_1\,(l_1+1)}{2\mu R^2}\delta_{l_1 l_2}\delta_{m_1 m_2} + \langle l_1\, m_1|\, V\,|\, l_2\, m_2\rangle. \qquad (5.245)
$$

Clearly, the first term is diagonal, but the second term may not be. For orbital angular momenta, we can use the position representation and write the matrix element of the potential as a double integral over the polar angles,

$$
\langle l_1\, m_1|\, V\,|\, l_2\, m_2\rangle = \int_0^{2\pi}\int_0^{\pi}\left[Y_{l_1}^{m_1}(\theta,\phi)\right]^* V(\theta,\phi)\, Y_{l_2}^{m_2}(\theta,\phi)\, \sin\theta\, d\theta\, d\phi.
$$

$$(5.246)$$

The vector coupling coefficients are used to find explicit expressions for the matrix elements of the potential whenever the potential energy can be obtained as an expansion over the spherical harmonics,

$$
V(\theta,\phi) = \sum_{\lambda=0}^{\infty}\sum_{\mu=-\lambda}^{\lambda} C_{\lambda\mu} Y_{\lambda}^{\mu}(\theta,\phi). \qquad (5.247)
$$

This expansion is legitimate on mathematical grounds, since the basis vectors $|l\, m\rangle$ are a complete set, i.e., they satisfy a closure relation and all the requirements for a genuine vector space, as discussed in Chapter 4. The centerpiece of our discussion is the following useful result:

$$
\int_0^{2\pi}\int_0^{\pi} Y_{l_1}^{m_1}(\theta,\phi)\, Y_{l_2}^{m_2}(\theta,\phi)\, Y_{l_3}^{m_3}(\theta,\phi)\, \sin\theta\, d\theta d\phi
$$

$$
= \left[\frac{(2l_1+1)(2l_2+1)(2l_3+1)}{4\pi}\right]^{1/2}\begin{pmatrix} l_1 & l_2 & l_3 \\ 0 & 0 & 0 \end{pmatrix}\begin{pmatrix} l_1 & l_2 & l_3 \\ m_1 & m_2 & m_3 \end{pmatrix},
$$

$$(5.248)$$

where the symbol $\begin{pmatrix} l_1 & l_2 & l_3 \\ m_1 & m_2 & m_3 \end{pmatrix}$ is a Wigner 3-j symbol, a symmetric form of the vector coupling coefficient,

$$
\begin{pmatrix} l_1 & l_2 & l_3 \\ m_1 & m_2 & m_3 \end{pmatrix}
$$

$$
= (-1)^{l_1-l_2-m_3}(2l_3+1)^{-1/2}\langle l_1\, m_1\, l_2\, m_2|l_1\, l_2\, l_3\, -m_3\rangle. \qquad (5.249)
$$

If we combine

$$
\left(Y_{l_1}^{m_1}\right)^* = (-1)^{m_1}\, Y_{l_1}^{-m_1},
$$

with Equations 5.247 and 5.248 into Equation 5.245, we obtain

$$\langle l_1 m_1 | \mathcal{H} | l_2 m_2 \rangle = \frac{\hbar^2 l_1 (l_1 + 1)}{2\mu R^2} \delta_{l_1 l_2} \delta_{m_1 m_2}$$

$$+ \sum_{\lambda=0}^{\infty} \sum_{\mu=-\lambda}^{\lambda} (-1)^{m_1} \, C_{\lambda\mu} \left[\frac{(2l_1 + 1)(2\lambda + 1)(2l_2 + 1)}{4\pi} \right]^{1/2}$$

$$\times \begin{pmatrix} l_1 & \lambda & l_2 \\ 0 & 0 & 0 \end{pmatrix} \begin{pmatrix} l_1 & \lambda & l_2 \\ -m_1 & \mu & m_2 \end{pmatrix}. \tag{5.250}$$

This equation is convenient for a couple of reasons. Firstly, for most applications, only a few terms in the sum over λ and μ (in the expansion of Equation 5.247) have to be considered. Secondly, the coefficients often have important physical meanings. The reader is familiar (perhaps without realizing it) with this kind of expansion from junior level electrostatics, and perhaps the only unfamiliarity is the angular momentum language that we are using here. Consider the scalar potential at a point $\mathbf{r}$ that a system of N charges located at $\mathbf{r}_1, \mathbf{r}_2, \mathbf{r}_3, \ldots$ produces. Coulomb's law gives

$$V = Q \sum_{i=1}^{N} \frac{q_i}{4\pi\epsilon_0 R_i}, \tag{5.251}$$

where Q is the test charge, q_i the charge at $\mathbf{r}_i$, and R_i is the distance between the point at $\mathbf{r}_i$ and $\mathbf{r}$. Using the law of cosines, we rewrite R_i as

$$R_i = \left(r^2 + r_i^2 - 2rr_i \cos \theta_i \right)^{1/2}. \tag{5.252}$$

It is well known that

$$\frac{1}{(1 + y^2 - 2xy)^{1/2}}, \tag{5.253}$$

is the generating function of the Legendre polynomials,

$$\frac{1}{(1 + y^2 - 2xy)^{1/2}} = \sum_{\lambda=0}^{\infty} P_\lambda(x) \, y^\lambda. \tag{5.254}$$

We use this theorem to rewrite R_i^{-1} as follows,

$$\frac{1}{R_i} = \frac{1}{(r^2 + r_i^2 - 2rr_i \cos \theta_i)^{1/2}} = \frac{1}{r \left[1 + \left(\dfrac{r_i}{r} \right)^2 - 2 \left(\dfrac{r_i}{r} \right) \cos \theta_i \right]^{1/2}},$$

$$\frac{1}{R_i} = \frac{1}{r} \sum_{\lambda=0}^{\infty} P_\lambda(\cos \theta_i) \left(\frac{r_i}{r} \right)^\lambda = \frac{1}{r} \sum_{\lambda=0}^{\infty} \left(\frac{r_i}{r} \right)^\lambda \frac{\sqrt{4\pi}}{\sqrt{2\lambda + 1}} Y_\lambda^0 (\theta_i, \phi_i),$$

where

$$Y_l^0 = \left(\frac{2l+1}{4\pi}\right)^{1/2} P_l(\cos\theta),$$

is used. The reader should recall that the spherical harmonics with a zero magnetic quantum number are the Legendre polynomials. The potential is in general,

$$V = Q \sum_{i=1}^{N} \frac{q_i}{4\pi\epsilon_0 r} \sum_{\lambda=0}^{\infty} \left(\frac{r_i}{r}\right)^{\lambda} \sqrt{\frac{4\pi}{2\lambda+1}} Y_\lambda^0(\theta_i,\phi_i). \tag{5.255}$$

Equation 5.255 is the multipole expansion of electrostatics. Note that the result in Equation 5.255 has nothing to do with quantum mechanics. The $\lambda = 0$ term is known as the monopole of a charge distribution, the $\lambda = 1$ term is its dipole moment, the $\lambda = 2$ term the quadrupole moment, and so on. As an application of the multipole expansion, and the vector coupling theorem to obtain matrix elements, consider a molecule in the electromagnetic field. The matrix elements of the dipole moment of a molecule are used to derive the selection rule, $\Delta j = \pm 1$. This selection rule can be understood by using j_1, j_2 as two rotational states of the molecule, $\lambda = 1$ from the dipole term of the electromagnetic field. The vector addition rules state that the allowed rotational state can only have the following values for the quantum number:

$$j_2 = j_1 + 1, j_1, |j_1 - 1|.$$

Otherwise, the vector coupling coefficient vanishes, i.e., the transition will be forbidden. The trivial $j_1 \leftarrow j_1$ transition is also forbidden, since the sum $j_1 + 1 + j_1$ is odd, and in that case

$$\begin{pmatrix} j_1 & 1 & j_1 \\ 0 & 0 & 0 \end{pmatrix}, \tag{5.256}$$

equals zero for any allowed value of j_1.

5.25   The Physical Rigid Dipole in a Constant Electric Field

The example in this section becomes important for the work in Chapter 11, when we test numerically the stereographic path integral simulation for a particle of mass μ on a two-sphere subjected to a dipole-moment-like potential,

$$\mathcal{H} = \frac{L^2}{2\mu R^2} + V_0 \cos\theta. \tag{5.257}$$

Using the equations for the matrix elements of anisotropic potentials, we can derive the following equation for this particular case:

$$\langle l_1\ m_1\,|\mathcal{H}|\ l_2\ m_2\rangle = \frac{\hbar^2 l_1\,(l_1+1)}{2\mu R^2}\delta_{l_1 l_2}\delta_{m_1 m_2}$$

$$+\,V_0\left(\frac{2l_1+1}{2l_2+1}\right)^{1/2}\langle l_1 010\,|l_1 1l_2 0\rangle\,\langle l_1-m_1 10\,|l_1 1l_2-m_2\rangle.$$

$$(5.258)$$

Note that the Hamiltonian matrix now has four subscripts. Before we can send this to a diagonalization routine, we have to reformat the indices of the matrix. A possibility is to arrange all the states in the following manner.

	$l_1=0$ $m_1=0$ $j=1$	$l_1=1$ $m_1=-1$ $j=2$	$l_1=1$ $m_1=0$ $j=3$	$l_1=1$ $m_1=1$ $j=4$	...
$l_2=0$ $m_2=0$ $i=1$	$H_{1,1}$	$H_{1,2}$	$H_{1,3}$	$H_{1,4}$	...
$l_2=1$ $m_2=-1$ $i=2$	$H_{2,1}$	$H_{2,2}$	$H_{2,3}$	$H_{2,4}$	...
$l_2=1$ $m_2=0$ $i=3$	$H_{3,1}$	$H_{3,2}$	$H_{3,3}$	$H_{3,4}$	...
...	...	...	...	...	...

The following snippet of code takes care of the index arrangement, calculates the elements of the potential energy matrix, the Hamiltonian matrix, and calls **eigen** for the diagonalization of real symmetric matrices.

```
      program exactamts2
implicit real*8(a-h,o-z)
parameter (mjmx = 14)
      parameter (N = (mjmx+1)*(mjmx+1))
real*8 h(N,N),v(N,N),z(N,N)
real*8 e(N), d(N)
      write(6,*) 'N = ',N
      i = 1
      j = 1
      do l1 = 0,mjmx
      do m1 = -l1,+l1
      do l2 = 0,mjmx
      do m2 = -l2,+l2
      h(i,j) = 0.d0
      v(i,j) = 0.d0
```

```
call racah(l1,1,l2,-m1,0,-m2,r1)
call racah(l1,1,l2,0,0,0,r2)
  f = sqrt(dfloat(2 *l1+1)/dfloat(2*l2+1))
  vij = f*r1*r2
  if (dabs(vij) .gt. 1.d-8) write(6,1000) l1,l2,m1,m2,i,j,vij
  if (j .eq. i) then
    h(i,j) = dfloat(l1*(l1+1))/(2.d0*207.d0)
  endif
    h(i,j) = h(i,j) + vij
    v(i,j) = vij
    i = i+1
  enddo
  enddo
  j = j+1
  i = 1
  enddo
  enddo
  call eigen(N,N,h,d,z,e,ierr)
  write(6,*) 'ierr =',ierr
  do k=1,25
  write (6,*) k,d(k)
  enddo
1000  format(6i6,f14.7)
  end
```

In the code `exactamts2`, `mjmx` is the maximum number that l_1 and l_2 are allowed. Setting `mjmx` $= 2$ produces the following results for the potential energy.

l_1	l_2	m_1	m_2	i	j	$V_{i,j}/V_0$
0	1	0	0	1	3	0.57735027
1	2	−1	−1	2	6	0.44721360
1	0	0	0	3	1	0.57735027
1	2	0	0	3	7	0.51639778
1	2	1	1	4	8	0.44721360
2	3	−2	−2	5	11	0.37796447
2	1	−1	−1	6	2	0.44721360
2	3	−1	−1	6	12	0.47809144
2	1	0	0	7	3	0.51639778
2	3	0	0	7	13	0.50709255
2	1	1	1	8	4	0.44721360
2	3	1	1	8	14	0.47809144
2	3	2	2	9	15	0.37796447

Choosing atomic units, setting $V_0 = 1$ a.u., the radius of the sphere to 1 bohr, and the mass to 207 a.u., and diagonalizing, yields the following eigenvalues:

degeneracy	E_n
1	-0.93170831
2	-0.86219228
1	-0.79515817
2	-0.79138313
2	-0.72686455
2	-0.71920695
$\vdots$	$\vdots$

The degeneracy of many excited states, indicated by the left column, is the result of the symmetry of the potential energy. But the pattern of degeneracy is nontrivial. Convergence, with respect to the size of the basis set (mjmx in this case), must be carefully checked. A value of 28 produces convergence in the first 25 eigenvalues with at least eight significant figures.

6

Time Evolution in Quantum Mechanics

6.1 Introduction

This is an important chapter because it introduces the temporal degree of freedom. Time evolution opens up opportunities to develop several powerful exact methods that circumvent the basis set exponential growth bottleneck. After introducing the time-dependent Schrödinger equation, the reader is guided to experience the time evolution using the most elementary model of quantum mechanics, the particle in a box. Then, we move on to the matrix representation of the time evolution operator, with the formal solution to the coupled equations and their first order Magnus expansion propagator. These methods are used to tackle the time evolution of quantum systems even when the Hamiltonian is time dependent and the equation is no longer separable. We explore the development of real-time path integrals. A separate chapter is dedicated to the development of computation methods for imaginary time path integrals. In this chapter, we develop two important stochastic ground state methods in imaginary time that are complementary to the path integral of statistical mechanics: the diffusion Monte Carlo (DMC) method and the variational Monte Carlo (VMC) method. The latter approach provides the approximate importance sampling function for the guided DMC algorithms. Two Importance Sampling Diffusion Monte Carlo (IS-DMC) methods are presented.

6.2 The Time-Dependent Schrödinger Equation

Time is a parameter of the equations of motion of classical mechanics; it is measured by comparing the motion of a system (e.g., the motions of planets or satellites) against a standard, which in modern time is the inverse of the frequency of electromagnetic radiation absorbed between two sharp states of the cesium atom. How does time enter into the laws of quantum mechanics? This is a complex question; however, if we neglect relativistic effects, then time-dependent processes in quantum mechanics can be described by,

$$i\hbar\frac{\partial}{\partial t}\Psi\left(x,t\right) = \widehat{H}\Psi\left(x,t\right), \qquad (6.1)$$

where $\widehat{H}$ is the familiar Hamiltonian operator. For bound states, and for cases where the Hamiltonian does not depend on time explicitly, the time dimension is separable from the spacial dimensions and the solution is written as a product of two functions, one a function of time $f(t)$ and the other a function of space $\psi(x)$. Inserting $\Psi(x,t) = f(t)\psi(x)$ into Equation 6.1, operating with respect to time on the left and with respect to space on the right, and dividing through by $f(t)\psi(x)$ produces the following result,

$$i\hbar \frac{1}{f(t)} \frac{\partial}{\partial t} f(t) = \frac{1}{\psi(x)} \widehat{H} \psi(x). \tag{6.2}$$

The L.H.S. is an expression that contains only time, whereas the expression on the right only contains space variables. Generally, the two expressions cannot equal one another unless they are constant. Let us call this constant E, for reasons that will become immediately obvious. Then, we can write two separate differential equations:

$$\frac{1}{\psi(x)} \widehat{H} \psi(x) = E \tag{6.3}$$

and

$$i\hbar \frac{1}{f(t)} \frac{\partial}{\partial t} f(t) = E. \tag{6.4}$$

Equation 6.3 is the familiar time-independent Schrödinger equation encountered in Chapter 5. Equation 6.4 is a simple one-dimensional differential equation that has the following solution:

$$f(t) = \exp\left(\pm i \frac{Et}{\hbar}\right). \tag{6.5}$$

Now demonstrate that states of the system that satisfy the Schrödinger equation have a time-independent probability density. Since, by assumption,

$$\widehat{H} \psi_n(x) = E_n \psi_n(x), \tag{6.6}$$

then

$$\Psi_n(x,t) = \exp\left(-i \frac{E_n t}{\hbar}\right) \psi_n(x), \tag{6.7}$$

and the probability density $\rho(x,t)$ constructed by squaring that overall wavefunction is

$$\rho(x,t) = |\Psi(x,t)|^2 = \exp\left(i \frac{E_n t}{\hbar}\right) \psi_n^*(x) \exp\left(-i \frac{E_n t}{\hbar}\right) \psi_n(x) = |\psi_n(x)|^2. \tag{6.8}$$

The eigenstates of the Hamiltonian are called stationary states because their probability density is time independent, as Equation 6.8 demonstrates.

6.3 Wavepackets, Measurements, and Time Propagation of Wavepackets

So far, the time degree of freedom seems pretty uninteresting. Nothing could be further from the truth, however. As the reader will see shortly, the temporal degree of freedom is connected intimately with the act of measuring properties in quantum mechanics. This connection has led the community to contemplate features of quantum theory that are not yet fully resolved at the time of writing. Consider an electron in a box, 4 bohr in length, for which we measure the position with some means. The details of the actual experiment are not relevant. Suppose the measurement yields a value of 3 bohr with some uncertainty, say ±0.1 bohr. The actual length of the time interval for the second measurement depends on the mass of the particle measured and the uncertainty of the measurement. Assuming we repeat, in a sufficiently small time interval, the measurement of its position, we get the same result, that is, 3 bohr within +0.1 bohr. Furthermore, every time we repeat the two successive measurements separated by a sufficiently small time interval anywhere in the box, we verify that the second measurement is, within the uncertainty, equal to the first. The only way that we can explain these observations is if we accept that the probability density (that is, the square of the wavefunction) looks sharply peaked at the place of the first measurement when the second measurement is carried out. For the first exercise, we are going to demonstrate how the superposition principle, and the mathematical structure of the Hilbert space of solutions of the particle in a box, support such wavefunctions.

```
      program position
      implicit real*8 (a-h,o-z)
      parameter (pi = 3.141592654d0)
      parameter (d_L = 4.d0) ! length of the box in bohr
c This program demonstrates the use of wavepackets and the superposition
c principle to represent the state of a particle in a box a short time
c after a position measurement
      write(6,*) '    Wavepacket for a particle in a box'
      write(6,*) ' Enter the position between 0 and L = ', d_L
      read(5,*) a
      n_max = 2
      write(6,*) ' Enter the number of states involved
                (a whole number)'
      read(5,*) n_max
c  Find the normalization constant
      sumn = 0.d0
      do i=1,n_max
       arg = dfloat(i)*pi*a/d_L
        sarg = sin(arg)
        sumn = sumn+ 2.d0*sarg*sarg/d_L
      enddo
```

```
c Produce a graph of the wavepacket
      x = 0.d0
      dx = d_L/1.0d4
      do j=1,10000
       sum = 0.d0
       do i=1,n_max
        arg = dfloat(i)*pi*a/d_L
        argx = dfloat(i)*pi*x/d_L
        sarg = sin(arg)
        sargx = sin(argx)
        sum = sum + sarg*sargx
       enddo
       psi = sum*sum/sumn
       write(8,*) x, n_max, psi
       x = x + dx
      enddo
      end
```

In the program `position`, we construct a linear superposition of the energy states of the particle in a box of length L that is sharply peaked at $x = a$. This is a perfectly good solution, and, in fact, the only solution that can explain the repeated position measurements. The form of the solution that reproduces the two consecutive measurements of position at a is

$$\Psi\left(x, t=0\right) = A \sum_{n=1}^{n_{max}} \sin\left(\frac{n\pi a}{L}\right) \sin\left(\frac{n\pi x}{L}\right), \tag{6.9}$$

where the normalization constant is

$$A = \left[\frac{L}{2} \sum_{n=1}^{n_{max}} \sin^2\left(\frac{n\pi a}{L}\right)\right]^{-1/2}. \tag{6.10}$$

In Figure 6.1, we graph Equation 6.9 for the $a = 3$ bohr case and for several values of n_{max}. Therefore, in order to reconcile the results of consecutive position measurements, we have to add up a number of eigenstates of the Hamiltonian. A wavefunction constructed with a linear superposition of eigenstates of the Hamiltonian is called a **wavepacket**. The interesting feature of wavepackets is that the probability density obtained by squaring the wavefunction at every value of x is no longer time independent. In order to account for the time dependence after some time $t > 0$, we have to include all the time dependences of each state:

$$\Psi\left(x, t>0\right) = A\left(t\right) \sum_{n=1}^{n_{max}} \exp\left(-i\frac{E_n t}{\hbar}\right) \sin\left(\frac{n\pi a}{L}\right) \sin\left(\frac{n\pi x}{L}\right) \tag{6.11}$$

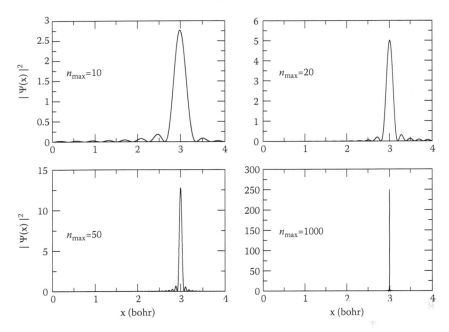

FIGURE 6.1
Particle in a box wavepacket placed at $x = 3$ bohr, with several total number of states included. The convergence to a Dirac delta distribution as $n_{max} \to \infty$ is evident.

$$|\Psi(x,t)|^2 = A(t)^2 \sum_{n=1}^{n_{max}} \sum_{m=1}^{n_{max}} \exp\left[-i\frac{(E_n - E_m)t}{\hbar}\right]$$
$$\times \sin\left(\frac{n\pi a}{L}\right) \sin\left(\frac{m\pi a}{L}\right) \sin\left(\frac{n\pi x}{L}\right) \sin\left(\frac{m\pi x}{L}\right). \quad (6.12)$$

It is still possible to normalize the wavefunction after some time t:

$$\int_0^L dx |\Psi(x,t)|^2 = A(t)^2 \sum_{n=1}^{n_{max}} \sum_{m=1}^{n_{max}} \exp\left[-i\frac{(E_n - E_m)t}{\hbar}\right]$$
$$\times \sin\left(\frac{n\pi a}{L}\right) \sin\left(\frac{m\pi a}{L}\right) \int_0^L dx \sin\left(\frac{n\pi x}{L}\right) \sin\left(\frac{m\pi x}{L}\right),$$
$$(6.13)$$

$$\int_0^L dx |\Psi(x,t)|^2 = A(t)^2 \frac{L}{2} \sum_{n=1}^{n_{max}} \sin^2\left(\frac{n\pi a}{L}\right). \quad (6.14)$$

Time evolution preserves the normalizable property of the wavefunction, and that is important, since we do not have to change the interpretation of the wavefunction to allow for time evolution. Furthermore, the analysis tells us that A is time independent and equal to the value in Equation 6.10. It is

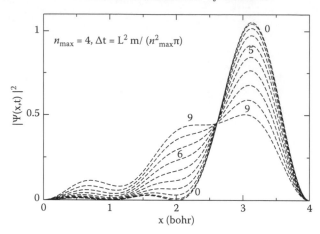

FIGURE 6.2
Time evolution of a 4-state wavepacket placed at 3 bohr and propagated for nine consecutive time steps.

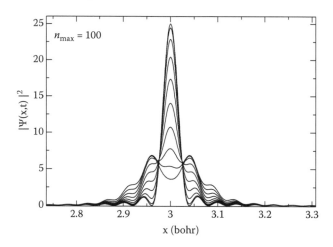

FIGURE 6.3
Time evolution of a 100-state wavepacket placed at 3 bohr and propagated for nine consecutive time steps.

instructive to inspect graphs of Equation 6.12 for several values of time, the initial position a, and the number of states involved. The program `wavepacket` allows the reader to study the behavior of several wavepackets. The results in Figures 6.2 and 6.3 are obtained with 4 and 100 states, respectively. The wavepackets are placed at $a = 3$ in both cases, and are graphed at $t = 0$ and for nine consecutive time steps. Δt is obtained from the maximum energy:

$$\Delta t = \frac{1}{2E_{n_{max}}}. \qquad (6.15)$$

```
      program wavepacket
      implicit real*8 (a-h,o-z)
      parameter (pi = 3.141592654d0)
      parameter (d_L = 4.d0) ! length of the box in bohr
c This program demonstrates the use of wavepackets and the superposition
c principle to represent the state of a particle in a box a short time
c after a position measurement
      write(6,*) '    Wavepacket for a particle in a box'
      write(6,*) '  Enter the initial position between 0 and L = ', d_L
      read(5,*) a
      n_max = 2
      write(6,*) ' Enter the number of states involved (a whole number)'
      read(5,*) n_max
c  Find the normalization constant
      sumn = 0.d0
      do i=1,n_max
       arg = dfloat(i)*pi*a/d_L
       sarg = sin(arg)
       sumn = sumn+ d_L*sarg*sarg/2.d0
      enddo
c Produce a graph of the wavepacket at time t
c tau is the shortest time
      tau = d_L*d_L/(pi*pi*n_max*n_max)
      t = 0.d0
      do k1 = 1, 10    ! ten time steps
      x = 0.d0
      dx = d_L/1.0d4
      do j=1,10000
       sumr = 0.d0
       sumi = 0.d0
       do i=1,n_max
c  mass of an electron (atomic units)
       en = dfloat(i*i)*pi*pi/(2.d0*d_L*d_L)
       arg = dfloat(i)*pi*a/d_L
       argx = dfloat(i)*pi*x/d_L
       sarg = sin(arg)
       sargx = sin(argx)
       sumr = sumr + cos(en*t)*sarg*sargx
       sumi = sumi - sin(en*t)*sarg*sargx
      enddo
       psi = (sumr*sumr + sumi*sumi)/sumn
       write(9,*) x, n_max, psi
       x = x + dx
      enddo
      t = t + tau
      enddo
      end
```

The interpretation of the wavepacket as a solution consistent with our measurements is not so simple. The first position measurement **prepares** a relatively sharp position state as a superposition of as many energy states as needed to reproduce the uncertainty in the measurement. The second rapidly following measurement confirms for us that, indeed, the state is in a sharply peaked position in the box. However, having constructed the position eigenstate, we lost sharpness in our knowledge of the energy. Another way of thinking about the physics of the position measurement and how we lose knowledge of the energy, is to consider that we had to put energy into the system (we had to illuminate it), in order to know where it is with some precision. Energy and position are incompatible observables, and the measurement of one requires the linear superposition of many eigenstates of the other. This is a manifestation of Heisenberg's uncertainty principle. We can compute the average energy for the wavepacket, and the standard deviation of the energy, and show that both are time independent,

$$\langle E\left(t\right)\rangle = A^2 \frac{L}{2} \sum_{n=1}^{n_{\max}} E_n \sin^2\left(\frac{n\pi a}{L}\right),\tag{6.16}$$

$$\langle E^2\left(t\right)\rangle = A^2 \frac{L}{2} \sum_{n=1}^{n_{\max}} E_n^2 \sin^2\left(\frac{n\pi a}{L}\right).\tag{6.17}$$

At the time of the measurement of the position, and for all the successive times, we are ignorant of the energy state of the system. Remarkably, the theory still describes the state of a single particle, not an ensemble; therefore, we cannot say that excited states would, say, in the case of a charged particle, couple with the electromagnetic field, radiate, and cascade to the ground state. In that case, we could collect those photons, measure their wavelengths, and gain knowledge of the energy. Our position measurement would no longer be ideal in that case, since the determination of the wavelengths of emitted radiation are sharp energy measurements. Furthermore, Figures 6.2 and 6.3 demonstrate that as time evolves, we lose sharpness in the position of the system as well.

Now, suppose that we prepare an electron in a box at 3 bohr, and we then measure the energy after some time, t. Only one of the allowed energy values, E_n, can be measured. The theory tells us that E_n will be measured with probability

$$P\left(E = E_n\right) = A^2 \frac{L}{2} \sin^2\left(\frac{n\pi a}{L}\right),\tag{6.18}$$

and all we can do is repeat the measurement many times to confirm Equation 6.18. Indeed, the theory matches the outcome of the experiments, and has survived almost a century of close scrutiny by generations of researchers. After an arbitrary amount of time following a position-energy pair of measurements, the energy measurement can be repeated, and the second energy measurement yields a sharp value identical to the first. Measuring the energy has **collapsed** our position wavepacket into one (and only one) of the energy eigenstates. In other words, measuring the energy prepares the system into a state of sharp energy, and measuring the position prepares the system into a sharply peaked position state.

A conundrum emerges from the quantum theory as it is used to interpret the outcome of measurements. We are only marginally interested in this outstanding problem in quantum mechanics; however, it is an important feature of the theory that any serious student in the field must know. There is nothing in Equation 6.1 that can duplicate the dynamics of the system as it evolves at the time of a measurement from a sharp energy state into a position state, or vice versa. It was Einstein's main contention that this feature of quantum theory violates the locality principle, which is the foundation of special and general relativity. Including the electromagnetic field (which is the medium by which position and energy measurements can be made in the laboratory) does explain why measurements interfere drastically with the states of matter. However, the time-dependent Schrödinger equation solved for the electromagnetic field coupled with the system, still cannot explain the collapsing of the wavepacket as the properties of the system are measured. The field and the original system become a new system that, if left unperturbed, evolves in time and can be prepared in a superposition of energy states, and nothing but an energy measurement will collapse it. Despite this issue, quantum theory as it stands is the most successful theory known to man, in terms of predictive power. Therefore, we will continue to adopt it until the day that a new all-encompassing theory is proposed, perhaps after studying phenomena at a scale several orders of magnitude smaller than the atomic scale. From this future theory, quantum theory should emerge in some limiting way, just as classical mechanics emerges at the macroscopic scale from quantum theory.

Exercises

1. Derive Equations 6.16 and 6.17, and then modify the program `position` to compute these quantities and obtain a value for $\Delta E = \sqrt{\langle E^2 \rangle - \langle E \rangle^2}$.

2. Change the program `position` so that the wavepacket is placed at $a = 1.0$ bohr instead, and verify that Equation 6.9 works as it is supposed to for this new value of a.

3. Modify the program `wavepacket` so that the average position $\langle x \rangle$ as a function of time is computed:

$$
\langle x(t) \rangle = \int_0^L dx\ \Psi^*(x,t)\ x\ \Psi(x,t)
$$

$$
= A^2 \sum_{n=1}^{n_{max}} \sum_{m=1}^{n_{max}} \exp\left[-i\frac{(E_n - E_m)t}{\hbar}\right] \sin\left(\frac{n\pi a}{L}\right) \sin\left(\frac{m\pi a}{L}\right)
$$

$$
\times \int_0^L dx\ x\ \sin\left(\frac{n\pi x}{L}\right) \sin\left(\frac{m\pi x}{L}\right),
$$

where the integral evaluates to,

$$\int_0^L dx\, x \sin\left(\frac{n\pi x}{L}\right) \sin\left(\frac{m\pi x}{L}\right)$$

$$= \begin{cases} \dfrac{L^2}{2\pi^2}\left[\dfrac{\cos\left[(n-m)\,\pi\right]-1}{(n-m)^2} - \dfrac{\cos\left[(n+m)\,\pi\right]-1}{(n+m)^2}\right] & n \neq m, \\[1.5em] \dfrac{L^2}{4} & n = m \end{cases}$$

,

for the $n_{\max} = 100$ case, and follow its value for 200 time steps. Explain what is happening physically, in particular, why is the particle moving at all? There is no force on it! Use your intuition to predict what happens if the wavepacket is placed at 2 bohr, and then run the experiment to verify your hypothesis.

4. Modify the program **wavepacket** so that the autocorrelation function,

$$\left| \int_0^L dx\, \Psi^*\left(x,t\right)\, \Psi\left(x,0\right) \right|,$$

is computed. Use 10 states and 10,000 time steps. You should see a periodic pattern. Determine the time it takes for the autocorrelation to return to one. This time is called the **Poincaré recurrence time**. Graph the wavepacket at the initial time and at the recurrence time. Change the number of states to 12, 13, 14, 15, and 16 and determine how the recurrence time changes with $n_{\max}$.

5. Modify the program **wavepacket** so that the average momentum $\langle p \rangle$ as a function of time is computed:

$$\langle p\left(t\right)\rangle = \int_0^L dx\, \Psi^*\left(x,t\right)\, \hat{p}\, \Psi\left(x,t\right)$$

$$= -i\hbar A^2 \sum_{n=1}^{n_{\max}} \sum_{m=1}^{n_{\max}} \exp\left[-i\frac{(E_n - E_m)\,t}{\hbar}\right] \sin\left(\frac{n\pi a}{L}\right) \sin\left(\frac{m\pi a}{L}\right)$$

$$\times \int_0^L dx\, \sin\left(\frac{m\pi x}{L}\right) \cos\left(\frac{n\pi x}{L}\right),$$

where the integral evaluates to

$$\int_0^L dx\, \sin\left(\frac{m\pi x}{L}\right) \cos\left(\frac{n\pi x}{L}\right)$$

$$= \begin{cases} \dfrac{n}{2}\left[\dfrac{\cos\left[(n+m)\,\pi\right]-1}{(n+m)} - \dfrac{\cos\left[(n-m)\,\pi\right]-1}{(n-m)}\right] & n \neq m \\[1.5em] 0 & n = m. \end{cases}$$

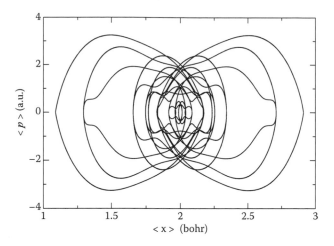

FIGURE 6.4
The average momentum plotted vs the average position for a 10-state wavepacket at 3 bohr.

Use the results from the previous exercise to produce phase space trajectories of the averages, i.e., $\langle p \rangle$ as a function of $\langle x \rangle$. The graph in Figure 6.4 is a plot of $\langle p \rangle$ as a function of $\langle x \rangle$ for a wavepacket created at $a = 3$ bohr, with 10 states and evolved for one recurrence time. After the first recurrence time, the "phase space trajectory" repeats itself.

6. Modify the program **wavepacket** to compute the product $\Delta p \, \Delta x$ as a function of time. The program that results from the previous two exercises is a good starting point for this exercise.

7. Modify the program **wavepacket** so that the propagation takes place in imaginary time, $t = i\tau$. Note that the wavepacket must be renormalized at every time step in this case:

$$\int_0^L dx \, |\Psi(x, \tau)|^2$$

$$= A(\tau)^2 \sum_{n=1}^{n_{\max}} \sum_{m=1}^{n_{\max}} \exp\left[-\frac{(E_n + E_m)\tau}{\hbar}\right] \sin\left(\frac{n\pi a}{L}\right) \sin\left(\frac{m\pi a}{L}\right)$$

$$\times \int_0^L dx \, \sin\left(\frac{n\pi x}{L}\right) \sin\left(\frac{m\pi x}{L}\right),$$

where E_m adds to E_n rather than subtracting from it, since the complex conjugate operation no longer changes the argument of the time evolution exponential function. The normalization constant is now a function of τ:

$$\int_0^L dx\, |\Psi(x,t)|^2 = A(\tau)^2\, \frac{L}{2} \sum_{n=1}^{n_{\max}} \exp\left(-\frac{2E_n\tau}{\hbar}\right) \sin^2\left(\frac{n\pi a}{L}\right),$$

$$A(\tau) = \left[\frac{L}{2} \sum_{n=1}^{n_{\max}} \exp\left(-\frac{2E_n\tau}{\hbar}\right) \sin^2\left(\frac{n\pi a}{L}\right)\right]^{-1/2}.$$

Place a 10-state wavepacket at 3 bohr and monitor the average energy,

$$\langle E(\tau)\rangle = A^2(\tau)\, \frac{L}{2} \sum_{n=1}^{n_{\max}} \exp\left[-\frac{2E_n\tau}{\hbar}\right] E_n \sin^2\left(\frac{n\pi a}{L}\right),$$

for the equivalent of 20, 100, 200, and 1000 recurrence times. Plot the wavepacket at these times. Note that the conservation of energy is violated in imaginary time and consequently, the system never returns to its initial state. The concept of imaginary time sounds more disturbing to beginner students than the collapse of the wavepacket in a measurement. However, the imaginary time evolution of a wavepacket is much simpler to interpret. In Chapter 7, we introduce imaginary time as $\beta\hbar = \hbar(k_{\mathrm{B}}T)^{-1}$, so the superposition of energy states now describes the state of an ensemble of systems that are in some sharp energy state. As τ becomes large, the temperature approaches zero and a large fraction of the ensemble is in the ground state, as you should have found with this exercise. Later in this chapter, we will take advantage of this feature of the imaginary time evolution to produce powerful stochastic algorithms with remarkably simple implementations.

6.4 The Time Evolution Operator

In the last section, we presented the solution to the time evolution problem in quantum mechanics for a simple system, for which the eigenstates are easy to obtain. For a more general case, we need to develop more powerful methods based on matrix representations and Lie algebras. Let $\mathbf{U}(t,0)$ be the matrix representation of the time evolution operator defined as

$$\mathbf{\Psi}_t = \mathbf{U}(t,0)\, \mathbf{\Psi}_0. \tag{6.19}$$

The time evolution operator is a member of a unitary group since, as we have seen, the normalization constant for a wavepacket is time independent. This translates to

$$\mathbf{U}^\dagger(t,0)\, \mathbf{U}(t,0) = \mathbf{1}. \tag{6.20}$$

In the following, the reader will find that the time evolution operator can be represented as the exponentiation of a Lie algebra, in all cases

$$\mathbf{U}(t,0) = \exp\left(-\frac{i}{\hbar}\mathbf{A}\right), \tag{6.21}$$

where, if $\mathbf{A}$ is in the space of Hermitian matrices, then all the members of the time evolution group derived by Equation 6.21 are unitary, as shown in Chapter 4.

6.5 The Dyson Series and the Time-Ordered Exponential Representation

There are, in general, four cases for the solution of the time evolution problem in quantum mechanics:

i. The Hamiltonian is time independent and diagonal. We have seen an example of this case in the previous section.

ii. The Hamiltonian is time independent but not diagonal. In this case, the exact solution of the problem (in matrix notation),

$$i\hbar\frac{d}{dt}\mathbf{\Psi} = \mathbf{H}\mathbf{\Psi}, \tag{6.22}$$

can be computed with,

$$\mathbf{\Psi}_t = \exp\left(-\frac{i}{\hbar}\mathbf{H}t\right)\mathbf{\Psi}_0. \tag{6.23}$$

Therefore, for cases i and ii, the matrix $\mathbf{A}$ in Equation 6.21 is given by

$$\mathbf{A} = \mathbf{H}t. \tag{6.24}$$

iii. The Hamiltonian is time dependent, not diagonal, and the following commutation rule holds for any time t and $t' \neq t$:

$$[\mathbf{H}(t), \mathbf{H}(t')] = 0. \tag{6.25}$$

In this case, the exact solution of Equation 6.22 can be computed with Equation 6.21, and with

$$\mathbf{A} = \int_0^t \mathbf{H}\,dt. \tag{6.26}$$

iv. The Hamiltonian is time dependent, not diagonal, and it does not commute with itself at a time $t' \neq t$:

$$[\mathbf{H}(t), \mathbf{H}(t')] \neq 0. \tag{6.27}$$

In this case, the exact solution of Equation 6.22 can be written formally in a number of interesting ways. Suppose we partition the interval $(0,t)$ into a number of small subintervals, $0 < t_1 < t_2 < \cdots < t_N$, with $t_{N+1} = t$. Then, as $N \to \infty$, we can approximate the time evolution of the wavepacket in the following manner:

$$\mathbf{\Psi}_{t_1} = \left(1 - \frac{i}{\hbar} \int_0^{t_1} \mathbf{H}(t)\, dt\right) \mathbf{\Psi}_0, \tag{6.28}$$

$$\mathbf{\Psi}_{t_2} = \left(1 - \frac{i}{\hbar} \int_{t_1}^{t_2} \mathbf{H}(t)\, dt\right) \mathbf{\Psi}_{t_1}, \tag{6.29}$$

$$\mathbf{\Psi}_{t_2} = \left(1 - \frac{i}{\hbar} \int_{t_1}^{t_2} \mathbf{H}(t)\, dt\right) \times \left(1 - \frac{i}{\hbar} \int_0^{t_1} \mathbf{H}(t)\, dt\right) \mathbf{\Psi}_0. \tag{6.30}$$

If we continue in this manner for all N time steps, we obtain the **Dyson series** expansion of the time evolution operator,

$$\mathbf{U}(t,0) = 1 - \frac{i}{\hbar} \int_0^t \mathbf{H}(t)\, dt + \left(-\frac{i}{\hbar}\right)^2 \int_0^t dt' \int_0^{t'} dt'' \mathbf{H}(t')\, \mathbf{H}(t'') + \cdots. \tag{6.31}$$

Most books rewrite the upper limit of integration in each of the n-tuple integrals using the abstract notation,

$$\int_0^t dt' \int_0^{t'} dt'' \ldots \int_0^{t^{(n-1)}} dt^{(n)} \mathbf{H}(t')\, \mathbf{H}(t'') \ldots \mathbf{H}\left(t^{(n)}\right)$$

$$= \frac{1}{n!} \left(\mathcal{P} \int_0^t dt\, \mathbf{H}(t)\right)^n, \tag{6.32}$$

where $\mathcal{P}$ is the **time-ordering operator**. The $n!$ term is the number of ways the order of integration needs to be changed to fill the time hypercube as the overall range of integration. The time-ordering operator is needed to order the argument of the integrals so that the integration variables $t^{(1)}, t^{(2)} \ldots > t^{(n)}$,

$$\mathcal{P}\left(\mathbf{H}(t^{(1)})\, \mathbf{H}(t^{(2)}) \cdots \mathbf{H}(t^{(n)})\right), \tag{6.33}$$

are relabeled in **causal** order, namely,

$$t^{(1)} > t^{(2)} > \cdots > t^{(n)}. \tag{6.34}$$

The causal ordering of the Hamiltonian matrix is not important if it is time independent, as in cases i and ii, or if it commutes with itself at two different times, as in case iii. The formal expression for the time evolution operator in the most general case can be written as

$$\mathbf{U}(t, 0) = \mathcal{P} \exp\left(-\frac{i}{\hbar} \int_0^t \mathbf{H}(t)\, dt\right). \tag{6.35}$$

Both the Dyson series and the time-ordering exponentiation are ways of expressing the causality of time evolution in quantum mechanics. However, neither expression demonstrates explicitly the unitary property of the time evolution operator, since $\mathbf{U}$ is not expressed as in Equation 6.21. Furthermore, neither expression is very convenient for computation. In the following, we present two other ways to express the time evolution operator; both are very instructive. One of these two ways is also amenable to numerical solutions, which is the main interest of this book.

6.6 The Magnus Expansion

The methods in the rest of the chapter are powerful enough to deal with the most complicated and more interesting cases of those considered in the last section. The Magnus expansion, the first order Trotter expansion and the propagator we derive from it, the path integral, and the DMC are all suitable for tackling problems of the complexity in case iv: $[\mathbf{H}(t), \mathbf{H}(t')] \neq 0$. The Magnus expansion gives an expression that makes explicit both the causality of the theory and the unitary property of the time evolution operator, since it is written as in Equation 6.21. In the Magnus expansion, $\mathbf{A}$ in Equation 6.21 is computed with the following recursive power series expansion:

$$\mathbf{A} = \sum_{n=0}^{\infty} \frac{B_n}{n!} \int_0^t dt^{(1)}\, \mathbf{\Omega}(\mathbf{A}^n, \mathbf{H}), \tag{6.36}$$

where $\mathbf{\Omega}$ stands for a sequence of nested commutators:

$$\mathbf{\Omega}(\mathbf{A}^0, \mathbf{H}) = \mathbf{H}$$

$$\mathbf{\Omega}(\mathbf{A}^1, \mathbf{H}) = [\mathbf{A}, \mathbf{H}]$$

$$\mathbf{\Omega}(\mathbf{A}^2, \mathbf{H}) = [\mathbf{A}, [\mathbf{A}, \mathbf{H}]]$$

$$\vdots \qquad \qquad \vdots$$

$$\mathbf{\Omega}(\mathbf{A}^k, \mathbf{H}) = [\mathbf{A}, \mathbf{\Omega}(\mathbf{A}^{k-1}, \mathbf{H})], \tag{6.37}$$

and the B_n are the Bernoulli's numbers generated by the following power series:

$$\frac{x}{e^x - 1} = \sum_{n=0}^{\infty} \frac{B_n}{n!} x^n, \qquad (6.38)$$

$B_0 = 1$, $B_1 = -1/2$, $B_2 = 1/6$, ... The nice feature of the Magnus expansion is that we can truncate the power series in Equation 6.36 at any desired power, and obtain an approximation to the time evolution operator, which is automatically unitary. Further, the approximation to the time evolution operator can be systematically improved, at least in principle. To the second order, i.e., $n = 1$, the matrix $\mathbf{A}$ is

$$\mathbf{A} = \int_0^t dt^{(1)} \, \mathbf{H}\left(t^{(1)}\right) - \frac{1}{2} \int_0^t dt^{(1)} \, \left[\mathbf{A}, \mathbf{H}\left(t^{(1)}\right)\right] + \dots. \qquad (6.39)$$

Inserting Equation 6.39 back into itself, and ignoring the third and higher order terms, yields the second order approximation of the Magnus expansion:

$$\mathbf{U}(t, 0) = \exp\left(-\frac{i}{\hbar} \int_0^t dt^{(1)} \, \mathbf{H}\left(t^{(1)}\right)\right.$$

$$\left. + \left(\frac{i}{\hbar}\right)^2 \frac{1}{2} \int_0^t dt^{(1)} \int_0^t dt^{(2)} \, \left[\mathbf{H}\left(t^{(2)}\right), \mathbf{H}\left(t^{(1)}\right)\right]\right). \qquad (6.40)$$

6.7 The Trotter Factorization

The Magnus expansion provides an expression that manifests in a concise way, both the causality of temporal evolution and the unitary property of the evolution operator. However, it suffers the same drawbacks as the Dyson series and the time-ordered exponential. All these representations of the time evolution operator are hard to compute numerically. Their implementation involves computing multidimensional integrals. Even without the multidimensional integrals, computing the global time evolution operator from cases ii or iii presents numerical challenges. For example, we could try to make use of the program cmatrix_exponentiation from Chapter 4 and compute $\mathbf{U}$ by exponentiation of $-i \int_0^t dt' \, \mathbf{H}(t')/\hbar$. We would soon discover, however, that depending on the size of t, the power series would require an enormous number of terms to converge properly. The careful reader may have realized that we have already solved a problem of the caliber of case iv for the time evolution operator in Chapter 4, where the split Liouville operator is integrated and exponentiated locally. The present case is similar. Instead of looking for a global solution to the time evolution problem, the idea to develop a stable and efficient algorithm is to derive a temporally local solution that preserves the unitary property of the operator and is easy to compute at the same time. If we partition the interval $(0,t)$ into a number of small subintervals, $0 < t_1 < t_2 < \cdots < t_N$, with $t_{N+1} = t$, then, as $N \to \infty$, we can neglect rigorously the commutators of the Hamiltonian. Simply inserting the first order Magnus expansion yields a stable first order integrator. Formally, this is written as

$$\mathbf{U}(t,0) = \lim_{N \to \infty} \prod_{i=1}^{N} \exp\left(-\frac{i}{\hbar} \int_{t_i}^{t_{i+1}} dt \, \mathbf{H}(t)\right). \qquad (6.41)$$

Equation 6.41 is general, in that it applies for all of the four cases that we have considered and the algorithms that result from it are generally applicable. In the following program, we solve the time evolution of a particle of 207 a.u. of mass in a harmonic well with unit force constant. The reader can easily modify the code to integrate time-dependent Hamiltonians, and even problems in scattering theory.

6.8 The `time_evolution_operator` Program

The program `time_evolution_operator` propagates a Gaussian wavepacket centered 1 bohr away from equilibrium, and with a variance of 0.06 bohr^2. The potential is harmonic and centered at zero. The program `time_evolution_operator` constructs a Discrete Variable Representation (DVR) of the Hamiltonian matrix, prepares an initial state

$$\psi_0 = A \exp\left[-(x-1)^2/0.12\right], \qquad (6.42)$$

and propagates it by implementing Equation 6.41. The time evolution takes place in the subroutine `cmexp`. Note that the time evolution propagator is always a complex matrix, and evolving a state vector in time produces a complex vector. The propagation takes place for 100 small time steps. Even when we begin with a real state vector to represent $\mathbf{\Psi}_0$, we need to propagate a complex vector after the first propagation step. Therefore, the overall operation is (in atomic units so $\hbar = 1$)

$$\exp(-i\mathbf{H}t)(\mathbf{b}_r + i\mathbf{b}_i), \qquad (6.43)$$

where $\mathbf{b}_r$ and $\mathbf{b}_i$ are the real and imaginary parts of a state vector after n propagations. In the program, we do not exponentiate the Lie algebra to get the unitary operator. Instead, a power series expansion for the propagated vector is developed using Taylor's theorem:

$$\exp(-i\mathbf{H}t)(\mathbf{b}_r + i\mathbf{b}_i) = \sum_{n=0}^{\infty} \frac{(-i\mathbf{H}t)^n (\mathbf{b}_r + i\mathbf{b}_i)}{n!}. \qquad (6.44)$$

Breaking up the real and imaginary part of i^n,

$$\exp\left(-i\mathbf{H}t\right)\left(\mathbf{b_r} + i\mathbf{b_i}\right) = \sum_{n=0}^{\infty}(-1)^n \frac{(-\mathbf{H}t)^{2n}\left(\mathbf{b_r} + i\mathbf{b_i}\right)}{(2n)!}$$

$$+ i\sum_{n=0}^{\infty}(-1)^n \frac{(-\mathbf{H}t)^{2n+1}\left(\mathbf{b_r} + i\mathbf{b_i}\right)}{(2n+1)!}, \qquad (6.45)$$

·ibuting through, we get two sums for the real part and two sums for
ıary part:

$$\mathbf{H}t)\left(\mathbf{b_r} + i\mathbf{b_i}\right)$$

$$\left. \sum_0 (-1)^n \frac{(-\mathbf{H}t)^{2n}\,\mathbf{b_r}}{(2n)!} - \sum_{n=0}^{\infty}(-1)^n \frac{(-\mathbf{H}t)^{2n+1}\,\mathbf{b_i}}{(2n+1)!} \right\}$$

$$\left. \sum_{n=0}^{\infty}(-1)^n \frac{(-\mathbf{H}t)^{2n}\,\mathbf{b_i}}{(2n)!} + \sum_{n=0}^{\infty}(-1)^n \frac{(-\mathbf{H}t)^{2n+1}\,\mathbf{b_r}}{(2n+1)!} \right\}. \qquad (6.46)$$

ـ.ue, these four sums are computed using the recursions of two
ـ.ors·

$$\mathbf{b}_{m+1} = -\frac{\mathbf{H}t}{m+1}\mathbf{b}_m, \qquad \mathbf{b}_0 = \mathbf{b_r}, \qquad (6.47)$$

$$\mathbf{c}_{m+1} = -\frac{\mathbf{H}t}{m+1}\mathbf{c}_m \qquad \mathbf{c}_0 = \mathbf{b_i}. \qquad (6.48)$$

```
      program time_evolution_operator
      implicit real*8 (a-h,o-z)
c Time evolution of a Gaussian wavepacket
c using a DVR Hamiltonian (1-d harmonic well)
      parameter (nbas = 1000)
      real*8 a(nbas,nbas)
      real*8 bn(nbas),bnp1(nbas)
      real*8 cn(nbas),cnp1(nbas)
      real*8 psi0(nbas),psi0i(nbas)
      real*8 rpsi(nbas),impsi(nbas)
      do i=1,nbas
c initialize all the arrays
        rpsi(i) = 0.d0
        impsi(i) = 0.d0
        do j=1,nbas
         a(i,j) = 0.d0
        enddo
      enddo
      n_max = 100
      dL = 5.D0
      epsilon = dL/dfloat(n_max)
```

```fortran
      do i=1,n_max
c  set up the Hamiltonian matrix
        x = -0.5*dL + dfloat(i-1)*epsilon
        do j=1,n_max
         if (i .eq. j) then
          a(i,j) = 3.14159**2/(6.*207*epsilon**2) + 0.5*x*x
         else
          sign = -1.d0
          if (mod(abs(i-j),2) .eq. 0) sign = +1.d0
          a(i,j) = sign/(207*epsilon**2*(i-j)**2)
         endif
        enddo
       enddo
c form the initial distribution
       do i=1,n_max
        x = -0.5*dL + dfloat(i-1)*epsilon
        psi0(i) =  exp(-(x - 1.d0)**2/0.12)
        psi0i(i) = 0.d0
        sum = sum + psi0(i)*psi0(i)
       enddo
       do i=1,n_max
        x = -0.5*dL + dfloat(i-1)*epsilon
        psi0(i) = psi0(i)/sqrt(sum)
        write(8,*) x,psi0(i)*psi0(i),0.5*x*x
       enddo
c get the real and imaginary part of the propagator
       t = 1.0d0
       do ii=1,50  !  time propagation loop
       call cmexp(nbas,n_max,a,psi0,psi0i,bn,bnp1,cn,cnp1,rpsi,impsi,t)
c  propagate the initial solution
        do i=1,n_max
         x = -0.5*dL + dfloat(i-1)*epsilon
         write(9,*) x,(rpsi(i)*rpsi(i)+ impsi(i)*impsi(i))
         psi0(i) = rpsi(i)
         psi0i(i) = impsi(i)
        enddo
       enddo    !  time propagation loop
c
       end
       subroutine cmexp(NMAX,N,a,ps,psi,bn,bnp1,cn,cnp1,sumr,sumi,t)
       implicit real*8 (a-h,o-z)
       real*8 a(NMAX,NMAX),ps(NMAX),psi(NMAX)
       real*8 bn(NMAX),bnp1(NMAX),cn(NMAX),cnp1(NMAX)
       real*8 sumr(NMAX),sumi(NMAX)
c
```

```
c This program computes exp(-iA*t) * b where i = sqrt(-1),
c and A is real and symmetric, b a real vector.
c  On input
c              NMAX = physical size of all arrays
c              N    = number of basis sets
c              ps   = input vector      b
c              bn   = scratch array used for the exponentiation
c              bnp1 =      "       "       "
c              t =  time in inverse hartree
c On output
c              sumr = real part of the exp(-iAt) * b    vector
c              sumi = imaginary part of the exp(-iAt) * b    vector
        do i=1,N    !initialize
          bnp1(i) = 0.0d0
          cnp1(i) = 0.0d0
          bn(i)   = ps(i)
          cn(i)   = psi(i)
          sumr(i) = bn(i)
          sumi(i) = cn(i)
        enddo
        mtop = 50
10      do m = 1,mtop !exponentiate
          do i=1,N        ! do no more than mtop iterations
            bnp1(i) = 0.d0
            cnp1(i) = 0.d0
            do k=1,N
            bnp1(i) = bnp1(i) - a(i,k)*t*bn(k)/dfloat(m)
            cnp1(i) = cnp1(i) - a(i,k)*t*cn(k)/dfloat(m)
            enddo
            if (mod(m,2) .eq. 0) then
             sign = -1.d0
             if (mod(m/2,2) .eq. 0) sign = +1.d0
              sumr(i) = sumr(i) + sign*bnp1(i)
              sumi(i) = sumi(i) + sign*cnp1(i)
             else
             sign = -1.d0
             if (mod((m-1)/2,2) .eq. 0) sign = +1.d0
              sumi(i) = sumi(i) + sign*bnp1(i)
              sumr(i) = sumr(i) - sign*cnp1(i)
            endif
          enddo
          do i=1,N     ! set bnp1 to bn
            bn(i) = bnp1(i)
            cn(i) = cnp1(i)
          enddo
```

```
        enddo       ! iteration loop ends here

        return
2000    format(10f12.6)
2010    format(2i4,2f12.6)
        return
        end
```

The program `time_evolution_operator` is far from a user-friendly black box that can be canned into a fail-safe routine. Changes in either the mass or the potential require careful testing. Smaller mass values require smaller values for the time increment. Without proper values for the time increment, n_max and `mtop`, the propagated vector can explode, causing floating point exceptions and sudden job crashes. A potential other than the quadratic form will cause significant incoherence of the wavepacket, and a much larger basis set is necessary to obtain a smooth graph of the wavefunction only after a handful of steps. The algorithm is stable when optimal values of t, n_max, and `mtop` are used, and can be run for extremely long times. However, the `write(9,*)` line should be commented out or an enormous amount of output will be generated onto the device file. This could fill the disk on your workstation and cause the operating system to crash. Real-time evolution of wavepackets requires some experience; the next set of exercises are designed to provide some hands-on learning and are sorted in order of increasing difficulty.

Exercises

1. Write out explicitly the $m = 0, 1, 2, 3$ terms of Equations 6.47 and 6.48, and insert these into Equation 6.46. In plugging terms into Equation 6.46, it is important to realize that for even values of m, only the first sum of the real and imaginary part of the evolved vector needs to be used. Whereas for odd values of m, the second sum of the real and imaginary part of the evolved vector are used. The resulting expression should coincide with the first four terms ($n = 0, 1, 2, 3$) from Equation 6.44. Verify all these statements, then consult the code in the subroutine `cmexp` and make sure the recursion procedure matches the algorithm made explicit by these equations.

2. The algorithm in the program `time_evolution_operator` has two different sources of truncation error, the basis set truncation error that governs the size of the matrix to be exponentiated, and the number of terms in the power series representation of the exponential. Tweak the variables n_max and `mtop`, and verify that the wavefunctions are converged at all times, by visual inspection of the wavefunction at several time snapshots.

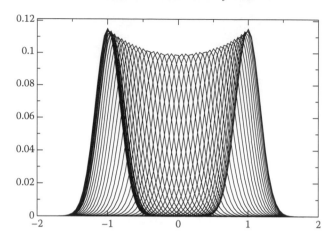

FIGURE 6.5
Time evolution of a Gaussian wavepacket (c.f. Equation (6.42)) evolving for 50 hatree^{-1}. The mass is 207 a.u. and is inside a harmonic well with unit force constant.

3. The wavepacket in Figure 6.5 is quite coherent, meaning that its shape remains nearly unchanged as it moves back and forth. Place the wavepacket at $x = 0$ and repeat the calculation for 100 steps. Does the wavepacket become less coherent? Does it become broader?

4. The Hamiltonian for the example used to test `time_evolution_operator` is time independent. Therefore, the propagated wavefunctions should converge for any time interval, provided `mtop` is adjusted to ensure that the exponentiation part of the algorithm is converged. Verify this statement by letting the solution evolve for 50 hartree^{-1} using 200, 100, 50, and 40 steps. This can be accomplished by changing the top value of the time propagation loop, and fixing the value of `t` accordingly on the code line immediately above the beginning of the time propagation loop.

5. Modify `time_evolution_operator` so that the average energy $\langle E \rangle$ and its fluctuation $\langle E^2 \rangle - \langle E \rangle^2$ are computed as a function of time. These quantities can be computed by two matrix-vector multiplications,

$$\langle E \rangle = (\mathbf{b_r} + i\mathbf{b_i})^\dagger \mathbf{H} \, (\mathbf{b_r} + i\mathbf{b_i}).$$

The expression contains real and imaginary parts,

$$\langle E \rangle = \mathbf{b_r^T H b_r} + \mathbf{b_i^T H b_i} + i \left(\mathbf{b_r^T H b_i} - \mathbf{b_i^T H b_r} \right).$$

Of course, the imaginary part of the average energy should be zero, but this needs to be verified numerically, and it is a good test to make sure everything is running as expected. The average energy and its fluctuation should be constant. With the mass as given, you should get $\langle E \rangle = 0.5351$ hartree and $\langle E^2 \rangle = 0.3164$ hartree^2.

6. Compute the average position $\langle x \rangle$ and its fluctuation $\langle x^2 \rangle - \langle x \rangle^2$,

$$\langle x \rangle = (\mathbf{b_r} + i\mathbf{b_i})^\dagger \, \mathbf{x} \, (\mathbf{b_r} + i\mathbf{b_i}),$$

where the matrix representation of the position in the DVR is diagonal with the eigenvalues equal to x_i. Comment out the `write(9,*)` line and with `t` set to 1, propagate for 5000 steps. Graph $\langle x \rangle$ as a function of time, then perform the Fourier transform (see Chapter 2). Interpret the physical meaning of your results. How are the frequencies related to the classical frequency $\omega - \sqrt{k/m}$ and to the eigenfrequencies of the quantum harmonic oscillator? Examine Equation 6.12, then use the expression for the energy levels of a harmonic oscillator in Chapter 5 to determine if the frequencies are what you expected.

7. Change the code for the DVR matrix portion of the code, so that the quartic double well potential with $\gamma = 0.9$ is used instead. At $t = 0$, prepare a particle with 207 a.u. of mass in a Gaussian wavepacket state in the upper well, symmetrically about $x = -\alpha$ and with a variance of ± 0.12 bohr^2. Check that the average energy is well below 1.0 hartree, (it should be about 0.3 hartree), then watch part of the wavepacket penetrate the barrier (a process called **tunneling**) and come out on the other well. Be sure to use a large enough basis set, and that a sufficient number of recursions are performed in the exponentiation. A time increment of 0.1 a.u. is better. The wavepacket can be written to the device file every 20 propagations by placing the `write(9,*)`...statement inside an `if (mod(ii,20) .eq. 0) then` branching statement. Repeat with smaller masses, making sure that `n_max` and `m_top` are optimized with every new mass value.

6.9   Feynman's Path Integral

With Equation 6.41, we are very close to developing the path integral representation of quantum mechanics proposed by R. P. Feynman. The path integral is an operator free representation of the time evolution of states, and it is based on the following two principles formulated by R. P. Feynman in his 1948 paper.

- *If a set of ideal position measurements is made to determine if a particle follows a certain path in a region R of space as time moves forward, then the probability $p(R)$ that the outcome of the position measurements* *is affirmative is the result of the sum of the contribution from all the possible paths in the region.*

- *Each path contributes the same amount in amplitude to the probability, but the phase contribution of each path is the action associated with the path as the time integral of the Lagrangian $\mathcal{L}(\dot{x}, x)$.*

$$S[\dot{x}(t), x(t)] = \frac{1}{\hbar} \int dt \, \mathcal{L}(\dot{x}, x). \tag{6.49}$$

The Lagrangian in Equation 6.49 is the classical Lagrangian introduced in Chapter 2.

Both of these principles are consistent with the postulates of quantum mechanics and give us a way of computing the probability $p(R)$. The path integral formulation of quantum mechanics is the most transparent way to see how classical mechanics emerges from the laws of quantum physics. In the second principle, the action taken by the system is not the least action that defines the classical trajectory, as shown in Chapter 2. Instead, all paths can contribute to $p(R)$. However, as we construct the probability $p(R)$, the paths that lie closest to the classical path will interfere constructively for the most part with one another, whereas paths that are far away from the classical path will interfere destructively for the most part with the other paths nearby. To construct the probability $p(R)$, Feynman partitions the time into N finite segments equally spaced by Δt, as we do to derive Equation 6.41. Then, the action can be broken up as a sum,

$$S[\dot{x}(t), x(t)] = \sum_i S(x_{i+1}, x_i) \approx \frac{\Delta t}{\hbar} \sum_i \mathcal{L}\left(\frac{x_{i+1} - x_i}{\Delta t}, \frac{x_{i+1} + x_i}{2}\right). \tag{6.50}$$

Note that as Δt becomes smaller, we can approximate the integral for the action by assuming that the Lagrangian is constant and expand it about the midpoint of the linear path, $x_i, t_i \to x_{i+1}, t_{i+1}$. The probability $p(R)$ is obtained in the limit $\Delta t \to 0$ of the following multidimensional integral,

$$p(R) = \lim_{\Delta t \to 0} A \int_R dx_1 \dots \int_R dx_N \exp\left[i \sum_i S(x_{i+1}, x_i)\right], \tag{6.51}$$

where A is a normalization constant obtained by setting the integral of $p(R)$ over all the space equal to one. The relation of the probability $p(R)$ with the time-dependent wavefunction is,

$$p(R) = \int_R dx \, \Psi^\dagger(x, t) \, \Psi(x, t). \tag{6.52}$$

Equation 6.51 provides a way to formulate the path integral expressions for the wavefunction and its complex conjugate

$$\Psi(x_N, t) = \lim_{\Delta t \to 0} A \int_R dx_{N-1} \int_R dx_{N-2} \dots \exp\left[i \sum_{i=-\infty}^{N-1} S(x_{i+1}, x_i)\right], \tag{6.53}$$

$$\Psi^\dagger(x_N, t) = \lim_{\Delta t \to 0} A \int_R dx_{N+1} \int_R dx_{N+2} \cdots \exp\left[i \sum_{i=N}^\infty S(x_{i+1}, x_i)\right].$$

(6.54)

These remarkable equations tell us that the probability that a position measurement at time t yields the result $x \in R$, is produced by the sum of all past histories (which produce Ψ) and all future histories (which produce $\Psi^\dagger$). The restrictions on the regions of integration R for Ψ and $\Psi^\dagger$ depend on all past (and future!) position measurements. The reference to future measurement impacting on the present measurement should be understood as nothing more than a manifestation of Heisenberg's uncertainty principle as it applies to the time-position pair, since these are incompatible properties. Note, however, that in nonrelativistic quantum mechanics time is a parameter, namely, there is no time operator that measures time. In Feynman's theory this difficulty is resolved because the formalism is operator free.

The derivation of the path integral representation of quantum mechanics is highly heuristic, and guided for the most part by Feynman's great intuitive abilities. Nevertheless, Feynman's path integral representation of quantum mechanics was immediately accepted, since he demonstrated in the same article that Equation 6.53 satisfies Equation 6.1, the time-dependent Schrödinger equation. It is possible to use Equation 6.53 and find a path integral representation of the matrix elements of the time evolution operator in the position representation

$$\langle x'|U|x\rangle = \lim_{\Delta t \to 0} A \int_{-\infty}^\infty dx_1 \cdots \int_{-\infty}^\infty dx_N \exp\left[i \sum_{i=1}^N S(x_{i+1}, x_i)\right], \quad (6.55)$$

where $x_0 = x, x_{N+1} = x'$, and where,

$$A = \left(\frac{i2\pi\hbar}{m}\right)^{-N/2}, \quad (6.56)$$

for a particle of mass m in $\mathbb{R}^1$.

Feynman's path integral formulation can be used for a number of interesting problems, including the fully relativistic quantum field theory and the scattering of elementary particles. One could propose using random numbers distributed according to the exponential in Equation 6.55 to compute average properties and overcome the exponential growth bottleneck that algorithms based on basis sets and matrix quantum mechanics suffer. Unfortunately, stochastic methods based on the real-time path integral are plagued with the infamous **sign problem**. Generating distributions of random numbers with $\exp\left[i\sum_{i=1}^N S(x_{i+1}, x_i)\right]$ as the importance sampling function is not possible since the function fluctuates from $+1$ to -1. Consequently, stochastic simulation of the real time path integral are challenging for any finite propagation time. The same oscillations that constructively and destructively interfere with path contributions, make any attempt to generate the correct distribution of random numbers futile. Without importance sampling, the statistical fluctuation of Monte Carlo integration for all but the smallest number of dimensions becomes overwhelming. Nevertheless, since the mid-1980s, chemists have found a large number of applications for Feynman's path integral. The majority of these are in statistical mechanics, and in this book we concentrate our attention on the imaginary time propagation by path integral because there, the sign problem does not exist. Chapter 7 is dedicated to that topic. Furthermore, it is possible to simply use a classical path, insert it into Feynman's path integral expression, and produce a semiclassical approximation to the time evolution propagator. The initial value representation and the centroid path integral methods are examples of these. More recently, other groups have used the coherent state path integral to formulate real-time stochastic algorithms for time propagation. Solutions to the time-dependent Schrödinger equation, in as many as 24 dimensions, have been reported with the Matching Pursuit Split Operator Fourier Transform (MP-SOFT) algorithm, for example [859, 860, 861, 862]. These novel techniques are promising, and it may be possible in the near future to implement them to simulate gas phase clusters, liquids, and other forms of condensed matter in real time.

6.10   Quantum Monte Carlo

Chapter 7 makes it clear that systems which are predominantly in the ground state are hard to simulate with path integrals. The DMC and the IS-DMC are methods that naturally produce the ground state of systems of spinless particles (nonfermionic in nature) without trouble. Therefore, atomic and molecular clusters composed of closed shell species of, in principle, any size can be investigated. The application to molecular clusters requires some additional work, if the molecules have rigid degrees of freedom. Amusingly, it is relatively less difficult to extend the DMC approach to alternate coordinate systems, and with holonomic constraints. The same exercise is considerably more challenging for the path integral, especially when not all internal degrees of freedom are justifiably "rigid" by a reasonable adiabatic approximation.

We begin the development of Quantum Monte Carlo methods by rewriting Equation 6.1 in imaginary time, $t = i\tau$. For a free particle in $\mathbb{R}^1$ for example, Equation 6.1 reads,

$$-\hbar \frac{\partial}{\partial \tau} \Psi(x, t) = \frac{\hbar^2}{2m} \frac{\partial^2}{\partial x^2} \Psi(x, t). \tag{6.57}$$

Equation 6.57 is isomorphic to the diffusion equation. Simulating diffusion with random numbers is a relatively simple process. Equation 6.57 with the initial condition,

$$\Psi(x, 0) = \Psi_0(x), \tag{6.58}$$

yields the following two linearly independent solutions,

$$\Psi(x, t) = \Psi_0(x) \times \exp\left(\pm \frac{mx^2}{2\hbar^2 \tau}\right). \tag{6.59}$$

The solution with the positive argument in the exponential is rejected on physical grounds, since the amount of diffusing "medium" cannot grow without bound. To simulate diffusion, all we have to do is generate random numbers that have a Gaussian distribution with a standard deviation

$$\sigma = \hbar \sqrt{\frac{\Delta \tau}{m}}, \tag{6.60}$$

where $\Delta \tau$ is the imaginary time increment. The proposal is to employ a population of pseudoparticles whose only purpose is to create a distribution that matches some starting state, $\Psi(x, 0)$. Each pseudoparticle is completely devoid of any physical meaning, only the distribution of position of the entire population has the physical interpretation of a ground state wavefunction once equilibrium is reached.

In order to reach equilibrium, it is necessary to consider a potential energy function, otherwise Equation 6.59 predicts that diffusion will continue without bounds in both directions indefinitely. Equation 6.57 becomes,

$$-\hbar \frac{\partial}{\partial \tau} \Psi(x, \tau) = \frac{\hbar^2}{2m} \frac{\partial^2}{\partial x^2} \Psi(x, \tau) - V \Psi(x, \tau). \tag{6.61}$$

Equation 6.61 is still isomorphic to a diffusion equation. The last term on the R.H.S. of Equation 6.61 acts as a source or sink, depending on the sign. Therefore, to guide the equilibration process toward the ground state, we need to perform a simulation that grows in areas "favored" by the potential energy, and decreases in regions that are unfavorable. To define favorable vs. unfavorable regions of space, it is necessary to introduce a reference energy value, V_{ref}, against which the potential energy of each pseudoparticle is measured before the decision to grow or annihilate is made. V_{ref} is a "guess" of the ground state energy that is refined at every diffusion step. The growth-annihilation step of the simulation is referred to as the **branching process**. After the population is propagated by a step Δx distributed according to Equation 6.59, and each pseudoparticle is grown or annihilated according to

$$w_i = \exp\left[-\left(V(x_i) - V_{\text{ref}}\right) \Delta \tau\right], \tag{6.62}$$

the value of V_{ref} is adjusted in order to maintain a constant population of pseudoparticles. Then, V_{ref} approaches the ground state energy, and the population distribution approaches the absolute ground state wavefunction. Usually, the weight, w_i, is taken to be an integer, with values of 0, 1, or 2.

To adjust the reference energy value that determines the branching process in order that the population remains approximately constant around a target value of N_0, Anderson proposed to fix V_{ref} after every step with

$$V_{ref} = \langle V \rangle - (\Delta\tau)^{-1} \left(\frac{N_\tau}{N_0} - 1 \right), \tag{6.63}$$

where N_τ is the population number at time, $\tau = M\Delta\tau$, after propagation and branching has taken place,

$$N_\tau = \sum_{i=1}^{N_{\tau-\Delta\tau}} w_i. \tag{6.64}$$

Recently, a more stable feedback strategy for the reference energy has been proposed by Umrigar *et al.* [402],

$$V_{ref} = \langle V \rangle - \log\left(\frac{N_\tau}{N}\right), \tag{6.65}$$

which suppresses the exponential explosion or implosion of the population if the estimate of V_{ref} is far off or fluctuates excessively.

A statistically better-behaved estimator for the ground state energy is the population average of V.

$$\langle E \rangle_\tau = \frac{\displaystyle\sum_{i=1}^{N_{\tau-\Delta\tau}} w_i \, V\left(x_i\right)}{\displaystyle\sum_{i=1}^{N_{\tau-\Delta\tau}} w_i}, \tag{6.66}$$

where x_i is the position of pseudoparticle i, after it is diffused and grown (or annihilated) with w_i. The upper limits of the sums is the number of pseudoparticles in the previous step of the algorithm. This number is updated with the denominator on the R.H.S. of Equation 6.66 before starting the next diffusion-branching cycle for the population. In the program dmc, we set up a population of 10,000 pseudoparticles distributed sharply at $x = 0$. The physical system is a particle in $\mathbb{R}^1$, with a mass of 1 a.u., subjected to a confining potential. The potential is computed in the subroutine pot, which must be supplied before compiling. The random number generator ransi from Chapter 3 must be appended as well. The diffusion step distribution is generated using the rejection technique introduced in Chapter 3. Adjustments to V_{ref} take place according to Equation 6.65. The graph in Figure 6.6 is obtained using a harmonic potential symmetric about $x = 0$ and with unit force constant.

```
program dmc
implicit real*8 (a-h,o-z)
```

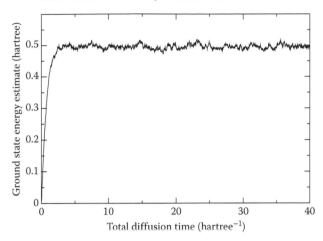

FIGURE 6.6
Imaginary time dependence of the population average potential for a particle of unit mass inside a harmonic well with unit force constant. The time increment is set at 0.02 a.u.

```
      parameter(MNX = 30000)
      parameter(gamma= 0.9d0)
      real*8 x(MNX),y(MNX),xr(2),xn(2)
      integer m(MNX)
c One dimensional diffusion monte Carlo method.
c unguided first order branching
      MN = 10000
      dm = 1.0d0
      dMN = dfloat(MN)
      iseed =  secnds(0.)
      call ransi(iseed)
      sume = 0.d0
      sume2 = 0.d0
      vref = 0.d0
      do i=1,MN
       x(i) = 0.d0
       m(i) = 0
      enddo
      deltax = 0.013
      ve = deltax
      ne = 0
      nrej = 0
      mv = 1000
      mw = 1000
      write(6,*) 'Enter dt'
```

```
      read(5,*) dt
c nw = number of "warm up steps",
c nv = number of steps to average properties
      do k=1,mv+mw
       sumv = 0.d0
       isum = 0        ! isum = population at step k
       do i=1,MN
c rejection method used to generate diffusion steps with a Gaussian
c distribution
         ne = ne + 1
         call ransr(xn,2)
         dx = deltax*(xn(1)-0.5d0)
         vet = ve + dx
         at = (dm*vet*vet)/(2.d0*dt)
         a = (dm*ve*ve)/(2.d0*dt)
         q = exp(-at+a)
         if (xn(2) .lt. q) then
          ve = vet
          a = at
         else
          nrej = nrej +1
         endif
c self adjusting step (used to generate the diffusion step)
         if (mod(ne,1000) .eq. 0 .and. k .lt. mw) then
          if (nrej .gt. 550) then
          deltax = deltax*0.9d0
          endif
          if (nrej .lt. 450) then
           deltax = deltax/0.9d0
          endif
          nrej = 0
         endif
c diffusion
         x(i) = x(i) + ve
         xin = x(i)
         call pot(xin,v)
c branching
         w = dexp(-(v - vref)*dt)
         if (w .gt. 10.d0) w = 10.d0
         call ransr(xr,1)
         m(i) = min(int(w + xr(1)),2)
         sumv = sumv + dfloat(m(i))*v
         isum = isum + m(i)
       enddo
       av = sumv/dfloat(isum)
```

```
c adjust vref [C. J. Umigar, et. al. JCP 99, 2865 (1993).]
c          vref = av -(dfloat(isum)/dMN - 1.d0)/dt
           vref = av -log(dfloat(isum)/dMN)
           write(8,*)k*dt, av,isum,vref
c correct the array for the creation - destruction process
           index = 0
           if (k .gt. mw) then
             sume = sume + av/dfloat(mv)
             sume2 = sume2 + av*av/dfloat(mv)
           endif
           do i=1,MN
             do j = 1, m(i)
               index = index + 1
               y(index) = x(i)
             enddo
           enddo
           MN = isum
           do i=1,MN
             x(i) = y(i)
             m(i) = 1
           enddo
         enddo                ! time step
       end
```

We could have used the Box–Muller algorithm in this program to generate the Gaussian random numbers. We have chosen the rejection technique instead, since this takes us a step closer toward developing the DMC and the IS-DMC extensions to curved manifolds. These developments are in Chapter 11.

Exercises

1. Verify that the function in Equation 6.59 satisfies the differential equation (Equation 6.57).

2. Modify and run **dmc** so that the variable **ve**, the diffusion step size, is written to a file. Collect one million values for the diffusion step size and compute its histogram. Verify that the distribution is Gaussian, centered at zero, and that the standard deviation is $\sqrt{\Delta\tau/m}$.

3. Obtain a histogram of the population for several values of the loop variable **k** during the first few diffusion steps. Let all the members of the population diffuse and branch first, then modify the program **histogram** of Chapter 3 so that it becomes a subroutine that prepares and writes histograms to a separate device file for every value of **k** from 1 to 50. Then, modify the program once more so that the distribution of the population of pseudoparticles is collected at the end of the simulation.

Compare your result with the actual ground state wavefunction of the harmonic oscillator,

$$\psi(x) = \left(\frac{\alpha}{\pi}\right)^{1/4} \exp\left(-\frac{\alpha x^2}{2}\right),$$

where

$$\alpha = \sqrt{\frac{m k}{\hbar^2}}.$$

Note that the program `histogram` of Chapter 3 does not normalize the distribution appropriately if the correct normalization constant for the histogram is the square root of the inverse of the **sum of the squares** of the count in each class.

4. The diffusion algorithm, as it stands, has several sources of errors.

- Statistical fluctuations
- Failure to reach the asymptotic distribution
- The time step bias
- Population bias
- Nonconfining potential energy surface bias

It is evident in Figure 6.6 that the population requires a number of steps to reach its steady state shape. This equilibration time can become very long when the potential energy surface is rugged, and the starting distribution is from the ground state wavefunction. Simulations that require exceedingly small values of the time increment to converge, aggravate the problem, and are inefficient. Molecular clusters simulated with Cartesian coordinates for all atoms have both very rugged surfaces and require very small time increments. Special methods to take care of these issues are developed in Chapters 11 and 12. The simplest approach to reduce the number of steps to reach the asymptotic distribution is to begin the DMC simulation from a good guess of the ground state wavefunction. In the next section, we develop one possible way to obtain a good initial guess of the wavefunction.

Statistical fluctuations can be estimated by averaging the energy over a number of independent runs. The ground state energy estimate in Figure 6.6 approaches the steady state in about 4 hartree^{-1}, and from there it fluctuates visibly about the correct value, 0.5 hartree. Change the population size from 10^5 to one million, then inspect the corresponding energy graph and verify that the fluctuations are approximately one-third smaller.

Change back to a population of 10^5 pseudoparticles. Estimate the time step bias by repeating the simulation with $\Delta\tau =$ 1.0, 0.4, 0.2, 0.1, 0.02

hartree^{-1}. Find out how the time step bias depends on the value of the time increment. Then, change the branching function to

$$w_i = \exp\left[-\frac{1}{2}\left(V\left(x_i\right) + V\left(x_i'\right) - 2V_{\text{ref}}\right)\Delta\tau\right],$$

where x_i' is the position of pseudoparticle i before the diffusion step. Repeat the simulation with $\Delta\tau = 1.0$, 0.4, 0.2, 0.1, 0.02 hartree^{-1} and verify that the time step bias is now second order in the time step.

5. Use the quartic double well potential with $\gamma = 0.9$ and simulate a particle with a mass of 10 a.u. Find the ground state energy and wavefunction. Begin the simulation by placing all the pseudoparticles at $x = 1$. Check the accuracy of the ground state energy and wavefunction by running the DVR algorithm from Chapter 5.

6.11 A Variational Monte Carlo Method for Importance Sampling Diffusion Monte Carlo (IS-DMC)

We mentioned in the last section that the time it takes for DMC simulations to reach a steady state can be shortened considerably if one begins with a "good guess" for the ground state wavefunction. In this section, we present the variational principle of quantum mechanics, and we propose a simple set of trial wavefunctions that promises to ameliorate simulations on rugged surfaces. Soon the reader will experience just how rugged potential energy surfaces can be, even for relatively simple mathematical models like the Lennard–Jones interaction. Having a good guess for the wavefunction can do more for DMC than setting up favorable starting distributions. It can be used to either guide the diffusion process with drift terms, or to modify the branching process. Guided DMC simulations have a significantly smaller statistical error, often smaller by orders of magnitude. Furthermore, guided DMC simulations have a significantly smaller bias from the nonconfining nature of potential energy surfaces. This latter source of error can be a serious problem in the simulation of clusters, since all the potential energy surfaces that describe the interactions among the constituents of a cluster are nonconfining.

The development in this section is based on the **variational principle** of quantum mechanics. The variational principle is one of the ways that one can solve differential equations exactly, provided the formalism is created so that systematic improvements to the solution can be performed. Let us attempt to solve the time-independent Schrödinger equation with a trial function $\psi_\alpha\left(x\right)$, where α is some parameter that we vary. If we assume that the trial solution can be constructed to be flexible, we can show that the estimate of the **local**

energy

$$\langle E \rangle_\alpha = \frac{\int dx \, \psi_\alpha^* (x) \, \hat{H} \psi_\alpha (x)}{\int dx \, \psi_\alpha^* (x) \, \psi_\alpha (x)}, \tag{6.67}$$

is stationary, meaning, there is a value of α for which

$$\frac{\partial \langle E \rangle_\alpha}{\partial \alpha} = 0. \tag{6.68}$$

Furthermore, one can show that if a sufficiently large set of parameters are used to construct a trial wavefunction, when these parameters are optimized, the "trial" wavefunction satisfies the time-independent Schrödinger equation. The proof of this statement can be found in many quantum mechanics books, and we do not reproduce it here. The easiest approach is to write $\psi_\alpha (x)$ as a linear superposition of basis sets in a Hilbert space. Then, the optimization of the parameter set $\{\alpha\}$ is carried out by diagonalization. The variation principle is equivalent to the Schrödinger equation. When a restricted number of parameters is used, Equation 6.68 still applies and yields an upper bound for the true value of the energy. It is also possible to approximate excited states with the variation principle. Our trial wavefunction contains one parameter and is designed so that the variational energy in Equation 6.67 can be computed in many dimensions and with systems that are characterized by rugged potential energy surfaces. The wavefunction in Equation 6.69 allows us to explore the rugged contour of the potential for a cluster by using parallel tempering. This trial wavefunction, while far from accurate, is ideal to provide the approximation to the true wavefunction and to guide to the important region of configuration space for DMC. Our proposal for the trial wavefunction is,

$$\psi_\alpha (x) = \exp \left(-\frac{\alpha V}{2} \right), \tag{6.69}$$

where V is the system's potential. Then, Equation 6.67 can be written as

$$\langle E \rangle_\alpha = \frac{\int dx \, E_{\mathrm{L}} (x) \exp (-\alpha V)}{\int dx \, \exp (-\alpha V)}, \tag{6.70}$$

where,

$$E_{\mathrm{L}} (x) = \psi_\alpha^{-1} (x) \, \hat{H} \psi_\alpha (x). \tag{6.71}$$

For problems in $\mathbb{R}^1$, this yields,

$$E_{\mathrm{L}} (x) = -\frac{\hbar^2}{2m} \left[-\frac{\alpha}{2} \frac{d^2 V}{dx^2} + \left(\frac{\alpha}{2} \frac{dV}{dx} \right)^2 \right] + V. \tag{6.72}$$

The program vmc implements the parallel tempering strategy to compute the variational theory integrals stochastically for the quartic double well. The subroutine pot must be supplied to run the program vmc. The local energy estimator in Equation 6.72 is evaluated in the subroutine elocal.

```
      program vmc
      implicit real*8 (a-h,o-z)
c program for variational Monte Carlo (in R^1) using parallel tempering
c the trial wavefunction is exp(- a V /2) a = parameter
      real*8 xn(1),y(100),ym1(100),vk(100),temp(100)
      real*8 sumv(100),sumv2(100),deltax(100)
      integer nrej(100),nrejs(100), iseed(10),nrejt(100)
      dm = 110.d0    ! mass of particle
      gamma = 9.0d-1
      tempm = 4.0d0
      tempx = 16.0d0
      nwalkers = 80
      nmoves = 1000000
      do nk = 1, nwalkers     ! initialize the arrays
       y(nk) = 1.d0
       ym1(nk) = 1.d0
       sumv(nk) = 0.d0
       sumv2(nk) = 0.d0
       deltax(nk) = 0.2
       x = y(nk)
       call pot(gamma, x,v,dv,d2v)
       vk(nk) = v
       temp(nk) = tempm + dfloat(nk-1)*(tempx-tempm)/dfloat(nwalkers-1)
       nrej(nk) = 0
       nrejs(nk) = 0
       nrejt(nk) = 0
      enddo
c seed the generator
        iseed(1) = secnds(0.)
      call random_seed(put=iseed)
      do moves=1, 2*nmoves
      do nk = 1, nwalkers        ! loop over all the walkers
       call random_number(xn)
       if (xn(1) .gt. 0.9 .and. nk .ne. nwalkers)  then
c swap
        dv = (temp(nk) - temp(nk+1))*(vk(nk+1) - vk(nk))
        q = exp(-dv)
        call random_number(xn)
        if (xn(1) .lt. q) then
          y(nk) = ym1(nk+1)          ! accept
          v = vk(nk+1)
          ym1(nk+1) = ym1(nk)
          vk(nk+1) = vk(nk)
          ym1(nk) = y(nk)
```

```
              vk(nk) = v
              else
              nrej(nk) = nrej(nk) + 1 ! reject
              nrejs(nk) = nrejs(nk) + 1
            endif
          else
c    METROPOLIS move
          call random_number(xn)
          y(nk) = ym1(nk) + deltax(nk)*(xn(1) - 0.5)
          x = y(nk)
          call pot(gamma,x,v,dv,d2v)
          dv = (v - vk(nk))*temp(nk)
          q = exp(-dv)
          call random_number(xn)
          if (xn(1) .lt. q) then
c   accept the move
            ym1(nk) = y(nk)
            vk(nk) = v
          else
c   reject the move
            nrej(nk) = nrej(nk) + 1
            nrejt(nk) = nrejt(nk) + 1
          endif
          endif
          if (moves .gt. nmoves) then
c         if (nk .eq. nwalkers) write(11,*) ym1(nk)
          x = ym1(nk)
          par = temp(nk)
          call elocal(gamma,dm,par, x,elv)
          sumv(nk) = sumv(nk) + elv/dfloat(nmoves)
          sumv2(nk) = sumv2(nk) + elv*elv/dfloat(nmoves)
          else
c adjust deltax automatically during the equilibration run
            if (mod(moves,1000) .eq. 0) then
            if (nrejt(nk) .gt. 550) deltax(nk) = deltax(nk)*0.9
            if (nrejt(nk) .lt. 450) deltax(nk) = deltax(nk)/0.9
             nrejt(nk) = 0
            endif
           endif
          enddo     ! walker loop
          enddo
          do nk=1,nwalkers
          e =  sumv(nk)
          write(9,1000) temp(nk),e,nrej(nk),nrejt(nk)
          enddo
1000    format(2f15.8,2i8)
          end
          subroutine elocal(gamma,dm,a, x,el)
c This subroutine calculates the local
```

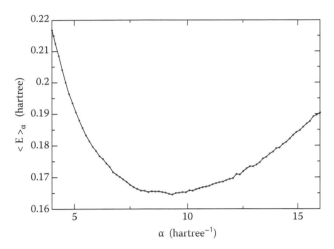

FIGURE 6.7
Variational energy for a particle with a mass of 110 a.u. trapped in the quartic
double well ($\gamma = 0.9$). The function $\langle E \rangle_\alpha$ vs α is smooth and quadratic about
the minimum ≈ 9.3.

```
c variational energy when the trial wavefunction is exp( -a V/2)
c On input:   x = coordinate
c             gamma = potential energy parameter parameter
c             dm    = mass
c             a     = wavefunction parameter
c On output: el =  local energy
      implicit real*8 (a-h,o-z)
      call pot(gamma, x,v,dv,d2v)
      el = v - (0.5d0/dm)*(0.25d0*a*a*dv*dv - 0.5d0*a*d2v)
      return
      end
```

The program **vmc** is used to produce the graph in Figure 6.7, where the
average energy in Equation 6.67 is computed using the trial wavefunction in
Equation 6.69, and the resulting integral in Equation 6.70 is computed with
parallel tempering using 80 walkers and one million moves. Figure 6.8 is a
histogram of configurations sampled by the thirty-sixth walker. Note that x
is distributed according to

$$p_\alpha\left(x\right) dx = \frac{\exp\left(-\alpha V\right)\, dx}{\displaystyle\int dx\, \exp\left(-\alpha V\right)}, \tag{6.73}$$

the normalized wavefunction squared.

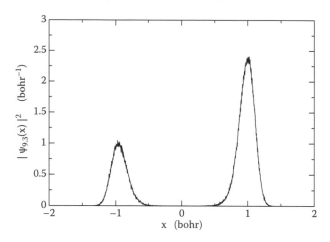

FIGURE 6.8
Optimal $\alpha \approx 9.3$ variational wavefunction (squared) for a particle with a mass of 110 a.u. trapped in the quartic double well ($\gamma = 0.9$).

Exercises

1. Assume the following trial wavefunction,

$$\psi_\alpha(x) = \exp\left(-\alpha x^2\right),$$

   for the harmonic oscillator,

$$\hat{H} = -\frac{\hbar^2}{2\,m}\frac{d^2}{d\,x^2} + \frac{1}{2}k\,x^2.$$

   Obtain an expression for $\langle E \rangle_\alpha$, then find the optimal value for α by setting the derivative of $\langle E \rangle_\alpha$ with respect to α to zero. Show that the wavefunction with the optimized parameter is the actual ground state of the harmonic oscillator. Repeat the whole procedure, using

$$\psi_\alpha(x) = x\exp\left(-\alpha x^2\right),$$

   as the trial wavefunction.

2. Derive Equation 6.72 from Equation 6.71.

3. Switch the potential in the program vmc, so that a particle of mass, m, inside the usual quartic potential ($\gamma = 0.9$) is simulated. Is parallel tempering necessary for this problem? Run the program as it is, then change the line

   ```
   if (xn(1) .gt. 0.9 .and. nk .ne. nwalkers) then
   ```

   to

```
if (xn(1) .gt. 1.1 .and. nk .ne. nwalkers) then
```

which turns off swapping moves, and find out by comparing energy curves. Several masses of increasing size from 100 up to several thousands atomic units need to be investigated.

6.12 Importance Sampling Diffusion Monte Carlo (IS-DMC) with Drift

Once we have a good approximation for the ground state wavefunction, we can reformulate the diffusion of pseudoparticles, producing in the process much more efficient algorithms. The optimized wavefunction produces the equivalent of importance sampling to DMC. The introduction of importance sampling into DMC can be accomplished in two ways, either it is inserted in the diffusion part, or into the branching part of the algorithm. If we insert

$$f(x, \tau) = \psi_T(x) \psi(x, \tau), \tag{6.74}$$

into the Schrödinger equation, we obtain a diffusion equation for $f(x, \tau)$,

$$-\hbar \frac{\partial f}{\partial \tau} = \frac{\hbar^2}{2\mu} \nabla^2 f - \frac{\hbar^2}{\mu} \nabla \left(\frac{f \nabla [\psi_T]}{\psi_T} \right) - (E_L - V_{\text{ref}}) f, \tag{6.75}$$

where E_L is the local energy of Equation 6.71, and

$$\frac{\hbar^2}{\mu} \nabla \left(\frac{f \nabla [\psi_T]}{\psi_T} \right), \tag{6.76}$$

is a drift term. Therefore, this process suggests that the diffusion moves should be adjusted by using the Gaussian distribution of steps η, and a drift term d,

$$x_i \rightarrow x_i + \eta + d, \tag{6.77}$$

where in $\mathbb{R}^1$, d is,

$$d = \frac{\hbar^2 \Delta \tau}{\mu} \psi_T(x)^{-1} \frac{d\psi_T}{dx}. \tag{6.78}$$

Given the trial wavefunction in Equation 6.69, one derives

$$d = -\frac{\hbar^2 \alpha \Delta \tau}{2\mu} \frac{dV}{dx}. \tag{6.79}$$

Therefore, the drift for the function f is proportional to the force felt by the particle at the position x. This method increases the efficiency of unguided DMC because the drift term drives particle replicas toward regions of the configuration space that are most important. The estimate of the ground state energy now, is

$$\langle E \rangle_\tau = \frac{\displaystyle\sum_{i=1}^{N_{\tau-\Delta\tau}} w_i \, E_{\mathrm{L}}\,(x_i)}{\displaystyle\sum_{i=1}^{N_{\tau-\Delta\tau}} w_i}, \tag{6.80}$$

and the local energy enters into the feedback and branching function as well,

$$w_i = \exp\left[-\frac{1}{2}\left(E_{\mathrm{L}}\,(x_i) + E_{\mathrm{L}}\,(x_i') - 2V_{\mathrm{ref}} \right) \Delta\tau \right]. \tag{6.81}$$

Finally, at equilibrium, the population distribution is the product of the "true" ground state wavefunction multiplied by the trial wavefunction.

The program dmc can be modified with the following code inside the population loop,

```
c diffusion
        xold = x(i)
        call pot(gamma, xold,vold,dvold,d2vold)
        call elocal(gamma,dm,alpha, xold ,elold)
        drift = -0.5d0*alpha*dvold*dt/dm
        x(i) = x(i) + ve + drift
        xin = x(i)
        call pot(gamma, xin,v,dv,d2v)
        call elocal(gamma,dm,alpha, xin ,el)
c branching (second order)
        w = dexp(-0.5d0*(el + elold - 2.d0*vref)*dt)
        if (w .gt. 10.d0) w = 10.d0
        call ransr(xr,1)
        m(i) = min(int(w + xr(1)),2)
        sumv = sumv + dfloat(m(i))*el
        isum = isum + m(i)
```

Note that elocal must be appended and alpha must be initialized (after having optimized its values with a vmc run).

6.13   Green's Function Diffusion Monte Carlo

Consider the continuous equivalent of Equation 6.23 in imaginary time, $\tau = it$. The sum that one writes for the multiplication of a vector by a matrix becomes an integral over all values of x,

$$\psi\,(x, \tau + \Delta\tau) = \int_{-\infty}^{\infty} G\,(x, x', \Delta\tau)\, \psi\,(x', \tau)\, dx. \tag{6.82}$$

Equation 6.82 is the integral representation of the time-dependent Schrödinger equation, and the function $G(x, x', \Delta\tau)$ can be shown to satisfy the Schrödinger equation subjected to the initial condition

$$\lim_{\Delta\tau \to 0} G(x, x', \Delta\tau) = \delta(x - x'). \tag{6.83}$$

The condition in Equation 6.83 guarantees that the wavefunction satisfies

$$\lim_{\Delta\tau \to 0} \psi(x', \tau + \Delta\tau) = \psi(x, \tau), \tag{6.84}$$

and is the continuous analog of the unit matrix. The function $G(x, x', \Delta\tau)$ is called the Green's function for the time-dependent Schrödinger equation. Equation 6.82 can be solved by approximating the local effects of the Green function. To first order in $\Delta\tau$, we have

$$G(x, x', \tau) \approx \exp\left(-\frac{m\,\Delta x^2}{2\Delta\tau}\right) \exp\left[-(V - V_{\text{ref}})\Delta\tau\right]. \tag{6.85}$$

We could choose to simulate the effects of the propagator in Equation 6.85 by making random moves for each replica with Δx distributed according to the first exponential on the L.H.S., and then subjecting each replica to a multiplication (annihilation) process guided by the second exponential on the R.H.S. The resulting algorithm would be identical to what we implement in dmc. To improve the process with a given approximation of the wavefunction, one inserts the trial wavefunction on both sides of Equation 6.82,

$$\psi_T(x)\,\psi(x, \tau + \Delta\tau) = \int_{\infty}^{\infty} G(x, x', \Delta\tau)\,\psi_T(x')\,\psi(x', \tau)\,dx'. \tag{6.86}$$

This equation can be rearranged to read

$$\psi(x, \tau + \Delta\tau) = \int_{-\infty}^{\infty} G(x, x', \Delta\tau)\,\frac{\psi_T(x')}{\psi_T(x)}\,\psi(x', \tau)\,dx', \tag{6.87}$$

where $\psi_T(x)$ is brought into the integral, since it is independent of the integration variable. The first order expansion for $G(x, x', \Delta\tau)$ equivalent to Equation 6.85 is,

$$G(x, x', \Delta\tau)\,\frac{\psi_T(x')}{\psi_T(x)} \approx \exp\left(-\frac{m\,\Delta x^2}{2\Delta\tau}\right) \frac{\psi_T(x + \Delta x)}{\psi_T(x)} \exp\left[-(V - V_{\text{ref}})\Delta\tau\right]. \tag{6.88}$$

Therefore, we choose to simulate the effects of the propagator in Equation 6.88 by making random moves for each replica with Δx distributed according to the first exponential on the L.H.S., and then subjecting each replica to a multiplication (annihilation) process guided by the following branching function,

$$w_i = \frac{\psi_T(x + \Delta x)}{\psi_T(x)} \exp\left[-(V - V_{\text{ref}})\Delta\tau\right]. \tag{6.89}$$

The average energy is still computed using Equation 6.80 with this method.

The original `dmc` program can be modified as follows,

```
c diffusion
          xold = x(i)
          call pot(gamma, xold,vold,dvold,d2vold)
          fxold = dexp(-0.5d0*alpha*vold)
          x(i) = x(i) + ve
          xin = x(i)
          call pot(gamma, xin,v,dv,d2v)
          fx = dexp(-0.5d0*alpha*v)
c branching
          w = dexp(-0.5d0*(vold + v - 2.d0*vref)*dt)*fx/fxold
          if (w .gt. 10.d0) w = 10.d0
          call ransr(xr,1)
          m(i) = min(int(w + xr(1)),2)
c compute the local energy
          call elocal(gamma,dm,alpha,xin,el)
          sumv = sumv + dfloat(m(i))*el
          isum = isum + m(i)
```

Once more, `elocal` must be appended and `alpha` must be initialized (after having optimized its values with a `vmc` run).

In Figure 6.9, the three different DMC schemes are compared for a particle of mass 110 a.u. trapped in the quartic double well with $\gamma = 0.9$.

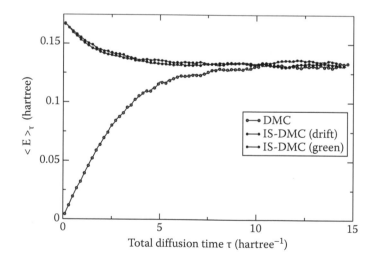

FIGURE 6.9
Ground state energy for a particle with a mass of 110 a.u. trapped in the quartic double well ($\gamma = 0.9$) estimated by three different DMC approaches.

The parameter α is 9.3, obtained from the VMC run, as graphed in Figures 6.7 and 6.8.

Exercises

1. Derive Equation 6.75. Proceed as follows. Let $\psi(x, \tau) = \psi_T^{-1}(x) f(x, \tau)$. Insert this expression into the imaginary-time Schrödinger equation,

$$-\hbar \frac{\partial \psi(x, \tau)}{\partial \tau} = \frac{\hbar^2}{2\mu} \frac{\partial^2}{\partial x^2} \psi(x, \tau) - (V - V_{\text{ref}}) \psi(x, \tau), \qquad (6.90)$$

then note that f depends on both x and τ, whereas ψ_T only depends on x.

2. Either IS-DMC method can reduce the statistical fluctuations by several orders of magnitude. In order to see this best, a nonharmonic but confining potential must be used. Use

$$V(x) = x^4,$$

to test the two IS-DMC methods. Begin by running the DVR of Colbert and Miller to find the ground state energy and wavefunction of the particle with 110 a.u. of mass trapped in the quartic potential. Then, modify the potential in dmc and vmc so that the optimal values of $\Delta\tau$ and α are found. Finally, run the two IS-DMC programs and measure the fluctuations by collecting data with 10 independent runs that are at "equilibrium." You will need to modify the program dmc to do this exercise.

3. Use the same mass, and potential energy function as in the previous exercise, and compare the convergence with respect to $\Delta\tau$ for both IS methods.

4. With the statistical error under control, it is possible to measure the population bias. Use the same system mass, potential energy function as in the previous two exercises, and use either one of the IS-DMC to study the impact of the population bias by running IS-DMC using the two programs you have written for the previous exercise. The statistical fluctuations and the correlations in the random numbers are best controlled with the following procedure. Use 10 independent runs of equal length (say 1000 moves at equilibrium) with a population of 10,000 and average the energy over the moves. Switch to a population of 1000 and collect 100 independent samples of equal length (identical to the previous value of the population), then drop to 100 for the target population and collect 1000 independent samples, always averaging and computing the standard deviation as a measure of the statistical fluctuations. You should discover that the population bias drops as N_0^{-1}.

5. To see how nonconfining potentials can produce error in unguided DMC, consider a particle of 500.0 a.u. subjected to the following potential,

$$V = \frac{4x}{x^2 + 4}.$$

Run the DVR of Colbert and Miller to find the ground state, modify the programs dmc and vmc so that the optimal values of $\Delta\tau$ and α are found, then use the programs for IS-DMC and verify that, in both cases, the nonconfining potential bias drops below the statistical fluctuations.

7

The Path Integral in Euclidean Spaces

7.1   Introduction

In this chapter, we use the analytical results of the harmonic oscillator (both classical and quantum mechanical) to develop and test three stochastic algorithms based on the Metropolis scheme. The classical canonical ensemble integral for the harmonic oscillator is calculated first. The results should be found within the equipartition values at all the temperatures within the error bars. The reader is encouraged to experiment with different walk lengths to verify that the error bars drop as the square root of the sample size and that the results remain within significance around the equipartition. The Metropolis algorithm is then applied to the Feynman path integral, using both the discretized approach and the Fourier path expansion. The partial averaging enhancement for the Fourier algorithm, and the reweighted Fourier series are introduced. All the programs in this chapter are for monodimensional problems. Later in the book, it is shown that the extension to multidimensional problems with the effort increasing linearly, is simple. The programs are sufficiently general, to allow for a substitution of the potential energy. Caution should be used, however, when introducing potential energy models that contain barriers greater than k_BT, as it is likely that quasiergodicity will arise.

7.2   The Harmonic Oscillator

The harmonic oscillator is one of the simplest systems for which classical and quantum mechanical properties can be expressed analytically with elementary functions. The classical Hamiltonian should be considered a function of both the momentum p and the position x.

$$\mathcal{H}\left(p, x\right) = \frac{1}{2m}p^2 + \frac{1}{2}kx^2. \tag{7.1}$$

In this equation, m is the mass of the particle and k is the force constant. In Chapter 2, we obtain the classical trajectory from a given initial position, x_0, and velocity, $\dot{x}_0$,

$$x(t) = x_0 \cos(\omega t) + \frac{\dot{x}_0}{\omega} \sin(\omega t), \qquad (7.2)$$

where ω is the characteristic frequency,

$$\omega = \sqrt{\frac{k}{m}}. \qquad (7.3)$$

The quantum mechanical equivalent of $\mathcal{H}(p, x)$ is the Hamiltonian operator $\hat{H}$,

$$\hat{H} = -\frac{\hbar^2}{2m} \frac{d^2}{dx^2} + \frac{1}{2} k x^2. \qquad (7.4)$$

The solution to the time-independent Schrödinger equation for the harmonic oscillator can be found in many introductory quantum mechanics texts. The set $\{\Psi_n(x)\}_{n=0}^{\infty}$ is the basis set for the Hilbert space of $\hat{H}$,

$$\Psi_n(x) = \frac{1}{(2^n n!)^{1/2}} \left(\frac{\alpha}{\pi}\right)^{1/4} H_n\left(\alpha^{1/2} x\right) e^{-\alpha x^2 / 2}, \qquad (7.5)$$

where

$$\alpha = \sqrt{\frac{km}{\hbar^2}}, \qquad n = 0, 1, 2, \ldots, \qquad (7.6)$$

and $H_n(x)$ are special functions, called Heremite's polynomials,

$$H_0(x) = 1, \qquad (7.7)$$

$$H_1(x) = 2x, \qquad (7.8)$$

$$H_{n+1}(x) = 2x\, H_n(x) - 2n\, H_{n-1}(x). \qquad (7.9)$$

The energy levels have a simple expression,

$$E_n = \left(n + \frac{1}{2}\right) \hbar\omega. \qquad (7.10)$$

Therefore, both the classical and quantum theory of the harmonic oscillator are solvable analytically. This makes the harmonic oscillator a prime candidate for the development and subsequent testing of theoretical and numerical tools aimed at solving similar problems with more complex potentials.

7.3 Classical Canonical Average Energy and Heat Capacity

Consider an ensemble (with constant N, V, T) of N noninteracting one-dimensional harmonic oscillators (e.g., an ideal gas of nonrotating diatomic molecules) at temperature T. The canonical partition function is,

$$Q = \left\{ \int_{-\infty}^{+\infty} dp \int_{-\infty}^{+\infty} dx \exp\left[-\beta \mathcal{H}\left(p, x\right)\right] \right\}^N, \tag{7.11}$$

where

$$\beta = \frac{1}{k_B T}, \tag{7.12}$$

and k_B is Boltzmann's constant. The resulting integrals are analytical,

$$\int_{-\infty}^{+\infty} dp \int_{-\infty}^{+\infty} dx \exp\left[-\beta\left(\frac{1}{2m}p^2 + \frac{1}{2}kx^2\right)\right] = \frac{2\pi}{\beta}\sqrt{\frac{m}{k}} = \frac{2\pi}{\beta\omega}. \tag{7.13}$$

Therefore, the partition function takes a rather simple form,

$$Q = \left(\frac{2\pi k_B T}{\omega}\right)^N. \tag{7.14}$$

The average energy and the heat capacity become simply,

$$U = \langle \mathcal{H} \rangle = k_B T^2 \frac{\partial \ln Q}{\partial T} = N k_B T, \tag{7.15}$$

$$C_V = \frac{\partial U}{\partial T} = N k_B, \tag{7.16}$$

a result that we could have obtained from the equipartition theorem, of course. Now that we have these nice analytical results, let's recast the expression for the average Hamiltonian using the general result in Equation 3.18.

Additionally, let us develop a convenient expression for the heat capacity. Using the postulate, $U = \langle \mathcal{H} \rangle$ and the chain rule, we derive,

$$C_V = -\frac{1}{k_B T^2}\frac{\partial \langle \mathcal{H} \rangle}{\partial \beta}, \tag{7.17}$$

and if we use Equation 3.18, we get,

$$\frac{\partial \langle \mathcal{H} \rangle}{\partial \beta} = -\frac{1}{2\beta^2} - \frac{\displaystyle\int_{-\infty}^{+\infty} dx \, [V\left(x\right)]^2 \exp\left[-\beta V\left(x\right)\right]}{\displaystyle\int_{-\infty}^{+\infty} dx \exp\left[-\beta V\left(x\right)\right]}$$

$$+ \left\{\frac{\displaystyle\int_{-\infty}^{+\infty} dx \, V\left(x\right) \exp\left[-\beta V\left(x\right)\right]}{\displaystyle\int_{-\infty}^{+\infty} dx \exp\left[-\beta V\left(x\right)\right]}\right\}^2, \tag{7.18}$$

i.e.,

$$\frac{\partial \langle \mathcal{H} \rangle}{\partial \beta} = -\frac{k_B^2 T^2}{2} - \langle V^2 \rangle + \langle V \rangle^2, \tag{7.19}$$

and the expression for C_V becomes,

$$C_V = \frac{k_B}{2} + \frac{\langle V^2 \rangle - \langle V \rangle^2}{k_B T^2}. \tag{7.20}$$

The program `classical_ho` computes the average energy and heat capacity for the harmonic oscillator using the Metropolis algorithm outlined in Chapter 3.

```
      program classical_ho
      implicit real*8 (a-h,o-z)
      real*8 xr(2)
      iseed = 1032
c   seed the random number generator (pseudo mode)
      call random_seed(iseed)
      write(6,1010)
      MW = 1000000
      M = 1000000
      temp = 0.02
      deltax = 1.0
      do k_t = 1,81
        nrej = 0.d0
        av = 0.d0
        av2 = 0.d0
        xim1 = 0.d0
        xi = 0.d0
c Calculate  V(x_0) where x_0 is set to zero
        call pot(xim1,vim1)
        do i=1,MW + M
c Draw two random numbers in  [0,1]
        call random_number(xr)
c   Calculate the new trial position
          xi = xim1 +  deltax *(xr(1)-0.5d0)
          call pot(xi,vxi)
c   Calculate the probability factor
          p = exp(-(vxi - vxim1)/temp)
          if (xr(2) .lt. p) then
            xim1 = xi
            vxim1 = vxi
          else
            nrej = nrej + 1
c   nrej is a counter for the number of rejections
            xi = xim1
            vxi = vxim1
          endif
          if (i.gt. MW) then
```

```
c    Accumulate the average V and V**2
               av = av + vxi/dfloat(M)
               av2 = av2 + vxi*vxi/dfloat(M)
             endif
           enddo
c    calculate the total energy and heat capacity
           u = 0.5*temp + av
           cv = 0.5 + (av2 - av*av)/(temp*temp)
c calculate the percent rejections
           pr = 100.*dfloat(nrej)/dfloat(M+MW)
           e_exact = temp
           cv_exact = 1.d0
           write(6,1000) u,e_exact,cv,cv_exact,pr
           temp = temp + 0.04
           deltax = deltax + 0.3
         enddo
1000   format(6f12.4)
1010   format('   < H > (K)   T   (K)      Cv/kB   exact value   %rj')
       end
c      POT
       subroutine pot(x,v)
       implicit real*8 (a-h,o-z)
       v = 0.5*x*x !   The force constant is 1.000 K/(bohr^2)
       return
       end
```

A partial output of classical_ho is,

< H > (K)	T (K)	Cv/kB	exact value	%rj
0.0200	0.0200	0.9966	1.0000	56.6481
0.0600	0.0600	1.0000	1.0000	46.3559
0.1002	0.1000	1.0067	1.0000	44.6343
0.1397	0.1400	0.9951	1.0000	44.8357
0.1796	0.1800	0.9940	1.0000	45.5339
0.2204	0.2200	1.0054	1.0000	46.5074
0.2601	0.2600	0.9994	1.0000	47.5902
0.3002	0.3000	0.9991	1.0000	48.6773
0.3400	0.3400	0.9982	1.0000	49.6491
0.3798	0.3800	0.9984	1.0000	50.7313
0.4204	0.4200	1.0005	1.0000	51.6908
0.4600	0.4600	1.0009	1.0000	52.7362
0.4996	0.5000	0.9939	1.0000	53.6046

Note that δ (in the program called deltax) is adjusted empirically by running the program a few times to generate a rejection rate between 40 and 60%. The program adjusts δ at every temperature using a linear function, however, for a larger temperature interval, one may in fact need a quadratic or cubic equation. Note also, that at every temperature, the walk is allowed to reach its asymptotic distribution by running NW $= 10^6$ moves without collecting data. Data sets are collected for an additional 10^6 moves.

7.4   Quantum Canonical Average Energy and Heat Capacity

Let's consider an ensemble (with constant N, V, T) of N noninteracting one-dimensional harmonic oscillators at temperature T. The canonical partition function is a sum over quantum state energies,

$$Q = \sum_{n=0}^{\infty} \exp\left(-\beta E_n\right), \qquad (7.21)$$

where β and k_B are as before, and N is still set to one for convenience. If the explicit form of the spectrum of the Hamiltonian operator is used, the canonical partition function can be re-expressed as a geometric series,

$$Q = \sum_{n=0}^{\infty} \exp\left[-\beta \hbar \omega \left(n + \frac{1}{2}\right)\right], \qquad (7.22)$$

$$Q = \exp\left(-\frac{1}{2}\beta \hbar \omega\right) \sum_{n=0}^{\infty} \left[\exp\left(-\beta \hbar \omega\right)\right]^n = \frac{\exp\left(-\frac{1}{2}\beta \hbar \omega\right)}{1 - \exp\left(-\beta \hbar \omega\right)}. \qquad (7.23)$$

The last derivation is identical to that in Chapter 5 in the context of the Planck distribution. It turns out that the harmonic oscillator plays a central role when Maxwell's equations for the electromagnetic fields are quantized.

The internal energy and heat capacity can be calculated analytically as well. The following results are found in many introductory statistical mechanics books. The reader is encouraged to work through the details.

$$U = -\frac{\partial \ln Q}{\partial \beta} = \hbar \omega \left[\frac{1}{2} + \frac{\exp\left(-\beta \hbar \omega\right)}{1 - \exp\left(-\beta \hbar \omega\right)}\right], \qquad (7.24)$$

$$C_V = -\frac{1}{k_B T^2} \frac{\partial U}{\partial \beta} = \frac{(\hbar \omega)^2}{k_B T^2} \left\{\frac{\exp\left(-\beta \hbar \omega\right)}{\left[1 - \exp\left(-\beta \hbar \omega\right)\right]^2}\right\}. \qquad (7.25)$$

Let us recast the expression for U and C_V into a more useful form for a general spectrum,

$$U = \langle E_n \rangle = -\frac{\partial \ln Q}{\partial \beta} = \frac{\displaystyle\sum_{n=0}^{\infty} E_n \exp\left(-\beta E_n\right)}{\displaystyle\sum_{n=0}^{\infty} \exp\left(-\beta E_n\right)}, \tag{7.26}$$

$$C_V = -\frac{1}{k_B T^2} \frac{\partial U}{\partial \beta}$$

$$= -\frac{1}{k_B T^2} \left[-\frac{\displaystyle\sum_{n=0}^{\infty} E_n^2 \exp\left(-\beta E_n\right)}{\displaystyle\sum_{n-0}^{\infty} \exp\left(-\beta E_n\right)} + \left(\frac{\displaystyle\sum_{n=0}^{\infty} E_n \exp\left(-\beta E_n\right)}{\displaystyle\sum_{n=0}^{\infty} \exp\left(-\beta E_n\right)} \right)^2 \right], \tag{7.27}$$

$$C_V = \frac{\langle E_n^2 \rangle - \langle E_n \rangle^2}{k_B T^2}. \tag{7.28}$$

This form, too, is amenable for computation using a stochastic algorithm. The trick here is to generate a set of random integers with a normalized discrete distribution,

$$p(n) = \frac{\exp\left(-\beta E_n\right)}{\displaystyle\sum_{n=0}^{\infty} \exp\left(-\beta E_n\right)}, \tag{7.29}$$

by using a rejection technique. In the program **quantum_ho**, the averages are calculated using

$$\langle E_n \rangle = \frac{1}{M} \sum_{i=1}^{M} E_{n(i)}, \tag{7.30}$$

and

$$\langle E_n \rangle^2 = \frac{1}{M} \sum_{i=1}^{M} E_{n(i)}^2, \tag{7.31}$$

where $\{n(i)\}_{i=1}^{M}$ is a random walk with trial moves,

$$n(i) = n_{\max}\xi, \tag{7.32}$$

and ξ is a random number in $[0, 1]$. Moves are accepted with probability,

$$p = \min\left\{1, \exp\left[-\beta\left(E_{n(i)} - E_{n(i-1)}\right)\right]\right\}. \tag{7.33}$$

The value of $n_{\max}$ is chosen to make the $n_{\max}$ term of the sum smaller than the machine precision ε.

$$n_{\max}^2 \exp\left[-\beta E_{n_{\max}}\right] < \varepsilon. \tag{7.34}$$

The units of temperature, energy, and heat capacity are reduced units,

$$T^* = \frac{k_B T}{\hbar \omega}, \qquad \langle E_n^* \rangle = \frac{\langle E_n \rangle}{\hbar \omega}, \qquad C_v^* = \frac{C_V}{k_B}. \qquad (7.35)$$

The exact expressions for the average energy and heat capacity in these reduced units are

$$\langle E_n^* \rangle = \frac{1}{2} + \frac{\exp\left(-\dfrac{1}{T^*}\right)}{1 - \exp\left(-\dfrac{1}{T^*}\right)}, \qquad \frac{C_V}{k_B} = \frac{\langle (E_n^*)^2 \rangle - \langle E_n^* \rangle^2}{(T^*)^2}. \qquad (7.36)$$

```
      program quantum_ho
      implicit real*8 (a-h,o-z)
      real*8 xr(2)
c this program calculates the average quantum energy
c at finite temperature (canonical ensemble) using
c a stochastic simulation from known energy levels
c Reduced units of temperature are used (T* = k_B T/ hbar omega)
c Reduced units of energy are used (E* = E/ hbar omega)
      iseed = 1109
      call random_seed(iseed)
      write(6,1010)
      nw = 1000000
      nmc = 1000000
      temp = 0.02
      delttemp = 0.04
      do kt = 1,80  !  loop over a range of temperatures
      nmax = max(1,int(temp*27.631021 - 0.5))
      n = 0
      en = 0.5
      sume = 0.d0
      sume2 = 0.d0
      irej = 0
      do i = 1,nw + nmc
       call random_number(xr)
       nt = int(nmax*xr(1))       ! make a move
       et = dfloat(nt) + 0.5d0
       det = (et - en)/temp
       prob = dexp(-det)
       if (xr(2) .lt. prob) then
         en  = et            !  accept
         n = nt
       else
         irej = irej + 1    !  reject
       endif
```

```
      if (i. gt. nw) then
        sume = sume + en/dfloat(nmc)        !  collect data
        sume2 = sume2 + en*en/dfloat(nmc)
      endif
     enddo
     cv = (sume2 - sume*sume)/(temp*temp)
     em1 = dexp(-1.d0/temp)
     eexact = 0.5d0+em1/(1.d0-em1)        !  the exact energy
c    the exact heat capacity
     cvexact = (em1/((1.d0 - em1)*(1.d0-em1)))/(temp*temp)
     pr = 100.*dfloat(irej)/dfloat(nmc+nw)
     write(6,1000) temp, sume,eexact,cv,cvexact,pr
     temp = temp +  delttemp
    enddo                    ! temperature loop
1000 format (6f12.4)
1010 format('      T*        <En>     exact <En>    Cv         exact
    &Cv  % rejection')
     end
```

The program `quantum_ho` should produce the following output, shown here only partially.

T*	<En>	exact <En>	Cv	exact Cv	% rejection
0.0200	0.5000	0.5000	0.0000	0.0000	0.0000
0.0600	0.5000	0.5000	0.0000	0.0000	0.0000
0.1000	0.5000	0.5000	0.0038	0.0045	50.0509
0.1400	0.5009	0.5008	0.0440	0.0404	66.5947
0.1800	0.5038	0.5039	0.1165	0.1202	74.8376
0.2200	0.5108	0.5107	0.2247	0.2241	79.5302
0.2600	0.5213	0.5218	0.3226	0.3299	82.5724
0.3000	0.5358	0.5370	0.4097	0.4262	84.6733
0.3400	0.5564	0.5557	0.5183	0.5091	86.0877
0.3800	0.5771	0.5775	0.5738	0.5787	87.1843

Note that there are no adjustable parameters that can reduce the percent rejection at high temperatures. Let's inspect two graphs for the quantum harmonic oscillator, that of the ensemble average energy, $\langle E_n^* \rangle$, and the heat capacity, C_V/k_B. These graphs are reproduced in Figures 7.1 and 7.2. In each graph, a comparison is made with the classical counterpart of the simulated physical property. The Metropolis algorithm for sums is not as useful for the calculation of quantum canonical properties for large dimensions. The problem with the algorithm in its present form is that one must first solve Schrödinger's equation in order to obtain all the quantum states. Unfortunately, such a task is prohibitive in general. The discrete rejection method, as outlined above, however, is useful for the solution of mixed classical-quantum problems in the adiabatic limit and when simulating metastable spin states for H_2 in the

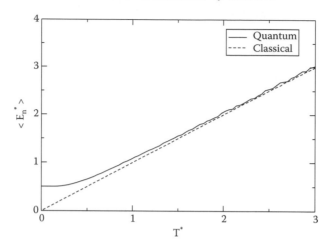

FIGURE 7.1
Quantum (—) and classical (– – –) average energy in the N, V, T ensemble for a single harmonic oscillator.

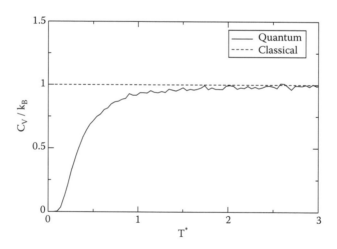

FIGURE 7.2
Heat capacity of the harmonic oscillator. The dashed line is the equipartition value obtained from the classical simulation.

condensed phase. Tools based on the Metropolis algorithm to obtain the finite temperature averages for fully quantum systems, without the knowledge of the energy states of a multidimensional Hamiltonian, have been developed. A powerful approach is the imaginary-time Feynman path integral. We develop two numerical methods for its computation in the following sections.

Exercises

1. Derive Equations 7.13, 7.15, 7.16, and 7.20.

2. Add code to the program `classical_ho` to produce a distribution of position at $T = 0.0200$ K, and compare it with the exact thermal distribution by analytically evaluating the integral in Equation 3.73.

3. In the program `classical_ho` change the force constant to 0.0010 $K/(bohr^2)$. How do you expect the results to change? You will need to readjust δ.

4. Derive Equations 7.24, 7.25, 7.28, and 7.36.

5. Run the program `quantum_ho` ten times, and calculate averages and standard deviations over the ten runs for both properties. Use the standard deviation obtained with the ten runs to estimate the statistical error (use twice the standard error in the mean), and show that the algorithm becomes less efficient (the statistical error is larger) at higher temperatures. A FORTRAN program that reads the ten lists and computes averages and standard deviations will be useful for further statistical analysis of estimators later.

7.5 The Path Integral in $\mathbb{R}^d$

In Euclidean manifolds, the derivation of the imaginary time $(t \rightarrow i\beta\hbar)$ path integral is straightforward. Schrödinger's equation is,

$$\hat{H} \, | \, n \rangle = E_n \, | \, n \rangle . \tag{7.37}$$

The familiar wavefunction, ψ_n, is represented as $\langle x \, | \, n \rangle$, and its complex conjugate, ψ_n^*, as $\langle n \, | \, x \rangle$. Matrix elements of the operator $\hat{O}$ are represented as $\langle i | \, \hat{O} \, | \, j \rangle$. The orthogonality of the eigenvectors of $\hat{H}$ is,

$$\langle m \, | \, n \rangle = \delta_{nm}. \tag{7.38}$$

We will require two mathematical theorems from the calculus on Hilbert spaces in Chapter 4. The first is the closure of the position basis set, which for Euclidean manifolds looks like

$$1 = \int_{-\infty}^{\infty} dx \, | \, x \rangle \, \langle x | . \tag{7.39}$$

This result is simple to prove, starting with

$$1 = \int_{-\infty}^{\infty} dx \, \psi_n^* \psi_n, \tag{7.40}$$

and using the bra-ket representation, we get

$$1 = \int_{-\infty}^{\infty} dx \, \langle n \mid x \rangle \, \langle x \mid n \rangle . \qquad (7.41)$$

The theorem now follows by bringing the integral sign inside the brackets (both $\mid n \rangle$ and $\langle n \mid$ are independent of x),

$$1 = \langle n | \left(\int_{-\infty}^{\infty} dx \mid x \rangle \, \langle x | \right) | n \rangle . \qquad (7.42)$$

Since $\langle n \mid n \rangle = 1$, the quantity inside the bracket must also be one. The second theorem is the invariance of the trace of a matrix under unitary transformation. For Hilbert spaces, N is infinite and the theorem is amended with the following caveat: *All the sums involved must be convergent.*

Let us express the quantum partition function in the following way,

$$Q = \sum_{n=0}^{\infty} \exp\left(-\beta E_n\right) = \sum_{n=0}^{\infty} \langle n | \exp\left(-\beta \hat{H}\right) | n \rangle = \int_{-\infty}^{\infty} dx \, \langle x | \exp\left(-\beta \hat{H}\right) | x \rangle ,$$
$$(7.43)$$

where the trace invariance theorem is used. The quantity $\langle x | \exp\left(-\beta \hat{H}\right) | x \rangle$ is called the **density matrix** element. The density matrix element can be shown to take a simple form for a free particle,

$$\rho_f\left(x, x', \beta\right) = \langle x' | \exp\left(-\beta \hat{K}\right) | x \rangle = \left[\frac{m}{2\pi \hbar^2 \beta}\right]^{1/2} \exp\left[-\frac{m}{2\beta \hbar^2}\left(x - x'\right)^2\right],$$
$$(7.44)$$

where $\hat{K}$ is the kinetic energy operator. The reader can verify that Equation 7.44 satisfies Bloch's equations,

$$-\frac{\hbar^2}{2m} \frac{\partial^2}{\partial x^2} \rho_f\left(x, x', \beta\right) = -\frac{\partial \rho_f\left(x, x', \beta\right)}{\partial \beta} . \qquad (7.45)$$

The imaginary-time interval, $0, \beta \hbar$, is subdivided into N equal subintervals. As N tends to infinity, one can write,

$$\exp\left(-\beta \hat{H}\right) = \lim_{N \to \infty} \prod_{i=1}^{N} \exp\left(-\frac{\beta}{N} \hat{H}\right) , \qquad (7.46)$$

where the limit is needed to eliminate commutation problems. Upon inserting the closure of the position basis $N - 1$ times, one arrives at,

$$Q = \lim_{N \to \infty} \int_{-\infty}^{\infty} dx \int_{-\infty}^{\infty} dx_1 \int_{-\infty}^{\infty} dx_2 \cdots \int_{-\infty}^{\infty} dx_{N-1}$$
$$\times \langle x | \exp\left(-\frac{\beta}{N} \hat{H}\right) | x_1 \rangle \langle x_1 | \exp\left(-\frac{\beta}{N} \hat{H}\right) | x_2 \rangle$$
$$\cdots \langle x_{N-1} | \exp\left(-\frac{\beta}{N} \hat{H}\right) | x \rangle . \qquad (7.47)$$

It may appear that we have complicated things with the last step. However, in subdividing the imaginary time, we have raised the temperature associated with each of the matrix elements, $\langle x_{i-1}| \exp\left(-\beta\hat{H}/N\right)|x_i\rangle$. Using a little intuition, we can approximate the matrix element by a classical expression. The reader will recall (see Figures 7.1 and 7.2) that as the temperature is raised for a particular system, the average energy and heat capacity approach their classical values. One approximates the high temperature density matrix element as,

$$\langle x_{i-1}| \exp\left(-\beta\hat{H}/N\right)|x_i\rangle \approx \left[\frac{m}{2\pi\hbar^2\beta}\right]^{1/2} \exp\left[-\frac{1}{2N}\beta m v^2 - \frac{1}{N}\beta V(x_i)\right],$$

$$(7.48)$$

where use is made of the classical Hamiltonian,

$$\mathcal{H} = \frac{1}{2m}p^2 + V(x),$$

$$(7.49)$$

and p is expressed in terms of v. This notation simplifies things,

$$v = \lim_{N\to\infty} N\frac{x_i - x_{i-1}}{\beta\hbar},$$

$$(7.50)$$

giving for the high temperature density matrix element,

$$\langle x_{i-1}| \exp\left(-\beta\hat{H}/N\right)|x_i\rangle \approx \left[\frac{m}{2\pi\hbar^2\beta}\right]^{1/2} \exp\left[-\frac{Nm}{2\beta\hbar^2}(x_i - x_{i-1})^2 - \frac{\beta}{N}V(x_i)\right].$$

$$(7.51)$$

The partition function then becomes,

$$Q = \lim_{N\to\infty} \left[\frac{m}{2\pi\hbar^2\beta}\right]^{N/2} \int_{-\infty}^{\infty} dx \int_{-\infty}^{\infty} dx_1 \dots \int_{-\infty}^{\infty} dx_{N-1}$$

$$\times \exp\left[-\frac{Nm}{2\beta\hbar^2}\sum_{i=1}^{N}(x_i - x_{i-1})^2 - \frac{\beta}{N}\sum_{i=1}^{N}V(x_i)\right],$$

$$(7.52)$$

where it is understood that $x_0 = x_N = x$. Note that upon setting $N = 1$, one obtains the classical partition function. The infinite-dimensional Riemann integral inside the expression of the partition function is usually represented by the following symbol:

$$\lim_{N\to\infty} \left[\frac{m}{2\pi\hbar^2\beta}\right]^{N/2} \int_{-\infty}^{\infty} dx_1 \cdots \int_{-\infty}^{\infty} dx_{N-1}$$

$$\times \exp\left[-\frac{Nm}{2\beta\hbar^2}\sum_{i=1}^{N}(x_i - x_{i-1})^2 - \frac{\beta}{N}\sum_{i=1}^{N}V(x_i)\right]$$

$$= \int_{x,0}^{x,\beta\hbar} \mathcal{D}[x(t)] \exp\left\{-\frac{1}{\hbar}\int_0^{\beta\hbar}\mathcal{H}[\dot{x}(t), x(t)]\,dt\right\}.$$

$$(7.53)$$

To simplify the expressions for the thermodynamic properties, the following notation is introduced.

$$W(x, x_1, \ldots, x_{N-1}, \beta) = \left(\frac{m}{2\pi\beta\hbar^2}\right)^{N/2} \exp\left[-S(x, x_1, \ldots, x_{N-1}, \beta)\right], \quad (7.54)$$

$$S(x, x_1, \ldots, x_{N-1}, \beta) = \frac{Nm}{2\beta\hbar^2} \sum_{i=1}^{N} (x_i - x_{i-1})^2 + \frac{\beta}{N} \sum_{i=1}^{N} V(x_i). \quad (7.55)$$

The average for a physical property estimated with the expression $A(x, x_1, \ldots, x_{N-1}, \beta)$ is,

$$
\langle A \rangle
$$
$$
= \frac{\displaystyle\int_{-\infty}^{\infty} dx \int_{-\infty}^{\infty} dx_1 \cdots \int_{-\infty}^{\infty} dx_{N-1} [A(x, x_1, \ldots, x_{N-1}, \beta)] W(x, x_1, \ldots, x_{N-1}, \beta)}{\displaystyle\int_{-\infty}^{\infty} dx \int_{-\infty}^{\infty} dx_1 \cdots \int_{-\infty}^{\infty} dx_{N-1} W(x, x_1, \ldots, x_{N-1}, \beta)}.
$$
$$(7.56)$$

Using $\langle E \rangle = -\partial \ln Q / \partial \beta$, one arrives at the following result,

$$\langle E \rangle = \frac{N}{2\beta} + \left\langle \frac{\partial S}{\partial \beta} \right\rangle, \quad (7.57)$$

$$\langle E \rangle = \frac{N}{2\beta} + \left\langle -\frac{Nm}{2\beta^2\hbar^2} \sum_{i=1}^{N} (x_i - x_{i-1})^2 + \frac{1}{N} \sum_{i=1}^{N} V(x_i) \right\rangle. \quad (7.58)$$

This energy estimator is called the T estimator in the literature. Its variance is known to grow with N, and in Euclidean space a much better estimator is the path average virial estimator (based on the virial theorem),

$$\langle E \rangle = \left\langle +\frac{1}{2N} \sum_{i=1}^{N} x_i \frac{dV(x_i)}{dx} + \frac{1}{N} \sum_{i=1}^{N} V(x_i) \right\rangle. \quad (7.59)$$

To calculate the heat capacity, one uses

$$\frac{C_V}{k_B} = -\frac{1}{(k_B T)^2} \frac{\partial \langle E \rangle}{\partial \beta}. \quad (7.60)$$

It is better to take the derivative of the path average virial estimator. The reader should prove the following relationship:

$$
\begin{aligned}
k_{\mathrm{B}} T^2 C_V &= \left\langle \left[+\frac{1}{2N} \sum_{i=1}^{N} x_i \frac{dV(x_i)}{dx} + \frac{1}{N} \sum_{i=1}^{N} V(x_i) \right] \right. \\
&\qquad \left. \times \left[-\frac{Nm}{2\beta^2 \hbar^2} \sum_{i=1}^{N} (x_i - x_{i-1})^2 + \frac{1}{N} \sum_{i=1}^{N} V(x_i) \right] \right\rangle \\
&\quad - \left\langle +\frac{1}{2N} \sum_{i=1}^{N} x_i \frac{dV(x_i)}{dx} + \frac{1}{N} \sum_{i=1}^{N} V(x_i) \right\rangle \\
&\qquad \times \left\langle -\frac{Nm}{2\beta^2 \hbar^2} \sum_{i=1}^{N} (x_i - x_{i-1})^2 + \frac{1}{N} \sum_{i=1}^{N} V(x_i) \right\rangle. \quad (7.61)
\end{aligned}
$$

All the averages are in a form suitable for computation with the Metropolis algorithm. The program **time_slice** calculates $\langle E \rangle$ (using both estimators) and C_V, for a particle of unit mass in a harmonic well with unit force constant, for several values of N.

```
      program time_slice
      implicit real*8 (a-h,o-z)
c This program calculates <E> and <Cv> using the
c discrete variable path integral in R^1. The T and path
c average virial estimator are compared.
      parameter (NMAX = 1000)
      real*8 wm1(0:NMAX),w(0:NMAX), xr(3)
      iseed = 1105  !  seed the random number generator
      call random_seed(iseed)
      N = 1 ! number of auxiliary variables
      do nQ = 1,20
      write(6,*) 'N =',N
      write(6,1001)
      MW = 1000000*N    !  number of warm up moves
      M = 1000000*N
      dM = dfloat(M)
      temp = 0.02
      deltax = 1.5
      do k_t = 1,42 ! loop through some temperatures
      nrej = 0.d0
      sumt = 0.d0
      sumv = 0.d0
      sumv2 = 0.d0
      xim1 = 0.d0
      xi = 0.d0
      do i=0,N
```

```
   w(i) = 0.d0
   wm1(i) = 0.d0
 enddo
 call getw(temp,w,xi,N,NMAX,wim1)   ! initiate the walk
 do i=1,MW + M
 call random_number(xr)
 imove = (N+1)*xr(1)      ! determine which variable to move
 if (imove .eq. 0 .or. imove .eq. N) then
  xi = xim1 +  deltax *(xr(2)-0.5d0)
 else
  w(imove) = wm1(imove) + deltax*(xr(2)-0.5d0)
 endif
 call getw(temp,w,xi,N,NMAX,wi)
 p = wi/wim1
 if (xr(3) .lt. p) then   !  accept or reject
  xim1 = xi    ! accept
  wim1 = wi
   do j=1,N-1
     wm1(j) = w(j)
   enddo
 else    ! reject
  nrej = nrej + 1
  xi = xim1
  wi = wim1
    do j=1,N-1
     w(j) = wm1(j)
     enddo
 endif
  if (i.gt. MW) then   ! accumulate data
   call testimator(temp,w,xi,N,NMAX,et)
   sumt = sumt + et/dM
   call vestimator(temp,w,xi,N,NMAX,ev)
   sumv = sumv + ev/dM
   sumv2 = sumv2 + ev*et/dM
  endif
 enddo
c the heat capacity with the virial estimator
 cvv =  (sumv2 - sumt*sumv)/(temp*temp)
c and calculate the percent rejections
 pr = 100.*dfloat(nrej)/dfloat(M+MW)
 write(6,1000) temp,sumt,sumv,cvv,pr
 temp = temp + 0.04
 deltax = deltax + 0.2
 enddo
  N = N  + 1    ! temperature loop
```

```
      enddo    ! quadrature loop
1000  format(6f12.4)
1001  format('        T        <E>T        <E>V       CV(V-T)   %r')
      end
c       VESTIMATOR
      subroutine vestimator(temp,w,xi,N,NMAX,sum)
      implicit real*8 (a-h,o-z)
      real*8 w(0:NMAX)
      sum = 0.d0
      dN = dfloat(N)
      do i=1,N
       x = xi + w(i)
       call pot(x,v,dv)
        sum = sum  + x*dv/(dN*2.d0) + v/(dN)
      enddo
      return
      end
c       TESTIMATOR
      subroutine testimator(temp,w,xi,N,NMAX,e)
      implicit real*8 (a-h,o-z)
      real*8 w(0:NMAX)
      sum = 0.d0
      dN = dfloat(N)
      do i=1,N
       x = xi + w(i)
       call pot(x,v,dv)
        sum = sum-(dN*temp*temp/(2.d0))*(w(i)-w(i-1))**2 + v/(dN)
      enddo
       e = sum + dN*temp/2.d0
      return
      end


c   GETW
      subroutine getw(temp,w,xi,N,NMAX,wi)
      implicit real*8 (a-h,o-z)
      real*8 w(0:NMAX)
      sum = 0.d0
      dN = dfloat(N)
      do i=1,N
       x = xi + w(i)
       call pot(x,v,dv)
        sum = sum+(dN*temp/(2.d0))*(w(i)-w(i-1))**2 + v/(dN*temp)
      enddo
       wi = exp(-sum)
      return
```

```
      end
c     POT
      subroutine pot(x,v,dv)
      implicit real*8 (a-h,o-z)
c The force constant is 1.000 hartree/(bohr^2)
      v = 0.5*x*x
      dv = x
      return
      end
```

The average energy $\langle E \rangle / (\hbar\omega)$ and heat capacity C_V / k_B in reduced units are calculated for several values of N. The reduced temperature is the same as before,

$$T^* = \frac{k_B T}{\hbar\omega}. \tag{7.62}$$

Note that $N+1$ auxiliary variables are introduced with the subscripted array, $w_i, i = 0, 1, 2, \ldots, N$ with $w_1 = w_N = 0$. Then, the position at time t_i is given by

$$x_i = x + w_i. \tag{7.63}$$

Using the units as chosen, and the auxiliary variables w_i, the quantity S in Equation 7.55 becomes,

$$S(x, x_1, \ldots, x_{N-1}, \beta) = \frac{NT^*}{2} \sum_{i=1}^{N} (w_i - w_{i-1})^2 + \frac{1}{NT^*} \sum_{i=1}^{N} V(x + w_i). \tag{7.64}$$

The quantity S is calculated in the subroutine **getw**. The T estimator is calculated in the subroutine **testimator**,

$$\langle E \rangle = \left\langle \frac{NT^*}{2} - \frac{N(T^*)^2}{2} \sum_{i=1}^{N} (w_i - w_{i-1})^2 + \frac{1}{N} \sum_{i=1}^{N} V(x + w_i) \right\rangle, \tag{7.65}$$

and the path average virial estimator is calculated in the subroutine **vestimator**. The output of the program for selected values of N is graphed in Figures 7.3 and 7.4.

Exercises

1. Prove that the expression for $\rho_f(x, x', \beta)$ in Equation 7.44 satisfies the differential equation in Equation 7.45.

2. Derive Equation 7.61.

3. Classical mechanics (or thermodynamics) emerges as a limit for large values of the mass and at high temperatures. In the limits, as β is small (or m is large), one term in the argument of Equation 7.52 dominates.

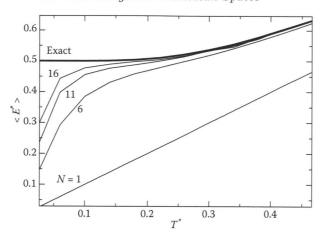

FIGURE 7.3
Average quantum canonical energy for the harmonic oscillator.

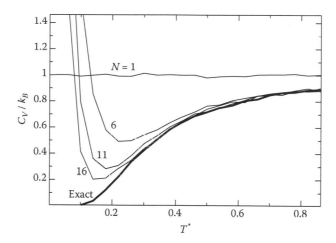

FIGURE 7.4
The heat capacity of the harmonic oscillator calculated with the $T - V$ estimator.

Use this to show that at any finite value of N, one derives the classical partition function. To see this, you will need to evaluate the partition function Q for $N = 2, 3, 4, \ldots$ and show that all these give the same result.

4. The partition function for a finite Trotter number N, as in Equation 7.52, can be computed analytically for the harmonic oscillator,

$$V = \frac{1}{2}kx^2, \tag{7.66}$$

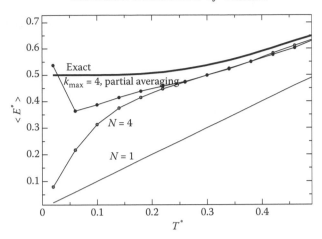

FIGURE 7.5
Comparison of the discretized PI and the FPI with partial averaging, with equal dimensionality.

since the resulting integrals can be recast as Gaussians. Evaluate the partition function Q for $1, 2, 3, 4, \ldots, N$, then use the resulting expression to obtain finite Trotter number expressions for the energy and the heat capacity. You can use the usual derivatives of Q with respect to β. Obtain graphs of these as a function of temperature for several values of N, to visualize the convergence pattern for E and C_V/k_B at low temperatures. These expressions are useful because they are free of statistical error.

5. Improve the program `time_slice` by developing a self-adjusting algorithm for δ (`deltax` in the code) as follows. During the warmup run, every 1000 moves, change δ by increasing its size slightly if the rejection number is smaller than 500, or decreasing its size slightly if the converse is true.

6. Run the computation ten times to estimate the random error for T and the virial estimator of the total energy as a function of N. Compare the efficiency of T and the virial estimator.

7. Use the program to compute the classical and the converged energy, and heat capacity for an atom of Argon confined in a quartic well,

$$V(x) = 0.01x^4, \tag{7.67}$$

where V is in hartree. Atomic units are the most convenient. Study the convergence pattern of the virial estimator and the T estimator, and compare them with the convergence for the harmonic oscillator. Use the

Discrete Variable Representation (DVR) to compute the energy levels analytically, then use these to carry out the estimate for the energy and heat capacity with Equations 7.26 and 7.28. Compare the DVR results with the path integral results.

8. It is often claimed in the literature that converged quantum algorithms allow particle tunneling to occur. If that were true, one could do without parallel tempering. Test this statement by running Metropolis simulations without configurational swaps for a particle in the quartic double well potential.

$$V(x) = V_0 \left(\frac{3}{2\alpha + 1} x^4 + \frac{4\alpha - 4}{2\alpha + 1} x^3 - \frac{6\alpha}{2\alpha + 1} x^2 + 1 \right), \qquad (7.68)$$

where the parameter α is the solution of

$$0.9 = \alpha^3 \left(\frac{\alpha + 2}{2\alpha + 1} \right). \qquad (7.69)$$

Look at the coldest temperatures for several carefully selected values of the mass of the particle. To verify that tunneling does indeed take place, begin the simulations in the global minimum, and after the warmup walk, collect position data to generate distributions. DVR solutions can be generated for any given value of the parameters, and the thermodynamic energy and heat capacity can be compared. Does quasiergodicity affect the convergence of the path integral?

9. To first order, the Trotter formula reads,

$$\exp\left[-\beta \left(\mathbf{A} + \mathbf{B}\right)\right] = \lim_{N \to \infty} \left\{ \exp\left[-\frac{\beta}{N}\left(\mathbf{A}\right)\right] \exp\left[-\frac{\beta}{N}\left(\mathbf{B}\right)\right] \right\}^N, \qquad (7.70)$$

where $\mathbf{A}$ and $\mathbf{B}$ are Hermitian operators. Inserting this into

$$Q = \int_{-\infty}^{\infty} dx \, \langle x | \exp\left(-\beta \hat{H}\right) | x \rangle, \qquad (7.71)$$

with $\hat{H} = \hat{T} + \hat{V}$, and $\hat{T}$ the kinetic energy operator, one derives Equation 7.52. Prove that if $[\mathbf{A}, \mathbf{B}] = 0$, then

$$\exp\left(\mathbf{A} + \mathbf{B}\right) = \exp\left(\mathbf{A}\right)\exp\left(\mathbf{B}\right). \qquad (7.72)$$

Then, prove that the symmetric product,

$$\exp\left(\mathbf{A} + \mathbf{B}\right) = \exp\left(\frac{1}{2}\mathbf{B}\right)\exp\left(\mathbf{A}\right)\exp\left(\frac{1}{2}\mathbf{B}\right), \qquad (7.73)$$

also holds if $[\mathbf{A}, \mathbf{B}] = 0$. This result leads to a symmetric Trotter formula,

$$\exp\left[-\beta\left(\mathbf{A} + \mathbf{B}\right)\right]$$

$$= \lim_{N \to \infty} \left\{ \exp\left[-\frac{\beta}{2N}\left(\mathbf{B}\right)\right] \exp\left[-\frac{\beta}{N}\left(\mathbf{A}\right)\right] \exp\left[-\frac{\beta}{2N}\left(\mathbf{B}\right)\right] \right\}^N, \tag{7.74}$$

where the limit is necessary since $[\mathbf{A}, \mathbf{B}] \neq 0$, in general. Show that if $\mathbf{A} = \hat{T}$ and $\mathbf{B} = \hat{V}$, inserting the symmetric Trotter formula into the expression for the partition function Q above, leads to the following variant of Equation 7.55:

$$\mathcal{S}\left(x, x_1, \ldots, x_{N-1}, \beta\right)$$

$$= \frac{Nm}{2\beta\hbar^2} \sum_{i=1}^{N} (x_i - x_{i-1})^2 + \frac{\beta}{2N} \sum_{i=1}^{N} \left[V\left(x_{i-1}\right) + V\left(x_i\right)\right]. \tag{7.75}$$

Modify the code for the path integral so that the symmetric Trotter expansion is used. You will need to modify the estimators as well. Which version has better convergence and statistical properties?

10. The method presented in this section is known as the primitive time-sliced algorithm. The primitive algorithm has a truncation error owing to the fact that N is taken as finite. The truncation error can be shown to be $O(N^{-2})$, meaning that the convergence of the primitive algorithm is linear. It is possible to modify the primitive algorithm [719, 669] by taking higher order terms in the Trotter expansion, e.g.,

$$\exp\left[-\beta\left(\mathbf{A} + \mathbf{B}\right)\right]$$

$$= \lim_{N \to \infty} \left\{ \exp\left[-\frac{\beta}{N}\left(\mathbf{A}\right)\right] \exp\left[-\frac{\beta}{2N}\left(\left[\mathbf{B}, \mathbf{A}\right]\right)\right] \exp\left[-\frac{\beta}{N}\left(\mathbf{B}\right)\right] \right\}^N, \tag{7.76}$$

and treating the resulting commutators perturbatively to obtain quadratic convergence. The truncation error scales as $O(N^{-3})$. The action in Equation 7.55 becomes,

$$\mathcal{S}\left(x, x_1, \ldots, x_{N-1}, \beta\right) = \frac{Nm}{2\beta\hbar^2} \sum_{i=1}^{N} (x_i - x_{i-1})^2 + \frac{\beta}{N} \sum_{i=1}^{N} V_{\text{eff}}\left(x_i\right), \tag{7.77}$$

where

$$V_{\text{eff}}\left(x_i\right) = V\left(x_i\right) + \frac{\beta^2\hbar^2}{24mN^2}\left(\frac{\partial V}{\partial x}\right)^2. \tag{7.78}$$

Higher order convergence implies a smaller value of N at a given temperature, and in turn, that may produce a more efficient algorithm if the gain in the reduced dimensionality of the integration is not offset by the added cost of calculating the gradient of the potential. Modify the program (including the expressions for the estimators) and compare the convergence properties of the energy estimators with the quartic potential.

7.6 The Canonical Fourier Path Integral

The Fourier expansion of the path allows one to partially integrate the action and obtain an alternative formulation. One starts by noting that the (imaginary) time-slice, Δt, is

$$\Delta t = \frac{\beta \hbar}{N}. \tag{7.79}$$

Then, in the kinetic energy part of the action, one can use the definition of the derivative

$$\frac{x_i - x_{i-1}}{\Delta t} = \dot{x}(t), \tag{7.80}$$

since $\Delta t \to dt$ as N approaches infinity. Noting further that

$$\frac{Nm}{2\beta \hbar^2} = \frac{m}{2\hbar \Delta t}, \tag{7.81}$$

and turning the sums from $i - 1$ to N into integrals from 0 to $\beta \hbar$, we can rewrite the action S as,

$$S = \frac{Nm}{2\beta \hbar^2} \sum_{i=1}^{N} (x_i - x_{i-1})^2 + \frac{\beta}{N} \sum_{i=1}^{N} V(x_i) = \int_0^\beta dt \left\{ \frac{m\dot{x}^2(t)}{2\hbar} + \frac{1}{\hbar} V[x(t)] \right\}. \tag{7.82}$$

To make any progress, we need an explicit formulation for the path. Continuous (but not necessarily differentiable everywhere) functions can be expanded using orthogonal functions. Perhaps the most convenient approach is to use sine functions. Let $x(t)$ be a function in imaginary time, $t = 0 \to \beta \hbar$, and let $x(0) = x_0$ and $x(\beta \hbar) = x_0'$. The Fourier expansion of the path is,

$$x(t) = x_0 + \frac{(x_0' - x_0) t}{\beta \hbar} + \sum_{k=1}^{\infty} a_k \sin \left(\frac{k\pi t}{\beta \hbar} \right). \tag{7.83}$$

The sum on the R.H.S. vanishes at the end points of the trajectory. Therefore, the Fourier expansion is the sum of the straight line connecting two points in space and time, and the quantum random fluctuations are accounted for by the randomness of a new set of auxiliary variables $\{a_k\}_{k=1}^{\infty}$. Normally, we are interested in the average of operators that are diagonal in the position representation, so the term linear in t in the path expansion is not necessary.

Noting that,

$$\dot{x}\left(t\right) = \sum_{k=1}^{\infty} a_k \frac{k\pi}{\beta\hbar} \cos\left(\frac{k\pi t}{\beta\hbar}\right), \tag{7.84}$$

we can carry out the kinetic energy integration without much trouble. Changing variables: $u = t/(\beta\hbar)$, and recognizing the orthogonality relation of the cosine function in the 0 to π range gives,

$$\int_0^{\beta\hbar} dt \frac{m\dot{x}^2\left(t\right)}{2\hbar} = \sum_{k=1}^{\infty} a_k^2 \frac{mk^2\pi^2}{4\hbar^2\beta}, \tag{7.85}$$

and the action becomes,

$$S\left(x, \{a_k\}, \beta\right) = \sum_{k=1}^{\infty} a_k^2 \frac{mk^2\pi^2}{4\hbar^2\beta} + \beta \int_0^1 du\, V\left[x\left(u\right)\right]. \tag{7.86}$$

To put this formulation into practice, one must consider a few points. Firstly, the path integral is recast into an infinite-dimensional Riemann integral, just as for the time-slice formalism in the previous section. In practice, we take a finite number ($k_{\max}$) of terms in the Fourier expansion and work toward convergence by systematically increasing it. Secondly, the potential energy, in general, needs to be integrated along a particular trajectory by using numerical quadratures. Researchers have found no real improvement in using a very high order quadrature on the potential. In fact, the composite trapezoid rule, with as many intervals as there are Fourier coefficients, is the most frequently used numerical strategy to calculate the potential contribution to the action $S\left(x, \{a_k\}, \beta\right)$,

$$\int_0^1 du\, V\left[x\left(u\right)\right]. \tag{7.87}$$

Therefore, as it stands, the Fourier path integral algorithm, combined with importance sampling, is no better than the discretized algorithm that was implemented earlier in this chapter. Why use the Fourier path expansion then? The real power of the Fourier formulation is that the convergence with respect to the number of auxiliary variables can be accelerated without the need to compute derivatives of the potential. The technique is called reweighted random series, and it is derived from partial averaging, a quadratically converging method that requires the Hessian of the potential [521].

$$S\left(x, \{a_k\}, \beta\right) = \sum_{k=1}^{k_{\max}} a_k^2 \frac{mk^2\pi^2}{4\hbar^2\beta} + \beta \int_0^1 du\, V_{\text{eff}}\left[x\left(u\right)\right], \tag{7.88}$$

where

$$V_{\text{eff}}\left[x\left(u\right)\right] = V\left[x\left(u\right)\right] + \frac{1}{2}V''\left[x\left(u\right)\right]\sigma^2\left(u\right), \tag{7.89}$$

and where

$$\sigma^2\left(u\right) = \frac{\beta\hbar^2}{m}\left[u\left(1-u\right)\right] - \sum_{k=1}^{k_{\max}} \frac{2\hbar^2\beta}{mk^2\pi^2} \sin^2\left(k\pi u\right). \qquad (7.90)$$

We begin by letting the reader explore partial averaging. The partial averaging expression is developed by treating the tail portion of the path by means of perturbation theory, the same way that the quadratically convergent time-sliced method is derived. The program fpi implements the Fourier path integral with the partial averaging strategy using, once more, the harmonic oscillator of unit mass in atomic units and with unit force constant. Note how the sine functions for the path expansion and $\sigma^2\left(u\right)/\beta$ are computed once outside all the loops and stored in a common block.

```
      program fpi
      implicit real*8 (a-h,o-z)
      parameter (NX = 1000)
      parameter (NMAX = 1000)
      real*8 wm1(0:NMAX),w(0:NMAX), xr(3)
      common /traj/sn(NX,NX),sigma2(NX)
c Program for the stochastic computation of the
c Fourier path integral.
c The array w() stores the path coefficients.
      see=secnds(0.)          !  seed the random number generator
      iseed = int(see/10.)
      call random_seed(iseed)
      pi = 3.1415926
      KMAX = 2       ! quadrature order
      call d2pot(x,d2v)
      if (d2v .ne. 0) write(6,997)
      if (dabs(d2v) .lt. 1.d-12) write(6,998)
      do nQ = 1,4
       N = KMAX
       dN = dfloat(N)
       write(6,999) KMAX, N
c Calculate the sines once for all
       do i=1,N
         sigma2(i) = (dfloat(i)/dN)*(-dfloat(i)/dN +1.d0)
         do k=1,KMAX
          sn(i,k) = sin(dfloat(k*i)*pi/dN)
          sigma2(i) = sigma2(i) - (2.d0)/((pi*dfloat(k))**2)*sn(i,k)**2
         enddo
c sigma2 is for the partial average method
      enddo
      write(6,1001)
4000  format(i10,f12.7)
```

```
MW = 1000000   !  number of warm up moves
M = 1000000
dM = dfloat(M)
temp = 0.02
deltax = 1.0
do k_t = 1,42 ! loop through some temperatures
nrej = 0.d0
sumv = 0.d0
sumt = 0.d0
sumv2 = 0.d0
xim1 = 0.d0
xi = 0.d0
do i=0,KMAX
 w(i) = 0.d0
enddo
call getw(temp,w,xi,N,KMAX,NMAX,wim1) ! initiate the walk
do i=1,MW + M
 call random_number(xr)    ! Draw three random numbers in [0,1].
 imove = (KMAX+1)*xr(1)    ! determine which variable to move
 if (imove .eq. 0 .or. imove .eq. KMAX) then
  xi = xim1 +  deltax *(xr(2)-0.5d0)
 else
  w(imove) = wm1(imove) + deltax*(xr(2)-0.5d0)
 endif
call getw(temp,w,xi,N,KMAX,NMAX,wi)
p = wi/wim1
if (xr(3) .lt. p) then   !  accept or reject
 xim1 = xi              ! accept
 wim1 = wi
  do j=1,KMAX-1
   wm1(j) = w(j)
  enddo
else               ! reject
 nrej = nrej + 1
 xi = xim1
 wi = wim1
 do j=1,KMAX-1
  w(j) = wm1(j)
 enddo
endif
 if (i.gt. MW) then                  ! accumulate data
  call testimator(temp,w,xi,N,KMAX,NMAX,et)
  sumt = sumt + et/dM
  call vestimator(temp,w,xi,N,KMAX,NMAX,ev)
  sumv = sumv + ev/dM
```

```
            sumv2 = sumv2 + ev*et/dM
          endif
        enddo
c the heat capacity with the virial estimator
          cvv =  (sumv2 - sumt*sumv)/(temp*temp)
c and calculate the percent rejections
        pr = 100.*dfloat(nrej)/dfloat(M+MW)
        write(6,1000) temp,sumt,sumv,cvv,pr
        temp = temp + 0.04
        deltax = deltax + 0.2
        enddo
         KMAX = KMAX  + 1                      ! temperature loop
        enddo              ! quadrature loop
997     format('# PARTIAL AVERAGING')
998     format('# NO PARTIAL AVERAGING')
999     format('# KMAX  N =',2i10)
1000    format(6f12.4)
1001    format('       T      <E>T       <E>V     CV(V-T)  %r')
        end
c       TESTIMATOR
        subroutine testimator(temp,w,xi,N,KMAX,NMAX,e)
        implicit real*8 (a-h,o-z)
        parameter (NX = 1000)
        common /traj/sn(NX,NX),sigma2(NX)
        real*8 w(0:NMAX)
        sum = 0.d0
        e = 0.d0
        pi = 3.1415926
        pi2 = pi*pi
        dK = dfloat(KMAX)
        dN = dfloat(N)
c the kinetic energy part
        do i=1,KMAX
          sum=sum-w(i)*w(i)*dfloat(i*i)*pi2*temp*temp/4.d0
        enddo
c the potential energy integral
        do i=1,N
         x = xi
          do k=1,KMAX
           x = x + w(k)*sn(i,k)
          enddo
         call pot(x,v,dv)
c add the partial average part
            call d2pot(x,d2v)
            veff = v + 0.5*d2v*sigma2(i)/temp
```

```
            sum = sum + veff/(dN)
          enddo
          e = sum + dK*temp/2.d0
          return
          end
c    VESTIMATOR
          subroutine vestimator(temp,w,xi,N,KMAX,NMAX,sum)
          implicit real*8 (a-h,o-z)
          parameter (NX = 1000)
          common /traj/sn(NX,NX),sigma2(NX)
          real*8 w(0:NMAX)
          sum = 0.d0
          dN = dfloat(N)
c the potential energy integral
          do i=1,N
           x = xi
            if (i .lt. N) then
            do k=1,KMAX
             x = x + w(k)*sn(i,k)
            enddo
            endif
            call pot(x,v,dv)
c add the partial average part
              call d2pot(x,d2v)
              veff = v + 0.5*d2v*sigma2(i)/temp
             sum = sum + x*dv/(dN*2.d0) + veff/(dN)
          enddo
          return
          end
c    GETW
          subroutine getw(temp,w,xi,N,KMAX,NMAX,wi)
          implicit real*8 (a-h,o-z)
          parameter (NX = 1000)
          common /traj/sn(NX,NX),sigma2(NX)
          real*8 w(0:NMAX)
          sum = 0.d0
          pi = 3.1415926
          pi2 = pi*pi
          dN = dfloat(N)
c the kinetic energy part
          do i=1,KMAX
             sum = sum + w(i)*w(i)*dfloat(i*i)*pi2*temp/4.d0
          enddo
c the potential energy integral
          do i=1,N
```

```
      x = xi
       if (i .lt. N) then
       do k=1,KMAX
       x = x + w(k)*sn(i,k)
       enddo
       endif
       call pot(x,v,dv)
c add the partial average part
          call d2pot(x,d2v)
          veff = v + 0.5*d2v*sigma2(i)/temp
       sum = sum + veff/(dN*temp)
      enddo
      wi = exp(-sum)
      return
      end
c      POT
      subroutine pot(x,v,dv)
      implicit real*8 (a-h,o-z)
c The force constant is 1.000 Hartreee/(bohr^2)
      v = 0.5*x*x
      dv = x
      return
      end
c      D2POT
      subroutine d2pot(x,d2v)
      implicit real*8 (a-h,o-z)
c The force constant is 1.000 Hartreee/(bohr^2)
      d2v = 1.d0   ! set this to 0 if you want to turn off
                   ! partial averaging
      return
      end
```

As evident in Figure 7.5, the partial averaging result is systematically better at all temperatures, compared with the primitive algorithm (obtained by setting the second derivative of the potential to zero, as suggested in the code).

Exercises

1. Derive the expressions for the energy estimators in **testimator** and **vestimator** inside the program **fpi**.

2. Note that **getw** computes the action $S(x, \{a_k\}, \beta)$. Follow the directions in the code to turn off partial averaging. Then, note that **getw**, is called with N, the number of steps in the composite trapezoid quadrature used

to compute

$$\int_0^1 du V\left[x\left(u\right)\right].$$

(7.91)

As it stands, the program `fpi` sets $N = k_{max}$, the total number of path coefficients. By studying the convergence pattern of the energy for the Argon atom in the quartic oscillator,

$$V = 0.01x^4,$$

(7.92)

determine if increasing the quadrature to $N = 2k_{max}, 3k_{max}, \ldots$ improves the convergence properties of the virial estimator. Compare the pattern of convergence of the canonical Fourier path integral with $N = k_{max}$ against the convergence pattern produced by the time-slice method.

3. Does increasing N, the number of steps in the composite trapezoid quadrature used to compute the integral over the potential, improve the statistical behavior of T and the virial estimator? Run the usual 10 blocks to determine statistical errors, and find out.

4. Repeat the simulations for the Argon atom in the quartic oscillator using $N = k_{max}$ as its quadrature, and turn on partial averaging. Note that you must modify the subroutine **d2pot** accordingly. Find out if the convergence pattern improves.

5. One can derive a quadratically convergent Fourier path integral by starting with the second order Trotter-Lie formula to develop a "time-sliced" $(N + 1)$-dimensional Riemann representation of the partition function, then change variables $w_k \rightarrow a_k$ by using

$$w_k = a_k \sin\left(\frac{k\pi t}{\beta\hbar}\right).$$

(7.93)

Verify these statements, and if true, modify the program accordingly to test them.

7.7   The Reweighted Fourier–Wiener Path Integral

Accelerating the convergence of path integral simulations requires either a first or a second order derivative of the potential. The gains obtained by reducing the number of slices N, or coefficients k_{max}, must be weighted against the additional cost of computing the derivatives. In multidimensional applications, especially when the potential energy models are sophisticated, the computation of the gradient, or the Hessian can, in fact, offset the gain at sufficiently cold temperatures. Additionally, the virial estimator requires that the first derivative of the effective potential be evaluated, therefore, estimating the energy requires additional derivatives of the potential one order higher than that is needed in the action. It is the combination of these two factors that made the reweighted random series, coupled with numerical derivative estimators, such powerful contributions [517, 518, 521]. With the reweighted random series, one gains quadratic convergence and statistically stable estimators without the need to evaluate any derivatives of V. For the reweighted Fourier–Wiener path integral, one begins by redefining the random path with $k'_m > k_m$ terms,

$$\tilde{x}\left(u\right) = x_0\left(u\right) + \sigma \sum_{k=1}^{k_m} a_k^\mu \Lambda_k\left(u\right) + \sigma \sum_{k=k_m+1}^{k'_m} a_k^\mu \widetilde{\Lambda}_k\left(u\right), \tag{7.94}$$

where

$$\sigma = \frac{\hbar \beta^{1/2}}{m^{1/2}}, \tag{7.95}$$

and constructs the functions $\widetilde{\Lambda}_k\left(u\right)$, so that the partial averaging expansion about the core path derived from Equation 7.94 is equal to the same derived with the infinite series. The requirement for $\widetilde{\Lambda}_k\left(u\right)$ is easily derived [521],

$$\sum_{k=k_m+1}^{k'_m} \widetilde{\Lambda}_k^2\left(u\right) = u\left(1 - u\right) - \sum_{k=1}^{k_m} \Lambda_k^2\left(u\right). \tag{7.96}$$

For the Fourier–Wiener path integral, the path functions are

$$\Lambda_k\left(u\right) = \sqrt{\frac{2}{\pi^2}} \frac{\sin\left(k\pi u\right)}{k}, \tag{7.97}$$

$$\widetilde{\Lambda}_k\left(u\right) = f\left(u\right) \sqrt{\frac{2}{\pi^2}} \frac{\sin\left(k\pi u\right)}{k}, \tag{7.98}$$

and

$$f\left(u\right) = \sqrt{\frac{u\left(1 - u\right) - \sum_{k=1}^{k_m} \frac{2}{\pi^2 k^2} \sin^2\left(k\pi u\right)}{\sum_{k=k_m}^{k'_m} \frac{2}{\pi^2 k^2} \sin^2\left(k\pi u\right)}}. \tag{7.99}$$

Then, for a N point quadrature, the density matrix ρ^{RW} becomes,

$$\rho^{\mathrm{RW}}\left(x, x', \beta\right)$$
$$= \left(\frac{1}{2\pi}\right)^{ND/2} \left(\hbar^2 \beta\right)^{-D/2} J_\Lambda \int d\left[a\right]_r \exp\left\{-\beta \int_0^1 du\, U\left(\tilde{x}\left(u\right)\right)\right\}, \tag{7.100}$$

where J_Λ is the Jacobian of the transformation $x_i \to a_i$, D is the dimension, and

$$U(\tilde{x}(u)) = \frac{1}{2}m\dot{x}^2 + V[\tilde{x}(u)]. \tag{7.101}$$

This algorithm, complete with the numerical difference estimators for the energy

$$\langle E \rangle_\beta = \frac{D}{2\beta} + \left\langle \frac{\partial}{\partial \beta} \left[\beta \int_0^1 du\, U(\tilde{q}^\mu(u)) \right] \right\rangle, \tag{7.102}$$

and heat capacity,

$$\begin{aligned}
\frac{C_V}{k_B} &= \frac{D}{2} + \frac{D^2}{4} + D\beta \left\langle \frac{\partial}{\partial \beta} \left[\beta \int_0^1 d\tau\, U(\tilde{q}^\mu(u)) \right] \right\rangle \\
&\quad + \beta^2 \left\langle \left\{ \frac{\partial}{\partial \beta} \left[\beta \int_0^1 du\, U(\tilde{q}^\mu(u)) \right] \right\}^2 \right\rangle \\
&\quad - \beta^2 \left\langle \frac{\partial^2}{\partial \beta^2} \left[\beta \int_0^1 du\, U(\tilde{q}^\mu(u)) \right] \right\rangle \\
&\quad - \left\{ -\frac{D}{2} - \beta \left\langle \frac{\partial}{\partial \beta} \left[\beta \int_0^1 du\, U(\tilde{q}^\mu(u)) \right] \right\rangle \right\}^2,
\end{aligned} \tag{7.103}$$

produces the desired convergence properties of the energy and heat capacity. The derivatives,

$$\frac{\partial}{\partial \beta} \left[\beta \int_0^1 du\, U(\tilde{q}^\mu(u)) \right], \tag{7.104}$$

are evaluated numerically. Here is an implementation of the Wiener–Fourier path integral for the harmonic oscillator with $D = 1$.

```
      program fwrrs
      implicit real*8 (a-h,o-z)
      parameter (NMAX = 300)
      parameter (NX = 300)
      parameter  (htok = 3.15773218d5)
      real*8 wm1(0:NMAX),w(0:NMAX), xr(10)
      common /traj/sn(NX,NX),sigma2(NX)
      see=secnds(0.)        !  seed the random number generator
      iseed = int(see)
      call init_random_seed
      do i=1,NX
        sigma2(i) = 0.d0
      do j=1,NX
        sn(i,j) = 0.d0
      enddo
      enddo
c         write(6,*) 'ENTER KMAX >'
```

```
         read(5,*) KMAX
c Calculate the sines ones and for all
         N = KMAX + 1
c         KMAX = N - 1
         do i=1,N
          do k=1,KMAX
           fk = sqrt(2.d0)/(3.1415926*dfloat(k))
           sn(i,k) = fk*sin(dfloat(k*i)*3.1415926/dfloat(N))
          enddo
         enddo
         do i=1,N - 1
            ui = dfloat(i)/dfloat(N)
            sigma2(i) = ui*(1.d0 - ui)
          do k=1,KMAX
             sigma2(i) = sigma2(i) - sn(i,k)*sn(i,k)
          enddo
         enddo
         do i=1,N
          sum00 = 0.d0
          do k=KMAX+1,4*KMAX
           fk = sqrt(2.d0)/(3.1415926*dfloat(k))
           sum00 =  sum00 + fk*sin(dfloat(k*i)*3.1415926/dfloat(N))*
     &         fk*sin(dfloat(k*i)*3.1415926/dfloat(N))
          enddo
          do k=KMAX+1,4*KMAX
           fk = sqrt(2.d0)/(3.1415926*dfloat(k))
           ru = sqrt(sigma2(i)/sum00)
           sn(i,k) = ru*fk*sin(dfloat(k*i)*3.1415926/dfloat(N))
          enddo
          enddo
         temp = 0.02
         do ntemp = 1,43  ! loop through some temperatures
         beta = 1/temp
         dbeta= 1.d-6*beta
c         write(6,*) '# N and KMAX =',N,KMAX
c         write(6,1001)
        MW = 1000000       !  number of warm up moves
        M =  1000000
         dM = dfloat(M)
        deltax = .00210
        nrej = 0.d0
        nrej2 = 0.d0
        sumt = 0.d0
        sumt2 = 0.d0
        sume = 0.d0
```

```
      sume2 = 0.d0
      sume3 = 0.d0
      xm1 = 0.d0
      x = 0.d0
c calculate sigma squared for the partial averaging potential
c initialize the coefficient array to zero
          do i=0,NMAX
           w(i) = 0.d0
           wm1(i) = 0.d0
          enddo
          call getw(beta,w,xi,N,KMAX,NMAX,wim1)    ! initiate the walk
      do i=1,MW + M
      call random_number(xr)
         imove = (4*KMAX)*xr(1) + 1
         xi = xim1 +   deltax *(xr(2)-0.5d0)
         w(imove) = wm1(imove) + deltax*(xr(3)-0.5d0)
         call getw(beta,w,xi,N,KMAX,NMAX,wi)
           ea = wi - wim1
           if (dabs(ea) .gt. 690) then
             p = 0.d0
             if (ea .lt. 0) p = 2.0d0
           else
           p = dexp(-ea)
           endif
      if (xr(4) .lt. p) then
       xim1 = xi                          ! accept
       wim1 = wi
         do j=1,4*KMAX
          wm1(j) = w(j)
         enddo
      else                               ! reject
       nrej = nrej + 1
       nrej2 = nrej2 + 1
       xi = xim1
       wi = wim1
         do j=1,4*KMAX
          w(j) = wm1(j)
         enddo
      endif
       if (i. lt. MW .and. mod(i,1000) .eq. 0) then
         deltax = deltax*(1.d0 - 0.1d0*(nrej2 - 500.d0)/1000.d0)
         nrej2 = 0
       endif
       if (i.gt. MW) then                      ! accumulate data
         call vestimator(beta,w,xi,N,KMAX,NMAX,pv)
```

```
         sumt = sumt + pv/dM
         sumt2 = sumt2 + pv*pv/dM
c calculate the energy and heat capacity using numerical
c derivatives
          betap = beta  + dbeta
          betam = beta - dbeta
         call getw(betap,w,xi,N,KMAX,NMAX,wip)
         call getw(betam,w,xi,N,KMAX,NMAX,wim)
         call getw(beta,w,xi,N,KMAX,NMAX,wi)
         dw = 0.5d0*(wip-wim)/dbeta
         d2w = (wip+wim-2.d0*wi)/(dbeta*dbeta)
         sume = sume + dw/dM
         sume2 = sume2 + (-0.5d0- beta*dw)/dM
         sume3 = sume3 + (0.75d0+ beta*dw+
     & beta*beta*(dw*dw - d2w))/dM
       endif
      enddo
c and calculate the percent rejections
      pr = 100.*dfloat(nrej)/dfloat(M+MW)
      e = sume + 0.5d0/beta
      cv = sume3 -sume2*sume2
      write(6,1000) KMAX,temp,sumt,e,cv,pr
       call flush(6)
      temp = temp + 0.04
      deltax = deltax + 0.02
      enddo            ! temperature loop
       N = N  + 1
c      enddo           ! quadrature loop
1000  format(i6,6f16.7)
1001  format('       Beta         <V>          error       %r')
      end
c        VESTIMATOR
         subroutine vestimator(beta,w,xi,N,KMAX,NMAX,e)
c path average potential energy estimator
         implicit real*8 (a-h,o-z)
         parameter (NX = 300)
         common /traj/sn(NX,NX),sigma2(NX)
         real*8 w(0:NMAX)
         sum = 0.d0
         dN = dfloat(N)
         t12 = sqrt(beta)
         do i=1,N
          x = xi
          do k=1,4*KMAX
```

```
            x = x  + t12*w(k)*sn(i,k) ! sin(dfloat(k*i)*3.14159826
                                      ! /dfloat(N))
         enddo
         call pot(x,v)
         sum = sum + v/dN
         enddo
          e = sum
         return
         end
c        GETW
         subroutine getw(beta,w,xi,N,KMAX,NMAX,wi)
         implicit real*8 (a-h,o-z)
         real*8 w(0:NMAX)
         parameter (NX = 300)
         common /traj/sn(NX,NX),sigma2(NX)
c note that temp is beta
         sum = 0.d0
         t12 = sqrt(beta)
         dN = dfloat(N)
         do i=1,N
c Get the path from the Fourier coefficients
            xm1 = xi
            x = xi
            do k=1,4*KMAX
               if (i .gt. 1) xm1 = xm1 + t12*w(k)*sn(i-1,k)
               x = x  + t12*w(k)*sn(i,k)
            enddo
            call pot(x,v)
            sum = sum + (dN/(beta*2.))*(x-xm1)**2
            sum = sum + beta*v/dN
         enddo
          wi = sum
         return
         end
c     POT
      subroutine pot(x,v)
      implicit real*8 (a-h,o-z)
      v = 0.5d0*x*x
      return
      end
```

The reader should note that the subroutine **vestimator** in the program **fwrrs** no longer computes the virial estimator for the energy; rather, it computes the path average potential energy.

Exercises

1. Beginning with reweighted random series partition function,

$$Q = \int dx \rho^{RW}(x, x, \beta), \qquad (7.105)$$

   and Equations 7.100, 7.25, 7.26, and 7.27, derive Equations 7.102 and 7.103.

2. Modify the program to study the convergence and statistical properties of the energy and heat capacity for the Argon atom in the quartic oscillator,

$$V = 0.01x^4. \qquad (7.106)$$

Part II

Atomic Clusters

8

Characterization of the Potential of Ar$_7$

8.1 Introduction

This chapter is dedicated to generalizing the characterization of potential energy surfaces into multidimensional spaces. Finding minima and transition states is something we can do in one or two dimensions with a simple graph. However, the exploration of multidimensional potential models for realistic simulations of clusters and other kinds of condensed matter presents formidable challenges. Not only can we not simply graph the potential energy surface and locate the points of interest, but the number of minima and transition states grows exponentially with size. Given the difficulty associated with the characterization of potential energy surfaces, the field remains a fertile ground for research and innovation.

The model system we use to train the reader, namely, the Lennard-Jones potential, has been employed extensively in the simulation of properties of liquefied noble gases [6, 7]. Of course, a very important question is, once a suitable potential model is chosen, what structure is the most favored energetically? This is a hard problem for most but the smallest clusters. Special techniques to find initial points for minimizations have been developed. Two of the most popular techniques are the genetic algorithm [816, 823, 824, 826], and basin hopping [806, 807, 810, 818].

The goal of the material in this chapter is to train the reader to use modern theoretical tools for the characterization of multidimensional potentials, however, the process brings the reader closer to the vast amount of literature on Lennard-Jones clusters as well. We begin with the most fundamental issues, such as, formats we use frequently for the storage of coordinates, and the manipulation of these. We go through the construction, the translation, and the rotation of configurations, since these are the basic building blocks for the structural comparison algorithm (SCA). SCA is especially useful during a minima search, to distinguish structures automatically.

After the fundamentals, the attention shifts to the energetics of configurations, and we lead the reader through the construction and the testing of a realistic potential energy model, its gradient, and its Hessian matrix. The reader is taken through the steps to carefully test all three calculations in detail. The Brownian dynamics algorithm is developed and explained in great detail as well. The $T = 0$ Brownian dynamics algorithm, developed in

Chapter 2, is the minimizer that we employ in both genetic algorithm and basin hopping searches of minima. As should be evident to the reader that completes the tutorials for the basin hopping and the genetic algorithm, the issue of finding minima is rather difficult. We train the reader with Ar_7, which is a "toy" problem for these powerful algorithms, since Ar_7 has only five minima. Even for such a problem, however, the reader will discover optical isomerism, the occurrence of which in homogeneous isotropic condensed matter requires lack of symmetry elements in the structure of the isomer.*

With the minima of Ar_7 in hand, we calculate the normal modes of vibration and the characteristic frequency associated with them. Normal modes, and instantaneous normal modes, have their own practical importance in theoretical investigation of the physical properties of condensed matter. The instantaneous normal modes from the diagonalization of the Hessian are the fundamental tools for finding transition states [812, 829, 832, 833, 847]. We introduce our own implementation of the Cerjan–Miller algorithm.

8.2   Cartesian Coordinates of Atomic Clusters

Atoms in a Lennard-Jones cluster are represented by point particles, so that only three degrees of freedom are needed to specify the configuration of each atom, and $3n$ total degrees of freedom are necessary to calculate the potential energy. A particular configuration is a $3n$ set of numbers, each representing a particular Cartesian coordinate. In most of our programs, we store configurations in two different ways. In **storage scheme 1**, a single array x() is filled with $3n$ Cartesian coordinates, where the x coordinates are stored in the entries x(1),...,x(n), the y coordinates in the entries x(n+1),...,x(2n), and the z coordinates in the x(2n+1),...,x(3n). In **storage scheme 2**, three separate arrays are filled with n values, where the x coordinates are stored in the array x(1),...,x(n), the y coordinates in the array y(1),...,y(n), and the z coordinates in the array z(1),...,z(n), or some modification of the identifiers thereof. In `construct7`, for example, we call these rx,ry,rz. Most of the rotations, translations, and computations of the center of mass are more transparent when storage scheme 2 is used for the configuration. However, storage scheme 1 is better suited for molecular dynamics computations, as it makes the storage of the gradient and the Hessian simpler. Therefore, we go back and forth between these two schemes, often within the same program.

In the subroutine `construct7`, we build the Cartesian coordinates of a pentagonal bipyramid. The structure is constructed so that the pentagonal ring, with one atom at each vertex, is in the xy plane. For a given atom to

---

*We refer to a minimum in the potential surface as an isomer interchangeably throughout.

atom distance d, the distance l from the center of the pentagon (placed at the origin) to the atom in the ring is obtained with simple trigonometry,

$$l = \frac{d}{2 \sin (\pi/5)}. \tag{8.1}$$

Once l is known, it is trivial to derive the locations of all five atoms in the xy plane. We use again simple trigonometry, and we place the first atom on the x axis at a distance l (dl in the code). Proceeding clockwise, the second atom is $2\pi/5$ degrees from the x axis and has coordinates, $x = l\cos(2\pi/5)$ and $y = l\sin(2\pi/5)$, etc. Atoms 6 and 7 are placed on the z axis at a distance d from all the atoms in the ring.

```
program buildar7
parameter (NMAX=55)
implicit real*8 (a-h,o-z)
real*8 rx(NMAX),ry(NMAX),rz(NMAX)
real*8 dis(NMAX,NMAX)
n = 7
do i=1,NMAX
 rx(i) = 0.d0 ! initialize all the coordinates to zero
 ry(i) = 0.d0
 rz(i) = 0.d0
 do j=1,NMAX
  dis(i,j) = 0.d0
 enddo
enddo
write(6,*) 'Enter the bond length'
read(5,*) d
call construct7(NMAX,rx,ry,rz,d)
write(6,*) n
write(6,*) 'Ar7 global minimum'
do  i=1,n
 write(6,1000) rx(i),ry(i),rz(i)
enddo
do i=1,n
 do j = i+1,n
 dx = rx(i) - rx(j)
 dy = ry(i) - ry(j)
 dz = rz(i) - rz(j)
 r2 = dx*dx + dy*dy + dz*dz
 dis(i,j) = dsqrt(r2)
 dis(j,i) = dis(i,j)
 enddo
enddo
write(6,*) 'DISTANCE MATRIX'
```

```
      do i=1,n
        write(6,1001) i,(dis(i,j),j=1,n)
      enddo
1000  format('Ar',3f12.5)
1001  format(i4,7f11.5)
      end
      subroutine construct7 (NMAX,rx,ry,rz,d)
      implicit real*8 (a-h,o-z)
      real*8 rx(NMAX),ry(NMAX),rz(NMAX)
          dl = d/1.175570505
          rx(1) = dl
          rx(2) = 0.309017*dl
          ry(2) = 0.951056*dl
          rx(3) = -0.80901699*dl
          ry(3) = 0.587785252*dl
          rx(4) = -0.80901699*dl
          ry(4) = -0.587785252*dl
          rx(5) = 0.309017*dl
          ry(5) = -0.951056*dl
          rz(6) = dsqrt(d**2 - dl**2)
          rz(7) = -dsqrt(d**2 - dl**2)
      return
      end
```

The program buildar7 with $d = 6$ produces the following Cartesian coordinates,

x	y	z
5.1039	0.0000	0.0000
1.5771	4.8541	0.0000
−4.1291	3.0000	0.0000
−4.1291	−3.0000	0.0000
1.5771	−4.8541	0.0000
0.0000	0.0000	3.1543
0.0000	0.0000	−3.1543

and the following distance matrix

	1	2	3	4	5	6	7
1	0.0000	6.0000	9.7082	9.7082	6.0000	6.0000	6.0000
2	6.0000	0.0000	6.0000	9.7082	9.7082	6.0000	6.0000
3	9.7082	6.0000	0.0000	6.0000	9.7082	6.0000	6.0000
4	9.7082	9.7082	6.0000	0.0000	6.0000	6.0000	6.0000
5	6.0000	9.7082	9.7082	6.0000	0.0000	6.0000	6.0000
6	6.0000	6.0000	6.0000	6.0000	6.0000	0.0000	6.3087
7	6.0000	6.0000	6.0000	6.0000	6.0000	6.3087	0.0000

Why do we calculate all the distances? The answer is simple, we are about to translate and rotate the structure produced by `construct7`. Both of these operations should preserve the bond lengths (and angles, for that matter).

8.3   Rotations and Translations

Consider a single three-vector $\mathbf{x} = (x, y, z)$. In Euclidean spaces, we can always rotate such a vector with a 3×3 matrix $\mathbf{U}$, $\mathbf{x}' = \mathbf{U}\mathbf{x}$, and do it in such a way that its size $|\mathbf{x}'|$ is conserved, $|\mathbf{x}'| = |\mathbf{x}|$. Since,

$$|\mathbf{x}'|^2 = (\mathbf{x}')^T \mathbf{x}', \tag{8.2}$$

using the definition of $\mathbf{x}'$ in terms of $\mathbf{U}$ on the right,

$$|\mathbf{x}'|^2 = (\mathbf{U}\mathbf{x})^T \mathbf{U}\mathbf{x}, \tag{8.3}$$

and recalling that $(\mathbf{U}\mathbf{x})^T = \mathbf{x}^T \mathbf{U}^T$, we derive the following equation,

$$|\mathbf{x}'|^2 = \mathbf{x}^T \mathbf{U}^T \mathbf{U}\mathbf{x} = |\mathbf{x}|^2. \tag{8.4}$$

The only way that Equation 8.4 can be true is if

$$\mathbf{U}^T \mathbf{U} = 1, \tag{8.5}$$

holds. That is, if the rotation matrix is orthogonal. We have explored the generators of rotation in Chapter 4 in detail. We return to the rotation of a rigid body and the three Euler angles in a later chapter. For now, we only need to consider rotations using two angles, as we do in Chapter 4. In that case, we have derived the following orthogonal matrix,

$$\mathbf{U} = \begin{pmatrix} \cos(\theta)\cos(\phi) & -\sin(\phi) & \sin(\theta)\cos(\phi) \\ \cos(\theta)\sin(\phi) & \cos(\phi) & \sin(\theta)\sin(\phi) \\ -\sin(\theta) & 0 & \cos(\theta) \end{pmatrix}. \tag{8.6}$$

The subroutine `rotate2` used a modification of the rotation matrix developed in Chapter 4. It rotates each of the (x, y, z) vectors of a configuration, one at the time. Since each vector is rotated with the same angle about the origin, the whole structure is rotated as well.

```
      subroutine rotate2(mna,nar,xv,yv,zv,ct,st,cp,sp)
      implicit real*8 (a-h,o-z)
      real*8 xv(mna),yv(mna),zv(mna)
      real*8 rot(3,3),un(3,3)
      integer*4 dblev
c     On entry:
c             Vectors xv,yv,zv are the coordinates of a configuration
```

```
c                      ct,st,cp,sp are the cosine and sine of the angles
c                      theta and phi. Right handed convention
c                      is used.
c      On return:
c                      Vectors xv,yv,zv are the coordinates of the configuration
c                      after the rotation has taken place
c Begin by constructing the rotation matrix rot(3,3)
       rot(1,1) = ct*cp
       rot(1,2) = -sp
       rot(1,3) = st*cp
       rot(2,1) = ct*sp
       rot(2,2) = cp
       rot(2,3) = st*sp
       rot(3,1) = -st
       rot(3,2) = 0.d0
       rot(3,3) = ct
c rotate  the whole thing...
       do i = 1,nar
          xp = rot(1,1)*xv(i) + rot(1,2)*yv(i) + rot(1,3)*zv(i)
          yp = rot(2,1)*xv(i) + rot(2,2)*yv(i) + rot(2,3)*zv(i)
          zp = rot(3,1)*xv(i) + rot(3,2)*yv(i) + rot(3,3)*zv(i)
          xv(i) = xp
          yv(i) = yp
          zv(i) = zp
       enddo
       return
       end
```

8.4   The Center of Mass

A useful concept we introduced briefly in Chapter 2 for the physics of a system consisting of n particles, is the center of mass. The center of mass is a vector computed with,

$$\mathbf{x}_{\text{Cm}} = \frac{\sum\limits_{i=1}^{n} m_i \mathbf{x}_i}{\sum\limits_{i=1}^{n} m_i}, \tag{8.7}$$

where m_i is the mass associated with the particle at point $\mathbf{x}_i$. The equation can be broken down to a component-by-component basis. For example, the equation for the x component is,

$$x_{\text{Cm}} = \frac{\sum\limits_{i=1}^{n} m_i x_i}{\sum\limits_{i=1}^{n} m_i}. \tag{8.8}$$

In the case of a homogeneous system of masses $(m_i = m \, \forall \, i)$, the equation for each component simplifies further,

$$x_{\mathrm{Cm}} = \frac{1}{n} \sum_{i=1}^{n} x_i. \tag{8.9}$$

The subroutine `centermass` calculates the center of mass vector $\mathbf{x}_{\mathrm{cm}}$ for a uniform system of n masses, and then translates the entire structure so that the center of mass is at the origin of the coordinates.

```
      subroutine centermass(NMAX,xv,yv,zv,n)
      implicit real*8 (a-h,o-z)
      real*8 xv(NMAX),yv(NMAX),zv(NMAX)
c     On entry:
c                    Vectors xv,yv,zv are the coordinates of a cluster
c                    in an arbitrary position in space
c     On return:
c                    Vectors xv,yv,zv are the coordinates of a cluster
c                    translated in the center of mass
c Get the center of mass
      sumx = 0.d0
      sumy = 0.d0
      sumz = 0.d0
      do i = 1,n
       sumx = sumx + xv(i)
       sumy = sumy + yv(i)
       sumz = sumz + zv(i)
      enddo
      xcm = sumx/float(n)
      ycm = sumy/float(n)
      zcm = sumz/float(n)
c translate the cluster
      do i = 1, n
      xv(i) = xv(i)-xcm
      yv(i) = yv(i)-ycm
      zv(i) = zv(i)-zcm
      enddo
      return
      end
```

The translation could be represented by a unitary matrix that belongs to a Lie group, and the Lie group has translation generators as a basis set of the translation Lie algebra. It should not be surprising to learn that these basis are the component of the momentum vector. This is another example of a classical Lie group. The basis set for the Lie algebra associated with translations is a more useful concept in quantum mechanics, therefore, we do not expand on it here.

8.5 The Inertia Tensor

An important physical property associated with the rotation of bodies, like clusters, is the moment of inertia tensor. We will need this concept later on when we deal with rigid body rotation. The elements of the inertia tensor can be used as structure descriptors during simulations. For N point particles, the inertia tensor $\mathbf{I}$ is,

$$I_{\mu\lambda} = \begin{bmatrix} \sum_{i=1}^{N} m_i \left(y_i^2 + z_i^2 \right) & -\sum_{i=1}^{N} m_i x_i y_i & -\sum_{i=1}^{N} m_i x_i z_i \\ -\sum_{i=1}^{N} m_i x_i y_i & \sum_{i=1}^{N} m_i \left(x_i^2 + z_i^2 \right) & -\sum_{i=1}^{N} m_i y_i z_i \\ -\sum_{i=1}^{N} m_i x_i z_i & -\sum_{i=1}^{N} m_i y_i z_i & \sum_{i=1}^{N} m_i \left(x_i^2 + y_i^2 \right) \end{bmatrix}. \quad (8.10)$$

This is calculated in units of mass by the following subroutine.

```
      subroutine inertiat(mna,nar,xv,yv,zv,ii)
      implicit real*8 (a-h,o-z)
      real*8 xv(mna),yv(mna),zv(mna)
      real*8 ii(3,3)
      integer*4 dblev
c     On entry:
c         Vectors xv,yv,zv are the coordinates of a configuration
c     On return:
c         ii(3,3) contains the moment of inertia tensor
c initialize
      ii(1,1) = 0.d0
      ii(1,2) = 0.d0
      ii(1,3) = 0.d0
      ii(2,1) = 0.d0
      ii(2,2) = 0.d0
      ii(2,3) = 0.d0
      ii(3,1) = 0.d0
      ii(3,2) = 0.d0
      ii(3,3) = 0.d0
      do i = 1,nar
       ii(1,1) = ii(1,1) + yv(i)*yv(i) +  zv(i)*zv(i)
       ii(2,2) = ii(2,2) + xv(i)*xv(i) +  zv(i)*zv(i)
       ii(3,3) = ii(3,3) + xv(i)*xv(i) +  yv(i)*yv(i)
       ii(1,2) = ii(1,2) - xv(i)*yv(i)
       ii(1,3) = ii(1,3) - xv(i)*zv(i)
       ii(2,3) = ii(2,3) - yv(i)*zv(i)
```

```
ii(2,1) = ii(2,1) - xv(i)*yv(i)
ii(3,1) = ii(3,1) - xv(i)*zv(i)
ii(3,2) = ii(3,2) - yv(i)*zv(i)
enddo
return
end
```

The moment of inertia tensor must be calculated in the center of mass frame (i.e., the center of mass is at the origin of the Cartesian coordinates). The elements of the tensor change with different orientation of the rigid body. The tensor can be diagonalized by a rotation. The most important frame of reference for a rigid body is one that diagonalizes the inertia tensor for obvious reasons. This frame of reference is call the **body fixed frame**. Even when the inertia tensor is diagonal, there are several cases that arise in practice:

1. $I_{11} = I_{22} = I_{33}$, spherical top

2. $I_{11} = I_{22} < I_{33}$, oblate top (disk-like)

3. $I_{11} = I_{22} > I_{33}$, prolate top (cigar-like)

4. $I_{11} \neq I_{22} \neq I_{33}$, asymmetric top

8.6 The Structural Comparison Algorithm

Now that we have covered some of the most basic transformations that can be performed on the Cartesian coordinates of a configuration without changing its shape, let us introduce the comparison algorithm. In the subroutine **compare** below, two distinct structures are processed according to the following steps.

a. Translate configuration A, so that atom 1 is at the origin.

b. Rotate configuration A, so that atom 2 is on the z axis.

c. Rotate configuration A, so that atom 3 is on the x-z plane.

d. Translate configuration B, so that atom i is at the origin.

e. Rotate configuration B, so that atom j is on the z axis.

f. Rotate configuration B, so that atom k is on the x-z plane.

g. Sort through the remaining atoms of configuration B to find the closest atom in configuration B to atom 4 of configuration A. Do the same for atom 5 of configuration A, then $6, 7, \ldots, n$.

   h. Calculate the distance between the atom of configuration A to the same
atom of configuration B, as permuted in step g, and sum up all such
distances from 1 to n.

   i. Repeat from step e for all $i, j, k = 1, \ldots, n$.

This process systematically goes through n^3 attempts to superimpose configuration B to configuration A. If it succeeds, then one or more of the sums evaluated in step h of the algorithm will be very close to zero. In the subroutine compare, the coordinate arrays are sized dynamically. That makes the subroutine portable and possibly a "canned" routine.

```
      subroutine compare(mna,nar,rx_1,ry_1,rz_1,rx_2,ry_2,rz_2,sigmamin)
c On entry:
c      rx_1,ry_1,rz_1
c      rx_2,ry_2,rz_2
c      contain the coordinates of the atoms of the two configurations to
c      be compared
c On return:
c      sigmamin is the smallest value of the sum
c      of the distances between atoms of the configurations
c      in rx_1,ry_1,rz_1 and those in rx_2,ry_2,rz_2 after these
c      are modified by translations and rotations in the attempt to
c      superimpose the two.
      implicit real*8 (a-h,o-z)
      real*8 rx_1(mna),ry_1(mna),rz_1(mna)
      real*8 rx_2(mna),ry_2(mna),rz_2(mna)
      integer*4 mp(mna)
      iflag = 0
      one = 1.d0
      zero = 0.d0
      sigmamin = 1.d20
c Prepare  configuration A, atom 1 is at the origin
      xi = rx_1(1)
      yi = ry_1(1)
      zi = rz_1(1)
      do j = 1,nar
         rx_1(j) = rx_1(j) - xi
         ry_1(j) = ry_1(j) - yi
         rz_1(j) = rz_1(j) - zi
      enddo  ! J loop
c Atom 2 is on the z axis
      k = 2
      d2_ik = (rx_1(k))**2 + (ry_1(k))**2 + (rz_1(k))**2
      d_ik = dsqrt(d2_ik)
      dxyi = dsqrt((rx_1(k))**2 + (ry_1(k))**2)
      if (d_ik .lt. 1.d-130) stop 'Rotation(1) about y NAN'
      ct = rz_1(k)/d_ik
      st = dxyi/d_ik
```

```
c If dxyi is zero atom 2 is already along the z axis, skip the rotation
      if (dxyi .lt. 1.d-130) goto 10
      cp = rx_1(k)/dxyi
      sp = ry_1(k)/dxyi
      call rotate2(mna,nar,rx_1,ry_1,rz_1,one,zero,cp,-sp)
      call rotate2(mna,nar,rx_1,ry_1,rz_1,ct,-st,one,zero)
10    l = 3
      d2_il = (rx_1(l))**2 + (ry_1(l))**2 + (rz_1(l))**2
      d_il = dsqrt(d2_il)
c Rotate configuration A so that atom 3 is in the x - z plane
      d2_il = (rx_1(l))**2 + (ry_1(l))**2 + (rz_1(l))**2
      d_il = dsqrt(d2_il)
      dxyl = dsqrt((rx_1(l))**2 + (ry_1(l))**2)
      if (dxyl .lt. 1.d-130) goto 11
      cp = rx_1(l)/dxyl
      sp = ry_1(l)/dxyl
      call rotate2(mna,nar,rx_1,ry_1,rz_1,one,zero,cp,-sp)
c Attempt to superimpose B
c translate B so that atom i in configuration 2 has the
c same coordinates (0,0,0) as atom i in configuration 1.
11      do i=1,nar
         xi = rx_2(i)
         yi = ry_2(i)
         zi = rz_2(i)
         do j = 1,nar
            rx_2(j) = rx_2(j) - xi
            ry_2(j) = ry_2(j) - yi
            rz_2(j) = rz_2(j) - zi
         enddo  ! J loop
c Configuration B has been translated now rotate B
c so that atom k is along the z axis
         do k = 1,nar
            if (k .ne. i) then
               d2_ik = (rx_2(k))**2 + (ry_2(k))**2 + (rz_2(k))**2
               d_ik = dsqrt(d2_ik)
               dxyi = dsqrt((rx_2(k))**2 + (ry_2(k))**2)
               if (d_ik .lt. 1.d-130) goto 20
               ct = rz_2(k)/d_ik
               st = dxyi/d_ik
c if dxyi is zero atom k is already along the z axis, skip the rotation
               if (dxyi .lt. 1.d-130) goto 20
               cp = rx_2(k)/dxyi
               sp = ry_2(k)/dxyi
c Rotate B
               call rotate2(mna,nar,rx_2,ry_2,rz_2,one,zero,cp,-sp)
               call rotate2(mna,nar,rx_2,ry_2,rz_2,ct,-st,one,zero)
20             do l = 1,nar
                 if (l .ne. i .and. l .ne. k) then
                    d2_il = (rx_2(l))**2 + (ry_2(l))**2 + (rz_2(l))**2
```

```
                        d_il = dsqrt(d2_il)
c now rotate B so that atom 1 is in the x - z plane
                        d2_il = (rx_2(1))**2 + (ry_2(1))**2 + (rz_2(1))**2
                        d_il = dsqrt(d2_il)
                        dxyl = dsqrt((rx_2(1))**2 + (ry_2(1))**2)
                        if (dxyl .lt. 1.d-130) goto 21
                        cp = rx_2(1)/dxyl
                        sp = ry_2(1)/dxyl
                        call rotate2(mna,nar,rx_2,ry_2,rz_2,one,zero,cp,-sp)
c Sort the remaining nar -2 atoms to minimize the atom of 1 to atom of 2
c distance.  This step reduces the scaling of the algorithm considerably.
21                      mp(1) = 1
                        mp(2) = 2
                        mp(3) = 1
                        do ni=1,nar
                          d2min = 1.d+20
                          do nj = 1,nar
                      iflaga = 0
                            do nk = 1,ni-1
                              if (nj .eq. mp(nk))  iflaga = 1
                            enddo
                            if (iflaga .eq. 0) then
                            d2 = (rx_1(ni) - rx_2(nj))**2 +
      & (ry_1(ni) - ry_2(nj))**2 + (rz_1(ni) - rz_2(nj))**2
                              if (d2 .lt. d2min) then
                               d2min = d2
                               mp(ni) = nj
                              endif
                            endif
22                        enddo
                        enddo
                        sum1 = 0.d0
                        do n = 1, nar
                          d2 = (rx_1(n) - rx_2(mp(n)))**2 +
      & (ry_1(n) - ry_2(mp(n)))**2 + (rz_1(n) - rz_2(mp(n)))**2
                  d = sqrt(d2)
                  sum1 = sum1 + d
                        enddo  ! n loop
                        sigma = sum1
c                        write(6,*) i,k,l,sigma
                        if (sigma .lt. sigmamin) then
                          sigmamin = sigma
c save this minimum sigma configuration....
                        endif
                    endif    ! l =\= k l=\= i
25                  enddo ! l loop    (third atom)
                endif    ! k =\= i
            enddo  ! k loop
        enddo  ! i loop
```

```
c        write(6,*) 'sigmamin = ', sigmamin
c        stop
30       return
         end
```

Albeit simple, the comparison algorithm is a useful and efficient structural analysis tool. We use it to identify distinct structures during minima searches, and to interpret simulations, both stochastic and deterministic. We provide examples of the latter application in Chapter 9. A number of groups around the world carry out simulations and use structural analysis to interpret trends in simulated physical properties. Some groups use inherent structures, which are the configurations obtained by beginning a minimization at every point of the simulation. The comparison algorithm can provide similar information at a fraction of the cost.

Exercises

1. Append `rotate2` to `buildar7`. After constructing the pentagonal bipyramid, call `rotate2` to transform the structure, using angles specified at run time by the user. Then, have the main program write out all the coordinates again using the same format. Note that the sines and cosines of the angles are the actual parameters used to call `rotate2`, unlike its version in Chapter 4. A run with $d = 6, \theta = 1.200$, and $\phi = 2.342$ rad, should give the following coordinates:

x	y	z
-1.28906	1.32618	-4.75704
-3.87908	-2.97349	-1.47001
-1.10835	-3.16390	3.84853
3.19409	1.01809	3.84853
3.08240	3.79311	-1.47001
2.04918	2.10820	1.14302
2.04918	-2.10820	-1.14302

2. Modify `buildar7` so that the atom to atom distance is checked after a rotation. Compare the atom to atom distance before and after the rotation, and convince yourself that all these are the same.

3. Perform two consecutive rotations of the coordinates generated by `buildar7` with two separate calls to `rotate2` as follows. First call: $\phi = 0, \theta = 1.200$ rad. Second call: $\phi = 2.342, \theta = 0.00$ rad. Compare the coordinates at the end of the two transformations with those obtained by reversing the rotation order. First call: $\phi = 2.342, \theta = 0.00$ rad. Second call: $\phi = 0.00, \theta = 1.200$ rad. Should these results be different?

4. Set $\phi = 0$ in the equations for the matrix elements of $\mathbf{U}$, and carry out, by hand, the rotation of a general vector (x, y, z). Find the linear combinations of $x, y,$ and z that determine the new components $x, y,$ and z, and use these equations to deduce the axis about which the rotation by θ degrees takes place.

5. Append `centermass` to `buildar7` and check that `construct7` builds a structure with the center of mass at the origin.

6. Translate the structure generated by `construct7` so that atom 1 is at the origin instead, then check that the distances are conserved after the translation.

7. In order to rotate a configuration correctly, must we make sure that we translate it to the center of mass first? Do an experiment by translating the structure so that atom 1 is at the origin, then, rotate the structure and find out if the distances are conserved after the rotation.

8. Rotate the structure generated by `construct7` so that the vector from atom 1 to atom 2 is on the z axis.

9. Write a program that constructs the coordinate of Ar$_7$ by calling `construct7`, and then computes the moment of inertia tensor. The output should be,

$$\mathbf{I} = \begin{pmatrix} 85.024871408 & 0 & 0 \\ 0 & 85.0249236840 & 0 \\ 0 & 0 & 130.249176912 \end{pmatrix}. \tag{8.11}$$

The inertia tensor is diagonal; note that the pentagonal bipyramid is an oblate top. It can be very hard to find the reference frame for an asymmetric top by guessing. In that case, the best approach is the numerical diagonalization of the inertia tensor.

10. Verify that an arbitrary rotation about the z axis produces an inertia tensor with small but nonzero values of I_{12} and I_{21} in the inertia tensor of the structure generated by `construct7`. Repeat the test with a rotation about the y axis. Then, try once more with an arbitrary rotation about both axis.

11. To test `compare`, modify `buildar7` and generate two configurations using `rotate2`. Then, append the subroutines `compare` and `rotate2`, above, and remove the comment marker at the line
```
c write(6,*) i,k,l,sigma.
```
To test `compare` further, first rotate and then exchange the coordinates between two atoms. Of course, the outcome, should be unchanged.

8.7 Gradients and Hessians of Multidimensional Potentials

Consider a two-dimensional function $V(x,y)$, where x and y are independent variables (degrees of freedom) and the whole set of possible pairs of x, y make up the domain of $V(x,y)$. The derivative of $V(x,y)$ is a vector with two components,

$$\frac{\partial V(x,y)}{\partial x}, \quad \frac{\partial V(x,y)}{\partial y}, \tag{8.12}$$

where the partial derivative is the multidimensional counterpart of the familiar derivative from calculus,

$$\frac{\partial V(x,y)}{\partial x} = \lim_{h \longrightarrow 0} \frac{V(x+h,y) - V(x,y)}{h}. \tag{8.13}$$

Note that the operation of taking a partial derivative with respect to x, simply treats y as a constant, and it is otherwise identical to its one-dimensional counterpart. For example, if $V(x,y) = x/y$, then

$$\frac{\partial V(x,y)}{\partial x} = \frac{1}{y}, \quad \frac{\partial V(x,y)}{\partial y} = -\frac{x}{y^2}. \tag{8.14}$$

The Hessian is a matrix that contains $n \times n$ elements for n degrees of freedom. For $n = 2$, as an example, we get a total of four second derivatives. The reader should verify that if $V(x,y) = x/y$, then the Hessian is

$$\begin{pmatrix} \dfrac{\partial^2 V(x,y)}{\partial x^2} & \dfrac{\partial^2 V(x,y)}{\partial x \partial y} \\ \dfrac{\partial^2 V(x,y)}{\partial y \partial x} & \dfrac{\partial^2 V(x,y)}{\partial y^2} \end{pmatrix} = \begin{pmatrix} 0 & -\dfrac{1}{y^2} \\ -\dfrac{1}{y^2} & \dfrac{2x}{\partial y^3} \end{pmatrix}. \tag{8.15}$$

The reader should note that the Hessian matrix is always symmetric. For n atoms, the number of degrees of freedom is $3n$, the gradient is a $3n$ vector, and the Hessian is a $3n \times 3n$ matrix.

Transition states are special configuration points at which, just like minima, the gradient is zero; no forces on the atoms are present at a transition state. Unlike minima, however, a transition state is a local maximum of the potential along one direction, and a minimum along all the other directions. Consider, for example a two-dimensional version of the quartic potential explored in Chapter 2.

$$V(x,y) = \frac{3}{2\alpha + 1}\left(x^4 + y^4\right) + \frac{4\alpha - 4}{2\alpha + 1}\left(x^3 + y^3\right) - \frac{6\alpha}{2\alpha + 1}\left(x^2 + y^2\right) + 2. \tag{8.16}$$

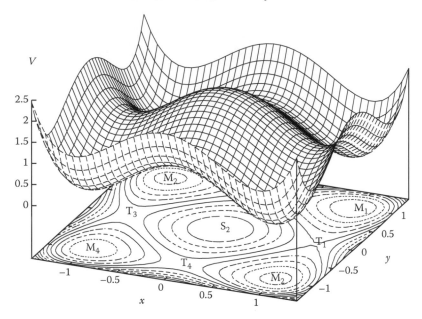

FIGURE 8.1

The two-dimensional potential energy surface for a quadratic potential, graphed with contour lines at $V = 0.5\,(- \cdot -)$, $V = 1.0\,(\cdots)$, and $V = 1.5\,(- - -)$. The points labeled M1, M2, M3 are the four minima of the potential, the points labeled T1, T3, T4 are the transition states (the point T2 at $x = 0, y = 1$ is not visible on the contour plot).

A graph of all the important features for $\gamma = 0.9$ is in Figure 8.1. We leave it to the reader to derive the gradient $\mathbf{D}$,

$$\mathbf{D} = \begin{pmatrix} \dfrac{12}{2\alpha + 1}x^3 + \dfrac{12\alpha - 12}{2\alpha + 1}x^2 - \dfrac{12\alpha}{2\alpha + 1}x \\[3mm] \dfrac{12}{2\alpha + 1}y^3 + \dfrac{12\alpha - 12}{2\alpha + 1}y^2 - \dfrac{12\alpha}{2\alpha + 1}y \end{pmatrix}, \tag{8.17}$$

and the Hessian matrix $\mathbf{H}$, which in this simple example is diagonal over the entire domain,

$$\mathbf{H} = \begin{pmatrix} \dfrac{36}{2\alpha + 1}x^2 + \dfrac{24\alpha - 24}{2\alpha + 1}x - \dfrac{12\alpha}{2\alpha + 1} & 0 \\[3mm] 0 & \dfrac{36}{2\alpha + 1}y^2 + \dfrac{24\alpha - 24}{2\alpha + 1}y - \dfrac{12\alpha}{2\alpha + 1} \end{pmatrix}. \tag{8.18}$$

The reader should set the gradient to zero and derive a total of nine critical points,

x	y	V	Description
1	1	0	Global minimum, M1
1	$-\alpha$	$1 - \gamma$	Second ranking minimum, M2
$-\alpha$	1	$1 - \gamma$	Second ranking minimum, M2
$-\alpha$	$-\alpha$	$2 - 2\gamma$	Third ranking minimum, M3
1	0	1	First order saddle, lowest energy, T1
0	1	1	First order saddle, lowest energy, T1
0	$-\alpha$	$2 - \gamma$	First order saddle, next lowest energy, T2
$-\alpha$	0	$2 - \gamma$	First order saddle, next lowest energy, T2
0	0	2	Second order saddle, S2

$$(8.19)$$

The best way to distinguish minima from transition states, or higher ranking saddles, is to inspect the eigenvalues of the Hessian, which, in this case, is easy to do. For example, the configuration point $(0,0)$ is a second order saddle. The Hessian at this point has two negative eigenvalues (recall that $\alpha > 0$),

$$\mathbf{H}(x, y \to 0, 0) = \begin{pmatrix} -\dfrac{12\alpha}{2\alpha + 1} & 0 \\ 0 & -\dfrac{12\alpha}{2\alpha + 1} \end{pmatrix}. \tag{8.20}$$

The negative eigenvalue indicates a negative curvature at that point along a particular direction. Therefore, the configuration point $(0,0)$ is a maximum along both the x and the y direction. By contrast, the configuration point $(1,1)$ is a minimum, since at this point both eigenvalues of the Hessian are positive,

$$\mathbf{H}(x, y \to 1, 1) = \begin{pmatrix} \dfrac{12\alpha + 12}{2\alpha + 1} & 0 \\ 0 & \dfrac{12\alpha + 12}{2\alpha + 1} \end{pmatrix}, \tag{8.21}$$

and the configuration point $(1,0)$ is a first order saddle, since one of the eigenvalues is positive and one is negative,

$$\mathbf{H}(x, y \to 1, 0) = \begin{pmatrix} \dfrac{12\alpha + 12}{2\alpha + 1} & 0 \\ 0 & -\dfrac{12\alpha}{2\alpha + 1} \end{pmatrix}. \tag{8.22}$$

We encourage the reader to find expressions for the Hessian at all the remaining critical points on the surface, and to verify that at those points the gradient is zero.

8.8   The Lennard-Jones Potential $V^{(\mathrm{LJ})}$

The Lennard-Jones potential for an n particle system is

$$V^{(\mathrm{LJ})} = 4\epsilon \sum_{i=1}^{n} \sum_{j=i+1}^{n} \left[\left(\frac{r_0}{r_{ij}} \right)^{12} - \left(\frac{r_0}{r_{ij}} \right)^{6} \right], \tag{8.23}$$

where r_{ij} is the size of the vector from the center of atom i to the center of atom j. The term proportional to r_{ij}^{-6} is the attractive branch of the interaction physically representing the dispersion part of the potential. The term proportional to r_{ij}^{-12} is the repulsive branch of the interaction that prevents the collapse of a structure down to a point. To see this, consider a single pair of atoms, then

$$V_{ij} = 4\epsilon \left[\left(\frac{r_0}{r_{ij}} \right)^{12} - \left(\frac{r_0}{r_{ij}} \right)^{6} \right]. \tag{8.24}$$

Let's find the equilibrium point, where the derivative of V_{ij} vanishes,

$$\frac{dV_{ij}}{dr_{ij}} = -\frac{4\epsilon}{r_{ij}} \left[12 \left(\frac{r_0}{r_{ij}} \right)^{12} - 6 \left(\frac{r_0}{r_{ij}} \right)^{6} \right]. \tag{8.25}$$

Setting dV_{ij}/dr_{ij} to zero and solving for r_{ij} gives,

$$r_{i,j} = 2^{1/6} r_0. \tag{8.26}$$

At this value of r_{ij}, the potential energy for the pair is at minimum,

$$V_{ij}(r_{ij} = 2^{1/6} r_0) = -\epsilon. \tag{8.27}$$

The fact that we can minimize, analytically, the Lennard-Jones potential for a pair, does not allow us to minimize the potential for 3 or more atoms in any way. It does, however, set a lower bound on the energy of the global minimum since, with n atoms there are at most $n(n-1)/2$ pairs, and these can contribute at most a total of $-\epsilon n(n-1)/2$ energy units. The graph in Figure 8.2 shows the trend of the energy of LJ_n global minima simulated with the Lennard-Jones pair potential, and compares the near linear trend with the curve $-n(n-1)/2$ as a function of n.

Clearly, for $n = 5$, the total number of pairs that are "in equilibrium" is somewhat less than 10. As n increases, the total number of "bonds" that a given structure can form is limited by the maximum possible coordination number that a single atom can have in three-dimensional space. The closest possible packing of a solid has a coordination number of 12, as the reader may recall from introductory chemistry.

A quick glance at the distance matrix for the pentagonal bipyramid (the global minimum for $n = 7$) we constructed earlier, will show a number of pairs with distances greater than the input parameter d. Much more about

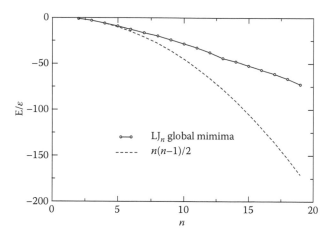

FIGURE 8.2

The energy of the global minima for the first 19 Lennard Jones clusters in units of ϵ compared with the maximum number of pairs.

the minimization problem will come later. For the moment, consider the subroutine pot below,

```
      subroutine pot(maxn,x,n,epsilon,r0,v)
      implicit real*8 (a-h,o-z)
      real*8 x(maxn)
c on entry: x = Cartesian coordinates x(1),...,x(3n)
c where the x coordinates are stored in the entries x(1),...,x(n)
c the y coordinates in the entries x(n+1),...,x(2n)
c and the z coordinates in the x(2n+1),...,x(3n)
c epsilon,r0 are the parameters for the Lennard-Jones interaction
      v = 0.d0
      do i=1,n
       do j = i+1,n
       dx = x(i) - x(j)
       dy = x(n+i) - x(n+j)
       dz = x(2*n+i) - x(2*n+j)
       r2 = dx*dx + dy*dy + dz*dz
       s2 = r0*r0/r2
       s4 = s2*s2
       s6 = s4*s2
       s12 = s6*s6
       v = v + 4.d0*epsilon*(s12 - s6)
       enddo
      enddo
      return
      end
```

First of all, the reader will note that we have used storage scheme 1 for the configuration. This change may seem unnecessary at this time, but it will help when we deal with dynamics later on. Appending `pot` to `builder7` and running with any value of d, should produce a potential equal to -16.474159 units of ϵ. The parameter ϵ is set to one so that the energy is calculated in units of ϵ (called reduced units in the literature), and the length scale parameter r_0 is set to $d/2^{1/6}$ so that the structure generated in `construct7` has optimal bond lengths. In actual simulations, the units of ϵ are going to be the units for the energy (typically, hartree or kelvin), whereas the units of r_0 are the units for all the coordinates, including the units of d in this example. A quick examination of the distance table we generated earlier, should reveal that about 16 pairs have distances in the proximity of the bond length d, therefore, a value of -16.474159 units of ϵ is reasonable, and the structure `construct7` builds is close to the actual minimum for $(LJ)_7$ (c.f. Figure 8.2).

8.9   The Gradient of $V^{(\mathrm{LJ})}$

Using the chain rule, it is not difficult to evaluate the gradient of the Lennard-Jones part of the potential,

$$V^{(\mathrm{LJ})} = \frac{1}{2} \sum_{j=1}^{n} \sum_{k=1,(k \neq j)}^{n} 4\epsilon \left[\left(\frac{r_0}{r_{jk}} \right)^{12} - \left(\frac{r_0}{r_{jk}} \right)^{6} \right]. \tag{8.28}$$

Note that we are representing the sum over pairs differently. The summation index in the innermost sum runs from 1 to n. This choice of range is more convenient for evaluating derivatives, but care must be used to avoid double counting (that is the reason why a factor of $1/2$ appears in front of the summation signs), and to avoid the "self-energy" term ($i = j$), which is, of course, infinite.

$$\frac{\partial V^{\mathrm{LJ}}}{\partial x_i} = \frac{1}{2} \sum_{j=1}^{n} \sum_{k=1,(k \neq j)}^{n} \left(\frac{\partial V_{jk}^{\mathrm{LJ}}}{\partial r_{jk}} \right) \left(\frac{\partial r_{jk}}{\partial x_i} \right). \tag{8.29}$$

Using,

$$\left(\frac{\partial r_{jk}}{\partial x_i} \right) = \frac{(\delta_{ji} - \delta_{ki})}{r_{jk}}, \tag{8.30}$$

we obtain,

$$\frac{\partial V^{\mathrm{LJ}}}{\partial x_i} = -\frac{1}{2} \sum_{j=1}^{n} \sum_{k=1,(k \neq j)}^{n} \frac{24\epsilon}{r_{jk}} \left[2 \left(\frac{r_0}{r_{jk}} \right)^{12} - \left(\frac{r_0}{r_{jk}} \right)^{6} \right] \frac{(x_j - x_k)(\delta_{ji} - \delta_{ki})}{r_{jk}}. \tag{8.31}$$

The Kronecker deltas collapse one of the two sums, since these symbols are zero unless the value of the subscript matches, giving a condition on j in the first delta and a condition on k in the second delta. Therefore, we end up with two terms,

$$\frac{\partial V^{LJ}}{\partial x_i} = -\frac{1}{2} \sum_{k=1,(k \neq i)}^{n} \frac{24\epsilon}{r_{ik}^2} \left[2 \left(\frac{r_0}{r_{ik}} \right)^{12} - \left(\frac{r_0}{r_{ik}} \right)^{6} \right] (x_i - x_k), \qquad (8.32)$$

$$-\frac{1}{2} \sum_{j=1,(j \neq i)}^{n} \frac{24\epsilon}{r_{ij}^2} \left[2 \left(\frac{r_0}{r_{ij}} \right)^{12} - \left(\frac{r_0}{r_{ij}} \right)^{6} \right] (x_i - x_j). \qquad (8.33)$$

The first term arises when $j = i$, the second one when $k = i$. The summands are identical, therefore, changing the summation index in the first term from k to j and combining with the second term, we end up with,

$$\frac{\partial V^{LJ}}{\partial x_i} = - \sum_{j=1,(j \neq i)}^{n} \frac{24\epsilon}{r_{ij}^2} \left[2 \left(\frac{r_0}{r_{ij}} \right)^{12} - \left(\frac{r_0}{r_{ij}} \right)^{6} \right] (x_i - x_j). \qquad (8.34)$$

The derivatives of V, with respect to y and z, can be obtained with an identical derivation, or directly from the last equation by substituting y (or z) where there is an x.

```
      subroutine potd(maxn,x,n,epsilon,r0,v,dv)
      implicit real*8 (a-h,o-z)
      real*8 x(maxn),dv(maxn)
c Potential energy and the gradient
c on entry: x = center of mass coordinates x(1),...,x(3n)
c where the x coordinates are stored in the entries x(1),...,x(n)
c the y coordinates in the entries x(n+1),...,x(2n)
c and the z coordinates in the x(2n+1),...,x(3n)
c epsilon,r0 are the parameters for the Lennard - Jones interaction{array}
c on return: v is the potential. The gradient is in the array
c dv. The arrangement of entries for dv is:
c dv(1),...,dv(n) are the elements corresponding
c                   to the x Cartesian coordinates of the centers in
c dv(n+1),...,dv(2n) are the elements corresponding
c                   to the y Cartesian coordinates
c dv(2n+1),...,dv(3n) are the elements corresponding
c                   to the z Cartesian coordinates
      v = 0.d0
      do i=1,n
       do j = i+1,n
        dx = x(i) - x(j)
        dy = x(n+i) - x(n+j)
        dz = x(2*n+i) - x(2*n+j)
        r2 = dx*dx + dy*dy + dz*dz
        s2 = r0*r0/r2
```

```
        s4 = s2*s2
        s6 = s4*s2
        s12 = s6*s6
        v = v + 4.d0*epsilon*(s12 - s6)
      enddo
    enddo
c in this loop get the gradient
    do i=1,n
      dv(i) = 0.d0
      dv(n+i) = 0.d0
      dv(2*n+i) = 0.d0
      do j = 1,n
       if (i .ne. j) then
        dx = x(i) - x(j)
        dy = x(n+i) - x(n+j)
        dz = x(2*n+i) - x(2*n+j)
        r2 = dx*dx + dy*dy + dz*dz
        s2 = r0*r0/r2
        s4 = s2*s2
        s6 = s4*s2
        s12 = s6*s6
        dvdr = -24.d0*epsilon*(2.d0*s12 - s6)/r2
        dv(i) = dv(i) + dvdr*dx
        dv(n+i) = dv(n+i) + dvdr*dy
        dv(2*n+i) = dv(2*n+i) + dvdr*dz
       endif
      enddo
    enddo
    return
    end
```

We have tested the subroutine `potd` using several techniques, some of which are left as exercise for the reader. Using the pentagonal bipyramid configuration generated by the subroutine `construct7`, with $d = 6$ produces $V = -16.474159$ in units of ϵ and the following x, y, z components of the force on the atoms

Atom number	F_x	F_y	F_z
1	−0.12372	0.00000	0.00000
2	−0.03823	−0.11766	0.00000
3	+0.10009	−0.07272	0.00000
4	+0.10009	+0.07272	0.00000
5	−0.03823	+0.11766	0.00000
6	0.00000	0.00000	−0.36595
7	0.00000	0.00000	+0.36595

These numbers were generated with the following code

```
call potd(NMAX,x,n,epsilon,r0,v,dv)
write(6,*) 'v =', v, ' in units of Epsilon'
write(6,*) 'force vector'
do i=1,n
  write(6,1002) -dv(i),-dv(n+i),-dv(2*n+i)
enddo
```

Not all the forces are zero, as the force vector indicates. That means that even with our best efforts, construct7 fails to produce the global minimum exactly. Next, we introduce a powerful dynamic algorithm for the minimization of this structure.

If the subroutine potd is free of errors, there should be no net force and no net torque about the center of mass. These two conditions are simple consequences of the fact that there are no external fields present in our potential energy model. In Figure 8.3, we plot the potential as a function of $x(1)$, the x coordinate for atom 1, whereas, in Figure 8.4, we compare the numerical and analytical partial derivative of the potential along the same degree of freedom. The numerical and analytical partial derivative of the potential with respect to $x(1)$ are in excellent agreement. The data in the two graphs were generated using a call from construct7 and the following code:

```
c generates a table of v,dv, central difference while traveling
c along the x(1) direction.
      x(1) = 3.58
      dx =  0.121
      h = 1.d-8     ! optimized for double precision (error around 10^{-7})
      do k=1,200
      call potd(NMAX,x,n,epsilon,r0,v,dv)
c calculate the partial derivative numerically
      x(1) = x(1) + h
      call pot(NMAX,x,n,epsilon,r0,vp)
      x(1) = x(1) - 2*h
      call pot(NMAX,x,n,epsilon,r0,vm)
      x(1) = x(1) + h
      dvn = 0.5d0*(vp - vm)/h
      write(8,*) x(1), v,dv(1),dvn,dabs(dv(1) - dvn)
      x(1) = x(1) + dx
      enddo
```

Here are some further suggestions for inquiry.

Exercises

1. Derive Equations 8.17 and 8.18 from Equation 8.16.

2. Set the gradient in Equation 8.17 to zero and derive all the critical points in Equation 8.19.

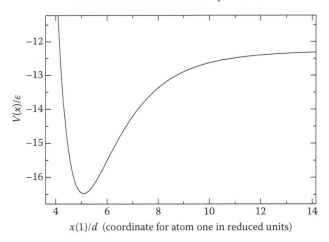

FIGURE 8.3

The energy of $(LJ)_7$ for several values of the x coordinate of atom 1. All the properties are expressed in reduced units.

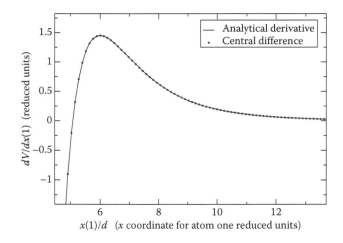

FIGURE 8.4

The partial derivative of the energy of $(LJ)_7$ for several values of the x coordinate of atom 1. All the properties are expressed in reduced units. The analytical derivative (line) is compared against the numerical one.

3. Evaluate the Hessian in Equation 8.18 at all the points in Equation 8.19.

4. Check that the potential energy for Ar_7 is rotationally and translationally invariant.

5. Investigate the dependence of the potential energy for the pentagonal bipyramid created in `construct7` as a function of the parameter d for a fixed value of the parameter r_0. Call `construct7` inside a loop and generate a graph of V vs d. What is the optimal d/r_0 ratio?

6. Test `potd`, first, by verifying that there is no net force on the center of mass. Use the pentagonal bipyramid created in `construct7`,

$$\sum_{i=1}^{n} F_{xi} = -\sum_{i=1}^{n} \frac{\partial V}{\partial x_i} = 0,$$

$$\sum_{i=1}^{n} F_{yi} = -\sum_{i=1}^{n} \frac{\partial V}{\partial y_i} = 0,$$

$$\sum_{i=1}^{n} F_{zi} = -\sum_{i=1}^{n} \frac{\partial V}{\partial z_i} = 0.$$

After a call to `potd`, this can be accomplished with the following code,

```
c verifies that there is not net force on the center of mass
      fx = 0.d0
      fy = 0.d0
      fz = 0.d0
      do i=1,n
       fx = fx -dv(i)
       fy = fy -dv(n+i)
       fz = fz -dv(2*n+i)
      enddo
      write(6,*) 'force on the center of mass'
      write(6,*) fx,fy,fz
```

7. Test `potd` by checking that there is no net torque on the center of mass.

$$\tau_x = \sum_{i=1}^{n} \left(y_i F_{zi} - z_i F_{yi} \right),$$

$$\tau_y = \sum_{i=1}^{n} \left(-z_i F_{xi} + x_i F_{zi} \right),$$

$$\tau_z = \sum_{i=1}^{n} \left(x_i F_{yi} - y_i F_{xi} \right).$$

Here is a suggestion for the code

```
c   verifies that there is not net torque
        tx = 0.d0
        ty = 0.d0
        tz = 0.d0
        do i=1,n
          tx = tx - x(n+i)*dv(2*n+i) + x(2*n+i)*dv(n+i)
          ty = ty + x(i)*dv(2*n+i) - x(2*n+i)*dv(i)
          tz = tz - x(i)*dv(n+i) + x(n+i)*dv(i)
        enddo
        write(6,*) 'new torque about the center of mass'
        write(6,*) tx,ty,tz
```

8. Compare the analytical derivatives obtained in potd against the numerical derivative obtained with the central difference formula at a convenient configuration.

8.10   Brownian Dynamics at 0 K

We have looked at dissipative systems in $\mathbb{R}^1$ space in Chapter 2. The extension to $\mathbb{R}^{3n}$ is straightforward, as we demonstrate. The equations of motion are a set of $3n$ coupled second order differential equations. Using a point slope algorithm similar to the one developed in Chapter 2, we decouple the set locally and use the resulting trivial integrals to update velocities and positions. In $\mathbb{R}^{3n}$ space, the equations of Brownian motion (at $T = 0$) are

$$\ddot{x}_i + \frac{1}{m}\partial_i V + \gamma \dot{x}_i = 0, \tag{8.35}$$

where $1 \leq i \leq 3n$, $\partial_i V$ is an abbreviation of the component i of the gradient of V, and γ is the drag coefficient. The second term on the left is the negative of a force component. The third is the phenomenological dissipative term we introduced in Chapter 2. In the present application, we use this term to reach the nearest minimum, since if there were no drag, the energy of the system would be conserved and the atoms would oscillate about the equilibrium position forever.

Once the acceleration for coordinate x_i is calculated, it can be assumed approximately constant over a sufficiently short time interval Δt, and the velocity component along that direction can be obtained. Using these simple assumptions, we obtain a two-part propagator. The first part for the velocities,

$$\dot{x}_i(t + \Delta t) = \dot{x}_i(t) + \left(\frac{1}{m}\partial_i V + \gamma \dot{x}_i^\mu(t)\right)\Delta t, \tag{8.36}$$

and the second part for the positions,

$$x_i(t + \Delta t) = x_i(t) + \dot{x}_i(t + \Delta t)\Delta t. \tag{8.37}$$

Essentially, these are the equations for rectilinear, uniformly accelerated motion in $3n$ dimension. The algorithm is stable provided the parameters γ and Δt are carefully selected.

It is possible to simulate a finite temperature bath by including a random number drawn from a Maxwell distribution at finite temperature to the right of Equation 8.36. This additional random contribution represents, phenomenologically, the multiple collisions with "particles" of a surrounding bath. With this more elaborate version of the Brownian algorithm, one can slowly decrease the temperature to simulate a real annealing process. Brownian simulated annealing was introduced as a global optimization method. If the annealing schedule is sufficiently slow, then the system has sufficient time to visit all the important configurations. Unfortunately, the likelihood that a single trajectory Brownian simulated annealing becomes trapped into minima other than the global minimum is still high, regardless of how high the initial temperature is chosen and how slow the annealing schedule is implemented. Much more efficient global optimization techniques exist. At the time of writing, two of the most popular approaches are the basin-hopping walk and the genetic algorithm. However, both basin hopping and the genetic algorithm require that a "candidate" configuration, obtained in the course of a random walk, is optimized to the nearest minimum. The minimizer is what we focus on at the moment. The subroutine **brown**, handles the integration by using the double recursion expressed in Equations 8.36 and 8.37.

```
      subroutine brown(maxn,x,dq,dqp,n,epsilon,r0,v)
      implicit real*8 (a-h,o-z)
      real*8 x(maxn),dv(maxn)
      real*8 dq(maxn),dqp(maxn)
c T=0 Brownian dynamics for n rigid centers mapped with 3n Cartesian
c coordinates
c On entry:
c x      the initial 3n Cartesian coordinates
c epsilon, r0, the parameters of the potential energy
c On return:
c x      the 3n Cartesian coordinates of the minimum
      dm = 39.95*1836.6  ! Mass of an argon atom
      fc = 0.0001  ! control from within fc = drag coefficient
      dt = 500  ! atomic units dt = time step
c initiate the velocity and acceleration components to zero
      do k=1,3*n
      dq(k) = 0.d0
      dqp(k) = 0.d0
      enddo
      rmsg = 100.00
c begin the integration
```

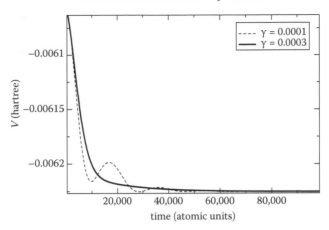

FIGURE 8.5

The energy of $(LJ)_7$ in hartree during two Brownian trajectories started from a pentagonal bipyramid with $d = 7$ bohr. Two values of the drag coefficient γ are tested.

```
c get the gradient and the metric tensor
      do kstep = 1,100000
10          call potd(maxn,x,n,epsilon,r0,v,dv)
            write(8,*) dfloat(kstep -1)*dt,v,rmsg
            if (sqrt(rmsg) .lt. 1.d-12) goto 999
c calculate the acceleration component
            rmsg = 0.d0
            do i=1,3*n
              rmsg = rmsg + dv(i)*dv(i)
              a = -dv(i)/dm -fc*dq(i)
c update the velocity components
              dqp(i) = a*dt
            enddo
c update the positions
            do i=1,3*n
              dq(i) = dq(i) + dqp(i)
              x(i) = x(i) + dq(i)*dt
            enddo
      enddo
999   return
1000  format(2f14.8)
1010  format(1p5e18.9)
      end
```

The graph of the energy vs. time generated inside the subroutine **brown**, reproduced in Figure 8.5, reveals oscillations about the equilibrium with the $\gamma = 0.0001$ a.u. Both trajectories are started from the structure generated by **construct7**, and end with the same value of the energy for the minimum.

The energy of the global minimum is in excellent agreement with the value reported in the literature. The transient nature of the energy with respect to time can be appreciated in Figure 8.5. As explained earlier, this energy loss trend arises from the $\gamma \dot{x}_i$ term in the equation of motion as energy is "dissipated into heat" from the drag term. Increasing γ has the effect of diminishing the intensity of some of the oscillations in the energy vs. time trace.

8.11   Basin Hopping

Brownian dynamics is a far more efficient method for minimizing a multidimensional function than the trial-and-error method of Chapter 3. As we have explained, the minimum that one finds depends on the starting positions and velocities provided to the Brownian dynamics algorithm. In a sufficiently large cluster, minima are very numerous and often close in energy. Systems where minima searches are highly likely to become trapped in local wells are called frustrated. The result of a single minimization by Brownian dynamics, even if it includes white noise that is very slowly annealed away, is unlikely to yield the global minimum in a frustrated system. Condensed matter seems to become a frustrated system after only as few as a dozen atoms. Over the years, many approaches have been developed to overcome this difficulty.

There exist several classes of unbiased methods that deal with the inherent difficulty of locating the global minimum for a particular system after a simulation. Some methods are designed to simulate an annealing schedule, and simulate physical properties in the process. Parallel tempering, a method designed to overcome quasiergodicity, can be used for such purpose, but parallel tempering is inefficient as a tool if all one is interested in is the global minimum. The broad class of phenomena that are classified as quasiergodicity are the same as those responsible for a minima search method to become trapped. However, we have seen systems for which parallel tempering does not find the global minimum. In this section, we introduce a method that has been shown to be powerful in some extreme examples of frustrated systems. Basin hopping is, in essence, a Metropolis walk performed on a transformation of the potential energy surface V,

$$W\left(\mathbf{x}\right) = \min\left\{V\left(\mathbf{x}\right)\right\}, \tag{8.38}$$

where $W\left(\mathbf{x}\right)$ is the transformed energy at a configuration point x, and $\min\left\{V\left(\mathbf{x}\right)\right\}$ is the potential energy obtained upon minimizing $V\left(\mathbf{x}\right)$, with the point $\mathbf{x}$ as the starting position. Here is a summary of the method to generate a basin-hopping random walk in pseudocode.

Step 1: Let $\mathbf{x_0}$ represent an arbitrary configuration of the cluster. Move $\mathbf{x_0}$ by a random amount. In the Metropolis algorithm of a multiatomic system, typically, one moves one atom (or molecule) selected randomly at the time. Instead, in the basin-hopping algorithm, all the atoms (or molecules) are moved at once, therefore, η in the following represents a $3n$-dimensional vector with random components between 0 and 1,

$$\mathbf{x}' = \mathbf{x} + \Delta \left(\eta - \frac{1}{2} \right). \tag{8.39}$$

Δ is the step of the random walk. This parameter is adjusted so that, on average, 50% of the moves are accepted.

Step 2: Submit the configuration $\mathbf{x}'$ to the Brownian algorithm. Let $\mathbf{x}''$ represent the resulting minima coordinates. Then,

$$W\left(\mathbf{x}'\right) = V\left(\mathbf{x}''\right), \tag{8.40}$$

is the transformed energy [818].

Step 3: Calculate the probability for acceptance p,

$$p = \exp \left\{ -\frac{W\left(\mathbf{x}'\right) - W\left(\mathbf{x}\right)}{kT} \right\}, \tag{8.41}$$

and draw a single random number χ between 0 and 1. If $p > \chi$, the move is accepted, otherwise it is rejected.

Step 4: Go back to step 1, until all the desired moves have been made.

All of this comes together in the program basin_hopping below. Only the main routine is shown. The program needs a number of subroutines: construct7, potd, the random number generator ransi, and the minimizer brown. Before running basin_hopping, be sure to comment out the line

```
write(9,*) dfloat(kstep -1)*dt,v,rmsg
```

inside of brown, otherwise a voluminous amount of output will be produced at run time.

```
program basin_hopping
implicit real*8 (a-h,o-z)
parameter (NMAX=55)
parameter (epsilon = 3.77169d-4)  ! hartree (for argon)
parameter (r0 = 6.424) ! bohr  (for argon)
real*8 rx(NMAX),ry(NMAX),rz(NMAX)
real*8 x(NMAX),dv(NMAX),ve(NMAX),acc(NMAX)
real*8 xt(NMAX),xtp(NMAX),xrand(20)
do i=1,NMAX      ! initialize all the arrays
  rx(i) = 0.d0
  ry(i) = 0.d0
  rz(i) = 0.d0
  x(i) = 0.d0
```

```
      xt(i) = 0.d0
      xtp(i) = 0.d0
      dv(i) = 0.d0
      ve(i) = 0.d0
      acc(i) = 0.d0
   enddo
c seed the random number generator
   see=secnds(0.)
   iseed1 = int(see/10.)
   write (6,*) 'ISEED1 ===>',iseed1
   call ransi(iseed1)
c construct the pentagonal bipyramid and quench it as a starting point
   write(6,*) 'Enter the bond length'
   read(5,*) d
   call construct7 (NMAX,rx,ry,rz,d)
   n = 7
   do i=1,n
     x(i) = rx(i)
     x(n+i) = ry(i)
     x(2*n|i) = rz(i)
   enddo
   call brown(NMAX,x,ve,acc,n,epsilon,r0,v)
   vsave = v
   write(6,*) 'Starting v = ', v, ' hartree'
     write(10,*) n    ! store this configuration
     write(10,1000) v
     do i=1,n
       write(10,1010) x(i),x(n+i),x(2*n+i)
     enddo
     icounter = 1
     step = 2.5
     nrej = 0
c basin hopping starts here
     do move = 1,500
       do i = 1,3*n
       xt(i) = x(i)
     enddo
c move all the coordinates at once
     do im=1,3*n
       call ransr(xrand,1)
       xt(im) = x(im) + step*(xrand(1) -0.5d0)
     enddo
     do i = 1,3*n
       xtp(i) = xt(i)     ! xtp contains the coordinates of the minimum
     enddo
c send the configuration to be minimized
     call brown(NMAX,xtp,ve,acc,n,epsilon,r0,v)
     argu = (v-vsave)/0.0001    ! KT is about 30 K
     p = exp(-argu)
```

```
        call ransr(xrand,1)
        if (p .gt. xrand(1)) then
c accept
          vsave = v
          do i = 1,3*n
            x(i) = xt(i)
          enddo
          write(10,*) n    ! store this configuration
          write(10,1000) v
          do i=1,n
            write(10,1010) xtp(i),xtp(n+i),xtp(2*n+i)
          enddo
          write(6,*) move,v
        else
c reject
          nrej = nrej + 1
        endif
80      enddo
        write(6,*) 'number of rejected moves', nrej
1000    format(f14.8)
1010    format(3f18.9)
        end
```

The random walk takes place at $k_B T = 1 \times 10^{-4}$ hartree, which corresponds to a temperature of 31.6 K (plenty to melt the system). If a move is accepted, the new configuration is accepted as well, and the structures are stored in `fort.10`. Many replicas of four different values of the energy should be scrolling on the screen. The format of `fort.10` can be imported into graphical software like XMAKEMOL, VMD, of RASMOL, so that the structures of the minima obtained can be displayed and rotated. Figures 8.6 through 8.9 were generated using XMAKEMOL. The seven-atom cluster is small enough to permit the visualization of the symmetric properties of these structures with still pictures. In Figure 8.9, only one of the two enantiomers is shown. The second can be obtained by reflecting the capping atom through either the horizontal plane of the structure (containing the four equatorial atoms) or the vertical plane. The energy of these structures is identical, but the two are not superimposable. It was surprising to us to encounter optical isomerism in homogeneous matter, since, unlike the typical organic compound, the chiral center does not coincide with any of the nuclei. The rediscovery of optical isomerism in homogeneous matter was made shortly after formulating the subroutine `compare` during the testing phase, precisely with Ar_7. Our observation was, at first, attributed to a bug in our program. A battery of tests revealed that, in fact, the occurrence is real.

In less than five minutes on a Pentium IV 2 GHz machine, basin hopping finds all the minima for Ar_7. If we save all the minima, and polish the list with `compare`, we also find the two enantiomers of the fourth minimum with 500 moves. The power of basin hopping can be understood if one considers

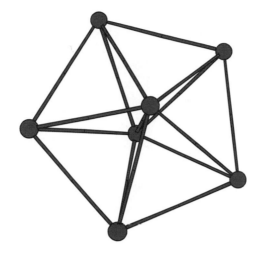

FIGURE 8.6
The global minimum of Ar$_7$, $V = -0.00622531$ hartree.

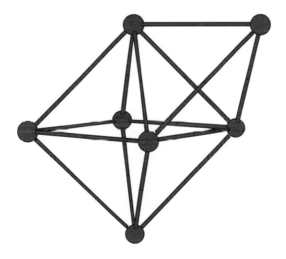

FIGURE 8.7
The second lowest minimum of Ar$_7$ (capped octahedron), $V = -0.0060102$ hartree.

the shape of $W(\mathbf{x})$. Since the transformation maps $W(\mathbf{x})$ to the value of the potential of the nearest minimum reached from $\mathbf{x}$, there are no barriers between minima in the same **funnel** of V, as this transformed function is, in essence, a set of steps. When the potential energy is sufficiently complicated to contain more than one funnel, $W(\mathbf{x})$ contains a set of steps that descend to one minimum, then one or more additional sets of steps exist that descend to another minimum. A sketch of a multifunneled potential energy surface

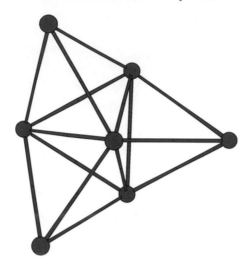

FIGURE 8.8
The third lowest minimum of Ar$_7$ (defective pentagonal bipyramid), $V = -0.0058812$ hartree.

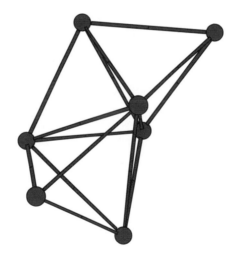

FIGURE 8.9
The fourth/fifth lowest minimum of Ar$_7$ (defective pentagonal bipyramid). $V = -0.0058585$ hartree. Only one of the two enantiomers is shown.

is shown in Figure 8.10. Three funnels are evident in the graph, and the lowest minimum in each funnel is labeled with A, B, and C. Funnels are a complicating feature of the potential energy surface. The barrier between two funnels is a kind of "super barrier." Without funnels, there would be no "global optimization problem."

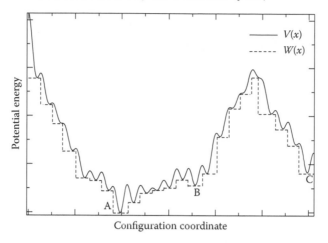

FIGURE 8.10
A sketch of the potential energy of a cluster sliced along a configuration coordinate, showing three separate funnels.

The program `basin_hopping` is not the complete implementation of the basin-hopping algorithm as it appears in the literature. Normally, the algorithm runs for 5000 moves as opposed to 500. More importantly, if the clusters are large enough, there will be more than one funnel, and the program `basin_hopping` will become trapped at the bottom of one of them. The basin-hopping walk has to be restarted frequently, to avoid sampling only a fraction of the important phase space. A number of techniques can be used to restart the basin-hopping walk. In the original implementation, the basin-hopping walk is restarted using angular moves. Angular moves on atoms are performed by selecting an atom i, drawing two random values for θ and ϕ, and setting its vector $\mathbf{r}_i$ relative to the center of mass equal to $r_{\max}(\cos\phi\sin\theta, \sin\phi\sin\theta, \cos\theta)$; $r_{\max}$ is the largest value in the configuration. Atoms are selected for angular moves that have a pair energy E_i,

$$E_i = \sum_{j \neq i} V_{ij}, \qquad (8.42)$$

greater by a fraction α than the lowest pair energy in the configuration; α is adjusted to produce 50% rejections. Furthermore, at the end of the run, we use programs that sort and distinguish structures by employing the Structural Comparison Algorithm in the subroutine `compare`. The basin-hopping algorithm with the implementation given here, can be used to purposely confine the walk to a funnel. In that case, the starting point for the walk are selected minima configurations obtained using intuitive assumptions (a process known as seeding), or obtained with the genetic algorithm discussed next. Learning about multiple funnels in potential energy surfaces, together with transition states, discussed later in the chapter, is indispensable for the proper interpretation of simulation results.

8.12   The Genetic Algorithm

However good one feels about the power of basin hopping, several things remain true about minimizing potential energy surfaces for clusters. Firstly, we never have definite assurance that the global minimum has been found because of the unfortunate presence of funnels. We can increase our confidence by running longer walks or restarting many times, and indeed the latter approach has been used extensively for a number of Lennard-Jones clusters. In the end, though, one is still left without a definite proof. Additionally, as the size of the cluster increases, the number of minima grows exponentially, and any hope of obtaining all the minima has to be abandoned. The explosion of minima also greatly complicates the search for the global minimum, since an increase in walk length continues to produce new distinct minima. Consequently, the possibility of finding a lower minimum remains.

As a result, it is not uncommon to find in the literature, claims that a "new" global minimum not previously reported, is found in the course of a search for a frustrated system. For Lennard-Jones systems, that possibility still exists, but for systems with many hundreds of atoms at this point. Given all these facts, it is good practice, when minimizing a new potential function, to employ an alternative method so that the results can be checked for consistency. If two separate methods find the same result, then one is confident that the true global minimum has been found.

The genetic algorithm is based on the principle of evolution for a population of configurations when this is subjected to an environmental stress. In the case of global optimization, members of a population in a given generation are selected randomly for "reproduction" with a distribution skewed to favor those configurations better fit with respect to the "stress of the environment," e.g., in our case, have lower energy. Through random mutations during the reproduction process, new configurations emerge with lower energy, i.e., better fitness. As the process continues, the growing population, in principle, marches toward the global minimum.

In the original version of the genetic algorithm, a number of operations on the configuration vector of one or two "parent" structures are performed to generate a "child." The reproduction takes place by either cutting and rejoining two sets of coordinates together a number of times (N-point crossover operator), arithmetic and geometric mean, or space inversion. The idea behind the reproduction step is to preserve most of the fitting character of the parent configurations and to introduce small random mutations through the generation process. The version of the genetic algorithm we employ here includes a minimization process for the child structure, and uses a relatively simple reproduction process. We use only a two-point crossover operator by selecting two parent configurations, translating them into the center of mass, rotating them by a random amount, then cutting and rejoining the Cartesian array in a random place to produce a single child. The crucial item for the genetic algorithm is the selection probability for the parents. The latter is implemented as follows. Suppose the population contains s configurations. Let E_i be the energy of configuration i in the population. Furthermore, let $E_{\min}$ be the lowest energy of the population,

$$E_{\min} = \min_i \{E_i\}_{i=1}^s, \tag{8.43}$$

then, the probability p_i that configuration i is selected for reproduction is,

$$p_i = \frac{e^{E_i/F_{\min}}}{\sum\limits_{i=1}^s e^{E_i/E_{\min}}}. \tag{8.44}$$

The program `genetic` is our implementation of the genetic algorithm, using the reproduction process we have described in the last paragraph, and using $T = 0$ Brownian dynamics to find the closest minimum from the children configurations as starting points.

```fortran
program genetic
implicit real*8 (a-h,o-z)
parameter (NMAX=55)
parameter (NS = 6000)  ! maximum number of structures
parameter (epsilon = 3.77169d-4)  ! hartree (for argon)
parameter (r0 = 6.424) ! bohr  (for argon)
real*8 x(NMAX),dv(NMAX),ve(NMAX),acc(NMAX)
real*8 xt(NMAX),xtp(NMAX),xrand(20)
real*8 xA(NMAX),yA(NMAX),zA(NMAX)  ! genetic algorithm arrays
real*8 xB(NMAX),yB(NMAX),zB(NMAX)
real*8 p(NS),vx(NS),xc(NMAX,NS)
do i=1,NMAX    ! initialize all the arrays
  xA(i) = 0.d0
  yA(i) = 0.d0
  zA(i) = 0.d0
  xB(i) = 0.d0
  yB(i) = 0.d0
  zB(i) = 0.d0
  x(i) = 0.d0
  xt(i) = 0.d0
  xtp(i) = 0.d0
  dv(i) = 0.d0
  ve(i) = 0.d0
  acc(i) = 0.d0
```

```
         enddo
         do i=1,NS
           p(i) = 0.d0
           vx(i) = 0.d0
           do j=1,NMAX
             xc(j,i) = 0.d0
           enddo
         enddo
c seed the random number generator
         see=secnds(0.)
         iseed1 = int(see/10.)
         write (6,*) 'ISEED1 ===>',iseed1
         call ransi(iseed1)
c get a starting structure and quench it
         write(6,*) 'Enter the bond length'
         read(5,*) d
         call construct7(NMAX,xA,yA,zA,d)
         n = 7
         do i=1,n
           x(i) = xA(i)
           x(n+i) = yA(i)
           x(2*n+i) = zA(i)
         enddo
         call brown(NMAX,x,ve,acc,n,epsilon,r0,v)
         write(6,*) 'Starting v = ', v, ' hartree'
c -------the genetic algorithm starts here -----
         do i=1,n
           xc(i,1) = x(i)
           xc(n+i,1) = x(n+i)
           xc(2*n+i,1) = x(2*n+i)
         enddo
         npts = 1     ! npts = number of parents
         vmin = v
         vx(1) = v
         ngen = 0
         write(6,*) 'CHILD  DAD    MOM  NCUT   RMIN              V'
c Calculate the selection probability vector
         do ngen = 1,200   ! generation loop
         sp = 0.0
         do j = 1,npts
           if (vx(j) .lt. vmin) vmin = v
         enddo
         do j = 1,npts
         sp = sp + exp(vx(j)/vmin)
         enddo
         do j = 1,npts
           p(j) = exp(vx(j)/vmin)/sp
         enddo
         n_child  = 0
```

```
c pick dad
5       do j = 1,npts
         call ransr(xrand,1)
         if (xrand(1) .lt. p(j)) then
          n_dad = j
          goto 10
         endif
        enddo
        goto 5
c pick mom
10      do j = 1,npts
         call ransr(xrand,1)
         if (xrand(1) .lt. p(j)) then
         n_mom = j
         goto 20
         endif
        enddo
20      do k=1,n
           xA(k) = xc(k, n_dad )
           yA(k) = xc(n+k, n_dad )
           zA(k) = xc(2*n+k, n_dad )
           xB(k) = xc(k,n_mom)
           yB(k) = xc(n+k,n_mom)
           zB(k) = xc(2*n+k,n_mom)
        enddo
c Put the two parent configurations in the center of mass
        call centermass(NMAX,xA,yA,zA,n)
        call centermass(NMAX,xB,yB,zB,n)
c Rotate the configurations by a random amount
        call ransr(xrand,4)
        theta = 2.d0*3.1415926*(xrand(1) - 0.5)
        phi = 4.d0*3.1415926*(xrand(2) - 0.5)
        ct = cos(theta)
        st = sin(theta)
        cp = cos(phi)
        sp = sin(phi)
        call rotate2(NMAX,n,xA,yA,zA,ct,st,cp,sp)
        theta = 2.d0*3.1415926*(xrand(3) - 0.5)
        phi = 4.d0*3.1415926*(xrand(4) - 0.5)
        ct = cos(theta)
        st = sin(theta)
        cp = cos(phi)
        sp = sin(phi)
        call rotate2(NMAX,n,xB,yB,zB,ct,st,cp,sp)
 c make the cut and rejoin in a place at random
        call ransr(xrand,1)
        n_cut = (n-4)*xrand(1) + 2
        do k= n_cut,n
         xA(k) = xB(k)
```

```
          yA(k) = yB(k)
          zA(k) = zB(k)
        enddo
c At this point the coordinates for the child configuration are stored
c in XA,YA,ZA.
c Make sure that no bad overlaps were produced
        rijmin = 1.d+4
        do i=1,n
         do j=i+1,n
          dx = xA(j) - xA(i)
          dy = yA(j) - yA(i)
          dz = zA(j) - zA(i)
          rij = sqrt(dx*dx + dy*dy + dz*dz)
          if (rij .lt. rijmin) rijmin = rij
         enddo
        enddo
        if (rijmin .lt. 4.8d0) goto 5
c quench the child
        do i=1,n
         x(i) = xA(i)
         x(n+i) = yA(i)
         x(2*n+i) = zA(i)
        enddo
        call brown(NMAX,x,ve,acc,n,epsilon,r0,v)
        npts = npts + 1
        do i=1,n
         xc(i,npts) = x(i)
         xc(n+i,npts) = x(n+i)
         xc(2*n+i,npts) = x(2*n+i)
        enddo
        vx(npts) = v
        write(6,1020) ngen, n_dad, n_mom, n_cut,rijmin,v
        write(10,*) n    ! store this configuration
        write(10,*) v
        do i=1,n
          write(10,1010) x(i),x(n+i),x(2*n+i)
        enddo
      enddo
1000  format(i8,2f12.6)
1001  format(i4,7f11.5)
1010  format(3f18.9)
1020  format(4i6,f10.3,f14.8)
      end
```

The genetic algorithm as implemented above, should have no trouble finding all five minima of Ar$_7$. To run **genetic**, a number of subroutines have to be appended, **brown,potd,rotate2**, the random number generator, and **centermass**. Again, many replicas of these minima will appear. The algorithm contained in **genetic** is rather crude. It makes use of only one reproduction strategy. Many more strategies have been studied. Additionally, it produces one child that is minimized and added to the population, even if it is a replica of structures already known. In practice, we apply the genetic algorithm on much more complicated systems than Ar$_7$, when a number of minima are already known, perhaps after running several seeded basin-hopping walks. We begin with about 100 distinct parent structures, and generate 100 children. These are all minimized at once at the end of the reproduction phase. The resulting configurations are added to the population, and the whole population is sorted by energy and pruned by the SCA. The sorting ensures that a new "gene" that drives the population to better fitness makes its way to the top 100 configurations. The pruning process ensures that all the replicas of a configuration in the population are eliminated. At the end of the generation-minimization-sorting-pruning cycle, the 100 distinct configurations with the lowest energy are selected to become the new parents for the next generation. The entire process is handled by two separate programs. Additionally, a shell script automates the calls to these two programs and can keep a single node busy for months with little or no supervision. Those readers that intend to use our programs `basin_hopping` and `genetic` for research purposes, are advised to consult the original literature [806, 807, 810, 818, 823, 824, 816, 826], then modify both programs accordingly. Different restarting strategies for the basin-hopping method and different generators for the genetic algorithm may work more efficiently than others, depending on the system.

Likely, there are two reasons why the genetic algorithm works well. Both have to do with the type of unconventional move made during the reproduction step. The move is designed to retain most of the character that makes the parent structures "fit." At the same time, the reproduction step can easily overcome both kinds of barriers, those between topologically adjacent minima and those that separate funnels.

8.13   The Hessian Matrix

Now, we are ready to move to the next phase of the characterization of Ar$_7$. We calculate the Hessian matrix, which is the matrix that contains all the second derivatives of the potential energy. While this matrix is well defined at every point of the potential energy surface, it takes a particularly special meaning when it is evaluated at a minimum. If we assume that very small displacements take place about a minimum, we can show rigorously that the leading term in the dynamics is harmonic. That makes the (classical) motion about the minimum, isomorphic to that of a set of coupled springs. The analysis of this coupled system of differential equations, derived directly from Newton's law, can be carried out numerically by diagonalization. The result is a set of characteristic frequencies, known as the normal frequencies, and a set of associated eigenvectors. We also need the Hessian to find transition states and classify them.

Deriving the equations for the second derivatives of a sum over pairs of isotropic interactions is a moderately difficult exercise with the regular vector notation we have been using. Tensor analysis significantly simplifies this derivation. It is exercises like these (and that of finding the Laplace operator in curvilinear coordinates) that can really make the user appreciate the power of tensor analysis and the efficiency of the Einstein sum convention. Tensor analysis will be introduced in Chapter 10, when molecular constrained motion and the path integral in curved spaces are introduced. Therefore, the derivation of the Hessian is carried out with the more cumbersome regular vector notation.

Let us begin by expressing the equations of the gradient,

$$\frac{\partial V}{\partial x_i} = \sum_{j=1,(j\neq i)}^{n} (x_i - x_j) V_{ij}', \tag{8.45}$$

and a similar expression applies for the y and z components. In these equations, we introduce the derivative of V_{ij} with respect to r_{ij},

$$V_{ij}' = -\frac{24\epsilon}{r_{ij}^2} \left[2 \left(\frac{r_0}{r_{ij}} \right)^{12} - \left(\frac{r_0}{r_{ij}} \right)^6 \right]. \tag{8.46}$$

These results were derived earlier. We need the second derivative of V_{ij} with respect to r_{ij},

$$\frac{\partial V_{ij}'}{\partial r_{ij}} = \frac{96\epsilon}{r_{ij}^3} \left[7 \left(\frac{r_0}{r_{ij}} \right)^{12} - 2 \left(\frac{r_0}{r_{ij}} \right)^6 \right]. \tag{8.47}$$

The reader is encouraged to derive this result. Now, we can begin to consider the multitude of cases that arise. For example, for $x - x$ partial derivatives, we obtain,

$$\frac{\partial^2 V}{\partial x_k \partial x_i} = \sum_{j=1,(j\neq i)}^{n} \left(\frac{\partial V_{ij}'}{\partial r_{ij}} \right) \left(\frac{\partial r_{ij}}{\partial x_k} \right) (x_i - x_j) + \sum_{j=1 \, (j\neq i)}^{n} V_{ij}' \left(\delta_{ik} - \delta_{jk} \right), \tag{8.48}$$

where, once again, we use,

$$\frac{\partial r_{ij}}{\partial x_k} = \frac{(x_i - x_j)(\delta_{ik} - \delta_{jk})}{r_{ij}}. \tag{8.49}$$

We still need to consider two cases. Case 1, $i = k$ (these are diagonal elements), the Kronecker deltas become $\delta_{ik} = 1$ and $\delta_{jk} = 0$, but we are still left with a sum, since there are no special conditions on j,

$$\frac{\partial^2 V}{\partial x_i^2} = \sum_{j=1,(j\neq i)}^{n} \left[V_{ij}' + \left(\frac{\partial V_{ij}'}{\partial r_{ij}} \right) \frac{(x_i - x_j)^2}{r_{ij}} \right]. \tag{8.50}$$

Case 2, when $k = j$, Equation 8.48 becomes,

$$\frac{\partial^2 V}{\partial x_k \partial x_i} = - \left(\frac{\partial V'_{ik}}{\partial r_{ik}} \right) \frac{(x_i - x_k)^2}{r_{ij}} - V'_{ik}, \tag{8.51}$$

as, this time, the Kronecker deltas become $\delta_{ik} = 0$ and $\delta_{jk} = -1$. Furthermore, the condition on j causes the sum to collapse to a single term. There are two other equations like Equation 8.50 and two like Equation 8.51. These can be obtained simply by replacing x with y and then, again with z.

At this point, there are two more cases we need to consider, when the variables themselves are mixed as in the $x - y$ case. Then,

$$\frac{\partial^2 V}{\partial x_k \partial y_i} = \sum_{j=1,(j\neq i)}^{n} \left(\frac{\partial V'_{ij}}{\partial r_{ij}} \right) \left(\frac{\partial r_{ij}}{\partial x_k} \right) (y_i - y_j). \tag{8.52}$$

These are missing a term compared to the $x - x$, $y - y$, and $z - z$ cases we considered earlier. Using the explicit form of the derivative of r_{ij} gives,

$$\frac{\partial^2 V}{\partial x_k \partial y_i} = \sum_{j=1,(j\neq i)}^{n} \left(\frac{\partial V'_{ij}}{\partial r_{ij}} \right) \frac{(x_i - x_k)(y_i - y_j)(\delta_{ik} - \delta_{jk})}{r_{ij}}. \tag{8.53}$$

The Kronecker deltas introduce two possibilities. Case 1, when $k = i$, we get a sum,

$$\frac{\partial^2 V}{\partial x_i \partial y_i} - \sum_{j=1,(j\neq i)}^{n} \left(\frac{\partial V'_{ij}}{\partial r_{ij}} \right) \frac{(x_i - x_j)(y_i - y_j)}{r_{ij}}, \tag{8.54}$$

whereas, when $k = j$, the condition on the summation index collapses the sum down to one term,

$$\frac{\partial^2 V}{\partial x_k \partial y_i} = - \left(\frac{\partial V'_{ik}}{\partial r_{ik}} \right) \frac{(x_i - x_k)(y_i - y_k)}{r_{ik}}. \tag{8.55}$$

A total of five equations like Equations 8.54 and 8.55 are derived for the $x - z$, $y - x$, $y - z$, $z - x$, and $z - y$ cases. The subroutine pot2d takes care of all these computations.

```
      subroutine pot2d(maxn,x,n,epsilon,r0,v,dv,hess)
        implicit real*8 (a-h,o-z)
        real*8 x(maxn),dv(maxn),hess(maxn,maxn)
c Potential energy and the gradient
c on entry: x = center of mass coordinates x(1),...,x(3n)
c   where the x coordinates are stored in the entries x(1),...,x(n)
c   the y coordinates in the entries x(n+1),...,x(2n)
c   and the z coordinates in the x(2n+1),...,x(3n)
c   epsilon,r0 are the parameters for the Lennard-Jones interaction
c on return: v is the potential. The gradient is in the array
```

```
c  dv. The arrangement of entries for dv is:
c  dv(1),...,dv(n) are the elements corresponding
c                        to the x Cartesian coordinates of the centers in
c  dv(n+1),...,dv(2n) are the elements corresponding
c                        to the y Cartesian coordinates
c  dv(2n+1),...,dv(3n) are the elements corresponding
c                        to the z Cartesian coordinates
c hess(i,j) is  the Hessian matrix
      v = 0.d0
      do i=1,n
       do j = i+1,n
       dx = x(i) - x(j)
       dy = x(n+i) - x(n+j)
       dz = x(2*n+i) - x(2*n+j)
       r2 = dx*dx + dy*dy + dz*dz
       s2 = r0*r0/r2
       s4 = s2*s2
       s6 = s4*s2
       s12 = s6*s6
       v = v + 4.d0*epsilon*(s12 - s6)
       enddo
      enddo
c in this loop get the gradient
      do i=1,n
        dv(i) = 0.d0
        dv(n+i) = 0.d0
        dv(2*n+i) = 0.d0
       do j = 1,n
       if (i .ne. j) then
       dx = x(i) - x(j)
       dy = x(n+i) - x(n+j)
       dz = x(2*n+i) - x(2*n+j)
       r2 = dx*dx + dy*dy + dz*dz
       s2 = r0*r0/r2
       s4 = s2*s2
       s6 = s4*s2
       s12 = s6*s6
       dvdr = -24.d0*epsilon*(2.d0*s12 - s6)/r2
       dv(i) = dv(i) + dvdr*dx
       dv(n+i) = dv(n+i) + dvdr*dy
       dv(2*n+i) = dv(2*n+i) + dvdr*dz
       endif
       enddo
      enddo
c done with the gradient, calculate the Hessian
      do k=1,n
       do i=1,n
c diagonal elements
         if (i .eq. k) then
```

```
          hess(i,i) = 0.d0
          hess(n+i,n+i) = 0.d0
          hess(2*n+i,2*n+i) = 0.d0
          do j=1,n
          if (j.ne. k) then
           dx = x(i) - x(j)
           dy = x(n+i) - x(n+j)
           dz = x(2*n+i) - x(2*n+j)
           r2 = dx*dx + dy*dy + dz*dz
           s2 = r0*r0/r2
           s4 = s2*s2
           s6 = s4*s2
           s12 = s6*s6
           dvdr = -24.d0*epsilon*(2.d0*s12 - s6)/r2
           d2vdr = 96.d0*epsilon*(7.d0*s12 - 2.d0*s6)/r2
c x-x terms
          hess(i,i) = hess(i,i) + dvdr + d2vdr*dx*dx/r2
c y-y terms
          hess(n+i,n+i) = hess(n+i,n+i) + dvdr + d2vdr*dy*dy/r2
c z-z terms
          hess(2*n+i,2*n+i) = hess(2*n+i,2*n+i) + dvdr +
     &       d2vdr*dz*dz/r2
c x-y terms (not exactly diagonal, but they still need a sum)
          hess(i,n+i) = hess(i,n+i) + d2vdr*dx*dy/r2
          hess(n+i,i) = hess(n+i,i) + d2vdr*dx*dy/r2
c x-z terms (not exactly diagonal, but they still need a sum)
          hess(i,2*n+i) = hess(i,2*n+i) + d2vdr*dx*dz/r2
          hess(2*n+i,i) = hess(2*n+i,i) + d2vdr*dx*dz/r2
c y-z terms (not exactly diagonal, but they still need a sum)
          hess(n+i,2*n+i) = hess(n+i,2*n+i) + d2vdr*dy*dz/r2
          hess(2*n+i,n+i) = hess(2*n+i,n+i) + d2vdr*dy*dz/r2
           endif
          enddo
          else
c off diagonal elements
              dx = x(i) - x(k)
              dy = x(n+i) - x(n+k)
              dz = x(2*n+i) - x(2*n+k)
              r2 = dx*dx + dy*dy + dz*dz
              s2 = r0*r0/r2
              s4 = s2*s2
              s6 = s4*s2
              s12 = s6*s6
              dvdr = -24.d0*epsilon*(2.d0*s12 - s6)/r2
              d2vdr = 96.d0*epsilon*(7.d0*s12 - 2.d0*s6)/r2
c x-x terms
              hess(i,k) = -(dvdr + d2vdr*dx*dx/r2)
c y-y terms
              hess(n+i,n+k) = -(dvdr + d2vdr*dy*dy/r2)
```

```
c z-z terms
        hess(2*n+i,2*n+k) = -(dvdr + d2vdr*dz*dz/r2)
c x-y terms
        hess(i,n+k)   =  -d2vdr*dx*dy/r2
        hess(i,2*n+k) = -d2vdr*dx*dz/r2
        hess(n+i,k) = -d2vdr*dy*dx/r2
        hess(n+i,2*n+k) =  -d2vdr*dy*dz/r2
        hess(2*n+i,k) = -d2vdr*dz*dx/r2
        hess(2*n+i,n+k) =  -d2vdr*dz*dy/r2
      endif
      enddo
    enddo
    return
    end
```

The program `hessian`, assumes that the a file named `input.dat` is present in the working directory and that it contains the coordinates of the minima of Ar_7. The Hessian is calculated at the global minimum and then checked by numerical differentiation. The reader should repeat the test with several configurations, even those generated by `construct7`.

```
    program hessian
    implicit real*8 (a-h,o-z)
    parameter (NMAX=55)
    parameter (NS = 6000)   ! maximum number of structures
    parameter (epsilon = 3.77169d-4)   ! hartree (for argon)
    parameter (r0 = 6.424) ! bohr  (for argon)
    real*8 rx(NMAX),ry(NMAX),rz(NMAX),dv(NMAX),hess(NMAX,NMAX)
    real*8 x(NMAX)
    do i=1,NMAX
      rx(i) = 0.d0
      ry(i) = 0.d0
      rx(i) = 0.d0
      dv(i) = 0.d0
      x(i) = 0.d0
      do j=1,NMAX
       hess(i,j) = 0.d0
      enddo
    enddo
    n = 7
c read the coordinates of the global minimum
    open(unit=2,file='input.dat',status='old')
    read(2,*) ene
    write(6,*) ene
    do i=1,n
     read(2,1000) rx(i),ry(i),rz(i)
     write(6,1000) rx(i),ry(i),rz(i)
    enddo
    do i=1,n
```

```
        x(i) = rx(i)
        x(n+i) = ry(i)
        x(2*n+i) = rz(i)
      enddo
      call  potd(NMAX,x,n,epsilon,r0,v,dv)
      write(6,*) 'FORCE VECTOR'
      do i=1,n
        write(6,1000) -dv(i),- dv(n+i),-dv(2*n+i)
      enddo
      call  pot2d(NMAX,x,n,epsilon,r0,v,dv,hess)
c check the Hessian numerically
      do i=1,3*n
        do j=1,3*n
          x(i) = x(i) + 1.d-8
          call  potd(NMAX,x,n,epsilon,r0,v,dv)
            dnhp = dv(j)
            x(i) = x(i) - 2.d-8
          call  potd(NMAX,x,n,epsilon,r0,v,dv)
            dnhm = dv(j)
            x(i) = x(i) + 1.d-8
            dnh  = 0.5d0*(dnhp - dnhm)/1.d-8
            write(6,1010) i,j,hess(i,j),dnh,(hess(i,j)-dnh)
        enddo
      enddo
      close(2)
1000  format(3f18.9)
1010  format(2i7,3f18.9)
      end
```

8.14   Normal Mode Analysis

Consider a minimum configuration, $\mathbf{x}_0$ (any minimum will do). In this section, it is convenient to introduce mass weighted coordinates, $\sqrt{m_i}x_i = q_i$, $1 \leq i \leq 3n$, where m_i is the mass associated with the degree of freedom i. Considering only small displacements from $\mathbf{q}_0$, we expand the potential energy surface into a Taylor series about $\mathbf{q}_0$,

$$V(\mathbf{q}) = V(\mathbf{q}_0) + \sum_{i=1}^{3n} \partial_i V(\mathbf{q}_0) \left(x_i - x_{i0}\right)$$

$$+ \frac{1}{2} \sum_{i=1}^{3n} \sum_{j=1}^{3n} \partial_i \partial_j V(\mathbf{q}_0) \left(q_i - q_{i0}\right) \left(q_j - q_{j0}\right), \qquad (8.56)$$

where $\partial_i V(\mathbf{q}_0)$ are the elements of the gradient evaluated at the minimum with respect to q_i. Of course, these are zero by definition. The elements of the

Hessian matrix, in terms q_i and q_j, are expressed with the symbol $\partial_i \partial_j V(\mathbf{q}_0)$. It is important to realize that both $V(\mathbf{q}_0)$ and the Hessian are evaluated at the minimum, and are therefore constants. If we take the negative of the derivative with respect to q_k on both sides of Equation 8.56, we obtain the force component along k,

$$F_k = -\frac{1}{2} \sum_{i=1}^{3n} \sum_{j=1}^{3n} \partial_i \partial_j V(\mathbf{q}_0) \left[(q_i - q_{i0}) \, \delta_{jk} + (q_j - q_{j0}) \, \delta_{ik} \right], \qquad (8.57)$$

and after we take care of the two cases (a condition on a summation index collapses one sum in both cases), we get

$$F_k = -\frac{1}{2} \sum_{i=1}^{3n} \partial_i \partial_k V(\mathbf{q}_0) \, (q_i - q_{i0}) - \frac{1}{2} \sum_{j=1}^{3n} \partial_k \partial_j V(\mathbf{q}_0) \, (q_j - q_{j0}). \qquad (8.58)$$

Since j is a dummy index and the Hessian is symmetric, we finally get,

$$F_k = -\sum_{i=1}^{3n} \partial_i \partial_k V(\mathbf{q}_0) \, (q_i - q_{i0}) = \ddot{q}_k. \qquad (8.59)$$

On the right of Equation 8.59, $\ddot{q}_k$ is the acceleration along the k degree of freedom from Newton's second law. (No mass enters here, the mass is hidden in the definition of q.) Equation 8.59 represents a set of $3n$ coupled, second order differential equations, since the second order time derivative of q_k is related to the value of all the other coordinates. These differential equations are easily decoupled and solved by diagonalization, as we now show. Assuming the motion is periodic, we write $q_k - q_{k0} = A_k \cos(\omega_k t)$, where A_k is an amplitude and t is time. Then, with two derivatives it is simple to derive $\ddot{q}_k = -\omega_k^2 A_k \cos(\omega_k t)$. We substitute these into Equation 8.59 and we get,

$$\sum_{i=1}^{3n} \partial_i \partial_k V(\mathbf{q}_0) A_i \cos(\omega_i t) = \omega_k^2 A_k \cos(\omega_k t). \qquad (8.60)$$

The L.H.S. is nothing more than the product of a matrix with a vector, therefore, we can simply write,

$$\mathbf{H} \mathbf{a} = \omega^2 \mathbf{a}, \qquad (8.61)$$

with $a_i = A_i \cos(\omega_i t)$, $H_{ij} = \partial_i \partial_j V(\mathbf{q}_0)$. This equation is in the form of an eigenvalue–eigenvector problem, which we can solve with a simple call to `eigen` to diagonalize $\mathbf{H}$. The eigenvalues of $\mathbf{H}$ are the squares of the natural frequencies associated with the configuration $\mathbf{q}_0$. The vector $\mathbf{a}_k$ contains the linear combination of the Cartesian coordinates that make up the collective motion of the mode k. Since the $3n$ vectors $\mathbf{a}$ make up a complete orthonormal basis set for the eigenvector space of $\mathbf{H}$, in principle, we can expand any trajectory using this set. In particular, the translation of the center of mass, and the three rotational degrees of freedom can be expressed as linear combinations of basis in this space. However, since there is no net force on the center of mass, and the potential produces no net torque on the configuration $\mathbf{q}_0$, there cannot be any frequency associated with those degrees of freedom (six in total when the configuration $\mathbf{q}_0$ is nonlinear, five otherwise). Therefore, when we diagonalize the Hessian, we should see six eigenvalues close to zero.

Since we have the Hessian with respect to x_i and x_j, we need to use the chain rule and derive the following result before we submit $\mathbf{H}$ to the diagonalizer,

$$\partial_i \partial_j V(\mathbf{q}_0) = \frac{\partial x_i}{\partial q_i} \frac{\partial x_j}{\partial q_j} \frac{\partial^2 V}{\partial x_i \partial x_j} = \frac{1}{\sqrt{m_j} \sqrt{m_i}} \frac{\partial^2 V}{\partial x_i \partial x_j}. \tag{8.62}$$

The eigenfrequencies of the Hessian about a minimum are important observables, as most of these can be directly measured with spectroscopy. The normal mode analysis is a very successful tool for the treatment of small covalent molecules and can be used to interpret infrared and Raman spectra. Additionally, a number of related approaches, such as the instantaneous normal mode analysis, have been used to treat liquids theoretically. Normal modes of minima can also be used to approximate partition functions and thermodynamic properties of clusters [799].

In associating the eigenfrequencies with infrared or Raman spectra, the wavenumber cm^{-1}, is a frequently used unit. The wavenumber is derived from the frequency ω in rad s^{-1} by dividing by $2\pi c$, where $c = 2.99724 \times 10^{10}$ cm s^{-1} is the speed of light. Our Hessian is in hartree bohr^{-2}. The square root of that unit is in hartree $\cdot$ rad (1 atomic time units = 1 hartree^{-1} = 2.4197×10^{-17} s). The factor 2.195×10^5 in the program normal_modes, converts from atomic units of angular frequency to wavenumbers.

```
program normal_modes
implicit real*8 (a-h,o-z)
parameter (NMAX=55)
parameter (NS = 6000)   ! maximum number of structures
parameter (epsilon = 3.77169d-4)   ! hartree (for argon)
parameter (r0 = 6.424) ! bohr  (for argon)
parameter (dm = 39.95*1822.8) ! mass of a single Ar atom
real*8 rx(NMAX),ry(NMAX),rz(NMAX),dv(NMAX),hess(NMAX,NMAX)
real*8 x(NMAX)
real*8 dia(NMAX),z(NMAX,NMAX),e(NMAX)
do i=1,NMAX
  rx(i) = 0.d0
  ry(i) = 0.d0
  rx(i) = 0.d0
  dv(i) = 0.d0
  x(i) = 0.d0
```

```
      do j=1,NMAX
       hess(i,j) = 0.d0
      enddo
     enddo
     n = 7
     open(unit=2,file='input.dat',status='old')
     read(2,*) ene
     write(6,*) ene
     do i=1,n
      read(2,1000) rx(i),ry(i),rz(i)
      write(6,1000) rx(i),ry(i),rz(i)
     enddo
     do i=1,n
       x(i) = rx(i)
       x(n+i) = ry(i)
       x(2*n+i) = rz(i)
     enddo
     call  potd(NMAX,x,n,epsilon,r0,v,dv)
     write(6,*) 'FORCE VECTOR'
     do i=1,n
      write(6,1000) -dv(i),- dv(n+i),-dv(2*n+i)
     enddo
     call  pot2d(NMAX,x,n,epsilon,r0,v,dv,hess)
     do i=1,3*n
      do j=1,3*n
       hess(i,j) = hess(i,j)/dm
      enddo
     enddo
     call eigen(NMAX,3*n,hess,dia,z,e,ierr)
c 2.195d+5 converts from atomic units of angular frequency to
c wavenumbers. (hbar = 1, so rad/atomic units of time = hartree
c then,  1 hartree = 4.360 \times 10^{-18} J.
c Divide the energy by hc (speed of light in cm/s) and you
c get wavenumbers.
     do i=1,3*n
      write(6,1020) i,dsqrt(dabs(dia(i)))*2.195d+5
     enddo
     close(2)
1000  format(3f18.9)
1010  format(2i7,3f18.9)
1020  format(i5,f12.5)
       end
```

The reader is encouraged to try diagonalizing the Hessian matrix at all the minima and compare the results with those in Table 8.1. The colums headed by M are the multiplicities (degeneracies) for a given frequency value.

TABLE 8.1

Characteristic Frequencies of the Four Minima of Ar$_7$ cm^{-1}

M	i = 1	M	i = 2	M	i = 3	M	i = 4
1	38.9	1	38.8	1	38.9	1	40.7
2	38.5	1	38.0	2	38.0	1	38.8
2	30.5	2	34.5	1	30.6	1	35.2
2	30.0	1	31.8	2	30.6	1	33.6
1	26.7	2	29.6	1	26.2	1	29.7
2	21.9	2	24.0	2	23.9	1	27.1
1	20.4	1	21.6	1	23.5	1	27.0
2	19.8	1	19.6	2	19.9	1	25.2
2	14.4	2	18.2	1	17.1	1	22.6
–		2	13.5	2	8.9	1	22.3
–		–		–		1	21.3
–		–		–		1	19.4
–		–		–		1	14.9
–		–		–		1	10.9
–		–		–		1	8.6

8.15   Transition States with the Cerjan–Miller Algorithm

The job of characterizing the potential energy surface through the structural analysis of minima yields a great deal of information about the sample of condensed matter under study. However, technically speaking, if we work with the global minimum (the most important minimum), then we neglect temperature effects, or we study the 0 K properties classically. While quite insightful, neither of these approaches can be a complete theoretical investigation of clusters, since at 0 K hardly anything behaves classically, and temperature effects should be studied carefully to establish their importance. Neglecting temperatures essentially implies that the global minimum is very deep, energetically, compared with the other minima, and that at some intermediate temperature, where quantum effects are negligible, there is still not enough thermal energy to reach other important minima of the potential. The combination of these two events only takes place in small covalent molecules. Neglecting temperature effects fails as an assumption in larger molecules, such as polymers and biopolymers, and in condensed matter.

Of course, one can simply include temperature effects by using stochastic simulations, like the classical parallel tempering simulation of Chapter 3, or at low temperatures, the path integral simulation in Chapter 7. These simulations are the subject of the next chapter. Nevertheless, it is often instructive to learn about the structure and the energetics of transition states of a potential energy surface, especially when studying how a surface morphology changes under the change of a parameter.

How does one find transition states (first order saddles) on a multidimensional potential energy surface, such as that of Ar$_7$? The task of finding minima is already a very hard problem, one would expect that finding transition states is harder. In fact, finding transition states, or higher order saddles, is not trivial. For example, we cannot use Brownian dynamics, because the chance that a trajectory reaches a metastable configuration with just the precise amount of energy to come to a stop at the top of a saddle, is remote. Minimizing the gradient, instead of the potential, is not efficient, since in the process, one will find far more minima than saddle points. A better approach, which we often use in our own research, has been developed by Cerjan and Miller [847]. Their algorithm is based on a local quadratic expansion of the potential (meaning, at every point along the walk, the gradient and Hessian are recalculated). To better understand how the method developed by Cerjan and Miller works, it is helpful to return to the simpler two-dimensional potential energy model in Equation 8.16. Let us assume that we are at point x_0, y_0 on that two-dimensional surface. By a local quadratic expansion, we mean,

$$V_L \approx V_0 + D_1 \Delta x + D_2 \Delta y + \frac{1}{2} H_{11} \Delta x^2 + \frac{1}{2} H_{22} \Delta y^2, \qquad (8.63)$$

where V_0 is the potential in Equation 8.16, D_1 and D_2 are the two elements of the gradient in Equation 8.17, and H_{11}, H_{22} are the two nonzero elements of the Hessian. All five of these are constants evaluated at x_0, y_0. Cerjan and Miller proposed to find the extrema of the potential using this local quadratic expansion subject to the constraint that the size of the step is fixed, $\Delta x^2 + \Delta y^2 = \Delta^2$. The constraint is introduced so that the best direction to walk uphill toward transition states can be found. The following Lagrangian:

$$\mathcal{L} = V_L + \frac{\lambda}{2} \left(\Delta^2 - \Delta x^2 - \Delta y^2 \right), \qquad (8.64)$$

governs the equations of motion along the desired direction once the best value for the Lagrange multiplier λ is obtained. The optimal values for λ, Δx, and Δy are obtained by setting the partial derivatives of $\mathcal{L}$ to zero,

$$\frac{\partial \mathcal{L}}{\partial \Delta x} = 0, \qquad (8.65)$$

$$\frac{\partial \mathcal{L}}{\partial \Delta y} = 0, \qquad (8.66)$$

$$\frac{\partial \mathcal{L}}{\partial \lambda} = 0. \qquad (8.67)$$

These yield,

$$D_1 + H_{11} \Delta x - \lambda \Delta x = 0, \qquad (8.68)$$

$$D_2 + H_{22} \Delta y - \lambda \Delta y = 0, \qquad (8.69)$$

$$\Delta^2 - \Delta x^2 - \Delta y^2 = 0. \qquad (8.70)$$

To find the best value for λ, one solves for Δx in Equation 8.68, for Δy in Equation 8.69, and plugs the two expressions into Equation 8.70,

$$\Delta^2 = \frac{D_1^2}{(\lambda - H_{11})^2} + \frac{D_2^2}{(\lambda - H_{22})^2}. \tag{8.71}$$

The local minimum of the function $\Delta^2(\lambda)$ is the best value of λ. The root of the equation,

$$\frac{d\Delta^2}{d\lambda} = -2\frac{D_1^2}{(\lambda_0 - H_{11})^3} - 2\frac{D_2^2}{(\lambda_0 - H_{22})^3} = 0, \tag{8.72}$$

is usually found numerically, and it is typically in between the two eigenvalues of the Hessian,

$$H_{11} < \lambda_0 < H_{22}. \tag{8.73}$$

Therefore, the best move uphill from the vicinity of a minimum toward the transition state is,

$$\Delta x = \frac{D_1}{\lambda_0 - H_{11}}, \tag{8.74}$$

$$\Delta y = \frac{D_2}{\lambda_0 - H_{22}}. \tag{8.75}$$

As the algorithm marches away from the vicinity of a minimum and toward the transition state, H_{11} becomes negative and this eventually makes the root, λ_0, negative as well. When λ_0 is negative, the best choice for the Lagrange multiplier is zero,

$$\Delta x = -\frac{D_1}{H_{11}}, \tag{8.76}$$

$$\Delta y = -\frac{D_2}{H_{22}}. \tag{8.77}$$

Equations 8.76 and 8.77 are the steps computed by the Newton–Raphson method for the root of the gradient of V.

Note that for this simple model, since the Hessian is diagonal at any value of x, y, these two coordinates are the local normal modes and the global normal modes. In more complicated models, the chosen coordinates will not necessarily be normal modes and no global normal modes may exist. Therefore, in the general case at every step of the walk, the Hessian has to be computed and diagonalized. Equations 8.71 through 8.77 are easily generalized once the normal modes are found. For example, Equation 8.71 becomes

$$\Delta^2 = \sum_{i=1}^{3n} \frac{D_i^2}{(\lambda - H_{ii})^2}, \tag{8.78}$$

and this equation will have $3n - 1$ local minima. Equations 8.74 through 8.77 now yield the optimal step toward the transition state in the normal mode representation. To find the actual coordinates, one must perform one additional similarity transformation. The Cerjan–Miller algorithm quickly marches to the nearest saddle. However, depending on the initial conditions, and the particular root one chooses to follow, it may find minima, or higher order saddles, as well. The program `cjlj` is our own implementation of the Cerjan–Miller algorithm. The program performs the following steps:

1. Read a minimum configuration from `input.dat` and calculate the Hessian and the normal modes.

2. Distort the configuration slightly along a normal mode.

3. Walk uphill following one of the $3n - 1$ roots λ_0.

4. Repeat step 3 for every possible root.

5. Repeat step 2 for all the modes, and in both the positive and the negative direction.

6. Repeat step 1 for all the minima in `input.dat`.

The termination criteria for the Cerjan–Miller walk is reached when the size of the gradient drops below 10^{-7} hartree/bohr. The configuration, the gradient, and the Hessian eigenvalues at the saddle are stored. Upon convergence, `cjlj` checks that the configuration is not a minimum. A minimum would have the first eigenvalue near zero, so that λ_7/λ_1 is a large number at a minimum. `eigen` sorts eigenvalues and eigenvectors, making λ_7 the first nonzero frequency. The code `if (dabs(eval(2)/eval(1)) .gt. 1.d-4) goto 200` checks that the saddle is a first order saddle before storing the information. That line should be commented out if one is interested in all the saddles of the potential energy surface. `cjlj` needs `pot2d` and `eigen` to run.

```
      program cjlj
      implicit real*8 (a-h,o-z)
c This program finds the transition state(s) for a multi-
c dimensional potential where both the gradient and the
c Hessian matrix are known. The Cerjan-Miller algorithm is used
c C. J. Cerjan and W. H. Miller, J. Chem. Phys. 75,2800, (1981).
      parameter (NMAX = 55)
      parameter (epsilon = 3.77169d-4)  ! hartree (for argon)
      parameter (r0 = 6.424) ! bohr  (for argon)
      parameter (dm = 39.95*1822.8) ! mass of a single Ar atom
      real*8 dv(NMAX),hesmat(NMAX,NMAX),ttr(NMAX,NMAX),eval(NMAX)
      real*8 sa1(NMAX),d(NMAX),dx(NMAX)
      real*8 x(NMAX),rx(NMAX),ry(NMAX),rz(NMAX)
      real*8 root(NMAX)
c initialization
```

```
      n = 7
      do nmin = 1,4
      open(unit=2,file='input.dat',status='old')
      read(2,*) ene
      write(6,*) ene
      do i=1,n
       read(2,999) x(i),x(n+i),x(2*n+i)
       rx(i) = x(i)
       ry(i) = x(n+i)
       rz(i) = x(2*n+i)
       write(6,999) x(i),x(n+i),x(2*n+i)
      enddo
      close(2)
      do nmode = 7,6*n
      do nroot=1, 3*n
c reload before you stretch again
      do i=1,n
       x(i) = rx(i)
       x(n+i) = ry(i)
       x(2*n+i) = rz(i)
      enddo
      do i=1,NMAX
      do j=1,NMAX
       hesmat(i,j) = 0.d0
       ttr(i,j) = 0.d0
      enddo
       root(i) = 0.d0
       dv(i) = 0.d0
       eval(i) = 0.d0
       d(i) = 0.d0
       dx(i) = 0.d0
       sa1(i) = 0.d0
      enddo
      call  pot2d(NMAX,x,n,epsilon,r0,v,dv,hesmat)
      call eigen(NMAX,3*n,hesmat,eval,ttr,sa1,ierl)
      if (ierl .gt. 0) stop 'diagonalization of the hessian failed'
c distort along a mode
        if (nmode .le. 3*n) then
         nm = nmode
         q = 0.1d0
        else
         nm = nmode - 3*n + 6
         q = - 0.1d0
        endif
      do i=1,3*n
       x(i) = x(i) + q*ttr(i,nm)
      enddo
      if (ierl .gt. 0) stop 'diagonalization of the hessian failed'
      ifrs = 1
```

```fortran
c The Cerjan Miller algorithm begin here
      do loop = 1,1000
c get the potential, the force and the Hessian
      do i=1,NMAX
      do j=1,NMAX
       hesmat(i,j) = 0.d0
       ttr(i,j) = 0.d0
      enddo
       root(i) = 0.d0
       dv(i) = 0.d0
       eval(i) = 0.d0
       d(i) = 0.d0
       dx(i) = 0.d0
       sa1(i) = 0.d0
      enddo
      call  pot2d(NMAX,x,n,epsilon,r0,v,dv,hesmat)
      call eigen(NMAX,3*n,hesmat,eval,ttr,sa1,ierl)
      if (ierl .gt. 0) stop 'diagonalization of the hessian failed'
      eige = 0.d0
c calculate the vector d
      do i=1,3*n
       d(i) = 0.d0
      do j=1,3*n
       d(i) = d(i) + ttr(j,i)*dv(j)   ! Gradient in  normal modes space
      enddo
      enddo
      if (ifrs .eq. 1) then
c map out the function Delta^2
      dlamda = 0.99*eval(1)
      ddlamda = (1.01*eval(3*n) - 0.999*eval(1))/2000.d0
      icountroot = 0
       do k=1,2001
         f2 = 0.d0
         f3 = 0.d0
         do i=1,3*n
         f2 = f2 -2.*d(i)*d(i)/((dlamda-eval(i))**3)
         enddo
         if (f2old*f2 .lt. 0) then
          icountroot = icountroot + 1
          root(icountroot) = dlamda
         endif
         f2old = f2
         dlamda = dlamda + ddlamda
       enddo
c follow all the roots
      if (nroot .le. icountroot) then
      gl = root(nroot)
      else
       exit
```

```
         endif
12       if (gl .lt. 0.d0) then
           gl = 0.d0
           ifrs = 0
         endif
         endif
c at this point the best value of lambda is  calculated
c plug this in to eq (2.12) of ref [1] in the normal mode
c space, and back transform to get deltax
c As termination criteria use the size of the gradient sdv
         sx = 0.d0
         sdv = 0.d0
         do i=1,3*n
          dx(i) = 0.d0
          do j=1,3*n
          dx(i) = dx(i) + ttr(i,j)*d(j)/(gl-eval(j))
          enddo
         sx = sx + dx(i)*dx(i)
         sdv = sdv + dv(i)*dv(i)
c        write(6,*)'i and dx(i)', i,dx(i)
         enddo
         if (sqrt(sdv) .lt. 1.d-7) then
c make a record of this saddle point
         do i=1,NMAX
         do j=1,NMAX
          hesmat(i,j) = 0.d0
          ttr(i,j) = 0.d0
         enddo
          root(i) = 0.d0
          dv(i) = 0.d0
          eval(i) = 0.d0
          d(i) = 0.d0
          dx(i) = 0.d0
          sa1(i) = 0.d0
         enddo
          call  pot2d(NMAX,x,n,epsilon,r0,v,dv,hesmat)
          do i=1,3*n
           do j=1,3*n
            hesmat(i,j) = hesmat(i,j)/dm
           enddo
          enddo
          call eigen(NMAX,3*n,hesmat,eval,ttr,sa1,ierl)
          if (ierl .gt. 0) stop 'diagonalization of the hessian failed'
c first make sure that it is not a minimum
         if (dabs(eval(7)/eval(1)) .gt. 1.d-4)  goto 200
c check that the saddle is first order only
         if (dabs(eval(2)/eval(1)) .gt. 1.d-4)  goto 200
           write(10,*) nmin,nmode,nroot
           write(10,*) n
```

```
         write(10,*) v, 'hartree,    Coordinates in bohr'
         write(6,*) v
         do i=1,n
            write(10,1010) x(i),x(n+i),x(2*n+i)
         enddo
         write(10,*) ' Force vector hartree/ bohr'
         do i=1,n
            write(10,999) -dv(i),-dv(n+i),-dv(2*n+i)
         enddo
         write(10,*) 'Frequencies in wavenumbers'
         do i=1,3*n
            write(10,1020) i,dsqrt(dabs(eval(i)))*2.195d+5
         enddo
         goto 200
      endif
c rescale the move to avoid  Newton-Rapson ballistic explosions
      if (sx .gt. 0.1d0) then
         do i=1,3*n
           dx(i) = 0.05*dx(i)/sqrt(sx)
         enddo
      endif
c move
      do i=1,3*n
        x(i) = x(i) + dx(i)
      enddo
      enddo  !      Step loop
200   enddo  ! loop over all the roots
      enddo ! loop over  the mode used to stretch the minimum
      enddo    ! loop over the minima
999   format(3f19.8)
1000  format(6i4,e14.6)
1001  format(i4,1p6e12.3)
1010  format('Ar',3f19.8)
1020  format(i4,f20.9)
      end
```

Running `cjlj` will create a large number of replica of several first order saddles.

8.16   Optical Activity

Just like some minima configurations can lack inversion symmetry, so can saddle point configurations. The program below, called `mirror`, makes use of `compare` and a reflection across the $y-z$ plane to identify configurations that are not superimposable to their mirror images.

```
      program mirror
      implicit real*8 (a-h,o-z)
      parameter (NMAX=55)
      parameter (epsilon = 3.77169d-4)   ! hartree (for argon)
      parameter (r0 = 6.424) ! bohr  (for argon)
      parameter (dm = 39.95*1836.6) ! mass of a single Ar atom
      real*8 rx(NMAX),ry(NMAX),rz(NMAX)
      real*8 rx2(NMAX),ry2(NMAX),rz2(NMAX)
c checks to see if a mirror image of a configuration
c is superimposable to the original configuration
      n = 7
      do isomer = 1,11
       open(2,file='input.dat',status='old')
       read(2,*) ene
       do i=1,n
        read(2,1000) rx(i),ry(i),rz(i)
       enddo
c generate the mirror image by translating and reflecting
c through the y-z plane
       xmin = 1.0d20
       do i=1,n
        if (rx(i) .lt. xmin) xmin = rx(i)
       enddo
       do i=1,n
        rx2(i) = -(rx(i) - xmin)
        ry2(i) = ry(i)
        rz2(i) = rz(i)
       enddo
       call compare(NMAX,n,rx,ry,rz,rx2,ry2,rz2,sigmamin)
       write(6,1000)isomer, sigmamin
      enddo
      close(2)
1000  format(i3,f10.5)
      end
```

Here are several suggested activities to deepen the reader's understanding of the algorithms we have showcased so far.

Exercises

1. Test **brown** by writing a program, which calls **construct7** and **potd**. To ensure that a minimum is reached, print the force vector on the screen before the call to **brown**, and then again after. Increase the number of decimals in the format statement to 12 so that more digits on the forces are displayed on the screen at run time. Set the parameters of the potential to simulate Ar$_7$ and run the minimization job with $d = 7$ twice,

once with $\gamma = 0.001$ a.u. and another (after recompiling) with $\gamma = 0.003$ a.u., and compare your results with those in Figure 8.5. The parameters of the Lennard-Jones potential for argon are $\epsilon = 3.77169 \times 10^{-4}$ hartree, and $r_0 = 6.424$ bohr. The potential energy for the pentagonal bipyramid generated in `construct7` with 7.0 bohr for the bond length is -0.00605783632 hartree. After you are done testing `brown`, you must add the comment flag on the leftmost column of the line,

```
c        write(8,*) dfloat(kstep -1)*dt,v,rmsg
```

otherwise successive uses of `brown` will create voluminous output that may fill the disk and cause your workstation to crash.

2. Run the basin hopping and the genetic algorithms for Ar_7, and compare your results with those in Figures 8.6 through 8.9.

3. Many copies of the minima are created in runs of the basin hopping and the genetic algorithms. The energy will be nearly identical for all the copies, but the coordinates will not. Write a program that reads the output of the basin hopping and the genetic algorithms, and compare structures that have nearly identical energies, using the SCA. Are all the minima of LJ_7 that have the same energy really superimposable?

4. Test the program `hessian` and make sure that the numerical and analytical second derivatives of the potential agree.

5. Test the program `normal_modes` for all `five` minima of the potential, and verify that the first six frequencies (generated by any net force and torque on the center of mass) are close to zero, and that the rest of the frequencies are in agreement with those in Table 8.1.

6. Modify the program `normal_modes` so that frames of moving Cartesian coordinates along a normal mode of choice are written to a file. These can be displayed using molecular graphic software. Here is a suggestion for the code lines to move the coordinates about normal mode number 18. The output is in a format compatible with XMAKEMOL.

```
c write out a normal mode of your choice
      fdm = 1.d0/sqrt(dm)
      t = 0.d0
      freq = dsqrt(dabs(dia(18)))
      dt = 1.d0/(60*freq)
      do j=1,360
      ct = cos(freq*t)
      write(10,*) n
      write(10,*) t
      do i=1,n
```

```
      rx(i) = x(i) + z(18,i)*ct
      ry(i) = x(n+i) + z(18,n+i)*ct
      rz(i) = x(2*n+i) + z(18,2*n+i)*ct
      write(10,*) 'Ar', rx(i),ry(i),rz(i)
    enddo
    t = t + dt
  enddo
```

7. Derive Equations 8.71 and 8.72.

8. Solve for Δx in Equation 8.68, for Δy in Equation 8.69, and plug the expressions into Equation 8.63. Show that the result is

$$V_{\mathrm{L}}(\lambda) - V_0 = \frac{D_1^2\left(\lambda - \frac{1}{2}H_{11}\right)}{\left(\lambda - H_{11}\right)^2} + \frac{D_2^2\left(\lambda - \frac{1}{2}H_{22}\right)}{\left(\lambda - H_{22}\right)^2}.$$

Write a program that computes V_0, D_1, D_2, H_{11}, H_{22}, for arbitrary values of x_0 and y_0 from the potential in Equation 8.16, its gradient, and its Hessian. Use $\gamma = 0.9$. Then, have the program use these constant parameters to produce two graphs, $\Delta^2(\lambda)$ and $V_{\mathrm{L}}(\lambda) - V_0$, as functions of λ. Produce these two graphs for two sets of points: (a) $(x_0, y_0 = 0.9, 0.9)$ and (b) $(x_0, y_0 = 0.9, 0.1)$. Inspection of these four graphs should shed additional light on how the optimal value of the Lagrange multiplier produces the best direction to locate the nearest transition state.

9. Run the program `cjlj` and locate all the transition states that it finds for Ar$_7$. Many copies of the transition states are created in runs of the Cerjan Miller algorithm. The energy will be nearly identical for all the copies, but the coordinates will not. Write a program that reads the output of the program `cjlj` and compares structures that have nearly identical energies, using the SCA. Are all the transition states of LJ$_7$ that have the same energy really superimposable? Confirm that the optically active minima and transition states that are found by the basin hopping, genetic, and Cerjan–Miller algorithms are not superimposable to their mirror images, using the program `mirror`.

10. Modify `brown` so that it integrates the dynamic equations, starting from a transition state structure with a relatively small amount of kinetic energy and propagates simultaneously in the positive and negative time direction. The motion along the **reaction path** will be somewhat complicated. Graph the values of the potential energy, and use the graph to find out which minima of the potential are connected with the particular transition state you decided to start with. The information obtained this way, for all the most important transition states, can be organized into a connectivity diagram for the potential energy surface that displays the funnel structure of the potential energy surface, as well as the energies of the most important configuration points.

11. The program `cjlj` begins by stretching the structure of a minimum along a normal mode. Change the program and test to see if a random move away from a minimum works just as well.

12. Log on to the Cambridge Cluster Data Base at

    `http:www-wales.ch.cam.ac.uk/CCD.html`

    and look through the minima for LJ_n, for $n = 2 \to 150$. Use the energy data to graph E_n/n, the average cohesive energy, as a function of n. The graph should be irregular, with several deep features that correspond to structures that complete icosahedral shells. Next, download the structures of each global minimum, and use the program `mirror` to find values of the size n for which the global minimum is optically active.

13. It is traditional to search for transition states by starting in the vicinity of a minimum, however, for more complicated systems it may be necessary to implement a combination of the basin hopping (or the genetic algorithm) that searches for transition states instead. Modify the programs `basin_hopping` and `genetic` so that searches for transition states are performed instead.

9

Classical and Quantum Simulations of Ar$_7$

9.1 Introduction

This chapter presents a number of classical and quantum simulation methods aimed at obtaining the basic physical properties of atomic clusters. So far, simulation techniques have been tested with systems of relatively small dimensionality. The stochastic techniques we have chosen in this book are powerful, and the reader will find in this chapter that all of these can be applied to multidimensional systems with only few modifications. The interpretation of simulation results can be accomplished only after an understanding of the features of the underlying potential energy surface is achieved. Since a number of structural characterization methods are presented in Chapter 8 using the potential energy surface of Ar$_7$, we are using Ar$_7$ to test the algorithms presented in this chapter as well. It has been an important goal to understand how clusters form in the first place. Cluster nucleation remains an important problem [848]. The community has made considerable progress toward understanding the thermodynamics of Lennard-Jones and similar isotropic systems. Numerous investigations have shown that edge effects deeply affect the thermodynamic behavior of finite systems. The peculiarity of melting in finite matter, and the discovery of the liquid–solid coexistence in relatively broad regions of temperature were first made in the process of simulating Lennard-Jones clusters [830, 837, 838, 841, 843, 844, 846].

On the other hand, the exploration of the thermodynamics and ground state properties of molecular clusters is only in its infancy. One of the purposes of this book is to demonstrate that the methods, in Chapter 8 and the present chapter, developed by the community can be relatively easily extended to molecular clusters. Another important purpose of this book is to train students and researchers in using these extensions. Molecular clusters are treated in later chapters, after the necessary tensor analysis is developed to handle the theoretical framework behind the extensions we propose. The simulation of atomic clusters provides an important benchmark against which closely related molecular simulations can be compared to gauge the influence of anisotropic interactions, geometric and dynamic couplings, and other interesting and complicating features pertinent to molecular clusters.

The study of phase changes in clusters requires powerful enhancements to the Metropolis algorithm [811, 821, 835, 839, 841], since, solid–solid and melting cluster phase changes are well-to-well hopping events, and quasier-godicity causes the Metropolis algorithm to become highly inefficient. The reader should already be familiar with the latest state-of-the-art in handling quasiergodicity, namely parallel tempering [3, 4].

In this chapter, we test the following simulation methods to find thermo-dynamic and ground state properties of Ar_7:

i. Classical parallel tempering in the canonical ensemble to compute the internal energy, average potential energy, and heat capacity in the clas-sical limit.

ii. Variational Monte Carlo (VMC) using the one parameter trial wave-function introduced in Chapter 6. An approximation to the ground state energy and average potential at $T = 0$ is obtained with these simula-tions. More importantly, these simulations optimize a trial wavefunction that is used to provide importance sampling for diffusion Monte Carlo (DMC), as explained in Chapter 6.

iii. DMC using an importance sampling scheme introduced in Chapter 6. The exact ground state energy (within the statistical fluctuations) and average potential at $T = 0$ are obtained with these simulations.

iv. Quantum parallel tempering in the canonical ensemble using the reweigh-ted random series path integral of Chapter 7. The path integral sim-ulations yield the internal energy, average potential energy, and heat capacity in the quantum limit for all but the coldest temperatures.

All these techniques have been introduced using "toy" models in $\mathbb{R}^1$ in earlier chapters. Their extension to $\mathbb{R}^{3n}$ spaces is straightforward. We could have included exercises that requested readers to simply modify the appro-priate programs to perform these simulations in $\mathbb{R}^{3n}$. Instead, we decided to provide code for them for the following reasons. Firstly, the path integral al-gorithm in Chapter 7 does not have the capability to handle quasiergodicity. The program we provide in this chapter is a combination of the Fourier–Wiener reweighted random series algorithm with the finite difference estima-tors discussed in Chapter 7, and the parallel tempering technique of Chapter 3. Designing the combination of these two techniques and testing the resulting code properly is a challenge for the beginner and, generally, a time-consuming task. Secondly, the goals of this chapter are as follows. (a) Demonstrate the most efficient way these methods can be generalized from their $\mathbb{R}^1$ version. (b) Lead the reader through the interpretation of the outcome of such sim-ulations. (c) It is highly likely that the reader of this chapter will need to modify the programs we present here. It is an equally important goal to pro-vide some general testing strategies for the programs. Therefore, we find it convenient to provide the code to direct the reader through some testing exercises.

Given the size of the three simulation programs, I place them in separate appendices at the end of the chapter. In the first few sections of this chapter, we provide the details of the functioning of the code, and reference the reader to those sections in earlier chapters for additional background information regarding the theory behind the simulations. The simulation techniques we have chosen are powerful, however, the results that one obtains still need to be interpreted carefully. Features in the graph of the heat capacity as a function of temperature, or features in the cohesive energy of the clusters, with or without quantum effects require the physical interpretation in terms of the structure of the important configurations of the potential energy, and the knowledge of distributions of structural descriptors. The following methods are introduced and discussed.

i. The Lindemann index, an average bond length fluctuation parameter

ii. Bond orientation parameters

iii. Distributions of inertia moments

iv. Distributions of structural distances from the global minimum achieved with the structural comparison algorithm (SCA) introduced in Chapter 8

v. Systematic quenching at all temperatures to find the percentage of well occupation in the random walk

vi. Using SCA to find the percentage of well occupation in the random walk in place of the systematic quenching method

9.2   Simulation Techniques: Parallel Tempering Revisited

In Chapter 3, we introduce the Metropolis algorithm as a rejection technique to produce random numbers (Markov chains) that have the desired distribution. With the Metropolis algorithm, one can simulate matter at a finite temperature by constructing a series of random numbers for the coordinates of configurations that are distributed according to the Boltzmann model. In Chapter 3, we introduce the parallel tempering "fix" for the Metropolis algorithm without deriving the acceptance probability for swaps. There, the goal is merely to demonstrate the quasiergodicity effect. In Chapter 7, we have explained how imaginary-time path integrals can be used to include quantum effects at finite temperatures into the Boltzmann distribution for the canonical ensemble. The toy models we use in Chapter 7 to develop the path integral are selected specifically to avoid quasiergodicity problems. We no longer have such luxury; path integral simulations of Ar_7 require measures to handle quasiergodicity. In order to merge parallel tempering and path integrals, we must provide a brief introduction to the theory of Markov chains, so that the swap moves can be generalized from the classical configuration integrals of Chapter 3.

The object of a random walk is to produce a Markov chain of states $\mathbf{x}_0, \mathbf{x}_1, \mathbf{x}_2, \ldots$ which approaches a desired distribution, $\rho(\mathbf{x})$, asymptotically. A sufficient condition for the creation of such a random process is the detailed balance relation. Let $\rho(\mathbf{x})$ and $\rho(\mathbf{x}')$ be the desired asymptotic distributions for the configurations $\mathbf{x}$ and $\mathbf{x}'$, and let $K(\mathbf{x}'|\mathbf{x})$ be the conditional probability that the system is found at $\mathbf{x}'$ if its previous position is $\mathbf{x}$. The detailed balance condition reads,

$$\rho(\mathbf{x}) K(\mathbf{x}'|\mathbf{x}) = \rho(\mathbf{x}') K(\mathbf{x}|\mathbf{x}'), \tag{9.1}$$

which implies that the overall probability that the state moving from $\mathbf{x}$ to $\mathbf{x}'$ is equal to that from $\mathbf{x}'$ to $\mathbf{x}$. Unlike many applications of Markov chains, in the Metropolis algorithm the conditional probability is unknown, but it can be replaced with the following product,

$$K(\mathbf{x}'|\mathbf{x}) = T(\mathbf{x}'|\mathbf{x}) \, p_a(\mathbf{x}'|\mathbf{x}). \tag{9.2}$$

$T(\mathbf{x}'|\mathbf{x})$ is the trial probability, a convenient starting distribution of random numbers to be modified by the rejection technique, and $p_a(\mathbf{x}'|\mathbf{x})$ is the acceptance probability for the move $\mathbf{x} \to \mathbf{x}'$. It is straightforward to show that,

$$p_a(\mathbf{x}'|\mathbf{x}) = \min\left\{1, \frac{T(\mathbf{x}|\mathbf{x}') \, \rho(\mathbf{x}')}{T(\mathbf{x}'|\mathbf{x}) \, \rho(\mathbf{x})}\right\}, \tag{9.3}$$

satisfies the detailed balance condition in Equation 9.1. In the Metropolis algorithm, the trial move is performed using a random number, η, uniform in $(-\Delta/2, \Delta/2)$. Δ is the move parameter introduced in Chapter 3, and optimized to produce 50% rejections. Consequently, the trial probability is a constant, namely,

$$T(\mathbf{x}|\mathbf{x}') = \frac{1}{\Delta}. \tag{9.4}$$

Inserting Equation 9.4 and $\rho(\mathbf{x}) = Q^{-1} \exp[-\beta V(\mathbf{x})]$ into Equation 9.3, one obtains

$$p_a(\mathbf{x}'|\mathbf{x}) = \min\{1, \exp\{-\beta [V(\mathbf{x}') - V(\mathbf{x})]\}\}. \tag{9.5}$$

This is exactly the acceptance probability for the Metropolis algorithm implemented to compute the configuration integrals in Chapter 3.

For parallel tempering simulations, 90% of the trial moves are Metropolis moves, and 10% are moves (called swaps) performed using a distribution of configurations from a walker at an adjacent temperature. The trial probabilities for swaps are

$$T\left(\mathbf{x}'\,|\mathbf{x}\right) = Q^{-1}\left(\beta'\right)\exp\left[-\beta'V\left(\mathbf{x}'\right)\right], \tag{9.6}$$

and,

$$T\left(\mathbf{x}\,|\mathbf{x}'\right) = Q^{-1}\left(\beta'\right)\exp\left[-\beta'V\left(\mathbf{x}\right)\right]. \tag{9.7}$$

Inserting Equations 9.7 and 9.6 into Equation 9.3 yields,

$$p_{\mathrm{a}}\left(\mathbf{x}'\,|\mathbf{x}\right) = \min\left\{1, \frac{\exp\left[-\beta V\left(\mathbf{x}'\right)\right] \times \exp\left[-\beta'V\left(\mathbf{x}\right)\right]}{\exp\left[-\beta V\left(\mathbf{x}\right)\right] \times \exp\left[-\beta'V\left(\mathbf{x}'\right)\right]}\right\}. \tag{9.8}$$

Equation 9.8 simplifies further into

$$p_{\mathrm{a}}\left(\mathbf{x}'\,|\mathbf{x}\right) = \min\left\{1, \exp\left\{-\left(\beta - \beta'\right)\left[V\left(\mathbf{x}'\right) - V\left(\mathbf{x}\right)\right]\right\}\right\}. \tag{9.9}$$

Equation 9.9 is the main result used for the parallel tempering example in Chapter 3, and in the program `parallel_tempering_r3n.f` in this chapter.

9.3 Thermodynamic Properties of a Cluster with n Atoms

In this section, we focus on computing the internal energy, average potential energy, and heat capacity of an n-atom cluster using the Metropolis algorithm enhanced by parallel tempering to compute the configuration integral,

$$\langle H \rangle = \frac{3n}{2\beta} + \frac{\displaystyle\int_{-\infty}^{\infty} dx_1 \cdots \int_{-\infty}^{\infty} dz_n\, V \exp\left(-\beta V\right)}{\displaystyle\int_{-\infty}^{\infty} dx_1 \cdots \int_{-\infty}^{\infty} dz_n \exp\left(-\beta V\right)}. \tag{9.10}$$

The main function of the program `parallel_tempering_r3n.f` is a modification of the program `parallel_tempering` of Chapter 3. The program `parallel_tempering_r3n.f` needs three subroutines to run: `construct7`, `pot` from Chapter 8, and the random number generator `ransi` from Chapter 3. The subroutine `pot` in Chapter 8 must be modified by the addition of a constraining term,

$$V_{\mathrm{c}} = \sum_{i=1}^{n}\left(\frac{|\mathbf{r}_i - \mathbf{r}_{\mathrm{CM}}|}{r_{\mathrm{LBA}}}\right)^{20}. \tag{9.11}$$

V_{c} is added to prevent dissociation events from taking place. In Equation 9.11, $\mathbf{r}_i$ is the position of atom i relative to the origin of the reference frame, r_C is the location of the center of mass of the cluster, and r_{LBA} is a parameter set equal to eight times the length scale parameter of the Lennard-Jones potential. The following code must be appended at the end of the subroutine **pot** before the **return** statement.

```
c Lee-Barker-Abraham continuous reflecting sphere (to prevent dissociations)
c compute the center of mass
      xm = 0.d0
      ym = 0.d0
      zm = 0.d0
      do i=1,n
       xm = xm + x(i)/dfloat(n)
       ym = ym + x(n+i)/dfloat(n)
       zm = zm + x(2*n+i)/dfloat(n)
      enddo
      vc = 0.d0
      rc = 8.d0*r0
      do i = 1,n
       dx = x(i) - xm
       dy = x(n+i) - ym
       dz = x(2*n+i) - zm
       r2 = dx*dx + dy*dy + dz*dz
       s2 = r2/(rc*rc)
       s4 = s2*s2
       s16 = s4*s4
       s20 = s16*s4
       vc = vc + s20
      enddo
      v = v + vc
```

Lee, Barker and Abraham [849] propose this type of constraining potential as they compute the free energy of formation of atomic clusters. Without V_c, a sufficiently long random walk at the highest temperatures begins to sample completely disconnected configurations, where a group of $1 \leq m < n$ atoms are sufficiently far from the rest, that neither of the two sets is subjected to the interactions from the other. In such a case, the simulation outcome cannot be interpreted as associated with a single n-atom cluster, rather the simulation is (in part) that of a "new," disconnected system,

$$\mathrm{Ar}_n \longrightarrow \mathrm{Ar}_m + \mathrm{Ar}_{n-m}. \tag{9.12}$$

There are also technical difficulties when attempting to simulate clusters without a confining feature of the overall potential energy surface. Since the Lennard-Jones and other types of realistic potentials are nonconfining, a fragment with one or more atoms can, and often does, wander so far away that computations of its distances and the powers of these necessary to compute interactions cause overflow exceptions in the course of simulations. These exceptions can be easily controlled with a few code segments, however, once a simulation has reached such critical distance scale, there is no easy way to recover from it and the simulation must be halted. Furthermore, a system that is allowed to wander in too large a confining volume has problems accepting swaps from dissociated configurations down to temperatures below the dissociation range. The result is that parallel tempering becomes inefficient against quasiergodicity in that situation. Occasionally, parallel tempering must draw configurations in the gaseous phase of the cluster, in order to overcome barriers among minima and funnels. Therefore, both V_c, and a fine tuning of the parameter r_C, are a must for successful parallel tempering simulations of clusters. The value of r_C should also be checked carefully to ensure that all the important minima fit inside the cavity and that V_c contributes a negligible amount of potential energy.

In a recent investigation, we were able to provide a plausible physical interpretation for this additional term of the potential: V_c can be interpreted as a phenomenological term describing the average interaction felt from collisions with an inert (and much lighter) carrier gas. Our studies demonstrated that an increase in the carrier gas density has the same impact on the thermodynamics as a decrease in the parameter r_C in Equation 9.11.

9.4 The Program `parallel_tempering_r3n.f`

In part 1 of the program `parallel_tempering_r3n.f`, the walk length (M) and the number of warm up moves (MW) are set to 10^6. The walkers are initialized with the configuration that is produced by `construct7` with a 7-bohr bond length. The array `px(maxn,100)` is a two-dimensional array that contains the running configuration of all the walkers. The second index is the walker pointer, whereas the first one is the coordinate index. The array `px(maxn,100)` stores configuration coordinates following the storage scheme 1 introduced in Chapter 8. The temperature stored in the array `pd0(100)` varies linearly from $T_1 = 1$ to $T_w = 100$ K, according to the following equation,

$$T_k = T_1 + T_w \frac{k-1}{w}, \tag{9.13}$$

where $k = 1, 2, \ldots w$, and w is the number of walkers set to 40. The array `pvsave(100)` contains the value of the potential energy that corresponds to the configuration stored in `px(maxn,100)`. Part 2 is the parallel tempering random walk itself. The outer loop is the move loop, therefore, the first move is attempted for all 50 temperature walkers, then the second move is attempted for all 50 walkers and so on. In part 2 (a), the running configuration is retrieved from the storage array `px()`. In part 2 (b), a random number is selected to decide if a swap or a random move should be performed. Part 2 (c) is where the swap takes place. The code in Section 2 (c) is nearly identical to its equivalent in Chapter 3. The array xt() is filled with the configuration of the walker at the adjacent higher temperature stored in px(). If the swap is accepted [Part 2 (e)], then px(i,k+1), $i = 1, \ldots, 3n$, is filled with the content of px(i,k), the lower temperature running configuration becomes the higher temperature configuration, and the corresponding potential energy [stored in the array pvsave(k+1)] is updated with the content of the memory location vtj. vtj is computed with the first call to pot in part 2 (c) using the configuration of the kth walker. In part 2 (d), the configuration in xt() is modified by a random move performed as follows. A total of four random numbers, η, between 0 and 1 are drawn. With the first random number, an atom of the configuration is selected. Since the distribution is adjusted to yield a uniform set of numerical outcomes from 1 to n, all atoms are selected with equal frequency. This fact is normally tested as a part of the preliminary assessment of the performance of the code. With the remaining three random numbers, the $x, y,$ and z coordinates of the randomly selected atom are moved according to

$$x \leftarrow x + \Delta_k \left(\eta - 0.5 \right), \tag{9.14}$$

where Δ_k is a set of parameters, one for each temperature, adjusted in part 2 (f) to produce a 50% rejection rate. In both the swap and the Metropolis move part, the boundaries of the exponents are checked prior to the call to the intrinsic function exp, to avoid underflow or overflow. In part 2 (e), the array px(i,k) , $i = 1, \ldots, 3n$, is updated if either move type is accepted, or else a number of counters to compute the rejection rates for swaps and moves are updated. Finally, in part 2 (g), the array x() is filled with the running configuration of the walker stored in px(), and this is used to call pot. The accumulation of V and V^2 is carried out after the first 10^6 moves. If a move or swap from a running configuration x to x' is rejected, the configuration x is used to call pot and accumulate V and V^2.

With the screen output from the program parallel_tempering_r3n.f as given, we produce the graphs of the total and potential energy in Figure 9.1, and the heat capacity in Figure 9.2. The following features of the two graphs deserve some attention, as they are regular tests to confirm that the code is working as it should. Firstly, both the total energy and the potential energy converge at the coldest temperatures to the value of the energy of the global minimum found in Chapter 8 (-0.006225319 hartree). The potential energy tends asymptotically toward zero as the temperature increases, after the dissociation range of temperatures, where the slopes of both energies increase abruptly. Eventually, $\langle V \rangle$ climbs slightly above zero, since at sufficiently high temperatures, with the cluster in the gaseous phase, the constraining potential V_c in Equation 9.11 begins to contribute significantly. The total energy climbs to higher values compared to the potential energy, since the average kinetic energy depends linearly on T. Needless to say, one has to confirm the reproducibility of these features, and determine the statistical error associated with the measured properties. The smooth graphs in Figures 9.1 and 9.2 indicate

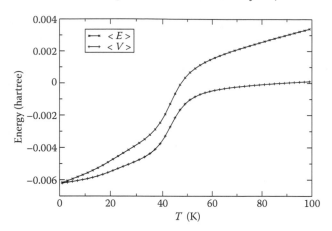

FIGURE 9.1

The average total and potential energy of Ar₇ obtained with a single 1,000,000 moves run of `parallel_tempering_r3n.f`.

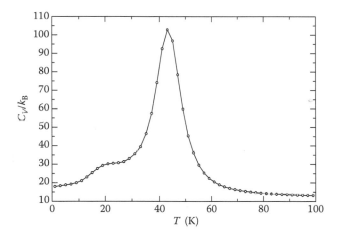

FIGURE 9.2

The heat capacity of Ar₇ obtained with a single 1,000,000 moves run of `parallel_tempering_r3n.f`.

that the statistical error is small, however, this must be carefully determined by running simulations with different seeds and performing block averages, as we do in Chapter 7.

In Chapter 3, we demonstrate that the average kinetic energy can be computed with the equipartition theorem,

$$\langle K \rangle = \frac{3n}{2} k_{\mathrm{B}} T. \tag{9.15}$$

If we also expand the potential energy surface quadratically about the global minimum, we introduce $(3n - 6)/2$ quadratic terms as potential energy into the Hamiltonian. As explained in Chapter 8, six normal modes (three translation of the center of mass, and the three rotations) do not contribute to the energy, since the associated frequencies vanish in the absence of external fields. Without external fields, there is no net force on the center of mass and no net torque, and the interaction is both translationally and rotationally invariant. Therefore, the heat capacity should approach the following equipartition limit,

$$\lim_{T \to 0} \frac{C_V}{k_B} = \frac{3n}{2} + \frac{3n - 6}{2} = 18. \tag{9.16}$$

Figure 9.2 demonstrates that the computed heat capacity does, in fact, approach 18 Boltzmann constant units.

On the high temperature side, the heat capacity should eventually approach the ideal gas limit,

$$\lim_{T \to \infty} \frac{C_V}{k_B} = \frac{3n}{2} = 11. \tag{9.17}$$

At 100 K, the heat capacity is approximately 13.2 units, but Figure 9.2 shows a gradual downward trend. The heat capacity approaches the "ideal gas" limit exponentially above the second peak at ≈ 42 K. We fit the ten highest temperature points to an exponential decay function, and estimating on the scale of the graph in Figure 9.2, the heat capacity is indistinguishable from the ideal gas limit at 200 K and above. Therefore, the sharp and intense peak around 42 K is a "boiling" peak. The contrast of the heat capacity of a cluster compared to the same for a bulk liquid is striking. There is a boiling temperature range. For a bulk liquid, the heat capacity graph displays an infinitely sharp and infinitely tall peak. Instead, in Figure 9.2, there is a broad boiling temperature range approximately 14 K wide at the peak half height. The broadening of the heat capacity is a finite size effect. In the thermodynamics limit $(n \to \infty)$, one recovers the proper infinitely sharp boiling (and melting) features of the thermodynamic variables. The finiteness effects on thermodynamic variables have been measured experimentally. However, the confining potential does play a significant role in how broad the boiling peak, and how steep (or gradual) the downward trend toward the ideal gas limit is.

Exercises

1. Show that Equation 9.3 satisfies the detailed balance condition in Equation 9.1.

2. Derive Equation 9.9 from Equation 9.8.

3. Modify the program `buildar7.f` so that `pot`, (the version that includes V_c in Equation 9.11) is called for a range of bond lengths. The range must

contain the minimum energy and extend far enough to see the effects of the constraining volume. The potential energy should rise above zero continuously, but rather sharply. Estimate the ratio of optimal length scale to the confining radius.

4. Perform several simulations of Ar7 using the program `parallel_tempering_r3n.f`, using different values of the parameter r_{LBA} in Equation 9.11. Begin with $r_C = 6r_0$, and increment its value by one unit of r_0 until the boiling peak becomes a shoulder. Monitor the error and the statistical fluctuation in the portion of the heat capacity below the boiling peak ($T < 41$ K).

5. Modify the program `parallel_tempering.f` from Chapter 3, so that the heat capacity for a particle trapped in the n-dimensional quartic potential,

$$V(x_1, x_2, \ldots, x_n) = \sum_{i=1}^{n} \left(\frac{3}{2\alpha + 1} x_i^4 + \frac{4\alpha - 4}{2\alpha + 1} x_i^3 - \frac{6\alpha}{2\alpha + 1} x_i^2 + 1 \right),$$

is obtained. Choose $\gamma = 0.9$. Note that the potential model is confining, therefore, there is no need to introduce terms like V_c in the potential. Use a uniform distribution to select a coordinate to move at the time. Compute the heat capacity at several temperatures for $n = 2, 3$, and 4. How does the heat capacity change with increasing dimension?

9.5   The Variational Ground State Energy

We move to quantum simulation methods for atomic clusters. The simplest place to continue is the VMC method. One of the benefits of using a trial wavefunction ψ_T as in Equation 9.18

$$\psi_T(\mathbf{x}) = \exp\left[-\frac{\beta}{2} V(\mathbf{x}) \right], \tag{9.18}$$

is that when we insert it into the variational integral,

$$\langle E_\beta \rangle = \frac{\displaystyle\int_{-\infty}^{\infty} dx_1 \cdots \int_{-\infty}^{\infty} dz_n \, E_L(\mathbf{x}) \, |\psi_T(\mathbf{x})|^2}{\displaystyle\int_{-\infty}^{\infty} dx_1 \cdots \int_{-\infty}^{\infty} dz_n \, |\psi_T(\mathbf{x})|^2}, \tag{9.19}$$

we obtain an expression similar to Equation 9.10,

$$\langle E_\beta \rangle = \frac{\displaystyle\int_{-\infty}^{\infty} dx_1 \cdots \int_{-\infty}^{\infty} dz_n \, E_L(\mathbf{x}) \exp(-\beta V)}{\displaystyle\int_{-\infty}^{\infty} dx_1 \cdots \int_{-\infty}^{\infty} dz_n \, \exp(-\beta V)}. \tag{9.20}$$

The local energy estimator is,

$$E_{\mathrm{L}}\left(\mathbf{x}\right) = \frac{1}{m} \sum_{i=1}^{3n} \left[\left(\frac{\beta}{2} \frac{\partial V}{\partial x_i} \right)^2 - \frac{\beta}{2} \frac{\partial^2 V}{\partial x_i^2} \right] + V\left(\mathbf{x}\right). \qquad (9.21)$$

Therefore, only a small modification to the code `parallel_tempering_r3n.f` is necessary to perform a random walk with the same distribution and accumulate the derivatives of the potential. Consider the following subroutine `elocal`.

```
          subroutine elocal(maxn,x,eloc)
          implicit real*8 (a-h,o-z)
          real*8 x(maxn),dv(maxn),hess(maxn,maxn)
          parameter(dmm =  39.96284*1822.89)   ! mass of Ar^{40} (99.6\% abundance)
c calculates the local energy from (H psi)/psi
             sumlb = 0.d0
             call pot2d(x,n,v,dv,hess)!  psit = exp(-0.5d0*temp*v)
          do i =1,3*n
             sumlb = sumlb + (temp*dv(i)/2.d0)*(temp*dv(i)/2.d0)
        &   - 0.5d0*temp*hess(i,i)
          enddo
          eloc = -0.5d0*sumlb/dmm + v
          return
          end
```

All one needs is to insert a call to `elocal` immediately after the array `x()` is assigned in part 2 (g) of `parallel_tempering_r3n.f`, and use the value in `eloc` returned by the subroutine to update an array such as `sume(k)` for all the walkers.

To produce the graph of $\langle E_\beta \rangle$ as a function of the parameter $T = (k_{\mathrm{B}}\beta)^{-1}$ in Figure 9.3, we change the parameters for the temperature schedule in Equation 9.13 to $w = 80$, $T_1 = 1$ K, and $T_w = 80$ K. A run with one million moves with 80 walkers takes less than one hour on a modern workstation. The minimum energy, $\min_\beta \langle E_\beta \rangle = -0.005184$ hartree, occurs for a value of the parameter $T = 15$ K. The function $\langle E_\beta \rangle$ is quadratic about the minimum, however, above 20 K, the melting and dissociation of the cluster have a visible impact on its shape. The type of trial wavefunction proposed in Chapter 6 and in Equation 9.18, has yet to undergo the most severe tests on clusters with extremely weak binding, and where the quantum effects are large, such as, for example, parahydrogen clusters or helium clusters. In those extreme cases, it may be possible that the optimal parameter T is above the dissociation temperature. How the resulting variational ground state wavefunction performs in improving the efficiency of DMC in those extreme cases is unknown. Recently, we have tested our variational method with Ne_{50}, and found that it works well, and that quasiergodicity affects the ground state energy significantly. As always, we caution those readers who plan on exploring new systems using our algorithms; unlike the methods one finds in tomes like *Numerical Recipes*, our algorithms are far from "canned routines" or black boxes. On the other hand, it is relatively easy to build additional flexibility into ψ_T, modify the parallel

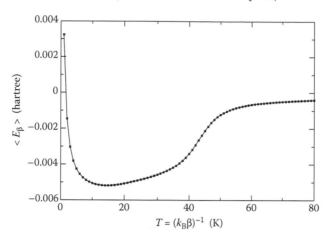

FIGURE 9.3

The average variational energy of Ar₇ obtained with a single 1,000,000 moves run of `parallel_tempering_r3n.f`.

tempering walk accordingly, and produce better estimates for the ground state wavefunctions than those obtainable with Equation 9.18.

From our variational simulation, we extract the starting configuration for DMC by reducing the number of moves and warm up moves to 100,000, and inserting the following snippet of code in part 2 (g) of the program `parallel_tempering_r3n.f`, immediately after the array `x()` is assigned:

```
c prepare a 100000 configuration distribution for DMC
      if (kw. eq. 15) then
        write(9,*) kw, move, v, pd0(kw), eloc
        do j=1,n
        write(9,*) x(j),x(n+j),x(2*n+j)
        enddo
      endif
```

The file `fort.9` produced in this second run constitutes the input for the program `gfis_dmc` in Appendix B.

9.6   Diffusion Monte Carlo (DMC) of Atomic Clusters

The program `gfis_dmc` is the implementation of the Green's function importance sampling DMC method that we introduce in Chapter 6. The program needs the subroutines, `elocal`, `pot`, `po2d`, and `ransi`. Before running `gfis_dmc`, the reader must produce a file that contains 100,000 configurations from a parallel tempering run at the optimized value of T, as described in the previous section.

The initialization of all the parameters for the Green's function importance sampling DMC simulation is carried out in part 1 of the program, gfis_dmc. The variable MN is the target size for the evolving population, and the parameter temp contains the optimal value of $\beta = (k_BT)^{-1}$, where $T = 15$ K. The code in part 2 (a) reads the file fort.9 produced by running parallel_tempering_r3n.f, as detailed earlier. There are several parts to the walk, but all of these are quite similar to those in the monodimensional example in Chapter 6. Part 2 (a) is the rejection technique to generate $3n$ random variables (the diffusion steps) that have a Gaussian distribution with a standard deviation equal to $\sqrt{\Delta t/m}$. The running values for the diffusion steps are stored in the $3n$ array vel(). Part 2 (a) also includes a self-adapting procedure for the Metropolis move parameter similar to that used in parallel_tempering_r3n.f. The coordinates are stored in the doubly subscripted array x(i,j), where the second index is the coordinate pointer (storage scheme 1) for the ith replica. The first index is the replica (or psip) pointer. The array x(i,j) is manipulated directly to affect the diffusion in part 2 (b), however, the array xnew(j) has to be loaded after diffusion before calling the potential subroutine and before calling elocal. The branching is a second order branching function and the integer weights are stored in the array m(), as we do in Chapter 6. The rest of the program is a straightforward extension of the code for DMC in Chapter 6.

In Figure 9.4, we compare the ground state energy computed by a DMC simulation (the noisy line labeled $< E >$ GFIS-DMC) with the variational estimate as detailed in the last section (the short dashed line labeled $< E_\beta >$ variational), the four energy minima (solid lines labeled M_1 through M_4), and some of the transition states (dotted lines labeled T_1 through T_6). The ground state energy at -0.00535 hartree is well above the four minima and the transition states plotted in Figure 9.4. Additionally, in Figure 9.4, we compare the average potential energy computed variationally, and by DMC. $< V_\beta >$ is the variational thermal average of the potential energy at the optimized value of β.

In Chapter 6, we make it clear that DMC provides a true estimate of the ground state energy. In the examples in Chapter 6, we are not as careful with the initial configuration of replicas as we are here. When simulating ground state properties of clusters, it becomes important to have a good starting distribution. Therefore, the variational computations are crucial to the success of the investigation of ground state properties of clusters. The trial wavefunction in Equation 9.18 was proposed in a recent article, where the impact of quasiergodicity was systematically investigated. We find that with a properly ergodic sampling of the potential landscape of a typical cluster, the trial wavefunction proposed in the last section, oversamples potential wells above the global minimum. However, DMC efficiently modifies the initial distribution with the

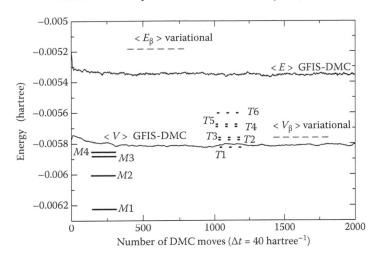

FIGURE 9.4

Important energy values for Ar₇.

branching process. On the other hand, if Metropolis is used, it is likely that the walk could be trapped in the wrong well. DMC is far less efficient in modifying the distribution if there are no replicas in the well that should be sampled by the starting distributions. DMC must first move a replica by diffusion into the right well. The time scale for this step is independent of the barrier between wells, and it is highly dependent on the mass of the system. A system as massive as Ar could remain trapped in the wrong minimum, or not sample other important wells for a very long simulation time. Unfortunately, the trend of the energy in time would look deceivingly flat.

9.7   Path Integral Simulations of Ar₇

The program `rewfpi` in Appendix C computes the canonical ensemble average of the internal energy using the reweighted random series path integral of Chapter 7, and the finite difference estimators for the energy and heat capacity. The program `rewfpi` requires the subroutine `getw`, in this section, the subroutines `construct7`, and `pot` from Chapter 8 (with the Lee–Barker–Abraham sphere introduced earlier), and the random number generator `ransi` from Chapter 3. In `rewfpi`, the path integral method is combined with parallel tempering, and much of its walk structure mimics the walk structure of the `parallel_tempering_r3n` program in Appendix A. The subroutine `getw` computes the exponent $\beta W_{k_{\max}}(\mathbf{a}, \mathbf{x}, \boldsymbol{\beta})$ of the importance sampling for the integral,

$$\langle A \rangle = \frac{\displaystyle\int_{-\infty}^{\infty} d\mathbf{x} \int_{-\infty}^{\infty} d\mathbf{a}\; A(\mathbf{a}, \mathbf{x}, \boldsymbol{\beta}) \exp\left[-\beta W_{k_{\max}}(\mathbf{a}, \mathbf{x}, \boldsymbol{\beta})\right]}{\displaystyle\int_{-\infty}^{\infty} d\mathbf{x} \int_{-\infty}^{\infty} d\mathbf{a}\; \exp\left[-\beta W_{k_{\max}}(\mathbf{a}, \mathbf{x}, \boldsymbol{\beta})\right]}. \tag{9.22}$$

The symbol $d\mathbf{x}$ is shorthand for the configuration space volume element $dx_1, \ldots, dx_{3n}$, and $d\mathbf{a}$ is shorthand for the differentials of the Wiener measure: $da_1^1, \ldots, da_{3n}^{4k_{\max}}$. The integration is over $3n(k_{\max}+1)$ dimensions.

The code for the subroutine `getw` is as follows.

```
      subroutine getw(temp,w,xin,N,ND,KMAX,wi)
      implicit real*8 (a-h,o-z)
      parameter (NMAX=200)
      parameter (NW=40)
      parameter (NX=200)
      parameter (epsilon = 3.77169d-4)  ! hartree (for argon)
      parameter (r0 = 6.424) ! bohr  (for argon)
       parameter(dmm =  39.96284*1822.89)  ! mass of Ar^40 (99.6% abundance)
      real*8 w(NMAX,NX)
      real*8 xin(NMAX)
      real*8 xm1(NMAX),x(NMAX)
      common /traj/sn(NX,NX),sigma2(NX)
c On input:
c  temp = beta in inverse hartree
c  w = array containing the path coefficients
c xin = Cartesian coordinates of the atoms
c N = quadrature for the action integral (Usually KMAX + 1)
c ND = number of HF molecules
c KMAX = number of core path variables
c On return:
c wi = the exponent for the importance sampling function ( beta V if kmax = 0)
      beta = temp
      sum = 0.d0
      t12 = sqrt(beta)
      dN = dfloat(N)
      do i=1,N
c Get the path from the Fourier coefficients
      do j=1,3*ND
       xm1(j) = xin(j)
       x(j) = xin(j)
       do k=1,4*KMAX
        if (i .gt. 1) then
         xm1(j) = xm1(j) + t12*w(j,k)*sn(i-1,k)
        endif
        x(j) = x(j) + t12*w(j,k)*sn(i,k)
       enddo
      enddo ! j loop
c the velocities  are calculated with the core path
      call pot(NMAX,x,ND,epsilon,r0,v)
```

```
      do j=1,3*ND
        sum = sum+(dN/(beta*2.d0))*dmm*(x(j)-xm1(j))*(x(j)-xm1(j))
      enddo
      sum = sum + beta*v/dN
    enddo  ! i loop
      wi = sum
      return
2001  format(' #',6f18.9)
      end
```

The exponent $\beta W_{k_{\max}}(\mathbf{a}, \mathbf{x}, \boldsymbol{\beta})$, is

$$\beta W_{k_{\max}}(\mathbf{a}, \mathbf{x}, \boldsymbol{\beta}) = \sum_{i=1}^{k_{\max}+1} \sum_{j=1}^{3n} \left\{ \frac{k_{\max}+1}{2\beta} m \left[\Delta x_j (\tau_i) \right]^2 + \frac{\beta}{k_{\max}+1} V \left[\mathbf{x}(\tau_i) \right] \right\},$$

(9.23)

where the path $x_j(\tau_i)$,

$$x_j(\tau_i) = \sum_{k=1}^{k_{\max}} \beta^{1/2} a_j^k \Lambda_k(\tau_i) + \sum_{k=k_{\max}+1}^{4k_{\max}} \beta^{1/2} a_j^k \widetilde{\Lambda}_k(\tau_i),$$

(9.24)

is expanded using the reweighted random series representation introduced in Chapter 7, and

$$\Delta x_j(\tau_i) = x_j(\tau_i) - x_j(\tau_{i-1}),$$

(9.25)

is the finite difference term for the velocities. The random series coefficients are stored in the doubly subscripted array w(j,k), with the first index as the pointer to the particular coordinate of the space $\mathbb{R}^{3n}$, and the second index is the pointer to the coefficient associated with the function $\Lambda_k(\tau_i)$.

The functions $\Lambda_k(\tau_i)$ and $\widetilde{\Lambda}_k(\tau_i)$ are the sinusoidal functions introduced in Chapter 7, and are computed in part 1 (b). Part (2) is similar in structure to the classical parallel tempering walk. The variables a_j^k are stored in the triply subscripted array pw(ii,kk,k), where the first index points to the particular coordinate $x_j \in \mathbb{R}^{3n}$, the second index is the pointer for the particular term of the Fourier expansion, and the third index is the temperature pointer. The configuration $(\mathbf{a}, \mathbf{x})$ must be loaded into arrays w(ii,kk) and xt(ii), respectively, before a call to getw can be made. In part 2 (b), the acceptance probability for a swap is a generalization of Equation 9.8,

$$p_a(\mathbf{x}' | \mathbf{x}) = \min \left\{ 1, \frac{\exp\left[-W(\mathbf{a}', \mathbf{x}', \boldsymbol{\beta}) \right] \times \exp\left[-W(\mathbf{a}, \mathbf{x}, \boldsymbol{\beta}') \right]}{\exp\left[-W(\mathbf{a}, \mathbf{x}, \boldsymbol{\beta}) \right] \times \exp\left[-W(\mathbf{a}', \mathbf{x}', \boldsymbol{\beta}') \right]} \right\}.$$

(9.26)

In part 2 (e), the data necessary to compute the finite difference estimators for the energy and heat capacity are accumulated. The finite difference estimators for the energy and heat capacity are introduced in Chapter 7 as well.

Test results for Ar$_7$ for $k_{\max} = 0, 2, 4, 6, 8, 10$, are compared in Figures 9.5 and 9.6 for the total energy and the heat capacity, respectively. The energy is

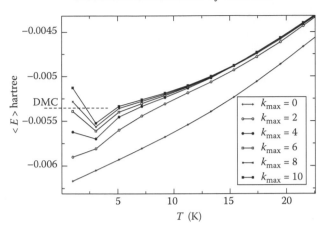

FIGURE 9.5
The total energy of Ar$_7$, as a function of temperature for several values of k_{max}.

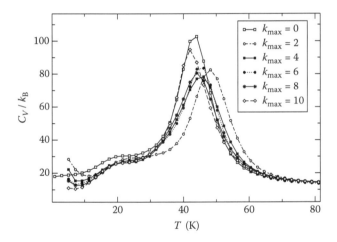

FIGURE 9.6
The heat capacity of Ar$_7$, as a function of temperature for several values of k_{max}.

converged at 12 K with $k_{\mathrm{max}} = 10$, and comparing the pattern of the energy at 5 K with the DMC value obtained in the previous section indicates that Ar$_7$ is predominantly in the ground state at or below 5 K. The convergence of the heat capacity is considerably less uniform, as Figure 9.6 indicates. The boiling peak shifts to higher temperatures at first, then drops to lower temperatures as k_{max} increases. For Ar$_7$, the heat capacity converges below the melting range for relatively elevated values of k_{max}, depending on how low the desired temperature is. At $k_{max} = 10$, the heat capacity in the boiling range is uniform and in the asymptotic regime. The amount of CPU time needed to simulate Ar$_7$ increases with k_{max}. We leave the readers to find the values of k_{max} for which the heat capacity is converging with the statistical error below the melting feature. A $k_{max} = 10$ simulation can require as much as 12 hours on a single modern workstation, therefore, it takes some time to find the answers to the convergence question, even for a "toy" problem like Ar$_7$.

Exercises

1. Modify the program `parallel_tempering_r3n` as suggested and run VMC and DMC simulations of Ar$_7$. Compare the results with those in Figures 9.3 and 9.4.

2. Verify that the DMC calculations are converged with the present choice of Δt for the ^{40}Ar$_7$ cluster. Run DMC with a value two, four, and eight times larger.

3. Repeat the VMC and DMC simulations for Ar$_7$ clusters composed of the two lighter isotopes of Ar, ^{38}Ar ($m = 37.96273$ amu. 0.63% abundance) and ^{36}Ar ($m = 35.96275$ amu. 0.34% abundance), to determine the isotope effect on the variational and exact ground states.

4. Turn off the guiding term by setting `fr = 1.d0` in part 2 (b) of the program `gfis_dmc`, and compare the value of the ground state and its fluctuations of the unguided DMC with the guided results.

5. Guided DMC simulations do not require confining potentials. Verify this statement by running `gfis_dmc` without the Lee-Barker-Abraham sphere in the potential.

6. Show that

$$\exp\left(-\beta V\right), \exp\left(-\beta V^2\right), \ldots, \exp\left(-\beta V^j\right),$$

is a valid basis set for a vector space of square integrable functions $\mathbb{R}^{3n} \to \mathbb{R}$. The vector space of square integrable functions, $\mathbb{R}^{3n} \to \mathbb{R}$, is usually denoted $\mathcal{L}^2\left[\mathbb{R}^{3n}\right]$. If

$$\int dx_1 \cdots \int dx_{3n} \left|f\left(x_1, \ldots, x_{3n}\right)\right|^2 < \infty,$$

then $f\left(x_1, \ldots, x_{3n}\right) \in \mathcal{L}^2\left[\mathbb{R}^{3n}\right]$. What property of the potential energy V is critical for $\exp\left(-\beta V\right) \in \mathcal{L}^2[\mathbb{R}^{3n}]$?

7. A trial wavefunction using two parameters can be constructed using the basis set in the last exercise,

$$\psi_T\left(\mathbf{x}, \beta, \lambda\right) = \lambda \exp\left(-\beta V\right) + (1 - \lambda) \exp\left(-\beta V^2\right).$$

To optimize the parameters, choose 10 values of λ between 0 and 1. For each value of λ, run a parallel tempering simulation with forty different values of β. You need to modify the code in `parallel_tempering_r3n.f` to reflect changes in the acceptance probability of regular moves and swaps. For Metropolis move, the trial probability is the same as Equation 9.4. Inserting Equation 9.4 and $\rho(\mathbf{x}) = A|\psi_T(\mathbf{x}, \beta, \lambda)|^2$ into Equation 9.3, derive an expression for p_a equivalent to its counterpart in Equation 9.5. A is a constant that should drop in the ratio you derive. For parallel tempering simulations, swaps are to be performed using a distribution of configurations from a walker at an adjacent value of the parameter β. The trial probabilities for swaps are:

$$T(\mathbf{x}'|\mathbf{x}) = A(\beta')|\psi_T(\mathbf{x}', \beta', \lambda)|^2,$$

and,

$$T(\mathbf{x}|\mathbf{x}') = A(\beta')|\psi_T(\mathbf{x}, \beta', \lambda)|^2.$$

You will need to derive a new equation for the local energy estimator as well.

8. After the optimal values of β and λ are found from the previous exercise, modify `gfis_dmc` so that it is guided by this more accurate trial wavefunction. Note that the subroutine `elocal` must be the one modified in the previous exercise. To a less extent, it is important to have a good starting configuration of walkers from a parallel tempering simulation at the optimal values for β and λ.

9. Compare the heat capacity from a path integral simulation of Ar_7 with $k_{max} = 14$ and $k_{max} = 18$ to a classical simulation $k_{max} = 0$. Determine the temperature range for which the heat capacity with $k_{max} = 14$ agrees with the the same computed with $k_{max} = 18$ within the statistical error.

9.8 Characterization Techniques: The Lindemann Index

The heat capacity of Ar_7 has a less intense shoulder centered around 20 K in Figures 9.2 and 9.6. In this section, we explore this feature, and other subtleties of the classical parallel tempering random walk, more closely. There are a number of useful structural identifiers and structural properties that are used to gain a deeper understanding of the thermodynamics of clusters and liquids. The radial distribution, for example, is a particularly useful concept for the simulation of liquids, but its definition is ambiguous for clusters. Over the years, the community has developed a number of translationally and rotationally invariant quantities that can be formulated unambiguously for clusters. The first one that we consider is the Lindemann index, which is a measure of the fluctuation of atom to atom distance,

$$\delta_r = \frac{2}{n(n+1)} \sum_{i=1}^{n} \sum_{j=i+1}^{n} \frac{\left(\langle r_{ij}^2 \rangle - \langle r_{ij} \rangle^2\right)^{1/2}}{\langle r_{ij} \rangle}. \tag{9.27}$$

In the course of a classical parallel tempering simulation with M set at 100,000, we write 100,000 configurations for every temperature to a set of fifty device files by inserting the following code at the end of part 2 (e) of `parallel_tempering_r3n`:

```
ifn = 6 + kw
write(ifn,*) n
do i=1,n
  write(ifn,*) x(i),x(n+i),x(2*n+i)
enddo
```

Therefore, the file `fort.7` contains 100,000 configurations visited by the walk at 1 K, `fort.56` contains 100,000 configurations visited by the walk at 100 K, etc. Filling the fifty device files with formatted data requires approximately 2.5 Gb of disk space and takes less than 15 minutes of CPU time on a modern workstation.

These files are used for a number of additional computations. The program `lindemann` reads these files, reproduces the value of the temperature associated with each file, and computes the index in Equation 9.27 for each temperature.

```
      program lindemann
      implicit real*8 (a-h,o-z)
      parameter (NW = 50)
      parameter (maxn = 100)
      real*8 x(maxn),y(maxn),z(maxn)
      real*8 rij(maxn,maxn),r2ij(maxn,maxn)
      real*8 xrand(10),tempw(100)
c this program analyzes groups of configurations and
c computes the Lindemann index
      do k = 1,NW
      tempw(k) = 1.00d0 + 100.d0*dfloat(k -1)/dfloat(NW)
c     initialize
      do i=1,maxn
       do j=1,maxn
         rij(i,j) = 0.d0
         r2ij(i,j) = 0.d0
      enddo
      enddo
      ifn = 6 + k
      icc = 0
```

```
      do number_of_configurations = 1,100000
      read(ifn,*,end=99) n
c       read(ifn,*,end=99) energy
      do kj=1,n
        read(ifn,*,end=99) x(kj),y(kj),z(kj)
      enddo
      do i=1,n
c calculate the distance between atom i and j
        do j=i+1,n
          dx = x(i) - x(j)
          dy = y(i) - y(j)
          dz = z(i) - z(j)
          r2 = dx*dx + dy*dy + dz*dz
          r = sqrt(r2)
          r2ij(i,j) = r2ij(i,j) + r2
          rij(i,j) = rij(i,j) + r
        enddo
      enddo
      icc = icc + 1
      enddo    ! configurations
c   normalize the sums
99    do i=1,n
        do j=i+1,n
         rij(i,j) = rij(i,j)/dfloat(icc)
         r2ij(i,j) = r2ij(i,j)/dfloat(icc)
        enddo
      enddo
c   compute the Lindemann index
      dlinr = 0.d0
      f = 2.d0/dfloat(n*(n-1))
      do i=1,n
        do j=i+1,n
         d1 = sqrt(r2ij(i,j) - rij(i,j)*rij(i,j))
         dlinr = dlinr + f*d1/rij(i,j)
        enddo
      enddo
      write(6,*) tempw(k), dlinr
      enddo    !sum over walkers
      stop
      end
```

The screen output from the program lindemann is graphed in Figure 9.7. The relative fluctuations in the radial distances should increase as the system transitions from a solid-like state into a liquid and then again into a gaseous state. The liquid to gas phase change matches nicely with the steep increase in the fluctuations the radial distances. The Lindemann index seems less

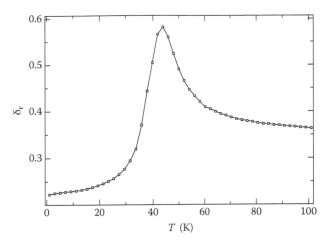

FIGURE 9.7
The Lindemann index from the classical parallel tempering simulation of Ar_7.

sensitive to the colder feature in the thermodynamics of Ar_7 observed in the heat capacity graphed in Figures 9.2 and 9.6.

9.9   Characterization Techniques: Bond Orientational Parameters

Fluctuations in the radial degrees of freedom yield useful information about the thermal processes that take place in both clusters and liquids. In thermodynamically stable clusters, the global minimum is deeper compared to all the other minima. The melting peak in the heat capacity for thermodynamically stable clusters is more pronounced and the fluctuations of the radial degrees of freedom increase in the melting range at first, and then again in the boiling region. For less stable clusters, more sensible structural descriptors have to be formulated. A number of rotationally and translationally invariant structural indicators have been developed by the community to inspect distributions and extract structural features of random walks. A class of such indicators is the class of **bond orientational parameters** denoted Q_l, and defined by,

$$Q_l = \left(\frac{4\pi}{2l+1} \sum_{m=-l}^{l} |Q_{l,m}|^2 \right)^{1/2}. \tag{9.28}$$

$Q_{l,m}$ are sums over spherical harmonic functions introduced in Chapter 4,

$$Q_{l,m} = \frac{1}{n_{\rm b}} \sum_{r_{ij}<r_b} Y_l^m \left(\theta_{ij}, \phi_{ij} \right). \tag{9.29}$$

The sum in Equation 9.29 is over all the atom pairs that have a distance less than an arbitrarily chosen cutoff value. For each of the n_b bonds, a vector is drawn from the center of mass of the cluster to the center of the bond. The angles θ_{ij}, ϕ_{ij} are the resulting polar and azimuthal angles, respectively, for the resulting vector. The cutoff is typically chosen as $r_b = 1.39 r_0$, where r_0 is the Lennard-Jones length scale parameter.

The lines of code of the program bopq1 are obtained by modifying the program lindemann. After reading the 100,000 configurations from a walker selected by the user at run time, the value of Q_1 is computed for each set of coordinates and written to a file (fort.1) for further processing into histograms.

The three spherical harmonic functions involved are:

$$Y_1^0 = \sqrt{\frac{3}{4\pi}} \cos\theta, \qquad (9.30)$$

$$Y_1^{\pm 1} = \mp \sqrt{\frac{3}{8\pi}} \sin\theta\, e^{\pm i\phi}, \qquad (9.31)$$

where $i = \sqrt{-1}$.

```
      program bopq1
      implicit real*8 (a-h,o-z)
      parameter (NW = 40)
      parameter (maxn = 100)
      parameter (r0 = 6.424) ! bohr  (for argon)
      parameter (rb = 1.39*r0) ! cutoff for a bond
      parameter (pi = 3.141592654d0)
      real*8 x(maxn),y(maxn),z(maxn)
c Analysis of groups of configurations:
c Bond "dipole"  distribution
         write(6,*) 'Enter the walker number'
         read(5,*) k
      ifn = 6 + k
      icc = 0
      do number_of_configurations = 1,100000
      read(ifn,*) n
      dn = dfloat(n)
      do kj=1,n
         read(ifn,*) x(kj),y(kj),z(kj)
      enddo
      call centermass(maxn,x,y,z,n)
      call jz(maxn,n,x,y,z,djz)
      nb = 0     ! count the number of bonds in a structure
c        initialize for q1
         f1 = sqrt(3.d0/(4.d0*pi))
         f2 = sqrt(3.d0/(8.d0*pi))
```

```
      f3 = 4.d0*pi/3.d0
      y10 = 0.d0
      y11r = 0.d0
      y11i = 0.d0
      y1m1r = 0.d0
      y1m1i = 0.d0
   do i=1,n
    do j=i+1,n
    dx = x(i) - x(j)
    dy = y(i) - y(j)
    dz = z(i) - z(j)
    r = sqrt(dx*dx + dy*dy + dz*dz)
c if r is less than the cutoff, then we have a bond
      if (r .lt. rb) then
      nb = nb + 1
c locate the middle of the bond and the distance to the CM
      xm = 0.5d0*(x(i) + x(j))
      ym = 0.5d0*(y(i) + y(j))
      zm = 0.5d0*(z(i) + z(j))
      rm = sqrt(xm*xm + ym*ym + zm*zm)
c now compute the orientation angles of the vector rm
      ct = zm/rm
      st = sqrt(xm*xm + ym*ym)/rm
      cp = 1.d0
      sp = 0.d0
      if ((dabs(st)) .gt. 1d-7) then
       cp = xm/(rm*st)
       sp = ym/(rm*st)
      endif
      y10 = y10 + f1*ct
      y11r = y11r + f2*st*cp
      y11i = y11i + f2*st*sp
      y1m1r = y1m1r + f2*st*cp
      y1m1i = y1m1i - f2*st*sp
c         write(6,*) i,j,r,rm,ct,st,cp,sp
     endif
    enddo
   enddo
      dn = dfloat(nb)
      y10 = y10/dn
      y11r = y11r/dn
      y11i = y11i/dn
      y1m1r = y1m1r/dn
      y1m1i = y1m1i/dn
     y102 = y10*y10
```

```
      y112 = y11r*y11r + y11i*y11i
      y1m12 = y1m1r*y1m1r + y1m1i*y1m1i
      q1 = sqrt(f3*(y1m12 + y102 + y112))
      write(1,*) q1
      write(2,*) djz
   enddo   ! configurations
   end
```

An inspection of Table 9.1 demonstrates that if distributions of Q_1 are not too broad, one could discriminate configurations that resemble M_1 from those that resemble M_2, M_3, and M_4. Figure 9.8 contains four graphs of the normalized distributions of Q_1 at 1, 7, 15, and 23 K. At 1 K, the distribution is sharply peaked around a value of 0.0625. Ar_7 seems to be in the global minimum and solid-like. The same peak broadens considerably at 7 and 15 K, and though it is asymmetric, there is no sufficient evidence that configurations near M_3 and M_4 are present. At 15 K, a new distinct peak at 0.01 is visible, indicating that the system is periodically visiting the two minima, M_1 and M_2. At 23 K, the system is liquid-like. The breath of the distributions of Q_1 gleaned in Figure 9.8 and another inspection of Table 9.1 tells us that the distributions of Q_3 probably cannot discriminate configurations as well. However, Q_4 could be used to detect $M_1 \rightarrow M_3$ events. The coding is left as an exercise, the nine spherical harmonic functions involved are,

$$Y_4^0 = \sqrt{\frac{9}{256\pi}} \left(35 \cos^4 \theta - 30 \cos^2 \theta + 3\right), \tag{9.32}$$

$$Y_4^{\pm 1} = \mp \sqrt{\frac{45}{64\pi}} \left(7 \cos^3 \theta - 3 \cos \theta\right) \sin \theta \, e^{\pm i\phi}, \tag{9.33}$$

$$Y_4^{\pm 2} = \sqrt{\frac{45}{128\pi}} \left(7 \cos^2 \theta - 1\right) \sin^2 \theta \, e^{\pm i2\phi}, \tag{9.34}$$

$$Y_4^{\pm 3} = \mp \sqrt{\frac{315}{64\pi}} \cos \theta \sin^3 \theta \, e^{\pm i3\phi}, \tag{9.35}$$

$$Y_4^{\pm 4} = \sqrt{\frac{315}{512\pi}} \sin^4 \theta \, e^{\pm i4\phi}. \tag{9.36}$$

TABLE 9.1

Bond Orientational Parameters for the Minima of Ar_7

Minimum i	n_b	Q_1	Q_2	Q_3	Q_4
1	16	0.062500	0.238851	0.062500	0.110714
2	15	0.003547	0.172634	0.121799	0.131807
3	15	0.047203	0.201953	0.125023	0.078516
4	15	0.058817	0.308391	0.131880	0.137021
5	15	0.058817	0.308391	0.131880	0.137021

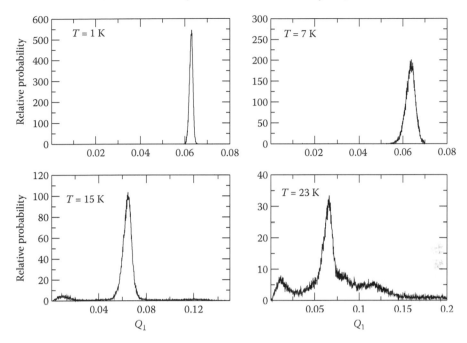

FIGURE 9.8

Distributions of the bond orientational parameter Q_1 from a classical simulation of Ar_7 for several values of temperature.

The results in Figure 9.9 are striking. The distribution of Q_4 is bimodal at 1 K and each peak can be assigned with the aid of the data in Table 9.1. However, looks can be deceiving! One must confirm that the two peaks at 1 K do, in fact, belong to different minima by selectively quenching configurations under the two peaks. At 15 K, there is a distinguishable shoulder from M_2-like structures and it indicates that the M_2-like configurations come to dominate the walk at 23 K. However, the Q_4 values for M_4 and M_5 are not sufficiently different from the value for M_2 to conclude that the shoulder growing around $Q_4 = 0.13$ is exclusively from the walker in the vicinity M_2. The data in Table 9.1 suggest that perhaps distributions of Q_2 can discriminate between M_2-like configurations and M_4, M_5 ones.

9.10 Characterization Techniques: Structural Comparison

It is often desirable to measure the relative importance of each minimum in random walks as a way of characterizing the phase behavior of clusters. It

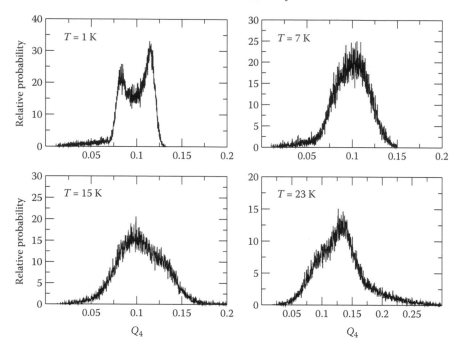

FIGURE 9.9

Distributions of the bond orientational parameter Q_4 from a classical simulation of Ar_7 for several values of temperature.

is hard to quantitatively extract percentages from distributions of structural descriptors like the bond orientational parameters, unless the peaks can be assigned unambiguously, are fully resolved, and can be integrated. It is rare that distributions of structural descriptors have resolved peaks, especially at temperatures of interest, such as, for example, in the liquid state. Furthermore, no single structural descriptor introduced so far has the ability to discriminate anantiomers. The relative importance of M_4 and M_5, however small, should be equal within the statistical error. In the present section, we introduce additional efficient ways to quantify random walks and detect phase changes beyond selective quenching under peaks of distributions.

There are two ways in which researchers have gauged the relative importance of minima in random walks of clusters.

i. Systematic quenching of all the configurations

ii. Analysis of structural distances with the SCA

The first one is straightforward to implement into a program. One could simply read all the configurations in each file, and submit all 100,000 of these to the subroutine **brown** introduced in Chapter 8. This brute force approach is likely to run out of steam for all but the smallest clusters. Even for a system as small as Ar₇, one would have to limit the investigation to a small range of temperatures. A little thought can improve the efficiency of systematic quenching, since every rejected move or swap in the course of simulations generates a consecutive pair (and occasionally consecutive triples, quadruples, etc.) of identical configurations. Identical consecutive configurations can be detected by inspecting consecutive potential energy values. The SCA can be helpful in the systematic quenching approach as well. Even the acceptance of a Metropolis move at cold temperatures, for which an atom changes its position relative to the others, does not cause the system to hop from one well to another, generally. It is much more likely that a swap induces a well-to-well hop. One can use SCA to discriminate configurations that have hopped into a different well by measuring structural distances consecutively. By choosing a cutoff distance sensibly, one can send only "sufficiently distant" structures to be quenched, considerably reducing the number of expensive minimizations.

The information gathered with the systematic quenching of all the configurations provide additional clues to the shape and size of the basin of attraction of the various minima of the surface. This type of information is important, however, there are advantages in pursuing a deeper analysis of structural distances. The analysis of structural distance goes as follows. After the first 10–20 minima are known for a particular cluster, one systematically compares the configurations generated in the course of the walk with those of the known minima. If the structural distance generated by SCA between the walker configuration and M_i falls below a sensibly chosen cutoff, then the configuration in the walk is labeled as "belonging to M_i." In the solid–solid coexistence region, the information gathered with the analysis of structural distances by SCA are identical to the data obtained by systematic quenches, assisted with SCA as detailed earlier with the same cutoff. Therefore, one advantage is practical in nature; the minimizations are eliminated altogether. Additionally, temperatures are found for which a significant number of configurations do not belong to any of the minima in the pool. In the case of Ar₇, if these structures are sent to be quenched, the resulting minima will be among the known ones. For a larger system, it is possible to find new higher energy (and therefore, less important) minima not included in the comparison pool. However, the onset of temperatures marked by significant distances from minima deemed "most important" is relevant information in itself. One could miss this information by systematically quenching without comparisons. The onset of such temperatures marks the beginning of a new phase coexistence. The work in previous sections has provided some evidence that between 1 and 15 K, for Ar₇, there may be a small but significant amount of well-to-well hopping. However, as the temperature reaches 20 K, the distributions of all the structural descriptors are too broad to provide a clear picture. The shoulder in the heat capacity indicates the onset of melting, and somehow, our structural description machinery must be able to identify such conspicuous phase changes unmistakably. SCA, with a sensibly chosen arbitrary cutoff, provides

TABLE 9.2

Structural Distances for the Minima of Ar_7

Minimum i	δ_1^i	δ_2^i	δ_3^i	δ_4^i	δ_5^i
1	0.000	21.75	15.33	17.49	17.49
2	13.60	0.00	17.06	15.66	15.66
3	14.37	10.95	0.00	16.15	16.15
4	14.37	11.90	7.98	0.00	14.62
5	14.37	10.74	8.07	14.62	0.00

TABLE 9.3

Relative Importance of the Minima of Ar_7

Temperature (K)	M_1	M_2	M_3	M_4	M_5	Liquid-like
1	100,000	0	0	0	0	0
3	100,000	0	0	0	0	0
6	100,000	0	0	0	0	0
7	99,956	44	0	0	0	0
9	99,509	395	38	58	0	0
11	98,287	1308	76	267	45	17
13	93,672	4280	476	815	406	351
15	86,581	153	1261	1403	1627	8975
17	76,112	1075	2340	2937	2493	15,043

a way to fine tune the structural analysis so that melting phases can be detected.

A glance at Table 9.2 tells us that the smallest minimum to minimum SCA distance is 7.98 bohr, roughly, the size of an Ar-Ar nearest neighbor distance in the minima. Note that $\delta_j^i \neq \delta_i^j$, because the atoms of miminum j, labeled as $1, 2$, and 3, that optimize the distance with i are, generally, not the same as atom $1, 2, 3$ of minimum j when the best labels for atoms $1, 2, 3$ of minimum i are optimized in defining δ_i^j.

Consider the program dmsca. The configurations of all walkers are read and systematically compared with those of the five known minima of the potential. The subroutines compare and rotate2 have to be appended. The partial screen output of dmsca is tabulated in Table 9.3. The column labeled "Liquid-like" is obtained by subtracting from 100,000 the sum of the tally for all the minima included. Assuming a 1% statistical fluctuation, the data in Table 9.3 indicate that at 11 K there is significant $M_1 \rightarrow M_2$ isomerization, M_3, M_4, M_5 are significant at 15 K, however, at that temperature there is a significant amount of melting as well.

```
program dmsca
implicit real*8 (a-h,o-z)
parameter (NW = 50)
```

```
      parameter (maxn = 100)
      parameter (pi = 3.141592654d0)
      real*8 x(maxn),y(maxn),z(maxn)
      real*8 rx(maxn),ry(maxn),rz(maxn)
      real*8 rxi(maxn,10),ryi(maxn,10),rzi(maxn,10)
      real*8 pd0(100)
      integer ic(10)
c SCA analysis of random walks. If the SCA distance (sigmanin)
c between a configuration and minimum i is less than 7.5 bohr, then
c the configuration is near minimum i.
      n = 7
c the coordinates of the global minimum
      open(unit=2,file='minima.dat',status='old')
      do k =1,5  ! read all five minima
      read(2,*) energy
      do i=1,n
       read(2,*) rxi(i,k), ryi(i,k), rzi(i,k)
      enddo
      enddo
      close(2)
      do j=1,40
       pd0(j) = 1.0d0  + 100.0d0*(dfloat(j-1)/dfloat(NW))  ! T in K
       ifn = 6 + j
       icc = 0
      do k=1,5
        ic(k) = 0.d0
      enddo
      do number_of_configurations = 1,100000
       read(ifn,*) n
       dn = dfloat(n)
       do kj=1,n
        read(ifn,*) x(kj),y(kj),z(kj)
       enddo
c compare wih minimum k
        do k=1,5
         do i=1,n
          rx(i) = rxi(i,k)
          ry(i) = ryi(i,k)
          rz(i) = rzi(i,k)
         enddo
         call compare(maxn,n,rx,ry,rz,x,y,z,sigmamin)
         if (sigmamin .lt. 7.5d0) then
          ic(k) = ic(k) + 1
         endif
        enddo
       enddo    ! configurations
       write(6,*) pd0(j), (ic(k),k=1,5)
      enddo   ! j
      end
```

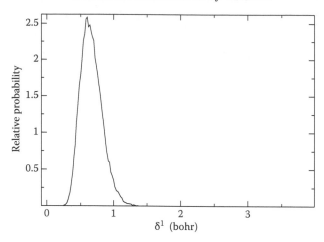

FIGURE 9.10
Distributions of the structural distance from the global minimum δ^1, at 1 K from a classical parallel tempering simulation.

The SCA can also be used to generate distributions of δ^i. To further confirm that Ar_7 is in the global minimum at 1 K, we collect the distribution of δ^1. In Figure 9.10, the distribution of δ^1, the structural distance from the global minimum, indicates that isomerizations are insignificant. The majority of configurations have a structural distance smaller than 1 bohr, and no value of δ^1 is observed above 2 bohr.

Exercises

1. Modify the program `parallel_tempering_r3n.f` so that it writes configurations to a file. Use the configurations to compute the Lindemann index, and the distributions of Q_1 at $1, 7, 15$, and 23 K. Compare the results with those in Figures 9.7 and 9.8.

2. Write a program that computes values of Q_2, Q_3, and Q_4, for input configurations. Then, have the program read the file `minima.dat` (generated in Chapter 8) that contains the Cartesian coordinates of all the atoms at the five minima of Ar_7 and reproduces the data in Table 9.1.

3. Use the program written in the previous exercise to duplicate the data in Figure 9.9. Select several configurations that have Q_4 values at the peaks, then quench them. Can you tell what structural feature gives rise to the peak at $Q_4 = 0.07$? A molecular graphics program is indispensable at this point.

4. Shah and Chakravarty [796] make use of third order rotationally invariant bond orientation parameters to study bulk Lennard-Jones systems.

$$W_l = \frac{\displaystyle\sum_{m_1,m_2=-l}^{+l} \begin{pmatrix} l & l & l \\ m_1 & m_2 & m_1+m_2 \end{pmatrix} Q_{l,-m_1} Q_{l,-m_2} Q_{l,m_1+m_2}}{\left(\displaystyle\sum_{m,=-l}^{+l} |Q_{l,m}|^2\right)^{3/2}},$$

where $Q_{l,m}$ is defined in Equation 9.29. The quantity inside () in the sum is a 3-j symbol introduced in Chapter 5. Modify the program bopq1.f so that distributions of W_1 are obtained instead.

5. Reproduce the data in Figure 9.10 and Tables 9.1 and 9.2.

6. The structural algorithm introduced in Chapter 8 performs on the order of n^5 operations to find the "best match" among two structures. Two improvements can be made to reduce the number of operations to n^2. Here is the suggested algorithm in pseudocode.

   a. Translate configuration A so that its center of mass is at the origin.

   b. Rotate configuration A so that atom 1 is on the z axis.

   c. Rotate configuration A so that atom 2 is on the $x - z$ plane.

   d. Translate configuration B so that its center of mass is at the origin.

   e. Rotate configuration B so that atom i is on the z axis.

   f. Rotate configuration B so that atom j is on the $x - z$ plane.

   g. Calculate the distance from atom 1 of configuration A to atom i of configuration B, and add the distance from atom 2 of configuration A to atom j of configuration B.

   h. Repeat from step e for all $i, j = 1, \ldots, n$.

   i. For the combination i and j that minimizes the sum computed in step g, sort the remaining atoms of configuration B to find the closest atom in configuration B to atom 3 of configuration A. Do the same for atom 4 of configuration A, then $5, 6, 7, \ldots, n$.

   j. Compute the distances from atom i of configuration A to atom i of configuration B as relabeled in step i. Sum all the distances for $i = 1$ to n. The result is δ_B^A.

Modify the subroutine compare and then repeat the SCA analysis for the classical simulation of Ar_7. Compare the execution times with those of the original version of SCA.

7. Perform the structural analysis with bond orientational parameters and SCA for the VMC (at the optimal temperature), the DMC, and the path integral walks at 40 K with $k_{max} = 20$ [201, 409]. Note that the interpretation of each distribution is slightly different. From VMC, the distribution pertains to the best wavefunction squared. For DMC, the distribution is the ground state wavefunction multiplied by the trial wavefunction, whereas for path integral simulations at finite temperature, the distribution is that of the diagonal elements of the density matrix. The latter approaches the square of the ground state wavefunction at cold temperatures, if it is converged.

9.11 Appendix A: `parallel_tempering_r3n`

```
program parallel_tempering_r3n
implicit real*8 (a-h,o-z)
parameter(maxn = 500)
parameter(NX = 350)
parameter (epsilon = 3.77169d-4)  ! hartree (for argon)
parameter (r0 = 6.424) ! bohr  (for argon)
parameter (htok = 3.15773218d+5) ! from  K/hatree
parameter (nwalk = 50)
real*8 x(maxn),xrand(maxn), xt(maxn)
real*8 px(maxn,100),pd0(100)
real*8 pvsave(100),deltax(100)
integer nrej(100),nrejj(100),najj(100)
real*8 sumv(100),sumv2(100)
real*8 rx(maxn),ry(maxn),rz(maxn)
c Classical Canonical ensemble parallel tempering simulation of Ar7
c Part 1: Initialization
      M = 1000000
      dM = dfloat(M)
      MW = M
      n = 7        ! 7 argon atoms
      do i=1,maxn
       rx(i) = 0.d0 ! initialize all the coordinates
       ry(i) = 0.d0
       rz(i) = 0.d0
      enddo
      see=secnds(0.)
      iseed1 = int(see/10.)
      write (6,*) 'ISEED1 ===>',iseed1
      call ransi(iseed1)
      d = 7.d0  ! bond length us ed to construct a pentagonal bipyramid
      call construct7(maxn,rx,ry,rz,d)
      write(6,*) n
```

```
      write(6,*) '# Ar7 starting structure'
      do  i=1,n
       write(6,*)"#", rx(i),ry(i),rz(i)
      enddo
        do i = 1,n
         x(i) = rx(i)
         x(i+n) = ry(i)
         x(i+2*n) = rz(i)
        enddo
      call pot(maxn,x,n,epsilon,r0,v)
      do kw=1, nwalk      ! nwalk walkers total
      pd0(kw) = 1.0d0  + 100.0d0*(dfloat(kw-1)/dfloat(nwalk))  ! T in K
      beta = htok/pd0(kw)
        do i = 1,n
         px(i,kw) = rx(i)
         px(n+i,kw) = ry(i)
         px(2*n+i,kw) = rz(i)
        enddo
      pvsave(kw) = v
      deltax(kw) = 0.1d0
      sumv(kw) = 0.d0
      sumv2(kw) = 0.d0
      nrej(kw) = 0
      nrejj(kw) = 0
      najj(kw) = 0
      write(6,*) '#',kw,pd0(kw),deltax(kw), pvsave(kw)
      enddo
c Part 2: Walk
100      do move = 1,M+MW
      do kw=1, nwalk    ! WALKER LOOP
c Part 2: (a) reset all the arrays
        do i = 1,n
         xt(i) = px(i,kw)
         xt(i+n) = px(i+n,kw)
         xt(i+2*n) - px(i+2*n,kw)
        enddo
      temp = htok/pd0(kw)    ! this is beta in hartree^{-1}
c Part 2: (b) decide wether to swap or to move
        call  ransr(xrand,1)
        if (xrand(1) .lt. 0.1 .and. kw .lt. nwalk) then
        najj(kw) =  najj(kw) + 1
        iflagjump = 1
c Part 2: (c) perform a swap
        tempj = htok/pd0(kw+1)
        call pot(maxn,xt,n,epsilon,r0,vtj)
        ea3 = tempj*vtj
        ea4 = tempj*pvsave(kw+1)
        do i = 1,n
         xt(i) = px(i,kw+1)
```

```
          xt(i+n) = px(i+n,kw+1)
          xt(i+2*n) = px(i+2*n,kw+1)
        enddo
        call pot(maxn,xt,n,epsilon,r0,v)
        ea1 = temp*v
        ea2 = temp*pvsave(kw)
        ea = -ea1 + ea2 - ea3 + ea4
        if (dabs(ea) .gt. 207) then
          if (ea .lt. 0) prob = 2.0d0
      if (ea .gt. 0) prob = 0.d0
        else
            prob = dexp(ea)
        endif
c Part 2: (d) perform a Metropolis move
        else
         iflagjump = 0
         call ransr(xrand,4)
c pick an atom at random
          k = n*xrand(1)+1
          xt(k) = xt(k) + deltax(kw)*(xrand(2)- 0.5d0)
          xt(n+k) = xt(n+k) + deltax(kw)*(xrand(3)- 0.5d0)
          xt(2*n+k) = xt(2*n+k) + deltax(kw)*(xrand(4)- 0.5d0)
          call pot(maxn,xt,n,epsilon,r0,v)
          ea = temp*(v-pvsave(kw))
        if (dabs(ea) .gt. 207) then
            if (ea .lt. 0) prob = 2.0d0
            if (ea .gt. 0) prob = 0.d0
        else
          prob = dexp(-ea) ! acceptance or rejection for  Metropolis
        endif
        endif
c Part 2: (e)  Now accept or reject
        call ransr(xrand,1)
        if (prob .gt. xrand(1)) then   ! accept !
          call flush(6)
        if (iflagjump .eq. 1) then  ! swap the kw+1 configuration
         do i = 1,n
          px(i,kw+1) = px(i,kw)
          px(i+n,kw+1) = px(i+n,kw)
          px(i+2*n,kw+1) = px(i+2*n,kw)
         enddo
         pvsave(kw+1) = vtj
        endif       !  swap branch ends here
        do i = 1,n   !   Store the new coordinates
         px(i,kw) = xt(i)
         px(i+n,kw) = xt(i+n)
         px(i+2*n,kw) = xt(i+2*n)
        enddo
        pvsave(kw) = v
```

```
      else !  reject and keep a  tally of rejected moves
        nrej(kw) = nrej(kw) + 1
        if (iflagjump .eq. 1) nrejj(kw) = nrejj(kw) + 1
      endif
c Part 2: (f) self adjusting step
        if (move .lt. MW .and. mod(move,1000) .eq. 0) then
            if (nrej(kw) .gt. 540) then
              deltax(kw) = deltax(kw)*0.7d0
              nrej(kw) = 0
            elseif (nrej(kw) .lt. 460) then
              deltax(kw) = deltax(kw)/0.7d0
              nrej(kw) = 0
            endif
            nrej(kw) = 0
          endif ! (mod (mm,1000) .eq. 0)
        if (move .eq. MW) then
            write(6,2005) move,kw,deltax(kw),nrej(kw),nrejj(kw),
     & najj(kw),pvsave(kw)
          endif
c Part 2: (g) collect data
      if (move .gt. MW) then
        do i = 1,n
          x(i) = px(i,kw)
          x(i+n) = px(i+n,kw)
          x(i+2*n) = px(i+2*n,kw)
        enddo
        call pot(maxn,x,n,epsilon,r0,v)
        sumv(kw) = sumv(kw) + v/dM
        sumv2(kw) = sumv2(kw) + v*v/dM
      endif
20      enddo    !  walker loop
        enddo   !  moves loop
      do kw=1,nwalk
        beta = htok/pd0(kw)
        e = 1.5d0*dfloat(n)/beta + sumv(kw)
        cv = 1.5d0*dfloat(n) + beta*beta*(sumv2(kw) - sumv(kw)*sumv(kw))
        write(6,1001) kw,deltax(kw),pd0(kw),nrej(kw),nrejj(kw),
     & najj(kw),e,sumv(kw),cv
      enddo
999     format(2i7,2f16.8)
1000    format(5e18.8)
1001    format(i4,2f12.6,3i8,3f12.6)
1002    format(1p4e18.7)
1010    format(1p5e24.9)
2005    format(' # ',i7,i4,f8.5,3i7,f16.8)
      end
```

9.12   Appendix B: gfis_dmc

```
      program gfis_dmc
      implicit real*8 (a-h,o-z)
      parameter(MNX = 200000)
      parameter(ms=100)
      parameter (epsilon = 3.77169d-4)  ! hartree (for argon)
      parameter (r0 = 6.424) ! bohr  (for argon)
      parameter (htok = 3.15773218d+5) ! from  K/hatree
      parameter(dmm =  39.96284*1822.89)  ! mass of Ar^40 (99.6% abundance)
      real*8 x(MNX,ms),y(MNX,ms),xr(ms),xn(ms),g(ms)

      real*8 ve(ms),vet(ms),xold(ms),xnew(ms),dv(ms),d2v(ms)
      real*8 xiold(ms),xinew(ms)
      integer m(MNX)
c Part 1: Initialization
      MN = 100000
      dMN = dfloat(MN)
      temp = htok/15.0d0   ! E = -0.005184 hartree with this  parameter
      dt = 10.d0
      n = 7
      sumfr = 0.d0
      sume = 0.d0
      sume2 = 0.d0
      vref = -0.005d0
      iseed =  secnds(0.)
      call ransi(iseed)
      nmoves = 1000
      ne = 0
      nrej = 0
      mv = nmoves
      mw = nmoves
      write(6,*) '# processing a ',n,' atom cluster'
      write(6,*) '# Number of moves = ',nmoves
      write(6,*) '# Delta t = ', dt
      write(6,*) '# seed = ',iseed
c Part 1 (a ):  input (from a variational calculation)
      do i=1,MN
       read(5,*) kw, nmm, v, tempk, eloc
       do j=1,n
        read(5,*) x(i,j),x(i,n+j),x(i,2*n+j)
       enddo
       m(i) = 1
      enddo
      deltax = 0.013
      do j=1,ms
        ve(j) = 0.d0
      enddo
```

```
c Part (2) Walk
        do k=1,mv+mw
         sumfr = 0.d0
         sumv = 0.d0
         sumav = 0.d0
         isum = 0
         do i=1,MN
          do j=1,3*n
           xold(j) = x(i,j)
          enddo
          call pot(ms,xold,n,epsilon,r0,vold)
c Part 2 (a):  acceptance - rejection scheme to generate
c the Gaussian distribution for the diffusion steps
          ne = ne + 1
          do j=1,3*n
          call ransr(xn,1)
           vet(j) = ve(j) + deltax*(xn(1)-0.5d0)
          enddo
c importance sampling function for the diffusion step
          a = 0.d0
          at = 0.d0
          do j=1,3*n
           a = a + 0.5d0*dmm*ve(j)*ve(j)/dt
           at = at + 0.5d0*dmm*vet(j)*vet(j)/dt
          enddo
c boundary check for exp
          if (dabs(a-at) .lt. 230.d0) then
            q = exp(-at+a)
          else
            if ((a-at) .lt. 0) then
              q = 0.0d0
            else
              q = 1.1d0
            endif
          endif
          call ransr(xn,1)
          if (xn(1) .lt. q) then
           do j=1,3*n
            ve(j) = vet(j)
           enddo
           a = at
          else
           nrej = nrej +1
          endif
c self adjusting random move step to get a 50% rejection
           if (mod(ne,1000) .eq. 0 .and. k .lt. mw) then
           if (nrej .gt. 550) then
           deltax = deltax*0.99d0
           endif
```

```
         if (nrej .lt. 450) then
          deltax = deltax/0.99d0
         endif
         nrej = 0
        endif
c Part 2 (b):  diffusion move
         do j=1,3*n
          x(i,j) = x(i,j) + ve(j)
         enddo
         do j=1,3*n
           xnew(j) = x(i,j)
         enddo
         call pot(ms,xnew,n,epsilon,r0,vnew)
         if (k .gt. 10) then
          fr = dexp(-temp*(vnew-vold)/2.d0)
         else
          fr = 1.d0
         endif
c Part 2(c) calculate eloc
         call elocal(ms,xnew,n,epsilon,r0,temp,eloc)
c Part 2(d) branch
         w = dexp(-0.5d0*(vnew + vold - 2.d0*vref)*dt)*fr
         if (w .gt. 10.d0) w = 10.d0
         call ransr(xr,1)
         m(i) = int(w+xr(1))
c Part 2 (e) average energy data
           sumv = sumv + dfloat(m(i))*eloc
           sumfr = sumfr + dfloat(m(i))*fr
           sumav = sumav + dfloat(m(i))*vnew
           isum = isum + m(i)
         enddo
         av = sumv/dfloat(isum)
         avfr = sumfr/dfloat(isum)
         avav = sumav/dfloat(isum)
c Part 2(f) adjust vref
         vref = av - log(dfloat(isum)/dMN)/dt
         write(6,*)k, av,isum,vref,avfr,avav
         call flush(6)
c Part 2(g)  correct the array for the creation - destruction process
         index = 0
         if (k .gt. mw) then
          sume = sume + av/dfloat(mv)
          sume2 = sume2 + av*av/dfloat(mv)
         endif
         do i=1,MN
          do kj=1,m(i)
           index = index + 1
           do j=1,3*n
           y(index,j) = x(i,j)
```

```
       enddo
      enddo
     enddo
     MN = isum    ! adjust the population number
     do i=1,MN
      do j=1,3*n
      x(i,j) = y(i,j)
      enddo
      m(i) = 1
     enddo
    enddo            ! time step
    end
```

9.13   Appendix C: rewfpi

```
    program rewfpi
    implicit real*8 (a-h,o-z)
    parameter (NMAX=200)
    parameter (NW=40)
    parameter (NX=200)
    parameter (epsilon = 3.77169d-4)  ! hartree (for argon)
    parameter (r0 = 6.424) ! bohr  (for argon)
    parameter (htok = 3.15773218d+5) ! From hartree to Kelvin
    parameter (twom12 = 2.44140625d-4) ! Second derivative formulas
    parameter (MW = 1000000) ! ''warm up'' moves
    parameter (M = 1000000) ! number of moves
    parameter (a = 0.1d0)
    real*8 rx(NMAX),ry(NMAX),rz(NMAX)
    real*8 x(NMAX),xt(NMAX)
    real*8 xi(3),xrand(20)
    real*8 px(NMAX,NW)
    real*8 deltax(NW),tempw(NW),sumv(NW),sumv2(NW)
    real*8 sumvv(NW),sumvv2(NW),vsave(NW)
    real*8 sume(NW),sume2(NW),sume3(NW)
    real*8 w(NMAX,NX),pw(NMAX,NX,NW)
    integer nrej(NW),nrejj(NW),najj(NW)
    common /traj/sn(NX,NX),sigma2(NX)
c Part 1 Initialization
c  Part 1 (a) seed the generator, produce a starting configuration
    iseed =  secnds(0.)
    call ransi(iseed)
    n = 7
    d = 7.1d0
    read(5,*) KMAX
    call construct7 (NMAX,rx,ry,rz,d)
    write(6,*) '# n = ',n
    write(6,*) '# kmax =',KMAX
```

```
      if (n .eq. 0) goto 999
    do i=1,7
     x(i) = rx(i)
     x(n+i) = ry(i)
     x(2*n+i) = rz(i)
    enddo
      write(6,*) '# starting configuration'
      do i=1,n
        write(6,*)'#', x(i),x(n+i),x(2*n+i)
      enddo
    call pot(NMAX,x,n,epsilon,r0,v)
    write(6,*) '# potential energy of the input configuration ',v
    dim = 3.d0*float(n)
    dM = dfloat(M)
    do i=1,3*n
       do kk = 1,4*KMAX
    w(i,kk) = 0.0d0
       enddo
      enddo
c Part 1 (b) compute the sines and the other path variables
      np = KMAX + 1
      do i=1,np
       do k=1,KMAX
        fk = sqrt(2.d0)/(3.1415926*dfloat(k))
        sn(i,k) = fk*sin(dfloat(k*i)*3.1415926/dfloat(np))
       enddo
      enddo
      do i=1,np - 1
        ui = dfloat(i)/dfloat(np)
        sigma2(i) = ui*(1.d0 - ui)
       do k=1,KMAX
          sigma2(i) = sigma2(i) - sn(i,k)*sn(i,k)
       enddo
      enddo
      do i=1,np
       sum00 = 0.d0
       do k=KMAX+1,4*KMAX
        fk = sqrt(2.d0)/(3.1415926*dfloat(k))
        sum00 =  sum00 + fk*sin(dfloat(k*i)*3.1415926/dfloat(np))*
     &      fk*sin(dfloat(k*i)*3.1415926/dfloat(np))
       enddo
       do k=KMAX+1,4*KMAX
        fk = sqrt(2.d0)/(3.1415926*dfloat(k))
        ru = sqrt(sigma2(i)/sum00)
        sn(i,k) = ru*fk*sin(dfloat(k*i)*3.1415926/dfloat(np))
       enddo
      enddo
c Part 1 (c) initialize all the walkers
      do k = 1,NW
```

```
      tempw(k) = 1.d0 + 80.d0*dfloat(k -1)/dfloat(NW - 1)
      beta = htok/tempw(k)
      call getw(beta,w,x,np,n,KMAX,wik)
      vsave(k) = wik
      sumv(k) = 0.d0
      sumv2(k) = 0.d0
      sumvv(k) = 0.d0
      sumvv2(k) = 0.d0
      sume(k) = 0.d0
      sume2(k) = 0.d0
      sume3(k) = 0.d0
      nrej(k) = 0
      nrejj(k) = 0
      najj(k) = 0
      deltax(k) = 3.0d-1
       do i=1,3*n
         px(i,k) = x(i)
         do kk = 1,4*KMAX
          pw(i,kk,k) = w(i,kk)
          enddo
         enddo
      write(6,2003) k,tempw(k),beta,vsave(k),nrej(k)
      enddo
c Part (2) Parallel Tempering walk
      do mm = 1,MW + M
      do k=1,NW     ! loop over all walkers
       do ii=1,3*n
        xt(ii) = px(ii,k)
        do kk=1,4*KMAX
         w(ii,kk) = pw(ii,kk,k)
         enddo
        enddo
       iflagjump = 0
       call ransr(xrand,1)
c Part 2 (b) swap configurations
       if (xrand(1) .lt. 0.1 .and. k .lt. NW) then
        najj(k) = najj(k) + 1
        iflagjump = 1
        betaj = htok/(tempw(k+1))
        call getw(betaj,w,xt,np,n,KMAX,wij)
        ea3 = wij
        ea4 = vsave(k+1)
        do ii=1,3*n
          xt(ii) = px(ii,k+1)
          do kk=1,4*KMAX
           w(ii,kk) = pw(ii,kk,k+1)
           enddo
          enddo
        beta = htok/(tempw(k))
```

```
       call getw(beta,w,xt,np,n,KMAX,wi)
       ea1 = wi
       ea2 = vsave(k)
       ea = -ea1 + ea2 - ea3 + ea4
       if (dabs(ea) .gt. 207) then
          if (ea .lt. 0) prob = 2.0d0
         if (ea .gt. 0) prob = 0.d0
       else
         prob = dexp(ea)
       endif
       else
c Part 2 (c) Metropolis move
        iflagjump = 0
       call ransr(xrand,8)
       j = n*xrand(1) + 1
       xt(j) = xt(j) +  deltax(k)*(xrand(2)- 0.5d0)
       xt(n+j) = xt(n+j) +  deltax(k)*(xrand(3)- 0.5d0)
       xt(2*n+j) = xt(2*n+j) +  deltax(k)*(xrand(4)- 0.5d0)
       kk = 4*KMAX*xrand(5)+1
       w(j,kk) = w(j+1,kk) + a*deltax(k)*(xrand(6)- 0.5d0)
       w(n+j,kk) = w(n+j,kk) + a*deltax(k)*(xrand(7)- 0.5d0)
       w(2*n+j,kk) = w(2*n+j,kk) + a*deltax(k)*(xrand(8)- 0.5d0)
       beta = htok/(tempw(k))
       call getw(beta,w,xt,np,n,KMAX,wi)
         ea = wi-vsave(k)
       if (dabs(ea) .gt. 207) then
          if (ea .lt. 0) prob = 2.0d0
          if (ea .gt. 0) prob = 0.d0
       else
         prob = dexp(-ea) ! probabilities for   Metropolis
       endif
       endif   ! Metropolis or swap
       call ransr(xrand,1)
        if (prob .gt. xrand(1)) then
c Part 2 (d) Accept or reject
        if (iflagjump .eq. 1) then   ! swap the kw+1 configuration
        do ii=1,3*n
          px(ii,k+1) = px(ii,k)
         do kk=1,4*KMAX
          pw(ii,kk,k+1) = pw(ii,kk,k)
         enddo
        enddo
        vsave(k+1) = ea3
        endif
        vsave(k) = wi
       do ii=1,3*n
        px(ii,k) = xt(ii)
        do kk=1,4*KMAX
         pw(ii,kk,k) = w(ii,kk)
```

```
          enddo
          enddo
          vsave(k) = wi
          else
           nrej(k) = nrej(k) + 1
c reject
          if (iflagjump .eq. 1)  nrejj(k) = nrejj(k) + 1
          endif
          if (mm .lt. MW .and. mod(mm,1000) .eq. 0) then
             if (nrej(k) .gt. 540) then
               deltax(k) = deltax(k)*0.9d0
               nrej(k) = 0
             elseif (nrej(k) .lt. 460) then
               deltax(k) = deltax(k)/0.9d0
               nrej(k) = 0
             endif
               nrej(k) = 0
           endif ! (mod (mm,1000) .eq. 0)
c Part 2 (e) Accumulate the data for the thermodynamic estimators
          if (mm .gt. MW) then
          do ii=1,3*n
           x(ii) = px(ii,k)
           do kk=1,4*KMAX
            w(ii,kk) = pw(ii,kk,k)
           enddo
          enddo
          call pot(NMAX,x,n,epsilon,r0,v)
          sumv(k) = sumv(k) + v/dM
          beta = htok/(tempw(k))
           call getw(beta,w,x,np,n,KMAX,wi)
           dbeta = twom12*beta
           betap = beta + dbeta
           betam = beta - dbeta
           call getw(betap,w,x,np,n,KMAX,wip)
           call getw(betam,w,x,np,n,KMAX,wim)
          dw = (-wim+wip)/(2.d0*dbeta)
          d2w = (wip+wim-2.d0*wi)/(dbeta*dbeta)
          sume(k) = sume(k) + dw/dM
          sume2(k) = sume2(k) + (-0.5d0*dim- beta*dw)/dM
          sume3(k) = sume3(k) + (0.5d0*dim + 0.25*dim*dim +
     & dim*beta*dw+ beta*beta*(dw*dw - d2w))/dM
           endif
           enddo  ! walkers
          enddo   ! moves
          do kw=1,NW
           beta = htok/tempw(kw)
           e = sume(kw) + 0.5*dim/beta
           cv = sume3(kw) -sume2(kw)*sume2(kw)
           write(6,1999) tempw(kw),e,cv,sumv(kw),
```

```
      & nrej(kw),nrejj(kw),najj(kw)
        enddo
999     stop 'DONE'
1999    format(4f15.9,3i7)
2001     format(' #',6f18.9)
2003    format(' #',i4,3f18.9,i4)
2005    format(' # ',i7,i4,f8.5,3i7,f16.8)
```

Part III

Methods in Curved Spaces

10

Introduction to Differential Geometry

10.1 Introduction

In this chapter, we are interested in making use of a selected number of results from differential geometry, which are necessary for the development of quantum methods for condensed molecular systems subjected to holonomic constraints. The objective is to introduce the concepts of covariant and contravariant vector (and tensor) spaces, the Einstein sum convention, the metric tensor, integration, differentiation, and classical dynamics in manifolds. In all the cases treated in this book, a **manifold** is simply a Euclidean space with n dimensions, or a subspace obtained generally by remapping the original n-dimensional Euclidean space with curvilinear coordinates, and then constraining one or more of these. We denote with the symbol $\mathbb{R}^n$, a Euclidean space with n dimensions when it is mapped with orthogonal Cartesian coordinates. Another possible source of manifolds is the use of a Lie group or semigroup of continuous transformations, introduced in Chapter 4. The Lie group approach is used for the treatment of rigid bodies composed of n point particles. These types of manifolds are most likely of intense interest to the molecular theorist. A typical example of a manifold of this type is the rigid rotor space, where the $\mathbb{R}^6$ space is remapped by transforming into center of mass coordinates (which we can remove from the dynamics), using spherical polar coordinates for the pseudo-one body problem that is left, and fixing the distance r to a constant r_0. The resulting manifold is the set of all points on the surface of a sphere of radius r_0, also known as a two-sphere and denoted with the symbol $\mathbb{S}^2$.

The development of path integrals and diffusion Monte Carlo (DMC) in manifolds like $\mathbb{S}^2$ is the subject of the next chapter. We have learned that these types of stochastic algorithms are considerably simpler to implement when curved manifolds like $\mathbb{S}^2$ are mapped projectively, meaning with coordinates that span the entire number line from $-\infty$ to $+\infty$, so that the vexing issue of imposing boundary conditions on the space, such as, e.g., $0 < \theta < \pi$, $0 < \phi < 2\pi$, is avoided. Boundary conditions in classical stochastic simulations are used to remove the edge effects in the simulations of liquids and solids. There are no difficulties in using open sets and boundary conditions in classical simulations. In fact, for classical simulations, angular variables may be the better choice. However, in quantum simulations, open sets and boundary conditions create technical difficulties. The resulting quantum simulations are challenging to first order and impossible to accelerate to second order. Therefore, a good portion of this chapter is dedicated to the introduction of stereographic projection coordinates for three important types of manifolds, the ring $\mathbb{S}^1$, the two-sphere $\mathbb{S}^2$, and the ellipsoid of inertia for the rotation of a rigid top $\mathbb{I}^3$. Dynamics in manifolds, the Hessian metric, the Cristoffel connections, and the curvature are essential tools for certain types of simulations discussed in the next chapter, therefore, a cursory introduction to these is provided as well.

10.2 Coordinate Changes, Einstein's Sum Convention, and the Metric Tensor

We must get acquainted with a new and very concise notation, known as Einstein's sum convention. Firstly, let us imagine a d-dimensional Euclidean space. We can build d mutually orthogonal axis and locate a point in space by using sets of d-tuples. The elements of the vector space are represented with the following notation:

$$\left(x^1, x^2, \ldots, x^{d-1}, x^d\right). \tag{10.1}$$

The superscripts in Equation 10.1 are not exponents, they are labels that identify each coordinate. For example, for $d = 3$, we identify x^1, x^2, x^3, with x, y, z. A special property of d-dimensional Euclidean spaces is reflected in the expression for the line element,

$$ds^2 = (dx^1)^2 + \cdots + (dx^d)^2. \tag{10.2}$$

For $d = 3$, Equation 10.2 is,

$$ds^2 = (dx^1)^2 + (dx^2)^2 + (dx^3)^2 = (dx)^2 + (dy)^2 + (dz)^2. \tag{10.3}$$

Spaces mapped with coordinates that satisfy Equation 10.2 are called **flat spaces**. In Einstein's notation, Equation 10.2 reads,

$$ds^2 = \delta_{\mu\upsilon}\, dx^\mu dx^\upsilon. \tag{10.4}$$

$\delta_{\mu\upsilon}$ is the $d \times d$ unit matrix. The operation implied in Equation 10.4 is a double summation, which is denoted by the repeated indices in both the lower and upper positions. In the notation of Chapter 4,

$$A_\mu B^\mu \equiv \sum_{i=1}^{d} A_i B^i.$$

10.3   Contravariant Tensors

With the tensor notation, vectors are no longer denoted with lowercase bold-face fonts. Rather, vectors are represented by their components, V^μ. The index in the upper position notates a particular way in which vectors change when coordinates are changed. Suppose a change of coordinates $x^\mu \rightarrow x^{\mu'}$ (like a finite or an infinitesimal rotation) takes place. Einstein's notation for the transformation of the vector V, reads,

$$V^{\mu'} = \frac{\partial x^{\mu'}}{\partial x^\mu} V^\mu. \tag{10.5}$$

Objects that transform according to Equation 10.5 are called **contravariant** vectors. Note that the primes are on the Greek indices. Before we generalize this notion to tensor spaces, let us work through an example. Consider the following transformation of variables, in the two-dimensional plane:

$$x^{1'} = x^1 \cos\phi + x^2 \sin\phi,$$
$$x^{2'} = -x^1 \sin\phi + x^2 \cos\phi. \tag{10.6}$$

This transformation can be written in the following form:

$$x^{\mu'} = \Lambda^{\mu'}_\mu x^\mu \qquad \mu, \mu' = 1, 2, \tag{10.7}$$

and the matrix $\Lambda^{\mu'}_\mu$ is,

$$\Lambda^{\mu'}_\mu = \begin{pmatrix} \cos\phi & \sin\phi \\ -\sin\phi & \cos\phi \end{pmatrix}, \tag{10.8}$$

a rotation. Now, the reader should note that a vector,

$$V^\alpha = \begin{pmatrix} V^1 \\ V^2 \end{pmatrix}, \tag{10.9}$$

in the original map transforms according to

$$V^{1'} = V^1 \cos\phi + V^2 \sin\phi$$
$$V^{2'} = -V^1 \sin\phi + V^2 \cos\phi. \tag{10.10}$$

The reader should be convinced that the last transformation satisfies Equation 10.5. Note that rotations of the axes and translations, are also legal transformations of variables.

10.4   Gradients as 1-Forms

Two of the most easily confused concepts by the beginner are the differential dx^μ and the partial derivative $\partial/\partial x^\mu$. Tensor notation helps to keep the two straight. The difference is immaterial in flat spaces. However, in curved or even flat spaces mapped with curvilinear coordinates, care must be taken to distinguish these two objects. Note that both quantities have a single index. This means that both the differential and the partial derivative are tensor quantities of rank 1. The difference is that the differential of x^μ has a raised index, whereas the partial derivative has the index in the lower position,

$$\frac{\partial}{\partial x^\mu} = \partial_\mu. \tag{10.11}$$

Note the abbreviation normally used for partial derivatives on the R.H.S. Tensor quantities of rank 1 with the index raised are known as vectors, these have been discussed earlier. What are the tensor quantities of rank 1 with the index down? They have a special name, they are called **1-forms**. The transformations of differentials and partial derivatives for a change of variables $x^\mu \rightarrow x^{\mu'}$ are,

$$\partial_{\mu'} = \frac{\partial x^\mu}{\partial x^{\mu'}} \partial_\mu. \tag{10.12}$$

All 1-forms transform the same way under change of variables.

$$\omega_{\mu'} = \frac{\partial x^\mu}{\partial x^{\mu'}} \omega_\mu. \tag{10.13}$$

Objects that transform according to Equation 10.13 are **covariant** tensors of rank 1. To see how Equation 10.12 works, consider the spherical polar coordinate set. Using the chain rule or Equation 10.12, we derive identically,

$$\frac{\partial}{\partial \theta} = \frac{\partial x}{\partial \theta} \frac{\partial}{\partial x} + \frac{\partial y}{\partial \theta} \frac{\partial}{\partial y} + \frac{\partial z}{\partial \theta} \frac{\partial}{\partial z}.$$

The Greek indices $\mu, \upsilon, \lambda, \rho, \alpha, \beta, \kappa$, are used throughout the chapter to imply either covariant or contravariant labels. The convention we use from here on, is that sums are implied only when the same Greek letter is in a covariant and contravariant position in the same term of a tensor expression. Roman indices and Arabic numbers are not used to abbreviate sums.

10.5   Tensors of Higher Ranks

We occasionally need tensors of rank greater than two. For the second order rank tensors, we have a total of three separate spaces, the space of 2-vectors (or a (2,0) tensor),

$$A^{\mu'\upsilon'} = \frac{\partial x^{\mu'}}{\partial x^{\mu}}\frac{\partial x^{\upsilon'}}{\partial x^{\upsilon}}A^{\mu\upsilon} \qquad \text{(contravariant).} \qquad (10.14)$$

The space of 2-forms (or a (0,2) tensor),

$$W_{\mu'\upsilon'} = \frac{\partial x^{\mu}}{\partial x^{\mu'}}\frac{\partial x^{\upsilon}}{\partial x^{\upsilon'}}W_{\mu\upsilon} \qquad \text{(covariant),} \qquad (10.15)$$

and the mixed case also denoted a (1,1) tensor,

$$Q^{\upsilon'}_{\mu'} = \frac{\partial x^{\mu}}{\partial x^{\mu'}}\frac{\partial x^{\upsilon'}}{\partial x^{\upsilon}}Q^{\upsilon}_{\mu} \qquad \text{(mixed).} \qquad (10.16)$$

A number of operations can change the rank of tensors. For instance, the product of 1-form with a vector produces a scalar a,

$$T_{\upsilon}V^{\upsilon} = a, \qquad (10.17)$$

an operation that resembles the vector dot product. A mixed (1,1) tensor can be contracted into a scalar w,

$$W^{\upsilon}_{\upsilon} = w. \qquad (10.18)$$

If the mixed (1,1) tensor is represented by a square matrix, then this operation is equivalent to evaluating its trace. As a last example, a (2,0) tensor multiplied by a (0,1) tensor produces a (1,0) tensor,

$$T_{\mu}W^{\mu\upsilon} = V^{\upsilon}. \qquad (10.19)$$

10.6   The Metric Tensor of a Space

Now we introduce a fundamental tensorial quantity (the metric g) that is central to the development of formulas for a number of coordinate transformations in a space, or for maps from one space to another. Let ds^2 represent the line element as defined in Equation 10.2, in a general space,

$$ds^2 = g_{\mu\upsilon}\,dx^{\mu}dx^{\upsilon}. \qquad (10.20)$$

Comparison with Equation 10.4 should make it obvious that for $\mathbb{R}^n$,

$$g_{\mu\upsilon} = \delta_{\mu\upsilon}. \qquad (10.21)$$

Note that the metric tensor is a 2-form, since it transforms under a remapping of the manifold by a map, $\Phi : x^{\mu} \to x^{\mu'}$, in the following way:

$$g_{\mu'\upsilon'} = \frac{\partial x^{\mu}}{\partial x^{\mu'}}\frac{\partial x^{\upsilon}}{\partial x^{\upsilon'}}g_{\mu\upsilon}. \qquad (10.22)$$

The metric tensor is a special type of quantity that contains all the geometric information of the space. For our purposes, the metric tensor will soon become the effective mass of a molecular system. The metric tensor is central to all we do from here on. In all cases in this book, we can compute the metric tensor by starting with some $\mathbb{R}^n$, for which the metric tensor is representable by a diagonal matrix with masses as eigenvalues, and transforming, according to Equation 10.22, into the manifold of interest. In all such cases, it is simple to show that the metric tensor is symmetric,

$$g_{\mu\upsilon} = g_{\upsilon\mu}. \tag{10.23}$$

There are a number of subtle mathematical properties that are tacitly assumed for the Φ and its inverse Φ^{-1} by which coordinates are transformed. A map, $\Phi : \mathbb{R}^n \to \mathbb{M}^d$, between the Euclidean space $(\mathbb{R}^n)$ mapped with Cartesian coordinates $(x^\mu, \mu = 1, \ldots, n)$, and a space $\mathbb{M}^d$ mapped with curvilinear coordinates $(q^\mu, \mu = 1, \ldots, d)$, must be such that, for every point in $\mathbb{R}^n$ there corresponds one and only one point in $\mathbb{M}^d$. In other words, Φ must be **onto**. The map must be invertible, i.e., $\Phi^{-1} : \mathbb{M}^d \to \mathbb{R}^n$ exists is unique and onto. Onto maps that poses a unique onto inverse map are called **one-to-one**, or **bijections**. Most of the maps in this book either satisfy these requirements fully, or if not, they satisfy them up to sets of points of **zero measure**. Furthermore, the following properties of the map are essential in this book.

$$\frac{\partial^2 q^\mu}{\partial x^\upsilon \partial x^\kappa} = \frac{\partial^2 q^\mu}{\partial x^\kappa \partial x^\upsilon}, \tag{10.24}$$

$$\frac{\partial^2 x^\mu}{\partial q^\upsilon \partial q^\kappa} = \frac{\partial^2 x^\mu}{\partial q^\kappa \partial q^\upsilon}. \tag{10.25}$$

In other words, the map must be integrable.
The Jacobian matrix $\mathbf{J}$ associated with $\Phi^{-1} : \mathbb{M}^d \to \mathbb{R}^n$ is,

$$J^\mu_\upsilon = \frac{\partial x^\mu}{\partial q^\upsilon}. \tag{10.26}$$

Note that the n total derivatives of x^μ can be abbreviated with,

$$dx^\mu = \frac{\partial x^\mu}{\partial q^\upsilon} dq^\upsilon, \tag{10.27}$$

where the sum is implied by the repeated index υ. The metric tensor elements can be calculated using the elements of the Jacobian matrix,

$$g_{\mu'\upsilon'} = J^\mu_{\mu'} J^\upsilon_{\upsilon'} \delta_{\mu\upsilon}. \tag{10.28}$$

More details can be found in a number of excellent books on the subject [852, 853]. Spaces for which the metric tensor does not satisfy Equation 10.23 are said to possess torsion. Spaces with torsions can be generated with differential mappings that do not satisfy Euler's exactness tests in Equations 10.24

and 10.25. While spaces with torsion have found important applications in physics, we do not make use of them in this book.

The metric tensor possesses an inverse with contravariant indices,

$$g^{\mu'\upsilon'} = \frac{\partial x^{\mu'}}{\partial x^{\mu}} \frac{\partial x^{\upsilon'}}{\partial x^{\upsilon}} g^{\mu\upsilon}. \tag{10.29}$$

The following relationship follows:

$$g^{\mu\alpha} g_{\alpha\upsilon} = \delta^{\mu}_{\upsilon}. \tag{10.30}$$

We devote a large portion of this chapter to the development of expressions of the metric tensor and its computation in a number of useful cases. The metric tensor is the tensorial quantity needed to lower the index of a vector,

$$V_{\mu} = g_{\mu\upsilon} V^{\upsilon}. \tag{10.31}$$

Similarly, we can use the inverse of the metric tensor to raise the index of a 1-form,

$$w^{\mu} = g^{\mu\upsilon} w_{\upsilon}. \tag{10.32}$$

Einstein's convention, when applied to a vector inner product, reads,

$$(\mathbf{u}, \mathbf{v}) = \mathbf{u}^T \mathbf{v} = g_{\mu\upsilon} u^{\mu} v^{\upsilon}. \tag{10.33}$$

Compare this expression with the normal notation that is typically taught in a linear algebra course for that particular product:

$$(\mathbf{u}, \mathbf{v}) = \sum_{i=1}^{N} u_i v_i. \tag{10.34}$$

Equation 10.34 is only valid if the basis vectors used to expand $\mathbf{u}$ and $\mathbf{v}$ are orthonormal. Equation 10.33 works even when this assumption is not true.

The following statements apply for a general curvilinear change of coordinates, or in a curved manifold resulting from holonomic constraints:

1. The axes corresponding to the new coordinate system may not be orthogonal. In that case, the metric tensor will have nonzero off-diagonal elements.

2. The orientation of the basis vectors $\hat{q}_1, \ldots, \hat{q}_n$, with respect to the Euclidean basis vectors $\hat{x}_1, \ldots, \hat{x}_n$, may change from point to point in space.

3. Vectors may not be transportable, as is normally done in Euclidean space mapped with Cartesian coordinates. In other words, moving vectors by preserving size and orientation is only possible with Cartesian coordinates, and not in general.

4. The smallest distance between two points may no longer be a straight line.

A few examples will help to clarify the derivation and application of the metric tensor with Equation 10.28.

10.6.1 Example: The Euclidean Plane in Polar Coordinates

The polar coordinates

$$\Phi^{-1}: \qquad x = r\cos\theta \qquad\qquad y = r\sin\theta, \qquad\qquad (10.35)$$

$$\Phi: \qquad r = \left[x^2 + y^2\right]^{1/2} \qquad \theta = \tan^{-1}\frac{y}{x}, \qquad (10.36)$$

can be used to locate points in the Euclidean plane $\mathbb{R}^2$. If we let $x^1 = x, x^2 = y$ and $q^1 = r, q^2 = \theta$, then the Jacobian becomes

$$J^{\mu}_{\mu'} = \begin{pmatrix} \dfrac{\partial x}{\partial r} & \dfrac{\partial y}{\partial r} \\[2mm] \dfrac{\partial x}{\partial \theta} & \dfrac{\partial y}{\partial \theta} \end{pmatrix}, \qquad (10.37)$$

$$J^{\mu}_{\mu'} = \begin{pmatrix} \cos\theta & \sin\theta \\ -r\sin\theta & r\cos\theta \end{pmatrix}. \qquad (10.38)$$

Using Equation 10.28, we get

$$g_{11} = \frac{\partial x}{\partial r}\frac{\partial x}{\partial r} + \frac{\partial y}{\partial r}\frac{\partial y}{\partial r} = \cos^2\theta + \sin^2\theta = 1, \qquad (10.39)$$

$$g_{22} = \frac{\partial x}{\partial \theta}\frac{\partial x}{\partial \theta} + \frac{\partial y}{\partial \theta}\frac{\partial y}{\partial \theta} = (-r\sin\theta)^2 + (r\cos\theta)^2 = r^2. \qquad (10.40)$$

The off-diagonal elements vanish, as we can easily prove.

$$g_{12} = g_{21} = \frac{\partial x}{\partial r}\frac{\partial x}{\partial \theta} + \frac{\partial y}{\partial r}\frac{\partial y}{\partial \theta} = -r\cos\theta\sin\theta + r\cos\theta\sin\theta = 0. \qquad (10.41)$$

Therefore,

$$g_{\mu\upsilon} = \begin{pmatrix} 1 & 0 \\ 0 & r^2 \end{pmatrix}. \qquad (10.42)$$

Even though we use a matrix to represent both the Jacobian and the metric tensor, these are quite different objects, as the placement of indices tells us. The inverse of the metric tensor, $g^{\mu\upsilon}$, can be used to raise indices on quantities, and the metric itself can be used to lower indices on quantities. The inverse of the metric is simple to derive in this case,

$$g^{\mu\upsilon} = \begin{pmatrix} 1 & 0 \\ 0 & 1/r^2 \end{pmatrix}. \qquad (10.43)$$

10.6.2 Example: The One-Sphere ($\mathbb{S}^1$)

The one-sphere is the set of all points on a circle of radius R. One of the possible constructions for a set of coordinates that can represent the points on the circle is the polar coordinates set for $r = R = \text{constant}$. Note, however, that not all the points in the circle can be expressed with polar coordinates. The point at $\theta = 0$ coincides with $\theta = 2\pi$. In fact, in the plane, the polar coordinates cannot represent points on the positive x semiaxis because there are two values of θ, and in every point on the positive x semiaxis, $\theta = 0$ coincides with $\theta = 2\pi$. Despite these complications, the polar coordinate θ is more convenient to describe points in a circle than Cartesian coordinates. Then, the points in $\mathbb{S}^1$ are represented by the open set $\{\theta \,|\, 0 < \theta < 2\pi\}$, the Jacobian matrix for this map is,

$$J_\upsilon^\mu = (-R\sin\theta \qquad R\cos\theta), \tag{10.44}$$

and the metric tensor is

$$g_{11} = R^2. \tag{10.45}$$

10.7 Integration on Manifolds

It is important to know that the metric tensor appears in the "volume" integration in generalized coordinates. We put volume in quotes because, of course, in one dimension what we call volume is actually length, in two is area, and so on. A more precise term for volume would be the **Riemann measure** of the set. The following result is of such great importance that it should be committed to memory. Let $x^1, x^2, \ldots, x^n$ be the Cartesian coordinates, $q^1, q^2, \ldots, q^n$ be a general set of coordinates, and let there be an invertible map, Φ, between the two sets that is onto and satisfies the integrability conditions. Then, the invariant infinitesimal volume element is given by,

$$dx^1 dx^2 \cdots dx^n = \sqrt{\det(g)}\, dq^1 dq^2 \cdots dq^n, \tag{10.46}$$

where $\det(g)$ is the determinant of the metric tensor. In Chapter 11, we sketch the derivation of Equation 10.46. To see how this result works, let's apply it to the two sets of coordinates we considered earlier. The measure of $\mathbb{S}^1$ is,

$$\int_{\mathbb{S}^1} \sqrt{\det(g)}\, dq^1 = \int_{\theta=0}^{2\pi} R\, d\theta = 2\pi R, \tag{10.47}$$

which is the length of the circumference, as one would expect.

Here is another example with polar coordinates in a plane. Consider the set A of points in the x, y plane that are inside a circle of radius R, the set is represented by $A = \{x, y \,|\, 0 < \sqrt{x^2 + y^2} < R\}$. If we use polar coordinates, we obtain the following measure for the set:

$$\int_A \sqrt{\det(g)}\, dq^1 dq^2 \cdots dq^n = \int_{r=0}^{R} \int_{\theta=0}^{2\pi} r\, dr\, d\theta = 2\pi \left(\frac{1}{2}R^2\right) = \pi R^2, \tag{10.48}$$

the area of the circle. Let's try it with spherical polar coordinates and the set A of points inside a sphere of radius R; $A = \{r, \theta, \phi \mid 0 < r < R; \ 0 < \phi < 2\pi; \ 0 < \theta < \pi\}$. Let $x^1 = x$, $x^2 = y$, $x^3 = z$, and let $q^1 = r$, $q^2 = \theta$, $q^3 = \phi$, then,

$$J^{\mu}_{\mu'} = \begin{pmatrix} \cos\phi\sin\theta & \sin\phi\sin\theta & \cos\theta \\ r\cos\phi\cos\theta & r\sin\phi\cos\theta & -r\sin\theta \\ -r\sin\phi\sin\theta & r\cos\phi\sin\theta & 0 \end{pmatrix}. \tag{10.49}$$

The diagonal elements of the metric tensor are,

$$g_{11} = \left(\frac{\partial x}{\partial r}\right)^2 + \left(\frac{\partial y}{\partial r}\right)^2 + \left(\frac{\partial z}{\partial r}\right)^2, \tag{10.50}$$

$$\begin{aligned} g_{11} &= \cos^2\phi\sin^2\theta + \sin^2\phi\sin^2\theta + \cos^2\theta \\ &= \sin^2\theta\left(\cos^2\phi + \sin^2\phi\right) + \cos^2\theta = 1. \end{aligned} \tag{10.51}$$

$$g_{22} = \left(\frac{\partial x}{\partial\theta}\right)^2 + \left(\frac{\partial y}{\partial\theta}\right)^2 + \left(\frac{\partial z}{\partial\theta}\right)^2, \tag{10.52}$$

$$\begin{aligned} g_{22} &= r^2\cos^2\phi\cos^2\theta + r^2\sin^2\phi\cos^2\theta + r^2\sin^2\theta \\ &= r^2\cos^2\theta\left(\cos^2\phi + \sin^2\phi\right) + r^2\sin^2\theta = r^2. \end{aligned} \tag{10.53}$$

Finally,

$$g_{33} = \left(\frac{\partial x}{\partial\phi}\right)^2 + \left(\frac{\partial y}{\partial\phi}\right)^2 + \left(\frac{\partial z}{\partial\phi}\right)^2, \tag{10.54}$$

$$\begin{aligned} g_{33} &= r^2\sin^2\phi\sin^2\theta + r^2\cos^2\phi\sin^2\theta \\ &= r^2\sin^2\theta\left(\cos^2\phi + \sin^2\phi\right) = r^2\sin^2\theta. \end{aligned} \tag{10.55}$$

The off-diagonal elements vanish,

$$g_{12} = g_{21} = \left(\frac{\partial x}{\partial r}\right)\left(\frac{\partial x}{\partial\theta}\right) + \left(\frac{\partial y}{\partial r}\right)\left(\frac{\partial y}{\partial\theta}\right) + \left(\frac{\partial z}{\partial r}\right)\left(\frac{\partial z}{\partial\theta}\right), \tag{10.56}$$

$$g_{12} = g_{21} = r\cos^2\phi\sin\theta\cos\theta + r\sin^2\phi\sin\theta\cos\theta - r\cos\theta\sin\theta, \tag{10.57}$$

$$g_{12} = g_{21} = r\sin\theta\cos\theta\left(\cos^2\phi + \sin^2\phi\right) - r\cos\theta\sin\theta = 0. \tag{10.58}$$

$$g_{13} = g_{31} = \left(\frac{\partial x}{\partial r}\right)\left(\frac{\partial x}{\partial\phi}\right) + \left(\frac{\partial y}{\partial r}\right)\left(\frac{\partial y}{\partial\phi}\right) + \left(\frac{\partial z}{\partial r}\right)\left(\frac{\partial z}{\partial\phi}\right), \tag{10.59}$$

$$g_{13} = g_{31} = -r\cos\phi\sin\phi\sin^2\theta + r\sin\phi\cos\phi\sin^2\theta = 0. \tag{10.60}$$

$$g_{23} = g_{32} = \left(\frac{\partial x}{\partial\theta}\right)\left(\frac{\partial x}{\partial\phi}\right) + \left(\frac{\partial y}{\partial\theta}\right)\left(\frac{\partial y}{\partial\phi}\right) + \left(\frac{\partial z}{\partial\theta}\right)\left(\frac{\partial z}{\partial\phi}\right), \tag{10.61}$$

$$g_{23} = g_{32} = -r^2\cos\phi\sin\phi\cos\theta\sin\theta + r^2\sin\phi\cos\phi\sin\theta\cos\theta = 0. \tag{10.62}$$

Therefore, the metric tensor is,

$$g_{\mu\upsilon} = \begin{pmatrix} 1 & 0 & 0 \\ 0 & r^2 & 0 \\ 0 & 0 & r^2\sin^2\theta \end{pmatrix}. \tag{10.63}$$

The determinant is simply the product of all the diagonal elements,

$$\sqrt{\det g} = \sqrt{g_{11}g_{22}g_{33}} = r^2 \sin\theta, \tag{10.64}$$

therefore,

$$dx\,dy\,dz = r^2 \sin\theta dr d\theta d\phi, \tag{10.65}$$

and the measure of A becomes,

$$\int_A \sqrt{\det g}\, dq^1 dq^2 dq^3 = \int_{\phi=0}^{2\pi}\int_{\theta=0}^{\pi}\int_{r=0}^{R} r^2 \sin\theta dr d\theta d\phi = \frac{4}{3}\pi R^3, \tag{10.66}$$

which is the familiar volume of a sphere of radius R.

10.8 Stereographic Projections

The simplest example to introduce stereographic projection coordinate(s) is the one-sphere (the ring). The stereographic projection coordinate map, Φ : $\mathbb{R}^2 \to \mathbb{S}^1$, is the projection onto a line at $y = -R$ made by the point at θ. The intersection of the line that passes through the $x = 0, y = R$ point and $x = R\cos\theta, y = R\sin\theta$ with the $y = -R$ line, is the stereographic projection coordinate, ξ. A sketch of the stereographic projection is given in Figure 10.1 In Figure 10.1, note that the two triangles $\widehat{OBC}$ and $\widehat{OAP}$ are equivalent. The $\overline{OB}$ segment is of length $2R$, whereas the $\overline{OA}$ and the $\overline{AP}$ segments correspond to $R - y$ and x for point P. The origin for x and y is the center of the circle. The segment $\overline{BC}$ is the stereographic projection. Now, since the two triangles

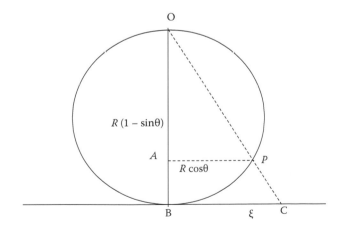

FIGURE 10.1
Sketch of the stereographic projection coordinate, ξ, to map the ring $\mathbb{S}^1$.

$\widehat{OBC}$ and $\widehat{OAP}$ are equivalent, the ratio of the length of the $\overline{BC}$ segment to the length of the $\overline{OB}$ segment is equal to the ratio of the length of the $\overline{AP}$ segment to the length of the $\overline{OA}$ segment. These simple geometric arguments lead to,

$$\xi = \frac{2R\cos\theta}{1 - \sin\theta}. \tag{10.67}$$

With trivial algebraic rearrangements, one can find the inverse of the map, $\Phi^{-1} : \mathbb{S}^1 \to \mathbb{R}^2$,

$$x = \frac{4R^2\xi}{\xi^2 + 4R^2}, \qquad y = \frac{R\left[\xi^2 - 4R^2\right]}{\xi^2 + 4R^2}. \tag{10.68}$$

Therefore, the metric tensor (which contains only one element), calculated using the transformation law for tensors, becomes,

$$g_{11} = \left(\frac{\partial x}{\partial \xi}\right)^2 + \left(\frac{\partial y}{\partial \xi}\right)^2 = \frac{\left(4R^2\right)^2}{\left(\xi^2 + 4R^2\right)^2}. \tag{10.69}$$

The stereographic projections for $\mathbb{S}^2$ can be defined in an equivalent manner. We let ξ^1 and ξ^2 be, respectively, the x and y coordinates, where the line from the north pole of the sphere through the point $P \in \mathbb{S}^2$ intersects the plane at $z = -R$. Using geometric arguments similar to those for the $\mathbb{S}^1$ case, one obtains,

$$\frac{\xi^1}{2R} = \frac{x}{R - z}, \qquad \frac{\xi^2}{2R} = \frac{y}{R - z}. \tag{10.70}$$

Squaring both sides of Equation 10.70, summing the two, and taking square roots yields a third relationship.

$$\frac{\sqrt{(\xi^1)^2 + (\xi^2)^2}}{2R} = \frac{\sqrt{x^2 + y^2}}{R - z}. \tag{10.71}$$

To avoid confusion, the square of the coordinate ξ^μ is represented using parenthesis $(\xi^\mu)^2$. From the equation of constraint, $x^2 + y^2 + z^2 = R^2$, and Equation 10.70, one can solve for z,

$$z = \frac{R\left[(\xi^1)^2 + (\xi^2)^2 - 4R^2\right]}{(\xi^1)^2 + (\xi^2)^2 + 4R^2}. \tag{10.72}$$

It becomes rather simple to solve for x and y with the last result.

$$x = \frac{4R^2\xi^1}{(\xi^1)^2 + (\xi^2)^2 + 4R^2}, \qquad y = \frac{4R^2\xi^2}{(\xi^1)^2 + (\xi^2)^2 + 4R^2}. \tag{10.73}$$

The metric tensor can be obtained readily by transforming from the metric for $\mathbb{R}^3$ once again,

$$g_{\mu\upsilon} = \frac{\left(4R^2\right)^2}{[(\xi^1)^2 + (\xi^2)^2 + 4R^2]^2} \begin{pmatrix} 1 & 0 \\ 0 & 1 \end{pmatrix}. \tag{10.74}$$

Therefore, the metric tensor is diagonal and has better symmetric properties than its version obtained with spherical polar coordinates in Equation 10.42.

Lastly, let us generalize the treatment to hyperspheres of arbitrary dimension $\mathbb{S}^n$. The $\mathbb{S}^n$ hypersphere (of unit radius) is defined by,

$$\sum_{\upsilon=1}^{n+1} \left(x^\upsilon\right)^2 = 1. \tag{10.75}$$

The set of generalized stereographic projections is a set of n independent coordinates defined by,

$$\frac{x^\mu}{1 - x^{n+1}} = \frac{\xi^\mu}{2}, \qquad \mu = 1, 2, \dots, n. \tag{10.76}$$

Solving for x^{n+1} in Equation 10.75 yields,

$$\left(x^{n+1}\right)^2 = 1 - \sum_{\mu=1}^{n} \left(x^\mu\right)^2, \tag{10.77}$$

$$\left(x^{n+1}\right)^2 = 1 - \frac{\left(1 - x^{n+1}\right)^2}{4} \sum_{\mu=1}^{n} \left(\xi^\mu\right)^2. \tag{10.78}$$

At this point, it is useful to introduce new notation that will shorten our expressions,

$$\sigma - \sum_{\mu=1}^{n} \left(\xi^\mu\right)^2, \tag{10.79}$$

then,

$$\left(x^{n+1}\right)^2 = 1 - \frac{\sigma}{4}\left(1 - x^{n+1}\right)^2. \tag{10.80}$$

Upon solving the quadratic equation and discarding the $x^{n+1} - 1$ solution, one produces the following relationship between the Cartesian coordinates of the $n + 1$-dimensional Euclidean space, inside which the hypersphere is embedded and the stereographic projections,

$$x^{n+1} = \frac{\sigma - 4}{\sigma + 4}, \qquad x^\mu = \frac{4\xi^\mu}{\sigma + 4}, \qquad i = 1, 2, \dots, n. \tag{10.81}$$

Therefore, a one-to-one mapping of points in the hypersphere is possible in general. We leave it as an exercise to show that the metric tensor for the hypersphere created by imposing one radial-like holonomic constraint on $\mathbb{R}^{n+1}$ is, in general,

$$g_{\mu\upsilon} = \frac{16}{\left(\sigma + 4\right)^2}\delta_{\mu\upsilon}. \tag{10.82}$$

We need the stereographic projections of $\mathbb{S}^3$ to treat the nonlinear rigid rotor case later in the chapter. However, the metric tensors for rigid bodies are generally not diagonal, since information about the effective mass (i.e., the moment of inertia) must enter into the metric as well. We elaborate in detail on this point later in the chapter.

Exercises

1. Use Equation 10.6 to show that Equation 10.10 satisfies Equation 10.5.

2. Consider, again, an example in the plane $\mathbb{R}^2$. The function $f\left(x^1, x^2\right) = \exp\left(x^1 \cdot x^2\right)$ is a scalar quantity. The gradient of f has the following elements:

$$\partial_\mu f\left(x\right) = \left(\partial_1 f, \quad \partial_2 f\right) = \left(x^2, \quad x^1\right) \exp\left(x^1 x^2\right). \tag{10.83}$$

   Now, let us affectuate the change of coordinates as in Equation 10.6. Note that functions, in general, transform by simply changing the arguments, and do not require any partial derivative term from the chain rule,

$$f\left(x^{1'}, x^{2'}\right) = \exp\left[\left(x^{1'} \cos\phi - x^{2'} \sin\phi\right)\left(x^{1'} \cos\phi + x^{2'} \sin\phi\right)\right].$$

   However, the chain rule leads immediately to Equation 10.13. Demonstrate the correctness of these statements by transforming Equation 10.83 according to Equation 10.13.

3. Derive Equations 10.38 and 10.42.

4. For the Euclidean plane in polar coordinates, check using the usual matrix product that,

$$g^{\mu\lambda} g_{\lambda\upsilon} = \delta^\mu_\upsilon = \begin{pmatrix} 1 & 0 \\ 0 & 1 \end{pmatrix}, \tag{10.84}$$

   holds.

5. Derive Equations 10.44 and 10.45.

6. Find the inverse of the metric tensor for three-dimensional space mapped by spherical polar coordinates, using the following transformation law for $(2,0)$ tensors:

$$g^{\mu'\upsilon'} = \frac{\partial q^{\mu'}}{\partial x^\mu} \frac{\partial q^{\upsilon'}}{\partial x^\upsilon} \delta^{\mu\upsilon},$$

   where q^1, q^2, q^3 are r, θ, ϕ. Then, check that $g^{\mu\upsilon'} g_{\upsilon'\upsilon} = \delta^\mu_\upsilon$.

7. Consider three-dimensional space mapped with spherical polar coordinates. Evaluate the gradient of $f(x, y, z) = \exp(-2r)$ first in spherical polar coordinates, then transform it to Cartesian coordinates using the chain rule.

$$\partial_\mu = \frac{\partial q^{\mu'}}{\partial x^\mu} \partial_\mu f.$$

8. With $f(x, y, z) = \exp(-2r)$, evaluate the gradient in spherical polar coordinates as in the previous exercise, then find an expression for the gradient with its index raised, in spherical polar coordinates,

$$\nabla^\mu f = g^{\mu\upsilon} \partial_\upsilon f.$$

9. Three-dimensional space can be mapped with elliptic cylindrical coordinates, u, v, z,

$$x^1 = a \cosh u \cos v,$$

$$x^2 - a \sinh u \sin v,$$

$$x^3 = z.$$

The family of curved manifolds are as follows:

   i. The constant u surfaces $0 \leq u < \infty$, are elliptic cylinders
   ii. The constant v surfaces $0 \leq v \leq 2\pi$, are hyperbolic cylinders
   iii. Constant z surfaces are planes parallel to the $x - y$ plane

Construct the metric tensor for three-dimensional space mapped with elliptic cylindrical coordinates, then for an elliptic cylinder, and a hyperbolic cylinder. Evaluate the volume of the set A,

$$A = \{a, u, v, z \,|\, a = 1, 0 \leq u < 1, 0 \leq v \leq 2\pi, 0 \leq z \leq 1\}.$$

10. The theoretical treatment of the hydrogen molecule ion, or any type of two-center problem, is simpler when three-dimensional space is mapped with prolate spheroidal coordinates, u, v, ϕ,

$$x^1 = a \sinh u \sin v \cos \phi,$$

$$x^2 = a \sinh u \sin v \sin \phi,$$

$$x^3 = a \cosh u \cosh v.$$

The family of curved manifolds are as follows:

   i. The constant u surfaces $0 \leq u < \infty$, are prolate spheroids
   ii. The constant v surfaces $0 \leq v \leq \pi$, are hyperboloids of two sheets
   iii. Constant ϕ surfaces are half-planes through the z axis

Construct the metric tensor for three-dimensional space mapped with prolate spheroidal coordinates. In this exercise, and the previous one, you will find that the metric tensor is diagonal. This is the result of the fact that the prolate spheroidal coordinates and the elliptic cylindrical coordinates are mutually orthogonal.

11. Verify Equation 10.68 by direct substitution of Equation 10.67 in the expressions for x and y.

12. Derive Equations 10.69, 10.74 and 10.82.

13. Derive Equation 10.70 using the same geometric analysis used for the ring space.

14. Evaluate the following Riemann measures:

   i. The subset $A \subset \mathbb{S}^1 = \left\{ \xi^1 \middle| -1 \leq \xi^1 \leq 1 \right\}$

   ii. The subset $A \subset \mathbb{S}^2 = \left\{ \xi^1, \xi^2 \middle| -1 \leq \xi^1, \xi^2 \leq 1 \right\}$.

15. In $\mathbb{S}^1$, transform the scalar quantity, $\cos\theta$, and its gradient into their respective expressions in terms of stereographic projection.

16. In $\mathbb{S}^2$, transform the scalar quantities, $\cos\theta$, $\sin\theta$, $\cos\phi$, $\sin\phi$, and their gradients into their respective expressions in terms of stereographic projection coordinates ξ^1 and ξ^2.

10.9 Dynamics in Manifolds

The first task in this section is to develop a multitude of ways of expressing the kinetic energy for a two and three body system under coordinate transformations that make it possible to introduce holonomic constraints. For the n body problem, we develop the necessary coordinate maps and associated differential geometry that are amenable for path integral simulations in a particularly challenging case: the asymmetric rigid body. This particular space forces us to work with nonorthogonal coordinates. Finding expressions and ways to efficiently compute the Jacobian terms is the second task. We have left a good deal of detail in our discussion so that the presentation may be used as a training tool for students interested in constrained molecular dynamics. We slowly develop the main results in the following order. We begin with a two body problem, where we remap the flat six-dimensional space, "remove" the center of mass, and remap the pseudo-one-dimensional problem stereographically. This case is relatively simple, if the stretch is constrained, the classical Jacobian for the infinitely stiff spring constant is a constant multiple of the root of the determinant of the metric tensor in the rotational subspace, and the curvature is constant and can be ignored. We then tackle the three body problem, and after remapping with relative coordinates and explaining how Jacobian coordinates can be built, we move on to the rigid n body case. We derive the Hessian metric tensor for a general case. When using Euler's angles for the rotations, and Cartesian coordinates for the translations, we find orthogonality among the translations and the rotations. The same orthogonality properties remain when we remap the ellipsoid of inertia submanifold with stereographic projections. These properties are not essential for quantum simulations, however, computer time costs when simulating molecular clusters can be reduced with orthogonalities. Furthermore, the orthogonality property among degrees of freedom plays a more important role when split operator methods are developed.

10.10 The Hessian Metric

We find it useful to include the mass of the system into the definition of the metric tensor. This definition is particularly handy when rotations of rigid bodies and internal degrees of freedom, such as bending and torsions, are considered. The Hessian metric for a particle of mass m in a three-dimensional space mapped with Cartesian coordinates is,

$$g_{\mu\upsilon} = m\delta_{\mu\upsilon}. \tag{10.85}$$

This definition allows us to write a very general form of the Lagrangian in a manifold,

$$\mathcal{L} = \frac{1}{2}g_{\mu\upsilon}\dot{q}^{\mu}\dot{q}^{\upsilon} - V, \tag{10.86}$$

and a general form for the Hamiltonian,

$$\mathcal{H} = \frac{1}{2}g^{\mu\upsilon}p_{\mu}p_{\upsilon} + V. \tag{10.87}$$

Note that the canonical momentum is a 1-form, since it is defined as a derivative of the Lagrangian,

$$p_{\upsilon} = \frac{\partial \mathcal{L}}{\partial \dot{q}^{\upsilon}}. \tag{10.88}$$

As stated in Chapter 2, the Euler–Lagrange equations apply, as written in Equation 10.89, for any manifold mapped with any independent set of coordinates q, and independent velocities $\dot{q}$, regardless of the orthogonality property among them.

$$\frac{d}{dt}\left(\frac{\partial \mathcal{L}}{\partial \dot{q}^{\mu}}\right) - \left(\frac{\partial \mathcal{L}}{\partial q^{\mu}}\right) = 0. \tag{10.89}$$

The direct application of Equation 10.89 to the general form of the Lagrangian yields the geodesic equations. Taking the derivative of Equation 10.86, with respect to the velocities, yields

$$\frac{\partial \mathcal{L}}{\partial \dot{q}^\mu} = g_{\mu\upsilon}\, \dot{q}^\upsilon. \tag{10.90}$$

Taking the time derivative of this last sum requires some care, since the metric tensor depends on time implicitly through its dependence on the coordinates. Using the chain rule, we obtain two terms,

$$\frac{d}{dt}\left(\frac{\partial \mathcal{L}}{\partial \dot{q}^\mu}\right) = g_{\mu\upsilon}\, \ddot{q}^\upsilon + (\partial_\mu g_{\upsilon\beta})\, \dot{q}^\upsilon \dot{q}^\beta. \tag{10.91}$$

Similarly, the partial derivative of the Lagrangian, with respect to q^μ, contains derivatives of the potential and the metric tensor,

$$\frac{\partial \mathcal{L}}{\partial q^\mu} = \frac{1}{2}\, (\partial_\mu g_{\upsilon\beta})\, \dot{q}^\upsilon \dot{q}^\beta - \partial_\mu V. \tag{10.92}$$

Combining these two partial derivative expressions into Equation 10.89 yields

$$g_{\mu\upsilon}\, \ddot{q}^\upsilon - \frac{1}{2}\, (\partial_\mu g_{\upsilon\beta})\, \dot{q}^\upsilon \dot{q}^\beta + \partial_\mu V = 0. \tag{10.93}$$

10.11   The Christofell Connections and the Geodesic Equations

It is customary to rearrange derivatives of the metric tensor in terms of the Christoffel connection coefficients, a symmetrized version of the derivative of the metric tensor. The Christoffel connection coefficients of the first kind read,

$$\Gamma_{\mu\upsilon\beta} = \frac{1}{2}\, (\partial_\mu g_{\upsilon\beta} + \partial_\upsilon g_{\beta\mu} - \partial_\beta g_{\mu\upsilon}), \tag{10.94}$$

and it is simple to show that,

$$\partial_\mu g_{\upsilon\beta} = \Gamma_{\mu\upsilon\beta} + \Gamma_{\mu\beta\upsilon}. \tag{10.95}$$

The set of coupled equations of motion on the manifold now becomes

$$g_{\mu\upsilon}\, \ddot{q}^\upsilon - \frac{1}{2}\, (\Gamma_{\mu\upsilon\beta} + \Gamma_{\mu\beta\upsilon})\, \dot{q}^\upsilon \dot{q}^\beta + \partial_\mu V = 0. \tag{10.96}$$

Multiplying through by the inverse of the metric tensor, produces the geodesic equations,

$$\ddot{q}^\upsilon - \Gamma^\upsilon_{\mu\beta}\, \dot{q}^\mu \dot{q}^\beta + g^{\upsilon\mu}\partial_\mu V = 0, \tag{10.97}$$

where the Christoffel connections of the second kind,

$$\Gamma_{\mu\beta}^{\upsilon} = \frac{1}{2} g^{\upsilon\rho} \left(\partial_\mu g_{\beta\rho} + \partial_\beta g_{\rho\mu} - \partial_\rho g_{\mu\beta} \right), \tag{10.98}$$

and their symmetric properties,

$$\Gamma_{\beta\mu}^{\upsilon} = \Gamma_{\mu\beta}^{\upsilon}, \tag{10.99}$$

are used.

It is worth noting that the Christoffel connections of either the first or second kind do not transform like tensors. However, the connection coefficients of the second kind are used to define a tensorial derivative in manifolds, since gradients of objects of rank one or greater generally do not transform as tensors either. The Christoffel connections of the second kind add the proper term and transform the gradients of objects of rank one or greater into **covariant derivatives**.

10.12   The Laplace–Beltrami Operator

The expression for the Laplacian in spherical polar coordinates is important for a number of problems in introductory quantum mechanics.

$$\nabla^2 = \frac{\partial^2}{\partial r^2} + \frac{2}{r} \frac{\partial}{\partial r} + \frac{\cos\theta}{r^2 \sin\theta} \frac{\partial}{\partial\theta} + \frac{1}{r^2} \frac{\partial^2}{\partial\theta^2} + \frac{1}{r^2 \sin^2\theta} \frac{\partial^2}{\partial\phi^2}. \tag{10.100}$$

This expression is more complicated than its equivalent in Cartesian coordinates. Nevertheless, every time the potential energy depends only on the distance from the origin (such as for the hydrogen atom in the center of mass frame) or when the constraints are spherical in nature (as in the particle in the spherical box, or the rotation of a rigid rotor in the center of mass frame), it is much simpler to work with spherical polar coordinates.

Let's begin by expressing the Laplace operator in Euclidean space. Using the tensor notation that should now be familiar, we write,

$$\nabla^2 = \partial_\mu \delta^{\mu\upsilon} \partial_\upsilon \quad \text{(Cartesian coordinates only).} \tag{10.101}$$

What happens to this equation when we apply a change of variables? The answer is the following expression:

$$\nabla^2 = g^{\mu\upsilon} \partial_\mu \partial_\upsilon + g^{\mu\upsilon} \left[\partial_\mu \ln \sqrt{\det(g)} \right] \partial_\upsilon + \left(\frac{\partial}{\partial x^\mu} g^{\mu\upsilon} \right) \partial_\upsilon. \tag{10.102}$$

The operator in Equation 10.102 is called the Laplace–Beltrami operator. Here is how this equation should be used. Let's take the spherical polar coordinate example; the inverse of the metric tensor is,

$$g^{\mu\upsilon} = \begin{pmatrix} 1 & 0 & 0 \\ 0 & 1/r^2 & 0 \\ 0 & 0 & 1/r^2\sin^2\theta \end{pmatrix}. \tag{10.103}$$

The first term of Equation 10.102 becomes,

$$g^{\mu\upsilon}\partial_\mu\partial_\upsilon = \frac{\partial^2}{\partial r^2} + \frac{1}{r^2}\frac{\partial^2}{\partial\theta^2} + \frac{1}{r^2\sin^2\theta}\frac{\partial^2}{\partial\phi^2}. \tag{10.104}$$

For the second term, remember that $\sqrt{\det(g)} = r^2\sin\theta$, therefore,

$$g^{\mu\upsilon}\left[\partial_\mu\ln\sqrt{\det(g)}\right]\partial_\upsilon = 1\left[\frac{\partial}{\partial r}\ln\left(r^2\sin\theta\right)\right]\frac{\partial}{\partial r} + \frac{1}{r^2}\left[\frac{\partial}{\partial\theta}\ln\left(r^2\sin\theta\right)\right]\frac{\partial}{\partial\theta}$$
$$+ \frac{1}{r^2\sin^2\theta}\left[\frac{\partial}{\partial\phi}\ln\left(r^2\sin\theta\right)\right]\frac{\partial}{\partial\phi}. \tag{10.105}$$

The derivatives are simple to evaluate,

$$g^{\mu\upsilon}\left[\partial_\mu\ln\sqrt{\det(g)}\right]\partial_\upsilon = 1\left[\frac{2r\sin\theta}{r^2\sin\theta}\right]\frac{\partial}{\partial r} + \frac{1}{r^2}\left[\frac{r^2\cos\theta}{r^2\sin\theta}\right]\frac{\partial}{\partial\theta}, \tag{10.106}$$

and a little simplification gives,

$$g^{\mu\upsilon}\left[\partial_\mu\ln\sqrt{\det(g)}\right]\partial_\upsilon = \frac{2}{r}\frac{\partial}{\partial r} + \frac{\cos\theta}{r^2\sin\theta}\frac{\partial}{\partial\theta}. \tag{10.107}$$

Finally, the third term becomes,

$$\left(\frac{\partial}{\partial x^\mu}g^{\mu\upsilon}\right)\partial_\upsilon = \left(\frac{\partial}{\partial r}g^{11}\right)\frac{\partial}{\partial r} + \left(\frac{\partial}{\partial\theta}g^{22}\right)\frac{\partial}{\partial\theta} + \left(\frac{\partial}{\partial\phi}g^{33}\right)\frac{\partial}{\partial\phi},$$

$$\left(\frac{\partial}{\partial x^\mu}g^{\mu\upsilon}\right)\partial_\upsilon = \left(\frac{\partial}{\partial r}1\right)\frac{\partial}{\partial r} + \left(\frac{\partial}{\partial\theta}\frac{1}{r^2}\right)\frac{\partial}{\partial\theta} + \left(\frac{\partial}{\partial\phi}\frac{1}{r^2\sin^2\theta}\right)\frac{\partial}{\partial\phi} = 0.$$

Upon collecting all the terms, one should obtain Equation 10.100. This derivation is much shorter than the direct use of the chain rule. One comes to appreciate the power that tensor analysis has to offer with these exercises.

10.13 The Riemann–Cartan Curvature Scalar

Quantum simulations with the discrete variable representation (DVR) and the path integral require a lattice definition. Choosing a set of lattice points in a generic manifold forces additional terms into the quantum dynamics. In curvilinear and curved manifolds, the choice of lattice point expansions (prepoint vs midpoint or postpoint) matters, since it changes the expressions for the quantum lattice corrections. We will provide more details about quantum lattice corrections in the next chapter. For now, we observe that the DeWitt prepoint, the Kleinert postpoint, and the Weyl midpoint expansion require certain combinations of the connection coefficients. The lattice corrections are scalar terms. These are typically interpreted as "quantum potentials." For example, the DeWitt prepoint lattice correction is,

$$V_{\text{eff}} = -\frac{\hbar^2}{6}\mathcal{R}, \tag{10.108}$$

where $\mathcal{R}$ is the Riemann–Cartan curvature scalar, a contraction of the Riemann curvature tensor,

$$R^{\rho}_{\sigma\mu\upsilon} = \partial_{\mu}\Gamma^{\rho}_{\upsilon\sigma} - \partial_{\upsilon}\Gamma^{\rho}_{\mu\sigma} + \Gamma^{\rho}_{\mu\lambda}\Gamma^{\lambda}_{\upsilon\sigma} - \Gamma^{\rho}_{\upsilon\lambda}\Gamma^{\lambda}_{\mu\sigma}. \tag{10.109}$$

The Riemann curvature tensor contracts to the Ricci tensor,

$$R_{\mu\upsilon} = R^{\lambda}_{\mu\lambda\upsilon}, \tag{10.110}$$

and the Riemann–Cartan curvature scalar is obtained by contracting the Ricci tensor with the inverse of the metric tensor,

$$\mathcal{R} = g^{\mu\upsilon}R_{\mu\upsilon}. \tag{10.111}$$

Here is an important example that we urge the reader to work through. The Christoffel connections for $\mathbb{S}^2$ mapped with stereographic projections are all nonzero, however, the expressions fall in a rather simple pattern,

$$\Gamma^1_{11} = -\frac{4\xi^1}{(\xi^1)^2 + (\xi^2)^2 + 4R^2}, \qquad \Gamma^2_{22} = -\frac{4\xi^2}{(\xi^1)^2 + (\xi^2)^2 + 4R^2}, \tag{10.112}$$

$$\Gamma^2_{11} = -\Gamma^1_{21} = -\Gamma^1_{12} = -\frac{1}{2}\Gamma^2_{22}, \tag{10.113}$$

$$\Gamma^1_{22} = -\Gamma^2_{12} = -\Gamma^2_{21} = -\frac{1}{2}\Gamma^1_{11}. \tag{10.114}$$

Given the high symmetry of the metric and the Christoffel connections, only four elements of the Riemann tensor are needed to obtain an expression of the quadrature. The following result:

$$R^1_{111} = R^2_{222} = 0, \tag{10.115}$$

is the consequence of the symmetric properties of the Riemann tensor,

$$R^{\rho}_{\mu\lambda\upsilon} = -R^{\rho}_{\upsilon\lambda\mu}. \tag{10.116}$$

Generally, the Riemann tensor with all lower indices, $R_{\mu\upsilon\lambda\rho} = g_{\mu\alpha}R^{\alpha}_{\upsilon\lambda\rho}$, has only one independent element in two-dimensional spaces, therefore, a number of simple relations similar to those obtained for the connections, can be obtained.

Furthermore, since the inverse of the metric tensor is diagonal, we need only two elements of the Riemann tensor,

$$R^1_{212} = 4\frac{-(\xi^1)^2 + (\xi^2)^2 + 4R^2}{[(\xi^1)^2 + (\xi^2)^2 + 4R^2]^2},\tag{10.117}$$

$$R^2_{121} = 4\frac{(\xi^1)^2 - (\xi^2)^2 + 4R^2}{[(\xi^1)^2 + (\xi^2)^2 + 4R^2]^2}.\tag{10.118}$$

The quadrature evaluates to (as can be checked),

$$\mathcal{R} = \frac{2}{R^2},\tag{10.119}$$

which is the same result one obtains by using the θ, ϕ coordinate map, of course. Since the curvature is a constant, the correction potential for the DeWitt lattice definition is also constant and can be ignored in quantum simulations with prepoint and postpoint expansions.

Exercises

1. Verify that in any manifold, the Hessian metric is,

$$g_{\mu\upsilon} = \frac{\partial^2 \mathcal{L}}{\partial \dot{q}^\mu \partial \dot{q}^\upsilon},$$

and that the inverse of the Hessian metric is,

$$g^{\mu\upsilon} = \frac{\partial^2 \mathcal{H}}{\partial \dot{p}_\mu \partial \dot{p}_\upsilon}.$$

Then, find the expression of the Hessian metric and its inverse for a symmetric spinning top in the body fixed frame. The Hamiltonian is,

$$\mathcal{H} = \frac{1}{2I_1}\left(p_1^2 + p_2^2\right) + \frac{1}{2I_3}p_3^2,$$

where p_1, p_2, p_3 are the components of the orbital angular momentum in the direction of the principal axis of inertia. $I_1 = I_2$ and I_3 are the eigenvalues of the inertia tensor.

2. Use the definition of the Hessian metric tensor to demonstrate that it is symmetric under index exchange,

$$g_{\mu\upsilon} = g_{\upsilon\mu}.$$

3. Prove that Equation 10.95 holds.

4. Use the definition of the Cristoffel connections of the second kind in Equation 10.98 to verify the symmetric property in Equation 10.95.

5. Verify that the connection coefficient, Γ^1_{11}, for the $\mathbb{S}^1$ mapped with the stereographic projection, ξ, is,

$$\Gamma^1_{11} = \frac{-2\xi}{\xi^2 + 4R^2}.$$

6. Verify that the Laplace–Beltrami operator in $\mathbb{S}^1$ mapped by the stereographic projection is,

$$\nabla^2_{\text{LB}} = \frac{\left(\xi^2 + 4R^2\right)^2}{\left(4R^2\right)^2} \frac{\partial^2}{\partial\xi^2} + \frac{2\xi\left(\xi^2 + 4R^2\right)}{\left(4R^2\right)^2} \frac{\partial}{\partial\xi}.$$

7. Derive the Laplace–Beltrami operator in $\mathbb{S}^2$ and $\mathbb{S}^3$ mapped by the stereographic projections.

8. Derive the connection coefficients for $\mathbb{S}^2$ mapped with stereographic projections. Show that Equations 10.112 and 10.114 hold.

9. The Riemann tensor in two dimensions has sixteen elements. Compute them for the $\mathbb{S}^2$ mapped with stereographic projections, then check that the curvature scalar is $2/R^2$.

10. Repeat the last two exercises for $\mathbb{S}^2$ mapped with the traditional polar angles.

11. Consider the spherical polar coordinates,

$$x = r\cos\phi\sin\theta,$$
$$y = r\sin\phi\sin\theta,$$
$$z = r\cos\theta. \tag{10.120}$$

The inverse of these equations can be found by using trigonometry principles,

$$r = \left(x^2 + y^2 + z^2\right)^{1/2},$$
$$\theta = \tan^{-1}\frac{\left(x^2 + y^2\right)^{1/2}}{z},$$
$$\phi = \tan^{-1}\frac{y}{x}. \tag{10.121}$$

Show that

$$\frac{\partial}{\partial x} = \cos\phi\sin\theta\frac{\partial}{\partial r} + \frac{1}{r}\cos\phi\cos\theta\frac{\partial}{\partial\theta} - \frac{\sin\phi}{r\sin\theta}\frac{\partial}{\partial\phi}, \tag{10.122}$$

$$\frac{\partial}{\partial y} = \sin\phi\sin\theta\frac{\partial}{\partial r} + \frac{1}{r}\sin\phi\cos\theta\frac{\partial}{\partial\theta} + \frac{\cos\phi}{r\sin\theta}\frac{\partial}{\partial\phi}, \tag{10.123}$$

$$\frac{\partial}{\partial z} = \cos\theta\frac{\partial}{\partial r} - \frac{1}{r}\sin\theta\frac{\partial}{\partial\theta}. \tag{10.124}$$

These are the components of the momentum operator in spherical polar coordinates. Appropriate combinations of these equations, with the components of the position operator, form the operators for the components of the angular momentum.

12. Use the results in the previous exercise to show that

i.

$$
\begin{aligned}
\frac{\partial^2}{\partial x^2} = \cos\phi\sin\theta &\left[\cos\phi\sin\theta\frac{\partial^2}{\partial r^2} - \frac{1}{r^2}\cos\theta\cos\phi\frac{\partial}{\partial\theta} \right.\\
&+ \frac{1}{r}\cos\phi\cos\theta\frac{\partial^2}{\partial r\partial\theta} + \frac{\sin\phi}{r^2\sin\theta}\frac{\partial}{\partial\phi} - \frac{\sin\phi}{r\sin\theta}\frac{\partial^2}{\partial r\partial\phi} \bigg]\\
&+ \frac{1}{r}\cos\phi\cos\theta\left[\cos\phi\cos\theta\frac{\partial}{\partial r} + \cos\phi\sin\theta\frac{\partial^2}{\partial\theta\partial r} \right.\\
&- \frac{1}{r}\cos\phi\sin\theta\frac{\partial}{\partial\theta} + \frac{1}{r}\cos\phi\cos\theta\frac{\partial^2}{\partial\theta^2}\\
&+ \frac{\sin\phi\cos\theta}{r\sin^2\theta}\frac{\partial}{\partial\phi} - \frac{\sin\phi}{r\sin\theta}\frac{\partial^2}{\partial\theta\partial\phi} \bigg]\\
&- \frac{\sin\phi}{r\sin\theta}\left[-\sin\phi\sin\theta\frac{\partial}{\partial r} + \cos\phi\sin\theta\frac{\partial^2}{\partial\phi\partial r} \right.\\
&- \frac{1}{r}\sin\phi\cos\theta\frac{\partial}{\partial\theta} + \frac{1}{r}\cos\phi\cos\theta\frac{\partial^2}{\partial\phi\partial\theta}\\
&- \frac{\cos\phi}{r\sin\theta}\frac{\partial}{\partial\phi} - \frac{\sin\phi}{r\sin\theta}\frac{\partial^2}{\partial\phi^2} \bigg]. \quad (10.125)
\end{aligned}
$$

ii.

$$
\begin{aligned}
\frac{\partial^2}{\partial y^2} = \sin\phi\sin\theta &\left[\sin\phi\sin\theta\frac{\partial^2}{\partial r^2} - \frac{1}{r^2}\cos\theta\sin\phi\frac{\partial}{\partial\theta} \right.\\
&+ \frac{1}{r}\sin\phi\cos\theta\frac{\partial^2}{\partial r\partial\theta} - \frac{\cos\phi}{r^2\sin\theta}\frac{\partial}{\partial\phi} + \frac{\cos\phi}{r\sin\theta}\frac{\partial^2}{\partial r\partial\phi} \bigg]\\
&+ \frac{1}{r}\sin\phi\cos\theta\left[\sin\phi\cos\theta\frac{\partial}{\partial r} + \sin\phi\sin\theta\frac{\partial^2}{\partial\theta\partial r} \right.\\
&- \frac{1}{r}\sin\phi\sin\theta\frac{\partial}{\partial\theta} + \frac{1}{r}\sin\phi\cos\theta\frac{\partial^2}{\partial\theta^2}\\
&- \frac{\cos\phi\cos\theta}{r\sin^2\theta}\frac{\partial}{\partial\phi} + \frac{\cos\phi}{r\sin\theta}\frac{\partial^2}{\partial\theta\partial\phi} \bigg]\\
&+ \frac{\cos\phi}{r\sin\theta}\left[\cos\phi\sin\theta\frac{\partial}{\partial r} + \sin\phi\sin\theta\frac{\partial^2}{\partial\phi\partial r} + \frac{1}{r}\cos\phi\cos\theta\frac{\partial}{\partial\theta} \right.\\
&+ \frac{1}{r}\sin\phi\cos\theta\frac{\partial^2}{\partial\phi\partial\theta} - \frac{\sin\phi}{r\sin\theta}\frac{\partial}{\partial\phi} + \frac{\cos\phi}{r\sin\theta}\frac{\partial^2}{\partial\phi^2} \bigg]. \quad (10.126)
\end{aligned}
$$

iii.

$$
\frac{\partial^2}{\partial z^2} = \cos\theta \left[\cos\theta \frac{\partial^2}{\partial r^2} + \frac{1}{r^2}\sin\theta\frac{\partial}{\partial\theta} - \frac{1}{r}\sin\theta\frac{\partial^2}{\partial r\partial\theta} \right]
$$
$$
- \frac{1}{r}\sin\theta \left[-\sin\theta\frac{\partial}{\partial r} + \cos\theta\frac{\partial^2}{\partial\theta\partial r} - \frac{1}{r}\cos\theta\frac{\partial}{\partial\theta} - \frac{1}{r}\sin\theta\frac{\partial^2}{\partial\theta^2} \right].
$$
$$(10.127)$$

These are the three expressions needed to transform the Laplacian in three-dimensional space from its expression in Cartesian coordinates to spherical polar coordinates.

13. Add the three results on the previous exercise and derive the Laplacian in three-dimensional space mapped with spherical polar coordinates. The three equations you derived in the last exercise can be simplified slightly. However, it is more convenient to consider the sum of the three, term by term, before simplifying. Look at each derivative operator coefficient, one at a time. The terms multiplying $\partial^2/\partial r^2$ for example, are:

$$\cos\phi\sin\theta\cos\phi\sin\theta + \sin\phi\sin\theta\sin\phi\sin\theta + \cos^2\theta = 1.$$

10.14 The Two-Body Problem Revisited

Most of the results in this section should be familiar to the reader from the work in Chapter 2. The two body problem is the simplest example of applications of the Hessian metric. In this section, the stereographic projections are used to express the classical Lagrangian for the two body problem and the rigid linear rotor. Let's consider a two point mass system with masses m_1 and m_2. Let x^μ ($\mu = 1, 2, \ldots, 6$) be the Cartesian coordinates of the system,

$$x^\mu = (x_1,\ y_1,\ z_1,\ x_2,\ y_2,\ z_2).$$
$$(10.128)$$

The Hessian metric tensor in these coordinates is representable by a diagonal matrix:

$$
g_{\mu\upsilon} = \begin{pmatrix} m_1 & 0 & 0 & 0 & 0 & 0 \\ 0 & m_1 & 0 & 0 & 0 & 0 \\ 0 & 0 & m_1 & 0 & 0 & 0 \\ 0 & 0 & 0 & m_2 & 0 & 0 \\ 0 & 0 & 0 & 0 & m_2 & 0 \\ 0 & 0 & 0 & 0 & 0 & m_2 \end{pmatrix}.
$$
$$(10.129)$$

The Lagrangian in Cartesian coordinates is,

$$\mathcal{L} = \frac{1}{2}m_1\left(\dot{x}_1^2 + \dot{y}_1^2 + \dot{z}_1^2\right) + \frac{1}{2}m_2\left(\dot{x}_2^2 + \dot{y}_2^2 + \dot{z}_2^2\right) - V.$$
$$(10.130)$$

Let us introduce a transformation of coordinates, $x^\mu \to q^\mu$,

$$q^\mu = (x_C, \ y_C, \ z_C, \ \delta_x, \ \delta_y, \ \delta_z), \tag{10.131}$$

where

$$x_C = \frac{m_1 x_1 + m_2 x_2}{m_t}, \quad y_C = \frac{m_1 y_1 + m_2 y_2}{m_t}, \quad z_C = \frac{m_1 z_1 + m_2 z_2}{m_t}, \tag{10.132}$$

are the coordinates of the center of mass, m_t is the total mass, and $\delta_x, \delta_y, \delta_z$ are relative coordinates,

$$\delta_x = x_2 - x_1, \qquad \delta_y = y_2 - y_1, \qquad \delta_z = z_2 - z_1. \tag{10.133}$$

The collection of these six equations is the map, Φ, from Cartesian to relative coordinates, $\Phi : x^\mu \to q^\mu$. We need to find the inverse of the map, $\Phi^{-1} : q^\mu \to x^\mu$, a simple task in this case,

$$x_1 = x_C - \frac{m_2}{m_t}\delta_x, \quad y_1 = y_C - \frac{m_2}{m_t}\delta_y, \quad z_1 = z_C - \frac{m_2}{m_t}\delta_z, \tag{10.134}$$

$$x_2 = x_C + \frac{m_1}{m_t}\delta_x, \quad y_2 = y_C + \frac{m_1}{m_t}\delta_y, \quad z_2 = z_C + \frac{m_1}{m_t}\delta_z. \tag{10.135}$$

If we agree to represent raised indices as the column labels and lower indices as row labels, then the Jacobian $\partial x^\nu / \partial q^\mu$ associated with Φ^{-1} can be represented by a 6×6 matrix.

$$\frac{\partial x^\mu}{\partial q^{\mu'}} = \begin{pmatrix} 1 & 0 & 0 & 1 & 0 & 0 \\ 0 & 1 & 0 & 0 & 1 & 0 \\ 0 & 0 & 1 & 0 & 0 & 1 \\ -\dfrac{m_2}{m_t} & 0 & 0 & \dfrac{m_1}{m_t} & 0 & 0 \\ 0 & -\dfrac{m_2}{m_t} & 0 & 0 & \dfrac{m_1}{m_t} & 0 \\ 0 & 0 & -\dfrac{m_2}{m_t} & 0 & 0 & \dfrac{m_1}{m_t} \end{pmatrix}. \tag{10.136}$$

The transformation of tensors law,

$$g_{\mu'\nu'} = \frac{\partial x^\nu}{\partial q^{\nu'}} \frac{\partial x^\mu}{\partial q^{\mu'}} g_{\mu\nu}, \tag{10.137}$$

with the row and column convention for the Jacobian that we introduced earlier, can be expressed as a matrix product, $\mathbf{G}' = \mathbf{J}\mathbf{G}\mathbf{J}^T$. The reader should go through this calculation by hand and prove that,

$$g_{\mu'\nu'} = \begin{pmatrix} m_t & 0 & 0 & 0 & 0 & 0 \\ 0 & m_t & 0 & 0 & 0 & 0 \\ 0 & 0 & m_t & 0 & 0 & 0 \\ 0 & 0 & 0 & \mu & 0 & 0 \\ 0 & 0 & 0 & 0 & \mu & 0 \\ 0 & 0 & 0 & 0 & 0 & \mu \end{pmatrix}, \tag{10.138}$$

where μ is the reduced mass,

$$\mu = \frac{m_1 m_2}{m_t}. \tag{10.139}$$

Remarkably, transforming the Hessian metric automatically produces the right mass for the "internal modes" of the rigid rotor. This result is general for all internal modes of linear and nonlinear tops. The last derivation should convince anyone of the power and conciseness of tensor analysis. The derivation of this result in Chapter 2 is much more, cumbersome, and error prone. Here is the Lagrangian:

$$\mathcal{L} = \frac{1}{2} m_t \left(\dot{x}_C^2 + \dot{y}_C^2 + \dot{z}_C^2 \right) + \frac{1}{2} \mu \left(\dot{\delta}_x^2 + \dot{\delta}_y^2 + \dot{\delta}_z^2 \right). \tag{10.140}$$

As explained in Chapter 2, the Cartesian to relative coordinate transformation for two body systems is particularly useful because a number of applications of two body theory involve interactions that depend on the relative coordinates, $\delta_x, \delta_y, \delta_z$ only. In Chapter 2, we have demonstrated that in the absence of external fields, the momenta associated with the center of mass are conserved and become trivial integrals of motion; these can be dropped altogether from the dynamics, their values are constant and equal to their initial values. Furthermore, a number of fundamental interactions depend only on the distance between the two bodies, which can be obtained from the relative coordinates, $r = \sqrt{\delta_x^2 + \delta_y^2 + \delta_z^2}$; in that case, one can use spherical polar coordinates. The conjugate momenta of the angles (or the components of the angular momentum) are conserved as well. The dynamics of the two body problem with a central potential are reducible to a single integral in that case.

Let us now consider the following change of variables:

$$\begin{pmatrix} x_C \\ y_C \\ z_C \\ \theta \\ \phi \\ r \end{pmatrix} \underset{\Phi_2^{-1}}{\overset{\Phi_2}{\rightleftharpoons}} \begin{pmatrix} x_C \\ y_C \\ z_C \\ \delta_x \\ \delta_y \\ \delta_z \end{pmatrix}, \tag{10.141}$$

where

$$r = \sqrt{\delta_x^2 + \delta_y^2 + \delta_z^2}. \tag{10.142}$$

Since we define these spherical polar coordinates with relative coordinates, we must find expressions for the inverse of Φ_2. These are relatively simple to obtain,

$$\delta_x = r \cos \phi \sin \theta,$$
$$\delta_y = r \sin \phi \sin \theta,$$
$$\delta_z = r \cos \theta, \tag{10.143}$$

with the center of mass coordinates unchanged. Now, we transform Equation 10.138 to get the new metric tensor. The Jacobian matrix associated with Φ_2^{-1} is straightforward to derive,

$$\frac{\partial x^\mu}{\partial q^{\mu'}} = \begin{pmatrix} 1 & 0 & 0 & 0 & 0 & 0 \\ 0 & 1 & 0 & 0 & 0 & 0 \\ 0 & 0 & 1 & 0 & 0 & 0 \\ 0 & 0 & 0 & r\cos\phi\cos\theta & r\sin\phi\cos\theta & -r\sin\theta \\ 0 & 0 & 0 & -r\sin\phi\sin\theta & r\cos\phi\sin\theta & 0 \\ 0 & 0 & 0 & \cos\phi\sin\theta & \sin\phi\sin\theta & \cos\theta \end{pmatrix}, \quad (10.144)$$

and the metric tensor becomes,

$$g_{\mu\nu} = \begin{pmatrix} m_t & 0 & 0 & 0 & 0 & 0 \\ 0 & m_t & 0 & 0 & 0 & 0 \\ 0 & 0 & m_t & 0 & 0 & 0 \\ 0 & 0 & 0 & \mu r^2 & 0 & 0 \\ 0 & 0 & 0 & 0 & \mu r^2 \sin^2\theta & 0 \\ 0 & 0 & 0 & 0 & 0 & \mu \end{pmatrix}. \quad (10.145)$$

Finally, the Lagrangian expressed with these coordinates is,

$$\mathcal{L} = \frac{1}{2} m_t \left(\dot{x}_C^2 + \dot{y}_C^2 + \dot{z}_C^2 \right) + \frac{1}{2}\mu \left(r^2\dot{\theta}^2 + r^2\sin^2\theta\dot{\phi}^2 + \dot{r}^2 \right) - V. \quad (10.146)$$

As stated earlier, for potentials that only depend on r (isotropic), this form of kinetic energy is the most useful; the conjugate momenta associated with the angular variables are known as the components of angular momentum. For isotropic potentials, the angular momentum is conserved, i.e., a constant of the motion, and one removes these coordinates from the equations of motion as well. Angular variables are not best-suited for quantum simulations of clusters of linear tops, as discussed at the beginning of this chapter. Some of the difficulties can be circumvented by using stereographic projections to map the rotations. This is the topic of the next section.

10.15　Stereographic Projections for the Two-Body Problem

Let us consider one additional change of variables,

$$\begin{pmatrix} x_C \\ y_C \\ z_C \\ \delta_x \\ \delta_y \\ \delta_z \end{pmatrix} \underset{\Phi_3^{-1}}{\overset{\Phi_3}{\rightleftharpoons}} \begin{pmatrix} x_C \\ y_C \\ z_C \\ \xi^1 \\ \xi^2 \\ r \end{pmatrix}, \quad (10.147)$$

where,

$$\xi^1 = \frac{2r\delta_x}{r - \delta_z}, \tag{10.148}$$

$$\xi^2 = \frac{2r\delta_y}{r - \delta_z}, \tag{10.149}$$

and r is the size of the relative distance as before.

$$r = \sqrt{\delta_x^2 + \delta_y^2 + \delta_z^2}. \tag{10.150}$$

It is possible to invert Φ_3. Let us introduce five auxiliary variables,

$$d^1 = \left(\xi^1\right)^2 + \left(\xi^2\right)^2 + 4r^2, \tag{10.151}$$

$$d^2 = \left(\xi^1\right)^2 + \left(\xi^2\right)^2 - 4r^2, \tag{10.152}$$

$$d^3 = -\left(\xi^1\right)^2 + \left(\xi^2\right)^2 + 4r^2, \tag{10.153}$$

$$d^4 = \left(\xi^1\right)^2 + \left(\xi^2\right)^2, \tag{10.154}$$

$$d^5 = \left(\xi^1\right)^2 - \left(\xi^2\right)^2 + 4r^2, \tag{10.155}$$

then, after some algebraic manipulations, one derives explicit expressions for Φ_3^{-1},

$$\delta_x = \frac{4r^2\xi^1}{d^1}, \tag{10.156}$$

$$\delta_y = \frac{4r^2\xi^2}{d^1}, \tag{10.157}$$

$$\delta_z = \frac{rd^2}{d^1}, \tag{10.158}$$

and the Jacobian matrix associated with Φ_3^{-1} becomes,

$$\frac{\partial x^\mu}{\partial q^{\mu'}} = \begin{pmatrix} 1 & 0 & 0 & 0 & 0 & 0 \\ 0 & 1 & 0 & 0 & 0 & 0 \\ 0 & 0 & 1 & 0 & 0 & 0 \\ 0 & 0 & 0 & \dfrac{4r^2d^3}{(d^1)^2} & -\dfrac{8r^2\xi^1\xi^2}{(d^1)^2} & \dfrac{16r^3\xi^1}{(d^1)^2} \\ 0 & 0 & 0 & -\dfrac{8r^2\xi^1\xi^2}{(d^1)^2} & \dfrac{4r^2d^5}{(d^1)^2} & \dfrac{16r^3\xi^2}{(d^1)^2} \\ 0 & 0 & 0 & \dfrac{8r\xi^1d^4}{(d^1)^2} & \dfrac{8r\xi^2d^4}{(d^1)^2} & \dfrac{d^2}{d^1} - \dfrac{16r^2d^4}{(d^1)^2} \end{pmatrix}. \tag{10.159}$$

The Jacobian matrix in Equation 10.159 is used to transform the metric tensor in Equation 10.138. The result is

$$
g_{\mu\upsilon} =
\begin{pmatrix}
m_{\mathrm{t}} & 0 & 0 & 0 & 0 & 0 \\
0 & m_{\mathrm{t}} & 0 & 0 & 0 & 0 \\
0 & 0 & m_{\mathrm{t}} & 0 & 0 & 0 \\
0 & 0 & 0 & \mu\dfrac{16r^4}{(d^1)^2} & 0 & -\mu\dfrac{16r^3\xi^1}{(d^1)^2} \\
0 & 0 & 0 & 0 & \mu\dfrac{16r^4}{(d^1)^2} & -\mu\dfrac{16r^3\xi^2}{(d^1)^2} \\
0 & 0 & 0 & -\mu\dfrac{16r^3\xi^1}{(d^1)^2} & -\mu\dfrac{16r^3\xi^2}{(d^1)^2} & \mu\dfrac{16r^2 d^4 + (d^1)^2}{(d^1)^2}
\end{pmatrix}.
\qquad (10.160)
$$

Therefore, it is possible to map, stereographically, the orientations of the two body system; however, there is a price to pay. The effective mass of the system is no longer constant, but depends on the configuration (as it happens with angular variables), and the coordinates ξ^1, ξ^2, are mutually orthogonal, but are not orthogonal to r. Consequently, the kinetic energy contains $\dot{\xi}^1\dot{r}$ terms and the Lagrangian of the system is more complicated.

$$
\mathcal{L} = \frac{1}{2}m_{\mathrm{t}}\left(\dot{x}_{\mathrm{C}}^2 + \dot{y}_{\mathrm{C}}^2 + \dot{z}_{\mathrm{C}}^2\right) + \frac{1}{2}\mu\frac{16r^4}{(d^1)^2}\left[\left(\dot{\xi}^1\right)^2 + \left(\dot{\xi}^2\right)^2\right]
$$

$$
+ \frac{1}{2}\mu\frac{16r^2 d^4 + (d^1)^2}{(d^1)^2}\dot{r}^2 - \mu\frac{16r^3\xi^1}{(d^1)^2}\dot{r}\dot{\xi}^1 - \mu\frac{16r^3\xi^2}{(d^1)^2}\dot{r}\dot{\xi}^2 - V. \qquad (10.161)
$$

Additionally, the radial coordinate is not convenient for quantum simulations. When angular variables are used to discretize paths for a two body problem with an isotropic interaction, the rotational barrier produces an infinitely attractive term in the $l = 0$ state. The presence of such abyss produces the well-known path collapse phenomenon [515], where a random walker along the radial degree of freedom, inexorably, is driven to values of r closer to zero at every step. The hydrogen atom problem, which contains the offending rotational barrier term and the Coulomb abyss, has been solved analytically with the Feynman quantization approach by clever alternative methods, which require nonholonomic remapping of the two body problem space. Difficulties with the Coulomb abyss have been reported even with simulations in flat spaces mapped by Cartesian coordinates, and in these cases, which are of importance in molecular physics, it has been found that the partial averaging technique eliminates the numerical problems.

10.16 The Rigid Rotor and the Infinitely Stiff Spring Constant Limit

Our interest in the projection coordinates becomes clear when one considers a two body problem with a relatively stiff central interaction. Then, one can approximate $r \approx r_{\mathrm{e}}, \dot{r} \approx 0$, and the Lagrangian simplifies substantially,

$$\mathcal{L} = \frac{1}{2} m_t \left(\dot{x}_C^2 + \dot{y}_C^2 + \dot{z}_C^2 \right) + \frac{1}{2} \mu \frac{16 r_e^4}{\left(d^1 \right)^2} \left[\left(\dot{\xi}^1 \right)^2 + \left(\dot{\xi}^2 \right)^2 \right] - V. \tag{10.162}$$

This is the Lagrangian that determines the dynamics. As we will see in the next chapter, the Jacobian $g^{1/2}$ affects the importance sampling expression for the rejection-acceptance of random moves in both the classical and quantum simulations.

The volume element in manifolds transforms according to the rules of differential geometry, as expressed in Equation 10.46. However, in the treatment of the physics of particles, the application of holonomic constraints must be carried out after the proper Jacobian ($g^{1/2}$) of the parent curvilinear space is at hand, otherwise, results could be in serious error. There is a difference that results from expressing the partition function first in the flat space, followed by the transformation of variables and the application of the infinite force constant(s) limits, as opposed to the alternative procedure of expressing the Boltzmann distribution or the density matrix expression after the transformation and the application of the holonomic constraints. Frenkel and Smit [7] have shown that the former procedure is the physically correct one. In curved manifolds, one evaluates the metric tensor in the $3n$- c-dimensional space (n atoms and c constraints), whereas in the c infinitely stiff springs model, one remaps the $\mathbb{R}^{3n}$ space with internal degrees of freedom, and then evaluates the Jacobian from the metric tensor in the $3n$ flat space. In the case of a linear rigid rotor, one evaluates the determinant of the appropriate submatrix of $g_{\mu\nu}$,

$$g^{1/2} - \left(\det g_{\mu\nu} \right)^{1/2} = \begin{vmatrix} m_t & 0 & 0 & 0 & 0 \\ 0 & m_t & 0 & 0 & 0 \\ 0 & 0 & m_t & 0 & 0 \\ 0 & 0 & 0 & \mu \dfrac{16 r_e^4}{\left(d^1 \right)^2} & 0 \\ 0 & 0 & 0 & 0 & \mu \dfrac{16 r_e^4}{\left(d^1 \right)^2} \end{vmatrix}^{1/2} = m_t^{3/2} \mu \frac{16 r_e^4}{\left(d^1 \right)^2}. \tag{10.163}$$

Whereas, for the infinitely stiff spring model of a linear molecule, one has to evaluate the determinant of the full metric tensor in Equation 10.160,

$$g^{1/2} = m_t^{3/2} \mu^{3/2} \frac{16 r_e^4}{\left(d^1 \right)^2}. \tag{10.164}$$

Clearly, the ratio of the two determinants is a constant independent of the configuration ($\mu^{1/2}$). Since shifting the importance sampling function by a constant cannot affect the dynamics, we conclude that for linear tops, and clusters of linear tops, the use of constrained spaces mapped by stereographic projections, and the infinitely stiff spring models yield the same answer at all temperatures. We return to this important point again in Chapter 11, where a much more general result regarding any type of rigid rotating body is obtained.

10.17   Relative Coordinates for the Three-Body Problem

Let's begin by developing a notation for the configuration of a three body system,

$$x^\mu = (x_1,\, y_1,\, z_1,\, x_2,\, y_2,\, z_2,\, x_3,\, y_3,\, z_3)\,. \tag{10.165}$$

In Cartesian coordinates, the Hessian metric tensor and the kinetic energy are simple.

$$g_{\mu\upsilon} = \mathrm{diag}\,(m_1,\, m_1,\, m_1,\, m_2,\, m_2,\, m_2,\, m_3,\, m_3,\, m_3)\,. \tag{10.166}$$

$$\mathcal{L} = \frac{1}{2}\sum_{i=1}^{3} m_i\left(\dot{x}_i^2 + \dot{y}_i^2 + \dot{z}_i^2\right) - V. \tag{10.167}$$

Unlike the two body problem, complications with the formulation of orthogonal coordinates begin when one attempts to use relative coordinates.

$$x^\mu \quad \underset{\Phi^{-1}}{\overset{\Phi}{\rightleftharpoons}} \quad \begin{pmatrix} x_C \\ y_C \\ z_C \\ \delta_{x1} \\ \delta_{y1} \\ \delta_{z1} \\ \delta_{x2} \\ \delta_{y2} \\ \delta_{z2} \end{pmatrix}. \tag{10.168}$$

Suppose Φ is defined by the following equations:

$$x_C = \frac{1}{m_t}\sum_{i=1}^{3} m_i x_i, \tag{10.169}$$

$$\delta_{x1} = x_1 - x_3, \tag{10.170}$$

$$\delta_{x2} = x_1 - x_2, \tag{10.171}$$

with identical definitions for the y and z part, and $m_t = m_1 + m_2 + m_3$. The Jacobian associated with Φ^{-1} becomes,

$$\frac{\partial x^\mu}{\partial q^{\mu'}} = \begin{pmatrix}
1 & 0 & 0 & 1 & 0 & 0 & 1 & 0 & 0 \\
0 & 1 & 0 & 0 & 1 & 0 & 0 & 1 & 0 \\
0 & 0 & 1 & 0 & 0 & 1 & 0 & 0 & 1 \\
\frac{m_3}{m_t} & 0 & 0 & \frac{m_3}{m_t} & 0 & 0 & \frac{m_3}{m_t}-1 & 0 & 0 \\
0 & \frac{m_3}{m_t} & 0 & 0 & \frac{m_3}{m_t} & 0 & 0 & \frac{m_3}{m_t}-1 & 0 \\
0 & 0 & \frac{m_3}{m_t} & 0 & 0 & \frac{m_3}{m_t} & 0 & 0 & \frac{m_3}{m_t}-1 \\
\frac{m_2}{m_t} & 0 & 0 & \frac{m_2}{m_t}-1 & 0 & 0 & \frac{m_2}{m_t} & 0 & 0 \\
0 & \frac{m_2}{m_t} & 0 & 0 & \frac{m_2}{m_t}-1 & 0 & 0 & \frac{m_2}{m_t} & 0 \\
0 & 0 & \frac{m_2}{m_t} & 0 & 0 & \frac{m_2}{m_t}-1 & 0 & 0 & \frac{m_2}{m_t}
\end{pmatrix}. \tag{10.172}$$

The reader should derive this last result, then show that the metric tensor takes the following form:

$$g_{\mu\nu} = \begin{pmatrix} m_t & 0 & 0 & 0 & 0 & 0 & 0 & 0 & 0 \\ 0 & m_t & 0 & 0 & 0 & 0 & 0 & 0 & 0 \\ 0 & 0 & m_t & 0 & 0 & 0 & 0 & 0 & 0 \\ 0 & 0 & 0 & \mu_1 & 0 & 0 & -\mu_2 & 0 & 0 \\ 0 & 0 & 0 & 0 & \mu_1 & 0 & 0 & -\mu_2 & 0 \\ 0 & 0 & 0 & 0 & 0 & \mu_1 & 0 & 0 & -\mu_2 \\ 0 & 0 & 0 & -\mu_2 & 0 & 0 & \mu_3 & 0 & 0 \\ 0 & 0 & 0 & 0 & -\mu_2 & 0 & 0 & \mu_3 & 0 \\ 0 & 0 & 0 & 0 & 0 & -\mu_2 & 0 & 0 & \mu_3 \end{pmatrix}, \quad (10.173)$$

where

$$\mu_1 = m_3 \left(1 - \frac{m_3}{m_t} \right), \tag{10.174}$$

$$\mu_2 = \frac{m_2 m_3}{m_t}, \tag{10.175}$$

$$\mu_3 = m_2 \left(1 - \frac{m_2}{m_t} \right). \tag{10.176}$$

Therefore, unlike the two body problem, relative coordinates are generally nonorthogonal in many body systems. However, the metric tensor just derived is very useful because the center of mass coordinates are removed, just as in the two body problem. The derivation of the metric tensor with relative coordinates is presented as evidence that one can still remove the center of mass from the dynamics. The center of mass coordinates are orthogonal to the relative ones we define here. The nine-dimensional problem has been reduced to a six-dimensional one. It is interesting to note that different choices for relative coordinates (there exist several possibilities) could be nonorthogonal to the center of mass coordinates.

The Hessian metric tensor is not dependent on configuration: one can diagonalize it once for all the points in the space. The resulting linear combinations of relative coordinates define mass weighted Jacobian coordinates, which are mutually orthogonal. The differential geometry procedure to obtain Jacobian coordinates is not limited to the three body problem, it is quite general. Jacobian coordinates are very useful, and are employed in theoretical chemistry quite frequently. Unfortunately, it is not simple to invoke constraints for the high-frequency degrees of freedom, such as stretching, of a small covalent molecule with mass weighted Jacobian coordinates.

10.18   The Rigid Body Problem and the Body Fixed Frame

In this section, we consider a series of maps, with the associated Jacobians, metric tensors, and kinetic energy expressions that are useful for the cases when the "internal degrees of freedom" are characterized by infinitely stiff spring constants and the resulting rigid top is not linear. We find that the best approach to arrive at a set of coordinates amenable for path integration is to develop the mapping in two stages. The first stage transforms from the Euclidean $\mathbb{R}^{3n}$ space mapped by Cartesian coordinates to the Euler angles. This is the standard treatment. The map, Φ_1, is defined by the transformation,

$$x^\mu \underset{\Phi_1^{-1}}{\overset{\Phi_1}{\rightleftharpoons}} \begin{pmatrix} x_C \\ y_C \\ z_C \\ \theta \\ \phi \\ \psi \end{pmatrix}. \tag{10.177}$$

The coordinates, θ, ψ, ϕ, in Equation 10.177 are the three Euler angles defined by the element $\mathbf{R}$ of the rotation group,

$$\mathbf{R} = \mathbf{R}_\phi \mathbf{R}_\theta \mathbf{R}_\psi, \tag{10.178}$$

where $\mathbf{R}_\psi$ is a rotation matrix about the z axis in the center of mass frame of Chapter 4,

$$\mathbf{R}_\psi = \begin{pmatrix} \cos\psi & -\sin\psi & 0 \\ \sin\psi & \cos\psi & 0 \\ 0 & 0 & 1 \end{pmatrix}, \tag{10.179}$$

$\mathbf{R}_\theta$ is a rotation about the x' axis in the center of mass frame after the rotation by ϕ has taken place,

$$\mathbf{R}_\theta = \begin{pmatrix} 1 & 0 & 0 \\ 0 & \cos\theta & -\sin\theta \\ 0 & \sin\theta & \cos\theta \end{pmatrix}, \tag{10.180}$$

and $\mathbf{R}_\phi$ is a rotation about the z'' axis in the center of mass frame after the rotations by ψ and θ,

$$\mathbf{R}_\phi = \begin{pmatrix} \cos\phi & -\sin\phi & 0 \\ \sin\phi & \cos\phi & 0 \\ 0 & 0 & 1 \end{pmatrix}. \tag{10.181}$$

We leave it to the reader to show that,

$$\mathbf{R} = \begin{pmatrix} \cos\psi\cos\phi - \cos\theta\sin\psi\sin\phi & -\cos\phi\sin\psi - \cos\theta\sin\phi\cos\psi & \sin\phi\sin\theta \\ \sin\phi\cos\psi + \cos\theta\cos\phi\sin\psi & -\sin\psi\sin\phi + \cos\theta\cos\psi\cos\phi & -\cos\phi\sin\theta \\ \sin\psi\sin\theta & \cos\psi\sin\theta & \cos\theta \end{pmatrix}. \tag{10.182}$$

Φ_1 is a map from a $3n$-dimensional Euclidean space mapped with Cartesian coordinates to a Cartesian product of two subspaces, $\Phi_1 : \mathbb{R}^{3n} \longrightarrow \mathbb{R}^3 \otimes \mathbb{I}^3$. The space $\mathbb{I}^3$ is the inertia ellipsoid, whereas $\mathbb{R}^3$ is the space of coordinates for the center of mass. There are an infinite number of possible ways of defining Φ_1, depending on a multitude of choices available for the definition of the body frame axis, and the choice between passive vs active rotations of the body fixed axis. Recall from Chapter 4 that the eigenbasis of a hermitian matrix are not unique. In the most general case, the body fixed coordinates can be obtained by diagonalization of the inertia tensor.

We denote the body frame coordinates with the following symbol: $x_{\mathrm{BF}}^{3(i-1)+\kappa}$, where the Roman index i is the atom label. In this section, the Greek letters $\kappa, \kappa', \kappa'', \ldots$ are used as the covariant and contravariant labels in the $\mathbb{R}^3$ space associated with atom i. Two items are important at this point. Firstly, the body fixed reference configuration is in the center of mass frame, i.e.,

$$\sum_{i=1}^{n} m_i x_{\mathrm{BF}}^{3(i-1)+\kappa} = 0, \quad \forall\kappa, \tag{10.183}$$

and is defined so that the inertia tensor is diagonal,

$$\sum_{i=1}^{n} m_i \left(x_{\mathrm{BF}}^{3(i-1)+k} x_{\mathrm{BF}}^{3(i-1)+k'} \right) = 0 \quad \text{if } \kappa \neq \kappa', \tag{10.184}$$

$$I_1 = \sum_{i=1}^{n} m_i \left[\left(x_{\mathrm{BF}}^{3(i-1)+2} \right)^2 + \left(x_{\mathrm{BF}}^{3(i-1)+3} \right)^2 \right], \tag{10.185}$$

$$I_2 = \sum_{i=1}^{n} m_i \left[\left(x_{\mathrm{BF}}^{3(i-1)+1} \right)^2 + \left(x_{\mathrm{BF}}^{3(i-1)+3} \right)^2 \right], \tag{10.186}$$

$$I_3 = \sum_{i=1}^{n} m_i \left[\left(x_{\mathrm{BF}}^{3(i-1)+1} \right)^2 + \left(x_{\mathrm{BF}}^{3(i-1)+2} \right)^2 \right]. \tag{10.187}$$

To transform from the body fixed to the laboratory frame, one rotates this rigid configuration and translates the center of mass. Let $\mathbf{r}_i$ represent the Cartesian vector for atom i $(1 \leq i \leq n)$ in the laboratory frame, and let $\mathbf{r}_i^{(\mathrm{BF})}$ represent the Cartesian vector for atom i in the body fixed frame. Then, Φ_1^{-1} can be represented with a single equation,

$$\mathbf{r}_i = \mathbf{r}_{\mathrm{C}} + \mathbf{R}\,\mathbf{r}_i^{(\mathrm{BF})}, \tag{10.188}$$

where $\mathbf{R}$ is the 3×3 matrix in Equation 10.182, and $\mathbf{r}_{\mathrm{C}} = (x_{\mathrm{C}}, y_{\mathrm{C}}, z_{\mathrm{C}})$. Equation 10.188 constitutes the representation of inhomogeneous Galilean group.

The Lie group of rotations is a subgroup of this. We have now linked the material in Chapter 4 on Lie algebra to the subject of conformal mapping and differential geometry. Naturally, we have only scratched the surface of this useful field of mathematics [51].

To find expressions for the element of the Jacobian, let us introduce some additional notation. Let $\partial_\mu \mathbf{R}$ represent a set of 3×3 matrices containing the derivative of the elements of the rotation matrix $\mathbf{R}$ with respect to q^μ. There are a number of useful properties for the set $\partial_\mu \mathbf{R}$. Some are immediately obvious,

$$\partial_\mu \mathbf{R} = 0 \qquad (1 \le \mu \le 3). \tag{10.189}$$

This result yields trivial elements of the Jacobian matrix associated with Φ_1^{-1},

$$J_\mu^{3(i-1)+\kappa} = \delta_{\mu\kappa} \qquad (1 \le \kappa \le 3, \, 1 \le i \le n, \, 1 \le \mu \le 3). \tag{10.190}$$

The second expression for the Jacobian marix elements is,

$$J_\mu^{3(i-1)+\kappa} = [\partial_\mu \mathbf{R}]_{\kappa'}^\kappa \, x_{(\mathrm{BF})}^{3(i-1)+\kappa'} \qquad (1 \le \kappa, \kappa' \le 3, \, 1 \le i \le n, \, 4 \le \mu \le 6). \tag{10.191}$$

Note that Einstein's notation is in full use here, as the upper and lower indices κ denote. With this notation, we can generate expressions for the transformation of a n body Hessian metric tensor in Equation 10.166,

$$g_{\mu\upsilon} = \sum_{i=1}^n m_i \delta_{\kappa\kappa'} J_\mu^{3(i-1)+\kappa} J_\upsilon^{3(i-1)+\kappa'}, \tag{10.192}$$

where the fact that the Cartesian n body metric tensor is diagonal has been used. We must inspect three cases separately:

I. $1 \le \mu \le 3, \, 1 \le \upsilon \le 3$. For this case, we insert Equation 10.190 into Equation 10.192 and get,

$$g_{\mu\upsilon} = \delta_{\mu\upsilon} \sum_{i=1}^n m_i. \tag{10.193}$$

II. $1 \le \mu \le 3, \, 4 \le \upsilon \le 6$. For this case, we make use of Equations 10.190 and 10.191. Inserting into Equation 10.192 as before, yields,

$$g_{\mu\upsilon} = \sum_{i=1}^n m_i \delta_{\kappa\upsilon} [\partial_\mu \mathbf{R}]_{\kappa'}^\kappa \, x_{(\mathrm{BF})}^{3(i-1)+\kappa'} = \delta_{\kappa\upsilon} [\partial_\mu \mathbf{R}]_{\kappa'}^\kappa \sum_{i=1}^n m_i \, x_{(\mathrm{BF})}^{3(i-1)+k'} = 0, \tag{10.194}$$

where we exchange the order of summation, and we use Equation 10.183. We conclude that, generally, the translations, and rotations, as expressed by the map in Equation 10.188, are orthogonal.

III. $4 \leq \mu \leq 6$, $4 \leq \upsilon \leq 6$. For this last case, we insert Equation 10.191 into Equation 10.192 and get,

$$g_{\mu\upsilon} = \sum_{i=1}^{n} m_i \delta_{\kappa\kappa'''} [\partial_\mu \mathbf{R}]^\kappa_{\kappa'} x^{3(i-1)+\kappa'}_{(BF)} [\partial_\upsilon \mathbf{R}]^{\kappa'''}_{\kappa''} x^{3(i-1)+k''}_{(BF)}. \tag{10.195}$$

Rearranging these terms slightly, we get

$$g_{\mu\upsilon} = \sum_{i=1}^{n} m_i \left(\delta_{\kappa\kappa'''} [\partial_\mu \mathbf{R}]^\kappa_{\kappa'} [\partial_\upsilon \mathbf{R}]^{\kappa'''}_{\kappa''} \right) x^{3(i-1)+\kappa'}_{(BF)} x^{3(i-1)+k''}_{(BF)}. \tag{10.196}$$

The term inside the parenthesis is independent of i and is a 2-form in $\mathbb{R}^3$ space for each atom,

$$g_{\mu\upsilon} = [\Gamma_{\mu\upsilon}]_{\kappa'\kappa''} \sum_{i=1}^{n} m_i x^{3(i-1)+\kappa'}_{(BF)} x^{3(i-1)+k''}_{(BF)}. \tag{10.197}$$

The symbol $\Gamma_{\mu\upsilon}$ represents a set of nine distinct 2-forms in the $\mathbb{R}^3$ space associated with the body fixed Cartesian coordinates for atom i,

$$[\Gamma_{\mu\upsilon}]_{\kappa'\kappa''} = \delta_{\kappa\kappa'''} [\partial_\mu \mathbf{R}]^\kappa_{\kappa'} [\partial_\upsilon \mathbf{R}]^{\kappa'''}_{\kappa''}. \tag{10.198}$$

This set of tensors is derived from the element of the rotation group, and its properties are of general importance. The reader should verify the following results ($q^4 = \theta$, $q^5 = \phi$, $q^6 = \psi$):

$$\Gamma_{44} = \begin{pmatrix} \sin^2 \psi & -\sin \psi \cos \psi & 0 \\ \sin \psi \cos \psi & \cos^2 \psi & 0 \\ 0 & 0 & 1 \end{pmatrix}, \tag{10.199}$$

$$\Gamma_{45} = \begin{pmatrix} -\sin \theta \sin \psi \cos \psi & \sin \theta \sin^2 \psi & 0 \\ -\sin \theta \cos^2 \psi & \sin \theta \sin \psi \cos \psi & 0 \\ -\cos \theta \cos \psi & \cos \theta \sin \psi & 0 \end{pmatrix}, \tag{10.200}$$

$$\Gamma_{46} = \begin{pmatrix} 0 & 0 & 0 \\ 0 & 0 & 0 \\ \cos \psi & \sin \psi & 0 \end{pmatrix}, \tag{10.201}$$

$$\Gamma_{55} = \begin{pmatrix} \cos^2 \theta \sin^2 \psi + \cos^2 \psi & \cos \psi \sin \psi (\cos^2 \theta - 1) & -\sin \theta \cos \theta \sin \psi \\ \cos \psi \sin \psi (\cos^2 \theta - 1) & \cos^2 \theta \cos^2 \psi + \sin^2 \psi & -\cos \theta \sin \theta \cos \psi \\ -\sin \theta \cos \theta \sin \psi & -\cos \theta \sin \theta \cos \psi & \sin^2 \theta \end{pmatrix}, \tag{10.202}$$

$$\Gamma_{56} = \begin{pmatrix} \cos \theta & 0 & 0 \\ 0 & \cos \theta & 0 \\ -\sin \theta \sin \psi & -\sin \theta \cos \psi & 0 \end{pmatrix}, \tag{10.203}$$

$$\Gamma_{66} = \begin{pmatrix} 1 & 0 & 0 \\ 0 & 1 & 0 \\ 0 & 0 & 0 \end{pmatrix}. \tag{10.204}$$

Equations 10.184 through 10.187, 10.197, and 10.198 yield,

$$g_{11} = g_{22} = g_{33} = m_t, \tag{10.205}$$

$$g_{44} = I_1 \cos^2 \psi + I_2 \sin^2 \psi, \tag{10.206}$$

$$g_{45} = (I_1 - I_2) \sin \theta \cos \psi \sin \psi, \tag{10.207}$$

$$g_{55} = I_1 \sin^2 \theta \sin^2 \psi + I_2 \sin^2 \theta \cos^2 \psi + I_3 \cos^2 \theta, \tag{10.208}$$

$$g_{56} = I_3 \cos \theta, \tag{10.209}$$

$$g_{66} = I_3. \tag{10.210}$$

With this result, we can conclude that the rigid n body problem can be treated with the three center of mass coordinates and three rotation angles. Furthermore, the center of mass Cartesian coordinates are orthogonal.

Exercises

1. Derive Equations 10.138, 10.145, 10.160, 10.156, 10.157, and 10.158. Prove that Φ_3 is onto and that it, and its inverse, satisfy the integrability conditions in Equations 10.24 and 10.25. δ_x, δ_y, and δ_z are the Cartesian coordinates of the $\mathbb{R}^3$ space of the pseudoparticle with mass μ.

2. For the two body problem, find the map, $\Phi : \theta, \phi \rightarrow \xi^1, \xi^2$, and its inverse, starting with Equations 10.156 through 10.158. Obtain the inverse $\Phi^{-1} : \xi^1, \xi^2 \rightarrow \theta\phi$. Then show that both Φ and Φ^{-1} are onto. If both Φ and Φ^{-1} are onto, the map Φ is called a bijection.

3. Obtain expressions for all the conjugate momenta associated with the Lagrangians in Equations 10.140, 10.146, and 10.161.

4. Show that the inverse of the relative coordinates map for the three body problem read,

$$x_1 = x_C + \frac{m_2}{m_t}\delta_{x2} + \frac{m_3}{m_t}\delta_{x1},$$

$$x_2 = x_C + \left(\frac{m_2}{m_t} - 1\right)\delta_{x2} + \frac{m_3}{m_t}\delta_{x1},$$

$$x_3 = x_C + \frac{m_2}{m_t}\delta_{x2} + \left(\frac{m_3}{m_t} - 1\right)\delta_{x1},$$

with identical equations for the y and z coordinates. Use this inverse map to derive the associated Jacobian $J^v_\mu = \partial x^v / \partial q^\mu$. Finally, derive Equation 10.173.

5. Consider the homogeneous three body problem, $m_1 = m_2 = m_3 = m$. Show that the following relative coordinates

$$x_C = \frac{1}{3}\left(x_1 + x_2 + x_3\right),$$

$$\delta_{x1} = 2x_1 - x_2 - x_3,$$

$$\delta_{x2} = x_2 - x_3,$$

(with identical equations for y and z) are orthogonal, then derive the following Lagrangian:

$$\mathcal{L} = \frac{m}{2}\left(3\dot{x}_C^2 + 3\dot{y}_C^2 + 3\dot{z}_C^2 + \frac{1}{6}\dot{\delta}_{x1}^2 + \frac{1}{6}\dot{\delta}_{y1}^2 + \frac{1}{6}\dot{\delta}_{z1}^2 + \frac{1}{2}\dot{\delta}_{x2}^2 \right.$$
$$\left. + \frac{1}{2}\dot{\delta}_{y2}^2 + \frac{1}{2}\dot{\delta}_{z2}^2\right) - V.$$

Begin by showing that the inverse map is,

$$x_1 = x_C + \frac{1}{3}\delta_{x1},$$

$$x_2 = x_C - \frac{1}{6}\delta_{x1} + \frac{1}{2}\delta_{x2},$$

$$x_3 = x_C - \frac{1}{6}\delta_{x1} - \frac{1}{2}\delta_{x2},$$

with identical relationships for the y and z coordinates.

6. Derive Equation 10.182 from Equations 10.179 through 10.181.

7. Derive Equations 10.199 through 10.210. Note that Equation 10.184 selects only diagonal elements of the tensors $[\Gamma_{\mu\upsilon}]_{\kappa'\kappa''}$ in Equation 10.197. For example, for g_{44}, one obtains,

$$g_{44} = \sin^2\psi \sum_{i=1}^{n} m_i \left(x^{3(i-1)+1}\right)^2 + \cos^2\psi \sum_{i=1}^{n} m_i \left(x^{3(i-1)+2}\right)^2$$
$$+ \sum_{i=1}^{n} m_i \left(x^{3(i-1)+3}\right)^2.$$

The substitution of $1 = \sin^2\psi + \cos^2\psi$ in the last term on the R.H.S. yields Equation 10.206.

10.19 Stereographic Projections for the Ellipsoid of Inertia

It is possible to map the three-dimensional rotation of a rigid body stereographically, and carry out the quantum simulations in a set of stereographic projection coordinates. The simplest strategy is to remap the inertia ellipsoid with two maps, Φ_1 and Φ_2, as follows:

$$
x^\mu \quad
\begin{matrix} \Phi_1 \\ \rightleftharpoons \\ \Phi_1^{-1} \end{matrix}
\quad
\begin{pmatrix} x_C \\ y_C \\ z_C \\ \theta \\ \phi \\ \psi \end{pmatrix}
\quad
\begin{matrix} \Phi_2 \\ \rightleftharpoons \\ \Phi_2^{-1} \end{matrix}
\quad
\begin{pmatrix} x_C \\ y_C \\ z_C \\ \xi^1 \\ \xi^2 \\ \xi^3 \end{pmatrix},
\tag{10.211}
$$

where x^μ is the configuration in $\mathbb{R}^{3n}$. We have investigated Φ_1 in the previous section, now we move on to examine Φ_2. It is possible to arrive at an expression for Φ_2 and its inverse by working through the quaternion parameter space associated with the three Euler angles. We turn our attention to the details of that particular process. The four quaternion parameters are defined by the following equations:

$$
x^1 = \cos\frac{\theta}{2} \cos\left(\frac{\psi + \phi}{2}\right),
\tag{10.212}
$$

$$
x^2 = \sin\frac{\theta}{2} \cos\left(\frac{\psi - \phi}{2}\right),
\tag{10.213}
$$

$$
x^3 = \sin\frac{\theta}{2} \sin\left(\frac{\psi - \phi}{2}\right),
\tag{10.214}
$$

$$
x^4 = \cos\frac{\theta}{2} \sin\left(\frac{\psi + \phi}{2}\right).
\tag{10.215}
$$

The ranges of the angles are $\theta : 0 \rightarrow \pi, \phi : 0 \rightarrow 2\pi, \psi : 0 \rightarrow 2\pi$. It is easy to show that in the range of the Euler angles, all four quaternion parameters take all possible values in $[-1, 1]$.

One can show that the four quaternion parameters are constrained to a three-dimensional sphere of unit radius,

$$
\left(x^1\right)^2 + \left(x^2\right)^2 + \left(x^3\right)^2 + \left(x^4\right)^2 = 1.
\tag{10.216}
$$

The stereographic projection definitions for a unit radius three-dimensional hypersphere can be extracted as a special case of Equation 10.75.

$$
\xi^\mu = \frac{2x^\mu}{1 - x^4}, \qquad \mu = 1, 2, 3.
\tag{10.217}
$$

It is now possible to invert the map Φ_2 and formulate the transformation of variables. The transformation of the metric tensor can be performed directly on Equations 10.205 through 10.210, if one has explicit forms for θ, ϕ, ψ as functions of ξ^1, ξ^2, ξ^3. These are obtained from the definitions of the quaternions.

The relationships can be inverted to obtain expressions for the three Euler angles,

$$\theta = 2\sin^{-1}\sqrt{(x^2)^2 + (x^3)^2},$$

(10.218)

$$\psi = \tan^{-1}\frac{x^4}{x^1} + \tan^{-1}\frac{x^3}{x^2},$$

(10.219)

$$\phi = \tan^{-1}\frac{x^4}{x^1} - \tan^{-1}\frac{x^3}{x^2}.$$

(10.220)

We get the following equations:

$$\theta = 2\sin^{-1}\frac{\sqrt{(4\xi^2)^2 + (4\xi^3)^2}}{\sigma + 4},$$

(10.221)

$$\psi = \tan^{-1}\left(\frac{\sigma - 4}{4\xi^1}\right) + \tan^{-1}\left(\frac{\xi^3}{\xi^2}\right),$$

(10.222)

$$\phi = \tan^{-1}\left(\frac{\sigma - 4}{4\xi^1}\right) - \tan^{-1}\left(\frac{\xi^3}{\xi^2}\right).$$

(10.223)

The Jacobian matrix for Φ_2^{-1} has a block diagonal structure,

$$J^\upsilon_\mu = \begin{pmatrix} \mathbf{1}_{3\times 3} & 0 \\ 0 & \mathbf{J}_{3\times 3} \end{pmatrix},$$

(10.224)

where the elements of the lower 3×3 block are developed in detail in this section. The Hessian metric tensor that is transformed has the following general form:

$$g_{\mu\upsilon} = \begin{pmatrix} m_t & 0 & 0 & 0 & 0 & 0 \\ 0 & m_t & 0 & 0 & 0 & 0 \\ 0 & 0 & m_t & 0 & 0 & 0 \\ 0 & 0 & 0 & g_{11}^{\mathbb{I}} & g_{12}^{\mathbb{I}} & g_{13}^{\mathbb{I}} \\ 0 & 0 & 0 & g_{12}^{\mathbb{I}} & g_{22}^{\mathbb{I}} & g_{23}^{\mathbb{I}} \\ 0 & 0 & 0 & g_{13}^{\mathbb{I}} & g_{23}^{\mathbb{I}} & g_{33}^{\mathbb{I}} \end{pmatrix}.$$

(10.225)

$g_{\mu\upsilon}^{\mathbb{I}}$ are the entries of the metric tensor in the rotation submanifold.

It is straightforward to show that the upper 3×3 block remains unchanged, thanks to the block diagonal structure of the metric tensor in Equation 10.225 and the Jacobian matrix (Equation 10.224). Therefore, for the ensuing tensor analysis, let us associate $q^1 = \theta$, $q^2 = \phi$, $q^3 = \psi$, and let us focus on the rotation 3×3 block on the bottom right of Equation 10.225. Additionally, we drop the $\mathbb{I}$ superscript in the following.

Let us define $h^\mu_{\mu'}$ as the partial derivatives involved in the transformation of covariant quantities, i.e.,

$$d\theta = h_1^1 d\xi^1 + h_2^1 d\xi^2 + h_3^1 d\xi^3,$$

(10.226)

$$d\phi = h_1^2 d\xi^1 + h_2^2 d\xi^2 + h_3^2 d\xi^3,$$

(10.227)

$$d\psi = h_1^3 d\xi^1 + h_2^3 d\xi^2 + h_3^3 d\xi^3.$$

(10.228)

The h_υ^μ are the elements of $\mathbf{J}_{3\times3}$ in Equation 10.224. Let us define the following five auxiliary variables:

$$d^1 = \sqrt{16\left(\xi^1\right)^2 + (\sigma - 4)^2}, \tag{10.229}$$

$$d^2 = \sqrt{\sigma - \left(\xi^1\right)^2}, \tag{10.230}$$

$$d^3 = \left(d^2\right)^2, \quad d^4 = 16\left(\xi^1\right)^2 + (\sigma - 4)^2 = \left(d^1\right)^2, \quad d^5 = 4\left[2\left(\xi^1\right)^2 - (\sigma - 4)\right], \tag{10.231}$$

then the entries of the Jacobian matrix are,

$$h_1^1 = -\frac{16d^2\xi^1}{d^1\left(\sigma + 4\right)}, \qquad h_2^1 = \frac{2d^5\xi^2}{d^1 d^2\left(\sigma + 4\right)}, \qquad h_3^1 = \frac{2d^5\xi^3}{d^1 d^2\left(\sigma + 4\right)}, \tag{10.232}$$

$$h_1^2 = h_1^3 = \frac{d^5}{d^4}, \qquad h_2^2 = \frac{8\xi^1\xi^2}{d^4} + \frac{\xi^3}{d^3}, \qquad h_3^2 = \frac{8\xi^1\xi^3}{d^4} - \frac{\xi^2}{d^3}, \tag{10.233}$$

$$h_2^3 = \frac{8\xi^1\xi^2}{d^4} - \frac{\xi^3}{d^3}, \qquad h_3^3 = \frac{8\xi^1\xi^3}{d^4} + \frac{\xi^2}{d^3}. \tag{10.234}$$

For simulations with nonspherical rotors, we translate Equations 10.232 through 10.234 directly into code, and use their values to transform the Hessian metric tensor from Euler angles into stereographic projections. However, the spherical case is both instructive and the results yield computational advantage. We consider this special case next.

10.20   The Spherical Top

The metric tensor for a spherical top is obtained by setting $I_2 = I_3 = I_1$ in Equations 10.206 through 10.210. The result is

$$g_{\mu\upsilon} = I_1 \begin{pmatrix} 1 & 0 & 0 \\ 0 & 1 & \cos\theta \\ 0 & \cos\theta & 1 \end{pmatrix}. \tag{10.235}$$

The following are important for the derivation of the metric tensor and the development of the theory that follows.

$$\cos\theta = \frac{2d^4}{(\sigma + 4)^2} - 1, \qquad \sin\theta = \frac{8d^1 d^2}{(\sigma + 4)^2}. \tag{10.236}$$

To prove these identities, begin by noting that,

$$\cos^2\frac{\theta}{2} = \left(x^1\right)^2 + \left(x^4\right)^2 = \frac{16\left(\xi^1\right)^2 + (\sigma - 4)^2}{(\sigma + 4)^2} = \frac{d^4}{(\sigma + 4)^2}. \tag{10.237}$$

Using the following trigonometric identities:

$$\cos\theta = \cos^2\frac{\theta}{2} - \sin^2\frac{\theta}{2}, \tag{10.238}$$

$$1 = \cos^2\frac{\theta}{2} + \sin^2\frac{\theta}{2}, \tag{10.239}$$

$$2\cos^2\frac{\theta}{2} = \cos\theta + 1, \tag{10.240}$$

we get,

$$\cos\theta = 2\cos^2\frac{\theta}{2} - 1 = \frac{2d^4}{(\sigma+4)^2} - 1.$$

Whereas,

$$\sin^2\frac{\theta}{2} = \left(x^2\right)^2 + \left(x^3\right)^2 = \frac{16\left[\left(\xi^2\right)^2 + \left(\xi^3\right)^2\right]}{(\sigma+4)^2} = \frac{16d^3}{(\sigma+4)^2},$$

and using

$$\sin\theta = 2\cos\frac{\theta}{2}\sin\frac{\theta}{2}, \tag{10.241}$$

$$\sin\theta = \left[\frac{2d^1}{(\sigma+4)}\right]\left[\frac{4d^2}{(\sigma+4)}\right] = \frac{8d^1d^2}{(\sigma+4)^2}. \tag{10.242}$$

The following three relationships prove crucial in the transformation of the metric tensor as well.

$$64\left(\xi^1\right)^2 d^3 + \left(d^5\right)^2 = 16d^4, \tag{10.243}$$

$$d^3\left(\xi^2\right)^2\left(d^5\right)^2 + d^4\left(\xi^3\right)^2(\sigma+4)^2 + \left(d^3\right)^2\left(8\xi^1\xi^2\right)^2$$
$$- \left(d^4\right)^2\left(\xi^3\right)^2 = 16\left(d^3\right)^2 d^4, \tag{10.244}$$

$$d^3\left(d^5\right)^2 - d^4(\sigma+4)^2 + 64\left(d^3\right)^2\left(\xi^1\right)^2 + \left(d^4\right)^2 = 0. \tag{10.245}$$

Finally, the last important relationship we need is,

$$d^3\left(d^5\right)^2\left(\xi^3\right)^2 + d^4\left(\xi^2\right)^2(\sigma+4)^2 + 64\left(d^3\right)^2\left(\xi^1\right)^2\left(\xi^3\right)^2$$
$$- \left(d^4\right)^2\left(\xi^2\right)^2 = 16d^4\left(d^3\right)^2. \tag{10.246}$$

These relationships can be proved by breaking down each term, replacing the auxiliary quantities d^1 through d^5, and simplifying.

After the zero entries in the metric tensor are considered, we get

$$I_1^{-1}g_{\mu'\upsilon'} = \delta_{\mu\upsilon}^3 h_{\mu'}^\mu h_{\upsilon'}^\upsilon + \left[\frac{2d^4}{(\sigma+4)^2} - 1\right]\left(h_{\mu'}^2 h_{\upsilon'}^3 + h_{\mu'}^3 h_{\upsilon'}^2\right). \tag{10.247}$$

The symmetry of the Hessian metric decreases to six the number of independent elements to be considered. Let us inspect all six terms. The following six relationships are obtained by substitution of the expressions for the elements of h^μ_ν derived earlier in Equation 10.247, followed by several algebraic steps. We begin with g_{11},

$$I_1^{-1}g_{11} = \frac{4}{d^4(\sigma+4)^2}\left\{64\left(\xi^1\right)^2 d^3 + \left(d^5\right)^2\right\} = \frac{64}{(\sigma+4)^2}. \qquad (10.248)$$

Therefore, considerable simplification takes place. Let us continue with g_{12},

$$I_1^{-1}g_{12} = -\frac{32\xi^1\xi^2 d^5}{d^4(\sigma+4)^2} + \frac{2d^5}{d^4}\left(\frac{8\xi^1\xi^2}{d^4}\right) + \frac{2d^5}{d^4}\left[\left(\frac{8\xi^1\xi^2}{d^4}\right)\right]\left[\frac{2d^4}{(\sigma+4)^2} - 1\right],$$

$$I_1^{-1}g_{12} = -\frac{32\xi^1\xi^2 d^5}{d^4(\sigma+4)^2} + 2d^5\left[\left(\frac{8\xi^1\xi^2}{d^4}\right)\right]\left[\frac{2}{(\sigma+4)^2}\right] = 0, \qquad (10.249)$$

and consider g_{13} next,

$$I_1^{-1}g_{13} = -\frac{32\xi^1\xi^3 d^5}{d^4(\sigma+4)^2} + \frac{2d^5}{d^4}\left(\frac{8\xi^1\xi^3}{d^4}\right) + \left[\frac{2d^4}{(\sigma+4)^2} - 1\right]\frac{2d^5}{d^4}\left(\frac{8\xi^1\xi^3}{d^4}\right),$$

$$I_1^{-1}g_{13} = -\frac{32\xi^1\xi^3 d^5}{d^4(\sigma+4)^2} + \left[\frac{4d^5}{(\sigma+4)^2}\right]\left(\frac{8\xi^1\xi^3}{d^4}\right) = 0. \qquad (10.250)$$

This last result follows the same way that it does for g_{12}. Next, consider the g_{22} term,

$$I_1^{-1}g_{22} = \frac{4}{d^4(d^3)^2(\sigma+4)^2}$$
$$\times \left[d^3(\xi^2)^2(d^5)^2 + d^4(\xi^3)^2(\sigma+4)^2 + (d^3)^2(8\xi^1\xi^2)^2 - (d^4)^2(\xi^3)^2\right].$$

Therefore,

$$I_1^{-1}g_{22} = \frac{64}{(\sigma+4)^2}. \qquad (10.251)$$

Let us move on to I_{23},

$$I_1^{-1}g_{23} = \frac{4\left(d^5\right)^2\xi^2\xi^3}{d^4 d^3(\sigma+4)^2} - \frac{4\xi^2\xi^3}{(d^3)^2} + \frac{4\xi^2\xi^3}{(\sigma+4)^2}\left[\frac{64\left(d^3\right)^2\left(\xi^1\right)^2 + \left(d^4\right)^2}{d^4(d^3)^2}\right],$$

$$I_1^{-1}g_{23} = \frac{4\xi^2\xi^3}{d^4\left(d^3\right)^2(\sigma+4)^2}$$
$$\times \left[d^3\left(d^5\right)^2 - d^4(\sigma+4)^2 + 64\left(d^3\right)^2\left(\xi^1\right)^2 + \left(d^4\right)^2\right] = 0. \qquad (10.252)$$

Finally, the last term gives,

$$I_1^{-1} g_{33} = \frac{4}{d^4 (d^3)^2 (\sigma+4)^2} \left[d^3 \left(d^5\right)^2 \left(\xi^3\right)^2 + d^4 \left(\xi^2\right)^2 (\sigma+4)^2 \right.$$

$$\left. + 64 \left(d^3\right)^2 \left(\xi^1\right)^2 \left(\xi^3\right)^2 - \left(d^4\right)^2 \left(\xi^2\right)^2 \right].$$

It follows that,

$$I_1^{-1} g_{33} = \frac{4}{d^4 (d^3)^2 (\sigma+4)^2} \left[16 d^4 \left(d^3\right)^2 \right] = \frac{64}{(\sigma+4)^2}. \tag{10.253}$$

To summarize our results, we begin with the Euler angles. By mapping the stereographic projections of the quaternion space associated with the rotation of the spherical top, we are able to show that,

$$g_{\mu\upsilon} = I_1 \begin{pmatrix} 1 & 0 & 0 \\ 0 & 1 & \cos\theta \\ 0 & \cos\theta & 1 \end{pmatrix} \Rightarrow \frac{64 I_1}{(\sigma+4)^2} \begin{pmatrix} 1 & 0 & 0 \\ 0 & 1 & 0 \\ 0 & 0 & 1 \end{pmatrix}. \tag{10.254}$$

Therefore, the treatment of the spherical top is simplified significantly by the stereographic projection approach.

10.21 The Riemann Curvature Scalar for a Spherical Top

In this section, we sketch the derivation of the curvature scalar for a spherical top. We work with Euler angles, since for the most general case, Euler angles are the most convenient coordinates. The transformation of the scalar does not require partial derivatives of the map. For the spherical top, we sketch the derivation with stereographic projections as well. These are relatively simple and instructive exercises to hone the reader's skills in differential geometry. Generally, the Riemann curvature scalar is constant, therefore quantum methods based on prepoint or postpoint lattice expansions have a constant lattice correction. Center point expansions, however, are sometimes desirable and the Weyl correction term must be computed carefully and included in the dynamics. The inverse of the metric tensor for the spherical top can be calculated readily by using cofactors and the determinant,

$$\det(g) = I_1 \sin^2\theta, \tag{10.255}$$

$$g^{\mu\upsilon} = \frac{1}{I_1 \sin^2\theta} \begin{pmatrix} \sin^2\theta & 0 & 0 \\ 0 & 1 & -\cos\theta \\ 0 & -\cos\theta & 1 \end{pmatrix}. \tag{10.256}$$

TABLE 10.1
Connection Coefficients for the Spherical Top Ellipsoid of Inertia

$\Gamma^1_{11} = 0$	$\Gamma^1_{12} = 0$	$\Gamma^1_{13} = 0$
$\Gamma^1_{21} = 0$	$\Gamma^1_{22} = 0$	$\Gamma^1_{23} = \dfrac{1}{2}\sin\theta$
$\Gamma^1_{31} = 0$	$\Gamma^1_{32} = \dfrac{1}{2}\sin\theta$	$\Gamma^1_{33} = 0$
$\Gamma^2_{11} = 0$	$\Gamma^2_{12} = \dfrac{\cos\theta}{2\sin\theta}$	$\Gamma^2_{13} = -\dfrac{1}{2\sin\theta}$
$\Gamma^2_{21} = \dfrac{\cos\theta}{2\sin\theta}$	$\Gamma^2_{22} = 0$	$\Gamma^2_{23} = 0$
$\Gamma^2_{31} = -\dfrac{1}{2\sin\theta}$	$\Gamma^2_{32} = 0$	$\Gamma^2_{33} = 0$
$\Gamma^3_{11} = 0$	$\Gamma^3_{12} = -\dfrac{1}{2\sin\theta}$	$\Gamma^3_{13} = \dfrac{\cos\theta}{2\sin\theta}$
$\Gamma^3_{21} = -\dfrac{1}{2\sin\theta}$	$\Gamma^3_{22} = 0$	$\Gamma^3_{23} = 0$
$\Gamma^3_{31} = \dfrac{\cos\theta}{2\sin\theta}$	$\Gamma^3_{32} = 0$	$\Gamma^3_{33} = 0$

The Christoffel connection coefficients (27 in total) are obtained from definition in Equation 10.98. The Γ^2_{12} coefficient, for example, becomes

$$\Gamma^2_{12} = \frac{1}{2}g^{21}\left(\partial_1 g_{21} + \partial_2 g_{11} - \partial_1 g_{12}\right) + \frac{1}{2}g^{22}\left(\partial_1 g_{22} + \partial_2 g_{21} - \partial_2 g_{12}\right)$$
$$+ \frac{1}{2}g^{23}\left(\partial_1 g_{23} + \partial_2 g_{31} - \partial_3 g_{12}\right) = \frac{1}{2}\left(-\frac{\cos\theta}{\sin^2\theta}\right)(-\sin\theta) = \frac{\cos\theta}{2\sin\theta}.$$
$$\tag{10.257}$$

Table 10.1 contains the expressions for all the connection coefficients. As a check, the connection is torsion free, as can be verified by checking the torsion tensor elements,

$$T^\lambda_{\mu\nu} = \Gamma^\lambda_{\mu\nu} - \Gamma^\lambda_{\nu\mu} = 0. \tag{10.258}$$

The following relationships are also easily verifiable.

$$\Gamma^\mu_{\mu\lambda} = \frac{1}{\sqrt{\det(g)}}\partial_\lambda\sqrt{\det(g)}, \tag{10.259}$$

$$\Gamma^\mu_{\mu 1} = \Gamma^1_{11} + \Gamma^2_{21} + \Gamma^3_{31} = \frac{1}{\sin\theta}\partial_1\sin\theta = \frac{\cos\theta}{\sin\theta}, \tag{10.260}$$

$$\Gamma^\mu_{\mu 2} = \Gamma^1_{12} + \Gamma^2_{22} + \Gamma^3_{32} = 0, \qquad \Gamma^\mu_{\mu 3} = \Gamma^1_{13} + \Gamma^2_{23} + \Gamma^3_{33} = 0. \tag{10.261}$$

The connection coefficients are needed to evaluate the elements of the Riemann tensor. The Riemann tensor is antisymmetric with respect to the last two lower indices, as can be easily seen by its definition; $R^\rho_{\alpha\mu\nu} = -R^\rho_{\alpha\nu\mu}$.

Therefore, a number of elements vanish; $R^\rho_{\alpha\mu\mu} = 0$. To calculate the curvature of the space, the Riemann tensor is contracted into the Ricci tensor, and this is contracted with the inverse of the Hessian metric to get the curvature. In the case of a three-dimensional space, one has nine terms,

$$\mathcal{R} = g^{11}R_{11} + g^{12}R_{12} + g^{13}R_{13}$$
$$+ g^{21}R_{21} + g^{22}R_{22} + g^{23}R_{23} + g^{31}R_{31} + g^{32}R_{32} + g^{33}R_{33}. \quad (10.262)$$

However, of these, only six terms are needed, since it can be shown that the Ricci tensor is symmetric $R_{\mu\nu} = R_{\nu\mu}$,

$$\mathcal{R} = g^{11}R_{11} + 2g^{12}R_{12} + 2g^{13}R_{13} + g^{22}R_{22} + 2g^{23}R_{23} + g^{33}R_{33}. \quad (10.263)$$

In the case of symmetric tops, inspection of the inverse metric eliminates the need to compute two additional off-diagonal elements of the Ricci tensor for the spherical top,

$$\mathcal{R} = g^{11}R_{11} + g^{22}R_{22} + 2g^{23}R_{23} + g^{33}R_{33}. \quad (10.264)$$

Therefore, only eight terms of the Riemann tensor are needed (there are three terms for each contraction into a Ricci tensor element), e.g.,

$$R_{11} = R^1_{111} + R^2_{121} + R^3_{131}, \quad (10.265)$$

and only eight of these are nonvanishing. Table 10.2 summarizes the results for the Riemann tensor elements needed to calculate the curvature. Plugging the relevant Ricci tensor and inverse metric elements gives,

$$I_1 \, \mathcal{R} = \frac{1}{2} + \frac{1}{\sin^2\theta}\left(\frac{1}{2}\right)$$
$$+ \left(-\frac{\cos\theta}{\sin^2\theta}\right)\frac{\cos\theta}{2} + \left(-\frac{\cos\theta}{\sin^2\theta}\right)\frac{\cos\theta}{2} + \frac{1}{\sin^2\theta}\left(\frac{1}{2}\right) = \frac{3}{2}. \quad (10.266)$$

TABLE 10.2
Relevant Elements of the Riemann and Ricci Tensors for the Spherical Top Ellipsoid of Inertia

$\mu\nu$	$\lambda = 1$	$\lambda = 2$	$\lambda = 3$	$R_{\mu\nu}$
11	$R^1_{111} = 0$	$R^2_{121} = \frac{1}{4}$	$R^3_{131} = \frac{1}{4}$	$\frac{1}{2}$
22	$R^1_{212} = \frac{1}{4}$	$R^2_{222} = 0$	$R^3_{232} = \frac{1}{4}$	$\frac{1}{2}$
23	$R^1_{213} = \frac{\cos\theta}{4}$	$R^2_{223} = \frac{\cos\theta}{4}$	$R^3_{133} = 0$	$\frac{\cos\theta}{2}$
33	$R^1_{313} = \frac{1}{4}$	$R^2_{323} = \frac{1}{4}$	$R^3_{333} = 0$	$\frac{1}{2}$

10.22   Coefficients and the Curvature for Spherical Tops with Stereographic Projection Coordinates (SPCs)

For spherical tops, it is much simpler to obtain the connections and the curvature tensors using the stereographic projections. For example, the inverse of the Hessian metric tensor, $g^{\sigma\lambda}$, which we find in a previous section is diagonal, is simpler to obtain, and its diagonal structure eliminates numerous Ricci tensor elements in the final contraction.

The high symmetry of the metric and its inverse yields only three independent connection coefficients,

$$\Gamma^1_{11} = -\frac{2\xi^1}{\sigma+4}, \tag{10.267}$$

$$\Gamma^1_{12} = -\frac{2\xi^2}{\sigma+4}, \tag{10.268}$$

$$\Gamma^1_{13} = -\frac{2\xi^3}{\sigma+4}. \tag{10.269}$$

A number of coefficients are related trivially to these:

$$\Gamma^1_{22} = \Gamma^1_{33} = -\Gamma^2_{12} = -\Gamma^3_{13} = -\Gamma^1_{11}, \tag{10.270}$$

$$\Gamma^2_{11} = -\Gamma^2_{22} = \Gamma^2_{33} = -\Gamma^3_{23} = -\Gamma^1_{12}, \tag{10.271}$$

while the rest are zero. Only one independent element of $R^\lambda_{\mu\sigma\upsilon}$ is required,

$$R^2_{121} = R^3_{131} = R^1_{212} = R^3_{232} = R^1_{313} = R^2_{323} = \frac{4}{\sigma+4} - \frac{4\sigma}{(\sigma+4)^2}. \tag{10.272}$$

Since the inverse of the Hessian metric is diagonal, we are only interested in diagonal elements of the Ricci tensor, and the high symmetry of the space yields one independent diagonal element of the Ricci tensor,

$$R_{11} = R_{22} = R_{33} = \frac{8}{\sigma+4} - \frac{8\sigma}{(\sigma+4)^2}. \tag{10.273}$$

The curvature follows immediately,

$$\mathcal{R} = g^{\mu\upsilon} R_{\mu\upsilon} = \frac{3}{2I_1}. \tag{10.274}$$

Therefore, while the space of a symmetric top is not a three-sphere, it is a maximally symmetric space.

10.23 The Riemann Curvature Scalar for a Symmetric Nonspherical Top

For the symmetric top, the eigenvalues of the inertia tensor satisfy the following relations:

$$I_1 = I_{xx} = I_{yy} \neq I_{zz} = I_2. \tag{10.275}$$

The Hessian metric tensor in the body frame is

$$g_{\mu\upsilon} = \begin{pmatrix} I_1 & 0 & 0 \\ 0 & I_1 \sin^2\theta + I_2 \cos^2\theta & I_2 \cos\theta \\ 0 & I_2 \cos\theta & I_2 \end{pmatrix}. \tag{10.276}$$

The determinant is

$$\det(g) = I_1^2 I_2 \sin^2\theta, \tag{10.277}$$

and $g^{\mu\upsilon}$ is easily obtained

$$g^{\mu\upsilon} = \begin{pmatrix} \dfrac{1}{I_1} & 0 & 0 \\ 0 & \dfrac{1}{I_1 \sin^2\theta} & \dfrac{-\cos\theta}{I_1 \sin^2\theta} \\ 0 & \dfrac{-\cos\theta}{I_1 \sin^2\theta} & \dfrac{I_1 \sin^2\theta + I_2 \cos^2\theta}{I_1 I_2 \sin^2\theta} \end{pmatrix}. \tag{10.278}$$

The connection coefficients are in Table 10.3. As a check for these expressions, we set $I_1 = I_2$, and we verify that all the connection coefficients for the symmetric top are reproduced. The symmetry of the coefficients is evident, and all the sum rules are easily verifiable. These are identical to the spherical top ones. The relevant Riemann tensor elements are as follows:

$$R^2_{121} = \frac{4I_1 - 3I_2}{4I_1}, \tag{10.279}$$

$$R^3_{131} = \frac{I_2}{4I_1}, \tag{10.280}$$

$$R^1_{212} = \left(\frac{I_2^2 - I_1 I_2}{4I_1^2}\right)\cos^2\theta - \left(\frac{I_2 - I_1}{I_1}\right)\sin^2\theta + \frac{I_2}{4I_1}, \tag{10.281}$$

$$R^3_{232} = \left(\frac{I_2^2 - I_1 I_2}{4I_1^2}\right)\cos^2\theta + \frac{I_2}{4I_1}, \tag{10.282}$$

$$R^1_{213} = R^2_{223} = \frac{I_2^2}{4I_1^2}\cos\theta, \tag{10.283}$$

TABLE 10.3

Connection Coefficients for the Symmetric Top Ellipsoid of Inertia

$\Gamma^1_{11} = 0$	$\Gamma^1_{12} = 0$	$\Gamma^1_{13} = 0$
$\Gamma^1_{21} = 0$	$\Gamma^1_{22} = \left(\dfrac{I_2 - I_1}{I_1}\right)\cos\theta\sin\theta$	$\Gamma^1_{23} = \dfrac{I_2}{2I_1}\sin\theta$
$\Gamma^1_{31} = 0$	$\Gamma^1_{32} = \dfrac{I_2}{2I_1}\sin\theta$	$\Gamma^1_{33} = 0$
$\Gamma^2_{11} = 0$	$\Gamma^2_{12} = \left(\dfrac{2I_1 - I_2}{2I_1}\right)\dfrac{\cos\theta}{\sin\theta}$	$\Gamma^2_{13} = -\dfrac{I_2}{2I_1\sin\theta}$
$\Gamma^2_{21} = \left(\dfrac{2I_1 - I_2}{2I_1}\right)\dfrac{\cos\theta}{\sin\theta}$	$\Gamma^2_{22} = 0$	$\Gamma^2_{23} = 0$
$\Gamma^2_{31} = -\dfrac{I_2}{2I_1\sin\theta}$	$\Gamma^2_{32} = 0$	$\Gamma^2_{33} = 0$
$\Gamma^3_{11} = 0$	$\Gamma^3_{12} = \dfrac{(I_2 - I_1)\cos^2\theta - I_1}{2I_1\sin\theta}$	$\Gamma^3_{13} = \dfrac{I_2\cos\theta}{2I_1\sin\theta}$
$\Gamma^3_{21} = \dfrac{(I_2 - I_1)\cos^2\theta - I_1}{2I_1\sin\theta}$	$\Gamma^3_{22} = 0$	$\Gamma^3_{23} = 0$
$\Gamma^3_{31} = \dfrac{I_2\cos\theta}{2I_1\sin\theta}$	$\Gamma^3_{32} = 0$	$\Gamma^3_{33} = 0$

and

$$R^1_{313} = R^2_{323} = \frac{I_2^2}{4I_1^2}. \tag{10.284}$$

The curvature for the space of a symmetric top is also a constant,

$$\mathcal{R} = \frac{4I_1 - I_2}{2I_1^2}. \tag{10.285}$$

Finally, the Riemann curvature for the general case when $I_1 \neq I_2 \neq I_3$, can be obtained in a similar way, though the work is considerably more tedious. The result is,

$$\mathcal{R} = \frac{(I_1 + I_2 + I_3)^2 - 2\left(I_1^2 + I_2^2 + I_3^2\right)}{2I_1 I_2 I_3}. \tag{10.286}$$

At this point, we have completed the necessary mathematical preliminaries that give us the ability to develop classical and quantum simulation methods of clusters comprised of any kind of rigid tops.

Exercises

1. Derive Equations 10.221 and 10.223.

2. Derive Equations 10.232 through 10.234.

3. Set $I_2 = I_3 = I_1$ in Equations 10.206 through 10.210, and derive Equation 10.235.

4. Work through the details to derive Equation 10.254.

5. Verify that Equations 10.258 and 10.259 apply for the connection coefficients in Tables 10.1 and 10.3.

6. Use the results in Table 10.2 to verify Equation 10.266.

7. Derive Equation 10.285.

8. Write a program that computes the Hessian metric for the asymmetric top, using Equations 10.205 through 10.210 for given values of $I_1, I_2, I_3, \theta, \phi$, and ψ. Then, using numerical analysis, add to the program subroutines that compute numerical values for all the connection coefficients, and all the Riemann and Ricci tensor entries. Finally, have the program compute the Riemann curvature scalar and use those to check Equation 10.286. Use the program to check all the results derived in the last three sections regarding the spherical top case, $I_1 = I_2 = I_3$, and the symmetric cases, $I_1 = I_2 \neq I_3$.

10.24   A Split Operator for Symplectic Integrators in Curved Manifolds

Consider the following Hamiltonian for a general manifold $\mathbb{M}^n$,

$$\mathcal{H} = \frac{1}{2} g^{\mu\upsilon} p_\mu p_\upsilon + V(q). \tag{10.287}$$

The one stage time evolution operator, derived for the Liouville operator is a $2n \times 2n$ matrix,

$$\exp\left(\int_t^{t+\Delta t} \mathbb{L}\, dt\right) = \begin{pmatrix} 1 & \dfrac{\partial \mathcal{H}}{p_\mu \partial p_\mu} \Delta t \\ 0 & 1 \end{pmatrix} \begin{pmatrix} 1 & 0 \\ -\dfrac{\partial \mathcal{H}}{q^\mu \partial q^\mu} \Delta t & 1 \end{pmatrix} \quad \text{(no sum)}, \tag{10.288}$$

where each $n \times n$ block is diagonal. The quantity,

$$\frac{\partial \mathcal{H}}{p_\mu \partial p_\mu} = \frac{g^{\mu\upsilon} p_\upsilon}{p_\mu}, \tag{10.289}$$

is the system's effective inverse mass. To see precisely how Equation 10.289 results, consider for example, the $n = 3$ most general case, which contains nonzero off-diagonal elements of the metric tensor,

$$\mathcal{H} = \frac{1}{2}g^{11}p_1p_1 + \frac{1}{2}g^{22}p_2p_2$$

$$+ \frac{1}{2}g^{33}p_3p_3 + g^{12}p_1p_2 + g^{13}p_1p_3 + g^{23}p_2p_3 + V(q), \qquad (10.290)$$

then,

$$\frac{\partial \mathcal{H}}{\partial p_\mu} = \frac{g^{11}p_1 + g^{12}p_2 + g^{13}p_3}{p_1} = \frac{g^{1\upsilon}p_\upsilon}{p_1}. \qquad (10.291)$$

Note that we assume a symmetric metric tensor, $g_{\mu\upsilon} = g_{\upsilon\mu}$.

In general, the metric tensor and its inverse are configuration dependent, therefore,

$$\partial_\mu \mathcal{H} = \frac{1}{2}\left(\partial_\mu g^{\rho\upsilon}\right)p_\rho p_\upsilon + \partial_\mu V(q). \qquad (10.292)$$

The Euler–Cromer algorithm for curved spaces is derived in a manner equivalent to its Euclidean space counterpart. Assuming the n positions are known after stepping in time by Δt, j times, and letting these be represented by $q^\mu(t_j)$ and $p_\mu(t_j)$, we can express the updating procedure with the following equations:

$$p_\mu(t_{j+1}) = p_\mu(t_j) - \partial_\mu \mathcal{H}\,\Delta t, \qquad (10.293)$$

$$q^\mu(t_{j+1}) = q^\mu(t_j) + p_\mu(t_{j+1})\frac{\partial \mathcal{H}}{p_\mu \partial p_\mu}\Delta t \qquad \text{(no sum).} \qquad (10.294)$$

An example should clarify the notation. Consider a particle of mass μ in a ring of unit radius mapped with stereographic projections and subjected to a potential V. We have introduced stereographic projection coordinates for $\mathbb{S}^1$ in the previous chapter. The Hamiltonian is,

$$\mathcal{H} = \frac{1}{32\mu}\left(\xi^2 + 4\right)^2 p_\xi^2 + V. \qquad (10.295)$$

The two relevant partial derivatives are,

$$\frac{\partial \mathcal{H}}{p_\mu \partial p_\mu} = \frac{\left(\xi^2 + 4\right)^2}{16\mu}, \qquad \partial_\mu \mathcal{H} = \frac{\xi\left(\xi^2 + 4\right)}{8\mu}p_\xi^2 + \frac{\partial V}{\partial \xi}. \qquad (10.296)$$

Inserting these expressions into the local split time evolution operator produces the following recursion:

$$\begin{pmatrix} \xi_{n+1} \\ p_{\xi,n+1} \end{pmatrix} = \begin{pmatrix} 1 & \dfrac{\left(\xi_n^2 + 4\right)^2}{16\mu}\Delta t \\ 0 & 1 \end{pmatrix} \begin{pmatrix} 1 & 0 \\ -\dfrac{\left(\xi_n^2 + 4\right)}{8\mu}p_{\xi,n}^2\,\Delta t - \dfrac{1}{\xi_n}\dfrac{\partial V}{\partial \xi}\Delta t & 1 \end{pmatrix} \begin{pmatrix} \xi_n \\ p_{\xi,n} \end{pmatrix}.$$

$$(10.297)$$

After the proper matrix-vector multiplications are carried out, one derives the following first order algorithm:

$$\xi_{n+1} = \xi_n + p_{\xi,n+1} \frac{\left(\xi_n^2 + 4\right)^2}{8\mu} \Delta t, \tag{10.298}$$

$$p_{\xi,n+1} = p_{\xi,n} - \left(\frac{\xi_n \left(\xi_n^2 + 4\right)}{8\mu} p_{\xi,n}^2 + \frac{\partial V}{\partial \xi} \right) \Delta t. \tag{10.299}$$

The integration can be performed using the following simple program, where a particle of unit mass trapped in a ring of unit radius is subjected to the sinusoidal potential in Equation 10.300,

$$V = \frac{4\xi}{\xi^2 + 4}. \tag{10.300}$$

The potential and its derivative are in Figure 10.2, whereas a graph of p_ξ vs ξ is in Figure 10.3

```
program Euler_Cromer_S1
implicit real*8 (a-h,o-z)
c The Euler - Cromer adapted for the integration of one dimensional
c manifolds.
dt = 1.d-3
c start the time evolution (the momentum has to be transported)
x = -1.50d0
p = 0.d0
call pot(x,v,vp)
call imetric(x,g,gp)
e - 0.5d0*g*p*p + v
write(6,*) ' initial energy =',e
do item = 1,100000
 call pot(x,v,vp)
 call imetric(x,g,gp)
 p = p - dt*vp - 0.5d0*dt*gp*p*p
 x = x + dt*g*p
 call pot(x,v,vp)
 call imetric(x,g,gp)
 e = 0.5d0*g*p*p + v
 t = t + dt
 write(8,*) t,x,p,e
enddo
end
subroutine imetric(x,g,gp)
implicit real*8 (a-h,o-z)
c computes the inverse of the metric tensor and its
c configuration derivative for a particle of unit
c mass in a ring of unit radius mapped stereographically.
```

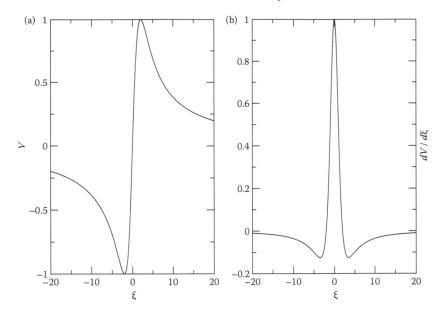

FIGURE 10.2
A graph of the potential $\cos\theta$ (a) and its derivative (b) expressed in terms of the stereographic projection coordinate, ξ, to map the ring $\mathbb{S}^1$.

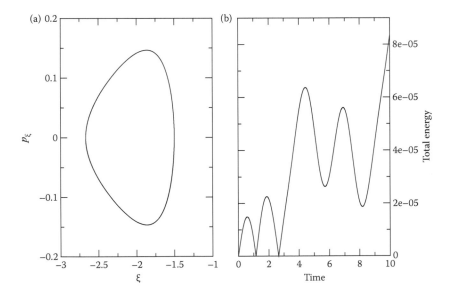

FIGURE 10.3
Phase space for a particle of unit mass in a ring of unit radius mapped by the stereographic projection coordinate ξ (a), and the conservation of energy property of the Euler–Cromer algorithm (b).

```
      g = (x*x + 4.d0)*(x*x + 4.d0)/16.d0
      gp = x*(x*x + 4)/4.d0
      return
      end
      subroutine pot(x,v,vp)
      implicit real*8 (a-h,o-z)
c computes the potential energy and its
c configuration derivative for a particle of unit
c mass in a ring of unit radius mapped stereographically.
c the potential transforms to cos \theta.
      v = 4.d0*x/(x*x + 4.d0)
      vp = 4.d0/(x*x + 4.d0) - 8.d0*x*x/(x*x + 4.d0)/(x*x + 4.d0)
      return
      end
```

The right panel of Figure 10.3 contains a graph of $|E - E_0|$, where E_0 is the initial energy. Unlike its counterpart in Euclidean space, this function drifts over time. Unfortunately, the Euler–Cromer is not as stable with stereographic projection coordinates, and the reader can easily demonstrate that the Candy-Rozmus algorithm does not perform any better in terms of stability and conservation of energy. Note that the angular variable is much simpler to integrate in this case for a number of reasons. Firstly, the metric tensor is simply a constant. Secondly, the potential we have used is nonconfining. If the energy is above the barrier, the motion of the particle, as it travels around the ring, produces values of ξ that approach minus infinity and return from positive infinity to repeat the process again, or vice versa, depending on the initial conditions. Boundaries on the values of the projections would have to be used to prevent numerical overflows. We urge the reader to modify the program and perform the integration with the angular variable.

10.25   The Verlet Algorithm for Manifolds

While the symplectic nature of Hamilton's equations, which are invariant under coordinate transformations, can be exploited to produce split operator integrators in manifolds, the stability of the resulting algorithms is not guaranteed. One should try several coordinate systems to find an efficient and stable implementation. There are good reasons that compel us to state this fact. For example, consider the split operator formalism of the last section. Inspection of Equation 10.289 reveals that when the metric tensor is not diagonal, one ends up with ratios of the component of the canonical momentum vector. These could accidentally go to zero during the trajectory run. That remote possibility becomes a certainty when the system integrated contains a dissipative source. At the end of a $T = 0$ Brownian integration, one ends up with the effective inverse mass in Equation 10.289, which is computed with several terms that are undetermined forms (0/0). These will spell numerical trouble.

It has been tempting for many researchers to find coordinates that can turn the metric tensor into a constant. This is possible to do in two cases, for the rotation in a ring, and for inertia ellipsoids mapped with angular variables. In the latter, the metric tensor is constant only approximately if the top is asymmetric. Consequently, one neglects the precession of the body fixed axes. Remarkably, the quantum propagator in the body fixed axes converges to first order to the right answer. Precession effects correct the propagation to second order. Therefore, to extend the accelerated methods, like the reweighted random series in inertia ellipsoids, it is necessary to use stereographic projections and to deal directly with the spacial dependence of the metric tensor from the outset.

One approach to fix the problem of undetermined forms in classical dynamic simulations with Equations 10.293 and 10.294, is to diagonalize the Hessian metric locally at every step. Here, we consider a simpler alternative. We begin with the Euler–Lagrange equations,

$$\frac{d}{dt}\left(\frac{\partial \mathcal{L}}{\partial \dot{q}^{\upsilon}}\right) - \left(\frac{\partial \mathcal{L}}{\partial q^{\upsilon}}\right) = 0. \tag{10.301}$$

We have shown that these yield,

$$g_{\mu\upsilon}\ddot{q}^{\mu} + \frac{1}{2}\left(\partial_{\upsilon}g_{\mu\beta}\right)\dot{q}^{\mu}\dot{q}^{\beta} + \partial_{\upsilon}V = 0. \tag{10.302}$$

The first two terms in Equation 10.302 result from the time derivative of the partial of $\mathcal{L}$ with respect to $\dot{q}^{\upsilon}$. At this point, rather than rearranging the derivatives of the metric in terms of the connection coefficients, we rearrange Equation 10.302 to read,

$$g_{\mu\upsilon}\ddot{q}^{\mu} = -\partial_{\upsilon}\mathcal{L}^{+}, \tag{10.303}$$

where $\mathcal{L}^{+}$ is the Euclidean Lagrangian,

$$\mathcal{L}^{+} = \frac{1}{2}g_{\mu\upsilon}\dot{q}^{\mu}\dot{q}^{\upsilon} + V. \tag{10.304}$$

Multiplication on both sides of Equation 10.303 by $g^{\mu\upsilon}$ produces the equivalent of Newton's equations in manifolds.

$$\ddot{q}^{\mu} = -g^{\mu\upsilon}\partial_{\upsilon}\mathcal{L}^{+}. \tag{10.305}$$

The simple form of Equation 10.305 suggests the following numerical procedure.

Step 1. Compute $\partial_{\upsilon}\mathcal{L}_{n}^{+}$, $g^{\mu\upsilon}$ and the product $\left(g^{\mu\upsilon}\partial_{\upsilon}\mathcal{L}^{+}\right)_{n}$ for $\upsilon = 1, 2, \ldots, d$ at $(p_{\mu})_{n}, (q^{\mu})_{n}$.

Step 2. For sufficiently small values of Δt, assume that $\left(g^{\mu\upsilon}\partial_\upsilon\mathcal{L}^+\right)_n$ is constant, integrate using the Verlet algorithm to find an update for the position and velocity vector,

$$\left(q^\mu\right)_{n+1} = 2\left(q^\mu\right)_n - \left(q^\mu\right)_{n-1} - \left(g^{\mu\upsilon}\partial_\upsilon\mathcal{L}^+\right)_n \Delta t^2, \qquad (10.306)$$

$$\left(\dot{q}^\mu\right)_{n+1} = \frac{\left(q^\mu\right)_{n+1} - \left(q^\mu\right)_{n-1}}{2\Delta t}. \qquad (10.307)$$

Step 3. Compute the energy $E\left(t\right) = \left(\mathcal{L}^+\right)_{n+1}$, update the simulation time, and repeat from step 1.

The algorithm does not require the explicit computation of the connection coefficients if $\partial_\upsilon\mathcal{L}^+$ is computed by finite differences. For asymmetric inertia ellipsoids mapped with stereographic projection coordinates, there are a large number of nonvanishing connection coefficients. Furthermore, when expressed with stereographic projection coordinates, the equations for all the connection coefficients are formidable. For the purpose of finding minima, the usual point slope approach is sufficient. The drag term can be added to produce a $T = 0$ Brownian dynamics algorithm,

$$\left(\dot{q}^\mu\right)_{n+1} = \left(\dot{q}^\mu\right)_n - \left(g^{\mu\upsilon}\partial_\upsilon\mathcal{L}^+\right)_n \Delta t - \gamma\left(\dot{q}^\mu\right)_n \Delta t. \qquad (10.308)$$

11

Simulations in Curved Manifolds

11.1 Introduction

This chapter deals with the development of stochastic simulation methods in manifolds. In particular, we explore the extension of the following computation techniques:

i. Classical statistical mechanics. The methods are straightforward extensions (for the most part) of the classical Metropolis algorithm and parallel tempering.

ii. Variational Monte Carlo (VMC). For the variational technique, we use the same approach developed in Chapters 6 and 9. Namely, we introduce the same trial wavefunction, which leads to a stochastic algorithm very similar to the classical Metropolis integrator enhanced by parallel tempering.

iii. Diffusion Monte Carlo (DMC) and Green's function propagation for the diffusion equation.

iv. Path integral (PI) simulations.

By now, the reader will have realized that this list closely mirrors the developments in Chapter 9. The order of presentation was thus conceived so for two reasons. Firstly, the list of methods in this chapter may contain objectives that expert researchers could be seeking for applications in their own research. For these seasoned readers, the material in previous chapters should be quite familiar. Other readers should have mastered the material in Chapters 6 through 10 before attempting to work on the material in this chapter. Here, the algorithms are designed to handle the additional complexities associated with the non-Euclidean nature of the spaces. The differential geometry for non-Euclidean spaces would unnecessarily complicate the first exposure to these methodologies. Secondly, the list is sorted in order of increasing difficulty as these are applied to non-Euclidean spaces.

Classical dynamics methods are the simplest to generalize to non-Euclidean spaces, owing to the fact that the equations of Euler–Lagrange, the equations of Hamilton, the symplectic nature of phase space, and the infinitesimal element of phase space are all invariant under changes of coordinates. For the integrals of classical statistical mechanics, however, one needs to include Jacobians, and here the subtleties begin. In Chapter 10, we have briefly discussed the fact that in the computation of Boltzmann distributions in non-Euclidean spaces, one must be careful about the order in which the transformation from Euclidean space, and the imposition of holonomic constraints is carried out. When dealing with internal degrees of freedom, namely atom–atom stretching, bending, and torsions, a different Jacobian is obtained if the order of transformation and application of holonomic constraints is reversed. It is clear that mathematical models of real physical systems obtained with holonomic constraints are nothing more than convenient approximations to the true system. Therefore, the physically meaningful procedure is to first transform the variables, and attain rigidity for stiff degrees of freedom with a limiting process, whereby the force constant is increased to infinite values. In the adiabatic approximation, the classical laws of physics for the unconstrained degrees of freedom are invariant under this limiting procedure. The Jacobian is the square root of the metric tensor for the original Euclidean space remapped by the curvilinear coordinates, even in the infinitely stiff degree of freedom limit. On the other hand, if the transformation of variables is performed from the Euclidean space to the manifold, ignoring the elements of the metric tensor along the stiff degrees of freedom from the outset, one obtains a different Jacobian. In Chapter 10, we have shown that for two-dimensional rotations, the difference between the two Jacobians can be ignored, since it is a mere multiplicative constant. However, when dealing with stretching, bending, and torsions, one must introduce the correct infinite spring constant Jacobian. The incorrect Jacobian would, clearly, impact the computation of our variational ground state approach as well. The same would be true for any property computed using the postulates of quantum mechanics to obtain averages of operators variationally.

A number of additional subtleties arise in the development of quantum methods in manifolds. The Green's function used for the propagation of the diffusion equation must satisfy the same initial condition as it does in any space, including Euclidean ones, namely, as t approaches zero, the kernel of the propagator must approach a Dirac delta distribution. Therefore, it follows that there cannot be Jacobians in the integral equations for propagation in real or imaginary times. This fact simplifies the extension of DMC schemes, introduced in Chapter 6 and 9. However, there are subtleties in the way the diffusion step distribution is generated. There remains spacial dependence in the kinetic part of the diffusion operator, which causes the distributions of diffusion steps to deviate from the standard Gaussian curve. Special methods to generate the correct step distributions have been generated and tested for the rotation of rigid bodies in one, two, and three dimensions.

The canonical quantization of classical variables, such as position and momentum, in non-Euclidean spaces does not yield the correct laws of physics. The proper procedure to derive the Hamiltonian operator in spaces generated by holonomic constraints is to quantize momentum and positions in the original Euclidean space mapped with Cartesian coordinates, then transform the Laplacian operator to curvilinear coordinates. This yields the Laplace-Beltrami operator, introduced in Chapter 10. Finally, the proper constraints are applied, and the rigid degrees of freedom are removed from the expression of the Laplace–Beltrami operator. The procedure is quite general, however, generating the discrete variable representation (DVR) matrix elements of the kinetic energy operator presents some interesting challenges. Generally, the kinetic energy operator contains both first and second order partial derivatives, and furthermore, it depends on space. Using finite differences, it is only possible to write symmetric second order derivative operators. Furthermore, to produce symmetric off-diagonal elements properly, one has to evaluate the space-dependent part of the matrix element at the midpoint between the two lattice points involved. The diagonalization step requires some thought as well. Diagonalizing the Hamiltonian matrix in the DVR produces eigenvalues and eigenvectors that are orthonormal. However, these cannot be the proper solutions to the Schrödinger equation, since the normalization of the eigenvectors is an integral over all space, generally, and this measure over the non-Euclidean space must contain a nontrivial Jacobian. The proper way to handle Jacobians within the DVR scheme is to introduce a generalized eigenvalue problem, which can be solved by first employing a Cholseky decomposition of a positive definite matrix. Thankfully, the Hamiltonian operators in manifolds we consider in this chapter can be easily constructed using angular momentum theory and, for the most part, that is going to be our approach. However, the DVR in manifolds can play an important future role when mixed theories with split operators aimed at dealing with internal high frequency degrees of freedom are developed.

The last topic, as mentioned earlier, is the most difficult. In this chapter, we develop a path integral approach by following the work of DeWitt. The resolution of the identity can be rigorously applied to derive the path integral expression using the same limiting procedure in non-Euclidean spaces, as we did in Chapter 7. However, the short time approximation has to be obtained by expanding the propagator into a power series and keeping all the terms that are first order in Δt. This produces a correction term that is proportional to the Riemann curvature scalar. Recently, Kleinert [515] has shown that the term proportional to the Riemann curvature scalar results from the proper transformation of the PI measure. Kleinert demonstrates that the proper way to produce correct PI formulas in spaces with curvature and torsion is to derive the time-sliced expression in the original Euclidean space, and then transform this into the curvilinear space. From the resulting expression, one simply removes the degrees of freedom that are constrained from the integration. The Jacobian of the transformation of the PI measure produces to first order in Δt, precisely the same curvature term (in the absence of torsion) that DeWitt derives. Unless the Riemann curvature scalar is constant, it cannot be omitted from the expression of the PI, or the latter will not converge.

The problems we treat in this book have spaces with constant or zero curvature, but once more, we caution the reader that releasing some internal degrees of freedom while keeping others constrained generates nonconstant curvatures and must be handled with care [442]. Our DMC methods are not unique, and, presently, only converge to the right answer to first order in manifolds [210]. The methods we present here should be carefully tested in any type of space that we do not include in this chapter. It is not known, for example, how a nonconstant curvature affects the convergence of our DMC schemes. Therefore, generalizations to other kinds of degrees of freedom should be done with the utmost care. Stretching degrees of freedom, for example, have only been tested in a simple instance, and it is not clear if the DMC or PI methods we present in this chapter, converge in general when radial degrees of freedom are involved. Before generalizing any of the methods proposed in this chapter to other types of manifolds, the reader should answer the following questions.

i. Can the manifold be covered with a single map? There are plenty of examples of physical applications requiring manifolds that cannot be covered with a single smooth map. The general approach is to use a set of maps (called an **atlas**) that can be smoothly sown together at the boundaries. As far as we know, there are no known ways to perform stochastic simulations over an atlas, though we have no evidence to exclude the feasibility of such approaches. Strictly speaking, rotations in one, two, and three dimensions, which concern us here, should be covered with atlases as well, since the maps commonly used do not cover all the points in the manifold. To see this, consider for example $\mathbb{R}^3$ covered with the spherical polar coordinate map. The point at the origin cannot be covered by the spherical polar coordinate map, since for $r \to 0$ there are an infinite number of angles, θ and ϕ. Consequently, the onto property of the map is violated at the origin, and that point cannot be covered. A similar argument applies for the $\mathbb{S}^1$ mapped with θ. The point $\theta = 0$, and $\theta = 2\pi$ coincide and must be excluded from the map. Clearly, the north pole for the n-sphere cannot be covered by the stereographic projection map either (see Chapter 10). In all the cases included in this chapter, we get away with a single map, since the Riemann measure over the set of points that are not covered, vanishes. For path integral methods in manifolds, it is necessary to verify that the same applies for the Wiener measure. Our approach has been to verify that the PI results in $\mathbb{S}^n$ agree with the results obtained by angular momentum theory. It would be an invaluable contribution to the chemical physics community to have a formal proof of this result.

ii. Is the manifold orientable? An example of a nonorientable manifold is the Möbius strip. In a nonorientable manifold, the formulas we use in this chapter and in Chapter 10 for the Jacobian included in the volume integral, may not apply.

iii. Is the metric tensor integrable in the Wiener sense? The potential cannot contain "bad" singularities for the path integral to converge in Cartesian coordinates. In manifolds, this requirement must be extended to the metric tensor and the logarithm of its determinant.

iv. Is the metric tensor positive definite everywhere in the manifold? Space-like manifolds all have positive definite metrics. Manifolds in special and general relativity are both space-like and time-like, and the Hessian metric is not positive definite. This fact is responsible for some of the complications that are encountered in extending quantum theory to general relativity. In some cases, these difficulties are limited to the classical Jacobians, and it has been possible to write a PI expression for exotic systems that do not have a classical counterpart. One example is the treatment of spin in quantum field theory with Grassman's algebra. All these complications are well outside the scope of this book. Nevertheless, special relativity, the Dirac equation, and quantum field theory in flat spaces do play a role in outstanding problems of chemical interest. Even the simpler Minkosky's space of special relativity does not contain a positive definite metric tensor. The relativistic limit of classical statistical mechanics remains a topic of active research in theoretical physics [5].

v. Is the space torsion free? If the map, Φ, violates the condition in Equation 10.24, or if Φ^{-1} violates the condition in Equation 10.25, then a number of assumptions we make in Chapter 5, 10, 11, and 12 break down. The metric tensor and the connection coefficients are also no longer symmetric, geodesics and autoparallels (lines of shortest distance) are no longer equivalent. Spaces with curvature and torsion find applications in physics, and path integrals in spaces with torsion are possible. The lattice corrections, the classical Jacobians, and a number of results developed in this chapter assume that the metric tensor is symmetric, and consequently, that the space is torsion free.

This list should be enough to convince anyone that extending typical classical and quantum simulation methods to non-Euclidean spaces, or even Euclidean spaces mapped curvilinearly, is not a trivial matter. Our equations may appear universal, but looks can be deceiving. Even if we write a Green's function or a PI propagator expression, their convergence is not guaranteed. Therefore, we introduce quantum applications of our algorithms for a number of simple systems, which can be solved exactly before we move on to molecular clusters. We begin with the particle in a two-sphere of radius R and we move on to ellipsoids of inertia. The space $\mathbb{S}^1$, though monodimensional, has additional complications not mentioned in the list above because it is not simply connected. Despite the number of difficulties that have been discussed, a number of items are missing from the list above. For example, it is not necessary that the manifolds have constant or zero curvature, that the metric tensor is constant, or that the coordinates are orthogonal. These exciting possibilities are the results of recent advances that have motivated a number of research endeavors of relevance to molecular condensed matter research. We write this chapter in the hope that others in the community find the training beneficial and the applications as important as we do.

11.2   The Invariance of the Phase Space Volume

We start this section by revisiting an important result from classical physics. The volume element of phase space is a conserved quantity under a regular change of variables. By "regular" change of variables, we mean those that satisfy the one-to-one and the integrability condition of the map, as elaborated in Chapter 10. In order to derive this result, we have to dig a little deeper into measure theory as it applies to manifolds. The result we need is that the integrand in a n-dimensional manifold is understood to be a n-form. The differential volume element, dq, is constructed as a n-form by taking the antisymmetrized tensor product (called a **wedge product**) among all the dq^μ. For example, the wedge product between two contravariant 1-tensors is

$$(V \wedge W)^{\mu\upsilon} = V^{[\mu}W^{\upsilon]} = V^\mu W^\upsilon - V^\upsilon W^\mu. \tag{11.1}$$

The wedge product between n vectors is constructed as a sum over all possible permutations of their indices, $\mu_1\mu_2\ldots,\mu_n$, with a $+$ for even permutations and a $-$ for odd permutations. The easiest way to simultaneously represent this sum and lower all the indices, is to employ the Levi–Civita symbol introduced in Chapter 4. In $\mathbb{R}^n$ mapped with Cartesian coordinates, we can write,

$$dx = \epsilon_{\mu_1\mu_2\cdots\mu_n}dx^{\mu_1}dx^{\mu_2}\cdots dx^{\mu_n}. \tag{11.2}$$

Now, let us change coordinates $x^{\mu'} \to x^\mu$ on the R.H.S.,

$$dx = \epsilon_{\mu_1\mu_2\cdots\mu_n}\frac{\partial x^{\mu_1}}{\partial x^{\mu'_1}}\frac{\partial x^{\mu_2}}{\partial x^{\mu'_2}}\cdots\frac{\partial x^{\mu_n}}{\partial x^{\mu'_n}}dx^{\mu'_1}dx^{\mu'_2}\cdots dx^{\mu'_n}. \tag{11.3}$$

This is usually abbreviated with,

$$dx = \left|\frac{\partial x^\mu}{\partial x^{\mu'}}\right|dx^{\mu'}, \tag{11.4}$$

where, as explained in Chapter 4, the determinant of the $n \times n$ matrix, J^μ_υ, is,

$$\det \mathbf{J} = \epsilon_{\mu_1\mu_2\cdots\mu_n}\frac{\partial x^{\mu_1}}{\partial x^{\mu'_1}}\frac{\partial x^{\mu_2}}{\partial x^{\mu'_2}}\cdots\frac{\partial x^{\mu_n}}{\partial x^{\mu'_n}} = \left|\frac{\partial x^\mu}{\partial x^{\mu'}}\right|. \tag{11.5}$$

This is the basic explanation for the appearance of the Jacobian in the volume elements when a coordinate change from Cartesian to curvilinear coordinates takes place. Generally, to construct an invariant volume element, it is necessary to multiply the wedge product of differentials by $\sqrt{|g|}$, the square root of the determinant of the metric tensor. The result is written formally as,

$$\sqrt{|g|}\; dx^{\mu_1} \wedge dx^{\mu_2} \wedge \cdots dx^{\mu_n} = \sqrt{|g|}\; dx^{\mu'_1} \wedge dx^{\mu'_2} \wedge \cdots dx^{\mu'_n}. \qquad (11.6)$$

Liouville phase space is the Cartesian product of configuration space, and the associated space of canonical momenta. The set of momenta associated with every point in a manifold is called the **cotangent bundle**. The integrand over a cotangent bundle is understood as a $(0,n)$ tensor. To see how the wedge product of the momenta volume differential transforms, we go back to $\mathbb{R}^n$ mapped with Cartesian coordinates, and we use the inverse of the Levi-Civita symbol,

$$dp - \epsilon^{\mu_1 \mu_2 \cdots \mu_n} dp_{\mu_1}\, dp_{\mu_2} \cdots dp_{\mu_n}. \qquad (11.7)$$

A change in coordinates yields,

$$dp = \epsilon^{\mu_1 \mu_2 \cdots \mu_n} \frac{\partial x^{\mu'_1}}{\partial x^{\mu_1}} \frac{\partial x^{\mu'_2}}{\partial x^{\mu_2}} \cdots \frac{\partial x^{\mu'_n}}{\partial x^{\mu_n}} dp_{\mu'_1}\, dp_{\mu'_2} \cdots dp_{\mu'_n}, \qquad (11.8)$$

since canonical momenta transform like 1-forms. If we combine Equation 11.3 with Equation 11.8, we obtain a profound and far-reaching result,

$$dx^{\mu_1}\, dx^{\mu_2} \cdots dx^{\mu_n} dp_{\mu_1}\, dp_{\mu_2} \cdots dp_{\mu_n} = dx^{\mu'_1}\, dx^{\mu'_2} \cdots dx^{\mu'_n} dp_{\mu'_1}\, dp_{\mu'_2} \cdots dp_{\mu'_n}. \qquad (11.9)$$

Equation 11.9 implies that the phase space volume element does not change under any type of valid transformation of variables, including canonical contact transformations along a trajectory that satisfies Hamilton's equations. This fact is usually expressed with the statement "phase space is incompressible." Note that we could derive Equation 11.6, by integrating the Boltzmann factor in a manifold $\mathbb{M}^d$, assume Equation 11.9 holds, and integrate over the conjugate momenta, p_μ. However, if holonomic constraints are used on the outset and a map is defined, $\Phi : \mathbb{R}^n \to \mathbb{M}^d$, for which $d < n$, then Equation 11.9 does not hold.

11.2.1 Soft and Hard Constraints and Parallel Tempering

The reason one invokes holonomic constraints is to handle degrees of freedom that have disparate time scales. In classical stochastic simulations, high frequency degrees of freedom are responsible for inefficiencies in the random walk. These manifest themselves with reproducibility problems. The typical symptoms are large fluctuations in sensitive quantities like the heat capacity. In some cases, as in clusters or liquids composed of small covalent molecules, the problem is sufficiently severe to warrant holonomic constraints. However, it is important to remember that constraining a degree of freedom is an approximation, which, in principle, can be approached in a controlled way by increasing the force constant along those degrees of freedom toward an infinite value. Let us introduce the Hamiltonian for a system represented with a curvilinear map, $\Phi : \mathbb{R}^n \to \mathbb{M}^d$, i.e., $d = n$.

$$\mathcal{H} = \frac{1}{2} g_{\mathbb{R}^n}^{\mu\upsilon}\, p_\mu p_\upsilon + V\left(q^1, \ldots, q^{n-c}\right) + \sum_{i=n-c}^{n} \frac{1}{2} k_i \left(q^i - q_{\text{eq.}}^i\right)^2. \tag{11.10}$$

In Equation 11.10, we have transformed the variables in $\mathbb{R}^n$ to curvilinear coordinates, we have assumed that the high frequency degrees of freedom can be separated adiabatically (i.e., we neglect **dynamic couplings**), and for convenience, we treat the latter harmonically. To emphasize that the metric tensor for $\mathbb{R}^n$ is mapped curvilinearly, we use the symbol $g_{\mathbb{R}^n}^{\mu\upsilon}$. We need to find the correct Jacobian for classical stochastic simulations in the N, V, T ensemble. Since the phase space is invariant under coordinate changes, for the expectation of a physical property A, we can write

$$\langle A\left(q\right)\rangle = \frac{\displaystyle\int dp \int_{\mathbb{R}^n} dq\, A\left(q\right) \exp\left(-\beta\mathcal{H}\right)}{\displaystyle\int dp \int_{\mathbb{R}^n} dq\, \exp\left(-\beta\mathcal{H}\right)}. \tag{11.11}$$

The integral over the momenta is analytical since, in general,

$$\int_{\mathbb{R}^n} dx\, \exp\left(-A_{\mu\upsilon} x^\mu x^\upsilon\right) = \pi^{n/2}\sqrt{|(\mathbf{A}^{\mu\upsilon})|}. \tag{11.12}$$

$$\int_{\mathbb{R}^n} dx\, \exp\left(-A^{\mu\upsilon} x_\mu x_\upsilon\right) = \pi^{n/2}\sqrt{|(\mathbf{A}_{\mu\upsilon})|}. \tag{11.13}$$

Therefore, we obtain,

$$\langle A\left(q\right)\rangle = \frac{\displaystyle\int dq\, \sqrt{|g_{\mathbb{R}^n}|}\, A\left(q\right) \exp\left(-\beta V - \beta \sum_{i=n-c}^{n} \frac{1}{2} k_i \left(q^i - q_{\text{eq.}}^i\right)^2\right)}{\displaystyle\int dq\, \sqrt{|g_{\mathbb{R}^n}|}\, \exp\left(-\beta V - \beta \sum_{i=n-c}^{n} \frac{1}{2} k_i \left(q^i - q_{\text{eq.}}^i\right)^2\right)}. \tag{11.14}$$

At this point, let us consider two acceptance probabilities. For the conventional Metropolis move, we have,

$$p_{\text{a}}\left(\mathbf{q}'\,|\mathbf{q}\right) = \min\left\{1, \sqrt{\frac{|g_{\mathbb{R}^n}'|}{|g_{\mathbb{R}^n}|}}\, \exp\left\{-\beta\left[V\left(\mathbf{q}'\right) - V\left(\mathbf{q}\right) - \delta\left(\mathbf{q}, \mathbf{q}'\right)\right]\right\}\right\}, \tag{11.15}$$

where,

$$\delta\left(\mathbf{q}, \mathbf{q}'\right) = \sum_{i=n-c}^{n} \frac{1}{2} k_i \left[\left(q^{i\,\prime} - q_{\text{eq.}}^{i\,\prime}\right)^2 - \left(q^i - q_{\text{eq.}}^i\right)^2\right]. \tag{11.16}$$

Following the development of the acceptance probability in Chapter 9, we obtain for swaps,

$$p_a\left(\mathbf{q}'\,|\mathbf{q}\right) = \min\left\{1, \exp\left\{-\left(\beta - \beta'\right)\left[V\left(\mathbf{x}'\right) - V\left(\mathbf{x}\right) + \delta\left(\mathbf{q},\mathbf{q}'\right)\right]\right\}\right\}. \quad (11.17)$$

The first important result is that there are no Jacobians in the acceptance probability for swaps. As the spring constants, k_i, increase linearly, the difference in the fluctuations of the internal degrees of freedom about the equilibrium drops to zero quadratically, and $\delta\left(\mathbf{q},\mathbf{q}'\right)$ tends to zero.

On the other hand, if we had chosen to separate the variables and apply the holonomic constraints first, we would have written the following Hamiltonian,

$$\mathcal{H} = \frac{1}{2}g_{\mathbb{M}^{n-c}}^{\mu\upsilon}p_\mu p_\upsilon + V\left(q^1,\dots,q^{n-c}\right), \quad (11.18)$$

reflecting the fact that we now deal with a subspace of $\mathbb{R}^n$, $\mathbb{M}^{n-c} \subset \mathbb{R}^n$. The momenta p_μ for $n - c < \mu \le n$ vanish, so there are no differences in the dynamics. However, the Boltzmann distribution is affected, since in general,

$$\sqrt{|g_{\mathbb{R}^n}|} \ne \sqrt{|g_{\mathbb{M}^{n-c}}|}. \quad (11.19)$$

The Jacobian on the right of Equation 11.19 is obtained if we assume that Equation 11.9 holds, and we repeat the integration process for Equations 11.11 through 11.14. Frenkel and Smit show [7], for example, that for a bending coordinate ψ, the R.H.S. contains an extra $\sqrt{1 - \left(\cos^2 \psi\right)/4}$ factor.

Now, let us demonstrate that for the rotation of rigid nonlinear tops, the ratio of $\sqrt{|g_{\mathbb{R}^n}|}$ and $\sqrt{|g_{\mathbb{M}^{n-c}}|}$ is a constant independent of the rotation angles and translation coordinates. We return to the map introduced in Chapter 10 between a rigid n-body system and the $\mathbb{R}^3 \otimes \mathbb{I}^3$ manifold,

$$\mathbf{r}_i = \mathbf{r}_C + \mathbf{R}\,\mathbf{r}_i^{(\mathrm{BF})}, \quad (11.20)$$

where $\mathbf{R}$ is the 3×3 matrix in Equation 10.182, and $\mathbf{r}_C = \left(x_C, y_C, z_C\right)$. The set $\mathbf{r}_i^{(\mathrm{BF})}$ is the set of coordinates for the reference body fixed frame. These coordinates do not depend on the orientation angles and the center of mass coordinates by definition. Therefore, they can only depend on the $3n - 6$ internal degrees of freedom. The Jacobian elements associated with such internal degrees of freedom are necessary to construct the expression for $g_{\mathbb{R}^n \mu\upsilon}$,

$$J_\mu^{3(i-1)+\kappa} = [\mathbf{R}]_{\kappa'}^\kappa\,\partial_\mu x_{(\mathrm{BF})}^{3(i-1)+\kappa'} \quad (1 \le \kappa, \kappa' \le 3,\ 1 \le i \le n,\ \mu > 6). \quad (11.21)$$

The Roman index i is the atom label, and the Greek letter κ, κ' are used as the covariant and contravariant labels in the $\mathbb{R}^3$ space associated with atom i. This notation is identical to that used in Chapter 10. The general form of the metric tensor elements for the $\mu > 6$, $\upsilon > 6$ case is,

$$g_{\mathbb{R}^n \mu\upsilon} = \sum_{i=1}^{n} m_i \left[\partial_\mu \mathbf{r}_i^{(\mathrm{BF})}\right]^T \partial_\upsilon \mathbf{r}_i^{(\mathrm{BF})}, \quad (11.22)$$

where the orthogonality of the rotation matrix $\mathbf{R}$ is used. The expression in Equation 11.22 does not depend on the Euler angles and $\mathbf{r}_C$. This implies that in the body fixed frame (where the upper 6×6 block of $g_{\mathbb{R}^n \ \mu\upsilon}$ is diagonal), $\sqrt{|g_{\mathbb{R}^n}|}$ is a constant if all the internal degrees of freedom are constrained. Therefore, for a rigid top, or clusters of rigid tops, the subtle difference between $\sqrt{|g_{\mathbb{R}^n}|}$ and $\sqrt{|g_{\mathbb{M}^{n-c}}|}$ can be ignored. This result parallels what we have shown in Chapter 10 for linear rotors.

11.3 Variational Ground States

Now, our attention shifts toward the development of a stochastic algorithm for quantum simulations on a general space-like orientable manifold $\mathbb{M}^n$ covered with a single smooth map. It is possible to write a generalization of the Schrödinger equation in manifolds of this kind,

$$\left(-\frac{\hbar^2}{2}\nabla^2 + V \right) \psi\left(q, t\right) = i\hbar \frac{\partial \psi\left(q, t\right)}{\partial t}, \tag{11.23}$$

where ∇^2 is the Laplace–Beltrami operator. To properly derive the Schrödinger equation in manifolds, it is necessary to begin with the Euclidean space that contains the manifold $\mathbb{R}^n$, remap it with curvilinear coordinates, and apply the holonomic constraints in that order. As mentioned earlier, our applications focus on manifolds created with the parameters of the rotation group. These manifolds can be covered by a single map up to a single point. Therefore, care must be taken to ensure that the Wiener measure vanishes over those points. For single rotors, we can make use of angular momentum theory and compare these diagonalization results against those obtained by stochastic simulations. This work is necessary before we can expand the applications of our methods to clusters, solids, and liquids composed of rigid rotors. The rest of this chapter is organized following the process explained above. We develop the necessary theoretical tools, and then we test each of them sequentially using a hierarchy of increasingly complex systems.

Once the issue of finding the correct classical Jacobian $\sqrt{g}$ is resolved, Metropolis and parallel tempering can be employed to perform variational simulations using any type of trial function and performing the configuration integral;

$$\langle A\left(q\right) \rangle = \frac{\displaystyle\int_{\mathbb{M}^n} dq \, \sqrt{g} \, A\left(q\right) \, |\Psi_{\mathrm{T}}\left(\beta, q\right)|^2}{\displaystyle\int_{\mathbb{M}^n} dq \, \sqrt{g} \, |\Psi_{\mathrm{T}}\left(\beta, q\right)|^2}, \tag{11.24}$$

where β represents the set of optimized parameters. We continue to use the simple trial wavefunction, $\Psi_{\mathrm{T}}\left(\beta, q\right) = \exp\left(-\beta V/2\right)$, introduced in Chapter 6.

The local energy estimator becomes,

$$
\Psi_T^{-1} \, \hat{\mathcal{H}} \, \Psi_T = V - \frac{\hbar^2}{2} \left\{ \frac{\beta^2}{4} g^{\mu\upsilon} \partial_\mu V \partial_\upsilon V - \frac{\beta}{2} g^{\mu\upsilon} \partial_\mu \partial_\upsilon V \right.
$$

$$
\left. - \frac{\beta}{2} \left\{ g^{\mu\upsilon} \left[\partial_\mu \ln \sqrt{\det(g)} \right] + \left(\frac{\partial}{\partial x^\mu} g^{\mu\upsilon} \right) \right\} \partial_\upsilon V \right\}. \tag{11.25}
$$

The mass is incorporated into the Hessian metric tensor in Equation 11.25. The estimator in Equation 11.25 replaces $A(q)$ in Equation 11.24, and the resulting integral yields an estimate of the ground state energy for the given parameter, $\beta \langle A(q) \rangle_\beta$. The optimization of β through the evaluation of $\langle A(q) \rangle_\beta$ over a set of parameter values proceeds exactly the same way as in Cartesian coordinates. As explained in Chapter 9, one can always refine Ψ_T by adding more parameters. Our main interest in the variational wavefunction is to guide DMC, which can correct the variational wavefunction with a relatively small amount of diffusions and branching steps.

11.4 Diffusion Monte Carlo (DMC) in Manifolds

In imaginary time, $t = i\tau$, the Schrödinger equation with a potential relative to some reference energy, E_{ref}, is,

$$
-\hbar \frac{\partial \psi}{\partial \tau} = \frac{\hbar^2}{2} \nabla^2 \psi - (V - E_{\text{ref}}) \, \psi. \tag{11.26}
$$

Equation 11.26 is the starting point for developing DMC methods in $\mathbb{M}^n$. However, we face two additional challenges when attempting to extend DMC to curved manifolds. The first challenge is abstract in nature. When we consider the integral representation of Equation 11.26,

$$
\psi(q, \tau + \Delta\tau) = \int_{\mathbb{M}^n} G(q, q', \Delta\tau) \, \psi(q', \tau) \, dq', \tag{11.27}
$$

we note that the integration extends over the entire manifold. However, this integral is of a different mathematical nature, since it does not require Jacobians. To see this, consider that $G(q, q', \Delta\tau)$ satisfies both the time-dependent Schrödinger equation in Equation 11.26, as well as the initial condition,

$$
\lim_{\Delta\tau \to 0} G(q, q', \Delta\tau) = \delta(q - q'). \tag{11.28}
$$

The Dirac delta distribution in a manifold must be defined in a manner similar to its Cartesian counterpart, namely, it must satisfy the following requirement:

$$
\int_{\mathbb{M}^n} \delta(q - q') \, \psi(q', \tau) \, dq' = \psi(q, \tau), \tag{11.29}
$$

as inserting Equation 11.28 into Equation 11.27 must yield back $\psi(q, \tau)$.

The second challenge is practical in nature. Equation 11.27 can be solved by approximating the kernel locally. To first order in $\Delta\tau$, we can write,

$$G\left(q, q', \tau\right) \approx \exp\left(-\frac{g_{\mu\upsilon}\, \Delta q^{\mu}\, \Delta q^{\upsilon}}{2\Delta\tau}\right) \exp\left[-\left(V - E_{\text{ref}}\right)\Delta\tau\right]. \qquad (11.30)$$

To simulate the population evolution according to Equation 11.30, one must create an algorithm that performs random moves for each replica with Δq distributed according to,

$$\exp\left(-\frac{g_{\mu\upsilon}\, \Delta q^{\mu}\, \Delta q^{\upsilon}}{2\Delta\tau}\right), \qquad (11.31)$$

and weight replicas according to the second exponential,

$$\exp\left[-\left(V - E_{\text{ref}}\right)\Delta\tau\right]. \qquad (11.32)$$

The dependence on configuration of the Hessian metric in Equation 11.31 complicates, only slightly, the method for generating steps that we use in Chapter 9. The following scheme is not unique, but has been thoroughly tested on $\mathbb{S}^n$, and has been found to converge to the right answer in all cases.

Step 1: Draw η with random numbers in $(0, 1)$ and compute $\Delta q'$ with

$$\Delta q' = \Delta q + \gamma\left(\eta - 0.5\right), \qquad (11.33)$$

where γ is a constant adjusted to produce a 50% rejection rate.

Step 2: Let q be the configuration of the present replica, compute the acceptance probability for moving the diffusion step from Δq to $\Delta q'$,

$$P(\Delta q \to \Delta q')$$
$$= \min\left\{1, \exp\left[-\frac{g_{\mu\upsilon}\left(q\right)\left(\Delta q^{\mu}\right)'\left(\Delta q^{\upsilon}\right)'}{2\Delta\tau} + \frac{g_{\mu\upsilon}\left(q\right)\Delta q^{\mu}\, \Delta q^{\upsilon}}{2\Delta\tau}\right]\right\}.$$
$$(11.34)$$

Step 3: Move the replica with $q + \Delta q' \to q'$ if the move is accepted, or with $q + \Delta q \to q'$ if it is rejected.

Step 4: Set $\Delta q' \to \Delta q$ if the move is accepted, and repeat from step 1 for all the replicas in the population.

Other possibilities to account for the spatial dependence of the metric tensor in step 2 exist, and a number of them have been tested. All tested approaches use different sensible assumptions about the values of q and q' inside the acceptance expression in step 2, and all yield the same result. The present approach is the simplest found in the course of our investigations. Recently, we have shown that it is relatively straightforward to derive the importance sampling enhancement in manifolds using Green's function propagators,

$$\exp\left(-\frac{g_{\mu\upsilon}\,\Delta q^\mu \Delta q^\upsilon}{2\Delta\tau}\right) \frac{\psi_T\left(q+\Delta q\right)}{\psi_T\left(q\right)} \exp\left[-\left(V-E_{\text{ref}}\right)\Delta\tau\right], \tag{11.35}$$

and the equivalent of Schmoluchowski's equation in $\mathbb{M}^n$,

$$-\hbar\frac{\partial f}{\partial\tau} = \frac{\hbar^2}{2}\left(\nabla^2+W_T\right)f - \hbar^2\nabla^\upsilon F_\upsilon^{(Q)} - \left(E_L-E_T\right)f, \tag{11.36}$$

where $\nabla^\upsilon = g^{\mu\upsilon}\partial_\mu$ is the gradient with the raised index, and

$$W_T = \frac{2}{\psi_T\sqrt{g}}\left(\partial_\mu\left[\psi_T\right]\right)\partial_\upsilon\left[g^{\mu\upsilon}\sqrt{g}\right], \tag{11.37}$$

is an added geometric modification of the diffusion operator. W_T is independent of the diffusion steps and vanishes when generating steps from the distribution, according to the strategy we implement here. The ground state energy for guided DMC using Green's function propagators or the Schmoluchowski equation in $\mathbb{M}^n$, is obtained by averaging the estimator in Equation 11.25 over the population.

11.5 The Path Integral in Space-Like Curved Manifolds

The earliest work that extended the Feynman path integral formula to non-Euclidean spaces is that of DeWitt. To derive the path integral in non-Euclidean manifolds, one needs to modify two theorems from Hilbert space theory in the position representation. The first is the closure of the position basis set. In non-Euclidean spaces, this is,

$$1 = \int_\infty^\infty dq\,[g\left(q\right)]^{1/2}\,|q\rangle\,\langle q|. \tag{11.38}$$

The second theorem is the invariance of the trace of a matrix under unitary transformation. This needs modification in the position representation also,

$$\sum_{n=0}^\infty \langle n|\,\hat{O}\,|n\rangle = \int_{-\infty}^\infty dq\,[g\left(q\right)]^{1/2}\left\langle q\left|\hat{O}\right|q\right\rangle. \tag{11.39}$$

The partition function is,

$$Q = \sum_{n=0}^\infty \exp\left(-\beta E_n\right)$$

$$= \sum_{n=0}^\infty \langle n|\exp\left(-\beta\hat{H}\right)|n\rangle = \int_{-\infty}^\infty dq\,[g\left(q\right)]^{1/2}\,\langle q|\exp\left(-\beta\hat{H}\right)|q\rangle. \tag{11.40}$$

If the imaginary-time interval $(0, \beta\hbar)$ is subdivided into N subintervals and the closure of the position basis is inserted $N - 1$ times then,

$$\langle q | \exp\left(-\beta\hat{H}\right) | q \rangle$$

$$= \int_{-\infty}^{\infty} dq_1 \, [g(q_1)]^{1/2} \int_{-\infty}^{\infty} dq_2 \, [g(q_2)]^{1/2} \cdots \int_{-\infty}^{\infty} dq_{N-1} \, [g(q_{N-1})]^{1/2}$$

$$\times \langle q | \exp\left(-\beta\hat{H}/N\right) | q_1 \rangle \langle q_1 | \exp\left(-\beta\hat{H}/N\right) | q_2 \rangle$$

$$\cdots \langle q_{N-1} | \exp\left(-\beta\hat{H}/N\right) | q \rangle . \tag{11.41}$$

The high temperature density matrix element is approximated by the expression,

$$\langle q_{i-1} | \exp\left(-\beta\hat{H}/N\right) | q_i \rangle \approx \left(\frac{m}{2\pi\hbar^2\beta}\right)^{D/2}$$

$$\times \exp\left\{-\frac{N}{2\beta\hbar^2} g_{\mu\upsilon}(q_{i-1}) \, \Delta q_i^\mu \, \Delta q_i^\upsilon - \frac{\beta}{N} V(q_{i-1}) - \frac{\beta}{N} \Delta V_{\text{DeWitt}}\right\}, \tag{11.42}$$

where,

$$\Delta q^\mu = q_i^\mu - q_{i-1}^\mu. \tag{11.43}$$

The last term in the argument of the exponential involves the curvature $\mathcal{R}$. In the formulation developed by DeWitt, the correction term,

$$\Delta V_{\text{DeWitt}} = \frac{\hbar^3}{6}\mathcal{R}(q_{i-1}), \tag{11.44}$$

is derived by forcing the action to agree through second order in the time increment. One expands $\langle q_{i-1} | \exp\left(-\beta\hat{H}\right) | q_i \rangle$ about q_{i-1}, in powers of imaginary time, keeping all terms that are first order in $\beta\hbar$. The resulting expression is shown to satisfy Bloch's equations up to that order, provided the DeWitt correction is included in the action [755].

Alternative lattice definitions (i.e., expanding the action at different points in the interval $[q_{i-1}, q_i]$) lead to different effective potentials. An alternative lattice expansion and effective potential is Weyl's [750]. The expansion of $\langle q_{i-1} | \exp\left(-\beta\hat{H}\right) | q_i \rangle$, is at the midpoint,

$$\bar{q}_i^\mu = \frac{q_{i-1}^\mu + q_i^\mu}{2}. \tag{11.45}$$

The expansion is shown to satisfy Bloch's equations up to first order if a different correction to the action is included. The following equation is used to obtain the correction term for midpoint expansions:

$$\Delta V_{\text{Weyl}} = \frac{\hbar^2}{2}\left(g^{\mu\upsilon}\Gamma_{\mu\lambda}^\kappa\Gamma_{\upsilon\kappa}^\lambda - \mathcal{R}\right). \tag{11.46}$$

It is temping to consider the terms in Equations 11.44 and 11.46 as effective potentials, however, Kleinert [515] has demonstrated that these correction terms are the result of transforming the measure of path integration from a parent flat space mapped with Cartesian coordinates to the manifold. Kleinert shows that the quantum corrections to the Wiener measure Jacobian for the prepoint q_{i-1} and the postpoint q_i are identical. Therefore, ΔV_{DeWitt} and ΔV_{Weyl} are quantum corrections to the Jacobian for the Wiener measure in manifolds. Without these terms, the path integral on the manifold does not converge, unless they happen to be constant.

The following notation,

$$W_N(q, q_1, \ldots, q_{N-1}, \beta)$$
$$= \left\{ \prod_{i=1}^{N} [g(q_i)]^{1/2} \right\} \left(\frac{m}{2\pi\hbar^2\beta} \right)^{Nd/2} \exp\left\{ -\mathcal{S}_N(q, q_1, \ldots, q_{N-1}, \beta) \right\}, \quad (11.47)$$

where d is the dimension of the manifold, and,

$$\mathcal{S}_N(q, q_1, \ldots, q_{N-1}, \beta) = \frac{Nm}{2\beta\hbar^2} \sum_{i=1}^{N} g_{\mu\upsilon} \Delta q_i^\mu \Delta q_i^\upsilon + \frac{\beta}{N} \sum_{i=1}^{N} V_{\text{eff}}(q_{i-1}), \quad (11.48)$$

is helpful to express the following results. V_{eff} is the system's potential with the proper lattice correction. We drop the dependence on the position for the metric tensor, to improve clarity in Equation 11.48, and in all that follows.

The canonical partition function becomes,

$$Q = \lim_{N \to \infty} \int_{-\infty}^{\infty} dq \int_{-\infty}^{\infty} dq_1 \int_{-\infty}^{\infty} dq_2 \cdots \int_{-\infty}^{\infty} dq_{N-1} W_N(q, q_1, \ldots, q_{N-1}, \beta).$$
$$(11.49)$$

The application of the Monte Carlo algorithm for the calculation of ensemble averages should be straightforward in principle. The quantity, $W_N(q, q_1, \ldots, q_{N-1}, \beta)$, functions as a convenient distribution that can be used in conjunction with the importance sampling technique. For estimators that are diagonal in the position representation, $\mathcal{A}(q)$, the ensemble average is

$$\langle \mathcal{A} \rangle =$$

$$\lim_{N \to \infty} \frac{\int_{-\infty}^{\infty} dq \int_{-\infty}^{\infty} dq_1 \int_{-\infty}^{\infty} dq_2 \cdots \int_{-\infty}^{\infty} dq_{N-1} \overline{\mathcal{A}}(q) W_N(q, q_1, \ldots, q_{N-1}, \beta)}{\int_{-\infty}^{\infty} dq \int_{-\infty}^{\infty} dq_1 \int_{-\infty}^{\infty} dq_2 \cdots \int_{-\infty}^{\infty} dq_{N-1} W_N(q, q_1, \ldots, q_{N-1}, \beta)},$$
$$(11.50)$$

where the line above the operator represents the path average,

$$\overline{\mathcal{A}}(q) = \frac{1}{N} \sum_{i=1}^{N} \mathcal{A}(q_i). \quad (11.51)$$

The average potential energy can be calculated directly using the approach just explained. The general form of the T estimator in curved spaces is derived using the usual approach.

$$\langle E \rangle = -\frac{\partial \ln Q}{\partial \beta},$$
(11.52)

$$\langle E \rangle = -\lim_{N \to \infty} \frac{\int_{-\infty}^{\infty} dq \int_{-\infty}^{\infty} dq_1 \int_{-\infty}^{\infty} dq_2 \cdots \int_{-\infty}^{\infty} dq_{N-1} \dfrac{\partial W_N(q, q_1, \ldots, q_{N-1}, \beta)}{\partial \beta}}{\int_{-\infty}^{\infty} dq \int_{-\infty}^{\infty} dq_1 \int_{-\infty}^{\infty} dq_2 \cdots \int_{-\infty}^{\infty} dq_{N-1} W_N(q, q_1, \ldots, q_{N-1}, \beta)},$$
(11.53)

where the derivative can be obtained for the general case,

$$\frac{\partial W_N (q, q_1, \ldots, q_{N-1}, \beta)}{\partial \beta}$$
$$= \left(-\frac{Nd}{2\beta} - \frac{\partial S_N (q, q_1, q_{N-1}, \beta)}{\partial \beta} \right) W_N (q, q_1, \ldots, q_{N-1}, \beta),$$
(11.54)

and the derivative of the action is,

$$\frac{\partial S_N (q, q_1, q_{N-1}, \beta)}{\partial \beta} = -\frac{Nm}{2\beta^2 \hbar^2} \sum_{i=1}^{N} g_{\mu\upsilon} \Delta q_i^{\mu} \Delta q_i^{\upsilon} + \frac{1}{N} \sum_{i=1}^{N} V_{\text{eff}} (q_{i-1}).$$
(11.55)

The final form of the T estimator can be expressed as,

$$\langle E \rangle = \frac{Nd}{2\beta} + \left\langle \frac{\partial S_N (q, q_1, q_{N-1}, \beta)}{\partial \beta} \right\rangle.$$
(11.56)

The T estimator has poor statistics in manifolds, and better energy estimators must be developed to tackle molecular systems. In the next section, we show how a general form for the energy estimator based on the virial theorem can be derived in manifolds.

11.6   The Virial Estimator for the Total Energy

The virial estimator for the path integral over a general manifold can be derived using a linear transformation of variables, which is introduced by Janke *et al.* [618].

$$s^k = \lambda q^k.$$
(11.57)

The canonical partition function in the rescaled system is,

$$Q = \lim_{N \to \infty} \int_{-\infty}^{\infty} ds \int_{-\infty}^{\infty} ds_1 \int_{-\infty}^{\infty} ds_2 \cdots \int_{-\infty}^{\infty} ds_{N-1} \lambda^{Nd} W_N (s, s_1, \ldots, s_{N-1}, \beta).$$
(11.58)

If we rearrange the expression for $W_N(s, s_1, \ldots, s_{N-1}, \beta)$, to bring the square root of the metric into the argument of the exponential function, and we set the parameter λ^2 equal to β/m, we can write,

$$\lambda^{Nd} W_N(s, s_1, \ldots, s_{N-1}, \beta) = \left(\frac{1}{2\pi\hbar^2}\right)^{Nd/2} \exp\left\{-\mathcal{S}'_N(s, s_1, \ldots, s_{N-1}, \beta)\right\}, \tag{11.59}$$

and the argument of the exponential is,

$$\mathcal{S}_N(s, s_1, \ldots, s_{N-1}, \beta)$$
$$= \frac{N}{2\hbar^2} \sum_{i=1}^{N} g_{\mu\upsilon} \Delta q_i^\mu \Delta q_i^\upsilon - \frac{1}{2} \sum_{i=1}^{N} \ln[g(s_i)] + \frac{\beta}{N} \sum_{i=1}^{N} V_{\text{eff}}(s_{i-1}). \tag{11.60}$$

Note that the β dependence of the partition function is contained in the potential if the metric tensor is constant, as happens with Cartesian coordinates. In general, the expression for the "virial" estimator contains derivatives of the metric tensor, as well as derivatives of the potential energy. If we take the derivative of the partition with respect to β in the rescaled coordinate system, we derive,

$$\langle E \rangle = \langle \partial_\beta \mathcal{S}_N \rangle. \tag{11.61}$$

The chain rule gives,

$$\partial_\beta V_{\text{eff}}(s_i) = \frac{\partial V_{\text{eff}}}{\partial s^\upsilon} \frac{\partial s^\upsilon}{\partial \beta} = \frac{1}{2\beta} s^\upsilon \partial_\upsilon V_{\text{eff}}. \tag{11.62}$$

Similar expressions are obtained by the two terms containing the metric tensor elements, and its determinant,

$$\partial_\beta g_{\mu\upsilon}(s_i) = \frac{1}{2\beta} s^\kappa \partial_\kappa g_{\mu\upsilon}, \tag{11.63}$$

$$\langle E \rangle = \left\langle \frac{1}{N} \sum_{i=1}^{N} V_{\text{eff}}(s_i) + \frac{1}{2N} \sum_{i=1}^{N} s_i^\upsilon \partial_\upsilon V_{\text{eff}}(s_i) + \mathcal{D}(s, s_1, \ldots, s_{N-1}, \beta) \right\rangle. \tag{11.64}$$

$\mathcal{D}$ is a geometric term,

$$\mathcal{D}(s, s_1, \ldots, s_{N-1}, \beta) = \frac{N}{4\beta\hbar^2} \sum_{i=1}^{N} s_{i-1}^\kappa \partial_\kappa g_{\mu\upsilon} \Delta q_i^\mu \Delta q_i^\upsilon - \frac{1}{4\beta} \sum_{i=1}^{N} g^{-1} s_i^\upsilon \partial_\upsilon g(s_i). \tag{11.65}$$

Using the chain rule, it can be shown that this expression is equivalent to that evaluated with the unscaled coordinates,

$$\langle E \rangle = \left\langle \frac{1}{N} \sum_{i=1}^{N} V_{\text{eff}}(q_i) + \frac{1}{2N} \sum_{i=1}^{N} q_i^\upsilon \partial_\upsilon V_{\text{eff}}(q_i) + \mathcal{D}(q, q_1, \ldots, q_{N-1}, \beta) \right\rangle. \tag{11.66}$$

$\mathcal{D}$ is now,

$$\mathcal{D}(q, q_1, \ldots, q_{N-1}, \beta) = \frac{N}{4\beta\hbar^2} \sum_{i=1}^{N} q_{i-1}^\kappa \partial_\kappa g_{\mu\upsilon} \, \Delta q_i^\mu \, \Delta q_i^\upsilon - \frac{1}{4\beta} \sum_{i=1}^{N} g^{-1} q_i^\upsilon \partial_\upsilon g(q_i).$$

$$(11.67)$$

To derive the virial theorem in Cartesian coordinates, it is assumed that the system remains bound. In that case, the second temporal derivative of the scalar moment of inertia, $m\delta_{\mu\upsilon}x^\mu x^\upsilon$, has a vanishing ensemble average. What constitutes a bound system in manifolds is less clear. Often, in molecular physics applications, one encounters nonconfining potentials. This is especially true for rotations. The virial theorem needs to be tested carefully in those cases, and the convergence of the scheme we present here is not guaranteed in general.

11.7   Angular Momentum Theory Solution for a Particle in $\mathbb{S}^2$

At this point, the task before us is to convince ourselves that the theoretical tools developed in the preceding sections of this chapter actually work. Thorough testing is necessary, since the theories are rich with subtleties, such as, e.g., points in the manifold excluded by maps, etc. As it turns out, the simplest non-Euclidean manifold to investigate is the two-sphere. This system is ideal for the developments in this chapter for several reasons. Firstly, it is a simply connected space, unlike its monodimensional counterpart, $\mathbb{S}^1$. Secondly, we can use angular momentum theory to express the Hamiltonian in a convenient matrix representation. In Chapter 5, we treat the physical rigid dipole in a constant electric field. We show that the Hamiltonian,

$$\mathcal{H} = \frac{L^2}{2\mu R^2} + V_0 \cos\theta, \qquad (11.68)$$

can be represented with the following matrix elements:

$$\langle l_1 \, m_1 \, |\mathcal{H}| \, l_2 \, m_2 \rangle = \frac{\hbar^2 l_1 (l_1 + 1)}{2\mu R^2} \delta_{l_1 l_2} \delta_{m_1 m_2}$$
$$+ V_0 \left(\frac{2l_1 + 1}{2l_2 + 1} \right)^{1/2} \langle l_1 0 1 0 | l_1 1 l_2 0 \rangle \langle l_1 - m_1 1 0 | l_1 1 l_2 - m_2 \rangle,$$

$$(11.69)$$

and we provide a program (`exactamts2`), that computes the Hamiltonian matrix and diagonalizes it.

With the eigenvalues (E_k) provided by running `exactamts2`, we have the energies and wavefunctions, and we can obtain all the thermodynamic properties of interest using well-established formulas. For the average energy, one uses,

$$\langle \mathcal{H} \rangle = \frac{\displaystyle\sum_{k=1}^{\infty} E_k \exp\left(-\beta E_k\right)}{\displaystyle\sum_{k=1}^{\infty} \exp\left(-\beta E_k\right)}, \tag{11.70}$$

the average potential is obtained with,

$$\langle V \rangle = \frac{\displaystyle\sum_{k=1}^{\infty} \langle k \,|V|\, k \rangle \exp\left(-\beta E_k\right)}{\displaystyle\sum_{k=1}^{\infty} \exp\left(-\beta E_k\right)}, \tag{11.71}$$

and the heat capacity is,

$$\frac{C_v}{k_\mathrm{B}} = \beta^2 \left[\langle \mathcal{H}^2 \rangle - \langle \mathcal{H} \rangle^2 \right]. \tag{11.72}$$

k is a state ordering quantum number. The matrix elements of V are computed by transforming the matrix,

$$V_{ij} = V_0 \left(\frac{2l_1 + 1}{2l_2 + 1} \right)^{1/2} \langle l_1 010 \,| l_1 1 l_2 0 \rangle \langle l_1 - m_1 10 \,| l_1 1 l_2 - m_2 \rangle, \tag{11.73}$$

into the energy representation

$$\langle k \,|V|\, k \rangle = \left(\mathbf{T}^\dagger \mathbf{V} \mathbf{T} \right)_{kk} = \sum_{i=1}^{N} \sum_{j=1}^{N} T_{ik} V_{ij} T_{jk}, \tag{11.74}$$

where $\mathbf{T}$ diagonalizes $\mathbf{H}$.

To `exactamts2` from Chapter 5, we append the following snippet of code (at the end of the main function) for the calculations of exact thermodynamic properties for this system.

```
c Calculation of  the potential energy diagonal elements
c in the energy representation
      do i=1,N
      vk(i) = 0.d0
       do k=1,N
       do l=1,N
       vk(i) = vk(i) + z(k,i)*v(k,l)*z(l,i)
       enddo
       enddo
      enddo
```

```
c Calculation of the thermal averages (beta = 1/KT in atomic units)
      beta = 1.0
      do i=1,80
        sumz = 0.d0
        sumh = 0.d0
        sumv = 0.d0
        sumh2 = 0.d0
        do j=1,N
          ej= exp(-beta*d(j))
          sumz = sumz + ej
          sumh = sumh + d(j)*ej
          sumv = sumv + vk(j)*ej
          sumh2 = sumh2 + d(j)*d(j)*ej
        enddo
        avh = sumh/sumz
        avv = sumv/sumz
        avh2 = sumh2/sumz
        cv = beta*beta*(avh2 - avh*avh)
        write(8,999) beta,avh,avv,cv
        beta = beta + 1.0
      enddo
999   format(4f20.7)
```

Convergence of the averages with basis size must be checked carefully, and it should be uniform with β. In Figure 11.1, we display the average potential $\langle V_0 \cos\theta \rangle$, for the case, $V_0 = 1$ hartree, $R = 1$ bohr, and a mass of

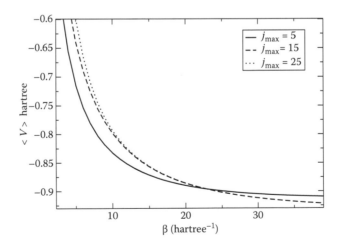

FIGURE 11.1
The average potential for a particle in $\mathbb{S}^2$ with $V_0, R = 1, m = 207$, all in atomic units computed with angular momentum theory.

207 a.u. As Figure 11.1 indicates, higher values of temperature (small values of β) require a larger number of bases. $j_{\max}$ is the largest value of the orbital angular momentum quantum number involved (`mjmx` in `exactamts2`). With diagonalization, the averages converge more slowly at smaller values of β. This numerical behavior is opposite to the PI convergence behavior, where larger and larger values of the Trotter number are needed to reach convergence for the larger values of β.

11.8 Variational Ground State for $\mathbb{S}^2$

Now, let us consider the stereographic projections for $\mathbb{S}^2$. In Chapter 10, we define the coordinate transformation,

$$x = \frac{4R^2\xi^1}{\sigma + 4R^2}, \qquad y = \frac{4R^2\xi^2}{\sigma + 4R^2}, \tag{11.75}$$

where $\sigma = \left(\xi^1\right)^2 + \left(\xi^2\right)^2$, and we derive the metric tensor,

$$g_{\mu\upsilon} = \frac{\left(4R^2\right)^2 m}{\left(\sigma + 4R^2\right)^2}\begin{pmatrix} 1 & 0 \\ 0 & 1 \end{pmatrix}. \tag{11.76}$$

The potential is simply z/R, therefore,

$$V\left(\xi^1, \xi^2\right) = V_0 \frac{\sigma - 4R^2}{\sigma + 4R^2}. \tag{11.77}$$

A plot of the potential surface is shown in Figure 11.2 for the $R = 1$ case. The following two subroutines compute V, the elements of the gradient,

$$\partial_\mu V\left(\xi^1, \xi^2\right) = 16V_0 \frac{\xi^\mu}{\left(\sigma + 4R^2\right)^2}, \tag{11.78}$$

and the diagonal elements of the Hessian for the $R = 1$ bohr, $V_0 = 1$ hartree case,

$$\partial_\mu^2 V\left(\xi^1, \xi^2\right) = 16V_0 \frac{\sigma + 4R^2 - 4\left(\xi^\mu\right)^2}{\left(\sigma + 4R^2\right)^3}. \tag{11.79}$$

```
subroutine pot(x,v)
implicit real*8 (a-h,o-z)
real*8 x(2)
sigma = x(1)*x(1) + x(2)*x(2)
v = (sigma - 4.d0)/(sigma + 4.d0)
return
end
```

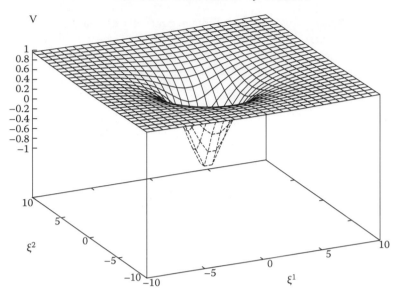

FIGURE 11.2
The potential $\cos\theta$ as a function of ξ^1, for $V_0 = 0, R = 1$.

c

```
      subroutine pot2(x,v,dv,d2v)
      implicit real*8 (a-h,o-z)
      real*8 x(2),dv(2),d2v(2)
      sigma = x(1)*x(1) + x(2)*x(2)
      den = sigma + 4.d0
      den2 = den*den
      den3 = den2*den
      v = (sigma - 4.d0)/den
      dv(1) = 16.d0*x(1)/den2
      dv(2) = 16.d0*x(2)/den2
      d2v(1) = 16.d0*(den - 4.d0*x(1)*x(1))/den3
      d2v(2) = 16.d0*(den - 4.d0*x(2)*x(2))/den3
      return
      end
```

Next, we note that for $\mathbb{S}^2$ space mapped by stereographic projections,

$$\sqrt{g}\, g^{\mu\upsilon} = \begin{pmatrix} 1 & 0 \\ 0 & 1 \end{pmatrix}. \tag{11.80}$$

This result has a number of consequences. If we write the Laplace-Beltrami operator as

$$\nabla^2 = \frac{1}{\sqrt{g}}\, \partial_\mu\, g^{\mu\upsilon}\sqrt{g}\, \partial_\upsilon, \tag{11.81}$$

we immediately derive an operator free of first order derivatives,

$$\nabla^2 = \frac{1}{\sqrt{g}} \delta^{\mu\upsilon} \partial_\mu \, \partial_\upsilon.$$

(11.82)

Equation 11.25 becomes,

$$\Psi_T^{-1} \, \hat{\mathcal{H}} \, \Psi_T = V - \frac{\hbar^2}{2} \left[\frac{\beta^2}{4} \frac{1}{\sqrt{g}} \delta^{\mu\upsilon} \partial_\mu V \partial_\upsilon V - \frac{\beta}{2} \frac{1}{\sqrt{g}} \delta^{\mu\upsilon} \partial_\mu \partial_\upsilon V \right].$$

(11.83)

The subroutine **vare** computes the estimator in Equation 11.83, for the $m = 207$ a.u. case.

```fortran
      subroutine vare(temp,x,ev,v)
      implicit real*8 (a-h,o-z)
      real*8 x(2),dv(2),d2v(2)
c local energy estimator from variational theory
      sum = 0.d0
      dlb = 0.d0
      call metric(x,g)
      call pot2(x,v,dv,d2v)
      do i=1,2
       dlb = dlb + temp*temp*dv(i)*dv(i)/(g*4.d0)
     & - temp*d2v(i)/(g*2.d0)
      enddo
      ev = v - dlb/(2.d0)
      return
      end
```

Its input is β (**temp**) and the configuration ξ^2, ξ^2. **vare** calls the subroutine **metric** that computes $\sqrt{g}$,

```fortran
      subroutine metric(x,g)
      implicit real*8 (a-h,o-z)
      real*8 x(2)
c Returns the square root of the determinant of the metric tensor
      den = (x(1)*x(1) + x(2)*x(2) + 4.d0)
      den2 = den*den
      g = 16.d0*207.d0/den2
      return
      end
```

The evaluation of Equation 11.24 carried out by Metropolis is performed by the program **varS2** given below.

```fortran
      program varS2
      implicit real*8(a-h,o-z)
      parameter (NMAX = 1000)
      real*8 xi(2),xim1(2),xr(4)
```

```fortran
      see=secnds(0.)              !  seed the random number generator
      iseed = int(see/10.)
      call ransi(iseed)
      MW = 1000000  !  number of warm up moves
      M = 1000000
      dM = dfloat(M)
      temp = 10.5d0 ! This is beta in inverse hartree
      deltax = 0.80d0
      do k_t = 1,50 ! loop through some values of beta
      nrej = 0.d0
      sumv = 0.d0
      sume = 0.d0
      do j = 1,2
       xim1(j) = 0.d0
       xi(j) = 0.d0
      enddo
      call pot(xi,vm1)    ! initiate the walk
      call metric(xi,gm1)
      do moves = 1,MW + M
       call ransr(xr,4)    ! Draw three random numbers in [0,1].
       xi(1) = xim1(1) +  deltax *(xr(2)-0.5d0)
       xi(2) = xim1(2) +  deltax *(xr(3)-0.5d0)
       call metric(xi,g)
       call pot(xi,v)
       p = (g/gm1)*exp(-temp*(v-vm1))
       if (xr(4) .lt. p) then  !  accept or reject
        do j=1,2
          xim1(j) = xi(j)   ! accept
        enddo
        vm1 = v
        gm1 = g
       else  ! reject
        nrej = nrej + 1   ! nrej is a counter for the number of rejections
        do j=1,2
         xi(j) = xim1(j)
        enddo
       endif
       if (moves .gt. MW) then ! accumulate data
        call pot(xi,v)
        call vare(temp,xi,ev,v)
        sumv = sumv + v/dM
   sume = sume + ev/dM
       endif
      enddo
c and calculate the percent rejections
      pr = 100.*dfloat(nrej)/dfloat(M+MW)
      write(6,1000) temp,sume,sumv,pr
       call flush(6)
       temp = temp + 1.0
```

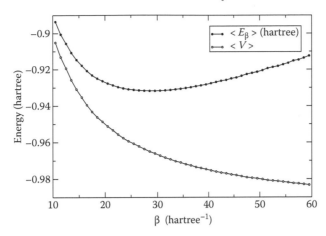

FIGURE 11.3
The variational energy for a body with 207 a.u. of mass, and the thermal average potential for the case $V_0 = 1$ hartree, $R = 1$ bohr.

```
        deltax = deltax - 0.003
        enddo
1000    format(6f14.7)
        end
```

The outermost loop is the loop over β. The linear adjustment in the parameter `deltax` is obtained empirically with a few short runs. The values of `sume`, the variational ground state as a function of the parameter β, and the average thermal potential (`sumv`) from a single run of `varS2`, are graphed in Figure 11.3. With two successive runs, we obtain an estimate of -0.93169 hartree, with the last digit indicating the uncertainty from statistical fluctuations. This ground state energy estimate is in excellent agreement with the ground state energy obtained with angular momentum theory -0.9317083 hartree. Both runs find the lowest value of the local energy for $\beta = 28.5$ hartree^{-1}.

11.9 Diffusion Monte Carlo (DMC) in $\mathbb{S}^2$

The agreement between the variational ground state and the energy obtained with vector space methods is so good that the optimized wavefunction would be a poor test for the DMC algorithm with importance sampling. Unguided DMC for a $V_0 \cos\theta$ potential, on the other hand, may not converge to the right answer because V is not confining, i.e.,

$$\lim_{\xi^1,\xi^2 \to \pm\infty} \frac{\sigma - 4}{\sigma + 4} \to 1 < \infty. \tag{11.84}$$

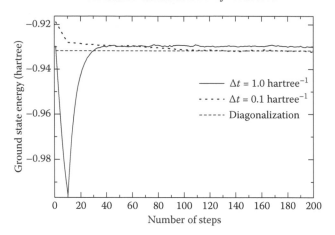

FIGURE 11.4
The ground state energy for a body with 207 a.u. of mass, and the thermal average potential for the case $V_0 = 1$ hartree, $R = 1$ bohr obtained with Diffusion Monte Carlo starting with a population distributed as $\exp(-53.5, V/2)$ and with a target population of 10^5.

To test DMC properly, we resort to using a trial wavefunction with the parameter β slightly off the optimal value of 28.5 hartree^{-1}. With a 10^4 variational run, we produce an initial distribution at $\beta = 53.5$ hartree^{-1}, at which value the local variational energy averages around -0.918 hartree. A program to carry out guided DMC simulations in $\mathbb{S}^2$ can be produced by making a small amount of modifications to `gfis_dmc.f` in Chapter 9. We leave this task as an exercise for the reader. Our results obtained by running DMC with two different values of $\Delta\tau$ are shown in Figure 11.4.

11.10   Stereographic Projection Path Integral in $\mathbb{S}^2$

As a further consequence of the symmetry of the metric tensor, the following relatively simple expression for the high temperature density matrix element is derived,

$$\langle \xi_{i-1}^1, \xi_{i-1}^2 | \exp\left(-\beta\hat{H}/N\right) | \xi_i^1, \xi_i^2 \rangle$$

$$= \left[\frac{m}{2\pi\hbar^2\beta}\right] \exp\left\{ -\left(\frac{Nm}{2\beta\hbar^2}\right) \left[\frac{4R^2}{(\xi_{i-1}^1)^2 + (\xi_{i-1}^2)^2 + 4R^2}\right]^2 \sum_{j=1}^2 \left(\xi_i^j - \xi_{i-1}^j\right)^2 \right.$$

$$\left. - \frac{\beta}{N} V\left(\xi_{i-1}^1, \xi_{i-1}^2\right) \right\}. \tag{11.85}$$

Using the same choices of units (atomic units), mass (207), radius (1.0), and well depth, $V_0 = 1$, as in the previous sections, we derive the following expression for the importance sampling function for the first order time-sliced imaginary-time path integral for $\mathbb{S}^2$ mapped by stereographic projections,

$$W_N = \left\{ \prod_{i=1}^{N} \frac{16}{\left[\left(\xi^1 + w_i^1\right)^2 + \left(\xi^2 + w_i^2\right)^2 + 4 \right]^2} \right\} \left(\frac{207}{2\pi\beta} \right)^N \exp\left(-S_N\right), \quad (11.86)$$

where,

$$S_N = \frac{207N}{2\beta} \sum_{i=1}^{N} \frac{16 \left[\left(w_i^1 - w_{i-1}^1\right)^2 + \left(w_i^2 - w_{i-1}^2\right)^2 \right]}{\left[\left(\xi^1 + w_{i-1}^1\right)^2 + \left(\xi^2 + w_{i-1}^2\right)^2 + 4 \right]^2} + \frac{\beta}{N} \sum_{i=1}^{N} V\left(\xi_{i-1}\right).$$

$$(11.87)$$

The curvature is constant and has been ignored.

The subroutine **getw**, given below, computes W_N from given values of β, the configuration ξ^1, ξ^2, and the auxiliary variables, $w_i^\mu \, i = 1, 2 \ldots, N - 1$, as input. The auxiliary variables, $w_i^\mu \, i = 1, 2 \ldots, N - 1$, define the path (for diagonal operators) as,

$$\xi^\mu\left(u_i\right) = \xi^\mu(0) + w_i^\mu. \quad (11.88)$$

Note that w_0^μ and w_N^μ are zero, and should remain such for any call to getw unless off-diagonal elements of the density matrix are desired.

```
subroutine getw(temp,w,xi,N,NMAX,wi)
implicit real*8 (a-h,o-z)
real*8 w(2,0:NMAX),xi(2),x(2),xm1(2)
sum = 0.d0
prod = 1.d0
dN = dfloat(N)
do i=1,N
 xm1(1) = xi(1) + w(1,i-1)
 xm1(2) = xi(2) + w(2,i-1)
 call metric(xm1,gm1)
 sumw2 = (w(1,i)-w(1,i-1))**2 + (w(2,i)-w(2,i-1))**2
 tkea = (dN/(temp*2.d0))*gm1*sumw2
 sum = sum + tkea
 x(1) = xi(1) + w(1,i)
 x(2) = xi(2) + w(2,i)
 call pot(xm1,v)
 sum = sum + temp*v/dN
 call metric(x,g)
 prod = prod*g
enddo
```

```
wi = prod*exp(-sum)
return
end
```

The subroutine `getw` calls `pot` and `metric`. These are identical to the ones used in the previous section. The program `pis2` performs the Metropolis walk for several values of β and N, using two outer loops.

```
program pis2
parameter (NMAX = 1000)
real*8 wm1(2,0:NMAX),w(2,0:NMAX),xi(2),xim1(2),xr(4)
see=secnds(0.)              !  seed the random number generator
iseed = int(see/10.)
call ransi(iseed)
read(5,*) N
do nQ = 1,1
write(6,*) '# N =',N
MW = 1000000  !  number of warm up moves
M = 1000000
dM = dfloat(M)
temp = 0.5     ! This is beta in inverse hartree
deltax = 0.80
do k_t = 1,50 ! loop through some temperatures
nrej = 0.d0
sumv = 0.d0
sumt = 0.d0
sumc1 = 0.d0
sumc2 = 0.d0
sumc3 = 0.d0
do j = 1,2
 xim1(j) = 0.d0
 xi(j) = 0.d0
 do i=0,N
  w(j,i) = 0.d0
  wm1(j,i) = 0.d0
 enddo
enddo
call getw(temp,w,xi,N,NMAX,wim1)    ! initiate the walk
do moves = 1,MW + M
 call ransr(xr,4)
 jp = 2
 if (xr(1) .lt. 0.5) jp = 1
 imove = (N+1)*xr(2)
 if (imove .eq. 0 .or. imove .eq. N) then
  xi(jp) = xim1(jp) +  deltax *(xr(3)-0.5d0)
 else
```

```
        w(jp,imove) = wm1(jp,imove) + deltax*(xr(3)-0.5d0)
        endif
        call getw(temp,w,xi,N,NMAX,wi)
        p = wi/wim1
        if (xr(4) .lt. p) then  !  accept or reject
        do j=1,2
         xim1(j) = xi(j)           ! accept
         do i=1,N-1
           wm1(j,i) = w(j,i)
         enddo
        enddo
        wim1 = wi
        else           ! reject
        nrej = nrej + 1
        wi = wim1
        do j=1,2
        xi(j) = xim1(j)
        do i=1,N-1
         w(j,i) = wm1(j,i)
        enddo
        enddo
        endif
        if (moves .gt. MW) then  ! accumulate data
         call vestimator(temp,w,xi,N,NMAX,ev)
         sumv = sumv + ev/dM
        endif
        enddo
c and calculate the percent rejections
        pr = 100.*dfloat(nrej)/dfloat(M+MW)
        write(6,1000) temp,sumv,pr
        call flush(6)
        temp = temp + 1.0
        deltax = deltax - 0.001
        enddo
        N = N  + 5
        enddo ! quadrature loop
1000    format(6f12.4)
        end
```

Figure 11.5 contains the path average potential energy computed using the subroutine `vestimator` listed here.

```
        subroutine vestimator(temp,w,xi,N,NMAX,ev)
        implicit real*8 (a-h,o-z)
c path averaged potential energy estimator
        real*8 w(2,0:NMAX),xi(2),x(2)
```

```
sum = 0.d0
ev = 0.d0
dN = dfloat(N)
do i=1,N
 x(1) = xi(1) + w(1,i)
 x(2) = xi(2) + w(2,i)
 call pot(x,v)
 sum = sum + v/dN
enddo
 ev = sum
return
end
```

The results in Figures 11.3 through 11.5 are encouraging. The simplicity of the programs involved in the computation of the ground state energy and canonical average potential energy for a particle in $\mathbb{S}^2$ should have convinced the reader that the PI quantization is attainable using a single stereographic projection map to cover $\mathbb{S}^2$. Nevertheless, much work still needs to be done. The present approaches converge to first order. Path integral applications to real clusters remain formidable. Secondly, the energy estimator remains an issue at this point. We leave it to the reader to discover that, since the potential is not confining, the virial estimator does not converge to the actual energy, and the T estimator is noisy, with the statistical error growing proportionally with the number of Trotter slices N. One possibility to overcome such formidable problems is to develop the random series expansion for the Brownian bridge, derive the reweighted tail approach, and implement the fi-

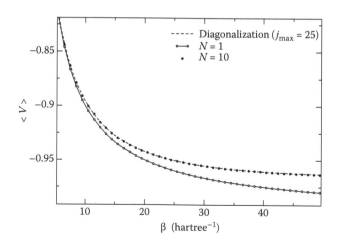

FIGURE 11.5
The canonical average potential energy for a particle in a $\mathbb{S}^2$ manifold. The dashed curve is obtained by diagonalization.

nite difference T estimator, as we did in Chapter 7 for Cartesian coordinates. For the moment, our main concern is to continue testing the approaches to ensure that the methods do converge for a set of manifolds that are of importance to molecular physics, especially as it applies to clusters of rigid bodies. Our next task is to investigate nonlinear rotors.

11.11   Higher Dimensional Tops

Fortunately, the solution of the Schrödinger equation for the symmetric top in a dipole or quadrupole-like field is only slightly more complicated than the solution for linear rotors. It can be shown, using hyperspherical harmonics, that the matrix elements of the Hamiltonian for a spherical top in the presence of a dipole field, $V_0 \cos\theta$, are given by,

$$H_{ll'm_1m_1'm_2m_2'} = \frac{\hbar^2\, l\,(l+1)}{2I}\, \delta_{ll'}\delta_{m_1m_1'}\delta_{m_2m_2'}$$

$$+ V_0\sqrt{\frac{(2l'+1)}{(2l+1)}}\, \langle l' - m_1'\, 1\, 0\, |\, l'\, 1\, l - m_1 \rangle \langle l' - m_2'\, 1\, 0\, |\, l'\, 1\, l - m_2 \rangle. \quad (11.89)$$

Note that there is a new pair of magnetic quantum numbers associated with the third Euler angle. More detail can be found in many angular momentum texts [856–858]. The following program constructs the Hamiltonian for a spherical top, carries out the diagonalization by calling eigen, and calculates the ensemble average of the potential and the energy. The moment of inertia tensor elements are, $I_{xx} = I_{yy} = I_{zz} = 207$ a.u.

```
program exactamts3
implicit real*8 (a-h,o-z)
parameter(MAXN=5000)
real*8 bs(MAXN,MAXN),h(MAXN,MAXN),v(MAXN,MAXN),d(MAXN)
real*8 e(MAXN),z(MAXN,MAXN),vk(MAXN)
i=1
j=1
write(6,*)'Input the maximum l value'
read(5,*)mjmx
N=((mjmx+1)*(2*mjmx+1)*((2.d0/3.d0)*mjmx+1))
write(6,*)'N=',N
do l1=0,mjmx
 do m1=-l1,l1
  do m2=-l1,l1
   do l2=0,mjmx
    do m1p=-l2,l2
     do m2p=-l2,l2
      h(i,j)=0.d0
```

```
         v(i,j)=0.d0
         call racah(l1,1,l2,-m1p,0,m1,r1)
         call racah(l1,1,l2,-m2p,0,m2,r2)
         f=sqrt(dfloat(2*l1+1)/dfloat(2*l2+1))
         vij=f*r1*r2
         if (i.eq.j)then
         h(i,j)=dfloat(l1*(l1+1))/(2.*207)
         endif
         h(i,j)=h(i,j)+vij
         v(i,j)=vij
         i=i+1
         enddo
        enddo
       enddo
      j=j+1
      i=1
     enddo
    enddo
   enddo

   write(6,1007)   (j, j=1,(10))
   do i=1,(10)
   write(6,1008) i, (h(i,j),j=1,(10))
   enddo
   call eigen(MAXN,N,h,d,z,e,ierr)
   write(6,*)'ierr=',ierr
   do k=1,25
   write(6,*)k,d(k)
   enddo
c   Calculation of the potential energy diagonal elements
c   in the energy representation
   do i=1,N
   vk(i)=0.d0
     do k=1,N
       do l=1,N
       vk(i)=vk(i)+z(k,i)*v(k,l)*z(l,i)
       enddo
     enddo
   enddo
c   Calculation of the thermal averages (beta=1/kt in atomic units)
   beta=8.5d0
   do i=1,80
   sumz=0.d0
   sumh=0.d0
   sumv=0.d0
   sumh2=0.d0
     do j=1,N
     ej=exp(-beta*d(j))
     sumz=sumz+ej
```

```
        sumh=sumh+d(j)*ej
        sumv=sumv+vk(j)*ej
        sumh2=sumh2+d(j)*d(j)*ej
        enddo
      avh=sumh/sumz
      avv=sumv/sumz
      avh2=sumh2/sumz
      cv=beta*beta*(avh2-avh*avh)
      write(9,999)beta,avh,avv,(avh-avv),cv
      beta=beta+1.d0
      enddo
999   format(4f20.7)
1006  format(I4,I4,I4,I4,I4,I4,'    ',f10.8)
1007  format('    ',20i10)
1008  format(i4,20f10.5)
      end
```

The program **exactamts3** requires the value of l_{max} as input. This value should be chosen carefully to ensure uniform convergence in the range of inverse temperature β. We find, empirically, that $l_{max} = 14$ creates the largest Hamiltonian matrix that can be handled by a typical desktop. Anything larger is difficult to store in memory as double precision, and the exact diagonalization requires the Lanczos algorithm and sparse matrix storage strategies. In the interest of brevity, we guide the reader through the testing of the time-sliced path integral with the space mapped by stereographic projections only, and we leave a number of other tests as exercises. In Chapter 10, we derive,

$$g_{\mu\upsilon} = \frac{64 l_1}{(\sigma + 4)^2} \begin{pmatrix} 1 & 0 & 0 \\ 0 & 1 & 0 \\ 0 & 0 & 1 \end{pmatrix}, \tag{11.90}$$

for spherical tops mapped by stereographic projection coordinates. Therefore, the treatment of the spherical top is simplified significantly by the stereographic projection approach. The metric tensor in Equation 11.90 can be handled with the following code.

```
      subroutine metric(x,g,det)
      implicit real*8 (a-h,o-z)
      real*8 x(3)
c on input x =   the vector containing the stereographic
c                  projection coordinates
c On output g =  the metric tensor element
c          det = the determinant of the metric tensor
      tiny = 1.d-300
      x12 = x(1)*x(1)
      x22 = x(2)*x(2)
      x32 = x(3)*x(3)
      sg = x12 + x22 + x32
```

```
sgp4 =  sg + 4.d0
sgp42 = sgp4*sgp4
g = 64.d0*207.d0/sgp42
det = g*g*g
return
end
```

Note that `metric` returns both $g_{\mu\upsilon}$ and g, the determinant of the Hessian metric tensor.

The subroutine `getw`, given below, computes the spherical top version of W_N,

$$W_N = \left\{ \prod_{i=1}^{N} \sqrt{g\left(\xi^1, \xi^2, \xi^3, w_1, \ldots, w_{N-1}\right)} \right\} \left(\frac{1}{2\pi\beta}\right)^{3N/2} \exp\left(-S_N\right), \quad (11.91)$$

$$S_N = \frac{N}{2\beta} \sum_{i=1}^{N} \frac{64I\left[\left(w_i^1 - w_{i-1}^1\right)^2 + \left(w_i^2 - w_{i-1}^2\right)^2 + \left(w_i^3 - w_{i-1}^3\right)^2\right]}{\left[\left(\xi^1 + w_{i-1}^1\right)^2 + \left(\xi^2 + w_{i-1}^2\right)^2 + \left(\xi^3 + w_{i-1}^3\right)^2 + 4\right]^2}$$

$$+ \frac{\beta}{N} \sum_{i=1}^{N} V\left(\xi_{i-1}\right), \quad (11.92)$$

and is a straightforward modification of the one used in the previous section. The input variables are β, the configuration ξ^1, ξ^2, ξ^3, and the auxiliary variables w_i^μ $i = 0, 1 \ldots, N$. In Chapter 11, we demonstrate that the curvature is constant, therefore, we ignore it in `getw`.

```
subroutine getw(temp,w,xi,N,NMAX,wi)
implicit real*8 (a-h,o-z)
real*8 w(3,0:NMAX),xi(3),x(3),xm1(3)
sum = 0.d0
prod = 1.d0
dN = dfloat(N)
do i=1,N
 xm1(1) = xi(1) + w(1,i-1)
 xm1(2) = xi(2) + w(2,i-1)
 xm1(3) = xi(3) + w(3,i-1)
 x(1) = xi(1) + w(1,i)
 x(2) = xi(2) + w(2,i)
 x(3) = xi(3) + w(3,i)
 call metric(xm1,gm1,det)
 call metric(x,g,det)
 call pot(x,v)
 c = 0.d0
 do k=1,3
  dk = w(k,i)-w(k,i-1)
```

```
      dk2 = dk*dk
      c = c + dk2*gm1
    enddo
    sum = sum + dN*c/(temp*2.d0) + temp*v/dN
    prod = prod*sqrt(det)
  enddo
  wi = prod*exp(-sum)
  return
  end
```

For the potential energy, using the map developed in Chapter 10 for non-linear tops, it is shown that,

$$\cos\theta = \frac{2\left[16\left(\xi^1\right)^2 + (\sigma - 4)^2\right]}{(\sigma + 4)^2} - 1, \tag{11.93}$$

which can be easily computed with the following subroutine.

```
  subroutine pot(x,v)
  implicit real*8 (a-h,o-z)
  real*8 x(3)
  x12 = x(1)*x(1)
  x22 = x(2)*x(2)
  x32 = x(3)*x(3)
  sg = x12 + x22 + x32
  sgp4 =  sg + 4.d0
  sgp42 = sgp4*sgp4
  sgm4 =  sg - 4.d0
  sgm42 = sgm4*sgm4
  v = 2.d0*(16.d0*x12 + sgm42)/sgp42 - 1.d0
  return
  end
```

The potential energy surface is rather complex when expressed in stereographic projection coordinates. The function V is clearly invariant under the exchange of ξ^2 and ξ^3. Therefore, to explore a projection of the surface, we set $\xi^2 = \xi^3$, and graph the two-dimensional function,

$$f\left(\xi^1, \xi^2\right) = \frac{2\left\{16\left(\xi^1\right)^2 + \left[\left(\xi^1\right)^2 + 2\left(\xi^2\right)^2 - 4\right]^2\right\}}{\left[\left(\xi^1\right)^2 + 2\left(\xi^2\right)^2 + 4\right]^2} - 1. \tag{11.94}$$

The surface in Figure 11.6 reveals two wells, $(0, -\sqrt{2}, -\sqrt{2})$ and $(0, +\sqrt{2}, +\sqrt{2})$, each with a minimum value of V equal to -1 hartree. The graph in Figure 11.6 also reveals that the point $(0, 0, 0)$ is a transition state, with a barrier of 1 hartree.

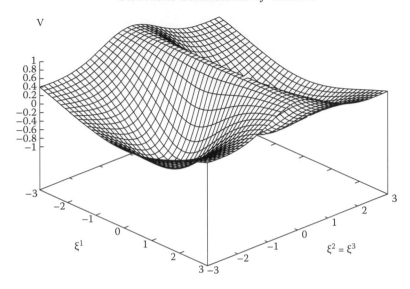

FIGURE 11.6

A slice of the surface $\cos\theta$ in $\mathbb{S}^3$ mapped with stereographic projection coordinates (c.f. Equation 11.94).

Then, program `pis3` is used to carry out the path integral simulation in an inertia ellipsoid space that has spherical symmetry and is mapped with the stereographic projection coordinates derived from the quaternion parameter space. To control possible ergodic sampling problems, parallel tempering is implemented with 30 walkers.

```
c The Stereographic projection path integral
c for spherical tops. The stereographic projection
c map is composed from the quaternions representation.
      program pis3
      implicit real*8 (a-h,o-z)
      parameter (NMAX = 1000)
      parameter (nw = 30)
      real*8 w(3,0:NMAX),xi(3),xr(4)
      real*8 pw(3,0:NMAX,100),pxi(3,100),deltax(100),ptemp(100)
      real*8 sumv(100),sumt(100)! parallel tempering runners
      integer nrej(100),nrejj(100),naj(100)
      see=secnds(0.)
      iseed = int(see/10.)
      call ransi(iseed)
        read(5,*) N
          write(6,*) '# N =',N
        write(6,1001)
      MW = 1000000    ! number of warm up moves
      M = 1000000
```

```
        dM = dfloat(M)
c initialization
      do k = 1,nw   ! loop through some temperatures
      nrej(k) = 0.d0
      nrejj(k) = 0.d0
      naj(k) = 0.d0
      sumv(k) = 0.d0
      sumt(k) = 0.d0
      ptemp(k) = 1.5 + 50.5*dfloat(k-1)/dfloat(nw -1) ! This is beta
      deltax(k) = 7.5d0 - 0.2*dfloat(k)
       do j = 1,3
        pxi(j,k) = 1.2d0
        do i=0,N
         pw(j,i,k) = 0.d0
        enddo
       enddo
       do j=1,3
       xi(j) = pxi(j,k)
         do i=1,N-1
          w(j,i) = pw(j,i,k)
         enddo
       enddo
      temp = ptemp(k)
      call getw(temp,w,xi,N,NMAX,wim1)
      enddo
c the random walk starts here
      do moves = 1,MW + M
      do k=1,nw
       call ransr(xr,1) ! Decide between swap or Metropolis
       if (xr(1) .lt. 0.1d0 .and. k .lt. nw) then
c Swap
           naj(k) = naj(k) + 1     ! number of attempted swaps
           do j=1,3
            xi(j) = pxi(j,k+1)
            do i=1,N-1
             w(j,i) = pw(j,i,k+1)
            enddo
           enddo
           temp2 = ptemp(k+1)
           call getw(temp2,w,xi,N,NMAX,wi4)
           temp1 = ptemp(k)
           call getw(temp1,w,xi,N,NMAX,wi1)
           do j=1,3
            xi(j) = pxi(j,k)
            do i=1,N-1
             w(j,i) = pw(j,i,k)
            enddo
           enddo
           temp2 = ptemp(k+1)
```

```
      call getw(temp2,w,xi,N,NMAX,wi2)
      temp1 = ptemp(k)
      call getw(temp1,w,xi,N,NMAX,wi3)
      p = wi1*wi2/(wi3*wi4)
  call ransr(xr,1)
      if (xr(1) .lt. p) then  !  accept or reject
      do j=1,3
       pxi(j,k) = pxi(j,k+1)                    ! accept
        do i=1,N-1
         pw(j,i,k) = pw(j,i,k+1)
        enddo
      enddo
      do j=1,3
       pxi(j,k+1) = xi(j)                       ! accept
        do i=1,N-1
         pw(j,i,k+1) = w(j,i)
        enddo
      enddo
      else                     ! reject
       nrej(k) = nrej(k) + 1
       nrejj(k) = nrejj(k) + 1
      endif
c Metropolis move
      else
        do j=1,3
          xi(j) = pxi(j,k)
          do i=1,N-1
           w(j,i) = pw(j,i,k)
          enddo
        enddo
        temp = ptemp(k)
        call getw(temp,w,xi,N,NMAX,wim1)
        call ransr(xr,4)
        jp = 3
        if (xr(1) .lt. 0.66666) jp = 2
        if (xr(1) .lt. 0.33333) jp = 1
        imove = (N+1)*xr(2)
        if (imove .eq. 0 .or. imove .eq. N) then
         xi(jp) = xi(jp) +  deltax(k)*(xr(3)-0.5d0)
        else
         w(jp,imove) = w(jp,imove) + deltax(k)*(xr(3)-0.5d0)
        endif
        call getw(temp,w,xi,N,NMAX,wi)
        p = wi/wim1
        if (xr(4) .lt. p) then  !  accept or reject
      do j=1,3
       pxi(j,k) = xi(j)                         ! accept
        do i=1,N-1
         pw(j,i,k) = w(j,i)
```

```
            enddo
          enddo
         else                        ! reject
           nrej(k) = nrej(k) + 1
         endif
        endif   ! Swap - Metropolis
c accumulate data
        if (moves .gt. MW) then
          do j=1,3
            xi(j) = pxi(j,k)
              do i=1,N-1
                w(j,i) = pw(j,i,k)
              enddo
            enddo
            temp = ptemp(k)
            call vestimator(temp,w,xi,N,NMAX,ev)
            sumv(k) = sumv(k) + ev/dM
            call testimator(temp,w,xi,N,NMAX,et)
            sumt(k) = sumt(k) + et/dM
          endif
        enddo
       enddo
c output
       do k=1,nw
         write(6,1000) ptemp(k),sumt(k),sumv(k),deltax(k),nrej(k),
     & nrejj(k),naj(k)
         call flush(6)
       enddo
1000  format(4f12.6,3i8)
1001  format(12(' '),'T',12(' '),'<E>',12(' '),'<V>',12(' '),'%r')
       end
```

The subroutines `testimator` and `vestimator` are used to compute the average energy and the average potential.

```
c  VESTIMATOR  path averaged potential energy estimator)
      subroutine vestimator(temp,w,xi,N,NMAX,ev)
      implicit real*8 (a-h,o-z)
      real*8 w(3,0:NMAX),xi(3),x(3)
      sum = 0.d0
      ev = 0.d0
      dN = dfloat(N)
      do i=1,N
        x(1) = xi(1) + w(1,i)
        x(2) = xi(2) + w(2,i)
        x(3) = xi(3) + w(3,i)
        call pot(x,v)
        sum = sum + v/dN
```

```
      enddo
      ev = sum
      return
      end
c TESTIMATOR the usual T estimator for the total ensemble energy
      subroutine testimator(temp,w,xi,N,NMAX,et)
      implicit real*8 (a-h,o-z)
      real*8 w(3,0:NMAX),xi(3),x(3),xm1(3)
      sum = 0.d0
      et = 0.d0
      dN = dfloat(N)
c the partial of SN with respect to beta is calculated here
      do i=1,N
       xm1(1) = xi(1) + w(1,i-1)
       xm1(2) = xi(2) + w(2,i-1)
       xm1(3) = xi(3) + w(3,i-1)
       x(1) = xi(1) + w(1,i)
       x(2) = xi(2) + w(2,i)
       x(3) = xi(3) + w(3,i)
       call metric(xm1,gm1,det)
       call pot(x,v)
       c = 0.d0
       do k=1,3
        dk = w(k,i)-w(k,i-1)
        dk2 = dk*dk
        c = c + gm1*dk2
       enddo
       sum = sum - dN*c/(temp*temp*2.d0) + v/dN
      enddo
      et = sum + 1.5*dN/temp   ! Now add ND/(2 beta)
      return
      end
```

The averages of the estimated properties in **testimator** and **vestimator** are graphed for several values of β in Figure 11.7. Once more, the agreement between angular momentum theory and the PI results is encouraging. The following will help corroborate the evidence that the methods outlined at the beginning of this chapter do work as expected in $\mathbb{S}^2$, and in inertia ellipsoids.

Exercises

1. Write a subroutine that estimates the energy for a particle in $\mathbb{S}^2$ during a PI simulation with the T estimator. Use the mass of 207 a.u. and $V_0 \cos \theta$ as the potential, with $V_0 = 1$ hartree.

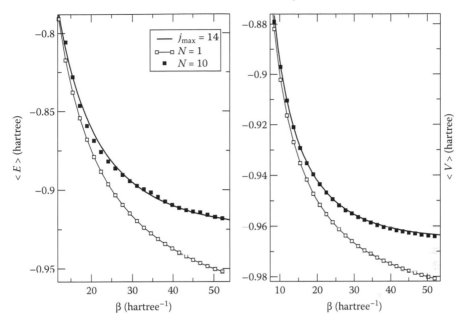

FIGURE 11.7
The energy of a spherical top with $I_{xx} = I_{yy} = I_{zz} = 207$ a.u. subjected to a $\cos\theta$ external field. Left: total energy computed with the T estimator. Right: the average potential computed with the path average estimator.

2. Write a subroutine that computes the energy for a particle in $\mathbb{S}^2$ during a PI simulation using the virial estimator. Compare the energy values with the T estimator. Use the mass of 207 a.u. and the $V_0\sigma$ potential, $V_0 = 1$ hartree.

3. The potential in the previous exercise is confining. Repeat the simulations in the previous exercise with the $V_0\cos\theta$ potential, $V_0 = 1$ hartree, and show that the virial estimator does not converge to the right answer.

4. Write a program that performs guided DMC simulations in $\mathbb{S}^2$, then use it to find the ground state for the particle of mass 207 a.u. subjected to the $V_0\cos\theta$ potential, $V_0 = 1$ hartree.

5. Modify the program varS2 and the guided DMC program in the previous exercise to find the ground state energy of a symmetric top with $I = 207$ a.u. The ground state energy for the system is -0.931708161 hartree. Repeat the variational calculation using parallel tempering to optimize the wavefunction parameter and to generate the initial distribution for the DMC simulation. Simulate for a number of values of I, to search for quasiergodic behavior.

6. For symmetric or asymmetric tops, we can calculate the inertia tensor and the determinant, numerically, using results obtained in Chapter 10.

```fortran
c calculates the moment of inertia
c tensor with the stereographic projection map.
      subroutine metric(x,g,det)
      implicit real*8 (a-h,o-z)
      real*8 x(3),g(3,3),h(3,3),mit(3),gu(3,3)
      integer tau
c on input x  =   the vector containing the stereographic
c                    projection coordinates
c On output g  =  the metric tensor and det its determinant
c          det = the determinant of the metric tensor
c the array h keeps all 9 partial derivative of the 3 Eulerian angles
c with respect to the three stereographic projection coordinates.
c I_x I_y and I_z are represented by the vector mit()
c This is the asymmetric case,
c gu is the unprimed metric tensor,
      mit(1) = 207.d0
      mit(2) = mit(1)    ! Symmetric top case
      mit(3) = 100.d0    ! Prolate
      det = 0.d0
      tiny = 1.d-300
      x12 = x(1)*x(1)
      x22 = x(2)*x(2)
      x32 = x(3)*x(3)
      sg = x12 + x22 + x32
      sgp4 =  sg + 4.d0
      sgp42 = sgp4*sgp4
      sgm4 =  sg - 4.d0
      sgm42 = sgm4*sgm4
      d1 = sqrt(sgp42 -16.d0*x22 -16.d0*x32)
      d2 = sqrt(x22+x32)
      d3 = x22+x32
      d4 = 16.d0*x12+ sgm42
      if (d1 .lt. tiny) STOP 'ERROR D1 is too small to handle'
      if (d2 .lt. tiny) STOP 'ERROR D2 is too small to handle'
      if (d3 .lt. tiny) STOP 'ERROR D3 is too small to handle'
      f1 = 8.d0/d1
      f2 = -2.d0*d2/sgp4
      h(1,1) = f1*x(1)*f2
      h(1,2) = f1*x(2)*(f2 + 1.d0/d2)
      h(1,3) = f1*x(3)*(f2 + 1.d0/d2)
      h(2,1) = (4.d0*(x12-x22-x32) + 16.d0)/d4
      h(2,2) = 8.d0*x(1)*x(2)/d4 + x(3)/d3
      h(2,3) = 8.d0*x(1)*x(3)/d4 - x(2)/d3
      h(3,1) = h(2,1)
      h(3,2) = 8.d0*x(1)*x(2)/d4 - x(3)/d3
      h(3,3) = 8.d0*x(1)*x(3)/d4 + x(2)/d3
```

```
ct = 2.d0*d4/sgp42 - 1.d0
st = 8.d0*d1*d2/sgp42
c2tps2 = mit(1)*st*st + mit(3)*ct*ct
spsi = (4.d0*x(1)*x(3)+x(2)*sgm4)/(d1*d2)
cpsi = (4.d0*x(1)*x(2)-x(3)*sgm4)/(d1*d2)
gu(1,1) = mit(1)*cpsi*cpsi + mit(2)*spsi*spsi
gu(1,2) = (mit(1)-mit(2))*st*spsi*cpsi
gu(1,3) = 0.d0
gu(2,1) = gu(1,2)
gu(2,2) = mit(1)*st*st*spsi*spsi +
& mit(2)*st*st*cpsi*cpsi + mit(3)*ct*ct
gu(2,3) = mit(3)*ct
gu(3,1) = 0.d0
gu(3,2) = gu(2,3)
gu(3,3) = mit(3)
do nup = 1,3
 do mup = 1,3
  g(mup,nup) = 0.d0
  do mu=1,3
   do nu = 1,3
    g(mup,nup) = g(mup,nup) + h(mu,mup)*h(nu,nup)*gu(mu,nu)
   enddo
  enddo
 enddo
enddo
det = g(1,1)*g(2,2)*g(3,3)          ! get the determinant
det = det - g(2,1)*g(1,2)*g(3,3)
det = det - g(3,1)*g(2,2)*g(1,3)
det = det - g(1,1)*g(3,2)*g(2,3)
det = det + g(3,1)*g(1,2)*g(2,3)
det = det + g(2,1)*g(3,2)*g(1,3)
return
end
```

Write a program that tests this version of `metric` as follows. Check that in the limit, $I_1 = I_2 = I_3$, the Hessian metric tensor is diagonal, with all its entries equal to,

$$\frac{64I}{(\sigma + 4)^2}.$$

Then, use the code to verify the following symmetry properties under coordinate exchange for the asymmetric case, $I_1 \neq I_2 \neq I_3$,

$$g_{11} \quad \overset{\xi^1 \leftrightarrow \xi^3}{\longleftrightarrow} \quad g_{11}, \qquad g_{22} \quad \overset{\xi^1 \leftrightarrow \xi^3}{\longleftrightarrow} \quad g_{22},$$

$$g_{33} \quad \overset{\xi^1 \leftrightarrow \xi^3}{\longleftrightarrow} \quad g_{33}, \qquad g_{12} \quad \overset{\xi^1 \leftrightarrow \xi^3}{\longleftrightarrow} \quad g_{12},$$

$$\xi^1 \leftrightarrow \xi^3 \qquad\qquad \xi^1 \leftrightarrow \xi^3$$
$$g_{13} \quad \longleftrightarrow \quad -g_{13}, \qquad g_{23} \quad \longleftrightarrow \quad -g_{23}.$$

7. For the symmetric rotor in the presence of a $\cos\theta$ field, the Hamiltonian matrix has a slightly different kinetic energy element,

$$
\begin{aligned}
H_{ll' m_1 m_1' m_2 m_2'} =\ & \frac{\hbar^2 \, l\,(l+1)}{2I_1} \delta_{ll'} \delta_{m_1 m_1'} \delta_{m_2 m_2'} \\
& + \frac{\hbar^2 \, (I_1 - I_2)\, m_2^2}{2I_1 I_2} \delta_{ll'} \delta_{m_1 m_1'} \delta_{m_2 m_2'} \\
& + V_0 \sqrt{\frac{(2l'+1)}{(2l+1)}}\, \langle\, l' - m_1' \, 1\,0 \,|\, l'\,1\,l - m_1 \,\rangle \\
& \times \langle\, l' - m_2' \, 1\,0 \,|\, l'\,1\,l - m_2 \,\rangle .
\end{aligned}
\tag{11.95}
$$

Choose sensible values for the inertia moments, I_1 and I_2, then modify `exactamts3.f` to compute the canonical average energy and the average potential for several values of β.

8. Modify `pis3` so that it performs a parallel tempering simulation for the same symmetric rotor considered in the previous exercise.

11.12   The Free Particle in a Ring

In this section, we consider some additional tests for the quantum methods we have outlined at the beginning of this chapter, to the $\mathbb{S}^1$ space. This manifold is the set of all points in a ring of radius R. These points are typically mapped using an angular variable $\theta \in (0, 2\pi)$. We will soon see that this set of spaces is more complicated, topologically, than its two- and three-dimensional counterparts. Additionally, it is important to make a distinction between the $\mathbb{S}^1$ space, the physical space for a rigid linear rotor confined to a plane, and other types of spaces for which one may use a similar type of map $\theta \in (0, 2\pi)$, but that have quite different Hessian metrics, like the bending and the torsional degrees of freedom. For the latter ones, the quantum dynamics are even more complicated than for the $\mathbb{S}^1$ case because the curvature (which is zero for $\mathbb{S}^1$) is not constant, even when the coupling between rotations and these degrees of freedom is ignored. Here, we limit our discussion to the $\mathbb{S}^1$ case.

For the particle of mass m in $\mathbb{S}^1$, the solution of Schrödinger's equation in the absence of the potential energy,

$$
-\frac{\hbar^2}{2mR^2} \frac{d^2}{d\theta^2} \psi\,(\theta) = E\psi\,(\theta),
\tag{11.96}
$$

subjected to the periodic boundary condition,

$$\psi(0) = \psi(2\pi), \tag{11.97}$$

is trivial,

$$\psi(\theta) = \frac{1}{\sqrt{2\pi}} \exp(in\theta), \qquad E_n = \frac{\hbar^2 n^2}{2mR^2}, \qquad n = 0, \pm 1, \pm 2, \ldots. \tag{11.98}$$

The canonical partition function is

$$Q = \sum_{n=-\infty}^{\infty} \exp\left(-\frac{\beta\hbar^2 n^2}{2mR^2}\right) = 1 + 2\sum_{n=1}^{\infty} \exp\left(-\frac{\beta\hbar^2 n^2}{2mR^2}\right). \tag{11.99}$$

Choosing the reduced units,

$$T^* = \frac{2mR^2 k_B T}{\hbar^2}, \qquad E_n^* = \frac{2mR^2 E_n}{\hbar^2} = n^2, \tag{11.100}$$

one derives for the ensemble average energy,

$$\langle E^* \rangle = \frac{2\sum_{n=1}^{\infty} n^2 \exp\left(-\frac{n^2}{T^*}\right)}{1 + 2\sum_{n=1}^{\infty} \exp\left(-\frac{n^2}{T^*}\right)}, \tag{11.101}$$

and the heat capacity,

$$\frac{C_V}{k_B} = (T^*)^{-2} \frac{2\sum_{n=1}^{\infty} n^4 \exp\left(-\frac{n^2}{T^*}\right)}{1 + 2\sum_{n=1}^{\infty} \exp\left(-\frac{n^2}{T^*}\right)} - (T^*)^{-2} \left\{ \frac{2\sum_{n=1}^{\infty} n^2 \exp\left(-\frac{n^2}{T^*}\right)}{1 + 2\sum_{n=1}^{\infty} \exp\left(-\frac{n^2}{T^*}\right)} \right\}^2. \tag{11.102}$$

These sums present no numerical difficulties when computed directly, as the following code demonstrates.

```
      program exact_freeS1
      implicit real*8 (a-h,o-z)
c This program calculates the average canonical energy
c and heat capacity of an (N,V,T) ensemble of particles
c of mass m constrained in a ring of radius R.
c Direct summation is used.
c loop over temperatures
c The unit or temperature T* are reduced units, T* = 2mR**2 kB T/hbar**2
c The average energy is simply <E*> = 2mR**2<E>/hbar**2
      tiny = 1.d-12
```

```
      temp = 0.02
      delta_t = 0.01
      do i=1,120
      n = 1
      sum1 = 0.d0
      sum2 = 0.d0
      sum3 = 0.d0
1     e1 = exp(-n*n/temp)
       sum1 = sum1 + e1
       sum2 = sum2 + n*n*e1
       sum3 = sum3 + n*n*n*n*e1
       n = n + 1
      if (n*n*n*n*e1 .gt. tiny) goto 1
      ae = 2.d0*sum2/(1.d0+2*sum1)
      cv = (2.d0*sum3/(1.d0+2*sum1) - ae*ae)/(temp*temp)
      write(6,1000) temp,ae,cv,n
      temp = temp + delta_t
      enddo    ! temperature loop
1000  format (3f12.4,i8)
      end
```

A graph of the energy $\langle E^* \rangle$ and C_V/k_B as a function of T^*, obtained with the program `exact_freeS1`, is in Figure 11.8. The properties of the particle in a ring approach the classical limit for reduced temperatures greater than 1. Surprisingly, as few as seven terms are necessary to converge within the machine precision at $T^* = 1$. The deviation of $\langle E^* \rangle$ from the classical estimate is one part in 5000, and drops to one part in 5000 by $T^* = 1.2$

Another interesting surprise is the heat capacity peak at $T^* = 0.38$. This is a feature of the topology of the space itself, since there are no forces at

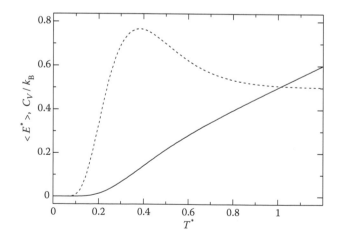

FIGURE 11.8

The energy (—) and heat capacity (···) of a particle in a ring in reduced units.

play other than the holonomic constraints. The heat capacity approaches the correct equipartition limit for temperatures greater than one, and approaches zero as $T^* \to 0$. The quest now, is to reproduce these results with the path integral. The free particle in a ring case has been solved analytically by Schulman [747] using the following arguments. The Lagrangian for the free particle in a ring mapped by the angular variable θ is,

$$\mathcal{L} = \frac{1}{2} m R^2 \dot{\theta}^2. \tag{11.103}$$

Paths in $\mathbb{S}^1$ can be represented by a continuous function, $0 < \theta(t) < 2\pi$. When summing over all contributions to the density matrix, one must take into account all paths, including those that, in going from θ to θ', wind around clockwise n times $(n > 0)$, and those that wind counterclockwise n times $(n < 0)$. Paths that have different winding numbers belong to different **homotopy** classes. Briefly, this means that a path with winding number n cannot be stretched and modified continuously into one with winding number n' without leaving the manifold. Schulman [747] proposes to map the ring from its **universal covering space**, $p : \mathbb{R} \to \mathbb{S}^1$, with the following projective map,

$$\begin{cases} \theta(x) = x - 2\pi \left[\dfrac{x}{2\pi} - 1\right] & x < 0 \\[2mm] \theta(x) = x - 2\pi \left[\dfrac{x}{2\pi}\right] & x > 0, \end{cases} \tag{11.104}$$

where the square brackets imply "the integer part" of what they contain. For example, $[x/2\pi]$ is equal to the integer part of $x/2\pi$. Note that if $x \in \mathbb{R}$, then $\theta(x) \in (0, 2\pi)$. Since the map $\theta(x)$ is smooth, one can solve for the density matrix in $\mathbb{R}^1$ with x, and then transport the solution into $\mathbb{S}^1$ for a given winding number. Since in $\mathbb{R}^1$ the Lagrangian is that of the free particle with "mass" mR^2, for a given winding number n, the density matrix takes the following form (c.f. Chapter 7),

$$\rho_n(\theta, \theta', \beta) = \left(\frac{mR^2}{2\pi\beta\hbar^2}\right)^{1/2} \exp\left[-\frac{mR^2(\theta' - \theta + 2n\pi)^2}{2\beta\hbar^2}\right], \tag{11.105}$$

and the partition function obtained by setting $\theta' = \theta$ and summing over all the winding numbers is,

$$Q = 2\pi \sum_{n=-\infty}^{\infty} \rho_n(\theta, \beta). \tag{11.106}$$

Homotopy phase factors are not necessary for Equation 11.106, since we are assuming the boundary conditions in Equation 11.97 hold [747]. Let us now use the same reduced units for energy and temperature that we introduce earlier, then we obtain,

$$Q = (\pi T^*)^{1/2} \left[1 + 2 \sum_{n=1}^{\infty} \exp\left(-n^2\pi^2 T^*\right) \right]. \tag{11.107}$$

Using the chain rule, it is straightforward to show that,

$$\langle E^* \rangle = \frac{(T^*)^2}{Q} \frac{\partial Q}{\partial T^*}, \tag{11.108}$$

and that

$$\frac{C_V}{k_B} = \frac{\partial \langle E^* \rangle}{\partial T^*}. \tag{11.109}$$

Using Equation 11.107, we obtain

$$\langle E^* \rangle = \frac{T^*}{2} - \frac{2(T^*)^2 \sum_{n=1}^{\infty} (\pi n)^2 \exp\left(-n^2\pi^2 T^*\right)}{1 + 2\sum_{n=1}^{\infty} \exp\left(-n^2\pi^2 T^*\right)}, \tag{11.110}$$

$$\frac{C_V}{k_B} = \frac{1}{2} - \frac{4T^* \sum_{n=1}^{\infty} (\pi n)^2 \exp\left(-n^2\pi^2 T^*\right)}{1 + 2\sum_{n=1}^{\infty} \exp\left(-n^2\pi^2 T^*\right)}$$

$$+ \frac{2(T^*)^2 \sum_{n=1}^{\infty} (\pi n)^4 \exp\left(-n^2\pi^2 T^*\right)}{1 + 2\sum_{n=1}^{\infty} \exp\left(-n^2\pi^2 T^*\right)}$$

$$+ 4(T^*)^2 \left\{ \frac{\sum_{n=1}^{\infty} (\pi n)^2 \exp\left(-n^2\pi^2 T^*\right)}{1 + 2\sum_{n=1}^{\infty} \exp\left(-n^2\pi^2 T^*\right)} \right\}^2. \tag{11.111}$$

Equations 11.110 and 11.111 are quite different from Equations 11.101 and 11.102, and it is not obvious that they should yield the same result. However, the reader can become convinced of this fact by running the program, `apiS1`, given below.

```
program apiS1
implicit real*8 (a-h,o-z)
pi = 3.141592654d0
```

```
        tiny = 1.d-12
        temp = 0.02
        delta_t = 0.01
        do i=1,120
        n = 1
        sum1 = 0.d0
        sum2 = 0.d0
        sum3 = 0.d0
1       e1 = exp(-n*n*pi*pi*temp)
         sum1 = sum1 + e1
         sum2 = sum2 + n*n*pi*pi*e1
         sum3 = sum3 + n*n*n*n*pi*pi*pi*pi*e1
         n = n + 1
        if (n*n*n*n*pi*pi*pi*pi*e1 .gt. tiny) goto 1
        ae = 0.5d0*temp - 2.d0*temp*temp*sum2/(1.d0+2*sum1)
        cv = 0.5d0 - 4.d0*temp*sum2/(1.d0+2*sum1)
       & + 2.d0*temp*temp*sum3/(1.d0+2*sum1)
       & - (2.d0*temp*sum2/(1.d0+2*sum1))**2
c         cv = (2.d0*sum3/(1.d0+2*sum1) - ae*ae)/(temp*temp)
        write(9,1000) temp,ae,cv,n
        temp = temp + delta_t
        enddo    ! temperature loop
1000    format (3f12.4,i8)
        end
```

No other example can display more clearly, the fundamental difference between the two methods of propagation. The energy representation corrects the thermodynamic properties from the ground state up. The higher the temperature (or real simulation time) is, the greater the effort to converge becomes. The PI contributes corrections from the classical limit down. Consequently, the lower the temperature is, the greater the effort to converge becomes. In Figure 11.9, we demonstrate this by graphing the values of n necessary to produce contributions smaller than the machine precision to the leading terms of the n^4 sum for both approaches. The PI solution requires a larger number of windings as T^* decreases, whereas the energy representation sums require a larger number of states as T^* increases.

11.13   The Particle in a Ring Subject to Smooth Potentials

The particle in a ring example brings into focus the advantage of using stereographic projections for quantum simulations, in $\mathbb{S}^n$. In order to carry out the

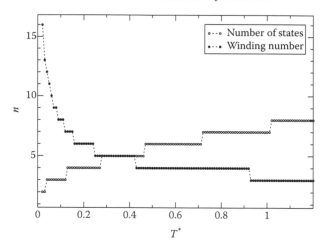

FIGURE 11.9

The number of windings and the number of states necessary to produce an error in the heat capacity for a particle in $\mathbb{S}^1$ below the machine precision.

stochastic simulations, we must map $\mathbb{S}^1$ from its universal covering space $\mathbb{R}^1$. For practical reasons, we must also introduce a potential to obtain an importance sampling function. Consider a simple constraining potential, such as $V\left[\theta\left(x\right)\right] = V_0\left(\theta - \pi\right)^2$. Mapping from $\mathbb{R}^1$ with $\theta\left(x\right)$ using Equation 11.104, would lead to a highly inefficient simulation, since $V\left[\theta\left(x\right)\right]$ has an infinite number of minima over the number line, all separated by finite barriers; however, this does not occur when mapping the space stereographically. Therefore, let's explore the DeWitt formulation with the stereographic projection coordinate by simulating the particle in a ring of 1.0 bohr and a mass of 207 a.u. subjected to a potential,

$$V = V_0 \cos\theta. \tag{11.112}$$

For simplicity, we set both V_0 and R to 1 hartree and 1 bohr, respectively, as we have in all the previous tests. The canonical ensemble average potential energy can be calculated by diagonalizing the Hamiltonian matrix (using the potential free ring basis),

$$H_{ij} = \frac{\hbar^2 i^2}{2mR^2}\delta_{ij} + \frac{V_0}{2}\left(\delta_{i(j-1)} + \delta_{i(j+1)}\right). \tag{11.113}$$

where the quantum numbers i and j are integers in

$$-n_{\max} < i, j < n_{\max}, \tag{11.114}$$

and $n_{\max}$ is systematically increased until results converge uniformly over all values of β under study. The canonical potential energy average is calculated using the resulting energy levels and vectors $E_k, |k\rangle$, and transforming the potential matrix in the energy representation. [c.f. Equation 11.71]. k is an energy ordering quantum number. The following code does the job of setting up the Hamiltonian matrix, sending it to the diagonalizer, producing the canonical average potential for several values of β, and producing the ground state wavefunction squared so that it can be compared with VMC and DMC distributions after appropriate transformations are made.

```
      exactamts1
      implicit real*8 (a-h,o-z)
      parameter(pi = 3.1415926)
      parameter(nbas = 401)
      real*8 h(nbas,nbas)    ! h() is the Hamiltonian matrix
      real*8 d(nbas), z(nbas,nbas),e(nbas),V(nbas,nbas),VK(nbas)
      common/box/d_l
      d_l = 1.0d0 ! the size of the well L
      n_max = 40
      tiny=1.d-10
      do k1 = 1,nbas
         e(k1)=0.d0
         d(k1)=0.d0
         VK(k1)=0.d0
        do k2 = 1, nbas
          h(k1,k2) = 0.d0
          V(k1,k2)=0.d0
          z(k1,k2)=0.d0
        enddo
      enddo
c R = 1 a.u. and V0 is in a.u.
c The mass is 207 a.u. (a muon)
      do n = (-n_max), n_max
      i=n+(n_max+1)
c the diagonal elements of the Hamiltonian
      h(i,i) = dfloat(n*n)*((1.0d0)**2)/(207.d0*2.d0*d_l*d_l)
      d(i) = 0.d0
      e(i) = 0.d0
      spec=0
      do m = n, n_max
      j=m+(n_max+1)
       spec=(i-j)**2
        if(spec.eq.1) V(i,j)=(.5d0)
        if(spec.ne.1) V(i,j)=0
        z(i,j) = 0.d0
        h(i,j) = h(i,j) + V(i,j)
        h(j,i) = h(i,j)
        V(j,i)=V(i,j)
```

```
   spec=0
 enddo
 enddo
n_size=2*n_max+1
    call eigen(nbas,(2*n_max+1),h,d,z,e,ierr)
    do i=1, n_size
     VK(i)=0.d0
      do k=1, n_size
       do kp=1,n_size
        VK(i)=VK(i)+ (z(k,i)*V(k,kp)*z(kp,i))
       enddo
      enddo
     enddo
     beta=0.4d0
     dbeta=0.2d0
     do ibeta=1,201
      sum1=0.d0
      sum2=0.d0
       do m=1,n_size
        sum1=sum1+exp(-beta*d(m))
        sum2=sum2+(VK(m))*(exp(-beta*d(m)))
       enddo
      ave=((sum2)/(sum1))
      write(14,*)beta,ave
      beta=beta+dbeta
     enddo
do x=0,2*pi,pi/100
 psi=0
 psi2=0
 psi3=0
 psi4=0
 psi5=0
 p15=0
 sum24=0
 sum25=0
  do i3=1,n_size
   i3_r=i3-(n_max+1)
    sum24=((z(i3,1))*(cos(i3_r*x)))+sum24
    sum25=((z(i3,1))*(sin(i3_r*x)))+sum25
  enddo
 psi=(1/(2*pi*d_1)*(sum24**2+sum25**2))
 write(12,*) x,psi
enddo
end
```

The content of `fort.14` is the value $\langle V \rangle_\beta$ for β from 0.4 to 40.4 hartree^{-1}. In Chapter 10, we derive the stereographic projection map for $\mathbb{S}^1$.

$$\xi = \frac{2R\cos\theta}{1 - \sin\theta}, \tag{11.115}$$

and the related metric tensor,

$$g = \frac{\left(4R^2\right)^2}{\left(\xi^2 + 4R^2\right)^2}, \tag{11.116}$$

which is both the tensor itself and its determinant. We have used the same potential for tests performed in Chapter 10,

$$V = V_0 \frac{4R\xi}{\xi^2 + 4R^2}. \tag{11.117}$$

Therefore, the following modifications to the subroutine pot and `metric` are sufficient. Note that `metric` returns the square root of the metric tensor.

```
c       POT
        subroutine pot(x,v)
        implicit real*8 (a-h,o-z)
        v = 4.d0*x/(x*x + 4.d0)
        return
        end
c       METRIC
        subroutine metric(x,g)
        implicit real*8 (a-h,o-z)
c Returns the square root of the determinant of the metric tensor
        g = 4.0d0/(x*x + 4.0d0)
        return
        end
```

The connection coefficient Γ_{11}^1 is nonvanishing,

$$\Gamma_{11}^1 = \frac{-2\xi}{\xi^2 + 4R^2}, \tag{11.118}$$

however, the curvature $\mathcal{R}$ for this case is zero. The following modification of `getw` takes care of the computation of the importance sampling function, W_N, using the usual set of auxiliary random variables $\{w\}_{i=0}^N$. The expressions for W_N and $\mathcal{S}_N$ for the $\mathbb{S}^1$ manifold become,

$$W_N\left(\xi, \{w\}, \beta\right)$$

$$= \left(\frac{207}{2\pi\beta}\right)^{N/2} \left\{ \prod_{i=1}^{N} \frac{4}{\left[\left(\xi + w_i\right)^2 + 4\right]} \right\} \exp\left[-\mathcal{S}_N\left(\xi, \{w\}, \beta\right)\right], \tag{11.119}$$

$$\mathcal{S}_N\left(\xi, \{w\}, \beta\right)$$

$$= \frac{207N}{2\beta} \sum_{i=1}^{N} \frac{4^2}{\left[\left(\xi + w_{i-1}\right)^2 + 4\right]^2} \left(w_i - w_{i-1}\right)^2 + \frac{\beta}{N} \sum_{i=1}^{N} V\left(\xi + w_{i-1}\right).$$

$$(11.120)$$

```
c       GETW
        subroutine getw(temp,w,xi,N,NMAX,wi)
        implicit real*8 (a-h,o-z)
        real*8 w(0:NMAX)
c note that temp is beta
        sum = 0.d0
        prod = 1.d0
        dN = dfloat(N)
        do i=1,N
         xm1 = xi + w(i-1)
        call metric(xm1,g)
         x = xi + w(i)
         call pot(x,v)
         sum = sum + (dN*207./(temp*2.))*g*g*(w(i)-w(i-1))**2 + temp*v/dN
        call metric(x,g)
        prod = prod*g
        enddo
         wi = prod*exp(-sum)
        return
        end
```

The output of `exactamts1` is compared with the DeWitt path integral calculation carried out with the program `pis1`. The subroutine `vestimator` calculates the path average potential estimator. The results are graphed in Figure 11.10.

```
        program pis1
        implicit real*8 (a-h,o-z)
c This program calculates <V>using the De Witt discretized path
c integral in S^1. A stereographic projection is used to map
c the points of the set S^1 -> R^1. The resulting  metric tensor
c and Christoffel symbol depend on the projection coordinate but
c the curvature vanishes leaving only one possible lattice definition.
c A cos(theta) potential is added. Atomic instead of reduced units
c are used throughout. The mass is 207 a.u. the ring radius is 1.00 bol
        parameter (NMAX = 1000)
        real*8 wm1(0:NMAX),w(0:NMAX), xr(3)
         see=secnds(0.)
        iseed = int(see/10.)
        call ransi(iseed)
         read(5,*) N
         do nQ = 1,1
```

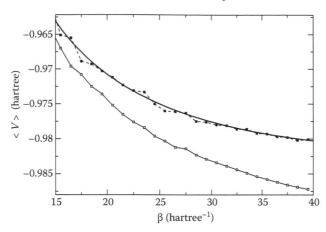

FIGURE 11.10

The canonical average potential for a particle in a ring subjected to a sinusoidal potential. The dark line is the diagonalization result, the white squares are the classical, and the black squares the quantum $N = 10$ results.

```
      write(6,*) '# N =',N
      write(6,1001)
MW = 1000000       !  number of warm up moves
M = 10000000
    dM = dfloat(M)
temp = 0. 5        ! This is beta in inverse hartree
deltax = .210
do k_t = 1,40      ! loop through some temperatures
nrej = 0.d0
sumt = 0.d0
sumt2 = 0.d0
xim1 = 0.d0
xi = 0.d0
  do i=0,N
    w(i) = 0.d0
    wm1(i) = 0.d0
  enddo
  call getw(temp,w,xi,N,NMAX,wim1)
do i=1,MW + M
call ransr(xr,3)
  imove = (N+1)*xr(1)
  if (imove .eq. 0 .or. imove .eq. N) then
 xi = xim1 +  deltax *(xr(2)-0.5d0)
  else
    w(imove) = wm1(imove) + deltax*(xr(2)-0.5d0)
```

```
            endif
            call getw(temp,w,xi,N,NMAX,wi)
        p = wi/wim1
        if (xr(3) .lt. p) then   !  accept or reject
         xim1 = xi                           ! accept
         wim1 = wi
           do j=1,N-1
             wm1(j) = w(j)
           enddo
        else                                 ! reject
         nrej = nrej + 1
         xi = xim1
         wi = wim1
           do j=1,N-1
             w(j) = wm1(j)
           enddo
        endif
         if (i.gt. MW) then         ! accumulate data
            call vestimator(temp,w,xi,N,NMAX,pv)
            sumt = sumt + pv/dM
            sumt2 = sumt2 + pv*pv/dM
         endif
        enddo
c and calculate the percent rejections
        pr = 100.*dfloat(nrej)/dfloat(M+MW)
          se = 2.*sqrt(sumt2 - sumt*sumt)
        write(6,1000) temp,sumt,se,pr
          call flush(6)
        temp = temp + 1.
        deltax = deltax + 0.02
        enddo
         N = N  + 1       ! temperature loop
         enddo              ! quadrature loop
1000    format(6f12.4)
1001    format('        Beta           <V>           error        %r')
        end
c       VESTIMATOR
        subroutine vestimator(temp,w,xi,N,NMAX,e)
c path average potential energy estimator
        implicit real*8 (a-h,o-z)
        real*8 w(0:NMAX)
        sum = 0.d0
        dN = dfloat(N)
        do i=1,N
         x = xi + w(i)
```

```
call pot(x,v)
sum = sum + v/dN
enddo
e = sum
return
end
```

Again, there is excellent agreement between the vector space solution and the PI simulation. However, the results in Figure 11.10 raise more questions than they provide answers. In particular, what happened to the winding numbers? Paths with different winding numbers in $\mathbb{S}^1$ are not homotopic, even under the stereographic projection map, and to understand homotopy, and what happens in our simulation a little more deeply, let us consider the following two convolutions,

$$\xi\{\theta[f_i(x)]\} = \frac{2R\cos\{\theta[f_i(x)]\}}{1 - \sin\{0[f_i(x)]\}}, \tag{11.121}$$

where,

$$f_1(x) = 1 - \frac{1}{2}\sin\left(\frac{\pi x}{3\pi + 1}\right), \tag{11.122}$$

for $x \in [1 : 3\pi + 1]$, and

$$f_2(x) = x - \left[\frac{x}{2\pi}\right]2\pi, \tag{11.123}$$

for $x \in [1, 3\pi + 1]$. A graph of both $\xi\{\theta[f_1(x)]\}$ and $\xi\{\theta[f_2(x)]\}$ can be found in Figure 11.11. $\xi\{\theta[f_1(x)]\}$ is in the $n = 0$ homotopy class, whereas $\xi\{\theta[f_2(x)]\}$ has three windings. If $\xi\{\theta[f_1(x)]\}$ and $\xi\{\theta[f_2(x)]\}$ were homotopic, we could transform smoothly from one path to the other, using the following linear combination,

$$(1 - \lambda)\xi\{\theta[f_1(x)]\} + \lambda\xi\{\theta[f_2(x)]\}, \tag{11.124}$$

with $\lambda \in [0, 1]$.

As Figure 11.11 demonstrates, $\xi\{\theta[f_2(x)]\}$ and the entire class of linear combinations containing any finite amount of $\xi\{\theta[f_2(x)]\}$, as in Equation 11.124, are discontinuous functions. From the practical point of view, the potential energy, albeit nonconfining, produces an attractor for the quantum systems that have one or more bound states. Consequently, for these, the wavefunction (and the $T \to 0$ limit of the density matrix) tend to zero for large values of ξ. We demonstrate this point in Figure 11.12, by graphing a histogram of values of ξ obtained during a ten million point run at $\beta = 40$ hartree^{-1} for $N = 10$ slices. A path in a $n > 0$ class is discontinuous at a point, and as Figure 11.11 demonstrates, must sample configurations with relatively high potential energy.

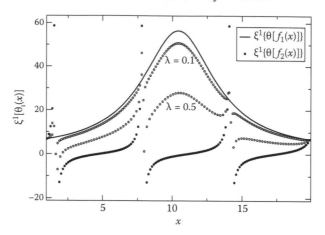

FIGURE 11.11

Two paths $\xi^1(\theta)$ mapped to $\mathbb{R}^1$ belonging to two different homotopy classes in $\mathbb{S}^1$, and their linear combinations.

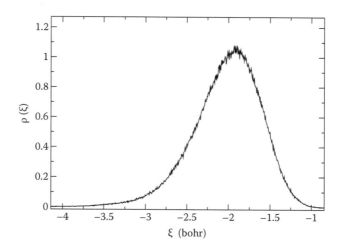

FIGURE 11.12

Distribution of ξ from a ten million point run at $\beta = 40.0$ hartree^{-1} for $N = 10$ slices.

Furthermore, as β increases, we approach the classical limit. Recall that as few as seven states produce a classical limit for this example for the potential free case, and that the number of windings needed to converge the PI expression of the density matrix for the same, drops to zero in the classical limit.

The combination of these two effects, the classical limit requiring only the $n = 0$ class, and the low distribution outside the well in the quantum limit, renders validity to the approximation obtained by dropping all homotopy classes with $n > 0$ in our PI simulation. Of course, this statement is hardly generalizable and any such approximation should be **carefully tested** for values of V_0, R, and mass m different from those used here. Molecular applications (for torsions) are likely to involve larger masses, but may require values of V_0 and R quite different from those tested here. Bending coordinates do not need more than the $n = 0$ homotopy class, since, presumably, as the bending angle sweeps to values close to zero or 2π, two of the three bodies involved approach each other, causing realistic energy models to approach infinity. In one of the exercises at the end of this section, we guide the reader to explore systems for which the $n = 0$ homotopy class might be insufficient to produce convergent path integrals in $\mathbb{S}^1$. Another compelling question that arises from the foregoing discussion is the following. If the potential has such drastic impact on the results compared to the topology of the space itself, is it a valid approximation to ignore the curved nature of the manifold altogether? The answer to this question can be provided with some additional experimentation, however, given that we have the means to easily include the proper Hessian metric into simulations, the additional work of controlling the approximation of $\mathbb{S}^1$ with $\mathbb{R}^1$ (meaning the use of a constant metric) is not justifiable.

Exercises

1. Modify the program `pis1` to compute the average of $V_0 \cos \theta$ for a particle with 207 a.u. of mass in a circle with a radius of 1 bohr, using several values for V_0 between 0.1 and 0.9 hartree. Modify the program `exactamts1`, and compare the outcome between the vector space solution and the PI simulation. Verify that for small values of β, the average potential has converged with respect to the basis size. Then, verify that the classical limit is always achieved by simulations with the $n = 0$ homotopy class.

2. Repeat the analysis of the previous exercise, but change the mass m instead of the value of V_0, which should be set back to 1 hartree. The quantum results should be carefully checked for convergence.

3. This exercise is designed to help the reader determine the impact of the topology of the ring on results, when a potential confines the particle to a region of configuration space. Modify program `pis1` to compare the results obtained for the particle with 207 atomic units of mass, in $\mathbb{S}^1$, with $R = 1$ bohr, $V_0 = 1$ hartree, V as in Equation 11.117 with those obtained when the metric tensor computed in `metric` is set to one.

4. For the Weyl ordering formulation, the high temperature approximation of the density matrix in $\mathbb{S}^1$ mapped stereographically becomes,

$$\rho_{\mathrm{f}} = \left(\frac{m}{2\pi\hbar^2\beta} \right)^{1/2} \exp\left[-\mathcal{S}\left(\xi, \xi', \beta \right) \right], \qquad (11.125)$$

with the following action,

$$S\left(\xi, \xi', \beta\right) = \frac{m}{2\beta\hbar^2} \frac{\left(4R^2\right)^2}{\left[\bar{\xi}^2 + 4R^2\right]^2} \left(\xi - \xi'\right)^2 + \frac{\beta\hbar^2\bar{\xi}^2}{8mR^4}, \tag{11.126}$$

where $\bar{\xi}$ is the midpoint,

$$\bar{\xi} = \frac{\xi + \xi'}{2}. \tag{11.127}$$

Derive the Weyl correction term in Equation 11.126 using,

$$\Delta V_{\text{Weyl}} = \frac{\hbar^2 \bar{\xi}^2}{2m\left(4R^4\right)} = \frac{\hbar^2}{2m} \left(g^{\mu\upsilon} \Gamma^\kappa_{\mu\lambda} \Gamma^\lambda_{\upsilon\kappa} - \mathcal{R}\right), \tag{11.128}$$

as the starting point. Note that the metric tensor is now evaluated at the midpoint between ξ and ξ'. Modify the program pis1 to compute the average of $V_0 \cos\theta$, $V_0 = 1$ hartree for a particle with 207 a.u. of mass in a circle with a radius of 1 bohr, using the midpoint expansion and the Weyl correction term. Verify that if the Weyl correction term is omitted, the PI simulation does not converge.

5. Repeat the analysis in the last exercise for $\mathbb{S}^2$ mapped stereographically, and then for the ellipsoid of inertia for spherical rotors. Use the same value for R, V_0, and inertia moment I (or mass for the two-sphere case) as we do in our tests.

6. Write the programs to carry out VMC and DMC in inertia ellipsoids for spherical tops, and symmetric tops, then compare the ground state energy and wavefunction with those obtained with angular momentum theory.

7. Repeat the previous exercise for $\mathbb{S}^2$.

8. Consider extending the reweighted Fourier–Wiener path integral to manifolds. To accomplish this, one begins by redefining the random path with $k'_m > k_m$ terms, as it is done with Cartesian coordinates,

$$\tilde{q}^\mu\left(u\right) = q^\mu\left(u = 0\right) + \sigma \sum_{k=1}^{k_m} a_k^\mu \Lambda_k\left(u\right) + \sigma \sum_{k=k_m+1}^{k'_m} a_k^\mu \widetilde{\Lambda}_k\left(u\right). \tag{11.129}$$

The difference is in the way the rescaling of the coefficients takes place through the definition of σ,

$$\sigma = \hbar\beta^{1/2}. \tag{11.130}$$

Note that the mass does not enter in the definition of σ. ρ^{RW} becomes,

$$
\rho^{RW}(q,q',\beta) = \left(\frac{1}{2\pi}\right)^{Nd/2}(\hbar^2\beta)^{-d/2}J_\Lambda
$$
$$
\times \int d[a]_r \exp\left\{-\beta\int_0^1 du\, U\left(\tilde{q}^\mu(u)\right)\right\}, \qquad (11.131)
$$

where

$$
U\left(\tilde{q}^\mu(u)\right) = -\frac{N}{2\beta}\ln\left[\det g_{\mu\upsilon}\left(\tilde{q}^\mu(u)\right)\right]
$$
$$
+\frac{1}{2}g_{\mu\upsilon}\left(\tilde{q}^\mu(u)\right)\dot{q}^\mu\dot{q}^\upsilon + V\left(\tilde{q}^\mu(u)\right). \qquad (11.132)
$$

$\sqrt{g}$ is computed at every slice endpoint, and has been included in the argument of the exponential with the first logarithmic term in $U\left(\tilde{q}^\mu(u)\right)$. Modify the program `fwrrs` in Chapter 7, so that it simulates a particle with 207 a.u. of mass in a ring with $R = 1$ bohr.

Part IV

Applications to Molecular Systems

12

Clusters of Rigid Tops

12.1 Introduction

In principle, this last chapter should serve as the capstone for all the material presented in the preceding parts of this book. The focus is on extending the methods of Chapters 5–9 that are adapted to curved manifolds in Chapter 11, to clusters of rigid linear and nonlinear molecules. The goal remains one of introducing pedagogical tools, rather than reviewing results from extensive simulations, or providing optimized tools ready for production runs. The quantum exploration of Lennard-Jones and other types of atomic clusters, is well on the way, with a large community still contributing heavily at the time of writing. We are unable to provide, let alone summarize, a complete list of references for the results from classical and quantum simulations of atomic clusters. However, the quantum simulations of molecular clusters beyond the dimers, and molecular liquids, are fields still in their infancies. Therefore, this chapter attempts to provide some of the wood and steel for the pioneering journey into this new frontier of knowledge. It is both humbling and awe inspiring to contemplate the vastness of what is left unexplored in this rich theoretical area. Our group has only just begun a few small scouting expeditions. These have already proved to require more methods, as, for example, the parameter space sweep technique for global minimization. This exciting area is likely to produce many more enhancements to the tools we have at our disposal, and it is likely to have a number of important applications.

The first section of this chapter is dedicated to the work on uniform Stockmayer clusters [773, 784]. The Stockmayer model is perhaps the simplest approach for the analytical representation of the interaction among polar linear molecules. The potential energy surface is obtained by summing over all the pairs of ideal (point-like) dipoles, and summing over all the pairs of Lennard-Jones interactions located at the same points. We abbreviate Stockmayer clusters with the symbol $(LJDD)_n$, representing n point particles interacting through a Lennard-Jones dipole-dipole potential energy surface. In recent works, we have characterized the surface of $(LJDD)_{13}$ using a range of strengths for the dipole moment at each center, and have investigated both the morphology of the potential energy surface as well as the classical thermodynamics as a function of the dipole moment size. We find a smooth transition from the icosahedron, characteristic of the Lennard-Jones aggregate at low values of the dipole moment, to a hexagonal antiprism, to a decahedral cage, and finally to a ring as the dipole moment increases. It is likely that these different shapes are in separate funnels of the potential. Global optimization is more involved at those parameter space boundaries, where two basins approach each other and switch rank energetically from the local to the global minimum. In order to surmount these difficulties, we develop a technique called the parameter space sweep. With one single basin hopping or genetic algorithm run over a grid of parameter values, we compile a library of distinct structures that are used to seed a second round of minimization sweeping through the desired range of parameters.

The classical thermodynamic behavior of $(LJDD)_{13}$ correlates quite well with the distribution of minima, as the systems transition from the stable icosahedron to a glassy surface with rings as the global minima. At intermediate values of the dipole moment, when the hexagonal antiprism and the decahedral cage coexist, we observe melting behavior. As far as we know, nothing else is known about the $(LJDD)_n$ system, with the exception of the ground state of the trimer with its parameters set to model $(HCl)_3$, obtained only very recently. Given its relative simplicity, and the possibility to obtain the potential surface and its gradient analytically, we find this model very important for a number of future applications of molecular aggregates. With the Stockmayer model, it may not be possible, perhaps, to obtain spectroscopic accuracy, but it does contain a sufficient number of parameters to achieve chemical accuracy for a relatively large set of homogeneous and heterogeneous condensed matter systems. Simulations with the Stockmayer model may be helpful in identifying the role played by masses, size, dipole strength, and dispersion strength on the structural, ground states, and thermodynamic properties of a vast number of important systems. These results can be used to guide future investigations with experiments and more sophisticated potential models.

Given the nature of the Stockmayer model, it is not possible to use Cartesian coordinates. Therefore, the developments in Chapters 10 and 11 are crucial for quantum simulations. The second section of this chapter is dedicated to extending the results for a rigid top, developed in Chapters 10 and 11, to clusters of these. At first, one might be tempted to run quantum simulations with all the atoms in each molecule free to move about. However, we have demonstrated that holonomic constraints, when used appropriately, yield very accurate models and provide massive efficiency gains in quantum simulations. For path integral simulations, in particular, internal degrees of freedom that, in the temperature of interest, are predominantly (99% or more) in the ground state, have to be frozen our of the simulation. If such a step is not taken, the path integral may never converge. Similar gains are obtained with diffusion Monte Carlo (DMC). Water clusters, for example, melt (for some sizes) at temperatures where the stretching and the bending degrees of freedom are predominantly in the ground state (below 200 K). Without holonomic constraints, it would be impossible to gauge the thermodynamic behavior of midsized molecular clusters, and the attainment of the ground state properties would be a formidable task. Therefore, the second part of this chapter includes a number of subroutines for the computation of the fundamental properties for clusters of rigid tops. These include the metric tensor, the reweighted random series path integral importance sampling function, the local energy for variational Monte Carlo (VMC), and guided diffusion [232, 461, 485, 508, 526]. The main programs that make use of these are straightforward modifications of the programs `parallel_tempering_r3n.f`, `gfis_dmc.f` and `rewfpi.f`, given in Chapter 9. Similarly, it is possible to modify the structural comparison algorithm, the computation of the Lindemann index, the bond orientational parameter, the basin hopping, and the genetic algorithm, presented in Chapters 8 and 9. Some of these modifications will undoubtedly need some additional experimentation, such as, for example, the implementation of the genetic operators, given two or more sets of coordinates. We leave these modifications to the reader.

12.2   The Stockmayer Model

In the Stockmayer model, each molecular top is represented as an ideal point dipole. The resulting potential model is a sum of pair interactions of two types. These are the dipole-dipole interaction, V^{DD}, and the Lennard-Jones interaction, V^{LJ},

$$V = \sum_{i=1}^{n} \sum_{j>i}^{n} \left(V_{ij}^{DD} + V_{ij}^{LJ} \right), \tag{12.1}$$

where n is the number of molecules in the aggregate. The dipole-dipole interaction between two linear molecules with a center to center vector, $\mathbf{r}_{ij}$, and with orientation unit vectors, $\mathbf{e}_i$ and $\mathbf{e}_j$, respectively, has the following form,

$$V_{ij}^{DD} = \frac{\mu_D^2}{4\pi\epsilon_0 r_{ij}^3} \left(\cos\gamma_{ij} - 3\cos\theta_i \cos\theta_j \right), \tag{12.2}$$

where ϵ_0 is the vacuum permittivity, γ_{ij} is the angle between the orientation unit vectors,

$$\cos\gamma_{ij} = \mathbf{e}_i \cdot \mathbf{e}_j, \tag{12.3}$$

and θ_i is the angle between $\mathbf{e}_i$ and $\mathbf{r}_{ij}$,

$$\cos\theta_i = \frac{\mathbf{r}_{ij} \cdot \mathbf{e}_i}{r_{ij}}, \qquad \cos\theta_j = \frac{\mathbf{r}_{ij} \cdot \mathbf{e}_j}{r_{ij}}. \tag{12.4}$$

The orientation of the dipoles is expressed in terms of stereographic projections rather than the traditional angles, for the reasons explained in Chapter 11. Each molecule, i, is characterized by three Cartesian coordinates, x_i, y_i, z_i, and two stereographic coordinates, ξ_i^μ, $\mu = 1, 2$. Superscript and subscript Greek letters are used as contravariant and covariant labels, whereas Roman subscripts are used to label particles.

Consider a diatomic molecule, $A - B$, with coordinates relative to the center of mass given as x_A, y_A, z_A and x_B, y_B, z_B. The distance between A and B is,

$$R = \sqrt{(x_B - x_A)^2 + (y_B - y_A)^2 + (z_B - z_A)^2}. \tag{12.5}$$

If R is fixed, namely, if the diatomic molecule is rigid, then the space spanned by the three relative variables after the center of mass is removed, is a sphere. Such a sphere can be mapped onto a plane by stereographic projections, as we demonstrate in Chapter 10,

$$x_B - x_A = \frac{4R^2 \xi^1}{\sigma + 4R^2}, \tag{12.6}$$

$$y_B - y_A = \frac{4R^2 \xi^2}{\sigma + 4R^2}, \tag{12.7}$$

$$z_B - z_A = R \frac{\sigma - 4R^2}{\sigma + 4R^2}, \tag{12.8}$$

where,

$$\sigma = \left(\xi^1\right)^2 + \left(\xi^2\right)^2. \tag{12.9}$$

Here, we have $x_A = x'_A - x$ and x is a coordinate of the center of mass of the diatomic.

To construct the potential, we reformulate the orientation vectors in terms of projection coordinates,

$$\mathbf{e}_i = \left(\frac{4R\xi_i^1}{\sigma_i + 4R^2}\right) \mathbf{i} + \left(\frac{4R\xi_i^2}{\sigma_i + 4R^2}\right) \mathbf{j} + \left(\frac{\sigma_i - 4R^2}{\sigma_i + 4R^2}\right) \mathbf{k}. \tag{12.10}$$

No tensor analysis is needed to derive the following results, since we are simply transforming scalar quantities.

$$\mathbf{e}_i \cdot \mathbf{e}_j = \frac{16R^2 \left(\delta_{\mu\nu}\xi_i^\mu \xi_j^\nu\right) + \left(\sigma_i - 4R^2\right)\left(\sigma_j - 4R^2\right)}{\left(\sigma_i + 4R^2\right)\left(\sigma_j + 4R^2\right)}, \tag{12.11}$$

$$\mathbf{r}_{ij} \cdot \mathbf{e}_i = \frac{4R\left(x_i - x_j\right)\xi_i^1 + 4R\left(y_i - y_j\right)\xi_i^2 + \left(z_i - z_j\right)\left(\sigma_i - 4R^2\right)}{\left(\sigma_i + 4R^2\right)}, \tag{12.12}$$

$$\mathbf{r}_{ij} \cdot \mathbf{e}_j = \frac{4R\left(x_i - x_j\right)\xi_j^1 + 4R\left(y_i - y_j\right)\xi_j^2 + \left(z_i - z_j\right)\left(\sigma_j - 4R^2\right)}{\left(\sigma_j + 4R^2\right)}. \tag{12.13}$$

The variable R can be held fixed if a cluster of rigid molecules is modeled, as we do here. Note that we have already assumed that the clusters are uniform, since we let $R_i = R_j = R$. For a linear molecule with charges $+\delta e$, $-\delta e$

separated by a distance R (e is the unit of charge, $e = 1.6021773(3) \times 10^{-19}$ C), the dipole moment, μ, is $\mu = \delta e R$ in Coulomb meters. If R is in bohr, then,

$$V_{ij}^{DD} = \frac{(\delta R)^2}{r_{ij}^3} \left[\mathbf{e}_i \cdot \mathbf{e}_j - 3 \frac{(\mathbf{r}_{ij} \cdot \mathbf{e}_i)(\mathbf{r}_{ij} \cdot \mathbf{e}_j)}{r_{ij}^2} \right], \qquad (12.14)$$

is in hartree.

At this point, it is important to remember that if two charges of equal magnitude and opposing sign are located a finite distance from each other, then the charge distribution has the dipole moment as the dominant term in the power expansion, but it also has finite quadrupole, octupole, hexadecapole moments, etc. (c.f. Chapter 5). In the Stockmayer model, these higher terms of the multipole expansion of the charge distribution are neglected.

The second part of the model is the Lennard-Jones potential, introduced in Chapter 8,

$$V_{ij}^{LJ} = 4\epsilon \left[\left(\frac{r_0}{r_{ij}} \right)^{12} - \left(\frac{r_0}{r_{ij}} \right)^6 \right], \qquad (12.15)$$

where r_{ij} is the same as in Equation 12.14. The following subroutine calculates the total $(LJDD)_n$ interaction for a given configuration, x_i, y_i, z_i, ξ_i^μ, $\mu = 1, 2$, $i = 1, n$.

```
      subroutine pot(x,xi,n,dmu,epsilon,r0,rd,v)
      implicit real*8 (a-h,o-z)
      parameter(maxn = 1000)
      real*8 x(maxn),xi(maxn)
c on entry: x = center of mass coordinates x(1),.....x(3n)
c where the x coordinates are stored in the entries x(1),....,x(n)
c the y coordinates in the entries x(n+1),...,x(2n)
c and the z coordinates in the x(2n+1),....,x(3n)
c epsilon,r0 are the parameters for the Lennard - Jones interaction
c   xi = stereographic projection coordinates xi(1),.....x(2n)
c where the xi^1 coordinates are stored in the entries xi(1),....,xi(n)
c the xi^2 coordinates in the entries xi(n+1),...,x(2n)
c dmu,rd are the parameters for the dipole-dipole interaction
c The Lennard - Jones interaction
      v = 0.d0
      do i=1,n
      do j = i+1,n
      dx = x(i) - x(j)
      dy = x(n+i) - x(n+j)
      dz = x(2*n+i) - x(2*n+j)
      r2 = dx*dx + dy*dy + dz*dz
      s2 = r0*r0/r2
      s4 = s2*s2
      s6 = s4*s2
      s12 = s6*s6
      v = v + 4.d0*epsilon*(s12 - s6)
```

```
      enddo
      enddo
c The dipole - dipole interaction
      do i=1,n
      si = xi(i)*xi(i) + xi(n+i)*xi(n+i)
      di = si + 4.d0*rd*rd
      do j = i+1,n
      sj = xi(j)*xi(j) + xi(n+j)*xi(n+j)
      dj = sj + 4.d0*rd*rd
      eiej = 16.d0*(xi(i)*xi(j) + xi(n+i)*xi(n+j))*rd*rd
      eiej = eiej + (si- 4.d0*rd*rd)*(sj- 4.d0*rd*rd)
      eiej = eiej/(di*dj)
      dx = x(i) - x(j)
      dy = x(n+i) - x(n+j)
      dz = x(2*n+i) - x(2*n+j)
      r2 = dx*dx + dy*dy + dz*dz
      r = sqrt(r2)
      r3 = r2*r
      rijei = 4.d0*rd*dx*xi(i) + 4.d0*rd*dy*xi(n+i)
      rijei = rijei + dz*(si - 4.d0*rd*rd)
      rijei = rijei/di
      rijej = 4.d0*rd*dx*xi(j) + 4.d0*rd*dy*xi(n+j)
      rijej = rijej + dz*(sj - 4.d0*rd*rd)
      rijej = rijej/dj
      v = v + dmu*dmu*(eiej - 3.d0*rijei*rijej/r2)/r3
      enddo
      enddo
      return
      end
```

The following array is the x array (filled with the center of mass coordinates, as explained in the introductory comments of pot),

$$x = \begin{pmatrix} 2.25005 \\ 0.426539 \\ -2.32268 \\ -0.353906 \\ -0.416782 \\ 2.31223 \end{pmatrix}, \tag{12.16}$$

and the stereographic projection coordinates in the ξ array are,

$$\xi = \begin{pmatrix} 0.169091 \\ -0.399769 \\ -0.180684 \\ 0.303733 \end{pmatrix}. \tag{12.17}$$

Testing pot with $n = 2, \mu_D = 10, \epsilon = 0, r_0 = 4.132, R = 1.70$, and with the coordinates equations (Equations 12.16 and 12.17) produces an energy equal to -0.46983 hartree.

12.3　The Map for $\mathbb{R}^{3n} \otimes \left(\mathbb{S}^2\right)^n$

Let there be n linear rigid molecules in a cluster. The space that contains the three center of mass coordinates and the two orientation degrees of freedom for each of the n moieties is represented with the symbol $\mathbb{R}^{3n} \otimes \left(\mathbb{S}^2\right)^n$. In order to construct the stereographic projection map, we convert the Cartesian coordinates of a moiety in the cluster by eliminating the center of mass, $x_A = x'_A - x$, where x is the coordinate of the center. The primed system gives the location of $2n$ atoms in Cartesian coordinates. From time to time, it is useful to convert from the x_i, y_i, z_i, ξ_i^μ, $\mu = 1, 2$, to the primed system. This is accomplished with the following steps. Firstly, there exists a relationship between x_A and x_B that can be easily derived by manipulating the definition of the center of mass coordinate,

$$x_A = x'_A - \frac{m_A x'_A + m_B x'_B}{m_A + m_B}, \tag{12.18}$$

$$x_B = x'_B - \frac{m_A x'_A + m_B x'_B}{m_A + m_B}. \tag{12.19}$$

Solving for x'_A in the first equation and plugging into the second yields,

$$x_B = -\frac{m_A}{m_B} x_A. \tag{12.20}$$

Similar relationships hold for the y and z coordinates. From the definition used to build the stereographic projection, we obtain,

$$x_B - x_A = \frac{4R^2 \xi^1}{\sigma + 4R^2}, \quad y_B - y_A = \frac{4R^2 \xi^2}{\sigma + 4R^2}, \quad z_B - z_A = R\frac{\sigma - 4R^2}{\sigma + 4R^2}, \tag{12.21}$$

where m_A and m_B are the masses of the atoms in the diatomic. Inserting the relationship between x_A and x_B into these,

$$-\frac{m_A}{m_B} x_A - x_A = \frac{4R^2 \xi^1}{\sigma + 4R^2}, \tag{12.22}$$

$$x_A = -\left(\frac{m_B}{m_A + m_B}\right) \frac{4R^2 \xi^1}{\sigma + 4R^2}, \tag{12.23}$$

and slightly rearranging $x_A = x'_A - x$, gives,

$$x'_A = x - \left(\frac{m_B}{m_A + m_B}\right) \frac{4R^2 \xi^1}{\sigma + 4R^2}. \tag{12.24}$$

A similar process yields the following results for the other two coordinates.

$$y'_A = y - \left(\frac{m_B}{m_A + m_B}\right) \frac{4R^2 \xi^2}{\sigma + 4R^2}, \tag{12.25}$$

$$z'_A = z - \left(\frac{m_B}{m_A + m_B}\right) R\frac{\sigma - 4R^2}{\sigma + 4R^2}. \tag{12.26}$$

The same substitution made in favor of x_B into $x'_B = x_B + x$, yields the coordinates for B,

$$x'_B = x + \left(\frac{m_A}{m_A + m_B}\right)\frac{4R^2\xi^1}{\sigma + 4R^2}, \qquad (12.27)$$

$$y'_B = y + \left(\frac{m_A}{m_A + m_B}\right)\frac{4R^2\xi^2}{\sigma + 4R^2}, \qquad (12.28)$$

$$z'_B = z + \left(\frac{m_A}{m_A + m_B}\right)R\frac{\sigma - 4R^2}{\sigma + 4R^2}. \qquad (12.29)$$

This procedure can be repeated for each center i; we have omitted the subscript in the preceding discussion for the sake of clarity. The subroutine xml that follows, handles the conversion.

```
      subroutine xml(x,xi,n,energy)
      implicit real*8 (a-h,o-z)
      parameter(maxn = 1000)
      real*8 x(maxn),xi(maxn)
c on entry: x = center of mass coordinates x(1),.....x(3n)
c where the x coordinates are stored in the entries x(1),....,x(n)
c the y coordinates in the entries x(n+1),...,x(2n)
c and the z coordinates in the x(2n+1),....,x(3n)
c  xi = stereographic projection coordinates xi(1),.....x(2n)
c where the xi^1 coordinates are stored in the entries xi(1),....,xi(n)
c the xi^2 coordinates in the entries xi(n+1),...,x(2n)
c This subroutine prepares a file for xmakemol using the center of mass
c data and the orientation data to put HI molecules with the correct
c orientation in space.
      dmt = 127.9d0
      dma = 1.d0/dmt
      dmb = 126.9d0/dmt
      rd = 1.7d0
      write(9,*) 2*n
      write(9,*) 'VSAVE = ',energy
      do i=1,n
       si = xi(i)*xi(i) + xi(n+i)*xi(n+i)
       dx = 4.d0*rd*rd*xi(i)/(si + 4.d0*rd*rd)
       dy = 4.d0*rd*rd*xi(n+i)/(si + 4.d0*rd*rd)
       dz = rd*(si - 4.d0*rd*rd)/(si + 4.d0*rd*rd)
       xa = x(i) - dmb*dx
       ya = x(n+i) - dmb*dy
       za = x(2*n+i) - dmb*dz
       xb = x(i) + dma*dx
       yb = x(n+i) + dma*dy
       zb = x(2*n+i) + dma*dz
       write(9,*) 'H', xa,ya,za
       write(9,*) 'I', xb,yb,zb
      enddo
      return
      end
```

12.4 The Gradient of the Lennard-Jones Dipole-Dipole (LJDD) Potential

Let us concern ourselves with the computation of the 1-form $\partial_\mu V$. This is required for quenching methods, such as the Brownian algorithm. Using the chain rule, it is not difficult to evaluate the gradient of the Lennard-Jones part of the potential. This is done in detail in Chapter 8. The result is,

$$\frac{\partial V^{LJ}}{\partial x_i} = -\sum_{j=1,(j\neq i)}^{n} \frac{24\epsilon}{r_{ij}^2}\left[2\left(\frac{r_0}{r_{ij}}\right)^{12} - \left(\frac{r_0}{r_{ij}}\right)^6\right](x_i - x_j). \tag{12.30}$$

The derivatives of the Lennard-Jones term with respect to the projection coordinates, vanish. The derivatives of the dipole-dipole interaction with respect to the coordinates of the centers, follow. Let us rewrite V^{DD} as,

$$V^{DD} = \frac{1}{2}\mu_D^2 \sum_{j=1}^{n} \sum_{k=1,(k\neq j)}^{n}\left[\frac{\mathbf{e}_j \cdot \mathbf{e}_k}{r_{jk}^3} - 3\frac{(\mathbf{r}_{jk}\cdot\mathbf{e}_k)(\mathbf{r}_{jk}\cdot\mathbf{e}_k)}{r_{jk}^5}\right]. \tag{12.31}$$

Note that the first term only depends on r_{jk} and the projection coordinates. The second term depends on r_{jk}^{-5} and on x_i in the numerator. The derivative with respect to x_i generates three terms,

$$\frac{\partial V^{DD}}{\partial x_i} = \frac{1}{2}\mu_D^2 \sum_{j-1}^{n} \sum_{k=1,(k\neq j)}^{n}\left[-3\frac{\mathbf{e}_j \cdot \mathbf{e}_k}{r_{jk}^4} + 15\frac{(\mathbf{r}_{jk}\cdot\mathbf{e}_j)(\mathbf{r}_{jk}\cdot\mathbf{e}_k)}{r_{jk}^6}\right]$$
$$\times \frac{(x_j - x_k)(\delta_{ji} - \delta_{ki})}{r_{jk}} + \frac{1}{2}\mu_D^2 \sum_{j=1}^{n} \sum_{k=1,(k\neq j)}^{n}$$
$$- \frac{3}{r_{jk}^5}\frac{\partial}{\partial x_i}\left[(\mathbf{r}_{jk}\cdot\mathbf{e}_j)(\mathbf{r}_{jk}\cdot\mathbf{e}_k)\right]. \tag{12.32}$$

The first two terms can be handled the same way as for the Lennard-Jones potential in Chapter 8,

$$\frac{\partial V^{DD}}{\partial x_i} = \mu_D^2 \sum_{j=1}^{n}\left[-3\frac{\mathbf{e}_i \cdot \mathbf{e}_j}{r_{ij}^5} + 15\frac{(\mathbf{r}_{ij}\cdot\mathbf{e}_j)(\mathbf{r}_{ij}\cdot\mathbf{e}_i)}{r_{ij}^7}\right](x_i - x_j)$$
$$+ \frac{1}{2}\mu_D^2 \sum_{j=1}^{n} \sum_{k=1,(k\neq j)}^{n} -\frac{3}{r_{jk}^5}\left[(\mathbf{r}_{jk}\cdot\mathbf{e}_j)\frac{\partial}{\partial x_i}(\mathbf{r}_{jk}\cdot\mathbf{e}_k)\right.$$
$$\left. + (\mathbf{r}_{jk}\cdot\mathbf{e}_k)\frac{\partial}{\partial x_i}(\mathbf{r}_{jk}\cdot\mathbf{e}_j)\right]. \tag{12.33}$$

Let's look at the last term in detail. Using the equations for the dot product, we arrive at,

$$\frac{1}{2}\mu_D^2 \sum_{j=1}^{n} \sum_{k=1,(k\neq j)}^{n} -\frac{3}{r_{jk}^5}\left[(\mathbf{r}_{jk}\cdot\mathbf{e}_j)\frac{\partial}{\partial x_i}(\mathbf{r}_{jk}\cdot\mathbf{e}_k) + (\mathbf{r}_{jk}\cdot\mathbf{e}_k)\frac{\partial}{\partial x_i}(\mathbf{r}_{jk}\cdot\mathbf{e}_j)\right]$$

$$=\frac{1}{2}\mu_D^2 \sum_{j=1}^{n} \sum_{k=1,(k\neq j)}^{n} -\frac{3}{r_{jk}^5}\left[(\mathbf{r}_{jk}\cdot\mathbf{e}_j)\frac{4R\xi_k^1}{(\sigma_k+4R^2)} + (\mathbf{r}_{jk}\cdot\mathbf{e}_k)\frac{4R\xi_j^1}{(\sigma_j+4R^2)}\right]$$

$$\times (\delta_{ji}-\delta_{ki}). \tag{12.34}$$

For the two Kronecker deltas, we get two cases. The $(j=i)$ case yields,

$$\frac{1}{2}\mu_D^2 \sum_{k=1,(k\neq j)}^{n} -\frac{3}{r_{ik}^5}\left[(\mathbf{r}_{ik}\cdot\mathbf{e}_i)\frac{4R\xi_k^1}{(\sigma_k+4R^2)} + (\mathbf{r}_{ik}\cdot\mathbf{e}_k)\frac{4R\xi_i^1}{(\sigma_i+4R^2)}\right], \tag{12.35}$$

whereas the $(k=i)$ case yields,

$$\frac{1}{2}\mu_D^2 \sum_{j=1,(j\neq i)}^{n} -\frac{3}{r_{ji}^5}\left[(\mathbf{r}_{ji}\cdot\mathbf{e}_j)\frac{4R\xi_i^1}{(\sigma_i+4R^2)} + (\mathbf{r}_{ji}\cdot\mathbf{e}_i)\frac{4R\xi_j^1}{(\sigma_j+4R^2)}\right](-1). \tag{12.36}$$

Note that the second case is identical to the first, since $r_{ji}=r_{ij}$. However, from the dot product, we get a sign change, $\mathbf{r}_{ji}\cdot\mathbf{e}_i = -\mathbf{r}_{ij}\cdot\mathbf{e}_i$ and $\mathbf{r}_{ji}\cdot\mathbf{e}_j = -\mathbf{r}_{ij}\cdot\mathbf{e}_j$. Switching i and j in the second case, changing the summation index into the sum for the first case from k to j, and putting everything together, we get the derivative of the dipole-dipole interaction with respect to the x coordinates of the centers,

$$\frac{\partial V^{DD}}{\partial x_i} = \mu_D^2 \sum_{j=1}^{n}\left[-3\frac{\mathbf{e}_i\cdot\mathbf{e}_j}{r_{ij}^5} + 15\frac{(\mathbf{r}_{ij}\cdot\mathbf{e}_j)(\mathbf{r}_{ij}\cdot\mathbf{e}_i)}{r_{ij}^7}\right](x_i-x_j)$$

$$+\mu_D^2 \sum_{j=1,(j\neq i)}^{n} -\frac{3}{r_{ij}^5}\left[(\mathbf{r}_{ij}\cdot\mathbf{e}_j)\frac{4R\xi_i^1}{(\sigma_i+4R^2)} + (\mathbf{r}_{ij}\cdot\mathbf{e}_i)\frac{4R\xi_j^1}{(\sigma_j+4R^2)}\right]. \tag{12.37}$$

With similar steps, one arrives at the expressions for the derivative of the dipole-dipole interaction, with respect to the y coordinates of the centers,

$$\frac{\partial V^{DD}}{\partial y_i} = \mu_D^2 \sum_{j=1}^{n}\left[-3\frac{\mathbf{e}_i\cdot\mathbf{e}_j}{r_{ij}^5} + 15\frac{(\mathbf{r}_{ij}\cdot\mathbf{e}_j)(\mathbf{r}_{ij}\cdot\mathbf{e}_i)}{r_{ij}^7}\right](y_i-y_j)$$

$$+\mu_D^2 \sum_{j=1,(j\neq i)}^{n} -\frac{3}{r_{ij}^5}\left[(\mathbf{r}_{ij}\cdot\mathbf{e}_j)\frac{4R\xi_i^2}{(\sigma_i+4R^2)} + (\mathbf{r}_{ij}\cdot\mathbf{e}_i)\frac{4R\xi_j^2}{(\sigma_j+4R^2)}\right], \tag{12.38}$$

and the z coordinates of the centers,

$$\frac{\partial V^{DD}}{\partial z_i} = \mu_D^2 \sum_{j=1}^{n} \left[-3\frac{\mathbf{e}_i \cdot \mathbf{e}_j}{r_{ij}^5} + 15\frac{(\mathbf{r}_{ij} \cdot \mathbf{e}_j)(\mathbf{r}_{ij} \cdot \mathbf{e}_i)}{r_{ij}^7} \right] (z_i - z_j)$$

$$+ \mu_D^2 \sum_{j=1,(j \neq i)}^{n} -\frac{3}{r_{ij}^5} \left[(\mathbf{r}_{ij} \cdot \mathbf{e}_j)\frac{(\sigma_i - 4R^2)}{(\sigma_i + 4R^2)} + (\mathbf{r}_{ij} \cdot \mathbf{e}_i)\frac{(\sigma_j - 4R^2)}{(\sigma_j + 4R^2)} \right].$$

$$(12.39)$$

The last piece is the expression for the gradient of the dipole-dipole potential, with respect to the projection coordinates. Let us begin with the same expression for V^{DD}, broken down into two parts for convenience,

$$V^{DD} = \frac{1}{2}\mu_D^2 \sum_{j=1}^{n} \sum_{k=1,(k \neq j)}^{n} \frac{\mathbf{e}_j \cdot \mathbf{e}_k}{r_{jk}^3} + \frac{1}{2}\mu_D^2 \sum_{j=1}^{n} \sum_{k=1,(k \neq j)}^{n} -3\frac{(\mathbf{r}_{jk} \cdot \mathbf{e}_k)(\mathbf{r}_{jk} \cdot \mathbf{e}_k)}{r_{jk}^5}.$$

$$(12.40)$$

For the first part, we get,

$$\frac{1}{2}\mu_D^2 \sum_{j=1}^{n} \sum_{k=1,(k \neq j)}^{n} \frac{\partial}{\partial \xi_i^\mu} \frac{\mathbf{e}_j \cdot \mathbf{e}_k}{r_{jk}^3}$$

$$= \frac{1}{2}\mu_D^2 \sum_{j=1}^{n} \sum_{k=1,(k \neq j)}^{n} \frac{1}{r_{jk}^3} \left[\frac{16R^2 \xi_k^\mu + 2\xi_j^\mu (\sigma_k - 4R^2)}{(\sigma_k + 4R^2)(\sigma_j + 4R^2)} - \frac{2\xi_j^\mu (\mathbf{e}_j \cdot \mathbf{e}_k)}{(\sigma_j + 4R^2)} \right] \delta_{ji}$$

$$+ \frac{1}{2}\mu_D^2 \sum_{j=1}^{n} \sum_{k=1,(k \neq j)}^{n} \frac{1}{r_{jk}^3} \left[\frac{16R^2 \xi_j^\mu + 2\xi_k^\mu (\sigma_j - 4R^2)}{(\sigma_k + 4R^2)(\sigma_j + 4R^2)} - \frac{2\xi_k^\mu (\mathbf{e}_j \cdot \mathbf{e}_k)}{(\sigma_k + 4R^2)} \right] \delta_{ki}.$$

$$(12.41)$$

Upon collapsing the appropriate sums, the first and second term are identical, and by replacing the index in the first term (as before) and adding the two terms, one arrives at

$$\frac{1}{2}\mu_D^2 \sum_{j=1}^{n} \sum_{k=1,(k \neq j)}^{n} \frac{\partial}{\partial \xi_i^\mu} \frac{\mathbf{e}_j \cdot \mathbf{e}_k}{r_{jk}^3}$$

$$= \frac{1}{2}\mu_D^2 \sum_{j=1,(j \neq i)}^{n} \frac{1}{r_{ij}^3} \left[\frac{16R^2 \xi_j^\mu + 2\xi_i^\mu (\sigma_j - 4R^2)}{(\sigma_i + 4R^2)(\sigma_j + 4R^2)} - \frac{2\xi_i^\mu (\mathbf{e}_i \cdot \mathbf{e}_j)}{(\sigma_i + 4R^2)} \right]. \quad (12.42)$$

In the derivative of the second part, we obtain,

$$\frac{1}{2}\mu_D^2 \sum_{j=1}^{n} \sum_{k=1,(k \neq j)}^{n} -\frac{3}{r_{jk}^5} \frac{\partial}{\partial \xi_i^\mu} [(\mathbf{r}_{jk} \cdot \mathbf{e}_j)(\mathbf{r}_{jk} \cdot \mathbf{e}_k)]$$

$$= \frac{1}{2}\mu_D^2 \sum_{j=1}^{n} \sum_{k=1,(k \neq j)}^{n} -\frac{3}{r_{jk}^5} \left[(\mathbf{r}_{jk} \cdot \mathbf{e}_j)\frac{\partial}{\partial \xi_i^\mu}(\mathbf{r}_{jk} \cdot \mathbf{e}_k) + (\mathbf{r}_{jk} \cdot \mathbf{e}_k)\frac{\partial}{\partial \xi_i^\mu}(\mathbf{r}_{jk}\mathbf{e}_j) \right],$$

$$(12.43)$$

where the derivatives of the dot product quantities are,

$$
\frac{\partial}{\partial \xi_i^1} \left(\mathbf{r}_{jk} \cdot \mathbf{e}_k \right) = \left[\frac{4R\left(x_j - x_k\right) + 2\xi_k^1\left(z_j - z_k\right)}{\left(\sigma_k + 4R^2\right)} - \frac{\left(\mathbf{r}_{jk} \cdot \mathbf{e}_k\right) 2\xi_k^1}{\left(\sigma_k + 4R^2\right)} \right] \delta_{ki}, \quad (12.44)
$$

$$
\frac{\partial}{\partial \xi_i^1} \left(\mathbf{r}_{jk} \cdot \mathbf{e}_j \right) = \left[\frac{4R\left(x_j - x_k\right) + 2\xi_j^1\left(z_j - z_k\right)}{\left(\sigma_j + 4R^2\right)} - \frac{\left(\mathbf{r}_{jk} \cdot \mathbf{e}_k\right) 2\xi_j^1}{\left(\sigma_j + 4R^2\right)} \right] \delta_{ji}. \quad (12.45)
$$

Two cases arise here as well. When $k = i$, the first term contributes and the second term vanishes. Conversely, when $j = i$, the second term contributes the exact same expression (aside from a different summation index as usual), while the first term vanishes. When the expressions for these two cases are added, one obtains,

$$
\frac{1}{2}\mu_D^2 \sum_{j=1}^{n} \sum_{k=1,(k \neq j)}^{n} -\frac{3}{r_{jk}^5} \frac{\partial}{\partial \xi_i^1} \left[\left(\mathbf{r}_{jk} \cdot \mathbf{e}_j\right)\left(\mathbf{r}_{jk} \cdot \mathbf{e}_k\right) \right]
$$

$$
= \mu_D^2 \sum_{j=1,(j \neq i)}^{n} -\frac{3\left(\mathbf{r}_{ij} \cdot \mathbf{e}_j\right)}{r_{ij}^5} \left[\frac{4R\left(x_i - x_j\right) + 2\xi_i^1\left(z_i - z_j\right)}{\left(\sigma_i + 4R^2\right)} - \frac{\left(\mathbf{r}_{ij} \cdot \mathbf{e}_i\right) 2\xi_i^1}{\left(\sigma_i + 4R^2\right)} \right],
$$
$$(12.46)$$

for the first projection coordinate and,

$$
\frac{1}{2}\mu_D^2 \sum_{j=1}^{n} \sum_{k=1,(k \neq j)}^{n} -\frac{3}{r_{jk}^5} \frac{\partial}{\partial \xi_i^2} \left[\left(\mathbf{r}_{jk} \cdot \mathbf{e}_j\right)\left(\mathbf{r}_{jk} \cdot \mathbf{e}_k\right) \right]
$$

$$
= \mu_D^2 \sum_{j=1,(j \neq i)}^{n} -\frac{3\left(\mathbf{r}_{ij} \cdot \mathbf{e}_j\right)}{r_{ij}^5} \left[\frac{4R\left(y_i - y_j\right) + 2\xi_j^2\left(z_i - z_j\right)}{\left(\sigma_i + 4R^2\right)} - \frac{\left(\mathbf{r}_{ij} \cdot \mathbf{e}_i\right) 2\xi_i^2}{\left(\sigma_i + 4R^2\right)} \right],
$$
$$(12.47)$$

for the second. All these equations present no problem for computation. The subroutine, potd, handles the computation of the gradient for the PES of $(LJDD)_n$ clusters.

```
      subroutine potd(x,xi,n,dmu,epsilon,r0,rd,v,dv)
      implicit real*8 (a-h,o-z)
      parameter(maxn = 1000)
      real*8 x(maxn),xi(maxn),dv(maxn)
c Potential energy and the gradient for LJDD_n clusters.
c on entry: x = center of mass coordinates x(1),.....x(3n)
c   where the x coordinates are stored in the entries x(1),....,x(n)
c   the y coordinates in the entries x(n+1),...,x(2n)
c   and the z coordinates in the x(2n+1),.....,x(3n)
c   epsilon,r0 are the parameters for the Lennard - Jones interaction
c   xi = stereographic projection coordinates xi(1),.....x(2n)
c   where the xi^1 coordinates are stored in the entries xi(1),....,xi(n)
c   the xi^2 coordinates in the entries xi(n+1),...,x(2n)
```

```
c  dmu,rd are the parameters for the dipole-dipole interaction
c  on return: v is the potential. The gradient is in the array
c  dv. The arrangement of entries for dv is:
c  dv(1),....,dv(n) are the elements corresponding
c                     to the x Cartesian coordinates of the centers in
c  dv(n+1),....,dv(2n) are the elements corresponding
c                     to the y Cartesian coordinates
c  dv(2n+1),....,dv(3n) are the elements corresponding
c                     to the z Cartesian coordinates
c  dv(3n+1),....,dv(4n) are the elements corresponding
c                     to the xi^1 Stereographic coordinates
c  dv(4n+1),....,dv(5n) are the elements corresponding
c                     to the xi^2 Stereographic coordinates
      v = 0.d0
      do i=1,n
       do j = i+1,n
       dx = x(i) - x(j)
       dy = x(n+i) - x(n+j)
       dz = x(2*n+i) - x(2*n+j)
       r2 = dx*dx + dy*dy + dz*dz
       s2 = r0*r0/r2
       s4 = s2*s2
       s6 = s4*s2
       s12 = s6*s6
       v = v + 4.d0*epsilon*(s12 - s6)
       enddo
      enddo
c  in this loop get the gradient
      do i=1,n
        dv(i) = 0.d0
      dv(n+i) = 0.d0
      dv(2*n+i) = 0.d0
      dv(3*n+i) = 0.d0    ! the first stereographic projection
      dv(4*n+i) = 0.d0    ! the second stereographic projection
       do j = 1,n
       if (i .ne. j) then
       dx = x(i) - x(j)
       dy = x(n+i) - x(n+j)
       dz = x(2*n+i) - x(2*n+j)
       r2 = dx*dx + dy*dy + dz*dz
       s2 = r0*r0/r2
       s4 = s2*s2
       s6 = s4*s2
       s12 = s6*s6
       dvdr = -24.d0*epsilon*(2.d0*s12 - s6)/r2
       dv(i) = dv(i) + dvdr*dx
       dv(n+i) = dv(n+i) + dvdr*dy
       dv(2*n+i) = dv(2*n+i) + dvdr*dz
       endif
```

```
      enddo
      enddo
c The dipole - dipole interaction
      do i=1,n
      si = xi(i)*xi(i) + xi(n+i)*xi(n+i)
      di = si + 4.d0*rd*rd
      do j = i+1,n
      sj = xi(j)*xi(j) + xi(n+j)*xi(n+j)
      dj = sj + 4.d0*rd*rd
      eiej = 16.d0*(xi(i)*xi(j) + xi(n+i)*xi(n+j))*rd*rd
      eiej = eiej + (si- 4.d0*rd*rd)*(sj- 4.d0*rd*rd)
      eiej = eiej/(di*dj)
      dx = x(i) - x(j)
      dy = x(n+i) - x(n+j)
      dz = x(2*n+i) - x(2*n+j)
      r2 = dx*dx + dy*dy + dz*dz
      r = sqrt(r2)
      r3 = r2*r
      rijei = 4.d0*rd*dx*xi(i) + 4.d0*rd*dy*xi(n+i)
      rijei = rijei + dz*(si - 4.d0*rd*rd)
      rijei = rijei/di
      rijej = 4.d0*rd*dx*xi(j) + 4.d0*rd*dy*xi(n+j)
      rijej = rijej + dz*(sj - 4.d0*rd*rd)
      rijej = rijej/dj
      v = v + dmu*dmu*(eiej - 3.d0*rijei*rijej/r2)/r3
      enddo
      enddo
c here is the gradient of the dipole - dipole interaction
      do i=1,n
      si = xi(i)*xi(i) + xi(n+i)*xi(n+i)
      di = si + 4.d0*rd*rd
      do j = 1,n
      if (j.ne. i ) then
      sj = xi(j)*xi(j) + xi(n+j)*xi(n+j)
      dj = sj + 4.d0*rd*rd
      eiej = 16.d0*(xi(i)*xi(j) + xi(n+i)*xi(n+j))*rd*rd
      eiej = eiej + (si- 4.d0*rd*rd)*(sj- 4.d0*rd*rd)
      eiej = eiej/(di*dj)
      dx = x(i) - x(j)
      dy = x(n+i) - x(n+j)
      dz = x(2*n+i) - x(2*n+j)
      r2 = dx*dx + dy*dy + dz*dz
      r = sqrt(r2)
      r3 = r2*r
      r5 = r2*r3
      r7 = r5*r2
      rijei = 4.d0*rd*dx*xi(i) + 4.d0*rd*dy*xi(n+i)
      rijei = rijei + dz*(si - 4.d0*rd*rd)
      rijei = rijei/di
```

```
      rijej = 4.d0*rd*dx*xi(j) + 4.d0*rd*dy*xi(n+j)
      rijej = rijej + dz*(sj - 4.d0*rd*rd)
      rijej = rijej/dj
      dvdr = dmu*dmu*(-3.d0*eiej/r5 +15.d0*rijei*rijej/r7)
      tx = rijei*4.d0*rd*xi(j)/dj+ rijej*4.d0*rd*xi(i)/di
      ty = rijei*4.d0*rd*xi(n+j)/dj+ rijej*4.d0*rd*xi(n+i)/di
      tz = rijei*(sj-4.d0*rd*rd)/dj+rijej*(si-4.d0*rd*rd)/di
      dv(i) = dv(i) + dvdr*dx   -3.d0*dmu*dmu*tx/r5
      dv(n+i) = dv(n+i) + dvdr*dy -3.d0*dmu*dmu*ty/r5
      dv(2*n+i) = dv(2*n+i) + dvdr*dz -3.d0*dmu*dmu*tz/r5
c gradient along the projection coordinates
      deiej = 16.d0*xi(j)*rd*rd + 2.d0*(sj- 4.d0*rd*rd)*xi(i)
      deiej = deiej/(di*dj)
      deiej = deiej - 2.d0*eiej*xi(i)/di
      dv(3*n+i) = dv(3*n+i) +   dmu*dmu*deiej/r3
      dre = rijej*(4.d0*rd*dx + 2.d0*xi(i)*dz)/di
      dre = dre - rijej*(2.d0*rijei*xi(i))/di
      dv(3*n+i) = dv(3*n+i) -3.d0*dmu*dmu*dre/r5
      deiej = 16.d0*xi(n+j)*rd*rd + 2.d0*(sj- 4.d0*rd*rd)*xi(n+i)
      deiej = deiej/(di*dj)
      deiej = deiej - 2.d0*eiej*xi(n+i)/di
      dv(4*n+i) = dv(4*n+i) +   dmu*dmu*deiej/r3
      dre = rijej*(4.d0*rd*dy + 2.d0*xi(n+i)*dz)/di
      dre = dre - rijej*(2.d0*rijei*xi(n+i))/di
      dv(4*n+i) = dv(4*n+i) -3.d0*dmu*dmu*dre/r5
      endif
      enddo
      enddo
      return
      end
```

Figures 12.1 and 12.2 are generated by computing the potential and the gradient of two centers. The calling parameters for potd are $n = 2, \mu_D = 10, \epsilon = 0, r_0 = 4.132, R = 1.70$, and the input coordinates are the same as in Equations 12.16 and 12.17.

12.5   Beyond the Stockmayer Model for Rigid Linear Tops

As mentioned earlier, linear charge distributions separated by finite distances are not ideal dipoles, they are sources of quadrupole fields, octupole fields, etc. The potential energy model for two sets of linear and nonlinear charge distributions, separated by a distance r greater than their sizes, can be constructed using a spherical tensors expansion. For example, for two rigid linear charge

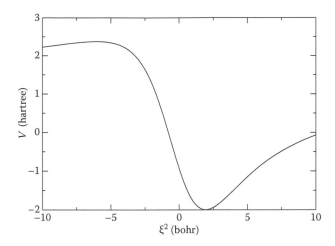

FIGURE 12.1

The potential energy surface for a uniform Stockmayer dimer as a function of the second projection coordinate for the second center. The input configuration is in Equations 12.16 and 12.17. The graph contains 4,000 points generated by spacing values of ξ_2^2 0.01 bohr apart.

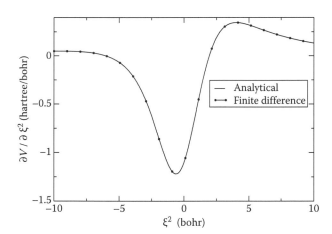

FIGURE 12.2

The gradient of the potential energy surface for a uniform Stockmayer dimer as a function of the second projection coordinate for the first center. The function in this figure is checked against that produced by finite differences.

distributions, we can expand the electrostatic energy using the following sum,

$$V\left(\theta_1, \phi_1, \theta_2, \phi_2, r\right) = \sum_{l_1, l_2, l_3} v_{l_1, l_2, l_3}\left(r\right) A_{l_1, l_2, l_3}\left(\theta_1, \phi_1, \theta_2, \phi_2\right), \qquad (12.48)$$

where the angular part is an expansion of spherical harmonics,

$$A_{l_1, l_2, l_3}\left(\theta_1, \phi_1, \theta_2, \phi_2\right)$$

$$= \sum_{m=-\min(l_1, l_2)}^{+\min(l_1, l_2)} \begin{pmatrix} l_1 & l_2 & l_3 \\ m & -m & 0 \end{pmatrix} Y_{l_1}^m\left(\theta_1, \phi_1\right) Y_{l_2}^m\left(\theta_2, \phi_2\right), \qquad (12.49)$$

and the electrostatic contribution to the radial part is equal to,

$$v_{l_1, l_2, l_3}^{(e)}\left(r\right) = \delta_{l_1 + l_2, l_3}\left(4\pi\right)\left(-1\right)^{l_2}\left[\frac{\left(2l_3 + 1\right)!}{\left(2l_3 + 1\right)\left(2l_1 + 1\right)!\left(2l_2 + 1\right)!}\right]^{1/2} \frac{Q_{l_1} Q_{l_2}}{r^{l_1 + l_2 + 1}}. \qquad (12.50)$$

The Q_l are the scalar spherical multipole moments (parameters of the potential that can be obtained from the charge distribution). It is possible to compute $Y_l^m\left(\xi^1, \xi^2\right)$, i.e., any order of the spherical harmonic function in terms of the two stereographic projection coordinates directly. The possibility to compute $Y_l^m\left(\xi^1, \xi^2\right)$ directly makes these expansions practical for the formulation of potential energy models of clusters of linear rigid tops. In Table 12.1, we have compiled all these functions up to $l = 3$. The functions in Table 12.1 can be obtained by normalizing the harmonic polynomials $H_l^m\left(\xi^1, \xi^2\right)$, obtained using the following generating function,

$$t^l \sum_{m=-l}^{l} H_l^m\left(\xi^1, \xi^2\right) t^m$$

$$= \frac{1}{\left(\sigma + 4R^2\right)^l}\left[4R\left(\xi^1 - i\xi^2\right) - 2\left(\sigma - 4R^2\right)t - 4R\left(\xi^1 + i\xi^2\right)t^2\right]^l. \qquad (12.51)$$

Equation 12.51, together with the Racah normalization for the spherical harmonics Y_l^m in Chapter 5, can be used to extend the entries in Table 12.1. It is recommended that the reader derives the expressions for the spherical harmonics, Y_l^m, with the normalizations, as given in Chapter 5, before attempting to extend Table 12.1. Expansions up to $l = 5$ are not uncommon in the literature. All moments up to $l = 5$ are usually necessary to achieve spectroscopic accuracy for dimers of cylindrically symmetric molecules.

It is also necessary to find additional modifications for the repulsive and dispersive part, which, unlike the simpler Stockmayer model, may have anisotropic contributions. Equation 12.48 allows for these contributions, using three or more terms in the radial part of V. For example, $v_{l_1, l_2, l_3}\left(r\right)$ can be constructed with three terms as follows:

$$v_{l_1, l_2, l_3}\left(r\right) = v_{l_1, l_2, l_3}^{(e)}\left(r\right) + v_{l_1, l_2, l_3}^{(r)}\left(r\right) + v_{l_1, l_2, l_3}^{(d)}\left(r\right). \qquad (12.52)$$

TABLE 12.1
Harmonic Polynomials

l	m	$Y_l^m \left(\xi^1, \xi^2 \right)$
0	0	$\dfrac{1}{2\sqrt{\pi}}$
1	0	$\dfrac{1}{2}\sqrt{\dfrac{3}{\pi}} \left(\dfrac{\sigma - 4R^2}{\sigma + 4R^2} \right)$
1	± 1	$\mp \dfrac{1}{2}\sqrt{\dfrac{3}{2\pi}} \left(4R\dfrac{\xi^1 \pm i\xi^2}{\sigma + 4R^2} \right)$
2	0	$\dfrac{1}{4}\sqrt{\dfrac{5}{\pi}} \left[\dfrac{2\left(\sigma - 4R^2\right)^2 - \left(4R\xi^1\right)^2 - \left(4R\xi^1\right)^2}{\left(\sigma + 4R^2\right)^2} \right]$
2	± 1	$\mp \dfrac{1}{2}\sqrt{\dfrac{15}{2\pi}} \left(\dfrac{\sigma - 4R^2}{\sigma + 4R^2} \right) \left(4R\dfrac{\xi^1 \pm i\xi^2}{\sigma + 4R^2} \right)$
2	± 2	$\dfrac{1}{4}\sqrt{\dfrac{15}{2\pi}} \left(4R\dfrac{\xi^1 \pm i\xi^2}{\sigma + 4R^2} \right)^2$
3	0	$\dfrac{1}{4}\sqrt{\dfrac{7}{\pi}} \left(\dfrac{\sigma - 4R^2}{\sigma + 4R^2} \right) \left[\dfrac{2\left(\sigma - 4R^2\right)^2 - 3\left(4R\xi^1\right)^2 - 3\left(4R\xi^1\right)^2}{\left(\sigma + 4R^2\right)^2} \right]$
3	± 1	$\mp \dfrac{1}{8}\sqrt{\dfrac{21}{\pi}} \left[\dfrac{4\left(\sigma - 4R^2\right)^2 - \left(4R\xi^1\right)^2 - \left(4R\xi^1\right)^2}{\left(\sigma + 4R^2\right)^2} \right] \left(4R\dfrac{\xi^1 \pm i\xi^2}{\sigma + 4R^2} \right)$
3	± 2	$\dfrac{1}{4}\sqrt{\dfrac{105}{2\pi}} \left(\dfrac{\sigma - 4R^2}{\sigma + 4R^2} \right) \left(4R\dfrac{\xi^1 \pm i\xi^2}{\sigma + 4R^2} \right)^2$
3	± 3	$\mp \dfrac{1}{8}\sqrt{\dfrac{35}{\pi}} \left(4R\dfrac{\xi^1 \pm i\xi^2}{\sigma + 4R^2} \right)^3$

The radial part of the electrostatic interaction, $v_{l_1,l_2,l_3}^{(e)}(r)$, is in Equation 12.50, whereas the repulsive portion is usually a Born-Mayer function,

$$v_{l_1,l_2,l_3}^{(r)}(r) = c_{l_1,l_2,l_3} \exp\left[-a_{l_1,l_2,l_3}\left(r - b_{l_1,l_2,l_3}\right)\right], \tag{12.53}$$

and the dispersion typically has several powers of r,

$$v_{l_1,l_2,l_3}^{(d)}(r) = \frac{d_6^{l_1,l_2,l_3}}{r^6} + \frac{d_7^{l_1,l_2,l_3}}{r^7} + \frac{d_8^{l_1,l_2,l_3}}{r^8} + \frac{d_9^{l_1,l_2,l_3}}{r^9} + \frac{d_{10}^{l_1,l_2,l_3}}{r^{10}}. \tag{12.54}$$

These more sophisticated potentials are used to invert high resolution spectroscopic data for dimers, and may contain as many as 30–60 parameters: a_{l_1,l_2,l_3}, b_{l_1,l_2,l_3}, c_{l_1,l_2,l_3}, $d_k^{l_1,l_2,l_3}$, Q_l, etc. The parameters are systematically adjusted to reproduce experimental data with high accuracy. There are several empirical potentials for the noble gas-HX, noble gas-CO, and HX-HY dimers (X,Y = halides, CN), which are used to guide additional high resolution spectroscopy experiments. It is possible to obtain analytical expressions for the gradient and the Hessian of V in Equation 12.48.

12.6 The Hessian Metric Tensor on $\mathbb{R}^{3n} \otimes (\mathbb{S}^2)^n$

Now that we have a set of reasonable model potential energy surfaces for linear tops, we concentrate on other essential parts needed for simulations. In this section, we provide one of the many ways that one can set up and compute a representation of the Hessian metric tensor. We are basing our algorithms, in this section, on the same choice of coordinate storage for the computation of the potential and the gradient of the Stockmayer model in the last section. We continue to assume that the clusters are uniform throughout.

The size squared of the infinitesimal line segment for an aggregate of n linear tops reads,

$$ds^2 = m \sum_{j=1}^n \delta_{\mu\upsilon} dx_j^\mu dx_j^\upsilon + \mu_{AB} \sum_{j=1}^n \left(\frac{4R^2}{\sigma_j + 4R^2} \right)^2 \delta_{\mu'\upsilon'} d\xi_j^{\mu'} d\xi_j^{\upsilon'}, \quad (12.55)$$

where the superscripts μ and υ on the Cartesian coordinates for the centers range from 1 to 3, and represent x for 1, y for 2 and z for 3; μ' and υ' range from 1 to 2; m is the total mass, and μ_{AB} is the reduced mass of AB for the linear top. The Hessian metric tensor in Equation 12.55 is diagonal,

$$g_{\mu\upsilon} = \text{diag} \left\{ \underbrace{m, \ldots}_{3n \text{ terms}}, \underbrace{\mu_{AB} \left(\frac{4R^2}{\sigma_1 + 4R^2} \right)^2, \ldots, \mu_{AB} \left(\frac{4R^2}{\sigma_n + 4R^2} \right)^2}_{n \text{ terms}}, \right.$$
$$\left. \underbrace{\mu_{AB} \left(\frac{4R^2}{\sigma_1 + 4R^2} \right)^2, \ldots, \mu_{AB} \left(\frac{4R^2}{\sigma_n + 4R^2} \right)^2}_{n \text{ terms}} \right\}. \quad (12.56)$$

The arrangement of entries we have chosen follows from having adopted a generalized coordinate set q^μ, where, on q, the Greek index ranges from 0 to $5n$. The first $3n$ entries on q^μ are the Cartesian coordinates of the centers.

The first orientation coordinate for each center are in the range $3n < \mu \leq 4n$, whereas the second orientation coordinate for each center are in the range $4n < \mu \leq 5n$. This ordering is consistent, as it must, with the ordering we have chosen for the array containing the gradient $\partial_\mu V$ in the previous section. If any other potential energy surface is used to model linear tops, the order in which the gradient is evaluated and stored must be preserved if any of the subroutines in this section are used without substantial modifications. The Hessian metric tensor is computed with the following version of the subroutine metric.

```
      subroutine metric(xi,n,g,rd)
      implicit real*8 (a-h,o-z)
      parameter(maxn = 1000)
      real*8 g(maxn),xi(maxn)
c  On entry:  xi = stereographic projection coordinates xi(1),.....x(2n)
c  where the xi^1 coordinates are stored in the entries xi(1),....,xi(n)
c  the xi^2 coordinates in the entries xi(n+1),...,x(2n)
c  rd is the radial distance between A and B in each center of the cluster
c  On return: the diagonal elements of the metric tensor are in the array
c  g. The arrangement of entries for g is: g(1),....,g(3n) are the
c  diagonal elements corresponding to the Cartesian Coordinates
c  of the centers. in  g(3n+1),....,g(5n) are the entries corresponding
c  to the stereographic projections for each center.
c  the masses used are 127 amu for A (converted to atomic units) and
c  1 amu  for B (converted to atomic units).
      dmb = 1.d0*1822.89d0
      dma = 126.9d0*1822.89  ! HI clusters
      dmt = dma + dmb
      dmu = dma*dmb/dmt
      do i=1,3*n
       g(i) = dmt
      enddo
      do i=3*n+1,4*n
      index1 = i - 3*n
      index2 = i - 2*n
      sigma = xi(index1)*xi(index1) + xi(index2)*xi(index2)
      write(6,*) i,index1,index2,sigma
      den = sigma + 4.d0*rd*rd
      sd = 4.d0*rd*rd/den
      g(i) = dmu*sd*sd
      g(n+i) = dmu*sd*sd
        enddo
      return
      end
```

Since the Hessian metric is diagonal and its elements are relatively simple functions, it is straightforward to obtain its gradient, which is necessary for dynamic simulations. Similarly, many of the $125n^3$ connection coefficients vanish, as the result of the high symmetry, diagonal structure, and relative simplicity of $g_{\mu\upsilon}$. The gradient of $g_{\mu\upsilon}$ is,

$$\partial_\mu g_{\lambda\upsilon} = \begin{cases} 0 & \text{if } \lambda \neq \upsilon \\[2ex] 0 & \text{if } \mu < 3n \text{ or } \lambda < 3n \\[2ex] -\dfrac{64\mu_{AB}R^4\xi_j^1}{\left(\sigma_j + 4R^2\right)^3} & \text{if } \lambda = \upsilon = \mu = 3n+j, \ (1 \leq j \leq n) \\[3ex] -\dfrac{64\mu_{AB}R^4\xi_j^1}{\left(\sigma_j + 4R^2\right)^3} & \text{if } \lambda = \upsilon = 4n+j, \ \mu = 3n+j, \ (1 \leq j \leq n) \\[3ex] -\dfrac{64\mu_{AB}R^4\xi_j^2}{\left(\sigma_j + 4R^2\right)^3} & \text{if } \lambda = \upsilon = 3n+j, \ \mu = 4n+j, \ (1 \leq j \leq n) \\[3ex] -\dfrac{64\mu_{AB}R^4\xi_j^2}{\left(\sigma_j + 4R^2\right)^3} & \text{if } \lambda = \upsilon = \mu = 4n+j, \ (1 \leq j \leq n). \end{cases}$$

$$(12.57)$$

The computations of the derivative of the metric tensor and the connection coefficients are carried out by the subroutine dmetric and connection, respectively.

```
      subroutine connection(xi,n,Gamma,rd,m1,m2,m3)
      implicit real*8 (a-h,o-z)
      parameter(maxn = 1000)
      real*8 g(maxn),xi(maxn)
c on entry:  xi = stereographic projection coordinates xi(1),.....x(2n)
c where the xi^1 coordinates are stored in the entries xi(1),....,xi(n)
c the xi^2 coordinates in the entries xi(n+1),...,x(2n)
c rd is the radial distance between A and B in each center of the cluster
c on return:  Gamma is the connection coefficient Gamma^{mu1}_{mu2 mu3}
      Gamma = 0.d0
      call metric(xi,n,g,rd)
c get the derivatives
      call dmetric(xi,n,dg1,rd,m2,m3,m1)
      call dmetric(xi,n,dg2,rd,m3,m1,m2)
      call dmetric(xi,n,dg3,rd,m1,m2,m3)
      Gamma = 0.5*(dg1 + dg2 - dg3)/g(m1)
      return
      end
      subroutine dmetric(xi,n,dg,rd,m1,m2,m3)
      implicit real*8 (a-h,o-z)
      parameter(maxn = 1000)
      real*8 g(maxn),xi(maxn)
c on entry:  xi = stereographic projection coordinates xi(1),.....x(2n)
c where the xi^1 coordinates are stored in the entries xi(1),.....,xi(n)
```

```
c the xi^2 coordinates in the entries xi(n+1),...,x(2n)
c rd is the radial distance between A and B in each center of the cluster
c on return:  dg is the derivative of g: partial_{mu1} g_{mu2 mu3}
      dg = 0.d0
      if (m2 .ne. m3) return   ! off diagonal elements are zero
      if (m1 .le. 3*n) return  ! g is independent of x,y,z
      if (m2 .le. 3*n) return   ! the first 3n elements of g are constant
      if (m2 .ne. m1) then
      if (abs(m2 - m1) .ne. n) return
      endif
      j = m1 - 3*n
      i = m2 - 3*n
      if (i .gt. n) i = i - n
      dmb = 1.d0*1822.89d0
      dma = 126.9d0*1822.89d0
      dmt = dma + dmb
      dmu = dma*dmb/dmt
      sigma = xi(i)*xi(i) + xi(n+i)*xi(n+i)
      den = sigma + 4.d0*rd*rd
      dg = -64.d0*dmu*rd*rd*rd*rd*xi(j)/(den*den*den)
      return
      end
```

The subroutines **dmetric** and **connection** can be used to compute the acceleration for coordinate q^μ, $1 \le \mu \le 5n$.

$$\ddot{q}^\mu + \Gamma^\mu_{\lambda\upsilon}\dot{q}^\upsilon\dot{q}^\lambda + g^{\mu\upsilon}\partial_\upsilon V + \gamma\dot{q}^\mu = 0. \tag{12.58}$$

The inverse of the metric tensor is trivial to obtain in this case. Note that we add a coefficient of drag so that we can derive an integrator for $T = 0$ Brownian dynamics. If we use the subscript i on coordinate q^μ to label a particular time, t_i, and we follow a point-slope approach, we derive a two-part propagator. The velocities are updated using,

$$\dot{q}^\mu_{i+1} = \dot{q}^\mu_i - \left(\Gamma^\mu_{\lambda\upsilon}\dot{q}^\upsilon_i\dot{q}^\lambda_i + g^{\mu\upsilon}_i\partial_\upsilon V_i + \gamma\dot{q}^\mu_i\right)\Delta t, \tag{12.59}$$

and the positions are updated with,

$$q^\mu_{i+1} = q^\mu_i + \dot{q}^\mu_i\Delta t - \left(\Gamma^\mu_{\lambda\upsilon}\dot{q}^\upsilon_i\dot{q}^\lambda_i + g^{\mu\upsilon}_i\partial_\upsilon V_i + \gamma\dot{q}^\mu_i\right)(\Delta t)^2. \tag{12.60}$$

A new version of the subroutine **brown**, adapted to propagate according to Equations 12.59 and 12.60, follows. The integration with values for γ and Δt equal to 0.0001, 0.0003, and 100 a.u., respectively, is adequate for the Stockmayer model for a relatively large range of dipole moments and for reasonable values of the masses inside **metric**. Needless to say, changing any of these values requires careful testing. Note that the subroutine does not return the minimum configuration, it returns only the potential energy value at the minimum configuration.

```
      subroutine brown(xn,xin,n,dmu,epsilon,r0,rd,v,d0,tdm)
      implicit real*8 (a-h,o-z)
      parameter(maxn = 1000)
      real*8 xn(maxn),xin(maxn)
      real*8 g(maxn),x(maxn),xi(maxn),dv(maxn)
      real*8 dq(maxn),dqp(maxn)
c T=0 Brownian dynamics for n rigid centers mapped with 3n Cartesian
c coordinates and 2*n stereographic projection coordinates
c On entry:
c xn,xin      the  3n Cartesian
c             coordinates and 2*n stereographic projection coordinates
c             (not modified by the subroutine)
c dmu,epsilon,
c r0,rd the parameters of the potential energy
c On return:
c v           the potential energy of the minimum
      fc1 = 0.0001  ! control from within fc = drag coefficient
      fc2 = 0.0003
      dt = 100  ! atomic units dt = time step
      do i = 1,3*n
        x(i) = xn(i)
        enddo
      do i = 1,2*n
        xi(i) = xin(i)
        enddo
c initiate the velocity components to zero
      do k=1,5*n
      dq(k) - 0.d0
      dqp(k) = 0.d0
      enddo
      rmsg = 100.00
c begin the integration
c get the gradient and the metric tensor
      do kstep = 1,100000
10    call potd(x,xi,n,dmu,epsilon,r0,rd,v,dv)
      call xmakemol(x,xi,n,v)
      write(6,*) dfloat(kstep -1)*dt,v
      call metric(xi,n,g,rd)
      if (sqrt(rmsg) .lt. 1.d-6) goto 999
c calculate the acceleration component
      rmsg = 0.d0
      do i=1,5*n
        rmsg = rmsg + dv(i)*dv(i)
        fc = fc1
        if (i .gt. 3*n) fc = fc2
        a = -dv(i)/g(i) -fc*dq(i)
        if (i .gt. 3*n .and. i .le. 4*n) then
          m1 = i  ! include the forces of constraint
          m2 = i  ! only call the coefficients you need
          m3 = i
```

```
      call connection(xi,n,giii,rd,m1,m2,m3)
      m1 = i
      m2 = n+i
      m3 = n+i
      call connection(xi,n,ginn,rd,m1,m2,m3)
      m1 = i
      m2 = i
      m3 = n+i
      call connection(xi,n,giin,rd,m1,m2,m3)
      a = a - giii*dq(i)*dq(i)
      a = a - 2.d0*giin*dq(i)*dq(n+i)
      a = a - ginn*dq(n+i)*dq(n+i)
      endif
      if (i .gt. 4*n) then
       m1 = i
       m2 = i
       m3 = i
      call connection(xi,n,giii,rd,m1,m2,m3)
       m1 = i
       m2 = -n+i
       m3 = -n+i
      call connection(xi,n,ginn,rd,m1,m2,m3)
       m1 = i
       m2 = i
       m3 = -n+i
      call connection(xi,n,giin,rd,m1,m2,m3)
      a = a - giii*dq(i)*dq(i)
      a = a - 2.d0*giin*dq(i)*dq(-n+i)
      a = a - ginn*dq(-n+i)*dq(-n+i)
      endif
c update the velocity components
      dqp(i) = a*dt
      enddo
c update the positions
      do i=1,3*n
      dq(i) = dq(i) + dqp(i)
      x(i) = x(i) + dq(i)*dt
      enddo
      do i=1,2*n
      if (dabs(dqp(3*n+i)) .lt. 1.d+20) then
       dq(3*n+i) = dq(3*n+i) + dqp(3*n+i)
      endif
      if (dabs(dq(3*n+i)) .lt. 1.d+20) then
       xi(i) = xi(i) + dq(3*n+i)*dt
      endif
      enddo
      enddo
999   vx = v/d0
      return
      end
```

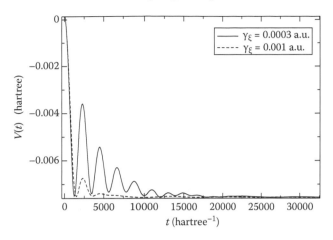

FIGURE 12.3
The time evolution of the potential energy surface for a uniform Stockmayer
dimer propagated with the subroutine **brown**.

We perform a test run initiated with the dimer configuration, as in Equations 12.16 and 12.17, the Stockmayer potential, and with parameters set at $\mu_D = 0.5, \epsilon = 0.00037685, r_0 = 4.132, R = 1.70$. The masses in **metric** represent the HI dimer. The total energy as a function of time for two runs with different values of γ_ξ (the drag coefficient associated with the orientations, **fc2** in the code), is graphed in Figure 12.3. The usual transient nature of the energy, with respect to time, can be seen. Increasing γ_t has the effect of diminishing the intensity of some of the peaks in the harmonic motion of the stereographic projection degrees of freedom. A value of 0.003 a.u. for the drag along the translations would cause the integration to be less efficient. A drag coefficient of 0.001 a.u. is too large for the rotations. With the masses used, the translations of the centers take place at a larger timescale.

12.7 Reweighted Random Series Action for Clusters of Linear Rigid Tops

The subroutine **getwlt** in this section is used to compute the importance sampling function for path integral simulations of n linear tops (**ND** in the code). The array **sn(NX,NX)** stores the values of the core,

$$\Lambda_k (u_i) \qquad 1 \leq k \leq k_{\max}, \qquad (12.61)$$

and the reweighted tail functions,

$$\widetilde{\Lambda}_k (u_i) \qquad k_{\max} < k \leq 4k_{\max}, \qquad (12.62)$$

for a number of values of u_i, $i = 1, 2, N - 1$,

$$u_i = \frac{i}{N}. \tag{12.63}$$

The values of the coordinates at two adjacent time points are computed at every time, u_i,

$$q^\mu (u_{i-1}) = q^\mu + \beta^{1/2} \sum_{k=1}^{k_m} a_k^\mu \Lambda_k (u_{i-1}) + \beta^{1/2} \sum_{k=k_m+1}^{k'_m} a_k^\mu \widetilde{\Lambda}_k (u_{i-1}), \tag{12.64}$$

$$q^\mu (u_i) = q^\mu + \beta^{1/2} \sum_{k=1}^{k_m} a_k^\mu \Lambda_k (u_i) + \beta^{1/2} \sum_{k=k_m+1}^{k'_m} a_k^\mu \widetilde{\Lambda}_k (u_i), \tag{12.65}$$

where $q^\mu (u_0) = q^\mu (u_N) = q^\mu$.

The variable sum contains U,

$$U (q^\mu (u_i)) = -\frac{N}{2\beta} \ln [\det g_{\mu\nu} (q^\mu (u_i))] + \frac{1}{2} g_{\mu\nu} (q^\mu (u_i)) \Delta q^\mu \Delta q^\upsilon + V (q^\mu (u_i)). \tag{12.66}$$

Note that the subroutine uses the postpoint expansion of Kleinert, as opposed to the prepoint expansion presented in Chapter 11, since $g_{\mu\nu}$ and V are evaluated at $q^\mu (u_i)$, as opposed to $q^\mu (u_{i-1})$. As explained in Chapter 11, the lattice correction is the same for both, and it is constant.

```
      subroutine getwlt(temp,w,xin,xiin,N,ND,KMAX,NMAX,wi,sv)
c On input
c   temp = beta in inverse hartree
c   w = array containing the path coefficients
c   xin,xiin =  configuration array
c   N = quadrature (usually KMAX +1)
c   ND = number of linear molecules
c   KMAX = number of core path coefficients
c   NMAX = parameters for sizing arrays
c On return
c wi = the quantum action
      implicit real*8 (a-h,o-z)
   parameter (rd = 2.412008)
      parameter (NX = 350)
      real*8 w(NMAX,0:NX)
   real*8 xin(NMAX),xiin(NMAX)
   real*8 xm1(NMAX),x(NMAX)
   real*8 xi(NMAX),xim1(NMAX)
   real*8 g(NMAX)
      common /traj/sn(NX,NX),sigma2(NX)
   sv = 0.d0
      beta = temp
      sum = 0.d0
   t12 = sqrt(beta)
      dN = dfloat(N)
```

```
       sum = 0.d0
    do i=1,N
c Get the path from the coefficients
       do j=1,ND
       xm1(j) = xin(j)
       x(j) = xin(j)
       xm1(ND+j) = xin(ND+j)
       x(ND+j) = xin(ND+j)
       xm1(2*ND+j) = xin(2*ND+j)
       x(2*ND+j) = xin(2*ND+j)
       xim1(j) = xiin(j)
       xi(j) = xiin(j)
       xim1(ND+j) = xiin(ND+j)
       xi(ND+j) = xiin(ND+j)
       do k=1,4*KMAX
        if (i .gt. 1) then
         xm1(j) = xm1(j) + t12*w(j,k)*sn(i-1,k)
         xm1(ND+j) = xm1(ND+j) + t12*w(ND+j,k)*sn(i-1,k)
         xm1(2*ND+j) = xm1(2*ND+j) + t12*w(2*ND+j,k)*sn(i-1,k)
         xim1(j) = xim1(j) + t12*w(3*ND+j,k)*sn(i-1,k)
         xim1(ND+j) = xim1(ND+j) + t12*w(4*ND+j,k)*sn(i-1,k)
      endif
         x(j) = x(j) + t12*w(j,k)*sn(i,k)
         x(ND+j) = x(ND+j) + t12*w(ND+j,k)*sn(i,k)
         x(2*ND+j) = x(2*ND+j) + t12*w(2*ND+j,k)*sn(i,k)
         xi(j) = xi(j) + t12*w(3*ND+j,k)*sn(i,k)
         xi(ND+j) = xi(ND+j) + t12*w(4*ND+j,k)*sn(i,k)
        enddo
       enddo ! j loop
c the metric tensor and the velocities  are calculated with the core path
       call metric(xi,ND,g,rd)
       call pot(x,xi,ND,v)    ! V is called with the reweighted path
       detg = 1.d0
       do j=1,ND
        sum = sum+(dN/(beta*2.))*g(j)*(x(j)-xm1(j))**2
        sum = sum+(dN/(beta*2.))*g(ND+j)*(x(ND+j)-xm1(ND+j))**2
        sum = sum+(dN/(beta*2.))*g(2*ND+j)*(x(2*ND+j)-xm1(2*ND+j))**2
        sum = sum+(dN/(beta*2.))*g(3*ND+j)*(xi(j)-xim1(j))**2
        sum = sum+(dN/(beta*2.))*g(4*ND+j)*(xi(ND+j)-xim1(ND+j))**2
        detg = detg*g(3*ND+j)*g(4*ND+j)
       enddo
       sv = sv + v
       sum = sum + beta*v/dN
       sum = sum - 0.5*log(detg)
      enddo   ! i loop
      wi = sum
      return
      end
```

12.8 The Local Energy Estimator for Clusters of Linear Rigid Tops

For DMC and VMC simulations, the local energy can be computed with the following code,

```
call d2pot(x,xi,n,v,dv,d2v)
call metric(xi,n,g,rd)
  sumlb = 0.d0
do j =1,5*n
  sumlb = sumlb + ((temp*dv(j)/2.d0)*(temp*dv(j)/2.d0)
& - 0.5d0*temp*d2v(j))/g(j)
  enddo
eloc = -0.5d0*sumlb + v
```

where the code takes advantage of the relatively simple form of the Laplace-Beltrami operator for linear tops,

$$\nabla^2 = \sum_{j=1}^{n} \frac{\partial^2}{\partial x_i^2} + \frac{\partial^2}{\partial y_i^2} + \frac{\partial^2}{\partial z_i^2} + \frac{(\sigma_i + 4R)^2}{(4R)^2} \frac{\partial^2}{\partial \xi_i^1 \partial \xi_i^1} + \frac{(\sigma_i + 4R)^2}{(4R)^2} \frac{\partial^2}{\partial \xi_i^2 \partial \xi_i^2}. \quad (12.67)$$

The subroutine d2pot (not shown) takes the configuration in the two arrays, x and xi, and returns the potential, the gradient dv, and the diagonal elements of the Hessian, $\partial_\mu \partial_\mu V$. The arrays, dv and d2v, must be organized the same way as in the subroutine potd for the Stockmayer model. The reader must supply his/her own pot and d2pot subroutines.

12.9 Clusters of Rigid Nonlinear Tops

It is tempting, after developing the dipole-dipole interaction and its gradient as a function of stereographic projections and distance, and extending it to higher order poles, to continue along the same path and derive interactions for nonlinear rigid charge distributions. However, the angular momentum theory treatment of the kinetic energy becomes more complicated. The tricky part is to have the proper "reduced" mass for simulations. The proper reduced mass is, in fact, the Hessian metric tensor. For nonlinear tops, the Hessian metric is different from that for linear tops, as demonstrated in Chapter 10. We will return to clusters of nonlinear tops and the resulting Hessian metric tensor later in the chapter. The quantum treatment of nonlinear rigid top with vector spaces is challenging, and requires perturbation techniques when the top is asymmetric. The expansions in powers of spherical tensors are useful to use angular momentum theory results and compute the matrix elements of the interactions in terms of vector coupling coefficients. Vector space methods are impractical for sizes beyond the dimer at this time. The easier course of action is to develop models that simply sum up the monopole-monopole energies for several partial charges distributed in sensible places around the rigid molecular framework. There is a multitude of such potential energy surface models in the literature. For water clusters and liquid water, for example, the TIP4P model 863–877, places a single isotropic Lennard-Jones source at each oxygen, and three partial charges distributed on and in the vicinity of the nuclei. The TIP4P assumes a rigid water molecule, and is capable of reproducing quantum properties of liquid water at and below room temperature with path integral simulations. The best way to compute the point-to-point electrostatic contributions for these models is by using the Cartesian coordinates of the charges. The majority of the point charges are located at the nuclei, therefore, it is convenient to have a subroutine that is called inside the potential energy subroutine, and that computes the Cartesian coordinates of all the atoms in a cluster of n rigid tops.

12.10   Coordinate Transformations for $\mathbb{R}^{3n} \otimes \mathbb{I}^n$

Consider the following configuration for a water molecule in the body fixed frame,

$$x_{BF}^{\mu} = \begin{array}{c|ccc} \mu & H & O & H \\ \hline 1 & 0 & 0 & 0 \\ 2 & \alpha & \beta & \alpha \\ 3 & \gamma & 0 & -\gamma \end{array}, \tag{12.68}$$

where the center of mass constraint along the y coordinates imposes the following relation,

$$\beta = -\frac{2m_H}{m_O}\alpha, \tag{12.69}$$

and m_H, m_O are the masses of hydrogen and oxygen, respectively. A rotation of x_{BF}^{μ} yields an intermediate set of Cartesian coordinates in the center of mass frame,

$$\mathbf{r}_i - \mathbf{r}_C = \mathbf{R}\,\mathbf{r}_{BF}, \tag{12.70}$$

where $\mathbf{R}$ is the full rotation matrix, $\mathbf{R}_\phi\mathbf{R}_\theta\mathbf{R}_\psi$, of Equation 10.182. Therefore, given the coordinates of the center of mass of the water $\mathbf{r}_C$, and the values of the three stereographic projections, ξ^μ, $\mu = 1, 2, 3$, as we introduce in Chapter 10, we find the Cartesian coordinates of all the atoms in a cluster of water molecules by implementing the following straightforward algorithm:

i. For center i, compute the three quaternion parameters,

$$q^\mu = \frac{4\xi^\mu}{\sigma + 4} \qquad (1 \le \mu \le 3), \tag{12.71}$$

$$q^\mu = \frac{\sigma - 4}{\sigma + 4} \qquad (\mu = 4), \tag{12.72}$$

where $\sigma = \left(\xi^1\right)^2 + \left(\xi^2\right)^2 + \left(\xi^3\right)^2$.

ii. Compute the Euler angles,

$$\theta = 2\tan^{-1}\sqrt{\frac{(q^2)^2 + (q^3)^2}{(q^1)^2 + (q^4)^2}}, \tag{12.73}$$

$$\psi - \phi = \tan^{-1}\frac{q^3}{q^2}, \tag{12.74}$$

$$\psi + \phi = \tan^{-1}\frac{q^4}{q^1}. \tag{12.75}$$

iii. Rotate x_{BF}^{μ} using the three angles, and translate the center of mass.

iv. Repeat step i for all the tops.

```
      subroutine mapr(n,xo,xh1,xh2,yo,yh1,yh2,zo,zh1,zh2,x)
      implicit real*8 (a-h,o-z)
      parameter (NMAX=100)
      real*8 xo(NMAX),xh1(NMAX),xh2(NMAX)
      real*8 yo(NMAX),yh1(NMAX),yh2(NMAX)
      real*8 zo(NMAX),zh1(NMAX),zh2(NMAX)
      real*8 xi(3),di(3),ii(3,3),xr(20),dm(3)
      real*8 cx(3),cy(3),cz(3),z(3,3),e(3)
      real*8 ax(3),ay(3),az(3)
      real*8 x(6*NMAX)
      parameter (PI = 3.141592654d0)
      parameter (TPI = 6.283185307d0)
      parameter (NTHREE = 3)
      parameter (dmh = 1.0079d0*1822.8d0)
      parameter (dmo = 15.9994d0*1822.8d0)
      parameter (dmt = dmo + 2.d0*dmh)
      parameter (alpha = 0.98327233 )! a reference frame coordinate (bohr)
      parameter (beta = -0.12388466)! a reference frame coordinate (bohr)
      parameter (gamma = 1.43042)! a reference frame coordinate (bohr)
c Transform back to the x,y,z coordinates of each atom for all the
c water molecules from the tree stereographic projections and the
c center of mass coordinates
c Input:   x (see the comments in mapf for the arrangement)
c x(6(i-1)+1),x(6(i-1)+2),x(6(i-1)+3),x(6(i-1)+4),x(6(i-1)+5), x(6i)
c XCM(i)       YCM(i)       ZCM(i)      Xi1(i)      Xi2(i)      Xi3(i)
c Output: jin                   = water molecule number
c          x0,xh1,xh2,y0,yh1,yh2,z0,zh1,zh2 = arrays containing the
c                              Cartesian configuration for the water cluster
c The reference frame that defines the space axis
c     x          y          z
c H  0        0.98327233  1.43042
c 0  0       -0.12388466  0.00000
c H  0        0.98327233 -1.43042
c
c
```

```
        do jin = 1,n
c start with the reference body frame
        cx(1) = 0.d0
        cy(1) = alpha
        cz(1) = gamma
        cx(2) = 0.d0
        cy(2) = beta
        cz(2) = 0.d0
        cx(3) = 0.d0
        cy(3) = alpha
        cz(3) = -gamma
        index1 = 6*(jin -1) + 1
        index2 = 6*(jin -1) + 2
        index3 = 6*(jin -1) + 3
        index4 = 6*(jin -1) + 4
        index5 = 6*(jin -1) + 5
        index6 = 6*jin
        xcm = x(index1)
        ycm = x(index2)
        zcm = x(index3)
        xi(1) = x(index4)
        xi(2) = x(index5)
        xi(3) = x(index6)
      eta2 = xi(1)*xi(1) + xi(2)*xi(2) + xi(3)*xi(3)
      q1 = 4.d0*xi(1)/(eta2 + 4.d0)
      q2 = 4.d0*xi(2)/(eta2 + 4.d0)
      q3 = 4.d0*xi(3)/(eta2 + 4.d0)
      q4 = (eta2 - 4.d0)/(eta2 + 4.d0)
c Now from the quaternions back to the Euler angles
      yyt = sqrt(q2*q2+q3*q3)
      xxt = sqrt(q1*q1+q4*q4)
      if (dabs(xxt) .lt. 1.d-10) then
       thetan = PI
      else
       thetan = 2.d0*atan(yyt/xxt)
      endif
      sp = atan2(q4,q1)
      if (sp .lt. 0) sp = sp + TPI
      dp = atan2(q3,q2)
      phin = sp - dp
      psin = sp + dp
      cps = cos(psin)
      sps = sin(psin)
      ct = cos(thetan)
      st = sin(thetan)
      cph = cos(phin)
      sph = sin(phin)
```

```
      call rotate3(NTHREE,NTHREE,cx,cy,cz,ct,st,cph,sph,cps,sps)
      do i=1,3
        cx(i) = cx(i) + xcm
        cy(i) = cy(i) + ycm
        cz(i) = cz(i) + zcm
      enddo
      xh1(jin) = cx(1)
      yh1(jin) = cy(1)
      zh1(jin) = cz(1)
      xo(jin) = cx(2)
      yo(jin) = cy(2)
      zo(jin) = cz(2)
      xh2(jin) = cx(3)
      yh2(jin) = cy(3)
      zh2(jin) = cz(3)
      enddo
      return
      end
```

The subroutine `rotate3` follows.

```
      subroutine rotate3(mna,nar,xv,yv,zv,ct,st,cph,sph,cps,sps)
      implicit real*8 (a-h,o-z)
      real*8 xv(3),yv(3),zv(3)
      real*8 rot(3,3),un(3,3)
c     On entry:
c                 Vectors xv,yv,zv are the coordinates of a configuration
c                 ct,st,cph,sph,cps,sps are the cosines and sines of
c                 Eulerian angles.
c                 to perform an inverse rotation use
c call rotate3(mna,nar,xv,yv,zv,ct,st,-cps,sps,-cph,sph)
c     On return:
c                 Vectors xv,yv,zv are the coordinates of the configuration
c                 after the rotation has taken place
c Begin by constructing the rotation matrix rot(3,3)
c H. Goldstein, (Classical Mechanics, Addison - Wesley 1965).
      rot(1,1) = cph*cps - ct*sph*sps
      rot(1,2) = -cph*sps -ct*sph*cps
      rot(1,3) = sph*st
      rot(2,1) = sph*cps + ct*cph*sps
      rot(2,2) = -sph*sps + ct*cph*cps
      rot(2,3) = -cph*st
      rot(3,1) = sps*st
      rot(3,2) = cps*st
      rot(3,3) = ct
c rotate  the whole thing...

      do i = 1,nar
         xp = rot(1,1)*xv(i) + rot(1,2)*yv(i) + rot(1,3)*zv(i)
         yp = rot(2,1)*xv(i) + rot(2,2)*yv(i) + rot(2,3)*zv(i)
```

```
      zp = rot(3,1)*xv(i) + rot(3,2)*yv(i) + rot(3,3)*zv(i)
      xv(i) = xp
      yv(i) = yp
      zv(i) = zp
   enddo
   return
   end
```

The subroutine `mapr` is the heart of simulations, since it is called inside the potential energy subroutine. All our subroutines are called with a single array y^μ (**x** in `mapr`) $\mu = 1, \ldots, 6n$, where the integers congruent to $1 \bmod (6)$, $2 \bmod (6)$, and $3 \bmod (6)$, respectively, are the pointers to the x, y, and z. Cartesian coordinate of the center of mass. Those integers congruent to $4 \bmod (6)$, $5 \bmod (6)$, and $0 \bmod (6)$ are the pointers to ξ^1, ξ^2, ξ^3. (See the comments in `mapr`.) The subroutine `mapr` can be easily modified to transform the coordinates of any cluster of rigid nonlinear tops.

At the beginning of a simulation, one needs to initiate the array y^μ, using perhaps a configuration of the cluster in Cartesian coordinates. In that case, the subroutine `mapf` is handy. The subroutine `mapf` is an implementation of the following algorithm:

i. For center i, compute the center of mass $\mathbf{r}_\mathrm{C}$, then translate to the center of mass frame, $\mathbf{r}'_i = \mathbf{r}_i - \mathbf{r}_\mathrm{C}$.

ii. Using Equation 12.70, the z coordinates of the two hydrogens in the prime system must be,

$$z'_{H2} = \alpha \sin \theta \cos \phi - \gamma \cos \theta,$$
$$x'_{H1} = \alpha \left(-\cos \phi \sin \psi - \cos \theta \sin \phi \cos \psi \right) + \gamma \sin \theta \sin \phi,$$
$$x'_{H2} = \alpha \left(-\cos \phi \sin \psi - \cos \theta \sin \phi \cos \psi \right) - \gamma \sin \theta \sin \phi,$$
$$y'_{H1} = \alpha \left(-\sin \phi \sin \psi + \cos \theta \cos \phi \cos \psi \right) - \gamma \sin \theta \cos \phi,$$
$$y'_{H2} = \alpha \left(-\sin \phi \sin \psi + \cos \theta \cos \phi \cos \psi \right) + \gamma \sin \theta \cos \phi,$$
$$z'_{H1} = \alpha \sin \theta \cos \phi + \gamma \cos \theta,$$
$$z'_{O} - \beta \sin \theta \cos \psi.$$

Compute the sines and cosines of the Euler angles in the following order,

$$\cos \theta = \frac{z_{H1} - z_{H2}}{2\gamma}, \tag{12.76}$$

$$\sin \theta = \frac{\sqrt{\left(x_{H1} - x_{H2} \right)^2 + \left(y_{H1} - y_{H2} \right)^2}}{2\gamma}, \tag{12.77}$$

$$\sin \phi = \frac{x_{H1} - x_{H2}}{2\gamma \cos \theta}, \tag{12.78}$$

$$\cos \phi = \frac{y_{H1} - y_{H2}}{2\gamma \cos \theta}, \tag{12.79}$$

$$\cos \psi = \frac{z_O}{\beta \sin \theta}, \tag{12.80}$$

$$\sin \psi = \frac{-\dfrac{x_{H1} + x_{H2}}{2\alpha} - \cos \theta \sin \phi \cos \psi}{\cos \phi}. \tag{12.81}$$

iii. Using sum difference and double angle trigonometric identities, compute the quaternion parameters

$$q^1 = \cos\frac{\theta}{2}\,\cos\left(\frac{\psi+\phi}{2}\right), \tag{12.82}$$

$$q^2 = \sin\frac{\theta}{2}\,\cos\left(\frac{\psi-\phi}{2}\right), \tag{12.83}$$

$$q^3 = \sin\frac{\theta}{2}\,\sin\left(\frac{\psi-\phi}{2}\right), \tag{12.84}$$

$$q^4 = \cos\frac{\theta}{2}\,\sin\left(\frac{\psi+\phi}{2}\right). \tag{12.85}$$

iv. Invert the definition and compute the stereographic projections,

$$\sigma = 4\frac{1+q^4}{1-q^4}, \tag{12.86}$$

$$\xi^\mu = \frac{1}{4}\left(\sigma+4\right)q^\mu \qquad (1\le\mu\le 3). \tag{12.87}$$

v. Repeat step i for all the tops.

In the event that $\theta \approx 0$, the inversions for the cosines and sines of the ψ and ϕ are invalid, and the equation for $\cos\psi$ can cause floating point exceptions. We trap the event and we set $\psi = 0$. Then, the prime coordinates are,

$$x'_{H1} = -\alpha\sin\phi,$$
$$y'_{H1} = \alpha\cos\phi,$$

and we use these relations to compute the sines and cosines of ϕ.

```fortran
      subroutine mapf(n,xo,xh1,xh2,yo,yh1,yh2,zo,zh1,zh2,x)
      implicit real*8 (a-h,o-z)
      parameter (NMAX=100)
      real*8 xo(NMAX),xh1(NMAX),xh2(NMAX)
      real*8 yo(NMAX),yh1(NMAX),yh2(NMAX)
      real*8 zo(NMAX),zh1(NMAX),zh2(NMAX)
      real*8 xi(3),di(3),ii(3,3),xr(20),dm(3)
      real*8 cx(3),cy(3),cz(3),z(3,3),e(3)
      real*8 ax(3),ay(3),az(3)
      real*8 x(6*NMAX)
      parameter (PI = 3.141592654d0)
      parameter (TPI = 6.283185307d0)
      parameter (NTHREE = 3)
      parameter (dmh = 1.0079d0*1822.8d0)
      parameter (dmo = 15.9994d0*1822.8d0)
      parameter (dmt = dmo + 2.d0*dmh)
      parameter (alpha = 0.98327233 )  ! a reference frame coordinate (bohr)
      parameter (beta = -0.12388466)   ! a reference frame coordinate (bohr)
      parameter (gamma = 1.43042)   ! a reference frame coordinate (bohr)
c Transform the x,y,z coordinates of each atom for all the water molecules
c to the tree stereographic projections and the center of mass coordinates
c Input: jin                 = water molecule number
```

```
c        NMAX                    = size of arrays
c        x0,xh1,xh2,y0,yh1,yh2,z0,zh1,zh2  = arrays containing the Cartesian
c                          configuration for the water cluster
c Output:  x  arranged as follows
c x(6(i-1)+1),x(6(i-1)+2),x(6(i-1)+3),x(6(i-1)+4),x(6(i-1)+5), x(6i)
c XCM(i)        YCM(i)       ZCM(i)      Xi1(i)      Xi2(i)      Xi3(i)
      do jin=1,n
    index1 = 6*(jin -1) + 1
    index2 = 6*(jin -1) + 2
    index3 = 6*(jin -1) + 3
    index4 = 6*(jin -1) + 4
    index5 = 6*(jin -1) + 5
    index6 = 6*jin
        cx(1) = xh1(jin)
        cy(1) = yh1(jin)
        cz(1) = zh1(jin)
        cx(2) = xo(jin)
        cy(2) = yo(jin)
        cz(2) = zo(jin)
        cx(3) = xh2(jin)
        cy(3) = yh2(jin)
        cz(3) = zh2(jin)
c get the center of mass
        xcm =  (dmh*cx(1) + dmo*cx(2)  + dmh*cx(3))/dmt
        ycm =  (dmh*cy(1) + dmo*cy(2)  + dmh*cy(3))/dmt
        zcm =  (dmh*cz(1) + dmo*cz(2)  + dmh*cz(3))/dmt
      x(index1) = xcm
      x(index2) = ycm
      x(index3) = zcm
c translate to the center of mass
      do i=1,3
        cx(i) = cx(i) - xcm
        cy(i) = cy(i) - ycm
        cz(i) = cz(i) - zcm
      enddo
c now invert the coordinates to find the angles
      ct = (cz(1) - cz(3))/(2.d0*gamma)
      dcx2 = (cx(1) - cx(3))*(cx(1) - cx(3))
      dcy2 = (cy(1) - cy(3))*(cy(1) - cy(3))
      st =  sqrt(dcx2 + dcy2)/(2.d0*gamma)
      if (st .gt. 0.d-12) then
       sph = (cx(1) - cx(3))/(2.d0*gamma*st)
       cph = (cy(3) - cy(1))/(2.d0*gamma*st)
       cps = cz(2)/(beta*st)
       sps = (-(cx(1) + cx(3))/(2.d0*alpha) -ct*sph*cps)/cph
      else
       cps = 1.d0
       sps = 0.d0
       sph = -cx(1)/(alpha)
       cph = cy(1)/(alpha)
       if (ct .lt. 0) cph = - cph
      endif
      psi = atan2(sps,cps)
      if (psi .lt. 0.d0) psi = psi + TPI
      phi = atan2(sph,cph)
```

```
      if (phi .lt. 0.d0) phi = phi + TPI
      ct2 = sqrt((ct+1.d0)/2.d0)
      st2 = sqrt((-ct+1.d0)/2.d0)
c quaternions
      q1 = ct2*cos(0.5d0*(psi+phi))
      q2 = st2*cos(0.5d0*(psi-phi))
      q3 = st2*sin(0.5d0*(psi-phi))
      q4 = ct2*sin(0.5d0*(psi+phi))
c Stereographic projections from the quaternions
      eta2 = 4.d0*(1.d0+q4)/(1.d0-q4)
      xi(1) = q1*(eta2 + 4.d0)/4.d0
      xi(2) = q2*(eta2 + 4.d0)/4.d0
      xi(3) = q3*(eta2 + 4.d0)/4.d0
      x(index4) = xi(1)
      x(index5) = xi(2)
      x(index6) = xi(3)
      enddo   ! loop over jin
      return
      end
```

One way to test the two subroutines, `mapf` and `mapr`, is to start with a Cartesian configuration of the TIP4 cluster, as, e.g., a dimer, call `mapf`, and use its output to call `mapr`. The output of `mapr` should be the starting Cartesian configuration of the TIP4 cluster.

12.11   The Hessian Metric Tensor for $\mathbb{R}^{3n} \otimes \mathbb{I}^n$

Let y^μ represent the $6n$-dimensional vector associated with a configuration point in the space of n nonlinear tops, defined as we have in the subroutine `mapf` and `mapr`. With the specific definition of y^μ in `mapf` and `mapr`, the metric tensor takes the following block-diagonal form,

$$
g_{\mu\upsilon} = \begin{pmatrix} \mathbf{g}^{(1)} & 0 & 0 \\ \cdot\cdot & \cdot\cdot & \cdot\cdot \\ 0 & 0 & \mathbf{g}^{(n)} \end{pmatrix}, \tag{12.88}
$$

where $\mathbf{g}^{(i)}$ is the molecular metric tensor represented by the following 6×6 block-diagonal matrix,

$$
\mathbf{g}^{(i)} = \left(\begin{array}{ccc|c} m & 0 & 0 & \\ 0 & m & 0 & \mathbf{0} \\ 0 & 0 & m & \\ \hline & \mathbf{0} & & \mathbf{g}^{(\mathbb{I},i)} \end{array} \right), \tag{12.89}
$$

with m representing the mass of the molecular top ($2m_{\mathrm{H}} + m_{\mathrm{O}}$ for the water example), and $\mathbf{g}^{(\mathbb{I},i)}$ representing the block (nondiagonal) associated with the orientations. The six independent elements of $\mathbf{g}^{(\mathbb{I},i)}$ are calculated using,

$$
g_{\mu\upsilon}^{(\mathbb{I},i)} = h_\mu^{\mu'} h_\upsilon^{\upsilon'} g_{\mu'\upsilon'}^{(\mathbb{I},i)}, \tag{12.90}
$$

where $h_\mu^{\mu'}$ is the Jacobian matrix element,

$$h_\mu^{\mu'} = \frac{\partial q^{\mu'}}{\partial x^\mu},$$

(12.91)

$q^{1'} = \theta, q^{2'} = \phi, q^{3'} = \psi$, $1 \leq \mu \leq 3$, and $g_{\mu'v'}^{(\mathbb{I},i)}$ is the metric tensor of the ellipsoid of inertia expressed in Euler angles, as in Chapter 10. The subroutine `mmetric` is a version of the subroutine introduced in Chapter 11 for the general asymmetric top. It computes the 3×3 block, $g_{\mu v}^{(\mathbb{I},i)}$, using I_1, I_2, and I_3 equal to 11,518.283, 7,518.195, 4,000.087 a.u., respectively, for the TIP4P water molecule. The metric tensor in Equation 12.89 is put together with the following code:

```
      subroutine metricnlt(n,x,g,det)
      implicit real*8 (a-h,o-z)
      parameter (maxn=100)
      real*8 g(6*maxn,6*maxn),x(6*maxn)
      real*8 xi(3),gm(3,3)
      parameter (dmh = 1.0079d0*1822.8d0)
      parameter (dmn = 15.9994d0*1822.8d0)
      parameter (dmt = dmn + 2.d0*dmh)
c note that for increased stability the natural logarithm of the
c determinant of the metric tensor is returned instead of the
c determinant itself.
      det = 0.d0
      do i=1,n
        ind1 = 6*(i-1) + 1
        ind2 = 6*(i-1) + 2
        ind3 = 6*(i-1) + 3
        ind4 = 6*(i-1) + 4
        ind5 = 6*(i-1) + 5
        ind6 = 6*i
        g(ind1,ind1) = dmt
        g(ind2,ind2) = dmt
        g(ind3,ind3) = dmt
        xi(1) = x(ind4)
        xi(2) = x(ind5)
        xi(3) = x(ind6)
        call mmetric(xi,gm,detm)
        det = log(detm) + det
        g(ind4,ind4) = gm(1,1)
        g(ind4,ind5) = gm(1,2)
        g(ind4,ind6) = gm(1,3)
        g(ind5,ind4) = g(ind4,ind5)
        g(ind6,ind4) = g(ind4,ind6)
        g(ind5,ind5) = gm(2,2)
        g(ind5,ind6) = gm(2,3)
        g(ind6,ind5) = g(ind5,ind6)
        g(ind6,ind6) = gm(3,3)
      enddo
      return
      end
```

12.12   Local Energy and Action for $\mathbb{R}^{3n} \otimes \mathbb{I}^n$

It is not difficult to design algorithms that take advantage of the sparse nature and block-diagonal structure of $g_{\mu\upsilon}$, to compute the kinetic energy block of the action,

$$g_{\mu\upsilon}\Delta y^\mu \Delta y^\upsilon = \sum_{i=1}^{n} g_{\rho\lambda}^{(i)} \Delta y^{6(i-1)+\rho} \, \Delta y^{6(i-1)+\lambda} \qquad \begin{array}{l} (1 \leq \mu, \upsilon \leq 6n) \\ (1 \leq \rho, \lambda \leq 6), \end{array} \qquad (12.92)$$

where y^μ is the same $6n$ vector associated with a configuration point in the space of n nonlinear tops, organized as explained in the subroutines mapf and mapr.

```
      subroutine getwnlt(temp,w,xin ,N,ND,KMAX,wi)
c On input:
c  temp = beta in inverse hartree
c  w = array containing the path coefficients
c xin = Cartesian coordinates of the center of mass of each molecule
c and stereographic projections of each molecule:
c x(6(i-1)+1),x(6(i-1)+2),x(6(i-1)+3),x(6(i-1)+4),x(6(i-1)+5), x(6i)
c XCM(i)       YCM(i)       ZCM(i)      Xi1(i)      Xi2(i)      Xi3(i)
c N = quadrature for action integral (Usually KMAX + 1)
c ND = number of HF molecules
c KMAX = number of core path variables
c On return:
c wi = the exponent of the statistical weight for a given configuration
      parameter (maxn=100)
      parameter (NW=40)
      parameter (NX=200)
      implicit real*8 (a-h,o-z)
      real*8 w(6*maxn,NX)
      real*8 xin(6*maxn),g02(6*maxn)
      real*8 xm1(6*maxn),x(6*maxn)
      real*8 g(6*maxn,6*maxn)
      common /traj/sn(NX,NX),sigma2(NX)
        beta = temp
        sum = 0.d0
        t12 = sqrt(beta)
        dN = dfloat(N)
        do i=1,N
c Get the path from the coefficients
        do j=1,6*ND
         xm1(j) = xin(j)
         x(j) = xin(j)
         do k=1,4*KMAX
          if (i .gt. 1) then
           xm1(j) = xm1(j) + t12*w(j,k)*sn(i-1,k)
          endif
          x(j) = x(j) + t12*w(j,k)*sn(i,k)
         enddo
```

```
      enddo ! j loop
c the metric tensor and the velocities  are calculated with the core path
      call metricnlt(ND,x,g,detg)
      call pot(ND,x,v)
        do j=1,ND
          do jj = 1,6
          jdxjj = 6*(j-1) + jj
c diagonal terms
            sum = sum+ (dN/(beta*2.d0))*
     & g(jdxjj,jdxjj)*(x(jdxjj)-xm1(jdxjj))*(x(jdxjj)-xm1(jdxjj))
          enddo
c off diagonal terms
          do jj = 4,6
           jdxjj = 6*(j-1) + jj
            do jjj = jj+1,6
             jdxjjj = 6*(j-1) + jjj
            sum = sum+ (dN/(beta))*
     & g(jdxjj,jdxjjj)*(x(jdxjj)-xm1(jdxjj))*(x(jdxjjj)-xm1(jdxjjj))
            enddo
          enddo
        enddo
        sum = sum + beta*v/dN
        sum = sum - 0.5*detg
      enddo  ! i loop
      wi = sum
      return
      end
```

Similarly, the local energy estimator,

$$\Psi_T^{-1}\,\hat{\mathcal{H}}\,\Psi_T = V - \frac{2}{2}\left\{\frac{\beta^2}{4}g^{\mu\upsilon}\partial_\mu V\partial_\upsilon V - \frac{\beta}{2}g^{\mu\upsilon}\partial_\mu\partial_\upsilon V\right.$$

$$\left. -\frac{\beta}{2}\left\{g^{\mu\upsilon}\left[\partial_\mu \ln\sqrt{\det(g)}\right] + \left(\frac{\partial}{\partial x^\mu}g^{\mu\upsilon}\right)\right\}\partial_\upsilon V\right\}, \qquad (12.93)$$

can be computed using the storage scheme for y^μ, using the subroutine `varenlt` in the following:

```
      subroutine varenlt(n,beta,xin,ve)
      implicit real*8 (a-h,o-z)
      parameter (dmh = 1.0079d0*1822.8d0)
      parameter (dmn = 15.9994d0*1822.8d0)
      parameter (dmt = dmn + 2.d0*dmh)
      parameter (maxn=100)
      real*8 xin(maxn)
      real*8 xi(3),g(3,3),gp(3,3),gm(3,3)
      real*8 gradv(6),gradd(3),hess(6,6),gradg(3,3,3)
      integer ind(6)
c variational local energy for a trial wavefunction
c   PSI = exp(-beta V/2)
```

```
c on input
c            n = number of water molecules
c            beta = (KT )^{-1}
c            xin = the six configuration coordinates
c x(6(i-1)+1),x(6(i-1)+2),x(6(i-1)+3),x(6(i-1)+4),x(6(i-1)+5), x(6i)
c XCM(i)        YCM(i)        ZCM(i)       Xi1(i)       Xi2(i)       Xi3(i)
      ve = 0.0d0
      do i=1,6
       do j=1,6
         hess(i,j) = 0.d0
       enddo
      enddo
      hh = 1.0d-5
      call pot(n,xin,v)
      ve = v    ! initialize
      do i=1,n    !  loop over all molecules
       do k=1,6
        ind(k) = 6*(i-1) + k   ! compute the pointers for
       enddo                    ! the coordinates of molecule i
       do k=1,3
        xi(k) = xin(ind(3+k)) ! xi() contains the ster. proj. for molecule i
       enddo
c get the inverse of the 3x3 rotation block
      call mmetrici(xi,g,detm)
c compute the gradient and the hessian of the potential numerically
      do k=1,6
         xin(ind(k)) = xin(ind(k)) + hh
         call pot(n,xin,vp)
         xin(ind(k)) = xin(ind(k)) - 2.0d0*hh
         call pot(n,xin,vm)
         xin(ind(k)) = xin(ind(k)) + hh
         gradv(k) = (vp - vm)/(2.0d0*hh)
         hess(k,k) = (vp + vm - 2.0d0*v)/(hh*hh)
         if (k .gt. 3) then
         do kk=4,6
          xin(ind(kk)) = xin(ind(kk)) + hh
          xin(ind(k)) = xin(ind(k)) + hh
          call pot(n,xin,vpp)
          xin(ind(kk)) = xin(ind(kk)) - 2.0d0*hh
          call pot(n,xin,vmp)
          xin(ind(k)) = xin(ind(k)) - 2.0d0*hh
          call pot(n,xin,vmm)
          xin(ind(kk)) = xin(ind(kk)) + 2.0d0*hh
          call pot(n,xin,vpm)
          xin(ind(kk)) = xin(ind(kk)) - hh
          xin(ind(k)) = xin(ind(k)) + hh
          hess(k,kk) = (vpp - vpm - vmp + vmm)/(4.0d0*hh*hh)
         enddo
         endif
```

```
          enddo
c compute the gradient of the inverse of the metric tensor numerically
      do k=1,3
         xi(k) = xi(k) + hh
         call  mmetrici(xi,gp,detmp)
         xi(k) = xi(k) - 2.0d0*hh
         call mmetrici(xi,gm,detmm)
         xi(k) = xi(k) + hh
         gradd(k) = (detmp - detmm)/(2.0d0*hh)
         do k1=1,3
         do k2=1,3
           gradg(k,k1,k2) = (gp(k1,k2) - gm(k1,k2))/(2.0d0*hh)
         enddo
         enddo
      enddo
c put it all together into the local energy
      sum1 = 0.d0
      sum2 = 0.d0
      sum3 = 0.d0
      sum4 = 0.d0
c first do the translations block
      do k=1,3
        sum1 = sum1 - 0.125d0*beta*beta*gradv(k)*gradv(k)/dmt
        sum2 = sum2 + 0.25d0*beta*hess(k,k)/dmt
      enddo
c now do the rotation block
      do k=1,3
        do k1=1,3
         sum1 = sum1 - 0.125d0*beta*beta*g(k,k1)*gradv(k+3)*gradv(k1+3)
         sum2 = sum2 + 0.25d0*beta*g(k,k1)*hess(k+3,k1+3)
         sum3 = sum3 + 0.25d0*beta*g(k,k1)*gradd(k)*gradv(k1+3)/detm
         sum4 = sum4 + 0.25d0*beta*gradg(k,k1,k)*gradv(k1+3)
        enddo
      enddo
      ve = ve + sum1 + sum2 + sum3 + sum4
      enddo          ! end of loop over all molecules
      return
      end
```

The subroutine `mmetrici` computes the inverse of $g_{\mu\upsilon}^{(\mathbb{I},i)}$. We simply modify the subroutine `mmetric` by adding the following lines at the end,

```
c compute the inverse of the 3x3 block
      d1 = g(2,2) - g(1,2)*g(1,2)/g(1,1)
      p23 = (g(2,3) - g(1,2)*g(1,3)/g(1,1))/d1
      d2 = g(3,3) - g(1,3)*g(1,3)/g(1,1) - d1*p23*p23
      p13 = g(1,3)/g(1,1) - g(1,2)*p23/g(1,1)
      gi(3,3) = 1.d0/d2
      gi(2,3) = -p23/d2
      gi(3,2) = gi(2,3)
```

```
      gi(1,3) = -p13/d2
      gi(3,1) = gi(1,3)
      gi(1,2) = -g(1,2)/(g(1,1)*d1) + p13*p23/d2
      gi(2,1) = gi(1,2)
      gi(1,1) = 1.d0/g(1,1) + g(1,2)*g(1,2)/(d1*g(1,1)*g(1,1))
  &  + p13*p13/d2
      gi(2,2) = 1.d0/d1 + p23*p23/d2
```

Upon return, `gi` contains the 3×3 block, $g^{\mu\upsilon}_{(\mathbb{I},i)}$. This is computed using the array `g`, which stores $g^{(\mathbb{I},i)}_{\mu\upsilon}$. The equations used for the inversion are derived using a few Gauss-Jordan eliminations.

12.13   Concluding Remarks

This book was designed to serve two main purposes. The first purpose was to develop a tool for training students in the fine art of stochastic simulations of clusters. The classical simulations of these, alone, can be challenging. The global minimization of Lennard-Jones clusters, for example, remains a formidable challenge if the size is sufficiently large 782, 787. I did not elaborate on new developments in the global optimization methods, because, for the moment, the community simulating molecular clusters is still focused on relatively small to medium-size systems. Global optimization is challenging there as well, and modifications to the basin hopping and the genetic algorithm have been proposed by us already. There are numerous goals for the trainee, including quantum stochastic simulations like those showcased in Chapter 9. In order to achieve these goals, and reach an undergraduate audience as well as a typical graduate student in statistical mechanics, I have included some background material in chapters such as Chapters 2 through 5. The second purpose of the textbook was to educate the community about a set of techniques designed to overcome the convergence difficulties of quantum methods implemented for molecular clusters; these are found when applying the methods of Chapter 6 and 7 to systems that contain multiple time scales. The adiabatic approximation, in these cases, is indispensable to achieve convergence with reasonable means.

It is clear that I have been influenced, not only by the voluminous literature cited in this work, but also by a number of outstanding books on the many subjects that this work builds upon. These include statistical mechanics [1], classical molecular simulations [6, 7], Lie algebras and Lie groups [51], sparse matrix technology [178], quantum Monte Carlo [427], differential geometry [855], and path integrals in flat and curved spaces [517, 744, 753]. These books are a constant source of ideas for me and my group, and it is my hope that this book serves those who read it in the same way as well. Certainly, our own ideas for possible future explorations, mentioned as humbling and awe inspiring at the beginning of this chapter, seem vast in scope to us. For example, the exploration of the Stockmayer clusters, with all their degrees of freedom $(n, \mu, \epsilon, \sigma)$, represent an effort that, alone, could take the community several years, if not decades.

Bibliography

[1] *Statistical Thermodynamics*, D. A. McQuarrie (University Science Books, Mill Valley, CA, 1973).

[2] *Handbook of Mathematical Functions*, M. Abramowitz and I. Stegun editors (Dover, New York, 1970).

[3] A biased Monte Carlo scheme for zeolite structure solution, M. Falcioni and M. W. Deem, *J. Chem. Phys.* **110**, 1754 (1999).

[4] Approach to ergodicity in Monte Carlo simulations, J. P. Neirotti, D. L. Freeman, and J. D. Doll, *Phys. Rev. E.* **62**, 7445 (2000).

[5] Thermal equilibrium and statistical thermometers in special relativity, D. Cubero, J. Casado-Pascual, J. Dunkel, P. Talkner, and P. Hänggi, *Phys. Rev. Lett.* **99**, 170610 (2007).

[6] *Computer Simulations of Liquids*, M. P. Allen and D. J. Tildesley (Claredon Press, Oxford, 1987).

[7] *Understanding Molecular Simulations*, D. Frenkel and B. Smit, Sec. Ed. (Academic Press, New York, 2002).

[8] *Numerical Recipes in FORTRAN*, W. H. Press, S. A. Teukolsky, W. T. Vetterling, and B. P. Flannery, Sec. Ed. (Cambridge University Press, New York, 1994).

Symplectic Integrators

[9] Molecular dynamics of surface-moving thermally driven nanocars, A. V. Akimov, A. V. Nemukhin, A. A. Moskovsky, A. B. Kolomeisky, and J. B. Tour, *J. Chem. Theory Comput.* **4**, 652 (2008).

[10] Explicit symplectic integrators of molecular dynamics algorithms for rigid-body molecules in the canonical, isobaric-isothermal, and related ensembles, H. Okumura, S. G. Itoh, and Y. Okamoto, *J. Chem. Phys.* **126**, 084103 (2007).

[11] Implementation of a symplectic multiple-time-step molecular dynamics algorithm, based on the united-residue mesoscopic potential energy function, F. Rakowski, P. Grochowski, B. Lesyng, A. Liwo, and H. A. Scheraga, *J. Chem. Phys.* **125**, 204107 (2006).

[12] Symplectic splitting operator methods for the time-dependent Schrödinger equation, S. Blanes, F. Casas, and A. Murua, *J. Chem. Phys.* **124**, 234105 (2006).

[13] Gradient symplectic algorithms for solving the radial Schrödinger equation, S. A. Chin and P. Anisimov, *J. Chem. Phys.* **124**, 054106 (2006).

[14] Time reversible and symplectic integrators for molecular dynamics simulations of rigid molecules, H. Kamberaj, R. J. Low, and M. P. Neal, *J. Chem. Phys.* **122**, 224114 (2005).

[15] Molecular dynamics integration and molecular vibrational theory. I. New symplectic integrators, D. Janezic, M. Praprotnik, and F. Merzel, *J. Chem. Phys.* **122**, 174101 (2005).

[16] Symplectic molecular dynamics simulations on specially designed parallel computers, U. Borstnik and D. Janezic, *J. Chem. Inf. Model.* **45**, 1600 (2005).

[17] Molecular dynamics integration meets standard theory of molecular vibrations, M. Praprotnik and D. Janezic, *J. Chem. Inf. Model.* **45**, 1571 (2005).

[18] Molecular dynamics with the united-residue model of polypeptide chains. I. Lagrange equations of motion and tests of numerical stability in the microcanonical mode, M. Khalili, A. Liwo, F. Rakowski, P. Grochowski, and H. A. Scheraga, *J. Phys. Chem.* B **109**, 13785 (2005).

[19] The canonical ensemble via symplectic integrators using Nosé and Nosé-Poincaré chains, B. J. Leimkuhler and C. R. Sweet, *J. Chem. Phys.* **121**, 108 (2004).

[20] Molecular dynamics integration time step dependence of the split integration symplectic method on system density, D. Janezic and M. Praprotnik, *J. Chem. Inf. Comput. Sci.* **43**, 1922 (2003).

[21] Gradient symplectic algorithms for solving the Schrödinger equation with time-dependent potentials, S. A. Chin and C. R. Chen, *J. Chem. Phys.* **117**, 1409 (2002).

[22] Symplectic quaternion scheme for biophysical molecular dynamics, T. F. Miller III, M. Eleftheriou, P. Pattnaik, A. Ndirango, D. Newns, and G. J. Martyna, *J. Chem. Phys.* **116**, 8649 (2002).

[23] Reaction cross sections and rate constants for the Cl + H$_2$ reaction from quasiclassical trajectory calculation on two new ab initio potential energy surfaces, C. Shen, T. Wu, G. Ju, and W. Bian, *J. Phys. Chem. A* **106**, 176 (2002).

[24] Fourth order gradient symplectic integrator methods for solving the time-dependent Schrödinger equation, S. A. Chin and C. R. Chen, *J. Chem. Phys.* **114**, 7338 (2001).

[25] Symplectic algorithm for constant-pressure molecular dynamics using a Nosé-Poincaré thermostat, J. B. Sturgeon and B. B. Laird, *J. Chem. Phys.* **112**, 3474 (2000).

[26] Symplectic integrators for the multilevel Redfield equation, C. Kalyanaraman and D. G. Evans, *Chem. Phys. Lett.* **324**, 459 (2000).

[27] Symplectic integration of closed chain rigid body dynamics with internal coordinate equations of motion, A. K. Mazur, *J. Chem. Phys.* **111**, 1407 (1999).

[28] Constant temperature molecular dynamics of a protein in water by high-order decomposition of the Liouville operator, H. Ishida and A. Kidera, *J. Chem. Phys.* **109**, 3276 (1998).

[29] Symplectic integrator for molecular dynamics of a protein in water, H. Ishida, Y. Nagai, and A. Kidera, *Chem. Phys. Lett.* **282**, 115 (1998).

[30] Symplectic integration of classical trajectories: A case study, C. Schlier and A. Seiter, *J. Phys. Chem. A* **102**, 9399 (1998).

[31] Open Newton-Cotes differential methods as multilayer symplectic integrators, J. C. Chiou and S. D. Wu, *J. Chem. Phys.* **107**, 6894 (1997).

[32] Symplectic splitting methods for rigid body molecular dynamics, A. Dullweber, B. Leimkuhler, and R. McLachlan, *J. Chem. Phys.* **107**, 5840 (1997).

[33] A symplectic method for rigid-body molecular simulation, A. Kol, B. B. Laird, and B. J. Leimkuhler, *J. Chem. Phys.* **107**, 2580 (1997).

[34] Application of symplectic integrator to stationary reactive-scattering problems: Inhomogeneous Schrödinger equation approach, K. Takahashi and K. S. Ikeda, *J. Chem. Phys.* **106**, 4463 (1997).

[35] Split integration symplectic method for molecular dynamics integration, D. Janezic and F. Merzel, *J. Chem. Inf. Comput. Sci.* **37**, 1048 (1997).

[36] The complexity of parallel symplectic molecular dynamics algorithms, R. Trobec, F. Merzel, and D. Janezic, *J. Chem. Inf. Comput. Sci.* **37**, 1055 (1997).

[37] Symplectic integrators tailored to the time-dependent Schrödinger equation, S. K. Gray and D. E. Manolopoulos, *J. Chem. Phys.* **104**, 7099 (1996).

[38] Symplectic integrator for molecular dynamics of a protein in water. An explicit and symplectic integrator for quantum-classical molecular dynamics, P. Nettesheim, F. A. Bornemann, B. Schmidt, and C. Schutte, *Chem. Phys. Lett.* **256**, 581 (1996).

[39] Accurate symplectic integrators via random sampling, W. G. Hoover, O. Kum, and N. E. Owens, *J. Chem. Phys.* **103**, 1530 (1995).

[40] Symplectic integrators for the multichannel Schrödinger equation, D. E. Manolopoulos and S. K. Gray, *J. Chem. Phys.* **102**, 9214 (1995).

[41] Qualitative study of the symplectic Störmer-Verlet integrator, D. J. Hardy and D. I. Okunbor, *J. Chem. Phys.* **102**, 8978 (1995).

[42] Symplectic reversible integrators: Predictor-corrector methods, G. J. Martyna and M. E. Tuckerman, *J. Chem. Phys.* **102**, 8071 (1995).

[43] An efficient symplectic integration algorithm for molecular dynamics simulations, D. Janezic and F. Merzel, *J. Chem. Inf. Comput. Sci.* **35**, 321 (1995).

[44] Symplectic integrators for large scale molecular dynamics simulations: A comparison of several explicit methods, S. K. Gray, D. W. Noid, and B. G. Sumpter, *J. Chem. Phys.* **101**, 4062 (1994).

[45] Applicability of symplectic integrator to classically unstable quantum dynamics, K. Takahashi and K. Ikeda, *J. Chem. Phys.* **99**, 8680 (1993).

[46] A symplectic integration algorithm for separable Hamiltonian functions, J. Candy and W. Rozmus, *J. Comp. Phys.* **92**, 230 (1991).

[47] Fourth-order symplectic integration, E. Forest and R. D. Ruth, *Physica D.* **43**, 105 (1990).

[48] Higher-order symplectic integrators, H. Yoshida, *Phys. Lett. A* **150**, 262 (1990).

[49] *Advanced Engineering Mathematics*, P. V. O'Neil, Sec. Ed. (Wadsworth, Belmont, CA, 1987).

[50] *Linear Algebra*, P. D. Lax (Wiley-Interscience, New York, 1997).

[51] *Lie Groups, Lie Algebras, and Some of Their Applications*, R. Gilmore (Dover, New York, 2006).

Discrete Variable Representation

[52] Using a nondirect product discrete variable representation for angular coordinates to compute vibrational levels of polyatomic molecules, X.-G. Wang and T. Carrington, Jr., *J. Chem. Phys.* **128**, 194109 (2008).

[53] Five-dimensional ab initio potential energy surface and predicted infrared spectra of H_2-CO_2 van der Waals complexes, H. Ran, Y. Zhou, and D. Xie, *J. Chem. Phys.* **126**, 204304 (2007).

[54] Optimal grids for generalized finite basis and discrete variable representations: Definition and method of calculation, V. Szalay, *J. Chem. Phys.* **125**, 154115 (2006).

[55] Application of Coulomb wave function discrete variable representation to atomic systems in strong laser fields, L.-Y. Peng and A. F. Starace, *J. Chem. Phys.* **125**, 154311 (2006).

[56] Structure of liquid water at ambient temperature from ab initio molecular dynamics performed in the complete basis set limit, H.-S. Lee and M. E. Tuckerman, *J. Chem. Phys.* **125**, 154507 (2006).

[57] Dynamical pruning of static localized basis sets in time-dependent quantum dynamics, D. A. McCormack, *J. Chem. Phys.* **124**, 204101 (2006).

[58] Finite basis representations with nondirect product basis functions having structure similar to that of spherical harmonics, G. Czakó, V. Szalay, and A. G. Császŕ, *J. Chem. Phys.* **124**, 014110 (2006).

[59] Three-dimensional ab initio potential-energy surface and rovibrational spectra of the H_2-Kr complex, Y. Zhou and D. Xie, *J. Chem. Phys.* **123**, 134323 (2005).

[60] Multidimensional time-dependent discrete variable representations in multiconfiguration Hartree calculations, R. van Harrevelt and U. Manthe, *J. Chem. Phys.* **123**, 064106 (2005).

[61] A coherent discrete variable representation method for multidimensional systems in physics, H.-G. Yu, *J. Chem. Phys.* **122**, 164107 (2005).

[62] How to choose one-dimensional basis functions so that a very efficient multidimensional basis may be extracted from a direct product of the one-dimensional functions: Energy levels of coupled systems with as many as 16 coordinates, R. Dawes and T. Carrington, Jr., *J. Chem. Phys.* **122**, 134101 (2005).

[63] Treating singularities present in the Sutcliffe-Tennyson vibrational Hamiltonian in orthogonal internal coordinates, G. Czakó, V. Szalay, A. G. Császár, and T. Furtenbacher, *J. Chem. Phys.* **122**, 024101 (2005).

[64] Degeneracy in discrete variable representations: General considerations and application to the multiconfigurational time-dependent Hartree approach, R. van Harrevelt and U. Manthe, *J. Chem. Phys.* **121**, 5623 (2004).

[65] A multidimensional discrete variable representation basis obtained by simultaneous diagonalization, R. Dawes and T. Carrington, Jr., *J. Chem. Phys.* **121**, 726 (2004).

[66] Quantum-classical dynamics of scattering processes in adiabatic and diabatic representations, P. Puzari, B. Sarkar, and S. Adhikari, *J. Chem. Phys.* **121**, 707 (2004).

[67] Spectral difference Lanczos method for efficient time propagation in quantum control theory, J. D. Farnum and D. A. Mazziotti, *J. Chem. Phys.* **120**, 5962 (2004).

[68] Full-dimensional quantum calculations of vibrational spectra of six-atom molecules. I. Theory and numerical results, H.-G. Yu, *J. Chem. Phys.* **120**, 2270 (2004).

[69] On one-dimensional discrete variable representations with general basis functions, V. Szalay, G. Czakó, Á. Nagy, T. Furtenbacher, and A. G. Császár, *J. Chem. Phys.* **119**, 10512 (2003).

[70] Symmetry-adapted direct product discrete variable representation for the coupled angular momentum operator: Application to the vibrations of $(CO2)_2$, H.-S. Lee, H. Chen, and J. C. Light, *J. Chem. Phys.* **119**, 4187 (2003).

[71] Molecular vibrations: Iterative solution with energy selected bases, H.-S. Lee and J. C. Light, *J. Chem. Phys.* **118**, 3458 (2003).

[72] Spin-orbit interactions, new spectral data, and deperturbation of the coupled $b^3 \prod_u$, and $A^1 \sum_u^+$ states of K2, M. R. Manaa, A. J. Ross, F. Martin, P. Crozet, A. M. Lyyra, L. Li, C. Amiot, and T. Bergeman, *J. Chem. Phys.* **117**, 11208 (2002).

[73] Clusters containing open-shell molecules. III. Quantum five-dimensional/two-surface bound-state calculations on $Ar_n OH$ van der Waals clusters ($X_2 \prod n = 4$ to 12), M. Xu, Z. Bačić, and J. M. Hutson, *J. Chem. Phys.* **117**, 4787 (2002).

[74] Spectral difference methods for solving the differential equations of chemical physics, D. A. Mazziotti, *J. Chem. Phys.* **117**, 2455 (2002).

[75] Tetrahedrally invariant discrete variable representation basis on the sphere, M. Cargo and R. G. Littlejohn, *J. Chem. Phys.* **117**, 59 (2002).

[76] An Airy discrete variable representation basis, R. G. Littlejohn and M. Cargo, *J. Chem. Phys.* **117**, 37 (2002).

[77] Bessel discrete variable representation bases, R. G. Littlejohn and M. Cargo, *J. Chem. Phys.* **117**, 27 (2002).

[78] A general framework for discrete variable representation basis sets, R. G. Littlejohn, M. Cargo, T. Carrington, Jr., K. A. Mitchell, and B. Poirier, *J. Chem. Phys.* **116**, 8691 (2002).

[79] Calculations of rotation-vibration states with the z axis perpendicular to the plane: High accuracy results for H_3^+, M. A. Kostin, O. L. Polyansky, and J. Tennyson, *J. Chem. Phys.* **116**, 7564 (2002).

[80] A theoretical study of the vibrational spectrum of the CS_2 molecule, J. Zúñiga, A. Bastida, A. Requena, and E. L. Sibert III, *J. Chem. Phys.* **116**, 7495 (2002).

[81] Multidimensional discrete variable representation bases: Sinc functions and group theory, R. G. Littlejohn and M. Cargo, *J. Chem. Phys.* **116**, 7350 (2002).

[82] Two Krylov space algorithms for repeated large scale sparse matrix diagonalization, J. H. Skone and E. Curotto, *J. Chem. Phys.* **116**, 3210 (2002).

[83] Quantum dressed classical mechanics, G. D. Billing, *J. Chem. Phys.* **114**, 6641 (2001).

[84] The structure of a weakly bound ionic trimer: Calculations for the $^4He_2H^-$ complex, F. A. Gianturco, F. Paesani, I. Baccarelli, G. Delgado-Barrio, T. Gonzalez-Lezana, S. Miret-Arts, P. Villarreal, G. B. Bendazzoli, and S. Evangelisti, *J. Chem. Phys.* **114**, 5520 (2001).

[85] Quasirandom distributed Gaussian bases for bound problems, S. Garashchuk and J. C. Light, *J. Chem. Phys.* **114**, 3929 (2001).

[86] A time-dependent discrete variable representation method, S. Adhikari and G. D. Billing, *J. Chem. Phys.* **113**, 1409 (2000).

[87] 6D vibrational quantum dynamics: Generalized coordinate discrete variable representation and (a)diabatic contraction, D. Luckhaus, *J. Chem. Phys.* **113**, 1329 (2000).

[88] Vibrations of the carbon dioxide dimer, H. Chen and J. C. Light, *J. Chem. Phys.* **112**, 5070 (2000).

[89] Six-dimensional calculation of the vibrational spectrum of the HFCO molecule, A. Viel and C. Leforestier, *J. Chem. Phys.* **112**, 1212 (2000).

[90] Phase space optimization of quantum representations: Direct-product basis sets, B. Poirier and J. C. Light, *J. Chem. Phys.* **111**, 4869 (1999).

[91] Ab initio calculations for the anharmonic vibrational resonance dynamics in the overtone spectra of the coupled OH and CH chromophores in CD_2H-OH, M. Quack and M. Willeke, *J. Chem. Phys.* **110**, 11958 (1999).

[92] Quantum calculations of highly excited vibrational spectrum of sulfur dioxide, I. Eigenenergies and assignments up to 15 000 cm^{-1}, G. Ma, R. Chen, and H. Guo, *J. Chem. Phys.* **110**, 8408 (1999).

[93] Hyperspherical surface functions for nonzero total angular momentum. I. Eckart singularities, B. K. Kendrick, R. T. Pack, R. B. Walker, and E. F. Hayes, *J. Chem. Phys.* **110**, 6673 (1999).

[94] Excited states of van der Waals clusters by projector Monte Carlo, with application to excitations of molecules in small ^4He_n, D. Blume, M. Mladenovic, M. Lewerenz, and K. B. Whaley, *J. Chem. Phys.* **110**, 5789 (1999).

[95] All the nonadiabatic $(J = 0)$ bound states of NO_2, R. F. Salzgeber, V. A. Mandelshtam, Ch. Schlier, and H. S. Taylor, *J. Chem. Phys.* **110**, 3756 (1999).

[96] Extended symmetry-adapted discrete variable representation and accelerated evaluation of $\hat{H}\psi$, R. Chen and H. Guo, *J. Chem. Phys.* **110**, 2771 (1999).

[97] Strong-field optical control of vibrational dynamics: Vibrational Stark effect in planar acetylene, L. Liu and J. T. Muckerman, *J. Chem. Phys.* **110**, 2446 (1999).

[98] A pseudospectral algorithm for the computation of transitional-mode eigenfunctions in loose transition states. II. Optimized primary and grid representations, A. J. Rasmussen, K. E. Gates, and S. C. Smith, *J. Chem. Phys.* **110**, 1354 (1999).

[99] Six-dimensional quantum calculation of the intermolecular bound states for water dimer, H. Chen, S. Liu, and J. C. Light, *J. Chem. Phys.* **110**, 168 (1999).

[100] Application of contracted distributed approximating functions to solving vibrational eigenvalue problems, V. Szalay and S. C. Smith, *J. Chem. Phys.* **110**, 72 (1999).

[101] Calculation of the rotation-vibration states of water up to dissociation, H. Y. Mussa and J. Tennyson, *J. Chem. Phys.* **109**, 10885 (1998).

[102] The effect of nonadiabatic coupling on the calculation of $N(E, J)$ for the methane association reaction, K. L. Mardis and E. L. Sibert III, *J. Chem. Phys.* **109**, 8897 (1998).

[103] Vibrational predissociation of the $I_2 \cdots Ne_2$ cluster: A molecular dynamics with quantum transitions study, A. Bastida, J. Zúñiga, A. Requena, N. Halberstadt, and J. A. Beswick, *J. Chem. Phys.* **109**, 6320 (1998).

[104] All the adiabatic bound states of NO_2, R. F. Salzgeber, V. Mandelshtam, Ch. Schlier, and H. S. Taylor, *J. Chem. Phys.* **109**, 937 (1998).

[105] An application of distributed approximating functional-wavelets to reactive scattering, G. W. Wei, S. C. Althorpe, D. J. Kouri, and D. K. Hoffman, *J. Chem. Phys.* **108**, 7065 (1998).

[106] Exterior complex dilation for grid methods: Application to the cumulative reaction probability, H. O. Karlsson, *J. Chem. Phys.* **108**, 3849 (1998).

[107] Hermiticity of the Hamiltonian matrix in a discrete variable representation, I. Tuvi and Y. B. Band, *J. Chem. Phys.* **107**, 9079 (1997).

[108] Empirical potential energy surface for ArSH/D and KrSH/D, P. P. Korambath, X. T. Wu, E. F. Hayes, C. C. Carter, and T. A. Miller, *J. Chem. Phys.* **107**, 3460 (1997).

[109] Vibrational eigenvalues and eigenfunctions for planar acetylene by wavepacket propagation, and its mode-selective infrared excitation, L. Liu and J. T. Muckerman, *J. Chem. Phys.* **107**, 3402 (1997).

[110] HO_2 rovibrational eigenvalue studies for nonzero angular momentum, X. T. Wu and E. F. Hayes, *J. Chem. Phys.* **107**, 2705 (1997).

[111] Six-dimensional quantum dynamics study for dissociative adsorption of H_2 on Cu(111) surface, J. Dai and J. C. Light, *J. Chem. Phys.* **107**, 1676 (1997).

[112] Ghost levels and near-variational forms of the discrete variable representation: Application to H_2O, H. Wei, *J. Chem. Phys.* **106**, 6885 (1997).

[113] Generalization of the multiconfigurational time-dependent Hartree method to nonadiabatic systems, K. Museth and G. D. Billing, *J. Chem. Phys.* **105**, 9191 (1996).

[114] Ab initio-discrete variable representation calculation of vibrational energy levels, E. Kauppi, *J. Chem. Phys.* **105**, 7986 (1996).

[115] Quantum analysis of absolute collision-induced scattering spectra from bound, metastable and free Ar diatoms, M. Chrysos, O. Gaye, and Y. Le Duff, *J. Chem. Phys.* **105**, 31 (1996).

[116] Calculation of vibrational ($J = 0$) excitation energies and band intensities of formaldehyde using the recursive residue generation method, N. M. Poulin, M. J. Bramley, T. Carrington, Jr., H. G. Kjaergaard, and B. R. Henry, *J. Chem. Phys.* **104**, 7807 (1996).

[117] Geometric phase effects in $H+O_2$ scattering. I. Surface function solutions in the presence of a conical intersection, B. Kendrick and R. T. Pack, *J. Chem. Phys.* **104**, 7475 (1996).

[118] On the direct complex scaling of matrix elements expressed in a discrete variable representation: Application to molecular resonances, K. Museth and C. Leforestier, *J. Chem. Phys.* **104**, 7008 (1996).

[119] Dynamics of the photodissociation of triplet ketene, J. D. Gezelter and W. H. Miller, *J. Chem. Phys.* **104**, 3546 (1996).

[120] Avoided resonance overlapping beyond the energy independent formalism. II. Electronic predissociation, V. Brems, M. Desouter-Lecomte, and J. Liévin, *J. Chem. Phys.* **104**, 2222 (1996).

[121] Avoided resonance overlapping beyond the energy independent formalism. I. Vibrational predissociation, M. Desouter-Lecomte, J. Liévin, and V. Brems, *J. Chem. Phys.* **103**, 4524 (1995).

[122] Three-dimensional study of predissociation resonances by the complex scaled discrete variable representation method: HCO/DCO, V. Ryaboy and N. Moiseyev, *J. Chem. Phys.* **103**, 4061 (1995).

[123] Variational discrete variable representation, G. C. Corey and J. W. Tromp, *J. Chem. Phys.* **103**, 1812 (1995).

[124] Time-dependent discrete variable representations for quantum wave packet propagation, E. Sim and N. Makri, *J. Chem. Phys.* **102**, 5616 (1995).

[125] A new method for calculating the rovibrational states of polyatomics with application to water dimer, S. C. Althorpe and D. C. Clary, *J. Chem. Phys.* **102**, 4390 (1995).

[126] "Pointwise" versus basis representations for two-dimensional spherical dynamics, O. A. Sharafeddin and J. C. Light, *J. Chem. Phys.* **102**, 3622 (1995).

[127] Calculation of triatomic vibrational eigenstates: Product or contracted basis sets, Lanczos or conventional eigensolvers? What is the most efficient combination? M. J. Bramley and T. Carrington, Jr., *J. Chem. Phys.* **101**, 8494 (1994).

[128] Initial state-selected reaction probabilities for OH+H$_2$ → H+H$_2$O and photodetachment intensities for HOH$_2^-$, W. H. Thompson and W. H. Miller, *J. Chem. Phys.* **101**, 8620 (1994).

[129] Large-scale ab initio calculations for C$_3$, M. Mladenović, S. Schmatz, and P. Botschwina, *J. Chem. Phys.* **101**, 5891 (1994).

[130] Resonance positions and widths by complex scaling and modified stabilization methods: van der Waals complex NeICl, V. Ryaboy, N. Moiseyev, V. A. Mandelshtam, and H. S. Taylor, *J. Chem. Phys.* **101**, 5677 (1994).

[131] Toeplitz matrices within DVR formulation: Application to quantum scattering problems, E. Eisenberg, S. Ron, and M. Baer, *J. Chem. Phys.* **101**, 3802 (1994).

[132] Discrete variable representations of complicated kinetic energy operators, H. Wei and T. Carrington, Jr., *J. Chem. Phys.* **101**, 1343 (1994).

[133] Exact quantum and time-dependent Hartree studies of the HBr/LiF(001) photodissociation dynamics, J. Y. Fang and H. Guo, *J. Chem. Phys.* **101**, 1231 (1994).

[134] Efficient calculation of highly excited vibrational energy levels of floppy molecules: The band origins of H$_3^+$ up to 35 000 cm^{-1}, M. J. Bramley, J. W. Tromp, T. Carrington, Jr., and G. C. Corey, *J. Chem. Phys.* **100**, 6175 (1994).

[135] Rotational resonance states of Ar-HCl ($\upsilon = 0$) by finite range scattering wave function method, H. W. Jang, S. E. Choi, and J. C. Light, *J. Chem. Phys.* **100**, 4188 (1994).

[136] Efficient polynomial expansion of the scattering Green's function: Application to the D+H$_2$ ($\upsilon = 1$) rate constant, S. M. Auerbach and W. H. Miller, *J. Chem. Phys.* **100**, 1103 (1994).

[137] A general discrete variable method to calculate vibrational energy levels of three- and four-atom molecules, M. J. Bramley and T. Carrington, Jr., *J. Chem. Phys.* **99**, 8519 (1993).

[138] Comparison of theoretical vibrational and rotational energies of the HCP molecule with experimental values, Y.-T. Chen and D. P. Chong, *J. Chem. Phys.* **99**, 8870 (1993).

[139] Discrete variable representations of differential operators, V. Szalay, *J. Chem. Phys.* **99**, 1978 (1993).

[140] Adiabatic pseudospectral methods for multidimensional vibrational potentials, R. A. Friesner, J. A. Bentley, M. Menou, and C. Leforestier, *J. Chem. Phys.* **99**, 324 (1993).

[141] Rotational excitation with pointwise vibrational wave functions, J. Tennyson, *J. Chem. Phys.* **98**, 9658 (1993).

[142] All the bound vibrational states of H_3^+: A reappraisal, J. R. Henderson, J. Tennyson, and B. T. Sutcliffe, *J. Chem. Phys.* **98**, 7191 (1993).

[143] van der Waals vibrational states of atom-large molecule complexes by a 3D discrete variable representation method: Naphthalene·Ar, M. Mandziuk and Z. Bačić, *J. Chem. Phys.* **98**, 7165 (1993).

[144] Highly vibrationally excited HCN/HNC: Eigenvalues, wave functions, and stimulated emission pumping spectra, J. A. Bentley, C.-M. Huang, and R. E. Wyatt, *J. Chem. Phys.* **98**, 5207 (1993).

[145] A three-dimensional study of NeICl predissociation resonances by the complex scaled discrete variable representation method, N. Lipkin, N. Moiseyev, and C. Leforestier, *J. Chem. Phys.* **98**, 1888 (1993).

[146] A new method for the calculation of photodissociation cross sections, T. Seideman, *J. Chem. Phys.* **98**, 1989 (1993).

[147] State-specific reaction probabilities from a DVR-ABC Green function, W. H. Thompson and W. H. Miller, *Chem. Phys. Lett.* **206**, 123 (1993).

[148] A DVR based time-dependent wave packet treatment for reactive scattering, S. Omar and J. Z. H. Zhang, *Chem. Phys. Lett.* **204**, 190 (1993).

[149] A DVR based time-dependent wave packet treatment for reactive scattering, O. Sharafeddin and J. Z. H. Zhang, *Chem. Phys. Lett.* **204**, 190 (1993).

[150] State-specific reaction probabilities from a DVR-ABC Green function, W. H. Thompson and W. H. Miller, *Chem. Phys. Lett.* **206**, 123 (1993).

[151] Highly excited vibrational eigenstates of nonlinear triatomic molecules. Application to H_2O, S. E. Choi and J. C. Light, *J. Chem. Phys.* **97**, 7031 (1992).

[152] A finite basis-discrete variable representation calculation of vibrational levels of planar acetylene, J. A. Bentley, R. E. Wyatt, M. Menou, and C. Leforestier, *J. Chem. Phys.* **97**, 4255 (1992).

[153] The discrete variable representation of a triatomic Hamiltonian in bond length-bond angle coordinates, H. Wei and T. Carrington, Jr., *J. Chem. Phys.* **97**, 3029 (1992).

[154] Quantum mechanical reaction probabilities via a discrete variable representation-absorbing boundary condition Green's function, T. Seideman and W. H. Miller, *J. Chem. Phys.* **97**, 2499 (1992).

[155] Calculation of the cumulative reaction probability via a discrete variable representation with absorbing boundary conditions, T. Seideman and W. H. Miller, *J. Chem. Phys.* **96**, 4412 (1992).

[156] A novel discrete variable representation for quantum mechanical reactive scattering via the S-matrix Kohn method, D. T. Colbert and W. H. Miller, *J. Chem. Phys.* **96**, 1982 (1992).

[157] Accurate calculation and assignment of highly excited vibrational levels of floppy triatomic molecules in a basis of adiabatic vibrational eigenstates, Z. Bačić, *J. Chem. Phys.* **95**, 3456 (1991).

[158] Dynamics of triatomic photodissociation in the interaction representation. I. Methodology, C. J. Williams, J. Qian, and D. J. Tannor, *J. Chem. Phys.* **95**, 1721 (1991).

[159] Quantum mechanics of small Ne, Ar, Kr, and Xe clusters, D. M. Leitner, J. D. Doll, and R. M. Whitnell, *J. Chem. Phys.* **94**, 6644 (1991).

[160] Grid representation of rotating triatomics, C. Leforestier, *J. Chem. Phys.* **94**, 6388 (1991).

[161] Three-dimensional analytic quantum theory for triatomic photodissociations. II. Angle dependent dissociative surfaces and rotational infinite order sudden approximation for bent triatomics, H. Grinberg, K. F. Freed, and C. J. Williams, *J. Chem. Phys.* **92**, 7283 (1990).

[162] Determination of the bound and quasibound states of Ar-HCl van der Waals complex: Discrete variable representation method, S. E. Choi and J. C. Light, *J. Chem. Phys.* **92**, 2129 (1990).

[163] Quantum reactive scattering in three dimensions using hyperspherical (APH) coordinates. IV. Discrete variable representation (DVR) basis functions and the analysis of accurate results for $F+H_2$, Z. Bačić, J. D. Kress, G. A. Parker, and R. T. Pack, *J. Chem. Phys.* **92**, 2344 (1990).

[164] Accurate quantum thermal rate constants for the three-dimensional $H+H_2$ reaction, T. J. Park and J. C. Light, *J. Chem. Phys.* **91**, 974 (1989).

[165] Use of the discrete variable representation in the quantum dynamics by a wave packet propagation: Predissociation of $NaI(^1\sum_0^+) \rightarrow NaI(0^+) \rightarrow Na(^2S)+I(^2P)$, S. E. Choi and J. C. Light, *J. Chem. Phys.* **90**, 2593 (1989).

[166] Efficient pointwise representations for vibrational wave functions: Eigenfunctions of H_3^+, R. M. Whitnell and J. C. Light, *J. Chem. Phys.* **90**, 1774 (1989).

[167] A variational localized representation calculation of the vibrational levels of the water molecule up to 27 000 cm^{-1}, Z. Bačić, D. Watt, and J. C. Light, *J. Chem. Phys.* **89**, 947 (1988).

[168] Adiabatic approximation and nonadiabatic corrections in the discrete variable representation: Highly excited vibrational states of triatomic molecules, J. C. Light and Z. Bačić, *J. Chem. Phys.* **87**, 4008 (1987).

[169] Accurate localized and delocalized vibrational states of HCN/HNC, Z. Bačić and J. C. Light, *J. Chem. Phys.* **86**, 3065 (1987).

[170] Use of the discrete variable representation in the infinite order sudden approximation for reactive scattering, R. M. Whitnell and J. C. Light, *J. Chem. Phys.* **86**, 2007 (1987).

[171] Highly excited vibrational levels of "floppy" triatomic molecules: A discrete variable representation-Distributed Gaussian basis approach, Z. Bačić and J. C. Light, *J. Chem. Phys.* **85**, 4594 (1986).

[172] Generalized discrete variable approximation in quantum mechanics, J. C. Light, I. P. Hamilton, and J. V. Lill, *J. Chem. Phys.* **82**, 1400 (1985).

[173] Direct determination of $\mu(R)$ from μ_{ij} – a simple DVR approach, I. P. Hamilton and J. C. Light, *Chem. Phys. Lett.* **116**, 169 (1985).

[174] Discrete variable theory of triatomic photodissociation, R. W. Heather and J. C. Light, *J. Chem. Phys.* **79**, 147 (1983).

[175] An iteration method for the solution of the eigenvalue problem of linear differential and integral operators, C. Lanczos, *J. Res. Natl. Bur. Stand.* **45**, 255 (1950).

[176] The Computation of Eigenvalues and Eigenvectors of Very Large Sparse Matrices, C. C. Paige, Ph.D. Dissertation, University of London (1971).

[177] Analysis of the symmetric Lanczos algorithm with reorthogonalization methods, H. D. Simon, *Linear Algebra Appl.* **61**, 101 (1984).

[178] *Sparse Matrix Technology*, S. Pissanetzky (Academic Press, New York, 1984).

[179] Conjugate gradient-like algorithms for solving nonsymmetric linear systems, Y. Saad and M. H. Schultz, *Math. Comp.* **44**, 417 (1985).

[180] QMR: A quasi-minimum residual method for non-Hermitian linear systems, R. W. Freund and N. M. Nachtigal, *Numer. Math.* **60**, 315 (1991).

[181] Methods of conjugate gradients for solving linear systems, M. R. Hestenes and E. Stiefel, *J. Res. Natl. Bur. Stand.* **49**, 409 (1952).

Variational Monte Carlo

[182] Quantum Monte Carlo study of first-row atoms using transcorrelated variational Monte Carlo trial functions, R. Prasad, N. Umezawa, D. Domin, R. Salomon-Ferrer, and W. A. Lester, Jr., *J. Chem. Phys.* **126**, 164109 (2007).

[183] An efficient sampling algorithm for variational Monte Carlo, A. Scemama, T. Lelièvre, G. Stoltz, E. Cancès, and M. Caffarel, *J. Chem. Phys.* **125**, 114105 (2006).

[184] A practical treatment for the three-body interactions in the transcorrelated variational Monte Carlo method: Application to atoms from lithium to neon, N. Umezawa, S. Tsuneyuki, T. Ohno, K. Shiraishi, and T. Chikyow, *J. Chem. Phys.* **122**, 224101 (2005).

[185] Excited electronic state calculations by the transcorrelated variational Monte Carlo method: Application to a helium atom, N. Umezawa and S. Tsuneyuki, *J. Chem. Phys.* **121**, 7070 (2004).

[186] Delayed rejection variational Monte Carlo, D. Bressanini, G. Morosi, S. Tarasco, and A. Mira, *J. Chem. Phys.* **121**, 3446 (2004).

[187] Properties of selected diatomics using variational Monte Carlo methods, S. Datta, S. A. Alexander, and R. L. Coldwell, *J. Chem. Phys.* **120**, 3642 (2004).

[188] Transcorrelated method for electronic systems coupled with variational Monte Carlo calculation, N. Umezawa and S. Tsuneyuki, *J. Chem. Phys.* **119**, 10015 (2003).

[189] A new variational Monte Carlo trial wavefunction with directional Jastrow functions, K. E. Riley and J. B. Anderson, *Chem. Phys. Lett.* **366**, 153 (2002).

[190] Improved efficiency with variational Monte Carlo using two level sampling, M. Dewing, *J. Chem. Phys.* **113**, 5123 (2000).

[191] Spatial-partitioning-based acceleration for variational Monte Carlo, D. Bressanini and P. J. Reynolds, *J. Chem. Phys.* **111**, 6180 (1999).

[192] A variational Monte Carlo study of the 2s-2p near degeneracy in beryllium, boron, and carbon atoms, A. Sarsa, F. J. Gálvez, and E. Buendía, *J. Chem. Phys.* **109**, 3346 (1998).

[193] Monte Carlo variational transition-state theory study of the unimolecular dissociation of RDX, D. V. Shalashilin and D. L. Thompson, *J. Phys. Chem. A* **101**, 961 (1997).

[194] An improved transition matrix for variational quantum Monte Carlo, M. Mella, A. Luchow, and J. B. Anderson, *Chem. Phys. Lett.* **265**, 467 (1997).

[195] Titrating polyelectrolytes-variational calculations and Monte Carlo simulations, B. Jonsson, M. Ullner, C. Peterson, O. Sommelius, and B. Soderberg, *J. Phys. Chem.* **100**, 409 (1996).

[196] Variational Monte Carlo and configuration interaction studies of C_{60} and its fragments, B. Srinivasan, S. Ramasesha, and H. R. Krishnamurthy, *J. Phys. Chem.* **100**, 11260 (1996).

[197] Calculating atomic properties using variational Monte Carlo, S. A. Alexander and R. L. Coldwell, *J. Chem. Phys.* **103**, 2572 (1995).

[198] Characteristics of electron movement in variational Monte Carlo simulations, Z. Sun, M. M. Soto, and W. A. Lester, Jr., *J. Chem. Phys.* **100**, 1278 (1994).

[199] A method for relativistic variational Monte Carlo calculations, H. Bueckert, S. M. Rothstein, and J. Vrbik, *Chem. Phys. Lett.* **190**, 413 (1992).

[200] Quantum and variational Monte Carlo interaction potentials for Li_2 $(X^1 \sum_g^+)$, Z. Sun, R. N. Barnett, and W. A. Lester, *Chem. Phys. Lett.* **195**, 365 (1992).

[201] Whither magic numbers: A variational Monte Carlo study of six and seven atom helium clusters, S. W. Rick and J. D. Doll, *Chem. Phys. Lett.* **188**, 149 (1992).

[202] A variational Monte Carlo study of argon, neon, and helium clusters, S. W. Rick, D. L. Lynch, and J. D. Doll, *J. Chem. Phys.* **95**, 3506 (1991).

[203] Variational Monte Carlo method in the connected moments expansion: H, H^-, Be, and Li_2, T. Yoshida and K. Iguchi, *J. Chem. Phys.* **91**, 4249 (1989).

[204] Connected moments expansion with variational Monte Carlo technique, T. Yoshida and K. Iguchi, *Chem. Phys. Lett.* **143**, 329 (1988).

Quantum Monte Carlo

[205] On the nodal structure of single-particle approximation based atomic wave functions, D. Bressanini and G. Morosi, *J. Chem. Phys.* **129**, 054103 (2008).

[206] Piris natural orbital functional study of the dissociation of the radical helium dimer, M. Piris, J. M. Matxain, and J. M. Ugalde, *J. Chem. Phys.* **129**, 014108 (2008).

[207] Full-dimensional quantum calculations of ground-state tunneling splitting of malonaldehyde using an accurate ab initio potential energy surface, Y. Wang, B. J. Braams, J. M. Bowman, S. Carter, and D. P. Tew, *J. Chem. Phys.* **128**, 224314 (2008).

[208] Quantum Monte Carlo study of the ground state and low-lying excited states of the scandium dimer, J. M. Matxain, E. Rezabal, X. Lopez, J. M. Ugalde, and L. Gagliardi, *J. Chem. Phys.* **128**, 194315 (2008).

[209] Full optimization of Jastrow-Slater wave functions with application to the first-row atoms and homonuclear diatomic molecules, J. Toulouse and C. J. Umrigar, *J. Chem. Phys.* **128**, 174101 (2008).

[210] Importance sampling for quantum Monte Carlo in manifolds: Addressing the time scale problem in simulations of molecular aggregates, T. Luan, E. Curotto, and M. Mella, *J. Chem. Phys.* **128**, 164102 (2008).

[211] Fixed-node diffusion Monte Carlo study of the structures of m-benzyne, W. A. Al-Saidi and C. J. Umrigar, *J. Chem. Phys.* **128**, 154324 (2008).

[212] Excitation levels and magic numbers of small parahydrogen clusters ($N \leq 40$), R. Guardiola and J. Navarro, *J. Chem. Phys.* **128**, 144303 (2008).

[213] Anionic microsolvation in helium droplets: $OH^-(He)_N$ structures from classical and quantum calculations, E. Coccia, F. Marinetti, E. Bodo, and F. A. Gianturco, *J. Chem. Phys.* **128**, 134511 (2008).

[214] Quantum Monte Carlo calculations of the potential energy curve of the helium dimer, R. Springall, M. C. Per, S. P. Russo, and I. K. Snook, *J. Chem. Phys.* **128**, 114308 (2008).

[215] Vibrational ground state properties of H_5^+ and its isotopomers from diffusion Monte Carlo calculations, P. H. Acioli, Z. Xie, B. J. Braams, and J. M. Bowman, *J. Chem. Phys.* **128**, 104318 (2008).

[216] A divide-and-conquer strategy to improve diffusion sampling in generalized ensemble simulations, D. Min and W. Yang, *J. Chem. Phys.* **128**, 094106 (2008).

[217] Quantum Monte Carlo study of small pure and mixed spin-polarized tritium clusters, I. Bešlić, L. V. Markić, and J. Boronat, *J. Chem. Phys.* **128**, 064302 (2008).

[218] Polymer chain dynamics at interfaces: Role of boundary conditions at solid interface, T. G. Desai, P. Keblinski, and S. K. Kumar, *J. Chem. Phys.* **128**, 044903 (2008).

[219] Quantum Monte Carlo for 3d transition-metal atoms, A. Sarsa, E. Buendia, F. J. Galvez, and P. Maldonado, *J. Phys. Chem.* A **112**, 2074 (2008).

[220] Bond dissocation and conformational energetics of tetrasulfur: A quantum Monte Carlo study, J. A. W. Harkless and J. S. Francisco, *J. Phys. Chem.* A **112**, 2088 (2008).

[221] Gamma distribution model to provide a direct assessment of the overall quality of quantum Monte Carlo-generated electron distributions, B. Coles, P. Vrbik, R. D. Giacometti, and S. M. Rothstein, *J. Phys. Chem.* A **112**, 2012 (2008).

[222] Toward the exact solution of the electronic Schrödinger equation for non-covalent molecular interactions: Worldwide distributed quantum Monte Carlo calculations, M. Korth, A. Luchow, and S. Grimme, *J. Phys. Chem.* A **112**, 2104 (2008).

[223] Blueshift and intramolecular tunneling of NH_3 umbrella mode in ^4He_n clusters, A. Viel, K. B. Whaley, and R. J. Wheatley, *J. Chem. Phys.* **127**, 194303 (2007).

[224] Full-dimensional (15-dimensional) quantum-dynamical simulation of the protonated water dimer. I. Hamiltonian setup and analysis of the ground vibrational state, O. Vendrell, F. Gatti, D. Lauvergnat, and H.-D. Meyer, *J. Chem. Phys.* **127**, 184302 (2007).

[225] Weak binding between two aromatic rings: Feeling the van der Waals attraction by quantum Monte Carlo methods, S. Sorella, M. Casula, and D. Rocca, *J. Chem. Phys.* **127**, 014105 (2007).

[226] Bosonic helium droplets with cationic impurities: Onset of electrostriction and snowball effects from quantum calculations, E. Coccia, E. Bodo, F. Marinetti, F. A. Gianturco, E. Yildrim, M. Yurtsever, and E. Yurtsever, *J. Chem. Phys.* **126**, 124319 (2007).

[227] Improved diffusion Monte Carlo for bosonic systems using time-step extrapolation "on the fly", P. Hakansson and M. Mella, *J. Chem. Phys.* **126**, 104106 (2007).

[228] Optimization of quantum Monte Carlo wave functions by energy minimization, J. Toulouse and C. J. Umrigar, *J. Chem. Phys.* **126**, 084102 (2007).

[229] Diffusion Monte Carlo study of the equation of state of solid ortho-D_2, F. Operetto and F. Pederiva, *J. Chem. Phys.* **126**, 074704 (2007).

[230] Energetics and dipole moment of transition metal monoxides by quantum Monte Carlo, L. K. Wagner and L. Mitas, *J. Chem. Phys.* **126**, 034105 (2007).

[231] The ground state tunneling splitting and the zero point energy of malonaldehyde: A quantum Monte Carlo determination, A. Viel, M. D. Coutinho-Neto, and U. Manthe, *J. Chem. Phys.* **126**, 024308 (2007).

[232] Stereographic projection diffusion Monte Carlo (SPDMC) algorithms for molecular condensed matter, M. Aviles and E. Curotto, *J. Phys. Chem. A* **111**, 2610 (2007).

[233] The electronic spectrum of Fe^{2+} ion in aqueous solution: A sequential Monte Carlo/quantum mechanical study, C. M. Aguilar, W. B. De Almeida, and W. R. Rocha, *Chem. Phys. Lett.* **449**, 144 (2007).

[234] Towards a field-free quantum Monte Carlo approach to polarizabilities of excited states: Application to the $n = 2$ hydrogen atom, Y. Li, J. Vrbik, and S. M. Rothstein, *Chem. Phys. Lett.* **445**, 345 (2007).

[235] Improved diffusion Monte Carlo propagators for bosonic systems using Itô calculus, P. Hakansson, M. Mella, D. Bressanini, G. Morosi, and M. Patrone, *J. Chem. Phys.* **125**, 184106 (2006).

[236] HF in clusters of molecular hydrogen: II. Quantum solvation by H_2 isotopomers, cluster rigidity, and comparison with CO-doped parahydrogen clusters, F. Sebastianelli, Y. S. Elmatad, H. Jiang, and Z. Bačić, *J. Chem. Phys.* **125**, 164313 (2006).

[237] Self-diffusion in submonolayer colloidal fluids near a wall, S. G. Anekal and M. A. Bevan, *J. Chem. Phys.* **125**, 034906 (2006).

[238] Quantum solvation dynamics of HCN in a helium-4 droplet, A. A. Mikosz, J. A. Ramilowski, and D. Farrelly, *J. Chem. Phys.* **125**, 014312 (2006).

[239] Magic numbers, excitation levels, and other properties of small neutral ^{4}He clusters ($N \leq 50$), R. Guardiola, O. Kornilov, J. Navarro, and J. P. Toennies, *J. Chem. Phys.* **124**, 084307 (2006).

[240] Excited states of weakly bound bosonic clusters: Discrete variable representation and quantum Monte Carlo, M. P. Nightingale and P.-N. Roy, *J. Phys. Chem. A* **110**, 5391 (2006).

[241] $(HCl)_2$ and $(HF)_2$ in small helium clusters: Quantum solvation of hydrogen-bonded dimers, H. Jiang, A. Sarsa, G. Murdachaew, K. Szalewicz, and Z. Bačić, *J. Chem. Phys.* **123**, 224313 (2005).

[242] An investigation of nodal structures and the construction of trial wave functions, D. Bressanini, G. Morosi, and S. Tarasco, *J. Chem. Phys.* **123**, 204109 (2005).

[243] Weak intermolecular interactions calculated with diffusion Monte Carlo, C. Diedrich, A. Lüchow, and S. Grimme, *J. Chem. Phys.* **123**, 184106 (2005).

[244] A diffusion quantum Monte Carlo study on the lowest singlet and triplet electronic states of BN molecule, S.-I. Lu, *J. Chem. Phys.* **123**, 174313 (2005).

[245] Towards accurate all-electron quantum Monte Carlo calculations of transition-metal systems: Spectroscopy of the copper atom, M. Caffarel, J.-P. Daudey, J.-L. Heully, and A. Ramírez-Solís, *J. Chem. Phys.* **123**, 094102 (2005).

[246] Quantum studies of the vibrations in $H_3O_2^-$ and $D_3O_2^-$, A. B. McCoy, X. Huang, S. Carter, and J. M. Bowman, *J. Chem. Phys.* **123**, 064317 (2005).

[247] Structure and stability of Ne^+He_n: Experiment and diffusion quantum Monte Carlo theory with "on the fly" electronic structure, C. A. Brindle, M. R. Prado, K. C. Janda, N. Halberstadt, and M. Lewerenz, *J. Chem. Phys.* **123**, 064312 (2005).

[248] Predicting atomic dopant solvation in helium clusters: The $MgHe_n$ case, M. Mella, G. Calderoni, and F. Cargnoni, *J. Chem. Phys.* **123**, 054328 (2005).

[249] Ar_nHF van der Waals clusters revisited: II. Energetics and HF vibrational frequency shifts from diffusion Monte Carlo calculations on additive and nonadditive potential-energy surfaces for $n = 1 - 12$, H. Jiang, M. Xu, J. M. Hutson, and Z. Bačić, *J. Chem. Phys.* **123**, 054305 (2005).

[250] The spectra of mixed ^{3}He-^{4}He droplets, S. Fantoni, R. Guardiola, J. Navarro, and A. Zuker, *J. Chem. Phys.* **123**, 054503 (2005).

[251] HF in clusters of molecular hydrogen. I. Size evolution of quantum solvation by parahydrogen molecules, H. Jiang and Z. Bačić, *J. Chem. Phys.* **122**, 244306 (2005).

[252] The vibrational predissociation spectra of the $H_5O_2^+ \cdot RG_n$ (RG = Ar,Ne) clusters: Correlation of the solvent perturbations in the free OH and shared proton transitions of the Zundel ion, N. I. Hammer, E. G. Diken, J. R. Roscioli, M. A. Johnson, E. M. Myshakin, K. D. Jordan, A. B. McCoy, X. Huang, J. M. Bowman, and S. Carter, *J. Chem. Phys.* **122**, 244301 (2005).

[253] A version of diffusion Monte Carlo method based on random grids of coherent states. II. Six-dimensional simulation of electronic states of H_2, D. V. Shalashilin and M. S. Child, *J. Chem. Phys.* **122**, 224109 (2005).

[254] Ab initio global potential-energy surface for $H_5^+ \rightarrow H_3^+ + H_2$, Z. Xie, B. J. Braams, and J. M. Bowman, *J. Chem. Phys.* **122**, 224307 (2005).

[255] Full-dimensional vibrational calculations for $H_5O_2^+$ using an ab initio potential energy surface, A. B. McCoy, X. Huang, S. Carter, M. Y. Landeweer, and J. M. Bowman, *J. Chem. Phys.* **122**, 061101 (2005).

[256] Interpolated potential energy surfaces: How accurate do the second derivatives have to be? D. L. Crittenden and M. J. T. Jordan, *J. Chem. Phys.* **122**, 044102 (2005).

[257] Performance of diffusion Monte Carlo for the first dissociation energies of transition metal carbonyls, C. Diedrich, A. Lüchow, and S. Grimme, *J. Chem. Phys.* **122**, 021101 (2005).

[258] Electronic polarization in liquid acetonitrile: A sequential Monte Carlo/quantum mechanics investigation, R. Rivelino, B. J. Costa Cabral, K. Coutiño, and S. Canuto, *Chem. Phys. Lett.* **407**, 13 (2005).

[259] Monte-Carlo simulations of photoinduced fluorescence enhancement in semiconductor quantum dot arrays, S. Maenosono, *Chem. Phys. Lett.* **405**, 182 (2005).

[260] Efficiency considerations in the construction of interpolated potential energy surfaces for the calculation of quantum observables by diffusion Monte Carlo, D. L. Crittenden, K. C. Thompson, M. Chebib, and M. J. T. Jordan, *J. Chem. Phys.* **121**, 9844 (2004).

[261] The ground state tunneling splitting of malonaldehyde: Accurate full dimensional quantum dynamics calculations, M. D. Coutinho-Neto, A. Viel, and U. Manthe, *J. Chem. Phys.* **121**, 9207 (2004).

[262] Thermochemistry of disputed soot formation intermediates C_4H_3 and C_4H_5, S. E. Wheeler, W. D. Allen, and H. F. Schaefer III, *J. Chem. Phys.* **121**, 8800 (2004).

[263] Rotational excitations of N_2O in small helium clusters and the role of Bose permutation symmetry, F. Paesani and K. B. Whaley, *J. Chem. Phys.* **121**, 5293 (2004).

[264] Calculating expectations with time-dependent perturbations in quantum Monte Carlo, M. H. Kalos and F. Arias de Saavedra, *J. Chem. Phys.* **121**, 5143 (2004).

[265] Unbiased expectation values from diffusion quantum Monte Carlo simulations with a fixed number of walkers, I. Bosá and S. M. Rothstein, *J. Chem. Phys.* **121**, 4486 (2004).

[266] Quantum and classical studies of vibrational motion of CH_5^+ on a global potential energy surface obtained from a novel ab initio direct dynamics approach, A. Brown, A. B. McCoy, B. J. Braams, Z. Jin, and J. M. Bowman, *J. Chem. Phys.* **121**, 4105 (2004).

[267] Electron pair localization function: A practical tool to visualize electron localization in molecules from quantum Monte Carlo data, A. Scemama, P. Chaquin, and M. Caffarel, *J. Chem. Phys.* **121**, 1725 (2004).

[268] Quantum Monte Carlo study of the reaction: $Cl + CH_3OH \rightarrow CH_2OH + HCl$, A. C. Kollias, O. Couronne, and W. A. Lester Jr., *J. Chem. Phys.* **121**, 1357 (2004).

[269] Irreversible adsorption of particles at random-site surfaces, Z. Adamczyk, K. Jaszczólt, B. Siwek, and P. Weronski, *J. Chem. Phys.* **120**, 11155 (2004).

[270] Properties of branched confined polymers, A. Sikorski and P. Romiszowski, *J. Chem. Phys.* **120**, 7206 (2004).

[271] Quantum Monte Carlo for electronic excitations of free-base porphyrin, A. Aspuru-Guzik, O. El Akramine, J. C. Grossman, and W. A. Lester, Jr., *J. Chem. Phys.* **120**, 3049 (2004).

[272] A diffusion quantum Monte Carlo study of geometries and harmonic frequencies of molecules, S.-I. Lu, *J. Chem. Phys.* **120**, 14 (2004).

[273] Ab initio diffusion Monte Carlo calculations of the quantum behavior of CH_5^+ in full dimensionality, A. B. McCoy, B. J. Braams, A. Brown, X. Huang, Z. Jin, and J. M. Bowman, *J. Phys. Chem.* A **108**, 4991 (2004).

[274] OCS in small para-hydrogen clusters: Energetics and structure with $N = 1 - 8$ complexed hydrogen molecules, F. Paesani, R. E. Zillich, and K. B. Whaley, *J. Chem. Phys.* **119**, 11682 (2003).

[275] Zero-variance zero-bias principle for observables in quantum Monte Carlo: Application to forces, R. Assaraf and M. Caffarel, *J. Chem. Phys.* **119**, 10536 (2003).

[276] Zero temperature quantum properties of small protonated water clusters $(H_2O)_n H^+$ $(n = 1 - 5)$, M. Mella and D. C. Clary, *J. Chem. Phys.* **119**, 10048 (2003).

[277] Coarse-grained stochastic processes and kinetic Monte Carlo simulators for the diffusion of interacting particles, M. A. Katsoulakis and D. G. Vlachos, *J. Chem. Phys.* **119**, 9412 (2003).

[278] Geometry optimization in quantum Monte Carlo with solution mapping: Application to formaldehyde, C. A. Schuetz, M. Frenklach, A. C. Kollias, and W. A. Lester, Jr., *J. Chem. Phys.* **119**, 9386 (2003).

[279] Analysis of the contributions of three-body potentials in the equation of state of ^{4}He, S. Ujevic and S. A. Vitiello, *J. Chem. Phys.* **119**, 8482 (2003).

[280] Intermolecular forces and fixed-node diffusion Monte Carlo: A brute force test of accuracies for He$_2$ and He-LiH, M. Mella and J. B. Anderson, *J. Chem. Phys.* **119**, 8225 (2003).

[281] Comparison of different propagators in diffusion Monte Carlo simulations of noble gas clusters, S. Chiesa, M. Mella, G. Morosi, and D. Bressanini, *J. Chem. Phys.* **119**, 5601 (2003).

[282] Librational modes of ice I, M. W. Severson, J. P. Devlin, and V. Buch, *J. Chem. Phys.* **119**, 4449 (2003).

[283] Quantum Monte Carlo study of singlet-triplet transition in ethylene, O. El Akramine, A. C. Kollias, and W. A. Lester, Jr., *J. Chem. Phys.* **119**, 1483 (2003).

[284] Linear scaling for the local energy in quantum Monte Carlo, S. Manten and A. Lüchow, *J. Chem. Phys.* **119**, 1307 (2003).

[285] Computing accurate forces in quantum Monte Carlo using Pulay's corrections and energy minimization, M. Casalegno, M. Mella, and A. M. Rappe, *J. Chem. Phys.* **118**, 7193 (2003).

[286] Performance of Ornstein-Uhlenbeck diffusion quantum Monte Carlo for first-row diatomic dissociation energies and dipole moments, S.-I. Lu, *J. Chem. Phys.* **118**, 6152 (2003).

[287] Effects of molecular rotation on densities in doped ^{4}He clusters, M. V. Patel, A. Viel, F. Paesani, P. Huang, and K. B. Whaley, *J. Chem. Phys.* **118**, 5011 (2003).

[288] A quantum Monte Carlo and density functional theory study of the electronic structure of peroxynitrite anion, J. A. W. Harkless, J. H. Rodriguez, L. Mitas, and W. A. Lester, Jr., *J. Chem. Phys.* **118**, 4987 (2003).

[289] Higher accuracy quantum Monte Carlo calculations of the barrier for the H + H$_2$ reaction, K. E. Riley and J. B. Anderson, *J. Chem. Phys.* **118**, 3437 (2003).

[290] Characterization of Ar$_n$O$^-$ clusters from ab initio and diffusion Monte Carlo calculations, J. Jakowski, G. Chalasinski, J. Gallegos, M. W. Severson, and M. M. Szczesniak, *J. Chem. Phys.* **118**, 2748 (2003).

[291] Computing potential energy curve for hydrogen fluoride in Ornstein-Uhlenbeck diffusion quantum Monte Carlo method, S.-I. Lu, *Chem. Phys. Lett.* **381**, 672 (2003).

[292] A quantum Monte Carlo study of electron correlation in transition metal oxygen molecules, L. Wagner and L. Mitas, *Chem. Phys. Lett.* **370**, 412 (2003).

[293] Electronic polarization of liquid water: Converged Monte Carlo-quantum mechanics results for the multipole moments, K. Coutinho, R. C. Guedes, B. J. Costa Cabral, and S. Canuto, *Chem. Phys. Lett.* **369**, 345 (2003).

[294] Density dependence of the hydrodynamic response to SF_6 rotation in superfluid helium, P. Huang and K. B. Whaley, *J. Chem. Phys.* **117**, 11244 (2002).

[295] Pulsed field gradient nuclear magnetic resonance study of long-range diffusion in beds of NaX zeolite: Evidence for different apparent tortuosity factors in the Knudsen, and bulk regimes, O. Geier, S. Vasenkov, and J. Kärger, *J. Chem. Phys.* **117**, 1935 (2002).

[296] Quantum hydrodynamic model for the enhanced moments of inertia of molecules in helium nanodroplets: Application to SF_6, K. K. Lehmann and C. Callegari, *J. Chem. Phys.* **117**, 1595 (2002).

[297] Positron, and positronium chemistry by quantum Monte Carlo. VI. The ground state of LiPs, NaPs, e+Be, and e+Mg, M. Mella, M. Casalegno, and G. Morosi, *J. Chem. Phys.* **117**, 1450 (2002).

[298] Benchmark quantum Monte Carlo calculations, J. C. Grossman, *J. Chem. Phys.* **117**, 1434 (2002).

[299] Cluster nucleation effects in $CO(Ar)_n$: A stochastic analysis, F. Paesani and F. A. Gianturco, *J. Chem. Phys.* **117**, 709 (2002).

[300] Vibrational effects in a weakly-interacting quantum solvent: The CO molecule in ^4He gas and in ^4He droplets, F. Paesani and F. A. Gianturco, *J. Chem. Phys.* **116**, 10170 (2002).

[301] Generic van der Waals equation of state and theory of diffusion coefficients: Binary mixtures of simple liquids, K. Rah and B. C. Eu, *J. Chem. Phys.* **116**, 7967 (2002).

[302] Quadratic diffusion Monte Carlo and pure estimators for atoms, A. Sarsa, J. Boronat, and J. Casulleras, *J. Chem. Phys.* **116**, 5956 (2002).

[303] Microsolvation and vibrational shifts of OCS in helium clusters, F. Paesani, F. A. Gianturco, and K. B. Whaley, *J. Chem. Phys.* **115**, 10225 (2001).

[304] Quantum structure and rotational dynamics of HCN in helium clusters, A. Viel and K. B. Whaley, *J. Chem. Phys.* **115**, 10186 (2001).

[305] Dynamics of 4He droplets, E. Krotscheck and R. Zillich, *J. Chem. Phys.* **115**, 10161 (2001).

[306] Stability and production of positron-diatomic molecule complexes, M. Mella, D. Bressanini, and G. Morosi, *J. Chem. Phys.* **114**, 10579 (2001).

[307] Relative stabilities of the two isomers of the methanol-water dimer: The effects of the internal rotations of the hydroxyl and methyl groups of methanol, J. W. Moskowitz, Z. Bačić, A. Sarsa, and K. E. Schmidt, *J. Chem. Phys.* **114**, 10294 (2001).

[308] Quantum Monte Carlo studies of the structure and spectroscopy of $Ne_n OH$ ($\tilde{A}^2 \sum^+$) van der Waals complexes, H.-S. Lee and A. B. McCoy, *J. Chem. Phys.* **114**, 10278 (2001).

[309] Torsional diffusion Monte Carlo: A method for quantum simulations of proteins, D. C. Clary, *J. Chem. Phys.* **114**, 9725 (2001).

[310] Quantum Monte Carlo characterization of small Cu-doped silicon clusters: $CuSi_4$ and $CuSi_6$, I. V. Ovcharenko, W. A. Lester, Jr., C. Xiao, and F. Hagelberg, *J. Chem. Phys.* **114**, 9028 (2001).

[311] Electronic excited-state wave functions for quantum Monte Carlo: Application to silane and methane, A. R. Porter, O. K. Al-Mushadani, M. D. Towler, and R. J. Needs, *J. Chem. Phys.* **114**, 7795 (2001).

[312] Density profile evolution and nonequilibrium effects in partial and full spreading measurements of surface diffusion, P. Nikunen, I. Vattulainen, and T. Ala-Nissila, *J. Chem. Phys.* **114**, 6335 (2001).

[313] Rotation in liquid ^{4}He: Lessons from a highly simplified model, K. K. Lehmann, *J. Chem. Phys.* **114**, 4643 (2001).

[314] A diffusion quantum Monte Carlo method based on floating spherical Gaussian orbitals and Gaussian geminals: Dipole moment of lithium hydride molecule, S.-I. Lu, *J. Chem. Phys.* **114**, 3898 (2001).

[315] Ground state of the quantum anisotropic planar rotor model: A finite size scaling study of the orientational order-disorder phase transition, B. Hetényi and B. J. Berne, *J. Chem. Phys.* **114**, 3674 (2001).

[316] Improved trial wave functions in quantum Monte Carlo: Application to acetylene and its dissociation fragments, R. N. Barnett, Z. Sun, and W. A. Lester, Jr., *J. Chem. Phys.* **114**, 2013 (2001).

[317] The electronic spectrum of N-methylacetamide in aqueous solution: A sequential Monte Carlo/quantum mechanical study, W. R. Rocha, K. J. De Almeida, K. Coutinho, and S. Canuto, *Chem. Phys. Lett.* **345**, 171 (2001).

[318] An efficient quantum mechanical/molecular mechanics Monte Carlo simulation of liquid water, W. R. Rocha, K. Coutinho, W. B. de Almeida, and S. Canuto, *Chem. Phys. Lett.* **335**, 127 (2001).

[319] Positron and positronium chemistry by quantum Monte Carlo. V. The ground state potential energy curve of e^+LiH, M. Mella, G. Morosi, D. Bressanini, and S. Elli, *J. Chem. Phys.* **113**, 6154 (2000).

[320] Quaternion formulation of diffusion quantum Monte Carlo for the rotation of rigid molecules in clusters, D. M. Benoit and D. C. Clary, *J. Chem. Phys.* **113**, 5193 (2000).

[321] Vibrationally excited states and fragmentation geometries of Ne_N and Ar_N clusters, $N = 3 - 6$, using hyperspherical coordinates, D. Blume and C. H. Greene, *J. Chem. Phys.* **113**, 4242 (2000).

[322] Quantum Monte Carlo determination of the atomization energy and heat of formation of propargyl radical, J. A. W. Harkless and W. A. Lester, Jr., *J. Chem. Phys.* **113**, 2680 (2000).

[323] Constraint dynamics for quantum Monte Carlo calculations, A. Sarsa, K. E. Schmidt, and J. W. Moskowitz, *J. Chem. Phys.* **113**, 44 (2000).

[324] Structure and energetics of Ar_nNO$^-$ clusters from ab initio calculations, J. Jakowski, J. Klos, G. Chalasinski, M. W. Severson, M. M. Szczesniak, and S. M. Cybulski, *J. Chem. Phys.* **112**, 10895 (2000).

[325] A stochastic study of microsolvation. II. Structures of CO in small helium clusters, F. A. Gianturco, M. Lewerenz, F. Paesani, and J. P. Toennies, *J. Chem. Phys.* **112**, 2239 (2000).

[326] A diffusion Monte Carlo accurate interaction potential between H and PsH, M. Mella, G. Morosi, and D. Bressanini, *J. Chem. Phys.* **112**, 1063 (2000).

[327] Quantum Monte Carlo and density functional theory characterization of 2-cyclopentenone and 3-cyclopentenone formation from $O(^3P)$ + cyclopentadiene, J. C. Grossman, W. A. Lester, and S. G. Louie, *J. Am. Chem. Soc.* **122**, 705 (2000).

[328] Quantum Monte Carlo study of the H^- impurity in small helium clusters, M. Casalegno, M. Mella, G. Morosi, and D. Bressanini, *J. Chem. Phys.* **112**, 69 (2000).

[329] Ab-initio molecular deformation barriers using auxiliary-field quantum Monte Carlo with application to the inversion barrier of water, R. Baer, *Chem. Phys. Lett.* **324**, 101 (2000).

[330] Energetics of carbon clusters C_{20} from all-electron quantum Monte Carlo calculations, S. Sokolova, A. Luchow, and J. B. Anderson, *Chem. Phys. Lett.* **323**, 229 (2000).

[331] Quantum Monte Carlo study of vibrational states of silanone, P. H. Acioli, L. S. Costa, and F. V. Prudente, *Chem. Phys. Lett.* **321**, 121 (2000).

[332] Diffusion quantum Monte Carlo on multiple-potential surfaces, A. B. McCoy, *Chem. Phys. Lett.* **321**, 71 (2000).

[333] An ab initio study of TiC with the diffusion quantum Monte Carlo method, S. Sokolova and A. Luchow, *Chem. Phys. Lett.* **320**, 421 (2000).

[334] Quantum Monte Carlo simulation of intermolecular excited vibrational states in the cage water hexamer, M. W. Severson and V. Buch, *J. Chem. Phys.* **111**, 10866 (1999).

[335] Diffusion Monte Carlo simulations of the dipole-bound state of the water dimer anion, D. C. Clary and D. M. Benoit, *J. Chem. Phys.* **111**, 10559 (1999).

[336] Structure and spectroscopy of Ne_nSH ($\tilde{A}^2 \sum^+$) complexes using adiabatic diffusion Monte Carlo (ADMC), H.-S. Lee, J. M. Herbert, and A. B. McCoy, *J. Chem. Phys.* **111**, 9203 (1999).

[337] A stochastic study of microsolvation. I. Structures of CO in small argon clusters, F. Paesani, F. A. Gianturco, M. Lewerenz, and J. P. Toennies, *J. Chem. Phys.* **111**, 6897 (1999).

[338] Quantum Monte Carlo calculations of molecular electron affinities: First-row hydrides, G. Morosi, M. Mella, and D. Bressanini, *J. Chem. Phys.* **111**, 6755 (1999).

[339] CO-oxidation model with superlattice ordering of adsorbed oxygen. I. Steady-state bifurcations, E. W. James, C. Song, and J. W. Evans, *J. Chem. Phys.* **111**, 6579 (1999).

[340] A spline approach to trial wave functions for variational and diffusion Monte Carlo, D. Bressanini, G. Fabbri, M. Mella, and G. Morosi, *J. Chem. Phys.* **111**, 6230 (1999).

[341] Structures and energetics of Ne_n-HN_2^+ clusters, M. Meuwly, *J. Chem. Phys.* **111**, 2633 (1999).

[342] From polypeptide sequences to structures using Monte Carlo simulations and an optimized potential, P. Derreumaux, *J. Chem. Phys.* **111**, 2301 (1999).

[343] Positron and positronium chemistry by quantum Monte Carlo. IV. Can this method accurately compute observables beyond energy? M. Mella, G. Morosi, and D. Bressanini, *J. Chem. Phys.* **111**, 108 (1999).

[344] Transport coefficients of electrolyte solutions from Smart Brownian dynamics simulations, M. Jardat, O. Bernard, P. Turq, and G. R. Kneller, *J. Chem. Phys.* **110**, 7993 (1999).

[345] Adiabatic diffusion Monte Carlo approaches for studies of ground and excited state properties of van der Waals complexes, H.-S. Lee, J. M. Herbert, and A. B. McCoy, *J. Chem. Phys.* **110**, 5481 (1999).

[346] Length and occupancy dependence of the tracer exchange in single-file systems, C. Rödenbeck and J. Kärger, *J. Chem. Phys.* **110**, 3970 (1999).

[347] Quantum-Monte Carlo study of rovibrational states of molecular systems, F. V. Prudente and P. H. Acioli, *Chem. Phys. Lett.* **302**, 249 (1999).

[348] Combined diffusion quantum Monte Carlo-vibrational self-consistent field (DQMC-VSCF) method for excited vibrational states of large polyatomic systems, S. Broude, J. O. Jung, and R. B. Gerber, *Chem. Phys. Lett.* **299**, 437 (1999).

[349] A classical dynamics study of the anisotropic interactions in NNO-Ar and NNO-Kr systems: Comparison with transport and relaxation data, M. A. ter Horst and C. J. Jameson, *J. Chem. Phys.* **109**, 10238 (1998).

[350] Diffusion quantum Monte Carlo calculation of the binding energy and annihilation rate of positronium hydride, PsH, N. Jiang and D. M. Schrader, *J. Chem. Phys.* **109**, 9430 (1998).

[351] Positron and positronium chemistry by quantum Monte Carlo. III. Ground state of [OH,Ps], [CH,Ps], and [NH₂,Ps] complexes, D. Bressanini, M. Mella, and G. Morosi, *J. Chem. Phys.* **109**, 5931 (1998).

[352] Temperature dependance of properties of star-branched polymers: A computer simulation study, P. Romiszowski and A. Sikorski, *J. Chem. Phys.* **109**, 2912 (1998).

[353] Positron chemistry by quantum Monte Carlo. II. Ground-state of positron-polar molecule complexes, D. Bressanini, M. Mella, and G. Morosi, *J. Chem. Phys.* **109**, 1716 (1998).

[354] A soft Hartree-Fock pseudopotential for carbon with application to quantum Monte Carlo, C. W. Greeff and W. A. Lester, Jr., *J. Chem. Phys.* **109**, 1607 (1998).

[355] Quantum Monte Carlo simulations of Ar_n-CO_2 clusters, M. W. Severson, *J. Chem. Phys.* **109**, 1343 (1998).

[356] Positronium chemistry by quantum Monte Carlo. I. Positronium-first row atom complexes, D. Bressanini, M. Mella, and G. Morosi, *J. Chem. Phys.* **108**, 4756 (1998).

[357] Modeling the concentration dependence of diffusion in zeolites. III. Testing mean field theory for benzene in Na-Y with simulation, C. Saravanan, F. Jousse, and S. M. Auerbach, *J. Chem. Phys.* **108**, 2162 (1998).

[358] Speed improvement of diffusion quantum Monte Carlo calculations on weakly bound clusters, D. M. Benoit, A. X. Chavagnac, and D. C. Clary, *Chem. Phys. Lett.* **283**, 269 (1998).

[359] Structural control of Ar-IIF complexes using dc electric fields: A diffusion quantum Monte Carlo study, R. J. Hinde, *Chem. Phys. Lett.* **283**, 125 (1998).

[360] Intramolecular excitations in the $H_2O \cdots CO$ complex studied by diffusion Monte Carlo and ab initio calculations, P. Sandler, J. Sadlej, T. Feldmann, and V. Buch, *J. Chem. Phys.* **107**, 5022 (1997).

[361] The ^{129}Xe nuclear shielding surfaces for Xe interacting with linear molecules CO_2, $N-2$, and CO, A. C. de Dios and C. J. Jameson, *J. Chem. Phys.* **107**, 4253 (1997).

[362] Accuracy of electronic wave functions in quantum Monte Carlo: The effect of high-order correlations, C.-J. Huang, C. J. Umrigar, and M. P. Nightingale, *J. Chem. Phys.* **107**, 3007 (1997).

[363] Folding a 20 amino acid alphabeta peptide with the diffusion process-controlled Monte Carlo method, P. Derreumaux, *J. Chem. Phys.* **107**, 1941 (1997).

[364] Instability scenarios for doped ^{4}He clusters, S. M. Gatica, E. S. Hernández, and M. Barranco, *J. Chem. Phys.* **107**, 927 (1997).

[365] Analytical theory of benzene diffusion in Na-Y zeolite, S. M. Auerbach, *J. Chem. Phys.* **106**, 7810 (1997).

[366] Ab initio study of the O_2 $(X^3 \sum_g^-)$+Ar(^{1}S) van der Waals interaction, S. M. Cybulski, R. A. Kendall, G. Chalasinski, M. W. Severson, and M. M. Szczesniak, *J. Chem. Phys.* **106**, 7731 (1997).

[367] Quantum Monte Carlo binding energies for silicon hydrides, C. W. Greeff and W. A. Lester, Jr., *J. Chem. Phys.* **106**, 6412 (1997).

[368] A collaborative theoretical and experimental study of the structure and electronic excitation spectrum of the Bar and Bar_2 complexes, M. H. Alexander, A. R. Walton, M. Yang, X. Yang, E. Hwang, and P. J. Dagdigian, *J. Chem. Phys.* **106**, 6320 (1997).

[369] The C_6H_6-$(H_2O)_2$ complex: Theoretical predictions of the structure, energetics, and tunneling dynamics, J. M. Sorenson, J. K. Gregory, and D. C. Clary, *J. Chem. Phys.* **106**, 849 (1997).

[370] Hybrid quantum and classical mechanical Monte Carlo simulations of the interaction of hydrogen chloride with solid water clusters, D. A. Estrin, J. Kohanoff, D. H. Laria, and R. O. Weht, *Chem. Phys. Lett.* **280**, 280 (1997).

[371] Fixed-sample optimization in quantum Monte Carlo using a probability density function, R. N. Barnett, Z. Sun, and W. A. Lester, *Chem. Phys. Lett.* **273**, 321 (1997).

[372] Nonadiabatic wavefunctions as linear expansions of correlated exponentials. A quantum Monte Carlo application to H_2^+ and Ps_2, D. Bressanini, M. Mella, and G. Morosi, *Chem. Phys. Lett.* **272**, 370 (1997).

[373] Quantum Monte Carlo study of the ground vibrational states of Ar_{2-6}-H_2S clusters. A case of microsolvation, G. de Oliveira and C. E. Dykstra, *Chem. Phys. Lett.* **264**, 85 (1997).

[374] Ground and excited states of the complex of CO with water: A diffusion Monte Carlo study, P. Sandler, V. Buch, and J. Sadlej, *J. Chem. Phys.* **105**, 10387 (1996).

[375] Combining ab initio computations, neural networks, and diffusion Monte Carlo: An efficient method to treat weakly bound molecules, D. F. R. Brown, M. N. Gibbs, and D. C. Clary, *J. Chem. Phys.* **105**, 7597 (1996).

[376] A comparative study of CO_2-Ar potential surfaces, M. A. ter Horst and C. J. Jameson, *J. Chem. Phys.* **105**, 6787 (1996).

[377] van der Waals clusters in the ultraquantum limit: A Monte Carlo study, M. Meierovich, A. Mushinski, and M. P. Nightingale, *J. Chem. Phys.* **105**, 6498 (1996).

[378] Multiconfiguration wave functions for quantum Monte Carlo calculations of first-row diatomic molecules, C. Filippi and C. J. Umrigar, *J. Chem. Phys.* **105**, 213 (1996).

[379] Diffusive dynamics of the reaction coordinate for protein folding funnels, N. D. Socci, J. N. Onuchic, and P. G. Wolynes, *J. Chem. Phys.* **104**, 5860 (1996).

[380] Dopant location in $SF_6He_{39,40}$, M. A. McMahon, R. N. Barnett, and K. B. Whaley, *J. Chem. Phys.* **104**, 5080 (1996).

[381] Theoretical calculations of zero-temperature absorption spectra of Li in solid H_2, E. Cheng and K. B. Whaley, *J. Chem. Phys.* **104**, 3155 (1996).

[382] Electronic states of Al and Al_2 using quantum Monte Carlo with an effective core potential, C. W. Greeff, W. A. Lester, Jr., and B. L. Hammond, *J. Chem. Phys.* **104**, 1973 (1996).

[383] Binding energy of $PsCH_3$ system by quantum Monte Carlo and ab initio molecular orbital calculations, T. Saito, M. Tachikawa, C. Ohe, K. Iguchi, and K. Suzuki, *J. Phys. Chem.* **100**, 6057 (1996).

[384] Ar_nHF ($n = 1 - 4$) van der Waals clusters: A quantum Monte Carlo study of ground state energies, structures and HF vibrational frequency shifts, P. Niyaz, Z. Bačić, J. W. Moskowitz, and K. E. Schmidt, *Chem. Phys. Lett.* **252**, 23 (1996).

[385] Three-body effects on molecular properties in the water trimer, J. K. Gregory and D. C. Clary, *J. Chem. Phys.* **103**, 8924 (1995).

[386] Quantum Monte Carlo studies of anisotropy and rotational states in $HeNCl_2$, M. A. McMahon and K. B. Whaley, *J. Chem. Phys.* **103**, 2561 (1995).

[387] Calculations of the tunneling splittings in water dimer and trimer using diffusion Monte Carlo, J. K. Gregory and D. C. Clary, *J. Chem. Phys.* **102**, 7817 (1995).

[388] Reaction path zero-point energy from diffusion Monte Carlo calculations, J. K. Gregory, D. J. Wales, and D. C. Clary, *J. Chem. Phys.* **102**, 1592 (1995).

[389] Many-electron correlated exponential wavefunctions. A quantum Monte Carlo application to H_2 and He_2^+, D. Bressanini, M. Mella, and G. Morosi, *Chem. Phys. Lett.* **240**, 566 (1995).

[390] Accurate quantum Monte Carlo calculations of the tunneling splitting in $(HF)_2$ on a six-dimensional potential hypersurface, M. Quack and M. A. Suhm, *Chem. Phys. Lett.* **234**, 71 (1995).

[391] A Monte Carlo model for simulating the behaviour of a quantum harmonic oscillator embedded in a classical cluster, liquid or solid, A. J. Stace, *Chem. Phys. Lett.* **232**, 283 (1995).

[392] Many-body trial wave functions for atomic systems and ground states of small noble gas clusters, A. Mushinski and M. P. Nightingale, *J. Chem. Phys.* **101**, 8831 (1994).

[393] Calculation of expectation values of molecular systems using diffusion Monte Carlo in conjunction with the finite field method, P. Sandler, V. Buch, and D. C. Clary, *J. Chem. Phys.* **101**, 6353 (1994).

[394] The complex of N_2 with H_2O, D_2O, and HDO: A combined ab initio and diffusion Monte Carlo study, P. Sandler, J. Jung, M. M. Szczesniak, and V. Buch, *J. Chem. Phys.* **101**, 1378 (1994).

[395] Generation of pseudopotentials from correlated wave functions, P. H. Acioli and D. M. Ceperley, *J. Chem. Phys.* **100**, 8169 (1994).

[396] Determination of surface diffusion coefficients by Monte Carlo methods: Comparison of fluctuation and Kubo-Green methods, C. Uebing and R. Gomer, *J. Chem. Phys.* **100**, 7759 (1994).

[397] Properties of ionic hydrogen clusters: A quantum Monte Carlo study, T. Pang, *Chem. Phys. Lett.* **228**, 555 (1994).

[398] A systematic study on the fixed-node and localization error in quantum Monte Carlo calculations with pseudopotentials for group III elements, H.-J. Flad, A. Savin, M. Schultheiss, A. Nicklass, and H. Preuss, *Chem. Phys. Lett.* **222**, 274 (1994).

[399] Quantum Monte Carlo vibrational analysis of the nitrogen-water complex, K. A. Franken and C. E. Dykstra, *Chem. Phys. Lett.* **220**, 161 (1994).

[400] Molecules in helium clusters: SF_6He_N, R. N. Barnett and K. B. Whaley, *J. Chem. Phys.* **99**, 9730 (1993).

[401] Rotational excitations of quantum liquid clusters: He_7 and $(H_2)_7$, M. A. McMahon, R. N. Barnett, and K. B. Whaley, *J. Chem. Phys.* **99**, 8816 (1993).

[402] A diffusion Monte Carlo algorithm with very small time-step errors, C. J. Umrigar, M. P. Nightingale, and K. J. Runge, *J. Chem. Phys.* **99**, 2865 (1993).

[403] Antisymmetry in the quantum Monte Carlo method with the A-function technique: H_2 $b^3 \sum_u^+$, H_2 $c^3 \prod_u$, He 1^3S, R. Bianchi, D. Bressanini, P. Cremaschi, and G. Morosi, *J. Chem. Phys.* **98**, 7204 (1993).

[404] Preferential adsorption of ortho-H_2 with respect to para-H_2 on the amorphous ice surface, V. Buch and J. P. Devlin, *J. Chem. Phys.* **98**, 4195 (1993).

[405] Multidimensional vibrational quantum Monte Carlo technique using robust interpolation from static or growing sets of discrete potential energy points, M. A. Suhm, *Chem. Phys. Lett.* **214**, 373 (1993).

[406] Treatment of rigid bodies by diffusion Monte Carlo: Application to the para-$H_2 \cdots H_2O$ and ortho-$H_2 \cdots H_2O$ clusters, V. Buch, *J. Chem. Phys.* **97**, 726 (1992).

[407] Glass transition of polymer melts: A two-dimensional Monte Carlo study in the framework of the bond fluctuation method, H.-P. Wittmann, K. Kremer, and K. Binder, *J. Chem. Phys.* **96**, 6291 (1992).

[408] Electronic and spin dynamics in one-dimensional chains studied by path-integral quantum Monte Carlo simulations, L. Utrera, J. Schulte, and M. C. Bohm, *Chem. Phys. Lett.* **191**, 299 (1992).

[409] The quantum mechanics of clusters: The low temperature equilibrium and dynamical behavior of rare-gas systems, S. W. Rick, D. L. Leitner, J. D. Doll, D. L. Freeman, and D. D. Frantz, *J. Chem. Phys.* **95**, 6658 (1991).

[410] Nonlocal pseudopotentials and diffusion Monte Carlo, L. Mitáš, E. L. Shirley, and D. M. Ceperley, *J. Chem. Phys.* **95**, 3467 (1991).

[411] Interdiffusion and self-diffusion in polymer mixtures: A Monte Carlo study, H. P. Deutsch and K. Binder, *J. Chem. Phys.* **94**, 2294 (1991).

[412] Antisymmetry in the quantum Monte Carlo method without a trial function, R. Bianchi, D. Bressanini, P. Cremaschi, and G. Morosi, *Chem. Phys. Lett.* **184**, 343 (1991).

[413] Instantaneous normal mode analysis as a probe of cluster dynamics, J. E. Adams and R. M. Stratt, *J. Chem. Phys.* **93**, 1332 (1990).

[414] Diffusion Monte Carlo simulations of hydrogen fluoride dimers, H. Sun and R. O. Watts, *J. Chem. Phys.* **92**, 603 (1990).

[415] Structure and energetics of Xe_n^- : Many-body polarization effects, G. J. Martyna and B. J. Berne, *J. Chem. Phys.* **90**, 3744 (1989).

[416] Escape probability of geminate electron-ion recombination in the limit of large electron mean free path, M. Tachiya and W. F. Schmidt, *J. Chem. Phys.* **90**, 2471 (1989).

[417] Effect of crystallite shape on exciton energy: Quantum Monte Carlo calculations, V. Mohan and J. B. Anderson, *Chem. Phys. Lett.* **156**, 520 (1989).

[418] Structure and energetics of Xe_n^-, G. J. Martyna and B. J. Berne, *J. Chem. Phys.* **88**, 4516 (1988).

[419] Phenomenological theory of the dynamics of polymer melts. I. Analytic treatment of self-diffusion, J. Skolnick, R. Yaris, and A. Kolinski, *J. Chem. Phys.* **88**, 1407 (1988).

[420] Time step, data correlation and sampling accuracy in the diffusion quantum Monte Carlo method, R. Bianchi, P. Cremaschi, G. Morosi, and C. Puppi, *Chem. Phys. Lett.* **148**, 86 (1988).

[421] Variable quadratic propagator for quantum Monte Carlo simulations, P. Zhang, R. M. Levy, and R. A. Friesner, *Chem. Phys. Lett.* **144**, 236 (1988).

[422] Local potential error in quantum Monte Carlo calculations of the Mg ionization potential, P. A. Christiansen and L. A. Lajohn, *Chem. Phys. Lett.* **146**, 162 (1988).

[423] A Green's function used in diffusion Monte Carlo, S. M. Rothstein and J. Vrbik, *J. Chem. Phys.* **87**, 1902 (1987).

[424] Diffusion Monte Carlo simulation of condensed systems, D. F. Coker and R. O. Watts, *J. Chem. Phys.* **86**, 5703 (1987).

[425] Is quantum Monte Carlo competitive? Lithium hydride test case, R. N. Barnett, P. J. Reynolds, and W. A. Lester, *J. Phys. Chem.* **91**, 2004 (1987).

[426] An efficient Monte Carlo method for calculating the equilibrium properties for a quantum system coupled strongly to a classical one, B. Carmeli and H. Metiu, *Chem. Phys. Lett.* **133**, 543 (1987).

[427] *Monte Carlo Methods*, M. H. Kalos and P. A. Whitlock (Wiley, New York, 1986).

[428] Quantum Monte Carlo calculations on Be and LiH, R. J. Harrison and N. C. Handy, *Chem. Phys. Lett.* **113**, 257 (1985).

[429] Surface self-diffusion constants at low temperature: Monte Carlo transition state theory with importance sampling, A. F. Voter and J. D. Doll, *J. Chem. Phys.* **80**, 5814 (1984).

[430] Extended Brownian dynamics. III. Three-dimensional diffusion, G. Lamm, *J. Chem. Phys.* **80**, 2845 (1984).

[431] Diffusion-controlled flocculation: The effects of attractive and repulsive interactions, P. Meakin, *J. Chem. Phys.* **79**, 2426 (1983).

[432] Extended Brownian dynamics. II. Reactive, nonlinear diffusion, G. Lamm and K. Schulten, *J. Chem. Phys.* **78**, 2713 (1983).

[433] Fixed-node quantum Monte Carlo for molecules, P. J. Reynolds, D. M. Ceperly, B. J. Alder, and W. A. Lester, Jr., *J. Chem. Phys.* **77**, 5593 (1982).

[434] Quantum Monte Carlo and the hydride ion, K. McDowell and J. D. Doll, *Chem. Phys. Lett.* **82**, 127 (1981).

[435] A Monte Carlo procedure for the study of solvent effects on quantum molecular degrees of freedom, M. F. Herman and B. J. Berne, *Chem. Phys. Lett.* **77**, 163 (1981).

[436] Rate constant and transmission coefficient in the diffusion theory of reaction rates, M. Mangel, *J. Chem. Phys.* **72**, 6606 (1980).

[437] Quantum chemistry by random walk: Higher accuracy, J. B. Anderson, *J. Chem. Phys.* **73**, 3897 (1980).

[438] A random-walk simulation of the Schrödinger equation: H_3^+, J. B. Anderson, *J. Chem. Phys.* **63**, 1499 (1975).

[439] Helium at zero temperature with hard-sphere and other forces, M. H. Kalos, D. Levesque, and L. Verlet, *Phys. Rev.* A **9**, 2178 (1974).

Path Integral

[440] An efficient ring polymer contraction scheme for imaginary time path integral simulations, T. E. Markland and D. E. Manolopoulos, *J. Chem. Phys.* **129**, 024105 (2008).

[441] Path integral Monte Carlo study of CO_2 solvation in ^4He clusters, Z. Li, L. Wang, H. Ran, D. Xie, N. Blinov, P.-N. Roy, and H. Guo, *J. Chem. Phys.* **128**, 224513 (2008).

[442] A stereographic projection path integral study of the coupling between the orientation and the bending degrees of freedom of water, E. Curotto, D. L. Freeman, and J. D. Doll, *J. Chem. Phys.* **128**, 204107 (2008).

[443] Stereographic projection path integral simulations of $(HCl)_n$ clusters ($n = 2 - 5$): Evidence of quantum induced melting in small hydrogen bonded networks, M. W. Avils, M. L. McCandless, and E. Curotto, *J. Chem. Phys.* **128**, 124517 (2008).

[444] A linearized path integral description of the collision process between a water molecule and a graphite surface, N. Markovic and J. A. Poulsen, *J. Phys. Chem.* A **112**, 1701 (2008).

[445] Ab initio path integral ring polymer molecular dynamics: Vibrational spectra of molecules, M. Shiga and A. Nakayama, *Chem. Phys. Lett.* **451**, 175 (2008).

[446] An automated integration-free path-integral method based on Kleinert's variational perturbation theory, K.-Y. Wong and J. Gao, *J. Chem. Phys.* **127**, 211103 (2007).

[447] Path-integral simulations beyond the adiabatic approximation, J. R. Schmidt and J. C. Tully, *J. Chem. Phys.* **127**, 094103 (2007).

[448] Car-Parrinello and path integral molecular dynamics study of the hydrogen bond in the chloroacetic acid dimer system, P. Durlak, C. A. Morrison, D. S. Middlemiss, and Z. Latajka, *J. Chem. Phys.* **127**, 064304 (2007).

[449] Proton momentum distribution in water: An open path integral molecular dynamics study, J. A. Morrone, V. Srinivasan, D. Sebastiani, and R. Car, *J. Chem. Phys.* **126**, 234504 (2007).

[450] Path integral formulation for quantum nonadiabatic dynamics and the mixed quantum classical limit, V. Krishna, *J. Chem. Phys.* **126**, 134107 (2007).

[451] Ground-state path integral Monte Carlo simulations of positive ions in ^{4}He clusters: Bubbles or snowballs? S. Paolini, F. Ancilotto, and F. Toigo, *J. Chem. Phys.* **126**, 124317 (2007).

[452] Rotational fluctuation of molecules in quantum clusters. I. Path integral hybrid Monte Carlo algorithm, S. Miura, *J. Chem. Phys.* **126**, 114308 (2007).

[453] An integrated path integral and free-energy perturbation-umbrella sampling method for computing kinetic isotope effects of chemical reactions in solution and in enzymes, D. T. Major and J. Gao, *J. Chem. Theory Comput.* **3**, 949 (2007).

[454] Thermodynamics of hydrogen adsorption in slit-like carbon nanopores at 77 K. Classical versus path-integral Monte Carlo simulations, P. Kowalczyk, P. A. Gauden, A. P. Terzyk, and S. K. Bhatia, *Langmuir* **23, 7**, 3666 (2007).

[455] Path integral molecular dynamics calculations of the H_6^+ and D_6^+ clusters on an ab initio potential energy surface, A. Kakizaki, T. Takayanagi, and M. Shiga, *Chem. Phys. Lett.* **449**, 28 (2007).

[456] Hydroxyl radical transfer between interface and bulk from transition path sampling, C. D. Wick and L. X. Dang, *Chem. Phys. Lett.* **444**, 66 (2007).

[457] H/D isotope effect on the dihydrogen bond of $NH_4^+ \cdots BeH_2$ by ab initio path integral molecular dynamics simulation, A. Hayashi, M. Shiga, and M. Tachikawa, *J. Chem. Phys.* **125**, 204310 (2006).

[458] Hybrid quantum/classical path integral approach for simulation of hydrogen transfer reactions in enzymes, Q. Wang and S. Hammes-Schiffer, *J. Chem. Phys.* **125**, 184102 (2006).

[459] Path integral ground state study of finite-size systems: Application to small (parahydrogen)$_N$ ($N = 2 - 20$) clusters, J. E. Cuervo and P.-N. Roy, *J. Chem. Phys.* **125**, 124314 (2006).

[460] Quantum mechanical single molecule partition function from path integral Monte Carlo simulations, S. Chempath, C. Predescu, and A. T. Bell, *J. Chem. Phys.* **124**, 234101 (2006).

[461] Stereographic projection path-integral simulations of $(HF)_n$ clusters, M. W. Avils, P. T. Gray, and E. Curotto, *J. Chem. Phys.* **124**, 174305 (2006).

[462] A comparative study of imaginary time path integral based methods for quantum dynamics, T. D. Hone, P. J. Rossky, and G. A. Voth, *J. Chem. Phys.* **124**, 154103 (2006).

[463] Structural properties of liquid N-methylacetamide via ab initio, path integral, and classical molecular dynamics, T. W. Whitfield, J. Crain, and G. J. Martyna, *J. Chem. Phys.* **124**, 094503 (2006).

[464] Path-integral virial estimator for reaction-rate calculation based on the quantum instanton approximation, S. Yang, T. Yamamoto, and W. H. Miller, *J. Chem. Phys.* **124**, 084102 (2006).

[465] Path-integral centroid dynamics for general initial conditions: A nonequilibrium projection operator formulation, S. Jang, *J. Chem. Phys.* **124**, 064107 (2006).

[466] Path-integral Monte Carlo simulation of the recombination of two Al atoms embedded in parahydrogen, Q. Wang and M. H. Alexander, *J. Chem. Phys.* **124**, 034502 (2006).

[467] Quantum simulation of a hydrated noradrenaline analog with the torsional path integral method, T. F. Miller and D. C. Clary, *J. Phys. Chem. A* **110**, 731 (2006).

[468] An adiabatic linearized path integral approach for quantum time-correlation functions II: A cumulant expansion method for improving convergence, M. S. Causo, G. Ciccotti, S. Bonella, and R. Vuilleumier, *J. Phys. Chem. B* **110**, 16026 (2006).

[469] High-precision quantum thermochemistry on nonquasiharmonic potentials: Converged path-integral free energies and a systematically convergent family of generalized Pitzer-Gwinn approximations, V. A. Lynch, S. L. Mielke, and D. G. Truhlar, *J. Phys. Chem.* A **110**, 5965 (2006).

[470] Highly optimized fourth-order short-time approximation for path integrals, C. Predescu, *J. Phys. Chem.* B **110**, 667 (2006).

[471] Condensed-phase relaxation of multilevel quantum systems. II. comparison of path integral calculations and second-order relaxation theory for a nondegenerate three-level system, S. Peter, D. G. Evans, and R. D. Coalson, *J. Phys. Chem.* B **110**, 18764 (2006).

[472] Feynman-Kleinert linearized path integral (FK-LPI) algorithms for quantum molecular dynamics, with application to water and He(4), J. A. Poulsen, G. Nyman, and P. J. Rossky, *J. Chem. Theory Comput.* **2**, 1482 (2006).

[473] Path integral simulations of proton transfer reactions in aqueous solution using combined QM/MM potentials, D. T. Major, M. Garcia-Viloca, and J. Gao, *J. Chem. Theory Comput.* **2**, 236 (2006).

[474] Variational path integral simulations using discretized coordinates, R. J. Hinde, *Chem. Phys. Lett.* **418**, 481 (2006).

[475] Path integral Monte Carlo simulation of the absorption spectra of an Al atom embedded in helium, Q. Wang and M. H. Alexander, *J. Chem. Phys.* **123**, 134319 (2005).

[476] Calculation of heat capacities of light and heavy water by path-integral molecular dynamics, M. Shiga and W. Shinoda, *J. Chem. Phys.* **123**, 134502 (2005).

[477] Path integral methods for rotating molecules in superfluids, R. E. Zillich, F. Paesani, Y. Kwon, and K. B. Whaley, *J. Chem. Phys.* **123**, 114301 (2005).

[478] Path-integral computations of tunneling processes, I. Benjamin and A. Nitzan, *J. Chem. Phys.* **123**, 104103 (2005).

[479] Path-integral virial estimator based on the scaling of fluctuation coordinates: Application to quantum clusters with fourth-order propagators, T. M. Yamamoto, *J. Chem. Phys.* **123**, 104101 (2005).

[480] Simulations of one- and two-electron systems by Bead-Fourier path integral molecular dynamics, S. D. Ivanov and A. P. Lyubartsev, *J. Chem. Phys.* **123**, 034105 (2005).

[481] Hybrid Monte Carlo implementation of the Fourier path integral algorithm, C. Chakravarty, *J. Chem. Phys.* **123**, 024104 (2005).

[482] Shape resonances as poles of the semiclassical Green's function obtained from path-integral theory: Application to the autodissociation of the $He_2^{++} 1\sum_g^+$ state, C. A. Nicolaides and T. G. Douvropoulos, *J. Chem. Phys.* **123**, 024309 (2005).

[483] Path-integral Monte Carlo simulations for electronic dynamics on molecular chains. II. Transport across impurities, L. Mühlbacher and J. Ankerhold, *J. Chem. Phys.* **122**, 184715 (2005).

[484] Simulation of electronic and geometric degrees of freedom using a kink-based path integral formulation: Application to molecular systems, R. W. Hall, *J. Chem. Phys.* **122**, 164112 (2005).

[485] Partial averaging and the centroid virial estimator for stereographic projection path-integral simulations in curved spaces, M. W. Avils and E. Curotto, *J. Chem. Phys.* **122**, 164109 (2005).

[486] Quantum dynamics in the highly discrete, commensurate Frenkel Kontorova model: A path-integral molecular dynamics study, F. R. Krajewski and M. H. Mser, *J. Chem. Phys.* **122**, 124711 (2005).

[487] Path integral ground state with a fourth-order propagator: Application to condensed helium, J. E. Cuervo, P.-N. Roy, and M. Boninsegni, *J. Chem. Phys.* **122**, 114504 (2005).

[488] Path integral evaluation of the quantum instanton rate constant for proton transfer in a polar solvent, T. Yamamoto and W. H. Miller, *J. Chem. Phys.* **122**, 044106 (2005).

[489] High-precision quantum thermochemistry on nonquasiharmonic potentials: Converged path-integral free energies and a systematically convergent family of generalized Pitzer-Gwinn approximations, V. A. Lynch, S. L. Mielke, and D. G. Truhlar, *J. Phys. Chem. A* **109**, 10092 (2005).

[490] Integrated light collimating system for extended optical-path-length absorbance detection in microchip-based capillary electrophoresis, K. W. Ro, K. Lim, B. C. Shim, and J. H. Hahn, *Anal. Chem.* **77**, 5160 (2005).

[491] An adiabatic linearized path integral approach for quantum time correlation functions: Electronic transport in metal-molten salt solutions, M. S. Causo, G. Ciccotti, D. Montemayor, S. Bonella, and D. F. Coker, *J. Phys. Chem. B* **109**, 6855 (2005).

[492] Ab initio path integral molecular dynamics simulation study on the dihydrogen bond of $NH_4^+...BeH_2$, A. Hayashi, M. Shiga, and M. Tachikawa, *Chem. Phys. Lett.* **410**, 54 (2005).

[493] Ab initio path integral simulation study on $^{16}O/^{16}O$ isotope effect in water and hydronium ion, M. Tachikawa and M. Shiga, *Chem. Phys. Lett.* **407**, 135 (2005).

[494] Quantum properties of many boson system: Path integral approach, S. Sanyal and A. Sethia, *Chem. Phys. Lett.* **404**, 192 (2005).

[495] Feynman path integral – ab initio study of the isotropic hyperfine coupling constants of the muonium substituted ethyl radical CH_2MuCH, R. Ramirez, J. Schulte, and M. C. Bohm, *Chem. Phys. Lett.* **402**, 346 (2005).

[496] Path-integral Monte Carlo simulations for electronic dynamics on molecular chains. I. Sequential hopping and super exchange, L. Mühlbacher, J. Ankerhold, and C. Escher, *J. Chem. Phys.* **121**, 12696 (2004).

[497] Path integral influence functional theory of dynamics of coherence between vibrational states of solute in condensed phase, T. Mikami and S. Okazaki, *J. Chem. Phys.* **121**, 10052 (2004).

[498] Theoretical study on isotope and temperature effect in hydronium ion using ab initio path integral simulation, M. Tachikawa and M. Shiga, *J. Chem. Phys.* **121**, 5985 (2004).

[499] Accurate vibrational-rotational partition functions and standard-state free energy values for H_2O_2 from Monte Carlo path-integral calculations, V. A. Lynch, S. L. Mielke, and D. G. Truhlar, *J. Chem. Phys.* **121**, 5148 (2004).

[500] Path integral approach to Brownian motion driven with an ac force, L. Y. Chen and P. L. Nash, *J. Chem. Phys.* **121**, 3984 (2004).

[501] Computation of the equation of state of the quantum hard-sphere fluid utilizing several path-integral strategies, L. M. Ses, *J. Chem. Phys.* **121**, 3702 (2004).

[502] Higher order and infinite Trotter-number extrapolations in path integral Monte Carlo, L. Brualla, K. Sakkos, J. Boronat, and J. Casulleras, *J. Chem. Phys.* **121**, 636 (2004).

[503] A path integral influence functional for excess electron in fluids: Density-functional formulation, T. Sumi and H. Sekino, *J. Chem. Phys.* **120**, 8157 (2004).

[504] Path integral Monte Carlo approach for weakly bound van der Waals complexes with rotations: Algorithm and benchmark calculations, N. Blinov, X. G. Song, and P.-N. Roy, *J. Chem. Phys.* **120**, 5916 (2004).

[505] Path integral calculation of thermal rate constants within the quantum instanton approximation: Application to the $H + CH_4 \rightarrow H_2 + CH_3$ hydrogen abstraction reaction in full Cartesian space, Y. Zhao, T. Yamamoto, and W. H. Miller, *J. Chem. Phys.* **120**, 3100 (2004).

[506] On the efficient path integral evaluation of thermal rate constants within the quantum instanton approximation, T. Yamamoto and W. H. Miller, *J. Chem. Phys.* **120**, 3086 (2004).

[507] Path integral hybrid Monte Carlo algorithm for correlated Bose fluids, S. Miura and J. Tanaka, *J. Chem. Phys.* **120**, 2160 (2004).

[508] Stereographic projections path integral for inertia ellipsoids: Applications to Ar_n-HF clusters, M. F. Russo, Jr. and E. Curotto, *J. Chem. Phys.* **120**, 2110 (2004).

[509] Ab initio centroid path integral molecular dynamics: Application to vibrational dynamics of diatomic molecular systems, Y. Ohta, K. Ohta, and K. Kinugawa, *J. Chem. Phys.* **120**, 312 (2004).

[510] Nonradiative electronic relaxation rate constants from approximations based on linearizing the path-integral forward-backward action, Q. Shi and E. Geva, *J. Phys. Chem.* A **108**, 6109 (2004).

[511] Determination of the van Hove spectrum of liquid He(4): An application of the Feynman-Kleinert linearized path integral methodology, J. A. Poulsen, G. Nyman, and P. J. Rossky, *J. Phys. Chem.* A **108**, 41, 8743 (2004).

[512] Quantum diffusion in liquid para-hydrogen: An application of the Feynman-Kleinert linearized path integral approximation, J. A. Poulsen, G. Nyman, and P. J. Rossky, *J. Phys. Chem.* B **108**, 19799 (2004).

[513] Local group velocity and path-delay: Semi-classical propagators for the time evolution of Wigner functions in deep tunneling and in dispersive media, S. Kallush, E. Tannenbaum, and B. Segev, *Chem. Phys. Lett.* **396**, 261 (2004).

[514] Nuclear degrees of freedom in calculated isotropic hyperfine coupling constants of the ethyl radical: A Feynman path integral-ab initio study, J. Schulte, M. C. Bohm, T. Lopez-Ciudad, and R. Ramirez, *Chem. Phys. Lett.* **389**, 367 (2004).

[515] *Path Integrals in Quantum Mechanics, Statistics and Polymer Physics*, H. Kleinert, 3rd ed. (World Scientific, River Edge, NJ, 2004).

[516] Practical evaluation of condensed phase quantum correlation functions: A Feynman-Kleinert variational linearized path integral method, J. A. Poulsen, G. Nyman, and P. J. Rossky, *J. Chem. Phys.* **119**, 12179 (2003).

[517] Heat capacity estimators for random series path-integral methods by finite-difference schemes, C. Predescu, D. Sabo, J. D. Doll, and D. L. Freeman, *J. Chem. Phys.* **119**, 12119 (2003).

[518] Energy estimators for random series path-integral methods, C. Predescu, D. Sabo, J. D. Doll, and D. L. Freeman, *J. Chem. Phys.* **119**, 10475 (2003).

[519] Transport properties of liquid para-hydrogen: The path integral centroid molecular dynamics approach, Y. Yonetani and K. Kinugawa, *J. Chem. Phys.* **119**, 9651 (2003).

[520] An analysis of molecular origin of vibrational energy transfer from solute to solvent based upon path integral influence functional theory, T. Mikami and S. Okazaki, *J. Chem. Phys.* **119**, 4790 (2003).

[521] Numerical implementation of some reweighted path integral methods, C. Predescu, D. Sabo, and J. D. Doll, *J. Chem. Phys.* **119**, 4641 (2003).

[522] Anharmonic effects on the structural and vibrational properties of the ethyl radical: A path integral Monte Carlo study, T. López-Ciudad, R. Ramírez, J. Schulte, and M. C. Böhm, *J. Chem. Phys.* **119**, 4328 (2003).

[523] Molecular collective dynamics in solid para-hydrogen and ortho-deuterium: The Parrinello-Rahman-type path integral centroid molecular dynamics approach, H. Saito, H. Nagao, K. Nishikawa, and K. Kinugawa, *J. Chem. Phys.* **119**, 953 (2003).

[524] Torsional path integral Monte Carlo method for calculating the absolute quantum free energy of large molecules, T. F. Miller III and D. C. Clary, *J. Chem. Phys.* **119**, 68 (2003).

[525] Path-integral Monte Carlo calculation of reaction-diffusion equation, K. Ando and H. Sumi, *J. Chem. Phys.* **118**, 8315 (2003).

[526] Stereographic projections path integral in S^1 and $(S^2)^m$ manifolds, M. F. Russo, Jr. and E. Curotto, *J. Chem. Phys.* **118**, 6806 (2003).

[527] Algorithms and novel applications based on the isokinetic ensemble. I. Biophysical and path integral molecular dynamics, P. Minary, G. J. Martyna, and M. E. Tuckerman, *J. Chem. Phys.* **118**, 2510 (2003).

[528] A path integral approach to molecular thermochemistry, K. R. Glaesemann and L. E. Fried, *J. Chem. Phys.* **118**, 1596 (2003).

[529] Trace gas measurement with an integrated porous tube collector/long-path absorbance detector, K. Toda, K.-I. Yoshioka, S.-I. Ohira, J. Li, and P. K. Dasgupta, *Anal. Chem.* **75**, 4050 (2003).

[530] Quantum path-integral molecular dynamics calculations of the dipole-bound state of the water dimer anion, M. Shiga and T. Takayanagi, *Chem. Phys. Lett.* **378**, 539 (2003).

[531] A. 'path-by-path' monotone extrapolation sequence for Feynman path integral calculations of quantum mechanical free energies, S. L. Mielke and D. G. Truhlar, *Chem. Phys. Lett.* **378**, 317 (2003).

[532] First-principle path integral study of DNA under hydrodynamic flows, S. Yang, J. B. Witkoskie, and J. Cao, *Chem. Phys. Lett.* **377**, 399 (2003).

[533] Fourier path integral Monte Carlo in the grand canonical ensemble, G. E. Lopez, *Chem. Phys. Lett.* **375**, 511 (2003).

[534] Ab initio path integral study of isotope effect of hydronium ion, M. Shiga and M. Tachikawa, *Chem. Phys. Lett.* **374**, 229 (2003).

[535] Photodissociation of Cl_2 in helium clusters: An application of hybrid method of quantum wavepacket dynamics and path integral centroid molecular dynamics, T. Takayanagi and M. Shiga, *Chem. Phys. Lett.* **372**, 90 (2003).

[536] Optimal series representations for numerical path integral simulations, C. Predescu and J. D. Doll, *J. Chem. Phys.* **117**, 7448 (2002).

[537] Wavelet formulation of path integral Monte Carlo, A. E. Cho, J. D. Doll, and D. L. Freeman, *J. Chem. Phys.* **117**, 5971 (2002).

[538] Improved heat capacity estimator for path integral simulations, K. R. Glaesemann and L. E. Fried, *J. Chem. Phys.* **117**, 3020 (2002).

[539] Path integral simulations of quantum Lennard-Jones solids, C. Chakravarty, *J. Chem. Phys.* **116**, 8938 (2002).

[540] Properties of the path-integral quantum hard-sphere fluid in k space, L. M. Ses, *J. Chem. Phys.* **116**, 8492 (2002).

[541] Torsional path integral Monte Carlo method for the quantum simulation of large molecules, T. F. Miller III and D. C. Clary, *J. Chem. Phys.* **116**, 8262 (2002).

[542] An improved thermodynamic energy estimator for path integral simulations, K. R. Glaesemann and L. E. Fried, *J. Chem. Phys.* **116**, 5951 (2002).

[543] Iterative path integral formulation of equilibrium correlation functions for quantum dissipative systems, J. Shao and N. Makri, *J. Chem. Phys.* **116**, 507 (2002).

[544] An adaptive, kink-based approach to path integral calculations, R. W. Hall, *J. Chem. Phys.* **116**, 1 (2002).

[545] Excited state properties of C_6H_6 and C_6D_6 studied by Feynman path integral-ab initio simulations, M. C. Bohm, J. Schulte, and R. Ramirez, *J. Phys. Chem. A* **106**, 3169 (2002).

[546] Reaction-path dynamics calculations using integrated methods. The CF_3CH_3 + OH hydrogen abstraction reaction, J. Espinosa-Garcia, *J. Phys. Chem. A* **106**, 5686 (2002).

[547] Path integral-molecular dynamics study of electronic states in supercritical water, D. Laria and M. S. Skaf, *J. Phys. Chem. A* **106**, 8066 (2002).

[548] Path integral molecular dynamics simulation of solid para-hydrogen with an aluminum impurity, D. T. Mirijanian, M. H. Alexander, and G. A. Voth, *Chem. Phys. Lett.* **365**, 487 (2002).

[549] Path integral molecular dynamics combined with discrete-variable-representation approach: The effect of solvation structures on vibrational spectra of Cl_2 in helium clusters, T. Takayanagi and M. Shiga, *Chem. Phys. Lett.* **362**, 504 (2002).

[550] Kink-based path integral calculations of atoms He-Ne, R. W. Hall, *Chem. Phys. Lett.* **362**, 549 (2002).

[551] A non-perturbative path-integral based thermal cluster expansion approach for grand partition function of quantum systems, S. H. Mandal, R. Ghosh, G. Sanyal, and D. Mukherjee, *Chem. Phys. Lett.* **352**, 63 (2002).

[552] Path integral Monte Carlo applications to quantum fluids in confined geometries, D. M. Ceperley and E. Manousakis, *J. Chem. Phys.* **115**, 10111 (2001).

[553] Quantum effect of solvent on molecular vibrational energy relaxation of solute based upon path integral influence functional theory, T. Mikami, M. Shiga, and S. Okazaki, *J. Chem. Phys.* **115**, 9797 (2001).

[554] A unified scheme for ab initio molecular orbital theory and path integral molecular dynamics, M. Shiga, M. Tachikawa, and S. Miura, *J. Chem. Phys.* **115**, 9149 (2001).

[555] Path integral centroid molecular-dynamics evaluation of vibrational energy relaxation in condensed phase, J. A. Poulsen and P. J. Rossky, *J. Chem. Phys.* **115**, 8024 (2001).

[556] An ansatz-based variational path integral centroid approach to vibrational energy relaxation in simple liquids, J. A. Poulsen and P. J. Rossky, *J. Chem. Phys.* **115**, 8014 (2001).

[557] Applications of higher order composite factorization schemes in imaginary time path integral simulations, S. Jang, S. Jang, and G. A. Voth, *J. Chem. Phys.* **115**, 7832 (2001).

[558] Quantum effects in liquid water: Path-integral simulations of a flexible and polarizable ab initio model, H. A. Stern and B. J. Berne, *J. Chem. Phys.* **115**, 7622 (2001).

[559] Uncertainty of path integral averages at low temperature, T. W. Whitfield and J. E. Straub, *J. Chem. Phys.* **115**, 6834 (2001).

[560] The asymptotic decay of pair correlations in the path-integral quantum hard-sphere fluid, L. E. Bailey and L. M. Ses, *J. Chem. Phys.* **115**, 6557 (2001).

[561] Path integral molecular dynamics method based on a pair density matrix approximation: An algorithm for distinguishable and identical particle systems, S. Miura and S. Okazaki, *J. Chem. Phys.* **115**, 5353 (2001).

[562] Path integral formulation of centroid dynamics for systems obeying Bose-Einstein statistics, N. V. Blinov, P.-N. Roy, and G. A. Voth, *J. Chem. Phys.* **115**, 4484 (2001).

[563] Displaced-points path integral method for including quantum effects in the Monte Carlo evaluation of free energies, S. L. Mielke and D. G. Truhlar, *J. Chem. Phys.* **115**, 652 (2001).

[564] Low lying vibrational excitation energies from equilibrium path integral simulations, R. Ramírez and T. López-Ciudad, *J. Chem. Phys.* **115**, 103 (2001).

[565] Simulation of material properties below the Debye temperature: A path-integral molecular dynamics case study of quartz, M. H. Müser, *J. Chem. Phys.* **114**, 6364 (2001).

[566] Quantum dynamics: Path integral approach to time correlation functions in finite temperature, A. Sethia, S. Sanyal, and F. Hirata, *J. Chem. Phys.* **114**, 5097 (2001).

[567] Effects of pressure on the trapping site structures and absorption spectra of Li in solid H_2: A path integral Monte Carlo study, Y. M. Ma, T. Cui, and G. T. Zou, *J. Chem. Phys.* **114**, 3092 (2001).

[568] Path-integral Monte Carlo study of the structural and mechanical properties of quantum fcc and bcc hard-sphere solids, L. M. Sesé, *J. Chem. Phys.* **114**, 1732 (2001).

[569] A semiclassical approach to the dynamics of many-body Bose/Fermi systems by the path integral centroid molecular dynamics, K. Kinugawa, H. Nagao, and K. Ohta, *J. Chem. Phys.* **114**, 1454 (2001).

[570] Path integral Monte Carlo study on the structure and absorption spectra of alkali atoms (Li, Na, K) attached to superfluid helium clusters, A. Nakayama and K. Yamashita, *J. Chem. Phys.* **114**, 780 (2001).

[571] A new Fourier path integral method, a more general scheme for extrapolation, and comparison of eight path integral methods for the quantum mechanical calculation of free energies, S. L. Mielke and D. G. Truhlar, *J. Chem. Phys.* **114**, 621 (2001).

[572] Path-integral simulation of finite-temperature properties of systems involving multiple, coupled electronic states, M. H. Alexander, *Chem. Phys. Lett.* **347**, 436 (2001).

[573] Simulation of quantum systems using path integrals in a generalized ensemble, I. Andricioaei, J. E. Straub, and M. Karplus, *Chem. Phys. Lett.* **346**, 274 (2001).

[574] Quantum path-integral simulation of poly(propylene oxide), M. Slabanja and G. Wahnstrom, *Chem. Phys. Lett.* **342**, 593 (2001).

[575] A two-chain path integral model of positronium, L. Larrimore, R. N. McFarland, P. A. Sterne, and A. L. R. Bug, *J. Chem. Phys.* **113**, 10642 (2000).

[576] A path integral ground state method, A. Sarsa, K. E. Schmidt, and W. R. Magro, *J. Chem. Phys.* **113**, 1366 (2000).

[577] Theoretical study on the structure of Na^+ -doped helium clusters: Path integral Monte Carlo calculations, A. Nakayama and K. Yamashita, *J. Chem. Phys.* **112**, 10966 (2000).

[578] Path integral molecular dynamics for Bose-Einstein and Fermi-Dirac statistics, S. Miura and S. Okazaki, *J. Chem. Phys.* **112**, 10116 (2000).

[579] Extrapolation and perturbation schemes for accelerating the convergence of quantum mechanical free energy calculations via the Fourier path-integral Monte Carlo method, S. L. Mielke, J. Srinivasan, and D. G. Truhlar, *J. Chem. Phys.* **112**, 8758 (2000).

[580] A relationship between centroid dynamics and path integral quantum transition state theory, S. Jang and G. A. Voth, *J. Chem. Phys.* **112**, 8747 (2000).

[581] A heat capacity estimator for Fourier path integral simulations, J. P. Neirotti, D. L. Freeman, and J. D. Doll, *J. Chem. Phys.* **112**, 3990 (2000).

[582] Constant pressure path integral molecular dynamics studies of quantum effects in the liquid state properties of n-alkanes, E. Balog, A. L. Hughes, and G. J. Martyna, *J. Chem. Phys.* **112**, 870 (2000).

[583] Global optimization: Quantum thermal annealing with path integral Monte Carlo, Y.-H. Lee and B. J. Berne, *J. Phys. Chem. A* **104**, 86 (2000).

[584] Ab initio molecular orbital calculation considering the quantum mechanical effect of nuclei by path integral molecular dynamics, M. Shiga, M. Tachikawa, and S. Miura, *Chem. Phys. Lett.* **332**, 396 (2000).

[585] Nuclear quantum effects in calculated NMR shieldings of ethylene; a Feynman path integral-ab initio study, M. C. Bohm, J. Schulte, and R. Ramrez, *Chem. Phys. Lett.* **332**, 117 (2000).

[586] Feynman path integral-ab initio investigation of the excited state properties of C_2H_4, J. Schulte, R. Ramrez, and M. C. Bohm, *Chem. Phys. Lett.* **322**, 527 (2000).

[587] Real time quantum correlation functions. II. Maximum entropy numerical analytic continuation of path integral Monte Carlo and centroid molecular dynamics data, G. Krilov and B. J. Berne, *J. Chem. Phys.* **111**, 9147 (1999).

[588] Iterative evaluation of the path integral for a system coupled to an anharmonic bath, N. Makri, *J. Chem. Phys.* **111**, 6164 (1999).

[589] Molecular dynamics study of vibrational energy relaxation of CN^- in H_2O and D_2O solutions: An application of path integral influence functional theory to multiphonon processes, M. Shiga and S. Okazaki, *J. Chem. Phys.* **111**, 5390 (1999).

[590] Extension of path integral quantum transition state theory to the case of nonadiabatic activated dynamics, C. D. Schwieters and G. A. Voth, *J. Chem. Phys.* **111**, 2869 (1999).

[591] A derivation of centroid molecular dynamics and other approximate time evolution methods for path integral centroid variables, S. Jang and G. A. Voth, *J. Chem. Phys.* **111**, 2371 (1999).

[592] Path integral centroid variables and the formulation of their exact real time dynamics, S. Jang and G. A. Voth, *J. Chem. Phys.* **111**, 2357 (1999).

[593] Simulation studies of liquid ammonia by classical ab initio, classical, and path-integral molecular dynamics, M. Diraison, G. J. Martyna, and M. E. Tuckerman, *J. Chem. Phys.* **111**, 1096 (1999).

[594] Molar excess volumes of liquid hydrogen and neon mixtures from path integral simulation, S. R. Challa and J. K. Johnson, *J. Chem. Phys.* **111**, 724 (1999).

[595] Asymptotic convergence rates of Fourier path integral methods, M. Eleftheriou, J. D. Doll, E. Curotto, and D. L. Freeman, *J. Chem. Phys.* **110**, 6657 (1999).

[596] Path-integral diffusion Monte Carlo: Calculation of observables of many-body systems in the ground state, B. Hetényi, E. Rabani, and B. J. Berne, *J. Chem. Phys.* **110**, 6143 (1999).

[597] Path integral approximation of state- and angle-resolved inelastic scattering, T. W. J. Whiteley and A. J. McCaffery, *J. Chem. Phys.* **110**, 5548 (1999).

[598] A path integral centroid molecular dynamics study of nonsuperfluid liquid helium-4, S. Miura, S. Okazaki, and K. Kinugawa, *J. Chem. Phys.* **110**, 4523 (1999).

[599] A modification of path integral quantum transition state theory for asymmetric and metastable potentials, S. Jang, C. D. Schwieters, and G. A. Voth, *J. Phys. Chem.* A **103**, 9527 (1999).

[600] Path-integral calculation of the mean number of overcrossings in an entangled polymer network, G. A. Arteca, *J. Chem. Inf. Comput. Sci.* **39**, 550 (1999).

[601] Eigenstates from the discretized path integral, A. Sethia, S. Sanyal, and F. Hirata, *Chem. Phys. Lett.* **315**, 299 (1999).

[602] On the Elber-Karplus reaction path-following method and related procedures, L. L. Stacho, G. Domotor, and M. I. Ban, *Chem. Phys. Lett.* **311**, 328 (1999).

[603] Path integral hybrid Monte Carlo for the bosonic many-body systems, S. Miura and S. Okazaki, *Chem. Phys. Lett.* **308**, 115 (1999).

[604] Path integral centroid molecular dynamics method for Bose and Fermi statistics: Formalism and simulation, K. Kinugawa, H. Nagao, and K. Ohta, *Chem. Phys. Lett.* **307**, 187 (1999).

[605] Theoretical studies on the magnetic quantum tunneling rates in Mn clusters by the path integral method, H. Nagao, S. Yamanaka, M. Nishino, Y. Yoshioka, and K. Yamaguchi, *Chem. Phys. Lett.* **302**, 418 (1999).

[606] An ab initio path integral molecular dynamics study of double proton transfer in the formic acid dimer, S. Miura, M. E. Tuckerman, and M. L. Klein, *J. Chem. Phys.* **109**, 5290 (1998).

[607] Centroid path integral molecular-dynamics studies of a para-hydrogen slab containing a lithium impurity, K. Kinugawa, P. B. Moore, and M. L. Klein, *J. Chem. Phys.* **109**, 610 (1998).

[608] Thermodynamic and structural properties of the path-integral quantum hard-sphere fluid, L. M. Sesé, *J. Chem. Phys.* **108**, 9086 (1998).

[609] An ab initio path integral Monte Carlo simulation method for molecules and clusters: Application to Li_4 and Li_5^+, R. O. Weht, J. Kohanoff, D. A. Estrin, and C. Chakravarty, *J. Chem. Phys.* **108**, 8848 (1998).

[610] Path integral study of hydrogen and deuterium diffusion in crystalline silicon, K. M. Forsythe and N. Makri, *J. Chem. Phys.* **108**, 6819 (1998).

[611] Improved Feynman's path integral method with a large time step: Formalism and applications, A. N. Drozdov, *J. Chem. Phys.* **108**, 6580 (1998).

[612] Lithium impurity recombination in solid para-hydrogen: A path integral quantum transition state theory study, S. Jang and G. A. Voth, *J. Chem. Phys.* **108**, 4098 (1998).

[613] Dynamic path integral methods: A maximum entropy approach based on the combined use of real and imaginary time quantum Monte Carlo data, D. Kim, J. D. Doll, and D. L. Freeman, *J. Chem. Phys.* **108**, 3871 (1998).

[614] Rate constants in quantum mechanical systems: A rigorous and practical path-integral formulation for computer simulations, N. Chakrabarti, T. Carrington, Jr., and B. Roux, *Chem. Phys. Lett.* **293**, 209 (1998).

[615] Path integral centroid molecular dynamics study of the dynamic structure factors of liquid para-hydrogen, K. Kinugawa, *Chem. Phys. Lett.* **292**, 454 (1998).

[616] Nuclear quantum effects in the electronic structure of C_2H_4: A combined Feynman path integral-ab initio approach, R. Ramrez, E. Hernandez, J. Schulte, and M. C. Bohm, *Chem. Phys. Lett.* **291**, 44 (1998).

[617] Path integral method by means of generalized coherent states and its numerical approach to molecular systems. I. Ensemble average of total energy, H. Nagao, Y. Shigeta, H. Kawabe, T. Kawakami, K. Nishikawa, and K. Yamaguchi, *J. Chem. Phys.* **107**, 6283 (1997).

[618] Optimal energy estimation in path-integral Monte Carlo simulations, W. Janke and T. Sauer, *J. Chem. Phys.* **107**, 5821 (1997).

[619] Efficient calculation of free-energy barriers in quantum activated processes. A path-integral centroid approach, R. Ramírez, *J. Chem. Phys.* **107**, 5748 (1997).

[620] Path integral grand canonical Monte Carlo, Q. Wang, J. K. Johnson, and J. Q. Broughton, *J. Chem. Phys.* **107**, 5108 (1997).

[621] Dynamics of quantum particles by path-integral centroid simulations: The symmetric Eckart barrier, R. Ramírez, *J. Chem. Phys.* **107**, 3550 (1997).

[622] High-accuracy discrete path integral solutions for stochastic processes with noninvertible diffusion matrices. II. Numerical evaluation, A. N. Drozdov, *J. Chem. Phys.* **107**, 3505 (1997).

[623] Circumventing the pathological behavior of path-integral Monte Carlo for systems with Coulomb potentials, M. H. Müser and B. J. Berne, *J. Chem. Phys.* **107**, 571 (1997).

[624] The quantum dynamics of interfacial hydrogen: Path integral maximum entropy calculation of adsorbate vibrational line shapes for the H/Ni(111) system, D. Kim, J. D. Doll, and J. E. Gubernatis, *J. Chem. Phys.* **106**, 1641 (1997).

[625] Centroid path integral molecular dynamics simulation of lithium parahydrogen clusters, K. Kinugawa, P. B. Moore, and M. L. Klein, *J. Chem. Phys.* **106**, 1154 (1997).

[626] Computation of the static structure factor of the path-integral quantum hard-sphere fluid, L. M. Sesé and R. Ledesma, *J. Chem. Phys.* **106**, 1134 (1997).

[627] Path integral simulation of charge transfer dynamics in photosynthetic reaction centers, E. Sim and N. Makri, *J. Phys. Chem.* B **101**, 5446 (1997).

[628] Quantum adsorbates: Path integral Monte Carlo simulations of helium in silicalite, C. Chakravarty, *J. Phys. Chem.* B **101**, 10, 1878 (1997).

[629] Generalized path integral based quantum transition state theory, G. Mills, G. K. Schenter, D. E. Makarov, and H. Jonsson, *Chem. Phys. Lett.* **278**, 91 (1997).

[630] A wavepacket-path integral method for curve-crossing problems: Application to resonance Raman spectra and photodissociation cross sections, A. E. Cardenas and R. D. Coalson, *Chem. Phys. Lett.* **265**, 71 (1997).

[631] Theoretical studies of the structure and dynamics of metal/hydrogen systems: Diffusion and path integral Monte Carlo investigations of nickel and palladium clusters, B. Chen, M. A. Gomez, M. Sehl, J. D. Doll, and D. L. Freeman, *J. Chem. Phys.* **105**, 9686 (1996).

[632] On the calculation of dynamical properties of solvated electrons by maximum entropy analytic continuation of path integral Monte Carlo data, E. Gallicchio and B. J. Berne, *J. Chem. Phys.* **105**, 7064 (1996).

[633] Cluster analogs of binary isotopic mixtures: Path integral Monte Carlo simulations, C. Chakravarty, *J. Chem. Phys.* **104**, 7223 (1996).

[634] Efficient and general algorithms for path integral Car-Parrinello molecular dynamics, M. E. Tuckerman, D. Marx, M. L. Klein, and M. Parrinello, *J. Chem. Phys.* **104**, 5579 (1996).

[635] Ab initio path integral molecular dynamics: Basic ideas, D. Marx and M. Parrinello, *J. Chem. Phys.* **104**, 4077 (1996).

[636] Path integral Monte Carlo study of SF_6-doped helium clusters, Y. Kwon, D. M. Ceperley, and K. B. Whaley, *J. Chem. Phys.* **104**, 2341 (1996).

[637] Adiabatic path integral molecular dynamics methods. I. Theory, G. J. Martyna, *J. Chem. Phys.* **104**, 2018 (1996).

[638] Adiabatic path integral molecular dynamics methods. II. Algorithms, J. Cao and G. J. Martyna, *J. Chem. Phys.* **104**, 2028 (1996).

[639] A wavepacket-path integral method for curve-crossing dynamics, R. D. Coalson, *J. Phys. Chem.* **100**, 7896 (1996).

[640] Path integral calculation of quantum nonadiabatic rates in model condensed phase reactions, M. Topaler and N. Makri, *J. Phys. Chem.* **100**, 4430 (1996).

[641] Path integral Monte Carlo simulations: Study of the efficiency of energy estimators, P. A. Fernandes, A. P. Carvalho, and J. P. Prates Ramalho, *J. Chem. Phys.* **103**, 5720 (1995).

[642] Path-integral treatment of multi-mode vibronic coupling. II. Correlation expansion of class averages, S. Krempl, M. Winterstetter, and W. Domcke, *J. Chem. Phys.* **102**, 6499 (1995).

[643] Variational upper and lower bounds on quantum free energy and energy differences via path integral Monte Carlo, G. J. Hogenson and W. P. Reinhardt, *J. Chem. Phys.* **102**, 4151 (1995).

[644] Path-integral Monte Carlo energy and structure of the quantum hardsphere system using efficient propagators, L. M. Sesé and R. Ledesma, *J. Chem. Phys.* **102**, 3776 (1995).

[645] Melting of neon clusters: Path integral Monte Carlo simulations, C. Chakravarty, *J. Chem. Phys.* **102**, 956 (1995).

[646] Dynamics of the carbon nuclei in C_{60} studied by Feynman path-integral quantum Monte Carlo simulations, M. C. Boehm and R. Ramirez, *J. Phys. Chem.* **99**, 12401 (1995).

[647] Path-integral simulations of zero-point effects for implanted muons in benzene, R. M. Valladares, A. J. Fisher, and W. Hayes, *Chem. Phys. Lett.* **242**, 1 (1995).

[648] Recursive evaluation of the real-time path integral for dissipative systems. The spin-boson model, M. Winterstetter and W. Domcke, *Chem. Phys. Lett.* **236**, 445 (1995).

[649] How to solve path integrals in quantum mechanics, C. Grosche and F. Steiner, *J. Math. Phys.* **36**, 2354 (1995).

[650] The absorption spectrum of the solvated electron in fluid helium by maximum entropy inversion of imaginary time correlation functions from path integral Monte Carlo simulations, E. Gallicchio and B. J. Berne, *J. Chem. Phys.* **101**, 9909 (1994).

[651] Quantum rates for a double well coupled to a dissipative bath: Accurate path integral results and comparison with approximate theories, M. Topaler and N. Makri, *J. Chem. Phys.* **101**, 7500 (1994).

[652] Real time path integral methods for a system coupled to an anharmonic bath, G. Ilk and N. Makri, *J. Chem. Phys.* **101**, 6708 (1994).

[653] Fourier path integral Monte Carlo method for the calculation of the microcanonical density of states, D. L. Freeman and J. D. Doll, *J. Chem. Phys.* **101**, 848 (1994).

[654] Path integral simulations of mixed para-D_2 and ortho-D_2 clusters: The orientational effects, V. Buch, *J. Chem. Phys.* **100**, 7610 (1994).

[655] A path integral study of electronic polarization and nonlinear coupling effects in condensed phase proton transfer reactions, J. Lobaugh and G. A. Voth, *J. Chem. Phys.* **100**, 3039 (1994).

[656] Path integral formulation of retardation effects in nonlinear optics, V. Chernyak and S. Mukamel, *J. Chem. Phys.* **100**, 2953 (1994).

[657] Path integral calculations of the free energies of hydration of hydrogen isotopes (H, D, and Mu), H. Gai and B. C. Garrett, *J. Phys. Chem.* **98**, 9642 (1994).

[658] Path-integral treatment of multi-mode vibronic coupling, S. Krempl, M. Winterstetter, H. Plöhn, and W. Domcke, *J. Chem. Phys.* **100**, 926 (1994).

[659] Tunneling mechanism in electron transfer. A view from the Feynman's path integral approach, J. Tang, *Chem. Phys. Lett.* **227**, 170 (1994).

[660] Path-integral Monte Carlo study of a lithium impurity in para-hydrogen: Clusters and the bulk liquid, D. Scharf, G. J. Martyna, and M. L. Klein, *J. Chem. Phys.* **99**, 8997 (1993).

[661] Path integral description of polymers using fractional Brownian walks, B. J. Cherayil and P. Biswas, *J. Chem. Phys.* **99**, 9230 (1993).

[662] The quantum dynamics of hydrogen and deuterium on the Pd(111) surface: A path integral transition state theory study, S. W. Rick, D. L. Lynch, and J. D. Doll, *J. Chem. Phys.* **99**, 8183 (1993).

[663] Particle exchange in the Fourier path-integral Monte Carlo technique, C. Chakravarty, *J. Chem. Phys.* **99**, 8038 (1993).

[664] Diatomic molecules, rotations, and path-integral Monte Carlo simulations: N_2 and H_2 on graphite, D. Marx, S. Sengupta, and P. Nielaba, *J. Chem. Phys.* **99**, 6031 (1993).

[665] Analysis of the statistical errors in conditioned real time path integral methods, A. M. Amini and M. F. Herman, *J. Chem. Phys.* **99**, 5087 (1993).

[666] Path integral calculation of hydrogen diffusion rates on metal surfaces, Y.-C. Sun and G. A. Voth, *J. Chem. Phys.* **98**, 7451 (1993).

[667] An approximate discretized real time path integral simulation method for nearly classical systems, A. M. Amini and M. F. Herman, *J. Chem. Phys.* **98**, 6975 (1993).

[668] Approximate path integral methods for partition functions, M. Messina, G. K. Schenter, and B. C. Garrett, *J. Chem. Phys.* **98**, 4120 (1993).

[669] Improved propagators for the path integral study of quantum systems, J. P. Prates Ramalho, B. J. Costa Cabral, and F. M. S. Silva Fernandes, *J. Chem. Phys.* **98**, 3300 (1993).

[670] Feynman path integral formulation of quantum mechanical transition-state theory, G. A. Voth, *J. Phys. Chem.* **97**, 8365 (1993).

[671] On smooth Feynman propagators for real time path integrals, N. Makri, *J. Phys. Chem.* **97**, 2417 (1993).

[672] The Huckel model by means of the path integral method. II. Application to the square lattice system, H. Nagao, K. Nishikawa, and S. Aono, *Chem. Phys. Lett.* **215**, 5 (1993).

[673] Quasi-adiabatic propagator path integral methods. Exact quantum rate constants for condensed phase reactions, M. Topaler and N. Makri, *Chem. Phys. Lett.* **210**, 285 (1993).

[674] Multidimensional path integral calculations with quasiadiabatic propagators: Quantum dynamics of vibrational relaxation in linear hydrocarbon chains, M. Topaler and N. Makri, *J. Chem. Phys.* **97**, 9001 (1992).

[675] Formally exact path integral Monte Carlo calculations using approximate projection operators, R. W. Hall, *J. Chem. Phys.* **97**, 6481 (1992).

[676] A partial averaging strategy for low temperature Fourier path integral Monte Carlo calculations, J. Lobaugh and G. A. Voth, *J. Chem. Phys.* **97**, 4205 (1992).

[677] Quantum free-energy calculations: Optimized Fourier path-integral Monte Carlo computation of coupled vibrational partition functions, R. Q. Topper and D. G. Truhlar, *J. Chem. Phys.* **97**, 3647 (1992).

[678] Path-integral Monte Carlo studies of para-hydrogen clusters, D. Scharf, M. L. Klein, and G. J. Martyna, *J. Chem. Phys.* **97**, 3590 (1992).

[679] Path-integral molecular-dynamics calculation of the conduction-band energy minimum V0 of excess electrons in fluid argon, J.-M. Lopez-Castillo, Y. Frongillo, B. Plenkiewicz, and J.-P. Jay-Gerin, *J. Chem. Phys.* **96**, 9092 (1992).

[680] Quantum path integral extension of Widom's test particle method for chemical potentials with application to isotope effects on hydrogen solubilities in model solids, T. L. Beck, *J. Chem. Phys.* **96**, 7175 (1992).

[681] Real-time path-integral simulation of vibrational transition probabilities of small molecules in clusters: Theory and application to Br_2 in Ar, A. M. Amini and M. F. Herman, *J. Chem. Phys.* **96**, 5999 (1992).

[682] A path integral Einstein model for characterizing the equilibrium states of low temperature solids, D. Li and G. A. Voth, *J. Chem. Phys.* **96**, 5340 (1992).

[683] Improved methods for path integral Monte Carlo integration in fermionic systems, W. H. Newman and A. Kuki, *J. Chem. Phys.* **96**, 1409 (1992).

[684] Path-integral study of magnetic response: Excitonic and biexcitonic diamagnetism in semiconductor quantum dots, E. L. Pollock and K. J. Runge, *J. Chem. Phys.* **96**, 674 (1992).

[685] Monte Carlo evaluation of real-time Feynman path integrals for quantal many-body dynamics: Distributed approximating functions and Gaussian sampling, B. M. Kouri, D. J. W. Zhu, X. Ma, B. M. Pettitt, and D. K. Hoffman, *J. Phys. Chem.* **96**, 9622 (1992).

[686] Toward a new time-dependent path integral formalism based on restricted quantum propagators for physically realizable systems, D. J. Kouri and D. K. Hoffman, *J. Phys. Chem.* **96**, 9631 (1992).

[687] The quantum mechanics of clusters: The low-temperature equilibrium and dynamical behavior of rare-gas systems, S. W. Rick, D. L. Leitner, J. D. Doll, D. L. Freeman, and D. D. Frantz, *J. Chem. Phys.* **95**, 6658 (1991).

[688] Development, justification, and use of a projection operator in path integral calculations in continuous space, R. W. Hall and M. R. Prince, *J. Chem. Phys.* **95**, 5999 (1991).

[689] Path-integral calculation of the tunnel splitting in aqueous ferrous-ferric electron transfer, M. Marchi and D. Chandler, *J. Chem. Phys.* **95**, 889 (1991).

[690] Path-integral study of excitons and biexcitons in semiconductor quantum dots, E. L. Pollock and S. W. Koch, *J. Chem. Phys.* **94**, 6776 (1991).

[691] Path integral studies of the 2D Hubbard model using a new projection operator, R. W. Hall, *J. Chem. Phys.* **94**, 1312 (1991).

[692] Feynman path integral approach for studying intramolecular effects in proton-transfer reactions, D. Li and G. A. Voth, *J. Phys. Chem.* **95**, 10425 (1991).

[693] Determination of the vertical ionization potentials of small sodium clusters using path integral Monte Carlo calculations, S. E. Hays and R. W. Hall, *J. Phys. Chem.* **95**, 8552 (1991).

[694] A Monte Carlo and transfer-matrix grid path-integral study of the vibrational structure of Br_2 in solid argon, J. P. Prates Ramalho, B. J. Costa Cabral, and F. M. S. Silva Fernandes, *Chem. Phys. Lett.* **184**, 53 (1991).

[695] Path integral study of the correlated electronic states of Na_4-Na_6, R. W. Hall, *J. Chem. Phys.* **93**, 8211 (1990).

[696] Discretized path integral method and properties of a quantum system, A. Sethia, S. Sanyal, and Y. Singh, *J. Chem. Phys.* **93**, 7268 (1990).

[697] Localization in a Lennard-Jones fluid: The path-integral approach, Y. Fan and B. N. Miller, *J. Chem. Phys.* **93**, 4322 (1990).

[698] Density functional theory of freezing for quantum systems. I. Path integral formulation of general theory, J. D. McCoy, S. W. Rick, and A. D. J. Haymet, *J. Chem. Phys.* **92**, 3034 (1990).

[699] The effect of discretization on a path integral expression for the one-electron density, G. G. Hoffman, *J. Chem. Phys.* **92**, 2966 (1990).

[700] A test of the maxentropic density matrix for hydrated electrons using the results of path-integral simulation, T. R. Tuttle and S. Golden, *Chem. Phys. Lett.* **170**, 69 (1990).

[701] Equilibrium and dynamical fourier path integral methods, J. D. Doll, D. L. Freeman, and T. L. Beck, *Adv. Chem. Phys.* **78**, 61 (1990).

[702] On energy estimators in path integral Monte Carlo simulations: Dependence of accuracy on algorithm, J. Cao and B. J. Berne, *J. Chem. Phys.* **91**, 6359 (1989).

[703] Path integral versus conventional formulation of equilibrium classical statistical mechanics, A. L. Kholodenko, *J. Chem. Phys.* **91**, 4849 (1989).

[704] The exchange potential in path integral studies: Analytical justification, R. W. Hall, *J. Chem. Phys.* **91**, 1926 (1989).

[705] Isotopic shift in the melting curve of helium: A path integral Monte Carlo study, J.-L. Barrat, P. Loubeyre, and M. L. Klein, *J. Chem. Phys.* **90**, 5644 (1989).

[706] Time correlation function and path integral analysis of quantum rate constants, G. A. Voth, D. Chandler, and W. H. Miller, *J. Phys. Chem.* **93**, 7009 (1989).

[707] Comparison of path integral and density functional techniques in a model two-electron system, R. W. Hall, *J. Phys. Chem.* **93**, **14**, 5628 (1989).

[708] Finite temperature effects in Na_3^+ and Na_3: A path integral Monte Carlo study, R. W. Hall, *Chem. Phys. Lett.* **160**, 520 (1989).

[709] Variance and correlation length of energy estimators in Metropolis path integral Monte Carlo, A. Giansanti and G. Jacucci, *J. Chem. Phys.* **89**, 7454 (1988).

[710] Quantum Monte Carlo dynamics: The stationary phase Monte Carlo path integral calculation of finite temperature time correlation functions, J. D. Doll, T. L. Beck, and D. L. Freeman, *J. Chem. Phys.* **89**, 5753 (1988).

[711] Electron attachment to ammonia clusters: A study using path integral Monte Carlo calculations, M. Marchi, M. Sprik, and M. L. Klein, *J. Chem. Phys.* **89**, 4918 (1988).

[712] The treatment of exchange in path integral simulations via an approximate pseudopotential, R. W. Hall, *J. Chem. Phys.* **89**, 4212 (1988).

[713] Extraction of ground state properties by discretized path integral formulations, H. Kono, A. Takasaka, and S. H. Lin, *J. Chem. Phys.* **89**, 3233 (1988).

[714] Monte Carlo calculation of the quantum partition function via path integral formulations, H. Kono, A. Takasaka, and S. H. Lin, *J. Chem. Phys.* **88**, 6390 (1988).

[715] Calculation of the free energy of electron solvation in liquid ammonia using a path integral quantum Monte Carlo simulation, M. Marchi, M. Sprik, and M. L. Klein, *J. Phys. Chem.* **92**, 12, 3625 (1988).

[716] Imaginary time path integral Monte Carlo route to rate coefficients for nonadiabatic barrier crossing, P. G. Wolynes, *J. Chem. Phys.* **87**, 6559 (1987).

[717] Path integral Monte Carlo study of the hydrated electron, A. Wallqvist, D. Thirumalai, and B. J. Berne, *J. Chem. Phys.* **86**, 6404 (1987).

[718] Path integral Monte Carlo studies of the behavior of excess electrons in simple fluids, D. F. Coker, B. J. Berne, and D. Thirumalai, *J. Chem. Phys.* **86**, 5689 (1987).

[719] High-order correction to the Trotter expansion for use in computer simulation, X.-P. Li and J. Q. Broughton, *J. Chem. Phys.* **86**, 5094 (1987).

[720] Path integral approach to electrostatic problems, J. I. Gersten and A. Nitzan, *J. Chem. Phys.* **86**, 3557 (1987).

[721] Statistical mechanics of solutions of semiflexible chains: A path integral formulation, S. M. Bhattacharjee and M. Muthukumar, *J. Chem. Phys.* **86**, 411 (1987).

[722] Electron-inert gas pseudopotentials for use in path integral simulations, X.-P. Li, J. Q. Broughton, and P. B. Allen, *J. Chem. Phys.* **85**, 3444 (1986).

[723] Atomic and molecular quantum mechanics by the path integral molecular dynamics method, D. Scharf, J. Jortner, and U. Landman, *Chem. Phys. Lett.* **130**, 504 (1986).

[724] Hydrated electron revisited via the Feynman path integral route, C. D. Jonah, C. Romero, and A. Rahman, *Chem. Phys. Lett.* **123**, 209 (1986).

[725] Study of electron solvation in liquid ammonia using quantum path integral Monte Carlo calculations, M. Sprik, R. W. Impey, and M. L. Klein, *J. Chem. Phys.* **83**, 5802 (1985).

[726] On quantum trajectories and an approximation to the Wigner path integral, F. McLafferty, *J. Chem. Phys.* **83**, 5043 (1985).

[727] Evaluation of microcanonical rate constants for bimolecular reactions by path integral techniques, D. Thirumalai, B. C. Garrett, and B. J. Berne, *J. Chem. Phys.* **83**, 2972 (1985).

[728] Path integral approach to multiparticle systems: The sudden representation, C. K. Chan and D. J. Kouri, *J. Chem. Phys.* **83**, 680 (1985).

[729] "Direct" calculation of quantum mechanical rate constants via path integral methods: Application to the reaction path Hamiltonian, with numerical test for the H+H$_2$ reaction in 3D, K. Yamashita and W. H. Miller, _J. Chem. Phys._ **82**, 5475 (1985).

[730] Quantum mechanical rate constants via path integrals: Diffusion of hydrogen atoms on a tungsten(100) surface, R. Jaquet and W. H. Miller, _J. Phys. Chem._ **89, 11**, 2139 (1985).

[731] Path integral methods for simulating electronic spectra, D. Thirumalai and B. J. Berne, _Chem. Phys. Lett._ **116**, 471 (1985).

[732] Path-integral simulation of pure water, A. Wallqvist and B. J. Berne, _Chem. Phys. Lett._ **117**, 214 (1985).

[733] Feynman path integral and the Hückel model, J. Marañon and H. Grinberg, _J. Chem. Phys._ **81**, 4537 (1984).

[734] Nonergodicity in path integral molecular dynamics, R. W. Hall and B. J. Berne, _J. Chem. Phys._ **81**, 3641 (1984).

[735] Monte Carlo Fourier path integral methods in chemical dynamics, J. D. Doll, _J. Chem. Phys._ **81**, 3536 (1984).

[736] A path integral Monte Carlo study of liquid neon and the quantum effective pair potential, D. Thirumalai, R. W. Hall, and B. J. Berne, _J. Chem. Phys._ **81**, 2523 (1984).

[737] Semiclassical collision theory within the Feynman path-integral formalism: The perturbed stationary state formulation, G. Jolicard, _J. Chem. Phys._ **80**, 2476 (1984).

[738] On the calculation of time correlation functions in quantum systems: Path integral techniques, D. Thirumalai and B. J. Berne, _J. Chem. Phys._ **79**, 5029 (1983).

[739] On nonadiabatic transitions and the Wigner path integral, F. McLafferty, _J. Chem. Phys._ **79**, 4922 (1983).

[740] Path integral solutions for Fokker-Planck conditional propagators in nonequilibrium systems: Catastrophic divergences of the Onsager-Machlup-Laplace approximation, P. M. Hunt, K. L. C. Hunt, and J. Ross, _J. Chem. Phys._ **79**, 3765 (1983).

[741] On classical paths and the Wigner path integral, F. McLafferty, _J. Chem. Phys._ **78**, 3253 (1983).

[742] Atomic orbitals of the nonrelativistic hydrogen atom in a four-dimensional Riemann space through the path integral formalism, H. Grinberg, J. Marañon, and H. Vucetich, _J. Chem. Phys._ **78**, 839 (1983).

[743] On path integral Monte Carlo simulations, M. F. Herman, E. J. Bruskin, and B. J. Berne, *J. Chem. Phys.* **76**, 5150 (1982).

[744] Path-integral method in the theory of the elementary act of the charge transfer in polar media, A. M. Kuznetsov, *Chem. Phys. Lett.* **91**, 34 (1982).

[745] Convenient and accurate discretized path integral methods for equilibrium quantum mechanical calculations, K. S. Schweizer, R. M. Stratt, D. Chandler, and P. G. Wolynes, *J. Chem. Phys.* **75**, 1347 (1981).

[746] Path integral solutions of stochastic equations for nonlinear irreversible processes: The uniqueness of the thermodynamic Lagrangian, K. L. C. Hunt and J. Ross, *J. Chem. Phys.* **75**, 976 (1981).

[747] *Techniques and Applications of Path Integration*, L. S. Schulman (John Wiley & Sons, New York, 1981).

[748] Recombination of radical pairs in high magnetic fields: A path integral-Monte Carlo treatment, K. Schulten and I. R. Epstein, *J. Chem. Phys.* **71**, 309 (1979).

[749] Path integral representation of the reaction rate constant in quantum mechanical transition state theory, W. H. Miller, *J. Chem. Phys.* **63**, 1166 (1975).

[750] The Weyl correspondence and path integrals, M. M. Mizrahi, *J. Math. Phys.* **16**, 2201 (1975).

[751] Wormlike chains near the rod limit: Path integral in the WKB approximation, H. Yamakawa and M. Fujii, *J. Chem. Phys.* **59**, 6641 (1973).

[752] Statistical mechanics of wormlike chains: Path integral and diagram methods, H. Yamakawa, *J. Chem. Phys.* **59**, 3811 (1973).

[753] Bounds for thermodynamic Green's functions from the application of path integral methods, L. W. Bruch and H. E. Revercomb, *J. Chem. Phys.* **58**, 751 (1973).

[754] *Quantum Mechanics and Path Integrals*, R. P. Feynman and A. R. Hibbs (McGraw-Hill, New York, 1965).

[755] Dynamical theory in curved spaces. I. A review of the classical and quantum action principles, B. S. DeWitt, *Rev. Mod. Phys.* **29**, 377 (1957).

[756] Space-time approach to non-relativistic quantum mechanics, R. P. Feynman, *Rev. Mod. Phys.* **20**, 367 (1948).

Lennard-Jones Systems

[757] Discretization error-free estimate of low temperature statistical dissociation rates in gas phase: Applications to Lennard-Jones clusters $X_{13-n}Y_n$ ($n = 0 - 3$), M. Mella, *J. Chem. Phys.* **128**, 244515 (2008).

[758] Morphing Lennard-Jones clusters to TIP4P water clusters: Why do water clusters look like they do? B. Hartke, *Chem. Phys.* **346**, 286 (2008).

[759] A dynamic lattice searching method with interior operation for unbiased optimization of large Lennard-Jones clusters, X. Shao, X. Yang, and W. Cai, *J. Comput. Chem.* **29**, 1772 (2008).

[760] Estimation of bulk liquid properties from Monte Carlo simulations of Lennard-Jones clusters, J. C. Barrett and A. P. Knight, *J. Chem. Phys.* **128**, 086101 (2008).

[761] Nonlinear scaling schemes for Lennard-Jones interactions in free energy calculations, T. Steinbrecher, D. L. Mobley, and D. A. Case, *J. Chem. Phys.* **127**, 214108 (2007).

[762] Understanding fragility in supercooled Lennard-Jones mixtures. I. Locally preferred structures, D. Coslovich and G. Pastore, *J. Chem. Phys.* **127**, 124504 (2007).

[763] Molecular dynamics of homogeneous nucleation in the vapor phase of Lennard-Jones. III. Effect of carrier gas pressure, K. Yasuoka and X. C. Zeng, *J. Chem. Phys.* **126**, 124320 (2007).

[764] Energy dependent decay rates of Lennard-Jones clusters for use in nucleation theory, J. C. Barrett, *J. Chem. Phys.* **126**, 074312 (2007).

[765] Solid-solid structural transformations in Lennard-Jones clusters: Accurate simulations versus the harmonic superposition approximation, V. A. Sharapov and V. A. Mandelshtam, *J. Phys. Chem.* A **111**, 10284 (2007).

[766] A dynamic lattice searching method with constructed core for optimization of large Lennard-Jones clusters, X. Yang, W. Cai, and X. Shao, *J. Comput. Chem.* **28**, 1427 (2007).

[767] Multiple structural transformations in Lennard-Jones clusters: Generic versus size-specific behavior, V. A. Mandelshtam and P. A. Frantsuzov, *J. Chem. Phys.* **124**, 204511 (2006).

[768] Structural transitions and melting in LJ_{74-78} Lennard-Jones clusters from adaptive, exchange Monte Carlo simulations, V. A. Mandelshtam, P. A. Frantsuzov, and F. Calvo, *J. Phys. Chem.* A **110**, 5326 (2006).

[769] Investigation of excess adsorption, solvation force, and plate-fluid interfacial tension for Lennard-Jones fluid confined in slit pores, D. Fu, *J. Chem. Phys.* **124**, 164701 (2006).

[770] Structural transitions in the 309-atom magic number Lennard-Jones cluster, E. G. Noya and J. P. K. Doye, *J. Chem. Phys.* **124**, 104503 (2006).

[771] High-symmetry global minimum geometries for small mixed Ar/Xe Lennard-Jones clusters, S. M. Cleary and H. R. Mayne, *Chem. Phys. Lett.* **418**, 79 (2006).

[772] Statistical-mechanical theory of rheology: Lennard-Jones fluids, R. Laghaei, A. E. Nasrabad, and B. C. Eu, *J. Chem. Phys.* **123**, 234507 (2005).

[773] Parallel tempering simulations of the 13-center Lennard-Jones dipole-dipole cluster ($\mu_D = 0 - 0.5$ a.u.), D. M. Pav and E. Curotto, *J. Chem. Phys.* **123**, 144301 (2005).

[774] Phase changes in Lennard-Jones mixed clusters with composition Ar_nXe_{6-n} ($n = 0, 1, 2$), R. P. White, S. M. Cleary, and H. R. Mayne, *J. Chem. Phys.* **123**, 094505 (2005).

[775] Tests of the homogeneous nucleation theory with molecular-dynamics simulations. 1. Lennard-Jones molecules, K. K. Tanaka, K. Kawamura, H. Tanaka, and K. Nakazawa, *J. Chem. Phys.* **122**, 184514 (2005).

[776] Combining smart darting with parallel tempering using Eckart space: Application to Lennard-Jones clusters, P. Nigra, D. L. Freeman, and J. D. Doll, *J. Chem. Phys.* **122**, 114113 (2005).

[777] Pressure dependent study of the solid-solid phase change in 38-atom Lennard-Jones cluster, D. Sabo, D. L. Freeman, and J. D. Doll, *J. Chem. Phys.* **122**, 094716 (2005).

[778] Extended-range order, diverging static length scales, and local structure formation in cold Lennard-Jones fluids, P. C. Whitford and G. D. J. Phillies, *J. Chem. Phys.* **122**, 044508 (2005).

[779] Minimization of the potential energy surface of Lennard-Jones clusters by quantum optimization, T. Gregora and R. Car, *Chem. Phys. Lett.* **412**, 125 (2005).

[780] Structural transition from icosahedra to decahedra of large Lennard-Jones clusters, X. Shao, Y. Xiang, and W. Cai, *J. Phys. Chem.* A **109**, 5193 (2005).

[781] Structural behavior and self-assembly of Lennard-Jones clusters on rigid surfaces, I. Paci, I. Szleifer, and M. A. Ratner, *J. Phys. Chem.* B **109**, 12935 (2005).

[782] An unbiased population-based search for the geometry optimization of Lennard-Jones clusters: $2 < N < 372$, W. Pullan, *J. Comp. Chem.* **26**, 899 (2005).

[783] Minimization of the potential energy surface of Lennard-Jones clusters by quantum optimization, T. Gregor and R. Car, *Chem. Phys. Lett.* **412**, 125 (2005).

[784] Parameter space minimization methods: Applications to Lennard-Jones-dipole-dipole clusters, C. A. Oppenheimer and E. Curotto, *J. Chem. Phys.* **121**, 6226 (2004).

[785] Phase changes in selected Lennard-Jones $X_{13-n}Y_n$ clusters, D. Sabo, C. Predescu, J. D. Doll, and D. L. Freeman, *J. Chem. Phys.* **121**, 856 (2004).

[786] Triple point of Lennard-Jones fluid in slit nanopore: Solidification of critical condensate, H. Kanda, M. Miyahara, and K. Higashitani, *J. Chem. Phys.* **120**, 6173 (2004).

[787] An efficient method based on lattice construction and the genetic algorithm for optimization of large Lennard-Jones clusters, Y. Xiang, H. Jiang, W. Cai, and X. Shao, *J. Phys. Chem.* A **108**, 3586 (2004).

[788] Evaporation dynamics of mixed Lennard-Jones atomic clusters, P. Parneix and P. Bréchignac, *J. Chem. Phys.* **118**, 8234 (2003).

[789] Density-dependent solvation dynamics in a simple Lennard-Jones fluid, M. M. Martins and H. Stassen, *J. Chem. Phys.* **118**, 5558 (2003).

[790] Computer modeling of the liquid-vapor interface of an associating Lennard-Jones fluid, J. Alejandre, Y. Duda, and S. Sokolowski, *J. Chem. Phys.* **118**, 329 (2003).

[791] *Energy Landscapes*, D. J. Wales (Cambridge Press, Cambridge, 2003).

[792] Entropic effects on the structure of Lennard-Jones clusters, J. P. K. Doye and F. Calvo, *J. Chem. Phys.* **116**, 8307 (2002).

[793] New lowest energy sequence of Marks' decahedral Lennard-Jones clusters containing up to 10000 atoms, H. Jiang, W. Cai, and X. Shao, *J. Phys. Chem.* A **107**, 4238 (2003).

[794] Saddle points and dynamics of Lennard-Jones clusters, solids, and supercooled liquids, J. P. K. Doye and D. J. Wales, *J. Chem. Phys.* **116**, 3777 (2002).

[795] Collapse of Lennard-Jones homopolymers: Size effects and energy landscapes, F. Calvo, J. P. K. Doye, and D. J. Wales, *J. Chem. Phys.* **116**, 2642 (2002).

[796] Comparison of inherent, instantaneous, and saddle configurations of the bulk Lennard-Jones system, P. Shah and C. Chakravarty, *J. Chem. Phys.* **115**, 8784 (2001).

[797] Magic number behavior for heat capacities of medium-sized classical Lennard-Jones clusters, D. D. Frantz, *J. Chem. Phys.* **115**, 6136 (2001).

[798] Supercooling in a two-dimensional Lennard-Jones mixture, E. Sim, A. Z. Patashinski, and M. A. Ratner, *J. Chem. Phys.* **114**, 9048 (2001).

[799] Quantum partition functions from classical distributions: Application to rare-gas clusters, F. Calvo, J. P. K. Doye, and D. J. Wales, *J. Chem. Phys.* **114**, 7312 (2001).

[800] Collapse transition of isolated Lennard-Jones chain molecules: Exact results for short chains, M. P. Taylor, *J. Chem. Phys.* **114**, 6472 (2001).

[801] Computer simulation of surface and adatom properties of Lennard-Jones solids: A comparison between face-centered-cubic and hexagonal-close-packed structures, S. Somasi, B. Khomami, and R. Lovett, *J. Chem. Phys.* **114**, 6315 (2001).

[802] Development of reference states for use in absolute free energy calculations of atomic clusters with application to 55-atom Lennard-Jones clusters in the solid and liquid states, L. M. Amon and W. P. Reinhardt, *J. Chem. Phys.* **113**, 3573 (2000).

[803] Phase changes in 38-atom Lennard-Jones clusters. II. A parallel tempering study of equilibrium and dynamic properties in the molecular dynamics and microcanonical ensembles, F. Calvo, J. P. Neirotti, D. L. Freeman, and J. D. Doll, *J. Chem. Phys.* **112**, 10350 (2000).

[804] Phase changes in 38-atom Lennard-Jones clusters. I. A parallel tempering study in the canonical ensemble, J. P. Neirotti, F. Calvo, D. L. Freeman, and J. D. Doll, *J. Chem. Phys.* **112**, 10340 (2000).

[805] Critical cluster size and droplet nucleation rate from growth and decay simulations of Lennard-Jones clusters, H. Vehkamäki and I. J. Ford, *J. Chem. Phys.* **112**, 4193 (2000).

[806] Evolution of the potential energy surface with size for Lennard-Jones clusters, J. P. K. Doye, M. A. Miller, and D. J. Wales, *J. Chem. Phys.* **111**, 8417 (1999).

[807] The double-funnel energy landscape of the 38-atom Lennard-Jones cluster, J. P. K. Doye, M. A. Miller, and D. J. Wales, *J. Chem. Phys.* **110**, 6896 (1999).

[808] Computer simulation study of gas-liquid nucleation in a Lennard-Jones system, P. Rein ten Wolde and D. Frenkel, *J. Chem. Phys.* **109**, 9901 (1998).

[809] Molecular dynamics of homogeneous nucleation in the vapor phase. I. Lennard-Jones fluid, K. Yasuoka and M. Matsumoto, *J. Chem. Phys.* **109**, 8451 (1998).

[810] Thermodynamics and the global optimization of Lennard-Jones clusters, J. P. K. Doye, D. J. Wales, and M. A. Miller, *J. Chem. Phys.* **109**, 8143 (1998).

[811] A j-walking algorithm for microcanonical simulations: Applications to Lennard-Jones clusters, E. Curotto, D. L. Freeman, and J. D. Doll, *J. Chem. Phys.* **109**, 1643 (1998).

[812] Efficient transition path sampling: Application to Lennard-Jones cluster rearrangements, C. Dellago, P. G. Bolhuis, and D. Chandler, *J. Chem. Phys.* **108**, 9236 (1998).

[813] Calculating free energies of Lennard-Jones clusters using the effective diffused potential, S. Schelstraete and H. Verschelde, *J. Chem. Phys.* **108**, 7152 (1998).

[814] Chaos and dynamical coexistence in Lennard-Jones clusters, F. Calvo, *J. Chem. Phys.* **108**, 6861 (1998).

[815] Microcanonical temperature and "heat capacity" computation of Lennard-Jones clusters under isoergic molecular dynamics simulation, U. A. Salian, *J. Chem. Phys.* **108**, 6342 (1998).

[816] An investigation of two approaches to basin hopping minimization for atomic and molecular clusters, R. P. White and H. R. Mayne, *Chem. Phys. Lett.* **289**, 463 (1998).

[817] A large-scale and long-time molecular dynamics study of supercritical Lennard-Jones fluid. An analysis of high temperature clusters, N. Yoshii and S. Okazaki, *J. Chem. Phys.* **107**, 2020 (1997).

[818] Global optimization by basin-hopping and the lowest energy structures of Lennard-Jones clusters containing up to 110 atoms, D. J. Wales and J. P. K. Doye, *J. Phys. Chem.* **101**, 5111 (1997).

[819] Molecular dynamics simulation for the cluster formation process of Lennard-Jones particles: Magic numbers and characteristic features, T. Ikeshoji, B. Hafskjold, Y. Hashi, and Y. Kawazoe, *J. Chem. Phys.* **105**, 5126 (1996).

[820] How do the properties of a glass depend on the cooling rate? A computer simulation study of a Lennard-Jones system, K. Vollmayr, W. Kob, and K. Binder, *J. Chem. Phys.* **105**, 4714 (1996).

[821] Study of the solid-liquid transition for Ar_{55} using the J-walking Monte Carlo method, G. E. López, *J. Chem. Phys.* **104**, 6650 (1996).

[822] Characterization of solvent clusters in a supercritical Lennard-Jones fluid, H. L. Martinez, R. Ravi, and S. C. Tucker, *J. Chem. Phys.* **104**, 1067 (1996).

[823] Global geometry optimization of atomic clusters using a modified genetic algorithm in space-fixed coordinates, J. A. Niesse and H. R. Mayne, *J. Chem. Phys.* **105**, 4700 (1996).

[824] Global geometry optimization of $(Ar)_n$ and $B(Ar)_n$ clusters using a mod ified genetic algorithm, S. K. Gregurick, M. H. Alexander, and B. Hartke, *J. Chem. Phys.* **104**, 2684 (1996).

[825] Structural optimization of Lennard-Jones clusters by a genetic algorithm, D. M. Deaven, N. Tit, J. R. Morris, and K. M. Ho, *Chem. Phys. Lett.* **256**, 195 (1996).

[826] Global optimization of atomic and molecular clusters using the space-fixed modified genetic algorithm method, J. A. Niesse and H. R. Mayne, *J. Comp. Chem.* **18**, 1233 (1996).

[827] Calculation of thermodynamic properties of small Lennard-Jones clusters incorporating anharmonicity, J. P. K. Doye and D. J. Wales, *J. Chem. Phys.* **102**, 9659 (1995).

[828] The kinetics of crystal growth and dissolution from the melt in Lennard-Jones systems, L. A. Bez and P. Clancy, *J. Chem. Phys.* **102**, 8138 (1995).

[829] Locating transition structures by mode following: A comparison of six methods on the Ar_8 Lennard-Jones potential, F. Jensen, *J. Chem. Phys.* **102**, 6706 (1995).

[830] Magic numbers for classical Lennard-Jones cluster heat capacities, D. D. Frantz, *J. Chem. Phys.* **102**, 3747 (1995).

[831] Rearrangements of 55-atom Lennard-Jones and $(C_{60})_{55}$ clusters, D. J. Wales, *J. Chem. Phys.* **101**, 3750 (1994).

[832] The construction of double-ended classical trajectories, A. E. Cho, J. D. Doll, and D. L. Freeman, *Chem. Phys. Lett.* **229**, 218 (1994).

[833] Use of an eigenmode method to locate the stationary points on the potential energy surfaces of selected argon and water clusters, C. J. Tsai and K. D. Jordan, *J. Phys. Chem.* **97**, 11227 (1993).

[834] A global optimization approach for Lennard-Jones microclusters, C. D. Maranas and C. A. Floudas, *J. Chem. Phys.* **97**, 7667 (1992).

[835] Extending J walking to quantum systems: Applications to atomic clusters, D. D. Frantz, D. L. Freeman, and J. D. Doll, *J. Chem. Phys.* **97**, 5713 (1992).

[836] Prediction of the thermodynamic properties of associating Lennard-Jones fluids: Theory and simulation, W. G. Chapman, *J. Chem. Phys.* **93**, 4299 (1990).

[837] New insight into experimental probes of cluster melting, J. E. Adams and R. M. Stratt, *J. Chem. Phys.* **93**, 1358 (1990).

[838] Melting and freezing of small argon clusters, D. J. Wales and R. S. Berry, *J. Chem. Phys.* **92**, 4283 (1990).

[839] Reducing quasi-ergodic behavior in Monte Carlo simulations by J-walking: Applications to atomic clusters, D. D. Frantz, D. L. Freeman, and J. D. Doll, *J. Chem. Phys.* **93**, 2769 (1990).

[840] Molecular dynamics studies of polar/nonpolar fluid mixtures. I. Mixtures of Lennard-Jones and Stockmayer fluids, S. W. de Leeuw, B. Smit, and C. P. Williams, *J. Chem. Phys.* **93**, 2704 (1990).

[841] Statistical thermodynamics of the cluster solid-liquid transition, P. Labastie and R. L. Whetten, *Phys. Rev. Lett.* **65**, 1567 (1990).

[842] Stability of face-centered cubic and icosahedral Lennard-Jones clusters, B. W. van de Waal, *J. Chem. Phys.* **90**, 3407 (1989).

[843] Rigid-fluid transition in specific-size argon clusters, M. Y. Hahn and R. L. Whetten, *Phys. Rev. Lett.* **61**, 1190 (1988).

[844] Melting and phase space transitions in small clusters: Spectral characteristics, dimensions, and K entropy, T. L. Beck, D. M. Leitner, and R. S. Berry, *J. Chem. Phys.* **89**, 1681 (1988).

[845] Structure and binding of Lennard-Jones clusters: $13 \leq N \leq 147$, J. A. Northby, *J. Chem. Phys.* **87**, 6166 (1987).

[846] Solid–liquid phase changes in simulated isoenergetic Ar_{13}, J. Jellineck, T. Beck, and R. S. Berry, *J. Chem. Phys.* **84**, 2783 (1986).

[847] On finding transition states, C. J. Cerjan and W. H. Miller, *J. Chem. Phys.* **75**, 2800 (1981).

[848] Monte Carlo calculation of argon clusters in homogeneous nucleation, N. G. Garcia and J. M. S. Torroja, *Phys. Rev. Lett.* **47**, 186 (1981).

[849] Theory and Monte Carlo simulation of physical clusters in the imperfect vapor, J. K. Lee, J. A. Barker, and F. F. Abraham, *J. Chem. Phys.* **58**, 3166 (1973).

[850] Third virial coefficients for mixtures of Lennard-Jones 6-12 molecules, D. E. Stogryn, *J. Chem. Phys.* **48**, 4474 (1968).

[851] Refinement of the Lennard-Jones and Devonshire theory of liquids and dense gases, W. J. Taylor, *J. Chem. Phys.* **24**, 454 (1956).

Differential Geometry

[852] *A First Course in General Relativity*, B. Schutz (Cambridge Press, New York, 1985).

[853] *Differential Manifolds and Theoretical Physics*, W. D. Curtis and F. R. Miller (Academic Press, New York, 1985).

FORTRAN and EISPACK

[854] *Introduction to FORTRAN 90 for Engineers and Scientists*, L. R. Nyhoff and S. C. Leestma (Prentice Hall, Upper Saddle River, NJ, 1997).

[855] *Matrix Eigensystem Routines-EISPACK Guide*, B. T. Smith, J. M. Boyle, and J. J. Dongarra, Sec. Ed. (Springer-Verlag, Berlin, 1976).

Angular Momentum Theory

[856] *Angular Momentum in Quantum Mechanics (Investigations in Physics)*, A. R. Edmonds (Princeton University Press, Princeton, NJ, 1996).

[857] *Angular Momentum: Understanding Spatial Aspects in Chemistry and Physics*, R. N. Zare (Wiley-Interscience, New York, 1988).

[858] *Elementary Theory of Angular Momentum*, M. E. Rose (John Wiley & Sons Inc., New York, 1957).

Matching-Pursuit Split-Operator

[859] Matching-pursuit/split-operator-Fourier-transform simulations of excited-state nonadiabatic quantum dynamics in pyrazine, X. Chen and V. S. Batista, *J. Chem. Phys.* **125**, 124313 (2006).

[860] Matching-pursuit split-operator Fourier-transform simulations of excited-state intramolecular proton transfer in 2-(2[prime]-hydroxyphenyl)-oxazole, Y. Wu and V. S. Batista, *J. Chem. Phys.* **124**, 224305 (2006).

[861] Matching-pursuit/split-operator Fourier-transform simulations of nonadiabatic quantum dynamics, Y. Wu, M. F. Herman, and V. S. Batista, *J. Chem. Phys.* **122**, 114114 (2005).

[862] Matching-pursuit/split-operator-Fourier-transform computations of thermal correlation functions, X. Chen, Y. Wu, and V. S. Batista, *J. Chem. Phys.* **122**, 064102 (2005).

Water TIP4P Clusters

[863] Hydration free energies of monovalent ions in transferable intermolecular potential four point fluctuating charge water: An assessment of simulation methodology and force field performance and transferability, G. L. Warren and S. Patel, *J. Chem. Phys.* **127**, 064509 (2007).

[864] Temperature and structural changes of water clusters in vacuum due to evaporation, C. Caleman and D. van der Spoel, *J. Chem. Phys.* **125**, 154508 (2006).

[865] Solvation free energies of amino acid side chain analogs for common molecular mechanics water models, M. R. Shirts and V. S. Pande, *J. Chem. Phys.* **122**, 134508 (2005).

[866] High-level ab initio calculations for the four low-lying families of minima of $(H_2O)_{20}$. I. Estimates of MP2/CBS binding energies and comparison with empirical potentials, G. S. Fanourgakis, E. Aprà, and S. S. Xantheas, *J. Chem. Phys.* **121**, 2655 (2004).

[867] Monte Carlo simulations of critical cluster sizes and nucleation rates of water, J. Merikanto, H. Vehkamäki, and E. Zapadinsky, *J. Chem. Phys.* **121**, 914 (2004).

[868] Effect of polarizability of halide anions on the ionic solvation in water clusters, S. Yoo, Y. A. Lei, and X. C. Zeng, *J. Chem. Phys.* **119**, 6083 (2003).

[869] Solvent-induced symmetry breaking of nitrate ion in aqueous clusters: A quantum-classical simulation study, M. C. G. Lebrero, D. E. Bikiel, M. D. Elola, D. A. Estrin, and A. E. Roitberg, *J. Chem. Phys.* **117**, 2718 (2002).

[870] Translational dynamics of a cold water cluster in the presence of an external uniform electric field, A. Vegiri, *J. Chem. Phys.* **116**, 8786 (2002).

[871] Exploring the ab initio/classical free energy perturbation method: The hydration free energy of water, S. Sakane, E. M. Yezdimer, W. Liu, J. A. Barriocanal, D. J. Doren, and R. H. Wood, *J. Chem. Phys.* **113**, 2583 (2000).

[872] Clustering of water in polyethylene: A molecular-dynamics simulation, M. Fukuda, *J. Chem. Phys.* **109**, 6476 (1998).

[873] A molecular dynamics study of the Cr^{3+} hydration based on a fully flexible hydrated ion model, J. M. Martnez, R. R. Pappalardo, and E. S. Marcos, *J. Chem. Phys.* **109**, 1445 (1998).

[874] Surface tension of water droplets: A molecular dynamics study of model and size dependencies, V. V. Zakharov, E. N. Brodskaya, and A. Laaksonen, *J. Chem. Phys.* **107**, 10675 (1997).

[875] Many-body effects in molecular dynamics simulations of $Na^+(H_2O)_n$ and $Cl^-(H_2O)_n$ clusters, L. Perera and M. L. Berkowitz, *J. Chem. Phys.* **95**, 1954 (1991).

[876] Dielectric relaxation in water. Computer simulations with the TIP4P potential, M. Neumann, *J. Chem. Phys.* **85**, 1567 (1986).

[877] Comparison of simple potential functions for simulating liquid water, W. L. Jorgensen, J. Chandrasekhar, J. D. Madura, R. W. Impey, and M. L. Klein, *J. Chem. Phys.* **79**, 926 (1983).

Index

Printed and bound by CPI Group (UK) Ltd, Croydon, CR0 4YY

18/10/2024

01776244-0019